WITHDRAWN

M. B. Hocking

# Modern Chemical Technology and Emission Control

With 152 Figures

Springer-Verlag
Berlin Heidelberg New York Tokyo

M. B. Hocking
Department of Chemistry
University of Victoria
Victoria, British Columbia
Canada V8W 2Y2

ISBN 3-540-13466-2 Springer-Verlag Berlin Heidelberg New York Tokyo
ISBN 0-387-13466-2 Springer-Verlag New York Heidelberg Berlin Tokyo

Library of Congress Cataloging in Publication Data
Hocking, M. B. (Martin Blake), 1938
Modern chemical technology and emission control.
Bibliography: p.
1. Chemistry, Technical. 2. Environmental chemistry. I. Title.
TP155.H58 1984 660.2 84-10604

Typesetting and printing: J. Kleindienst, Berlin; Brüder Hartmann, Berlin
Bookbinding: Lüderitz & Bauer, Berlin
2152/3020 − 5 4 3 2 1 0

# Preface

This text of applied chemistry considers the interface between chemistry and chemical engineering, using examples of some of the important process industries. Integrated with this is detailed consideration of measures which may be taken for avoidance or control of potential emissions. This new emphasis in applied chemistry has been developed through eight years of experience gained from working in industry in research, development and environmental control fields, plus twelve years of teaching here using this approach. It is aimed primarily towards science and engineering students as well as to environmentalists and practising professionals with responsibilities or an interest in this interface.

By providing the appropriate process information back to back with emissions and control data, the potential for process fine-tuning is improved for both raw material efficiency and emission control objectives. This approach also emphasizes integral process changes rather than add-on units for emission control. Add-on units have their place, when rapid action on an urgent emission problem is required, or when control simply is not feasible by process integral changes alone. Obviously fundamental process changes for emission containment are best conceived at the design stage. However, at whatever stage process modifications are installed, this approach to control should appeal to the industrialist in particular, in that something more substantial than decreased emissions may be gained.

This book may also be used as a general source of information and further leads to the details of process chemistry, or as a source of information relating to air and water pollution chemistry. Many references are cited to provide easy access to additional background material. The dominant sources cited may generally be recognized by the number of direct citations given in the chapter. Article titles are given with the citation for any anonymous material to aid in retrieval and consultation. Sources of further information on the subject of each chapter, but generally not cited in the text, are also given in a short Relevant Bibliography list immediately following the text. Tradenames have been recognized by capitalization, when known, and sufficient detail is mentioned or referenced to each of these to enable them to be followed up, if desired. It would be appreciated if any unrecognized tradename usage is brought to the author's attention.

# Acknowledgements

I am grateful to numerous contacts in industry and environmental laboratories who have willingly contributed and exchanged technical information included in this book. I would particularly like to thank the following who have materially assisted in this way: B. R. Buchanan, Dow Chemical Inc.; W. Cary, Suncor; R. G. M. Cosgrove, Imperial Oil Enterprises Ltd.; F. G. Colladay, Morton Salt Co.; J. F. C. Dixon, Canadian Industries Ltd.; R. W. Ford, Dow Chemical Inc.; T. Gibson, B. C. Cement Co.; G. J. Gurnon, Alcan Smelters and Chemicals Ltd.; D. Hill, B. C. Forest Products; J. A. McCoubrey, Lambton Industrial Society; R. D. McInerney, Canadian Industries Ltd.; R. C. Merrett, Canoxy, Canadian Occidental Petroleum; S. E. Moschopedis, Alberta Research Council; J. C. Mueller, B. C. Research; J. A. Paquette, Kalium Chemicals; J. N. Pitts, Jr., Air Pollution Research Centre, University of California; J. R. Prough, Kamyr Inc.; J. G. Sanderson, McMillan-Bloedel Ltd.; A. D. Shendrikar, The Oil Shale Corp.; J. G. Speight, Exxon; A. Stelzig, Environmental Protection Service; H. E. Worster, MacMillan-Bloedel Ltd. They have been credited wherever possible through their own recent publications. These contacts are especially valuable because of the notorious slowness of new industrial practice to appear in print.

I also thank all of the following individuals, each of whom read sections of the text in manuscript form, and C. G. Carlson who read all of it, for their valuable comments and suggestions that have contributed significantly to the authenticity of this presentation:

R. D. Barer, Metallurgical Division, Defence Research Establishment Pacific
G. Bonser, Husky Oil Limited
R. A. Brown, formerly of Shell Canada
M. J. R. Clark, Environmental Chemistry, Waste Management Branch, B. C. Government
H. Dotti, Mission Hill Vineyards
M. Kotthuri, Meteorology Section, Waste Management Branch, B. C. Government
J. Leja, Dept. of Mining and Mineral Process Engineering, University of British Columbia
L. J. Macaulay, Labatt Breweries of B. C., Ltd.
D. J. MacLaurin, formerly of MacMillan-Bloedel Ltd.
R. N. O'Brien, Department of Chemistry, University of Victoria
M. E. D. Raymont, Sulphur Development Institute of Canada
W. G. Wallace, Alcan Smelters and Chemicals, Ltd.
R. F. Wilson, Dow Chemical Canada Inc.
M. D. Winning, Shell Canada Resources Ltd.

But without the support of the University of Victoria, the Department of Chemistry, and my family to work within, this book would never have been completed. I owe a debt of gratitude to the inexhaustible patience of my wife, Diana, who handled the whole of the initial inputting of the manuscript into the computer, corrected several drafts, and executed all of the original line drawings. Thanks also go to K. Hartman who did the photographic work, to B. J. Hiscock and L. J. Proctor, who were unfailingly encouraging and helpful to adoption of the computer for manuscript preparation, even when the occasional seemingly hopeless scrambles occurred, and to L. G. Charron and M. Cormack, who completed the final manuscript.

Some of the line drawings and one photograph are borrowed courtesy of other publishers and authors, as acknowledged with each of these illustrations. To all of these I extend my thanks.

I am also grateful to the personnel at Springer-Verlag, in particular F. L. Boschke for his initial invitation and encouragement, and R. Stumpe, R. Sakolowski, and H. Schoenefeldt for their care and attention to detail during production.

# Table of Contents

# 1 Background and Technical Aspects of the Chemical Industry

## 1.1. Important General Characteristics

The business niche occupied by the chemical industry is of primary importance to the developed world in its ability to provide components of all the food, clothing, transportation, accommodation, and employment enjoyed by modern man. Most material goods are either chemical in origin or have involved one or more chemicals during the course of their manufacture. In some cases the chemical interactions involved in the generation of final products are relatively simple ones. In others, for instance for the fabrication of some of the more complex petrochemicals and drugs, more complicated and lengthy procedures are involved. But by far the bulk of all modern chemical processing uses raw materials naturally occurring on or near the earth's crust, as the raw materials for producing the commodities of interest.

It is instructive to consider, in an overview fashion, the sources of some of the common chemical raw materials and to relate these to the kinds of products that are accessible via one or two simple chemical transformations in a typical chemical complex. Even starting with the relatively few simple components such as air, water, salt (NaCl), and ethane, and some external source of energy which can be derived from coal, oil, natural gas, or hydroelectric power, quite a range of finished products is possible (Figure 1.1). While it is unlikely that all of these will be produced at any one location many will be, and all are based on commercially feasible and practised processes [1]. Thus, a company which is basic in the electrolytic production of chlorine and sodium hydroxide from salt can conveniently site itself on or near natural salt beds, which can provide a secure source of this raw material. Preferably this operation should also be sited near a large source of fresh water, such as a river or a lake, to provide for feedstock and cooling water requirements. Quite often an oil refinery forms a part of the matrix of companies which find it mutually advantageous to locate together. This can provide a supply of ethane or other hydrocarbon feedstocks. In this manner all the simple raw material requirements of the complex can flow smoothly into the production of more than a dozen products for sale (Figure 1.1).

A rapid rise in the numbers of chemicals produced commercially, and a steady growth in the uses and consumption of these chemicals historically (since the 1930-1940 period), has given the chemical industry a high growth rate relative to other industrial activities. In current dollars the average annual growth rate in the U.S.A. was about 11 % per year in the 1940's and just over 14 % per year through the 70's, seldom dropping below 6 % in the intervening period [2]. The production of plastics and basic organic chemicals have generally been the stronger performing sectors of the chemical industry as far as growth rate is concerned. Basic inorganic chemicals production, a generally "mature" area of the industry, has showed slower growth [3]. World chemical export growth has been strong too, having averaged just over a 17 % annual growth rate over the 1968-1978 interval [4]. But growth rates based on current dollar values, such as these, fail to recognize the salutary influence of inflation on these apparent rates of growth. Using a constant value dollar, and smoothing the values over a 10-year running average basis gives the maximum for the real growth rate of about 9 % per year occurring in 1959, tapering down to about 1-3 % per year for 1982 [2]. The slowing of the real growth rate in recent years may be because the chemical industry generally is maturing or stabilizing. There may also have been a contribution over the more recent short term from the global business recession of that period.

Most of the machinery and containment vessels required for chemical processing are costly, in part because of the high degree of automation used by this industry. This in turn means that, based on the value of products, the labour requirement is relatively low. Put in another way, in the U.S.A. the investment in chemical plant per employee has amounted to about $ 30,000 per worker at the time when the average for all manufacturing stood at $ 14,000 per worker. In the U.K. this ratio of capital investment per employee in the chemical industry versus the investment by all manufacturing is

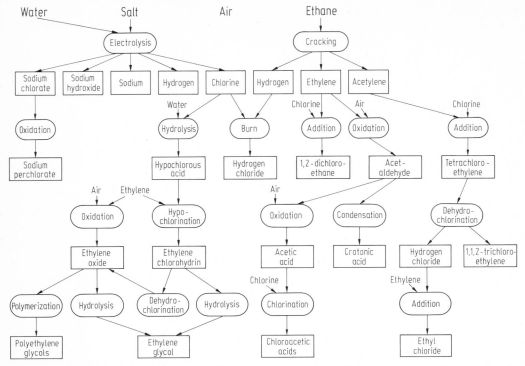

**Figure 1.1.** Flow sheet of a hypothetical though credible chemical complex based on only air, water, salt, and ethane raw materials

very similar to the experience in the U.S.A. In 1963 these figures stood at 7,000, and 3,000 pounds, and in 1972, 17,000 and 7,000 pounds respectively [5].

Yet another way of considering the relationship of investment to the number of employees is in terms of the "value added per employee". The value added, which is defined as the market price of a good minus the cost of raw materials required to produce that good [6, 7], is a measure of the worth of processing a chemical in terms of its new (usually greater) value after processing than before. When the gross increase in value of the products of a chemical complex is divided by the numbers of employees operating the complex, one arrives at a "value added per employee", a sort of productivity index. Considered on this basis the productivity of a worker in the chemical industry is at the high end of the range, in comparison with the productivity of all workers within any particular country. There are, however, quite significant differences in relative productivity when the value added per em-

ployee of one country is compared with the same figure from other countries. In 1978, the value added for the U.S.A. stood at $ 58,820 per employee per year, as compared to a value of $ 17,800 for Spain, the extremes of the range among the countries compared (Table 1.1). This comparison also amounts to a reflection of the much higher investment per employee and higher degree of automation generally used by American chemical companies versus their Spanish counterparts. The range of values given here is, however, also dependent on a number of other factors such as scale, and capacity usage rates etc. which have not yet been discussed. Relative positions may also change in a ten year span, such as shown by Canada and West Germany, from differing investment rates and other factors.

The products of the chemical industry, because of the more or less steady stream of better performing replacement products, tend to be subjected to a high rate of obsolescence. During the 50's and 60's most of the sales volume increase noted by chemi-

**Table 1.1** Numbers employed in chemicals production, and value added per
employee for selected countries[a]

| Country | Thousands employed, 1978 | Value added per employee, US$ 1968 | 1978 | Decade productivity growth factor |
|---|---|---|---|---|
| Austria | 61 | — | 19 670 | — |
| Belgium | 62.6 | 8 410 | 37 100 | 4.4 |
| Canada | 84.7 | 18 020 | 39 130 | 2.2 |
| France | 305 | 9 540 | 32 540 | 3.4 |
| West Germany | 548 | 11 350 | 46 220 | 4.1 |
| Italy | 292 | 7 890 | 20 000 | 2.5 |
| Japan | 470 | 9 460 | 33 600 | 3.6 |
| Netherlands | 87.1 | 9 810 | 44 100 | 4.5 |
| Norway | 17.4 | 8 290 | 23 800 | 2.9 |
| Spain | 144.1 | 5 780 | 17 800 | 3.1 |
| Sweden | 39.7 | 12 180 | 36 500 | 3.0 |
| United Kingdom | 467 | 8 110 | 19 800 | 2.4 |
| U.S.A. | 1088 | 24 760 | 58 820 | 2.4 |

[a] Data from [48], and data from The Chemical Industry, OECD, Paris, 1970 and 1981.

cal companies for the period was from products developed in the preceding 15 years [8]. But to provide the steady stream of improved products required to maintain this record and still grow, requires a significant commitment to research in order for a company to keep up with its competition. This requirement also provides much of the incentive for a chemical company to employ and provide facilities for the employment of chemists, engineers, biologists, and other professionals to ensure the continuing discovery and development of new products.

From 2.5 to 3.5 % of the value of sales of U.S. chemical companies is spent on research and development activity, about the same proportion of sales as spent by all industry [9]. West German companies tend to place a somewhat greater emphasis on research and development, showing a range of investment of 4 to 5 % of sales in this activity in the 1969-1979 interval [10]. Drug (pharmaceutical) companies represent the portion of the chemical sector which spends the largest fraction of sales, about 6 %, on research and development programs [11]. This is probably a reflection of both the generally higher rate of obsolescence of the products of this sector as compared to chemical industry as a whole, as well as of the greater costs involved in bringing new, human use drugs to market, as opposed to new commodity chemicals.

## 1.2 Types and Significance of Information

With the still existing moderately high growth rate of the chemical industry and its high rate of obsolescence of both the products themselves and of processes leading to an existing product, the competition in this industrial area is vigorous. Technological and market success of a company in this industry area is a composite of the financial resources, raw material position, capabilities and outlook of staff, and information resources that the company has at its disposal. The last of these factors, the information resource, is a particularly important one for the chemical processing industry. Information, or "know-how", may be derived from prior experience of all kinds. It may be generated from self-funded and practised research or process development, or it may be purchased from appropriate other companies if it is already available in this form. In this way, even if the results of research by a company are not used by that company itself to produce a product, they may still produce an income for the originating company in the form of licensing agreements, royalties per unit of product sold and the like. In many ways this is a highly desirable component of a company's earnings since it does not require any capital investment or raw material(s) and product(s) inventory as ordinarily required to generate an income from chemical processing.

Patents, and the patenting system generally represent the orderly system of public documents used in most parts of the world to handle much of this kind of information which may be of interest to chemical as well as other companies. They must be applied for in each country in which protection is desired, since the subject of a patent may be practised and the product sold without license in any country in which this precaution was not taken. "Composition of matter" patents, which relate primarily to newly-discovered chemical compounds, are issued on successful application by an inventor (individual or company). Utility, i.e. some type of useful function of the compound must be demonstrated before a patent application of this type can be filed. In return this class of patent provides the best kind of protection for a new compound because the compound itself is protected from its sale by others, regardless of the synthetic route developed to produce it.

"Process" patents are used to protect a new process, or refinements to an established process which is employed to produce an existing compound. This type of patent also provides useful protection against the commercial use by others of an improved, completely distinct process which may be developed by a company. The benefit provided by all types of process development may be lower costs achieved from higher conversion rates, or from better conversions, more moderate operating conditions and the like and in one or more of these ways provides the company with an economic advantage to practise this improvement. There are other patent areas that are used by chemical and other companies, such as those covering machines and registered designs, trademarks and symbols, and copyrights, but these are generally less fundamental to the operations of chemical companies than the composition of matter and process patent areas [12]. These areas are generally of more importance for sales, product recognition being a significant marketing factor.

The patent, in printed form, comprises a brief description of the prior art (the narrow segment of technology) in areas related to the subject of the patent. Usually this is followed by a brief summary of what is being patented. A more detailed description of what is involved in the invention is then given, accompanied by descriptions of some detailed examples which illustrate the application of the invention. Usually at least one of the examples described is a description of an experiment which

was actually carried out, but they need not all have been actually tested. Differentiation between actually tested examples and hypothetical examples described in the body of the patent is made on the basis of the tense used in the description. If it is described in the past tense, i.e. "was" used throughout, then it is a description of a tested example. If it is given in the present tense then it describes a hypothetical example. To be able to differentiate the two types of examples is of particular interest to synthetic chemists, for example, who are likely to be more successful if they follow a procedure of a tested rather than a hypothetical example. The last, and most important part of a patent is the claims section. Here, numbered paragraphs, each of which by custom is written all in one sentence [13], cover the one or more novel areas to be protected by the patent in order of importance. In the case of any contest of the patent by other parties, these claims must be disproved in reverse order, i.e. the last and least important claim first followed by the others if the last claim is successfully contested.

The granting of a patent confers on the holder a time-limited monopoly in the country of issue, for a period of 17 years, re the novel composition of matter or advance in the art that is claimed by the patent. During this time the company or individual may construct a plant using the patented principles, which may take 6 or 7 years. Once production has started a product can be marketed from this plant at a sufficiently high price that the research and development costs involved prior to patenting, as well as reasonable plant write-off expenses and the like, may be met. This stage of marketing is managed without competition from others for the 10-11 years remaining from the original patent interval. Or it may choose to license the technology and collect product royalties from another interested company. Or it may follow both options simultaneously, if it reasons that the market will be large enough. For these reasons the patent system encourages a company to carry out its own research since it provides a reasonable prospect of the company being able to recover its early development costs while it is using the new art protected from competition.

Seventeen years (20 years in European countries) [14] from the date of issue of a patent, however, the subject matter of the patent comes into the public domain, that is it becomes open to any other person or company who wishes to practise the art described in the patent and *sell* a product based on this technology. At this time, the price of the product will normally fall somewhat as the product starts to be produced competitively by others. But the originating company still has some production and marketing advantages from its longer experience in using the technology, from having one or more producing units which may be largely paid for by this time, and from having already developed some customer confidence and loyalty.

The increase in new regulatory requirements which must be met before marketing new drugs and pesticides are now taking extremely long to satisfy, up to 7 years in some instances. This has increased the new product development costs, at the same time as decreasing the period of time available for monopoly marketing to allow recovery of development costs. Realization of this has led to moves in the U.K. [15] and in the U.S.A. [16, 17] to extend the period of monopoly protection granted by the patent by the length of time required by a company to obtain regulatory clearance. These moves should at least encourage maintenance of the current level of research and development effort by companies even if it does not increase innovation [18].

In all cases patent protection for an idea is for a limited time, but even during the protected time the information in the patent becomes public knowledge. There may be some technological developments which a company wishes to keep completely to itself, or which are so close to prior art (already practised) that there is some doubt of the likelihood of success of a patent application. Information falling into these categories may be simply filed in company records for reference purposes and not be patented or otherwise publicized at all. This type of "know-how" is classed as proprietary information, useful to the company but kept within the company only. Agreements signed by all new employees working for a company ensure that this proprietary information does not become public knowledge. In return for risking possible eventual leakage and use of this information by others, the company gains the advantageous use of the information in the meantime, it saves patenting costs (even if feasible), and it avoids the certainty of public disclosure on issuance of a patent covering this information. But the ideas involved are not directly protected from use by others, whether or not the knowledge is lost via "leaks" or via independent discovery by a second company working on the same common knowledge premises as the first

company. Hence the value of the patent system in providing this assurance of protection.

## 1.3 The Value of Integration

Integration, as means of consolidation by which a company may improve its competitive position, can take a number of forms. Vertical integration, one of these, can be "forward" to carry an existing product of the company one or more stages closer to the final consumer. For instance a company producing polyethylene resin may decide to also produce film from this resin, for sale, or it might decide to produce both film, and garbage bags from the film. By doing this, more "value-added" manufacturing stages are undertaken within the company, and if these developments are compatible with the existing activities and markets of the company can significantly enhance the profitability of its operations.

Vertical integration may also be "backward" in the sense that the company endeavours to improve its raw material position by new resource discoveries and acquisition, or by purchase of resource-based companies strong in the particular raw materials of interest. Thus, it can explore for oil, or purchase an oil refinery to put itself into a secure position for ethane and ethylene. Or it can purchase land overlying beds of sodium chloride or potassium chloride, or near sodium sulfate rich waters and develop these to use for the preparation of existing product lines. Either of these routes of backward integration can help to secure for a company an assured source of supply and stable raw material pricing, both helpful in strengthening the reliability of longer term profit projections.

Horizontal integration is the further type, in which the technological, or information base of the company is applied to improve its competitive position in this and related areas. When a particular area of expertise has been discovered and developed this can be more fully exploited if a number of different product or service lines are put on the market using this technology. As examples, Procter and Gamble and Unilever, among their areas of activity, have both capitalized on surfactant technology in their development of a range of washing and cleaning products. Surface activity of different types has also been exploited by the Dow Chemical Company with its wide range of ion exchange resins, and by Union Carbide with its molecular sieve-based technology. It can be seen from these examples that judicious application of one or more of these forms of integration can significantly strengthen the market position and profitability of a chemicals based company.

## 1.4 The Economy of Scale

The size or scale of operation of a chemical processing unit is an important competitive factor since, as a general rule, a large scale plant operating at full capacity can produce a lower unit cost product. This is the so-called "economy of scale" factor. How does this lower cost product from a larger plant arise? Firstly, the labour cost per unit of product is lower for a very large than for a small plant. This comes about because proportionally less men are required per unit of product to run a 1,000 tonne per day plant than a 100 tonne per day plant. Secondly, the capital cost of the plant per unit of product is lower if the plant is operating at full capacity.

Reduced labour costs result from the fact that if one man is required to control for example raw material flows into a reactor in a 100 tonne per day plant, in all likelihood one man can still control these flows in a 1,000 tonne per day plant. In fact an empirical expression has been derived by correlation of more than 50 types of chemical operations which, knowing the labour requirement for one size of plant, allows one to estimate with reasonable assurance the labour requirement for another capacity [19] (Equation 1.1).

$$M = M_0 (Q/Q_0)^n, \text{ where} \tag{1.1}$$

M   is the labour requirement for plant capacity Q of interest,

$M_0$   is the known labour requirement for a plant capacity $Q_0$, and

n   is the exponent factor, normally about 0.25, for the estimation of labour requirements.

If 16 men are required to operate a 200 tonne/day sulfuric acid plant, this expression allows us to determine that only about 24 men $(16 \times (1,000/200)^{0.25})$ should be needed to operate a 1,000 tonne/day plant. Thus, when operating at full capacity, the larger plant would only have three tenths the labour charge of the smaller plant, per unit of product.

The lowered plant capital cost per unit of product comes about because of the relationship of capital costs of construction to plant capacity, which is an exponential, not a linear relation (Equation 1.2).

$$\text{capital cost } \alpha \text{ (plant capacity)}^{2/3} \tag{1.2}$$

The approximate size of the fractional exponent of this expression results from the fact that the cost to build a plant varies directly as the area (or weight) of metal used, resulting in a square exponent term [20]. At the same time the capacity of the various components of the processing units built increases in relation to the volume enclosed, or a cube root term. Hence, the logic of this approximate relationship.

In actual fact, a skilful design engineer is generally able to shave just a bit off this descriptively derived exponent, making capital cost relate to scale more closely in accord with Equation 1.3 for whole chemical plants.

$$\text{capital cost } \alpha \text{ (plant capacity)}^{0.60} \tag{1.3}$$

In order to use Equation 1.3 to estimate the capital cost of a larger or smaller plant, when one knows the capital cost of a particular size of plant, one has to insert a proportionality constant [21] (Equation 1.4).

$$C = C_0(Q/Q_0)^n \tag{1.4}$$

C   is the capital cost for the production capacity Q of the plant to be determined,

$C_0$   is the known capital cost for production capacity $Q_0$, given in the same units as C, and

n   is the scale exponent, which is usually in the 0.60 to 0.70 range for whole chemical plants.

Thus, if it is known that the capital cost for a 200 tonne/day sulfuric acid plant is $1.2 million ($1.2 mm) then using this relationship it is possible to estimate that the capital cost of an 1,800 tonne/day plant will be somewhere in the range $4.49 mm to $5.59 mm (Equation 1.5).

$$C = \$ 1.2 \text{ mm } (1800/200)^{0.60} \tag{1.5}$$
$$= \$ 1.2 \text{ mm } (3.7372)$$
$$= \$ 4.485 \text{ mm}$$
$$C = \$ 1.2 \text{ mm } (1800/200)^{0.70}$$
$$= \$ 1.2 \text{ mm } (4.6555)$$
$$= \$ 5.587 \text{ mm}$$

From construction cost figures the actual capital cost of construction of an 1,800 tonne/day sulfuric plant is about $5.4 mm, when taken at the time of these estimates [1]. This figure agrees quite well with the two values estimated from the known cost of the smaller sized plant.

Of course, if one has recent capital cost information on two different sizes of plant for producing the same product, this can enable a closer capital cost estimate to be made by determination of the value of exponent n from the slope of the capital cost versus production volume line plotted for the two sizes of plant. For the particular example given, this experimentally determined exponent value would be 0.685. It should also be noted that this capital cost estimation method is less reliable for plant sizes more than a decade larger or smaller than the plant size for which current costs are available. [22].

From a comparison of the foregoing capital cost figures, it can be seen that 9 times as much sulfuric acid can be made for a capital cost of only 3.7 to 4.7 times as much as that of a 200 tonne per day plant. Obviously if the large plant is operated at full capacity, the charge (or interest) on the capital which has to be carried by the product for sale by the larger plant is only about half (4.7/9.0), or even less than half (3.7/9.0) of the capital cost required to be borne by the 200 tonne/day plant, per unit of product.

To make the decision regarding the size of plant to build in any particular situation careful consideration has to be given to product pricing, market size and elasticity, and market growth trends. Also it is a useful precaution to survey the immediate geographical area and the local public construction announcements for any other plans for a plant to produce the same product. The final decision should be based on a scale of operation which,

within a period of 5 to 7 years can reasonably be expected, in the face of any existing or projected competition, to be running at full capacity. That is, it should be possible to stimulate a sufficient market, within this period of time, to sell all of the product that the plant can produce. If the final size of plant built is too small, not only are sales restricted from inadequate production capacity but also the profit margin per unit of product is smaller than it potentially could have been if the product were being produced in a somewhat larger plant. If the final result is too large, and even after 10 years or so the plant is only required to operate at 30 % of capacity to provide for the whole market, then the capital and generally also labour costs per unit of product become higher than they would have been with say $1/2$ or even $1/4$ of the plant size. In this event, planning too optimistically can significantly *decrease* the profitability of the operation [23]. It is the significance of decisions such as these as to the financial health of a chemical company that justify the handsome salaries of its senior executives.

One remaining point to consider regarding scale is that the capital cost exponential factor of 0.60 to 0.70 relates to most whole plants. If considering individual processing units this factor can vary quite widely (Table 1.2). With a jaw crusher, for example, a unit with three times the capacity costs 3.7 times as much. Obviously here scaling up imposes greater capital costs per unit of product for a larger than for a smaller unit. But other associated costs may still be reduced. A steel vent stack of three times the height costs about three times as

**Table 1.2** Typical values for the exponent scale factor and how these relate to the cost of the unit for particular types of chemical processing equipment [a]

| Type of equipment | Typical value of exponent n | Cost factor for three times scale |
|---|---|---|
| jaw crusher | 1.2 | 3.74 |
| fractionating column, bubble cap | 1.2 | 3.74 |
| steel stack | 1.0 | 3.00 |
| fractionating column, sieve tray | 0.86 | 2.57 |
| forced circulation evaporator | 0.70 | 2.16 |
| shell and tube heat exchanger | 0.65 | 2.04 |
| jacketed vessel evaporator | 0.60 | 1.93 |
| stainless steel pressure reactor, 20 bar | 0.56 | 1.85 |
| industrial boiler | 0.50 | 1.73 |
| drum dryer, atmosperic pressure | 0.40 | 1.55 |
| storage tank | 0.30 | 1.39 |

[a] Exponent values for use with Equation 1.4, $C = Co(Q/Qo)^n$, and selected from those of [21 and 22].

much, i.e. there is no capital cost economy of scale here, but these capital cost increases with height may still have to be borne by the plant. However for very simple components of processing units such as storage tanks, the value of this exponent is small, about 0.30, which allows a tank of three times the capacity to be built for only about 1.4 times the price. Thus, a composite of the scale-up exponent factors for individual units averages out to the 0.60 to 0.70 range for a whole chemical plant.

## 1.5  Chemical Processing

The chemical side of the chemical process industries is concerned with the change of raw materials into products by means of *chemical conversions*. Since it is very seldom that a single reacted starting material gives only pure product it is also usually necessary to use any of a number of physical separations, such as crystallization, filtration, distillation, phase separation etc. to recover product(s) from the unreacted starting materials and byproducts. Byproducts are materials other than product which are obtained from reacted starting materials. These physical separation processes are often called *unit operations* to distinguish these definable steps, similar features of which may be compared from process to process, from the chemical conversion aspect of a process [24]. The combination of the chemical conversion step, with all of the unit operations (physical separations) that are required to recover product resulting from the chemical conversion, is collectively referred to as a *unit process*.

Unit processes may be carried out in single-use (dedicated) equipment, which is used solely for generating the particular product for which it was designed. Or, they may be carried out in multi-use equipment that is used in sequence to produce first one product, followed by the production of one or more related products that have similar unit process requirements, in a series after this. Single use equipment is invariably used for large scale production, when 90 % or so of full time usage rates are required to obtain sufficient product to satisfy the market requirements. Multi-use equipment more often is chosen for small scale production, and particularly for more complicated processes such as required for the manufacture of many drugs, dyes, and some specialty chemicals.

Proper materials of construction particularly with regard to strength and toughness, corrosion resistance, and cost must all be kept in mind at the design stage for construction of a new chemical plant [25]. Early experiments during the conception of the process will have generally been conducted in laboratory glassware.Even though glass is almost universally corrosion resistant (and transparent, and thus useful in the lab) it is unduly fragile for most full scale process use. Mild steel is used wherever possible, because of its lowest cost and ease of fabrication [26]. But it is not, however, resistant to attack by many kinds of process fluids or gases. In these cases any of titanium, nickel, stainless steel, brass, Teflon, polyvinylchloride (PVC), wood, cement, and even glass (usually as a lining) among other materials may be used to construct components of a chemical plant. The final choice of construction material is based on a combination of experience and accelerated laboratory tests [27, 28]. Small coupons of the short-listed candidate materials are suspended in mock prepared process mixtures which are then heated to simulate anticipated plant conditions. These preliminary tests will be followed by further tests during small scale process test runs in a pilot plant, wherever possible. Even when the final full scale plant is completed, there may still be recurring corrosion failures of a particular component which may require construction material changes even at this stage of development [29, 30].

### 1.5.1  Types of Reactor

In considering industrial reactor types again the analogy between laboratory manipulations and a full scale production plant can be usefully applied. Very often in the laboratory a synthesis will be carried out by placing all the required reactants in the flask and then imposing the right conditions, heating, cooling, light etc. on the contents until the desired extent of reaction has been achieved. At this stage the contents of the flask are emptied into another vessel for the product recovery steps to be carried out. Operating an industrial process in this fashion, which can be done, is termed a *batch* process or batch operation. Essentially this situation is obtained when all starting materials are placed in the reactor at the beginning of the reaction, and all materials remain in the vessel until the reaction is over, when the contents are removed. This mode of operation is the one generally favoured for smaller scale processes, for multiple use equipment, or for new and untried, or some types of more hazardous reactions.

On the other hand an industrial process may be operated in a *continuous,* rather than in a batch mode. To achieve this, either a single or a series of interconnected vessels may be used. The required raw materials are continuously fed into this vessel or vessels and the reaction products continuously removed so that the volume of material in the reactor stays constant as the reaction proceeds. The concentrations of starting materials and products in the reactor eventually reach a steady state. Instead of one or more tanks being used to conduct the continuous process, a pipe or tube reactor may be used, in which case the reaction time is determined by the rate of flow of materials into the tube divided by the length of the pipe or tube.

Since, in general, the labour costs of operating a large scale continuous process are lower than for a batch process, most large scale industrial processes are eventually worked in a continuous mode [31]. However, because of the more complicated control equipment required for continuous operation, the capital cost of the plant is usually higher than for the same scale batch process. Thus, the final choice of the mode of operation to be used for a process will often depend on the relative cost of capital versus labour in the operating area in which the plant is to be constructed. Most developed countries opt for a high degree of automation and higher capital costs in new plant construction decisions. For Third World nations, however, where capital is generally scarce and labour is low cost and readily available, more manual, simpler, batch-type operations will often be the most appropriate [32-36]. Smaller scales of operation will generally be more than adequate to supply the generally smaller market requirements in these economies. Also the more straightforward operating, maintenance, and repair operations for such a plant are more easily accomplished under these circumstances than would be possible with the more complex control systems of a continuous reactor configuration.

There are several common combinations within this broad division into batch, and continuous types of reactor which use minor variations of the main theme. The simplest and least expensive of these subdivisions is represented by the straight batch reactor, which is frequently just a single stirred tank. All the raw materials are placed in the tank at the start of the process. There is no flow of materials into or out of the tank during the course of the reaction, i.e. the volume of the tank contents is fixed during the reaction (Table 1.3). There also usually will be some provision for heating or cooling of the reacting mixture, either via a metal jacket around the outside of the reactor or via coils placed inside the reactor, through which water, steam or heat exchange fluid may be passed for

**Table 1.3** A qualitative comparison of some of the main batch and continuous types of liquid phase

| Type of reactor | Illustration of concept | Uniformity of | | |
|---|---|---|---|---|
| | | Composition | | Temperature throughout process[b] |
| | | with time | within reactor[a] | |
| a. Batch | | no | yes | no |
| b. Semi-batch | feeds | no | yes | yes |
| c. Continuous stirred tank(CSTR) sometimes "tank flow reactor" | feeds / product | yes | yes | yes |
| d. Multi-stage CSTR | feeds / product | yes | partly | partly |
| e. Tubular flow, sometimes "pipe reactor", or "plug flow reactor" | feeds / product | yes | no | no |

[a] Meaning the composition within the reactor at any particular point in time.

[b] Referring to temperature constancy during the whole of the reaction phase.

temperature control. However, the temperature is not usually uniform in this situation since initially the concentrations and reaction rate of the two (or more) reactants are at a maximum, which will more or less rapidly taper off as the reaction proceeds. Thus, heat evolution (or uptake) is going to be high initially and then gradually subside coinciding with a slowing of the reaction rate. At the end of the reaction the whole of the reactor contents is pumped out for product recovery.

A semi-batch reactor is a type of batch configuration used particularly for processes which employ very reactive starting materials. Only one reactant, plus solvent if required, is present in the reactor at the start of the reaction. The other reactant(s) is then added gradually to the first, while continuing stirring and controlling the temperature. By controlling the rate of addition of one reactant in this way, the temperature of the reacting mixture may be kept uniform as the reaction proceeds.

Continuous reactor configurations are generally favoured for very large scale industrial processes. If the process is required to produce only 2 mm (two million) kg/year or less, generally the economics of construction will dictate that a batch process be used [37]. If, however, the process is called upon to produce 9 mm kg/year or more there is usually a strong incentive to apply some type of continuous reactor configuration in the design of the production unit.

The stirred tank is used as the main element of the simplest type of continuous reactor, the Continuous Stirred Tank Reactor (CSTR). Continuity of the process is maintained by continuous metering in of the starting materials in the correct proportions, and continuous withdrawal of the product from the same, well-stirred vessel. In this way the concentration and temperature gradients shown by simple batch reactors are entirely eliminated (Table 1.3). This type of continuous reactor is good for slow reactions, in particular, since it is a large, simple, and cheap unit to construct. However, reactors of this CSTR type are inefficient at large conversions [37]. For a process proceeding via first order kinetics and requiring a 99 % conversion a 7 times larger reactor volume is needed than if only 50 % conversion is desired. A solution to this volume requirement to obtain high conversions is to use two or more continuous stirred tank reactors (CSTR's) in series [37]. Using two CSTR's in series allows the first reactor to operate at some intermediate degree of conversion, the product of which is then used as the feed to the second reactor to obtain the final extent of reaction desired (Figure 1.2). From the diagrams it can be readily seen that the total reactor volume required to achieve the desired final degree of conversion is significantly reduced over the volume required to achieve the same degree of conversion in a single reactor. Carrying this idea further, it can also be seen that increasing the number of CSTR's operating in series to three or more units contributes further to the space-time yields and allows further reductions in reactor volume to be made and yet still obtains the same final degree of conversion. Therefore multiple CSTR's operating in series allow either a reduction in the total reactor volume used to obtain the same degree of conversion as with a single CSTR, or a higher degree of conversion for the same total reactor volume, or of course a suitable mix of both of these attributes. In either case the engineering cost to achieve these changes is in the additional

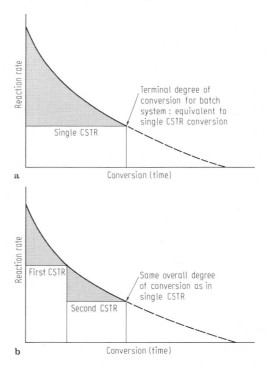

**Figure 1.2.** A comparison of the "space-time" yields, or saving in reactor volume, achieved by carrying out a continuous process in a single stirred tank reactor, versus two continuous stirred tank reactors (CSTR's) operating in series, to the same degree of conversion [37]

connecting piping, and fluid and heat control systems required for multiple units, over a single reactor. This generally is the factor that limits the extent to which multiple reactors are economic to use for improvement of process conversions.

Taking the multiple CSTR concept to its logical extreme, a very large number of series-connected and small tank reactors can be likened to carrying out the same process in a very long, narrow-bore tube, referred to as a *tubular flow,* or *pipe reactor.* By placing sections, or all of this tube in externally heated (or cooled) sections, any desired processing temperature of the fluid mixture flowing inside the pipe can be obtained. If a high flow rate is used, or if in-line mixers (streamline flow splitters employing an interacting series of baffles) are used, components of even immiscible mixtures may be made to intimately interact from turbulent flow as they move down the tube. So usually no external mixing is necessary to obtain good contact between the reacting materials moving through a pipe reactor. Turbulent flow also helps to ensure good heat transfer between all parts of the liquid flowing inside the tube, and the tube wall. If the temperature control fluid or gas flowing outside the tube is also moving vigorously, the temperature difference between this external heat transfer medium and the contents of the tube is also kept to a desirable minimum.

The concentrations of starting materials flowing in a tube reactor decreases, and the concentration of product increases, as the mixture flows down the tube and the reaction proceeds. Thus reaction times for the raw materials flowing into a tube reactor can be calculated from the relation (distance from the inlet)/(reactant velocity). The induced turbulence in the tube occurs mostly in the cross sectional dimension and very little along the length of the tube, i.e. there is little or no "backmixing", or mixing of newly entering raw materials, with raw materials that have already reacted for some time. This feature has led to the names *plug,* or *plug flow* reactor as other descriptive synonyms for tubular flow or pipe reactor.

Most industrial processes using the interaction of fluids to obtain chemical changes can be classified into one, or sometimes more of the preceding five liquid reactor types. There are variations on these themes which are used for gas-gas, gas-liquid, or gas-solid reactions, but parallels of many of the processing ideas used for liquid-liquid reactors are also applicable in these situations [38, 39].

## 1.5.2 Fluid Flow Through Pipes

To understand the mechanism of the turbulent mixing process occurring in pipe reactors we have to consider first some of the more general properties of fluid flow in pipes. The resistance to fluid flow in a pipe has two components, the viscous friction of the fluid itself within the pipe, which increases as the fluid viscosity increases, and the pressure differential caused by either a liquid level and vessel height difference or a pressure difference between the two vessels.

At relatively low fluid velocities, and particularly for a viscous fluid (where turbulence is damped) in a small pipe one will normally obtain streamline flow of the fluid within the pipe (Figure 1.3 a). Under these conditions the fluid is in a continuous state of shear with the fastest flow in the center of the pipe and low to zero flow right at the wall. The fluid velocity profile, along a longitudinal section of the pipe, is parabolic in shape.

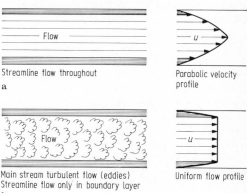

Figure 1.3. Fluid flow characteristics and profiles of fluid flow in pipes [40]. **a** At low Reynolds numbers, where streamline flow is obtained throughout cross section. **b** At high Reynolds numbers, where turbulent flow is obtained for most of pipe volume. Streamline flow is only obtained in a thin boundary layer adjacent to the pipe wall where the influence of the wall and viscous forces prevent turbulence

At high fluid velocities, and particularly for low viscosity fluids in large diameter pipes, small flow disturbances create eddies in the fluid stream which fill the whole of the cross sectional area of the pipe (Figure 1.3 b). Only a residual boundary layer against the inside wall of the pipe will maintain streamline flow under these conditions. This turbulent condition is in fact the more usual one for fluid movement in pipes since smaller pipes, which cost much less, may be used when high fluid velocities are used. The cost saving obtained by using smaller pipe generally far exceeds the small pumping cost increase required to achieve the higher fluid velocities. There are also other reasons for this [40].

The development of turbulent flow depends on the ratio of viscous to inertial (density and velocity) forces, a ratio known as the Reynolds Number, $R_e$ (Equation 1.6).

$$R_e = \frac{u\varrho d}{\mu}, \text{ where }\quad \begin{aligned} u &= \text{fluid velocity} \\ \rho &= \text{fluid density} \\ d &= \text{pipe diameter} \\ \mu &= \text{fluid viscosity} \end{aligned} \qquad (1.6)$$

Either metric or English units, e.g. g/cm/sec or lb/ft/sec may be used for substitution, as long as the usage is consistent. In either case, as long as consistency of units has been maintained, the result comes out to the same, dimensionless (unitless) Reynolds number. The lack of dimensionality of this value is expected for any pure ratio. In practice, whenever the Reynolds number for fluid flow in a pipe exceeds about 2,100, one obtains turbulent flow [40]. However, the division between streamline and turbulent flow situations is also somewhat dependent on other factors than those included in the Reynolds number calculation, such as the proximity of bends and flow-obstructing fittings, and the surface roughness of the interior of the pipe. So normally a Reynolds number range is given for the dividing line between streamline and turbulent flow. If it is 2,000 or less, this is indicative of a streamline flow situation. If it is 3,000 or more, turbulence is highly likely [41].

Tubular or pipe reactors are designed to take advantage of this phenomenon to obtain good mixing. In effect, this means that relatively small bore tubes, and relatively high flow rates are used for this type of continuous reactor. It should also be kept in mind that this dependence of good mixing on high flow rates may in some instances set a lower limit on the fraction of the design production rate at which the plant can operate. Turbulent flow of the raw materials in the pipe not only contributes good mixing, but also assists in maintaining good heat transfer conditions through the pipe wall separating the reactants flowing in the pipe from the jacketing fluid.

## 1.5.3  Controlling and Recording Instrumentation

Sensors of many kinds are needed to measure the process parameters important for effective operation of any kind of chemical conversion. The principles of manual or automated process control require first of all, a determination of the appropriate variable or variables which need to be measured, temperature, pressure, pH, viscosity, water content, etc. in order to know the progress of the reaction or separation process. For the measured variable to be significant in the control of the process it must represent a control parameter, such as steam flow, pump speed, acid addition rate and the like which, when altered will cause a response in the measured variable. Finally, there must be some actuating mechanism between the variable which is sensed and the process condition which requires adjustment, in order for the measured variable to be useful for the control of the process. In relatively simple processes and where labour costs are low the actuation may be a person who reads a dial or gauge, decides whether the parameter is high, low, or within normal range, and if necessary adjusts a steam valve or pump power flow to correct the condition. For automated plants the means of actuation may be a mechanical, pneumatic, electric, or hydraulic link between the sensor and the controlled parameter.

The amounts of materials fed to a chemical reaction are usually sensed by various types of flowmeters (Figure 1.4). The proportions of raw materials reacting are known from metered flow rates, the altering of which, in turn, may be used to obtain the correct ratio of raw materials moving into the reactor. Control of liquid flow rates is usually achieved via valves, in the early days by means of an on or off option. Today, flow control as well as many other process variables are designed to be proportional, that is there is control of the degree to which a variable may be altered, not simply an all or nothing situation. Valves may be set to a variety of different flow rates, pump speeds may be altered, conveyors moving solids may be made to feed at different rates and heat input as electrical energy or steam, or cooling water flows may by varied at will. The development of proportional controls of this kind for more process variables has greatly improved the degree of refinement of process operation now possible. Occasionally raw material quantity measurement is

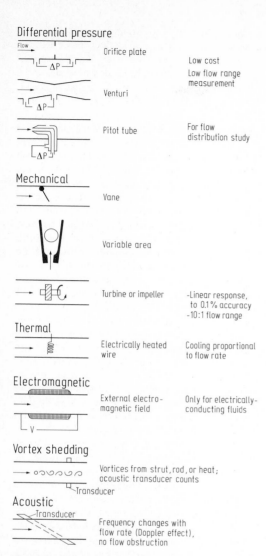

Differential pressure

Flow

Orifice plate

Low cost
Low flow range
measurement

Venturi

Pitot tube

For flow
distribution study

Mechanical

Vane

Variable area

Turbine or impeller

-Linear response,
 to 0.1% accuracy
-10:1 flow range

Thermal

Electrically heated
wire

Cooling proportional
to flow rate

Electromagnetic

External electro-
magnetic field

Only for electrically-
conducting fluids

Vortex shedding

Vortices from strut, rod, or heat;
acoustic transducer counts

Transducer

Acoustic

Transducer

Frequency changes with
flow rate (Doppler effect),
no flow obstruction

**Figure 1.4.** Some types of flow measuring devices used in chemical processing [59-61]

carried out using measuring tanks or bins, with a float, sight tube, electrically activated tuning fork or the like as level controls, or even mass measurement of the bin plus contents. These devices are more like the usual laboratory methods used for mass proportioning of reactions.

Actuation of process controls in response to the measured process variables has commonly been, and still is to a significant extent pneumatic (low pressure air) because of the reliability and inherent

ignition safety aspect of this system [42]. However, the greater ease of computer interfacing possible with electric or electronic actuation, and the now improved reliability and safety of these systems have contributed to the growth in the use of these actuating methods. Increasing use of computer-based technology for plant automation has stimulated this trend and at the same time has helped further to smooth process operations and improve yields and product quality while providing a savings in labour costs.

Automation of plant control using a computer to match ideal process parameters to the readings being taken from measured process variables, allows close refinement of the operating process to the ideal conditions (Figure 1.5). Manually, a process reading may be compared to the ideal condition every hour or half hour as a reasonable operating procedure when being run under human control. However, it is possible to program the operating system so that 20 (or more) variables are monitored, compared to their ideal ranges, and actuators motivated to adjust process parameters if necessary, every minute or at shorter or longer intervals as required, under computer control. The computer is given override management of the main process loop. With the very short monitoring intervals that computer control makes possible, while process control is still not easy [43, 44], no process using this system needs to deviate far from ideal operating conditions before parameter corrections are made.

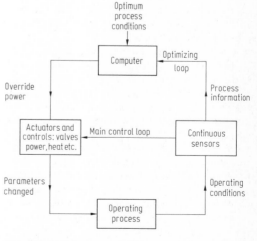

**Figure 1.5.** Scheme used for computer control of a chemical processing unit [40]

The frequent monitoring capability of the computer has also provided a stimulus to the design and implementation of rugged, on-line process analyzers such as infra-reds, mass spectrometers and the like, that can provide more frequent composition information on the progress of the reaction than is possible from conventional laboratory analysis. Without on-line analysis, frequently an interval of 2 hours or more had to elapse for completion and reporting of a laboratory analysis before adjustment of appropriate process parameters and optimizing of the process condition could be carried out. Under manual analysis conditions the extent of deviation of the process from ideal was sometimes quite significant, causing a reduction in product yields and quality [45].

Whether equipped with on-line analyzers or not, all chemical processes rely heavily on the information obtained from periodic manual laboratory analyses. These not only provide a necessary means of checking the performance of the on-line analyzers but more importantly provide quality control checks on all raw materials moving into the plant site and all products and waste streams (if required) moving out. Sampling frequency and methods of anlysis used for raw materials will be geared to the size of the shipment lot and the method of delivery. Process and product quality monitoring will be carried out at a frequency depending on the operating stability of the process and the rigour of the quality control requirements (allowed impurity concentrations) of the product. A process that requires frequent checks and parameter adjustments to keep operating smoothly is of course going to need more frequent sampling and manual analysis to ensure consistent product quality.

A part of the analytical effort of operating a chemical complex is the setting up and maintenance of a "sample library", where analyzed samples from each tank car load or reactor lot may be stored for reference purposes. The retention time for these samples is set to be appropriate so as to normally exceed the probable delay between time of product shipment and time of final consumption. A sample retention program of this kind enables the company to render rapid technical assistance to a customer who may be having difficulty with a particular batch of product. It also provides an independent means, by re-analysis of the sample, for the company to accept or reject claims regarding the quality of any particular product shipment. Thus, the whole area of proportional control, analysis, sample retention, and careful record-keeping all comprise important integrated parts of an operating chemical complex necessary for the maintenance of product quality and customer satisfaction.

## 1.5.4 Costs of Operation

Since the primary purpose for the existence of a chemical company is to make a profit for its owners and shareholders, it is vital that is be possible to accurately determine operating costs so that product pricing can achieve this objective. Today, of course, this profit picture is complicated by the fact that it has to be achieved while other company obligations relating to the welfare of its employees, to safety, to environmental quality, and to the community and country of operation as a corporate good citizen are also met. It is of course possible to produce an uneconomic chemical commodity, such as when required under the exigencies of war or with particular political objectives, but this requires the encouragement of artificial inducements such as tax concessions, subsidies, or the economics of an organized economy to achieve this.

As a rough rule of thumb, the chemical process industries aim at an 8 to 10 % after tax profit (earnings), stated as percent of sales [46]. In general, if the financial decision re construction of a new plant relates to a process for a new product, the economic projections required for construction to proceed will require a slightly higher profit, or higher rate of return on investment than this, for construction to proceed. If, however, the plant is to produce a standard chemical commodity that is already in large volume sales (i.e. a less risky venture) then lower profit projections may still be acceptable. This is particularly true if the company is intending to use a substantial fraction of the product in its own operations, quite a favourable "value added" practice known as providing a "captive market" for this product. However, these are just profit projections. The ever variable nature of business cycles places significant perturbations onto the realization of these projections so that actual, after tax profit margins more usually range from 4 or 5 % up to 12 % in any particular year, and sometimes outside both of these extremes.

What should be remembered, however, is that a profitable company not only earns an income in its own right, part of which goes to the investors who put up the money to construct the plant, but it also provides jobs and salaries to its employees who spend much of their earnings locally, stimulating further business activity. The company also usually pays local and federal taxes on a corporate basis and through its individual employees providing direct and indirect sources of income to different levels of government. Coupled with the sales taxes levied against many types of chemical commodities, because of this "multiplier effect", the local and country-wide benefit of a profitably operating commodity-based company is enormous.

In order to accurately determine the costs of operation for any particular process the chemical reactions being used must be known precisely. It is the effective stoichiometry, or quantities of product(s) which are to be expected from particular quantities of raw materials which must be known. It is useful to know something about the mechanism, or chemical pathway to the materials being produced since this knowledge can be used to intelligently exercise process changes which will have a high probability of increasing reaction rates, or of raising process yields. However, it is also quite possible and frequently happens that we have a process which operates profitably, and produces saleable product, long before anything significant is known about the mechanism.

As examples, the early facilities to produce phenol by chlorobenzene hydrolysis, and by cumene oxidation [47] were both constructed when the stoichiometry demonstrated acceptable economics. It was long after the product had already been on the market for some years before anything was known about the respective mechanisms involved. Other secondary aspects of the process, such as the capital costs, heat requirements, electric power, labour, and water needs all must also be known with some degree of certainty to determine the product pricing structure necessary for profitable operation.

## 1.5.5 Conversion and Yield

There are several important ways in which the description of the quantitative aspects of a reaction differ, using the same facts, when discussed in a research or academic context as opposed to in an applied or industrial context. Probably the best way to understand these distinctions is to define the various terms used, employing a general example (Equation 1.7).

The yield which would be reported in a research or academic setting would be based on the yield definition given in the word Equation 1.8.

Calculating a yield from Equation 1.7 using this definition gives a value of 60 % (0.60/1.00 x 100) for the result to be reported using this system. This result takes no account of any starting A which did not react. This is in keeping with what is frequently the primary objective in the academic setting, the synthesized product of interest. The amount of any unreacted starting material in the residues from a reaction is seldom bothered with, and in fact is usually disposed of with any byproducts etc. once the product of interest has been isolated.

In contrast to this, in an applied or industrial situation the definition of yield differs somewhat from that described above. Here, the amount of unreacted starting material remaining after a reaction is carried out is nearly as important as the amount of product obtained, and is taken care of in the yield definition used (Equation 1.9).

The reason for this being an important consideration, commercially, is that any unreacted starting material in an industrial process is usually recovered from product(s) and byproducts. It is then recycled to the front end of the process, with the fresh starting material just entering the process (Figure 1.6), and in this way decreases the amount of fresh starting material required. Again, using

| A | → B | + A | + byproducts | |
|---|---|---|---|---|
| 1.00 moles starting material | 0.60 mol product | 0.30 mol unreacted starting material | 0.10 mol from reacted A | (1.7) |

$$\% \text{ yield} = \frac{(\text{moles of B formed}) \times 100}{(\text{theoretical moles of B which could be formed from moles of A charged})} \quad (1.8)$$

$$\% \text{ yield} = \frac{(\text{moles B formed}) \times 100}{(\text{theoretical moles of B which could be formed from moles of A consumed})} \quad (1.9)$$

the example just given, the industrial yield works out to 85.7 % (0.60/(1.00-0.30) x 100) or, rounded, to 86 %. Thus, when taking into account the unreacted starting material a more favourable yield picture is presented.

However, to an applied chemist or an engineer at least one further piece of information relating to the performance of the reaction is vital in order to estimate quantitative results of a process, and that is the "conversion". A good working definition of this term is given by Equation 1.10, which takes into account the fact that not all reactions yield one mole of product for each mole of a starting material charged, or placed in the reactor. Cracking reactions, for example, may yield two (or more) moles of product per mole of raw material, while condensation reactions and polymerizations can yield less (and sometimes much less) than a mole of product per mole of raw material. Using this definition for the example at hand gives the working Equation 1.11. Using this equation, a value of 70 % ((1.00-0.30)/(1.00) x 100) is obtained for the conversion. What this means, in practical terms, is that 70 % of the raw material started with is no longer starting material and has been converted to something. This number alone does not tell us how much of what was converted or chemically changed is product.

One needs to use the definitions of both the industrial conversion and the yield in order to determine how much of product B we can expect. This can be determined when having only the amount of starting A and the conversion and yield data for the process of interest. The industrial yield essentially amounts to the fraction of converted starting material which ends up as product, a sort

of "selectivity" or "efficiency" term for the process. For this reason "selectivity" or "efficiency" are terms which are often used synonymously with industrial yield. As such, by multiplying the two fractions together one obtains the fractional yield of product to be expected from a batch process, or the "yield per pass" (yield on one passage of raw materials through the process) for a continuous process (Equation 1.12).

Again using the example, when the fractional product recovery is multiplied by the number of moles of A started with, one obtains a value of 0.60 ((0.70 x 0.86) x 1.00 mole of A) moles of B, as the amount of product to be expected from a batch operation, or from one pass of a continuous process on this scale. This is in agreement with the quantities specified in the original example, and, it will be noted, is the same as the academic yield specified on a fractional basis. Thus we can write down the form of an additional relationship, specified in fractional terms, which is often useful in quantitative calculations which relate to industrial processes (Equation 1.13). This can be further rearranged to another useful expression for calculations involving process efficiencies (Equation 1.14).

It is useful to consider the significance of having a facility with these yield and conversion manipulations at this point. It is of course desirable to have any industrial process operate with high yields and high conversions. If both of these conditions prevail simultaneously, more product will be obtained from each passage of raw materials through a given size of reactor, and there will be less starting material to separate and recycle, than if this were not true. Many industrial processes do in fact operate with this favourable situation.

It is possible, however, to make a success of an industrial process which only achieves low conversions, as long as high yields are maintained. Very few industrial processes operate with yields or selectivities of less than 90 %, and many operate with yields of 95 % or better. Yet some of these, for example the processes for the vapour phase hydration of ethylene to ethanol, and the ammonia synthesis reaction, have conversions which run in the 5 to 15 % range [1] and yet are operating on a large scale very competitively because both reliably achieve selectivities to the desired product which exceed 95 %. If one only had "academic" yield information about these processes, which would be 4 to 5 % and 15 to 20 % respectively, neither would appear to be promising candidates for commercia-

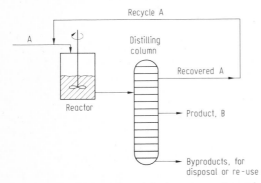

**Figure 1.6.** An illustration of the importance of recycle of recovered starting material in the industrial definition of yield

$$\% \text{ yield} = \frac{(\text{moles of main product}) \times 100}{(\text{moles of main product equivalent to the moles of main reactant charged})} \quad (1.10)$$

$$\% \text{ conversion} = \frac{(\text{initial moles of A}) - (\text{final moles of A}) \times 100}{(\text{initial moles of A})} \quad (1.11)$$

$$
\begin{array}{lll}
(\text{fractional industrial} & \times (\text{fractional} & = (\text{fractional product} \\
\text{yield}) & \text{conversion}) & \text{recovery}) \\
0.70 & 0.86 & 0.60
\end{array}
\quad (1.12)
$$

$$(\text{industrial yield}) \times (\text{conversion}) = (\text{``academic'' yield}) \quad (1.13)$$

$$\% \text{ selectivity} = \frac{\text{``academic'' yield (fractional)} \times 100}{(\text{fractional conversion})} \quad (1.14)$$

lization. Thus, while it is desirable for an industrial process to obtain high conversions with high yields (selectivity), it is *vital* for a successful industrial process to have high yields.

### 1.5.6 Importance of Reaction Rate

Fast reactions, in general, are conducive to obtaining a large output from a relatively small volume of chemical processing equipment. For example the ammonia oxidation reaction, the first stage of producing nitric acid from ammonia, is essentially complete in $3 \times 10^{-4}$ seconds at $750\,°C$ [24]. This is sufficiently rapid that the catalytic burner required to do this only occupies about half the volume of a large office desk for the production of some 250 tonnes of nitric acid daily. Except for the cost of the catalyst inventory, the fabrication cost of the ammonia burner itself is relatively low. Follow-up reactions for the process are much slower than this so that the volume of equipment required to contain them is much larger and more costly.

Ammonia oxidation represents a process with which it was realized, early in the design stage, that carrying out this step at 600-700 °C instead of at near ambient temperatures speeded up the process sufficiently to allow large conversion volumes to take place in a relatively small reactor. This same philosophy is followed, wherever feasible, with all chemical processes, i.e. a reactor volume saving is a capital cost saving. But with some processes, such as the esterification of glycerin with nitric acid, technical complications exist that put upper limits on feasible reaction temperatures, which effectively prevents the use of temperature elevation as a means of raising the rate of the reaction. Consequently this process, under the normal operating temperature of about 5 °C, has a 60 to 90 minute reaction time requirement [1]. So to produce even 20 tonnes of nitroglycerin per day would require a batch reactor of 2 tonnes or so capacity, much larger than the ammonia oxidation unit required for a 250 tonne per day nitric acid plant.

These examples illustrate the principle that, wherever feasible, reaction conditions, catalysts etc. are selected and developed such that the rate of a commercial process is maximized. In so doing the size of the processing units required for a given volume of production is reduced, in this way decreasing the costs of construction. Reducing the capital costs also reduces the capital charge per unit of product, decreasing the price required from the product to operate at a profit. In these ways improvement of the rate of a chemical process becomes a further contributing factor in the market competitiveness of the chemical industry.

## 1.6 Chemical Volume Perspectives

The chemicals listed in Table 1.4 are presently in the largest scale production and are examples of the so-called commodity chemicals, heavy, or bulk chemicals. Sodium chloride is not always classified as a *produced* chemical since most salt production is basically extractive in nature. Sulfur, too, is sometimes ranked with produced chemicals and sometimes with extracted minerals, depending on the origin of the sulfur. Leaving these two components aside sulfuric acid emerges as the leading volume chemical product, both in the U.S.A. and worldwide. The availability of world production data for many chemicals is somewhat sporadic, but the world production of those chemicals not on this

**Table 1.4** World and American production of large scale chemicals [a]

| U.S. ranking | | World, million tonnes | | | U.S.A., million tonnes | | | |
|---|---|---|---|---|---|---|---|---|
| 1980 | Chemical | 1978 | 1980 | 1981 | 1978 | 1980 | 1981 | 1982 |
| 1 | sodium chloride | 167.0 | 164.8 | 164.0 | 42.9 | 40.3 | 35.8 | |
| 2 | sulfuric acid | 121.7 | 143.0 | 139.3 | 37.3 | 40.1 | 36.3 | 29.4 |
| 3 | ammonia | (84.5) | | | 16.4 | 17.2 | 17.3 | 14.1 |
| 4 | lime | 118.7 | 119.4 | 116.9 | 17.6 | 16.0 | 17.1 | 12.9 |
| 5 | oxygen | | | | 16.0 | 15.6 | 12.2 | 10.1 |
| 6 | nitrogen | | | | 13.6 | 15.5 | 13.9 | 14.0 |
| 7 | ethylene | (45) | 50 | | 13.2 | 12.5 | 13.1 | 11.1 |
| 8 | sulfur | 53.9 | 54.5 | 53.6 | 11.2 | 11.8 | 10.6 | 8.6 |
| 9 | sodium hydroxide | 29.6 | 33.1 | | 11.2 | 10.3 | 9.4 | 8.3 |
| 10 | chlorine | 29.1[e] | 31.3 | | 11.0 | 10.2 | 9.8 | 8.3 |
| 11 | phosphoric acid | 125.4[b] | 134.9[b] | | 9.2 | 9.9 | 9.0 | 7.7 |
| 12 | ammonium nitrate | | | | 7.1 | 7.8 | 8.1 | 6.7 |
| 13 | nitric acid | 28.2 | | | 7.8 | 8.1 | 8.2 | 6.9 |
| 14 | sodium carbonate | 27.3 | 28.8 | 28.3 | 7.5 | 7.5 | 7.5 | 7.2 |
| | sugar | 92.0 | | | 5.1 | 4.3 | 4.7 | |

[a] Figures for the U.S.A. can be used as a means of estimating world figures, in the absence of world data for many categories (blanks). Parenthesized values are production *capacities*. Table compiled from the data of [46, 49-53], and from that supplied by Verband der Chemischen Industrie, e.V. (VCI).

[b] Stated as phosphate rock, for lack of phosphoric acid data.

[e] Calculated from known sodium hydroxide production figures.

list probably ranks somewhere near the U.S. ranking. Since the U.S. production of most, if not all of the chemicals on this list represents the largest single component of the world figure, and recent statistics on the production of American commodity chemicals are generally quite readily available and complete, the U.S. ranking data can often be usefully used to estimate world rankings and production levels when this data is not available directly.

There are a number of other interesting observations which can be made concerning these particular bulk chemicals. First, they are all not far removed, in terms of processing steps, from the natural raw materials from which they are derived. Virtually all of the oxygen and nitrogen and significant proportions of the salt, sulfur, and sodium carbonate are obtained relatively directly from natural sources. Also, these commodities inter-relate quite closely to one another, in chemical terms. Thus sulfuric acid, largely produced from sulfur, in turn is used to a significant extent to produce phosphoric acid from phosphate rock. A large fraction of the nitrogen produced goes into the production of synthetic ammonia. Ammonia, in turn, is used for nitric acid production and also is com-

bined with much of the nitric acid produced in an acid-base reaction for the preparation of ammonium nitrate. Chlorine and sodium hydroxide, too, are largely obtained from the electrolysis of a solution of sodium chloride in water. It is interrelationships of this kind, coupled with very large world fertilizer markets for some of the secondary and tertiary products of these sequences, in particular ammonia, ammonium nitrate, and phosphoric acid (as salts), that keeps many of these chemicals on this large volume production list.

If one attempts to rank all classes of industrial companies, worldwide, that are active in the production of chemicals in the broad sense, one obtains a list that is dominated by oil companies (Table 1.5). In fact this exercise shows that fully 11 of the world's 15 largest companies and 20 of the largest 32 companies are engaged in some aspect of what can be broadly considered to be chemical processing. However, of this list of twenty, only two, Unilever and E.I. du Pont de Nemours (Dupont), are substantially chemical companies rather than having oil production and refining as their primary areas of activity. Creating a separate listing which specifically omits the oil processing area, gives greater representation to the world's thirteen major

**Table 1.5** The world's twenty largest chemical processing companies ranked on the basis of gross sales relative to all industrial companies [a]

| Overall rank 1981 | 1979 | Company | Headquarters | 1981 Sales,[c] $10^9$ US$ |
|---|---|---|---|---|
| 1 | 1 | Exxon | New York | 108 |
| 2 | 3 | Royal Dutch/Shell | The Hague/London | 82 |
| 3 | 4 | Mobil | New York | 64 |
| 5 | 7 | Texaco | Harrison, N.Y. | 58 |
| 6 | 6 | British Petroleum | London | 52 |
| 7 | 8 | Standard Oil of Calif. | San Francisco | 44 |
| 9 | 10 | Standard Oil (Indiana) | Chicago | 30 |
| 10 | 13 | ENI (Ente Nationale Idrocarburi (oil)) | Rome | 29 |
| 12 | 9 | Gulf Oil | Pittsburg | 28 |
| 13 | 21 | Atlantic Richfield | Los Angeles | 28 |
| 15 | 12 | Unilever | London/Rotterdam | 24 |
| 16 | 39 | E.I. du Pont de Nemours | Wilmington, Delaware | 23 |
| 17 | 16 | Francaise des Petroles | Paris | 23 |
| 18 | 26 | Shell Oil | Houston | 22 |
| 19 | d | Kuwait Petroleum | Safat, Kuwait | 21 |
| 20 | 32 | Elf-Aquitaine | Paris | 20 |
| 21 | 29 | Petroleos de Venezuela | Caracas | 20 |
| 23 | 47 | Petrobas (Petroleo Brasiliero) | Rio de Janeiro | 19 |
| 24 | e | Pemex (Petroleos Mexicanos) | Mexico City | 19 |
| 32 | 49 | Phillips Petroleum | Bartlesville, Okla. | 16 |

[a] Compiled from [54, 55]

[b] The gaps in ranking are filled by companies which are not involved in chemical processing, e.g. ranked 4th, General Motors; 8th, Ford Motors; 11th, International Business Machines; 14th, General Electric; and 22nd, Fiat.

[c] Rounded upwards.

[d] Due to mergers, on list for first time this year.

[e] Annual sales value was less than those of the top fifty companies, therefore was not on the list previously.

very large international corporations that can be classed as companies whose principal business is the manufacturing of chemicals (Table 1.6). The dominant positions of the U.S.A., West Germany, and the United Kingdom in chemicals production also becomes evident from this treatment. Other companies which appear here, such as in the metals processing area (principally iron and steel) and food processing, also peripherally relate to the process areas discussed, and so are included. In terms of the scale of operations, just the first company of each of these lists has a gross annual sales volume which exceeds the gross national product of the whole economies of many countries.

**Table 1.6** The largest chemical and related companies whose primary business area is not oil, based on 1980 and 1981 sales [a]

| Company | Country | Business area | Annual Sales[b] $10^9$ US$ |
|---|---|---|---|
| Du Pont de Nemours | U.S.A. | chemicals | 33.3 |
| Unilever | Britain/ Netherlands | food products, soaps cosmetics | 24.1 |
| U.S. Steel | U.S.A. | metal refining | 18.4 |
| Hoechst | West Germany | chemicals, dyes | 15.3 |
| Bayer | West Germany | chemicals drugs | 15.0 |
| Nippon Steel | Japan | metal refining | 15.2 |
| Nestle | Switzerland | food products | 14.2 |
| BASF (Badische Anilin und Soda Fabrik) | West Germany | chemicals, plastics | 13.7 |
| Thyssen | West Germany | steel, manuf. products | 13.1 |
| ICI (Imperial Chemical Industries) | Great Britain | chemicals, plastics | 13.3 |
| Procter and Gamble | U.S.A. | soap, detergents | 12.0 |
| Dow Chemical | U.S.A. | chemicals, plastics | 10.6 |
| Canadian Pacific | Canada | metal refining | 10.3 |
| Union Carbide | U.S.A. | chemicals | 9.1 |
| Montedison | Italy | chemicals, drugs | 7.9 |
| Pechiney Ugine Kuhlmann | France | chemicals, metal refining (Al) | 7.6 |
| DSM (Dutch State Mines) | Netherlands | chemicals, fertilizers | 7.4 |
| Ciba-Geigy | Switzerland | chemicals, drugs | 7.1 |
| British Steel | Great Britain | metal refining | 6.9 |
| Rhone Poulenc | France | chemicals, drugs | 6.6 |
| 3M (Minnesota Mining and Manufacturing) | U.S.A. | chemicals, plastics | 6.6 |
| Friedrich Krupp | West Germany | metal refining | 6.6 |
| Monsanto | U.S.A. | chemicals | 6.3 |

[a] Compiled from [56, 57 and 58]
[b] Sales figures for U.S. Corporations are given for 1982, the others are for 1980.

## Relevant Bibliography

1. R.W. Thomas and P. Farago, Industrial Chemistry, Heinemann, London, 1973
2. H.P. Meissner, Processes and Systems in Industrial Chemistry, Prentice-Hall, Inc., Englewood Cliffs, N.J. 1971
3. The Modern Inorganic Chemicals Industry, R. Thompson, editor, The Chemical Society, London, 1977
4. D.F. Rudd, G.J. Powers, and J.J. Siirola, Process Synthesis, Prentice-Hall, Inc., Englewood Cliffs, N.J., 1973
5. M.S. Peters and K.D. Timmerhaus, Plant Design and Economics for Engineers, 3rd edition, McGraw-Hill, New York, 1980
6. R.F. Goldstein and A.L. Waddams, The Petroleum Chemicals Industry, 3rd edition, E., and F.N. Spon, Ltd., London, 1967
7. K. Weissermel and H.-J. Arpe, Industrial Organic Chemistry, Verlag Chemie, Weinheim, 1978
8. H.A. Wittcoff and B.G. Reuben, Industrial Organic Chemicals in Perspective, Part One: Raw Materials and Manufacture, John Wiley and Sons, New York, 1980
9. Chemistry in the Economy, American Chemical Society, Washington, 1973
10. B.G. Reuben and M.L. Burstall, The Chemical Economy, Longman, London, 1973
11. W.S. Howe, Industrial Economics: An Applied Approach, MacMillan, London, 1978
12. D.F. Ball, Some Aspects of Technological Economics, The Chemical Society, London, 1974
13. Applied Science and Technology Index, H.W. Wilson Company, Bronx, New York, since 1958; formerly called The Industrial Arts Index published since 1913
14. Chemical Abstracts, American Chemical Society, Columbus, Ohio, published since January, 1907. Chemistry, Technology, and patent indices
15. British Technology Index, The Library Association, London, published since January, 1962
16. Patent Office Record, Patent Office, Ottawa, Ontario, published since 1872. U.S. Patent Office, Official Gazette, Washington, published weekly since 1872, replacing the previous "Patent Office Reports". Also the parallel records of other countries of interest

17. Business Periodicals Index, H.W. Wilson Company, New York, published since January, 1958. Current economic data

# References

1. F.A. Lowenheim and M.K. Moran, Faith, Keyes and Clark's Industrial Chemicals, 4th edition, Wiley-Interscience, New York, 1975
2. Chemical Producers Face Yet Slower Growth, Chem. Eng. News 61 (15), 11, April 11, 1983
3. J.W. Barrett, Nature, 5, Dec. 31, 1971
4. World Chemical Exports Continue Growing, Chem. Eng. News 59 (36), 21, Sept. 7, 1981
5. M. Trowbridge, Chem. In Brit. 11 (1), 15, Jan. 1975
6. P.A.S. Taylor, A New Dictionary of Economics, Routledge and Kegan Paul, London, 1966, page 290
7. H.S. Sloan and A.J. Zurcher, Dictionary of Economics, 5th edition, Barnes and Noble, New York, 1971, page 459
8. Riegel's Industrial Chemistry, J.A. Kent, editor, Reinhold Book Corp., New York, 1962
9. W.L. Fallwell, Chem. Eng. News 60 (3), 33, Jan. 18, 1983
10. D.A. O'Sullivan, Chem. Eng. News 58 (7), 15, Feb. 1980
11. Facts and Figures for Chemical R and D, Chem. Eng. News 60 (30), 38, July 26, 1982
12. Most Patents Continue to be in Chemical Field, Chem. Eng. News 60 (14), 24, April 5, 1982
13. T.E.R. Singer and J.F. Smith, J. Chem. Educ. 44 (2), 111, Feb. 1967
14. L.L. Annett, Chem. In Can. 34 (6), 4, June 1982
15. Little Neddies Call For Longer Patent Protection, Chem. In Brit. 17 (11), 504, Nov. 1981
16. Committee Approves Patent Restoration Bill, Chem. Eng. News 59 (26), 14, June 29, 1981
17. Senate Approves Patent Term Restoration Bill, Chem. Eng. News 59 (29), 21, July 20, 1981
18. Patent Extension May Not Increase Innovation, Chem. Eng. News 59 (36), 28, Sept. 7, 1981
19. F.P. O'Connell, Chem. Eng. 69 (4), 150, Feb. 19, 1962
20. J.P. Stern and E.S. Stern, Petrochemicals Today, Edward Arnold, London, 1971
21. D.H. Allen, Chem. Ind. (London) 98, Feb. 5, 1977
22. M.S. Peters and K.D. Timmerhaus, Plant Design and Economics for Chemical Engineers, 3rd edition, McGraw-Hill, New York, 1980
23. Big Scale Sometimes Means Big Loss, Can. Chem. Proc. 55 (10), 60, Oct., 1971
24. R.N. Shreve and J.A. Brink, Jr., Chemical Process Industries, 4th edition, McGraw-Hill, New York, 1977
25. W.P. Long and E.D. Montrone, Chem. Eng. 77 (22), 41, Oct. 12, 1970
26. W.C. Benzer, Chem. Eng. 77 (22), 101, Oct. 12, 1970
27. A. Krisher, Chem. Eng. 77 (22), 47, Oct. 12, 1970
28. O.H. Fenner, Chem. Eng. 77 (22), 53, Oct. 12, 1970
29. R.J. Landrum, Chem. Eng. 77 (22), 75, Oct. 12, 1970
30. G. Sorel, Chem. Eng. 77 (22), 83, Oct. 12, 1970
31. K. Denbigh, Chemical Reactor Theory, The University Press, Cambridge, 1965
32. G.F. Reynolds, Chem. In Brit. 8 (12), 534, Dec. 1972
33. E. Simon, Chemtech. 5 (10), 582, Oct. 1975
34. W.E. Cowley, Chem. and Ind. (London), 284, April 3, 1976
35. R.P. Morgan, Chem. Eng. News 55 (11), 31, Nov. 14, 1977
36. W. Lepkowski, Chem. Eng. News 58 (24), 31, June 16, 1980
37. M.D. Wynne, Chemical Processing in Industry, Royal Institute of Chemistry, London, 1970
38. Chemistry and Industry, D.G. Jones, editor, Clarendon Press, Oxford, 1967
39. Chemical Reactors, H.S. Fogler, editor, ACS Symposium Series 167, American Chemical Society, Washington, 1981
40. Chemical Engineering; a Special Study, J.G. Raith, editor, Nuffield Foundation, Penguin, Harmondsworth, England, 1971
41. Chemical Engineer's Handbook, 4th edition, J.H. Perry, editor, McGraw-Hill, New York, 1963
42. S.C. Stinson, Chem. Eng. News 59 (49), 15, Dec. 7, 1981
43. F.G. Shinskey, Process Control Systems, McGraw-Hill, New York, 1967
44. G.D. Shilling, Process Dynamics and Control, Holt, Rinehart and Winston, New York, 1963
45. R. Douglas, Can. Chem. Proc. 66 (11), 24, Nov. 1982
46. Facts and Figures for the Chemical Industry, Chem. Eng. News 60 (24), 31, June 14, 1982, and 61 (24), 26, June 13, 1983
47. P. Wiseman, Industrial Organic Chemistry, Wiley Interscience, New York, 1972
48. A.C.H. Cairns, Roy. Inst. Chem. Revs. 2 (1), 41, Feb. 1969
49. Top Fifty Chemical Products and Producers, Chem. Eng. News 58 (18), 33, May 5, 1980
50. Top Fifty Chemical Products and Producers, Chem. Eng. News 59 (18), 35, May 4, 1981
51. Kirk-Othmer Encyclopedia of Chemical Technology, 3rd edition, John Wiley and Sons, New York, 1978, volumes 2 and 9
52. 1979/80 UN Statistical Yearbook, United Nations, New York, 1981
53. Minerals Yearbook, Volume 1, Metals and Minerals, U.S. Bureau of Mines, Washington, 1981

54. The Largest Industrial Companies of the World, Fortune, *106,* 182, Aug. 23, 1982
55. The Largest Industrial Companies of the World, Fortune, *104,* 206, Aug. 10, 1981
56. The Foreign 500, Fortune *104,* 206, Aug. 10, 1981
57. The 500, Fortune, *107,* 226, May 2, 1983
58. The Times 1000 1982-83, M. Allen, editor, Times Books Ltd., London, 1982
59. J.L. McShane and F.G. Geil, Research/Development *26* (2), 30, Feb. 1975
60. Non-intrusive Meters: New in Fluid Flow, Can. Chem. Proc. *62* (6), 32, June 1978
61. New Flowmeters Put Squeeze on Orifice Plate, Chem. Eng. News *55* (51), 20, Dec. 19. 1977

# 2 Air Quality and Emission Control

## 2.1 Significance of Man's Activity on Atmospheric Quality

In the early days of habitation of this planet, when the human population was small and its per capita consumption of energy was primarily as food (8,400-12,600 kJ/day; 2-3,000 kcal/day), total human demands on the biosphere were consequently small. Early requirements of goods were minimal and quite close to direct (requiring little fashioning) so that this early society's total demands and wastes were easily accommodated and assimilable by the biosphere, with little impact.

Today, with advances in technology providing an ever increasing range of appreciated and expected goods and services, it has been estimated that the individual consumption of energy of all forms has risen some 100-fold from the requirements of primitive man [1]. Of the one million kJ per person per day (230,000 kcal/person/day) that this collective consumption now represents, more than half, or 645,000 kJ (154,000 kcal) is estimated to be consumed by collective industrial, agricultural, and transportation needs. About a quarter is produced and consumed per person through the medium of electrical generation and consumption, either directly, for individual domestic use, or by the prorated industrial power consumption on their behalf [1]. All of the major fields of human endeavour, high technology agriculture, industrial production, and thermal electricity generation, use combustion processes to provide a major fraction of their energy requirements. When this gross increase in per capita energy consumption is coupled to a global population growth increase of a thousand or more times the population of primitive societies, it becomes easy to see how the total demand placed on the biosphere by modern industrial societies has now become so significant and evident.

Most of the world's population growth has taken place since the Middle Ages, and the industrial development has accelerated since about 1850, so that most of the increase in demand for energy and materials has occurred during the period when these two component increases coincided. Man's activities now contribute similar volumes of the minor gaseous constituents of the atmosphere as do natural processes. From a world fossil fuel consumption of about 5 billion tonnes per year, we now contribute about $1.5 \times 10^{10}$ tonnes of carbon dioxide to the atmosphere annually [2], a significant fraction of the natural contribution, now estimated to be about $7 \times 10^{10}$ tonnes/year [3, 4]. Of man's annual atmospheric contributions of sulfur-containing gases of about $10^{10}$ tonnes [5], of oxidized and reduced nitrogen compounds of about 50 million tonnes [6], and of carbon monoxide of about 200 million tonnes, all are estimated to be ranging in a similar order of magnitude to the natural contributions of these constituents. The old jingle "The solution to pollution is dilution" no longer holds true, particularly for atmospheric discharges. There is just too large a total mass of atmospheric contaminants being discharged, or too small an atmosphere to accept these, to still be able to discharge at the same rate and obtain sufficient dilution. This is particularly the case when natural air movement is sluggish for any reason so that little pollutant mixing occurs. Under these circumstances discharged pollutants simply remain in the same local air mass, greatly exaggerating the immediate effects of the discharge.

### 2.1.1 Natural Contaminants

Natural sources of many common air contaminants also make a contribution to this overall atmospheric pollutant loading. The seas contribute large masses of saltwater spray droplets to the air as a result of wave action. As the water evaporates from these droplets very fine particles of salts are left suspended in the moving air, one of the contributing factors to the sea "smell". It has been estimated that some 13 million tonnes of sulfate ion alone is contributed to the atmosphere annually from the oceans of the world [5].

Volatile organic compounds are contributed to the atmosphere by many forms of plant life, conifers such as cedar and pine, eucalyptus, and herbs such as mint. Extensive tracts of such plants, such as the

$$CH_2=C(CH_3)-CH=CH_2 \qquad \qquad \qquad \qquad \text{limonene} \qquad (2.1)$$

isoprene                                 α-pinene

pine forests of New England, contribute a large mass of terpenes to the air above them. Terpenes, plant products biosynthetically derived from isoprene, have a type formula $(C_5H_8)_n$ based on the isoprene multiplicity of the compound (Equation 2.1).

In general all terpenes have one or more reactive double bonds so that interaction of these natural terpene vapours with sunlight and air causes a development of a photochemically-promoted blue haze over these forests during periods of little air movement [7]. Estimates of the global atmospheric contribution by plants of terpenes and oxygenated terpenes alone range from $2 \times 10^8$ to $10^9$ tonnes per year [7].

The world's deserts contribute significant masses of dust and particulate to the atmosphere, some of which is transported a considerable distance. Meteoritic dust is estimated to be collected by the atmosphere to the extent of more than 900 tonnes annually [8]. However, an active volcano is a far more significant particulate contributor, some 90 million tonnes having been estimated to have been discharged into the atmosphere by the recent eruption of Mount St. Helens in Washington State [9, 10]. When several eruptions take place at one time the gross dust contribution to the global atmosphere is massive. This dust loading is generally long-lived since it is forcefully injected as high as 40 km into the atmosphere [11]. Therefore, the global influence of these events has more significance on atmospheric quality than more localized dusting events.

Sulfur dioxide and hydrogen sulfide are also contributed to the atmosphere on the scale of 1 to 2 million tonnes per year by volcanic activity [5, 12]. Other contaminating gases [13], as well as metal vapours, are also discharged in significant quantities during volcanic eruptions [14]. Emanations of the vapours of more volatile metallic elements such as mercury also occur continuously over ore bodies. Sensitive vapour detection instruments have been used as a prospecting method to converge on the ore body using this mercury vapour "halo" [15, 16].

Biological materials of many types do not represent a large mass of contamination but nevertheless comprise an important component of atmospheric pollution because of their potency. For example, very small quantities of the pollens from many types of wild flowers and grasses such as golden rod and ragweed, are sufficient to severely affect many people [17]. Bacteria, viruses, and the living spores of some of the common moulds can also cause problems, particularly when situations develop which promote a rapid, localized rise in numbers of the organism [18-20]. When this occurs in sea water, such as near sewer outfalls, dispersal of micro-organisms in the aerosols released by vigorous wave action can aggravate the problem [8, 21, 22].

Little can be done about controlling most of these natural sources of atmospheric contaminants. However, it is still important to catalogue and quantify them to be able to quantitatively relate these components of atmospheric quality to the contributions from our own activities. Only in this way is it possible to compare the relative significance of the two sources of contaminants. This is a necessary preliminary to provide informed and appropriate guidance for the modification of our own activities to decrease their negative impact on atmospheric quality where this is necessary.

## 2.2 Classification of Air Pollutants

Air pollutants can generally be classified into one of three main categories. By doing so, potential emissions may be grouped as to their appropriateness for one or more types of emission control devices, based on their mode of action.

The coarse particles, or the particulate class of air pollutants, comprise solid particles or liquid droplets which have an average diameter greater than about $10\,\mu m$ ($10^{-3}$ mm). Particles or droplets of this class of air contaminant are large enough that they fall out of the air of their own accord, and more or less rapidly.

Air pollutants of the aerosol class can also comprise solid particles or liquid droplets, but this time of a size range generally less than about $10\,\mu m$ average diameter. The important consideration for this class is that the particles or droplets are small enough in size that there is a strong tendency for the substance to stay in suspension in air [23]. Thus for the powders of denser solids, such as magnetite, the average particle size would have to be smaller than this $10\,\mu m$ guideline for the particle to stay in suspension. A suspension of a finely divided solid in air is referred to as a "fume", and that of a finely divided liquid as a "fog", each term being only a more specific representation within the aerosol classification.

The gases comprise the third major classification of air pollutants, which includes any contaminant that is in the gaseous or vapour state. This includes the more ordinary "permanent" gases, such as sulfur dioxide, hydrogen sulfide, nitric oxide (NO), nitrogen dioxide ($NO_2$), ozone, carbon monoxide, carbon dioxide (pollutant?) etc., as well as the less common ones such as hydrogen chloride, chlorine, tritium ($_1^3H$) and the like. It also includes materials which are not ordinarily gases, such as hydrocarbon vapours, volatile non-metal or metal vapours (e.g. arsenic, mercury, zinc) when these are in the vapour state.

The dividing line between the particulate/aerosol classes, and the gaseous classes is clear enough because of the phase difference. But the position of the dividing line between the particulate and the aerosol classes, being based on whether or not a second phase stays in suspension in air, is far less obvious. However, consideration of the terminal velocities for particles of differing diameters helps to clarify this distinction. It can be seen from Table 2.1 that, for a substance with a density of 1 $g/cm^3$, a significant terminal velocity begins to be observed at a particle diameter of about $10\,\mu m$. Hence the approximate dividing line between these two classes. Figure 2.1 gives examples of typical particle size ranges for some common industrial and domestic substances.

## 2.2.1 Quantification and Identification of Particulates

The particulate class of air pollutant, since it represents particles or droplets which more or less rapidly settle out of air, is also one of the easiest classes to measure. If a source particulate deter-

**Table 2.1** Gravitational settling velocity for spheres of unit density in air at 20°C[a]

| Particle diameter, μm | Terminal velocity, mm/sec |
|---|---|
| 0.1 | $8.5 \times 10^{-4}$ |
| 0.5 | $1.0 \times 10^{-2}$ |
| 1.0 | $3.5 \times 10^{-2}$ |
| 5.0 | 0.78 |
| 10 | 3.0 |
| 20 | 12 |
| 50[b] | 72 |
| 100[c] | 250 |

[a] From [174 and 275].
[b] Roughly equivalent to a powder which would pass through a No. 325 sieve (325-mesh, 54 μm average opening size) and be retained on a No. 400 sieve (400-mesh, 45 μm average opening size).
[c] Roughly equivalent to a powder which would pass through a No. 170 sieve (170-mesh, 103 μm averge opening size) and be retained on a No. 200 sieve (200-mesh, 86 μm average opening size).

mination is to be carried out, that is if the particulates in the flue gases of a chimney or the exhaust gases of a vent stack are to be sampled, then special holes are required in the ductwork. Probes with associated equipment and a means of reaching the sampling holes with this equipment are necessary. More details of this procedure are given in connection with aerosol determination.

It is also possible, however, to gain a useful quantitative record of source information by observation of the opacity of a plume. The Ringelmann system was set up by Maximilian Ringelmann in about 1890 to accomplish this for black smokes [24]. According to this system, Ringelmann numbers 1, 2, 3, 4, and 5 represent 20, 40, 60, 80, and 100 % opaque plumes respectively [8]. A circular card with shaded opacities corresponding to these Ringelmann numbers which surround a central viewing hole is available to assist in visual assessments. But a trained observer is able to reproduceably estimate plume opacity to within half a Ringelmann Number without this assistance. The obscuring properties of white smokes are specified simply on a percent opacity basis.

Another common type of particulate determination method uses the particulate fallout from ambient air as a measure of the particle loading. This method can be as simple as a series of glass jars placed in suitable areas for which dustfall measure-

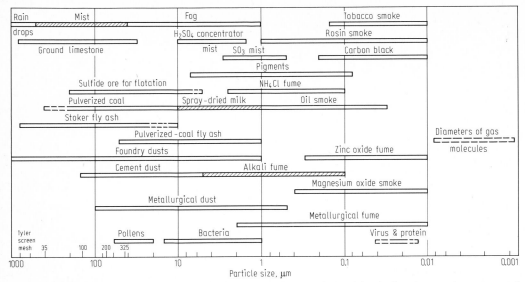

**Figure 2.1.** Typical particle size ranges of some types of precipitation, industrial and mineral processing streams and ambient air biologicals. From L.C. McCabe, ed., Air Pollution, Chap. 16 by H.P. Munger, p. 159, McGraw-Hill Book Co., New York, 1952, by courtesy of the publisher

ments are desired. The jars may be used dry, or may contain a liquid collecting agent to assist in trapping the fallout as it is collected. After a suitable interval, usually 30 days, the collected material is filtered (if wet collection was used) dried, and weighed. The weight obtained may be used to calculate a fallout value for each of the areas in which the collections were carried out. The result is specified as short tons per square mile per month, or more usually today as $mg/m^2/day$.

Essentially the same procedure is used with a plastic dustfall canister on a stanchion (upright stand) specially designed for the purpose, except that a few collection refinements are used to increase the reliability of the results. The stanchion, which places the top of the canister 4 feet above the base, is also sited a reasonable distance away from trees, buildings, or roof fixtures so as to minimize the interference of fallout by anomalous air currents [25]. It is also usually fitted with a bird ring to discourage the contamination of the canister contents by birds resting on the edge of the canister itself. A further modification, a series of vertical spikes welded to the bird ring, is necessary in areas frequented by larger species such as gulls and crows, to prevent their roosting on the edge of the canister *or* the bird ring.

The amount of fallout normally obtained with either device in relatively undisturbed open country is often a surprise to those who have not previously worked in the area of air pollution studies. For example, if 100 mg of dustfall is collected during a 30 day period by a glass container having a 10 cm diameter opening, this does not sound like much. But in conventional fallout units this amounts to $424 mg/m^2$ day, or more impressively as 36.4 $tons/(mile)^2$ month (Equations 2.2, 2.3), actually quite a heavy fallout. Table 2.2 gives a range of some recent fallout values for the U.K. and Canada, to give a basis for comparison. The Canadian figures are somewhat high since they were obtained for an industrialized, relatively high population density area.

An early Canadian particulate guideline stipulated that no discharge was to result in a fallout exceeding 20 tons per square mile per month for any area [26].

$$(100 \text{ mg} \times 10\,000 \text{ cm}^2/\text{m}^2) \div (3.1416 \times (5 \text{ cm})^2 \times 30 \text{ days}) \tag{2.2}$$

$$(0.10 \text{ g} \times (2.54 \text{ cm/in})^2 \div (1; \text{ in/ft})^2 \times (5280 \text{ ft/mile})^2 \div (3.1416 \times (5 \text{ cm})^2 \times 9.07 \times 10^5 \text{ g/ton} \tag{2.3}$$

**Table 2.2** Average values for the particulate fallout experienced by areas of differing urban activity in the United Kingdom and Canada [a]

| Type of area | United Kingdom, 1962-3 Particulate fallout | | Canada, 1968 Particulate fallout | |
|---|---|---|---|---|
| | mg/m² day | ton/mi² month[b] | mg/m² day² | ton/mi² month |
| industrial | 159 | 13.6 | 233 – 350 | 20 – 30 |
| high density housing | 116 | 9.95 | – | – |
| town center | 112 | 9.60 | 210 – 256 | 18 – 22 |
| town park | 90 | 7.7 | – | – |
| low density housing | 82 | 7.0 | 117 – 163 | 10 – 14 |
| suburbs, town edge | 70 | 6.0 | – | – |
| open country | 38 | 3.3 | 82 – 140 | 7 – 12 |

[a] Compiled from [174 and 276].
[b] Calculated using the factor $(\text{tons/mi}^2/\text{month}) \div (\text{mg/m}^2/\text{day}) = 0.08575$.

## 2.2.2 Quantification and Identification of Aerosols

Since the aerosol class of air pollutants represent particles and droplets too small to fall out of their own accord, sampling and measurement of this class requires dynamic sampling equipment. Work has to be done on the gas to force it through the recovery or analyzing equipment so that the suspended matter can be captured. Again the analysis can be a source test, where a stack or waste vent is sampled directly, or it can be an ambient air survey, where the general condition of the outside air is determined.

Source testing requires the placing of a probe into the stack or waste vent through a sampling port in the side of the stack. A vacuum pump at the exhaust end of the sampling train provides the means for drawing the test gas sample into the train (Figure 2.2). Also important for good source

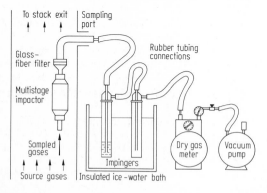

**Figure 2.2.** Simplified layout of an impactor sampling train [31]

sampling technique is that the gas flow rate into the end of the probe is the same as the gas flow rate in the stack at the point being sampled, referred to as iso-kinetic sampling [27]. Otherwise some sorting of the particles sampled inevitably occurs making the results unrepresentative and unreliable [28, 29]. Also different points in the cross section of the stack must be sampled to correct for any variation in particle concentrations across a section, so-called "equal area sampling" [30]. This is necessary because, even in a long straight vertical duct, gas flow rates close to the wall of the duct will be lower, due to frictional effects. This flow rate difference, in turn, affects the relative particle loadings in the gases moving at the center, and at the edge of the duct.

As the sampled gases flow into the first series of large jets in the impactor used for collection, large particles with significant momentum strike the impactor plate, many of which remain stuck there for later analysis [31]. Smaller particles with less momentum are diverted away from this plate, still carried by the diverted current of gases. As the gases proceed further into the impactor they reach smaller and smaller jet sizes forcing the gas to move at higher and higher velocities. As the gas velocities increase, smaller and smaller particles receive sufficient kinetic energy to impinge on the collector plates, and stick there. For some kinds of test the sticking tendency may be increased by application of a thin wipe of petroleum jelly [8]. In this manner particles are collected roughly classified as to size. On completion of the test the impactor is carefully disassembled. The sorted collected material will give some information about the particle size distribution of the tested source (or

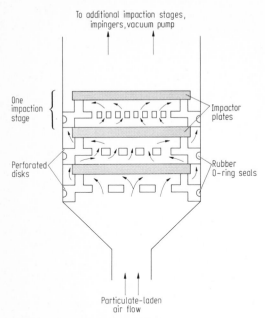

To additional impaction stages, impingers, vacuum pump

One impaction stage

Impactor plates

Perforated disks

Rubber O-ring seals

Particulate-laden air flow

**Figure 2.3.** Enlarged longitudinal section of a portion of a cascade impactor for source testing [31]

ambient air). The samples are also available for other tests, such as microscopic examination [32, 33] and wet chemical or instrumental tests if desired. This system does not permit good quantitative information to be obtained, unless it is fitted with a fine pore filter (membrane or the like) for final passage of the sampled gas. A filter would, however, impose gas flow rate restrictions onto the system which in turn could limit its use for isokinetic sampling.

Filters are, however, separately useful in the sampling of ambient air for suspended matter. The high volume "high vol" sampler is one of the commonest and simplest devices used for this purpose. An air turbine driven by a vacuum cleaner motor is used to provide the suction to pull an air sample of 2,000 m³ or more through a thick 12 x 15 cm filter sheet during a predetermined test period, usually 24 hours. A recording flowmeter is used to keep track of the volume of air filtered, since the flow rate gradually declines during the test period as material collected on the filter gradually impedes air flow. The quantities (flow rate) x (time) give the volume of air filtered, and the gross weight minus the tare weight of the filter, dried to standard conditions, gives the mass collected. A quantitative

mass per unit volume result for ambient air is thus obtained, limited only at the lower end of the particle size range by the pore size of the filter used. The large particles, plus the medium to larger size range of the aerosol classification will be collected. The result is usually specified in units of mass per unit volume, e.g. mg/m³ or μg/m³.

A refinement of the high vol sampling technique enables separation of the collected sample into coarse (2.5 to 15 μm) and fine (>2.5 μm) particle fractions with the help of different Teflon filter elements [34]. Ambient air sampling rates of about 1 m³ per hour are claimed for this device.

None of the preceding methods is very efficient at retaining particles of 1 μm diameter or less [8]. For information on this particle size range, ultrafiltration on dense cellulose or molecular membrane filters of cellulose acetate (e.g. Millipore, Isopore) is necessary [35]. Cellulose is fine for qualitative filtration work but is difficult to use in quantitative studies because of weight variations caused by the variable extent of moisture adsorption. Cellulose acetate, which is much less affected by atmospheric moisture, is also close to 100 % efficient at retaining particles larger than about 0.1 μm, and has the further advantage that the filter element can be used for direct examination under a light transmission microscope for size distribution or particle identification studies [8]. The fine pore size of this filter requires a pressure differential of nearly an atmosphere to obtain adequate air filtration rates, and a good filter element support to enable the membrane to tolerate these pressures without rupture. Even using these conditions only about 0.03 m³ of air per minute can be filtered through the standard 5 cm disk [8], one of the limitations of the method.

A Mine Safety Appliance development, an analytical scale electrostatic precipitator, provides one further technique for collection of fine aerosol material, particularly useful under low loading conditions. By prior taring of the removable glass tube of this device it is possible to determine particle loadings of as low as 10 μg/m³ after the passage of 10 m³ of air [8]. This, or even larger volumes of air may easily be sampled with this device since there is negligible resistance to air flow through the glass collection tube of this design.

Identification of particles and aerosols can frequently be made using direct microscopic examination coupled with a knowledge of the collection circumstances. Particle features such as transparent/opaque, colourless/coloured, rounded/angular, isotropic/anisotropic, density, and refractive index all may be determined in this way. When these properties are combined with the general visual features such as are catalogued in one of the particle atlases [36], this examination can very often be unequivocal for identification. Particles below about $0.2\,\mu m$ diameter, the limit for the best light microscopes, may be studied using electron microscopy [8, 33, 37]. Various destructive wet chemical, or instrumental techniques such as ion chromatography [38], neutron activation analysis [39, 40], or atomic absorption [41] may also be used to provide confirmation of the identities provisionally established by the visual process. Details are available from a variety of standard texts and review articles [42-46].

## 2.2.3  Analysis of Gaseous Air Pollutants

A wide variety of methods exists for the analysis of gaseous air pollutants within the realms of wet chemical, instrumental, and biological procedures. The analytical method of choice will depend on a number of factors among which the purpose (accuracy and number of results required), budget, equipment available, and level of training of available staff will all have a bearing. As an example of the spread of possibilities within these choices, an air (and water) pollution survey was conducted in the United Kingdom during 1972-73 which was publicized by the Sunday Times Magazine. Hardware and methods literature were made available at cost through the Advisory Centre for Education, Cambridge. The unique aspect of this survey was that it was conducted predominantly by school children. Yet it produced a set of results that has been classed as one of the most comprehensive overviews of air quality yet conducted in the British Isles.

The chief difference between the two "particle classes" of air pollutants and the gaseous air pollutants, as far as analysis is concerned, is that for the latter normal filtration methods can not be applied. Physical adsorption, such as on activated charcoal [47], or silica, or absorption into such matrices as silica gel, rubber, zeolites or the like can be used to capture many gaseous air pollutants [48]. Recovery of the gases or vapours required for later analysis, or for recycle depending on the scale, is frequently achieved by warming the bed of adsorbent or absorbent while passing air or an inert gas through

the bed. Sometimes this desorption is assisted by a pressure reduction. Wet absorption methods are also used, with a reagent dissolved in the liquid (usually water) to obtain reaction with and hence retain the pollutant of interest in the liquid as the air containing it is sparged through it. Sulfur dioxide, which is not well retained when contaminated air is sparged through water alone [49], is efficiently trapped when a solution of hydrogen peroxide or sodium hydroxide in water is used (Equations 2.4-2.6).

$$SO_2 + H_2O \rightarrow H_2SO_3 \qquad (2.4)$$

$$H_2SO_3 + H_2O_2 \rightarrow H_2SO_4 + H_2O \qquad (2.5)$$

$$H_2SO_3 + 2\,NaOH \rightarrow Na_2SO_3 + 2\,H_2O \qquad (2.6)$$

### 2.2.3.1  Wet Chemical Analysis of Gases

Sometimes the chemical capture reagent actually forms a part of the wet chemical analytical scheme to be used. For example, if hydrogen peroxide is used as the sulfur dioxide capture reagent, then titration using standard alkali gives the final sulfuric acid concentration obtained in the capture solution (Equation 2.5). Relating the measured acid concentration back to the original volume of air passed through the absorbing solution then gives the sulfur dioxide concentration originally present in the air. The answer obtained by this method is subject to errors when other acidic or basic gases (e.g. NO, $NO_2$, $NH_3$) or aqueous aerosols containing sulfuric acid, nitric acid, ammonium hydroxide or the like are present in the air being sampled. The influence of these interferences can be greatly decreased if the determination is carried out gravimetrically by the addition of barium chloride rather than by titration (Equation 2.7).

$$H_2SO_4 + BaCl_2 \rightarrow 2\,HCl + BaSO_4 \downarrow \qquad (2.7)$$

When sodium hydroxide in water is used as the medium to trap sulfur dioxide it allows a separate result to be obtained for the sulfur dioxide content, independent of any sulfur trioxide or sulfuric acid present in the sampled gas. This distinction is not possible when using hydrogen peroxide. To obtain this separate result requires addition of an involatile acid to the collection medium after collection. This is followed by heating, or sparging of the acidified solution with inert gas while heating, to release the sulfur dioxide from solution (Equation 2.8).

$$Na_2SO_3 + H_2SO_4 \rightarrow Na_2SO_4 + H_2O + SO_2 \uparrow \quad (2.8)$$

Capture of this sulfur dioxide in fresh aqueous base, followed by titration with standard iodine solution then gives the concentration of sulfur dioxide present in this second solution [50] (Equations 2.9, 2.10).

$$SO_2 + 2\,NaOH \rightarrow Na_2SO_3 + H_2O \quad (2.9)$$

$$SO_3^{2-} + I_2 \rightarrow SO_3 + 2\,I^- \quad (2.10)$$
$$\text{brown} \qquad \text{colourless}$$

Similar wet chemical techniques have been developed for many other polluting gases, but are beyond the scope of this presentation. The general features of wet chemical analysis are such that the investment in equipment and materials required is generally low, but the skill level required of the analyst, particularly for some of the tests, is high. While it is possible to streamline some of the wet chemical methods so as to enable many results to be obtained in a working day, these methods are generally slow relative to other alternatives. Comprehensive accounts of suitable methods are available from many sources [50-52].

There is one gas analysis technique which borrows from the colourimetry of manual procedures but with which the technology of the determination is prepackaged in sealed adsorbent tubes. Draeger, and now also the Gastec and the Matheson-Kitagawa systems for gas analysis all use glass tubes packed with a solid support coated with an appropriate colourimetric reagent for the gas of interest. Company literature assesses the coefficient of variation for the tubes with most readily discernable colour changes as 10 %, and for the less efficient tubes as 20 to 30 % [53] (Equations 2.11-2.13).

This is certainly sufficient level of precision for the occasional determination of the less common

Standard deviation =
$$[(x_1{}^2 + x_2{}^2 + \ldots . x_n{}^2)/(n-1)]^{0.5}, \quad (2.11)$$
where $x_1, x_2, \ldots x_n$ are the deviations of individual determinations from the mean of all determinations in the series being measured, and n represents the number of individual determinations in the series.

Mean deviation = $(x_1 + x_2 + \ldots x_n)/n$,   (2.12)
absolute values of the individual deviations are used, without regard to sign. It is also referred to as the average deviation.

**Table 2.3** Some examples of the gases which may be determined by prepacked colourimetric tubes [a]

| Gas or vapour | Measurement range[b] | |
| --- | --- | --- |
| | Minimum | Maximum |
| ammonia | 5 ppm | 10% |
| benzene vapour | 5 ppm | 420 ppm |
| carbon monoxide | 5 ppm | 7% |
| chlorine | 0.2 ppm | 500 ppm |
| ethylene oxide | 5 ppm | 3.5% |
| formaldehyde | 0.5 ppm | 40 ppm |
| hydrogen sulfide | 1 ppm | 7% |
| mercury vapour | 0.1 mg/m$^3$ | 2 mg/m$^3$ |
| methane thiol (methyl mercaptan) | 2 ppm | 100 ppm |
| nitrogen dioxide | 0.5 ppm | 50 ppm |
| oxygen | 5% | 23% |
| ozone | 0.05 ppm | 300 ppm |
| perchloroethylene | 5 ppm | 1.4% |
| phenol | 5 ppm | c |
| sulfur dioxide | 0.1 ppm | 2000 ppm |
| tetrahydrofuran | 100 ppm | 2500 ppm |
| vinyl chloride | 0.5 ppm | 50 ppm |
| xylene | 5 ppm | 1000 ppm |

[a] Compiled from [53, 277 and 278].
[b] Concentrations are specified by volume (see text). Wider concentration ranges are covered with two to five detector tubes of differing sensitivities.
[c] Not specified.

$$\text{Coefficient of variation} = \frac{\text{(standard deviation)}}{\text{mean}}$$
$$(2.13)$$

gases, and provides a much more convenient method for this purpose than most alternatives (Table 2.3). However, it should be noted that in the early days of the development of these systems a tube for carbon tetrachloride vapour was questioned as to its capability to achieve a precision within 50 % of stated values [54]. Nevertheless, even this level of accuracy is still adequate for many regulatory and industrial hygiene requirements.

### 2.2.3.2 Instrumental Methods for Gas Analysis

Instrumental analysis of air samples can frequently be conducted directly on the air or gas sample, particularly when either the concentration of the contaminant is high or when the analytical method being used is sensitive. Otherwise, prior concentration of the component of interest is necessary.

This may be accomplished by adsorption onto a substrate from low concentrations, to be later released at higher concentrations by heating, evacuation or the like. Or the concentration step may be accomplished by absorption from the air, as fine bubbles discharged under the surface of an entrapping liquid. The sampling method will be selected on the basis of the type of information required from the sample and the instrumentation that is to be used for analysis.

Spectrophotometric methods, both ultraviolet and infrared, are useful for air analysis in a wide variety of formats. Either a solution or a gas sample cell may be used. For good ultraviolet sensitivity the component of interest must have two or more double bonds present, preferably conjugated. Thus sulfur dioxide, acrolein ($CH_2 = CH - CHO$), and benzene all give very good ultraviolet sensitivity, by measurement at suitable wavelengths near their absorption maxima. In contrast acetone, phosgene, and gasoline all have direct atmospheric measurement detection limits of only about 5 ppm, too high to be useful for many situations [8]. However, even if direct measurement sensitivities are poor, a colourimetric method coupled with ultraviolet spectrophotometry may be used to obtain good sensitivity. For example, ultraviolet absorption is sufficiently more sensitive for the purple complex formed by reaction of formaldehyde with chromotropic acid than it is for formaldehyde itself to enable solution detection limits of $1 \mu m$ ($0.1 \mu g/mL$) to be achieved [55] (Equation 2.14). For a 10 L air sample this corresponds to an ambient air sensitivity of $0.1 mg/m^3$.

Ultraviolet absorption by sulfur dioxide may also be usefully applied in the field for direct plume observation. In a novel technique developed for the observation of the colourless discharge of sulfur dioxide and water vapour from a natural gas cleaning plant vent stack, use of silica optics and a special film allow the clear photographic differentiation and observation of plume behaviour [56, 57].

The infrared spectrum of many air pollutants is generally more complex than its ultraviolet spectrum, showing several absorption maxima. But for many single components the pattern of maxima obtained can permit positive identification of the material at the same time as obtaining the concentration from the absorbance. Any compound absorbs infrared energy although the intensity of the absorption varies greatly from compound to compound. This affects the relative sensitivity of the method for different compounds. However, recently developed multiple path cells compensate for this somewhat by providing effective absorption path lengths of 20 m or more in a complex but compact gas cell which uses internal reflections (Figure 2.4). Typical infrared detection limits quoted for this system are 1.2 ppm for carbon monoxide, 1.5 ppm and 0.08 ppm for nitric (NO) and nitrous ($N_2O$) oxides, respectively, 0.1 ppm for sulfur dioxide and 0.05 ppm for carbon tetrachloride [58]. Some of these infrared instruments are conveniently portable and fitted for 12 volt power, suitable for operating from an ordinary car battery. These field capabilities avoid many of the problems faced by sample containment systems required for field collection and then transport to the laboratory for analysis.

Rapid multiple scans, signal storage, plus fourier transform capabilities have combined to push infrared detection limits to about 1/100 of the concentrations possible with these more conventional infrared instruments, but at about 10 times the cost [59-62].

The really significant spectroscopic developments that are moving air pollutant analysis into the forefront of new capabilities are closely linked to developments in laser technology [63]. The coherence and high resolution possible with a laser source, coupled with the high power available with some configurations makes inherently high sensitivities possible, particularly where the match between laser and absorbing frequencies is good. For example, it has been estimated that sensitivities of the order of 1 ppb should be possible with nitrous oxide, using a 100 m path length. Even lower sensitivities are reasonable for more strongly absorbing gases [64, 65]. The high source powers available also allow the use of lasers for remote sensing of air pollutants, such as in smog situations, or for discharge plumes, or even for ambient air pollutant concentrations [64, 66-68].

Various laser instrument arrangements are possible depending on the particular remote sensing application (Figure 2.5). The direct absorption mode, $a$, is the simplest, but suffers from the inconvenience that source and detector units must be sited separately, and aligned for each particular test situa-

$H_2CO$ (colourless) + chromotropic acid $\rightarrow$ purple complex     (2.14)

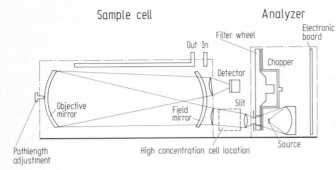

**Figure 2.4.** Diagram of the operating details of a portable infrared gas analyzer which uses a long path, multiple internal reflection gas cell. Adjusting the angle of the objective mirror alters the number of internal reflections obtained, before the beam enters the detector [58] (Foxboro Co.)

tion. Retroreflection, either from a surface-silvered mirror or from a geographical feature such as a tree or rock, *b,* at least allows source and receiver to be operated from the same site. In the backscatter mode, *c,* the reflections from aerosol particles which have passed through the target air sample are picked up by the receiver-detector operating at the same location. In this situation the laser is tuned to a high absorption frequency of interest and then to a low absorption frequency near this. The concentration of the component of interest may be computed from a comparison of the relative absorption signals, the basis of the so-called DIAL (*d*ifferential *a*bsorption *l*idar) system [64, 67].

One other non-spectroscopic instrumental technique which should be mentioned as being particularly appropriate to the analysis of gaseous air pollutants is gas chromatography, particularly when this is coupled to an efficient mass spectrometer. Gas chromatographic techniques allow efficient separation and detection of many common and less common air pollutants. With the more sensitive columns and an electron-capture or similar sensitivity detector it can give a measurable signal with as little as $10^{-12}$ g of separated component [69]. But gas chromatography itself does not permit unequivocal identification of the separated component. Coupling a quadrupole mass spectrometer to the exit of a gas chromatograph allows both functions to be performed, the gas chromatograph to carry out separations of constituents, and the rapid acting mass spectrometer to identify them from their fragmentation patterns or molecular ions [70, 71]. The value of this combination to provide rapid answers for the real time tracking of gases, vapours, and fumes released from a train derailment emergency involving released chemicals has been well demonstrated [72-74].

Many other instrumental methods are also used for gas analysis, the details of which are available elsewhere [69, 75-77]. Development, standardization and validation of the various methods used for particular pollutants of interest are continuing and on-going processes [78, 79].

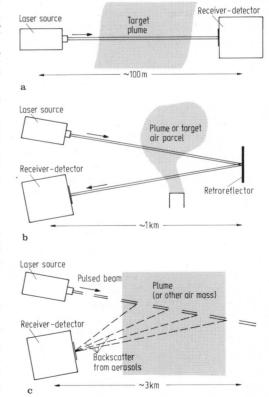

**Figure 2.5.** Some possible configurations of laser source and receiver detectors for use in remote sensing of air pollutants. **a** Simple direct absorption mode **b** Retroreflective direct absorption mode **c** Backscatter from aerosols mode

### 2.2.3.3 Concentration Units for Gases in Air

The concentration of a gas or vapour in air or any other gas phase, may be specified on a volume for volume basis, a weight for volume basis or on a partial pressure basis. Each of these concentration units has one or more features which makes its use convenient in particular situations.

Concentrations of one gas in another, specified on a volume for volume basis, are normally corrected to 25 °C and 1 atmosphere (1.01325 bar, 760 mm Hg) pressure for reporting [80]. Under this system relatively high concentrations are specified in percent (%) so that 3 % sulfur dioxide in air would correspond to 3 mL of sulfur dioxide mixed with 97 mL of air, both specified at 25 °C and 1 atmosphere. In more general terms this also corresponds to 3 parts by volume of sulfur dioxide in 100 parts by volume of the sulfur dioxide/air mixture, both specified in the same volume units. Lower concentrations are specified in smaller units, ppm for parts per million (1 in $10^6$), ppb for parts per billion (1 in $10^9$), and now that adequate sensitivity has been developed for the analysis of some air pollutants at these low concentrations occasionally even ppt for parts per trillion (1 in $10^{12}$) [7].

This volume for volume system of specifying gas gas concentrations is in essence unitless, i.e. it represents a pure ratio of quantities given in the same volume units which thus allows the units to divide out. To a chemist this system has the further advantage that comparisons made on a volume for volume basis are also on a molecular, or molar basis, i.e. a mole of any gas at normal temperature and pressure (N.T.P. or S.T.P.; 0°C and 1 atmosphere) occupies 22.4. L. This volume for volume comparison is true for most gases and vapours when existing as mixtures at or near ambient conditions of pressure and temperature, and only deviates significantly from this molar equivalency when the conditions (pressure and/or temperature) become extreme. But it must be remembered that volume for volume data are corrected to 25 °C and 1 atmosphere, under which conditions a molar volume corresponds to 24.5 (24.46) L (298 K/273 K x 22.41 L).

These combined properties of this volume for volume concentration unit are such that, if the temperature or pressure is changed, there is no change in the concentration specified in this manner, within reasonable limits. Any correction which may be necessary as a result of a minor variation of the measuring conditions is generally smaller than the experimental error of the analysis. Since many analyses of air samples are conducted at temperatures close to 25 °C and normal atmospheric pressure, corrections are only seldom necessary. Because of this feature the temperature or pressure does not normally need to be specified when using volume for volume units for quoting concentrations of a gas in air or another gas.

Increasingly often, however, the concentration of a gas (or vapour) in air is being specified on a weight per unit volume basis, common units for which are mg/L, mg/m$^3$, and $\mu$g/m$^3$ (Equation 2.14).

$$1 \text{ mg/L} = 1000 \text{ mg/m}^3 = 10^6 \text{ } \mu\text{g/m}^3 \qquad (2.14)$$

Knowing the concentration of an air pollutant specified in this way allows easier determination of mass rates of emission, which are important for regulatory purposes and for exposure hazard calculations. But using this system, there is no direct molecular basis for comparison of the concentrations of any two gases, which makes it more difficult to visualize chemical relationships with this system. Also the concentration of a gas in a gas, specified in these units, changes with changes in temperature or pressure. This happens because the mass (or weight) of a minor component per unit volume in this system becomes smaller as the temperature of the gas mixture is raised, whereas the volume of the air (or other main component) in the mixture becomes larger. Both influences tend to make the specified concentration smaller with a rise in temperature. Since 0 °C has been used as "standard temperature" by some authorities [81, 82] and 25 °C by others [7, 83] for specifying weight per unit volume gas concentrations, to avoid ambiguity in specifying results in these units the temperature or the standard temperature basis used *must* also be specified. Data presented in this format are normally corrected to one atmosphere pressure.

The common usage of both volume for volume, and weight for volume units in the quoting of air pollution results frequently makes it necessary to be able to inter-convert between these units. To obtain a value in mg/m$^3$ from a value in ppm one has to multiply the ppm value by the molecular weight in grams, of the component of interest and divide by 24.46, the molar volume at 25 °C (Equation 2.15).

The conversion process is similar for $\mu$g/m$^3$ and

$$mg/m^3 = \frac{1/10^6 \, (ppm) \times mol.wgt. \, (g/mol) \times 10^3 \, (mg/g)}{24.46 \, (L/mol) \times 10^{-3} \, (m^3/L)} \tag{2.15}$$

$$= (ppm \times mol.wgt. \, (g/mol))/24.46 \, (L/mol), \text{ at } 25\,°C.$$

$$\mu g/m^3 = (ppm \times mol.wgt. \, (g/mol) \times 10^3)/24.46 \, (L/mol) \tag{2.16}$$

$$mg/L = (ppm \times mol.wgt. \, (g/mol) \times 10^{-3})/24.46 \, (L/mol) \tag{2.17}$$

$$ppm \text{ of constituent} = \frac{(partial \text{ pressure of constituent}) \times 10^6}{(total \text{ barometric pressure})} \tag{2.18}$$

mg/L from ppm, again both at 25 °C (Equations 2.16, 2.17).

For weight for unit volume values at temperatures other that 25 °C the molar volume for the temperature of interest must be used in place of 24.46 L/mole.

The partial pressure system is also occasionally used to specify the concentration of a gas in a gas (e.g. Equation 2.18).

Common pressure units are used for the partial pressure of the component of interest and total barometric pressure readings, so that the units divide out. Thus, in all important respects the partial pressure system is equivalent to the volume per unit volume system for specifying concentrations of a gas in a gas. This system is of particular value, for instance, when making up synthetic mixtures of gases on a vacuum line, where the partial pressure of each component is known, and can be used to determine the relative concentrations.

### 2.2.3.4 Biological Methods for Air Pollution Assessment

High concentrations of air pollutants are known to kill many annual plants and trees. This amounts to a biological indicator, albeit a coarse one. Less severe exposures can cause premature senescence (early leaf drop) of sensitive species of trees and shrubs [84, 85]. For example aspen poplar, *Populus tremuloides* drops its leaves after exposure to as little as 0.34 ppm of sulfur dioxide for 1 hour [86]. Early leaf drop slows tree growth, a measurable result which can be qualitatively determined by examining growth ring widths from tree cores [87]. The olefactory membranes, as another kind of biological indicator, also provide an odour warning of air pollutant exposure to man and many animals which in some cases rivals the sensitivity of the most sophisticated instrumentation [88]. Since many toxic gases also possess a more or less intense odour the avoidance response that a smell activates is also useful in this way [7].

However the lichens, as a group, have been far more generally useful as indicators of pollution levels over time and as measures of air pollution trends for both city [89] and rural study areas [90]. Lichens represent a slow growing symbiosis between an alga, which photosynthetically manufactures carbohydrate, and a fungus, which aids the alga in water storage in an otherwise inhospitable site and uses the manufactured carbohydrate for its own metabolism and growth. Their very slow growth, and their high reliance on the air for most of the lichens' nutrient requirements, makes for efficient uptake and retention of many air pollutants. Hence, the level of air pollution more significantly affects the growth and reproduction of lichens than that of most other, faster-growing forms of plants which rely on a soil substrate [91, 92]. Also, heavy metal fumes or other particulate air pollutant exposures tend to accumulate to higher concentrations in lichens than in most other forms of plants [87]. In the hands of skilled botanists a lichen diversity and distribution study of an area can give useful long term exposure information regarding continuous or intermittent fumigation by pollutants such as sulfur dioxide which would be difficult to obtain in any other way [93, 94].

Even in the hands of the relatively uninitiated but careful observer it is possible to determine up to 6 or 7 zones of relative air pollutant exposures by recording the diversity and types of lichens found growing in an area [95]. Thus, in a relatively polluted area few, or no healthy lichen varieties will be found. Those more resistant varieties that do occur in this exposed situation, such as *Pleurococcus* species, will be forms which are closely associated with their support and recumbent, a habit that minimizes exposure of the lichen to air pollutants. At the other extreme, in areas little exposed to air pollutants, both the species diversity and the number of specimens of each species found will be larger. There will also be a better representation of the more pollution sensitive foliose (leafy) forms such as *Parmelia* and *Letharia* species, or the fruiticose (shrubby) forms such as *Usnea* or *Alectoria* species. Biological methods based on lichen surveys such as this represent an important complement to the information obtained from wet chemical and instrumental analysis fot determining the integrated levels of air pollution over time for an area.

## 2.3 Effects of Air Pollutants

Much has been said and written about the effects of air pollutants on plants and animals so that only a brief summary needs to be presented here. The health effects on man probably comprise the most direct, even if somewhat anthropocentric, concern with air pollutants. The primarily nuisance effect of the smell emanating from a fish packing plant represents an example at the minor end of the effects range. However, since the average person takes in some 22,000 breaths per day, amounting to some 14 to 16 kg (30 to 35 lb) of air per day [8], even a relatively low concentration of most air pollutants is sufficient in this large mass of inspired air to begin to have a noticeable effect. For this reason alone the air environment has to be the most important of the biospheric elements to be worth striving to maintain or improve in quality.

The elevated air pollutant levels which occur during localized smog episodes tend to severely impact on the ill, and the young and the old, causing effects all the way from excessive lachrymation and restricted breathing, to an aggravation of respiratory illnesses in particular, even to a noticeable rise in the death rate recorded for the affected area in the most severe episodes [96, 97]. London, England, the Donora valley, Pennsylvania, and the Meuse valley, Belgium, are just a few of the fairly well documented older examples of occurrences of this last level of severity [7, 98-100].

Today, two clearly differentiated types of smogs are recognized. In the classical variety of smog, represented by the pollution episodes mentioned above, the problem was caused by the accumulation of primary air pollutants such as sulfur dioxide, particulates, and carbon monoxide contributed by smoke, complicated by the presence of fog (Table 2.4). The fog also tended to slow air pollutant dispersal by cutting off sunlight, preventing the normal mixing process which is stimulated via the warming of air close to the earth's surface by the sun. The characteristics and time of occurrence of a classical smog differ markedly from the more recent phenomenon of photochemical smog which was first described for Los Angeles, but which is also experienced by Tokyo, Mexico City and other major centers with similar circumstances [101].

In photochemical smog episodes, secondary air pollutants such as ozone, nitrogen dioxide, and

**Table 2.4** Distinguishing features of classical and photochemical smogs [a]

|  | Classical smog | Photochemical smog |
|---|---|---|
| Location example | London, early | Los Angeles |
| Peak time of occurence | winter | summer |
| Conditions | early a.m., 0 to 5 °C, high humidity plus fog | around noon, 22-35 °C, low humidity, clear sky |
| Atmospheric chemistry | primarily reducing, $SO_2$, particulates, carbon monoxide, moisture | oxidizing, nitrogen dioxide, ozone, peroxyacetylinitrate |
| Human effects | chest, bronchial irritation | primarily eye irritation |
| Underlying causes | fog plus stable high, surface inversion, emission dispersal of primary pollutants is prevented | sheltered basin, frequent stable highs, accumulation of secondary polutants from photochemical oxidation |

[a] Compiled from [7 and 101].

peroxyacetylnitrate play the dominant role in the effects felt by plants and man [101-105]. These are formed as a result of the chemical interaction of the primary air pollutants, principally nitric oxide and hydrocarbon vapours, with sunlight and air [106]. The interesting recent development is that the authorities of the London area have been successful in reducing emissions of sulfur dioxide and particulates sufficiently via the regulations of their Clean Air Act that classical smogs are a thing of the past in this area [101]. But the more frequent and intense sunshine experienced as a result of this improvement coupled with the recent increase in automobile traffic in this city now means that London, too, occasionally experiences the irritating effects of photochemical smog occurrences [107, 108].

Air pollutant concentrations such as occur in either type of smog episode are also sufficient to affect plants more or less severely from a combination of interactions [84]. When this occurs in conjunction with mist or rain, the pH of wetted surfaces is dropped low enough to seriously damage statuary and stone or masonry buildings from a simple solution reaction of the stonework with the acidic water (e.g. Equation 2.19).

$$CaCO_3 + 2\,HNO_3 \rightarrow \underset{\text{soluble}}{Ca(NO_3)_2} + H_2O \quad (2.19)$$

Similar processes cause the attack of ironwork and other metalwork, and accelerate the decay of exposed wood. Even aluminum, which is highly favoured for use in exterior metalwork because of its resistance to corrosion, is now showing the classic pits and erosion scars of corrosion when used in areas which have severe air pollution.

**Table 2.5** Contributions of sulfuric and nitric acidity of acid rain [a]

| Substance | Concentration (mg/L) | Contribution to total acidity[b] |
|---|---|---|
| $H_2SO_4$ | 5.10 | 57 |
| $HNO_3$ | 4.40 | 39 |
| $NH_4{}^+$ | 0.92 | 51 |
| $H_2CO_3$ | 0.62 | 20 |
| all others | ca. 0.4 | ca. 12 |

[a] Data selected from that of [110]. Determined for a sample of rain of pH 4.01 collected at Ithaca, N.Y. in October 1975.
[b] Microequivalents per liter. Ammonium by titration to pH 9.0.

The lowered pH caused by rainout and washout of nitrate and sulfate from the atmosphere has also resulted in the now well known "acid rain" phenomenon (Table 2.5) [109-111]. When it is realized that a pH of as low as 2.4, about the acidity of lemon juice, or vinegar, has been measured for the rainwater of individual storms (Pitlochry, Scotland), and an annual pH value for precipitation of 3.78 has also been recorded in the Netherlands [112, 113], it is evident that this trend can only serve to further aggravate structural damage to buildings. This lowered pH of precipitation is also having a particularly serious effect on the biota of those lakes which have a limited carbonate-bicarbonate natural buffering capacity [113, 114]. When the ionic exchange capacity of soils becomes exhausted it can result in reduced growth, or actual die-back of forests [115].

Related to smogs but only relatively recently noticed by the scientific community is the phenomenon of Arctic haze, a brownish turbidity occurring in the atmosphere of Arctic regions from late fall to March or April of each year [116]. A suspended aerosol of primarily sulfates, $2\,\mu g/m^3$, organic carbon, $1\,\mu g/m^3$ and black carbon, 0.3-0.5 $\mu g/m^3$ reduces the visibility of the region to 3 to 8 km, over an area of several thousand square kilometers and up to an altitude of about 3,000 m during this period. The novelty here is that this represents an accumulation of pollutants, remote from the original points of discharge, which by the idiosyncracies of atmospheric movements and conditions resides over and adversely influences a site not responsible for the emissions [117].

This tendency of aerosols to reduce visibility during Arctic haze episodes is also a very general effect of atmospheric particles, caused by the scattering of light by particles or droplets in the aerosol size range [7]. Apart from aesthetic considerations, high loadings of solid and liquid (fogs) atmospheric aerosols also influence flight conditions at airports, and have been implicated in rainfall promotion patterns for St. Louis [118].

Another important consideration related to atmospheric loadings and particulate size distributions concerns the potential for human effects on inhalation. The body's defences in the upper respiratory system are adequate to trap more than 50 % of particles larger than about 2 $\mu m$ in diameter which are present in the air breathed. However, particles smaller than this that are not efficiently captured by the upper respiratory tract, the "respirable particulates", penetrate the lungs to the alveolar level [119]. Here, the more vigorous Brownian motion of these small aerosols increases their collision frequency with the moist walls of the alveoli, thus trapping a large proportion of them. This occurs in a region of the respiratory system poorly equipped to degrade or flush out accumulated material, particularly if the aerosol is inert or insoluble. Hence, any physiological effect of the presence of these foreign substances is aggravated, increasing the incidence of respiratory illness experienced in areas with high concentrations of polluting aerosols. The control efficiencies of many types of air pollution control equipment for the lower aerosol size range is poor, an important consideration when the decision is made re process emission control options [120].

Proceeding from the local and regional effects of air pollutants to the global scale, there is no doubt that our escalating use of fossil fuels coupled with the widespread cutting of forests have contributed to a steady rise in the atmospheric carbon dioxide concentration, now averaging about 0.7 ppm per year (Figure 2.6) [121-124]. Carbon dioxide is virtually transparent to the short wavelength (ca. 1 $\mu$m) maximum in the incoming solar radiation, but has a substantial absorption band in the region of the longer wavelength infrared irradiation emanating from the relatively low surface temperature of the earth. Thus, an increase in the concentration of atmospheric carbon dioxide does not significantly affect the energy gained by the earth's surface from incoming solar radiation, because of the atmospheric absorption "window" in this re-

gion. But the carbon dioxide of the atmosphere does absorb a significant fraction of the energy in the outgoing long wavelengths, a fraction which is increasing with the increasing concentration of carbon dioxide. This net energy gain, coupled with other related influences which are sometimes collectively dubbed the "greenhouse effect", is predicted to cause a warming trend in the earth's climate [7, 125-129].

While the increase in average global temperatures that this thermal effect brings about may be only a Celsius degree or less, it could still cause a gradual decrease in water storage in the earth's ice caps. This in turn could raise the world's ocean levels, causing changes to low lying coastlines [122]. Also, an increase in average global temperatures would be likely to increase the rate and extent of desertification of marginally arable land, reducing the already limited agricultural resource. Recently, however, there appears to have been a global downturn in surface temperatures, thought possibly to arise from an increased loading of sunlight-obscuring aerosols in the atmosphere [128, 130]. Thus, consideration of atmospheric carbon dioxide concentrations in isolation is inadequate to predict overall climatic trends, even though it may be possible to predict the climatic influence of this component. Much further study and more detailed modelling of the global energy balance is necessary before the conclusions from carbon dioxide concentration trends will be any more than speculative (e.g. reference 131).

The other potential global problem from increased atmospheric pollutant levels, and of halocarbons and fixed nitrogen in particular, concerns the perceived negative influence of these gases on the stratospheric concentrations of ozone [132, 133]. This potential problem relates to the interference by these gases in the ozone-forming and degrading processes which normally contribute to the natural steady state concentration of ozone in the stratosphere. When first recognized, it was predicted that this interference would cause a decrease in the equilibrium ozone concentration present [132, 134-136]. The ozone content of the stratosphere serves as an important atmospheric filter for damaging short wavelength ultraviolet. So the net effect of a decrease in the ozone concentration was predicted to cause an increase in the global exposure to short ultraviolet which, in turn, was anticipated could cause in increase in the incidence of human skin melanomas [137]. Fortunately the apparent down-

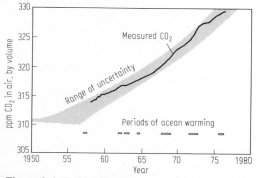

**Figure 2.6.** Estimated past perspectives and present trends of atmospheric carbon dioxide concentrations [122]

ward trend in atmospheric fluorocarbon concentrations in recent years resulting from voluntary use curtailment by the industry promises to minimize the influence of these effects [138, 139].

## 2.4  Air Pollutant Inventories, and Pollutant Weighting

In order to best apply pollution control strategies, it is an important preliminary, from a cost-benefit standpoint, to start with an inventory of the sources, mass emission rates, and types of pollutants being discharged in the area of concern. This is accomplished by a detailed analysis of the emissions from each point source of the inventory area. Data obtained by actual measurements are used wherever possible. But where measured data are not available, chemical and engineering theory is applied to come up with the best possible emission estimates within the time frame required for the inventory. Moving sources, such as automobiles, aircraft and the like are tallied by averages per unit, times the number of miles per unit, times the number of units to come up with good estimates for the contributions from these sources.

The final pollutant inventory obtained gives an overview of the total current emission picture for the study area. Table 2.6 presents this data for Los Angeles, where the inventory was conducted to assist in determining the origins/causes of the photochemical smog problem. For comparison, parallel data is presented for the whole of Canada which also has incipient smog problems in some of the major centers [140]. The best foundation for mean-

ingful planning of abatement measures, with regard to the cost and ease of control of the various alternatives relative to the extent of improvement likely to be seen from instituting these measures, is obtained from inventories of this kind.

From the tabulated data it is evident that automobiles rather than industrial activity are the source of by far the largest mass of pollutants overall (Table 2.6). Road transport, consisting of multiple moving sources under the control of individuals makes emission abatement for this classification difficult to achieve. Nevertheless, by fundamental engine design changes coupled to accessory control units installed at the manufacturing stage, significant reductions have been achieved.

### 2.4.1  Automotive Emission Control

As much as 70 % of the hydrocarbons, 98 % of the carbon monoxide, and 60 % of the nitrogen oxides ($NO_x$), have originated from the operation of automobiles (Table 2.6). Hence, these are the principal automotive emissions of concern. The hydrocarbons and carbon monoxide arise from the fact that the most power for a given engine capacity (displacement) is obtained with enough air for only 70 to 80 % complete combustion. Nitrogen oxides are formed from the interactions of atomic oxygen and atomic nitrogen, which are formed against hot metal surfaces at high temperatures, with the corresponding elements [141] (Equations 2.20, 2.21).

$$N_2 + [O] \rightarrow NO + [N] \qquad (2.20)$$
$$O_2 + [N] \rightarrow NO + [O] \qquad (2.21)$$

These emission problems are not easy to solve. To run with a leaner tuned engine (a higher air to fuel

**Table 2.6** Air emission inventory for Los Angeles in November 1973 compared on an percentage basis to the annual inventory for Canada, 1974 [a]

| Emission Source | Emissions to the atmosphere, tonnes/day | | | | | Percent of total | |
|---|---|---|---|---|---|---|---|
| | Hydro-carbons | Nitrogen oxides | Sulfur dioxide | Carbon monoxide | Totals, by source | Los Angeles | Canada |
| Automobiles | 1 750 | 445 | 27 | 9 375 | 11 597 | 86.9 | 47.9 |
| Organic solvent use | 450 | | | | 450 | 3.4 | |
| Oil refining | 200 | 41 | 41 | 154 | 436 | 3.2 | 27.1 |
| Chemicals production | 50 | | 60 | | 110 | 0.8 | |
| Combustion of fuels | 13 | 245 | 280 | 1 | 539 | 4.1 | 9.1 |
| Miscellaneous | 37 | 27 | 5 | 144 | 213 | 1.6 | 15.9 |
| Totals, by pollutant | 2 500 | 758 | 413 | 9 674 | 13 345 | 100.0 | 100.0 [b] |

[a] From [142 and 279].
[b] Canadian air pollutant inventory total for the listed categories for 1974 was 25.8 million tonnes.

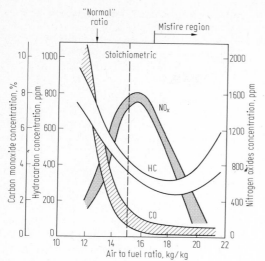

**Figure 2.7.** Plot of the approximate effect of various air to fuel ratios on the composition of automotive exhaust gas [142]

ratio) does obtain more complete fuel combustion and in so doing decreases the hydrocarbon and carbon monoxide concentrations in the exhaust, as would be expected. But at the same time the higher combustion temperatures obtained in the process raises the concentrations of nitrogen oxides obtained in the exhaust, and decreases the power output per liter of engine displacement [142] (Fig-

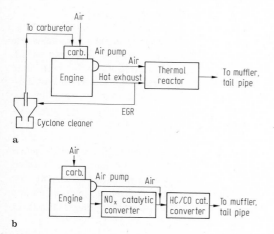

a

b

**Figure 2.8.** Block diagrams of two automotive emission control configurations. **a** Thermal reactor plus exhaust gas recirculation (EGR). **b** Catalytic exhaust purification

ure 2.7). It also affects driveability or smooth engine response at various throttle settings.

Early emission control methods were based on the use of a thermal reactor, for hydrocarbon and carbon monoxide oxidation, combined with exhaust gas recirculation (EGR) for reduction of nitrogen oxide emissions [143] (Figure 2.8 a). Hydrocarbons and carbon monoxide in the hot exhaust fed to the reactor, once heated, were rapidly oxidized to carbon dioxide and water by the additionally pumped air which was fed to the reactor (representative Equations 2.22, 2.23).

$$HC (= hydrocarbon) + O_2 \rightarrow CO_2 + H_2O \quad (2.22)$$

$$2\,CO + O_2 \rightarrow 2\,CO_2 \quad (2.23)$$

Some of the relatively inert exhaust gas, before it entered the reactor, was used to dilute the air fed to the carburettor by some 15 %, in so doing decreasing the peak combustion temperatures during normal engine operation [141]. This effectively decreased the concentrations of nitrogen oxides formed. These combined measures were effective in achieving significant emission reductions (Table 2.7), but at the same time caused a noticeable increase in fuel consumption and loss of performance [144, 145].

While thermal reactors have been used extensively [146], two-stage catalytic exhaust purification approached these emission problems in a different way. Here the first stage of control, $NO_x$ reduction, is achieved catalytically using the reserve chemical reducing capacity of the residual hydrocarbons and carbon monoxide already present in the exhaust [147] (Figure 2.8 b), (e.g. Equations 2.24-2.27).

$$2\,NO + 2\,CO \rightarrow N_2 + 2\,CO_2 \quad (2.24)$$

$$2\,NO_2 + 4\,CO \rightarrow N_2 + 4\,CO_2 \quad (2.25)$$

$$2\,NO + HC (= hydrocarbon) \rightarrow N_2 + CO + H_2O \quad (2.26)$$

$$2\,NO_2 + HC \rightarrow N_2 + CO_2 + H_2O \quad (2.27)$$

All of these $NO_x$-reducing reactions are much more rapid and effective than even the catalyzed direct redox reaction (Equation 2.28),

$$2\,NO \overset{cat.}{\rightarrow} N_2 + O_2 \quad (2.28)$$

and hence accomplish most of the $NO_x$ conversion to nitrogen. Subsequent to $NO_x$ reduction, secondary air injection and passage of the hot mixture into

**Table 2.7** Typical automotive emissions before and after regulation, related to U.S. legislated standards [a]

| | Grams per mile | | | | Gasoline evapn, g/test |
|---|---|---|---|---|---|
| | Carbon monoxide | Hydro-carbons | Nitrogen oxides | Particulates | |
| Before control | 80 | 11 | 4 | 0.3 | 60 |
| 1970 average | 23 | 2.2 | 4 – 5 | – | – |
| 1975 standard | 15 | 1.5 | 3.1 | 0.1 | 2 |
| 1980 standard | 7.0 | 0.41 | 3.1 | – | – |
| Thermal reactor + EGR[b] | 4.5 | 0.2 | 1.3 | | |
| Catalytic, + cyclone | 0.85 | 0.11 | 1.65 | 0.1 | 1 – 2 |

[a] Compiled from [142, 143, 150 and 151].
[b] EGR short for exhaust gas recirculation, referring to the practice of recycling of about 10 to 15% of exhaust gases to the intake of the engine to decrease peak combustion temperatures.

a second catalytic converter accomplish complete oxidation of the residual hydrocarbons and carbon monoxide not used for $NO_x$ reduction in the first converter [142, 148] (Equations 2.22, 2.23). One or two grams of platinum or palladium finely distributed on a ceramic matrix provides the catalytic activity of both converters [149, 150]. This combination exhaust treatment achieves better control of the hydrocarbons and carbon monoxide, and almost as good control of nitrogen oxides as achieved with the thermal reactor, and at the same time restores much of the lost fuel economy and performance [151] (Table 2.7). If a limited exhaust gas recirculation circuit is integrated with this dual converter system, the $NO_x$ emission rate can be further improved.

Since lead-containing antiknock additives of leaded gasolines rapidly destroy the activity of the catalytic emission control system, non-leaded gasolines have to be used in automobiles fitted with this or related systems. However, since using this fuel modification also eliminates lead emissions this trend also has a highly favourable impact on the major source of lead particulate discharges to the atmosphere [152-157]. Before non-leaded gasolines were on the market it was common to find high lead levels in the core areas of busy cities [158, 159] and alongside freeways [160-163], a trend which is gradually reversing as the proportion of automobiles on the road equipped to burn non-leaded gasolines increases.

Whilst it is possible to significantly reduce particulate lead discharges from automobiles using leaded gasolines by employing cyclone-type particle collectors in the exhaust train [164-166], these methods only achieve incremental reduction, not elimination, of lead dispersal from this source.

Today, to maintain or further improve the gaseous emission characteristics of internal combustion engines the trend is towards integral improvements in engine design, such as fuel injection and various stratified charge modifications which accomplishes the chemistry of the add-on units but as an integral part of engine operation [146, 167-169]. These developments are leading to more driveable, fuel efficient automobiles.

Combinations of these control measures have resulted in a decrease in ambient carbon monoxide concentrations, and in the severity and frequency of photochemical smog incidences, where they have been applied, even though there has been a concurrent increase in the number of automobiles on the road.

### 2.4.2 Air Pollutant Weighting

Accumulation of emission data on a mass basis is a required step to assess the overall impact of air pollutant discharges of a region on this basis. But for development of the most effective control strategy for a control area, the contributing pollutants should be considered on the basis of their relative significance in terms of health effects, smog occurrences, and the like, and not solely on a mass basis. Two of the various weighting factors which have been proposed to do this are given in Table 2.8. While the weightings of these two systems differ significantly from each other, particularly for hydrocarbon vapours, they both assign the lowest weighting to carbon monoxide [170, 171].

If the weighting factors based on California standards are used to recalculate the relative significance of the Los Angeles emission sources on this basis, transportation still remains as the largest

**Table 2.8** Examples of weighting factors used for primary air pollutants

| | Weighting factors based on: | |
|---|---|---|
| | California ambient air standards[a] | U.S. federal secondary air quality standards[b] |
| carbon monoxide | 1.00 | 1.00 |
| hydrocarbons | 2.07 | 125 |
| sulfur oxides, $SO_x$ | 28.0 | 21.5 |
| nitrogen oxides, $NO_x$ | 77.8 | 22.4 |
| particulates | 106.7 | 37.3 |
| oxidants | 186.9 | – |

[a] Used as the basis of the proposed Pindex air pollutant rating system [170].
[b] Based primarily on federal standards with health effect considerations [171].

single sector, but now only just larger than the combustion sources sector (Table 2.9). This treatment thus has the effect of redistributing the significance of emission control strategies among the transportation, oil refining and chemicals production, and combustion source sectors to obtain maximum effectiveness of ambient air improvement for this control area. It can be seen that a similar treatment of the Canadian data would give even higher emphasis to the industrial and power production sectors than found for Los Angeles.

# 2.5 Methods and Limitations of Air Pollutant Dispersal

Chimneys and vent stacks have been, and still are popular for waste gas discharge from all types of stationary sources, large and small [172]. Thermal power stations, smelters, refineries and even domestic heating appliances all use discharge and dispersal methods to dispose of their waste gases. Serious problems can arise, however, if the source is a very large one (i.e. has a high mass rate of emission) discharging into a stable air mass, particularly in the absence of control of the more noxious components of its waste gas. With the occurrence of a long term inversion a temporary shutdown of the facility may be necessary to protect public health.

Simple dispersal of the flue gases of large sources, when operated without emission controls, is becoming less and less acceptable today because of the limited capability of the atmosphere of highly industrialized regions of the globe to accept further loading. Particulate matter discharged in this way only falls out again on the immediate area, and more or less rapidly [173]. Discharged fumes, fogs, and mists too, although they are more widely dispersed, also eventually return to the earth's surface as they coagulate or coalesce into larger agglomerated particles, or are washed out of the atmosphere in precipitation. Discharged gases too, which have the best prospect of efficient dispersal, eventually return adsorbed onto, or reacted with other gases, particles, or water and also return in

**Table 2.9** Comparison of actual versus weighted [a] daily air pollution inventory for Los Angeles, November 1973.

| | Total mass of gaseous primary air pollutants[b] | | Estimated mass of particulates and aerosols tonnes/day | | Weighted mass of primary pollutant emissions[c] | |
|---|---|---|---|---|---|---|
| Emission source | tonnes/day | % of total | Actual | Weighted | tonnes/day | % of total |
| Automobiles | 11 597 | 86.9 | 74 | 7 896 | 56 271 | 41.5 |
| Organic solvent use | 450 | 3.4 | 0 | 0 | 930 | 0.7 |
| Oil refining | 436 | 3.2 | 123 | 13 124 | 19 714 | 14.6 |
| Chemicals production | 110 | 0.8 | | | | |
| Combustion of fuels | 539 | 4.1 | 245 | 26 142 | 53 108 | 39.2 |
| Miscellaneous | 213 | 1.6 | 27 | 2 881 | 5 371 | 4.0 |
| | 13 345 | 100.0 | 469 | 50 043 | 135 394 | 100.0 |

[a] Weightings from California Standards, Table 2.6. See also [170].
[b] Includes hydrocarbon vapours, nitrogen oxides, sulfur oxides and carbon monoxide.
[c] Includes the primary gaseous air pollutants, plus particulates and aerosols.

precipitation. So a higher chimney merely spreads the combined fallout from these processes over a wider area; it does not decrease the gross atmospheric loading or fallout rate.

Multiple high chimneys in a single area tend to produce plume overlap some distance downwind of the original discharge points. This effect begins to negate the reduction in ground level concentrations of air pollutants, the original objective of the high stack (Figure 2.9 a).

If stack dispersal is to be used, it is important that the stack be sited in such a way that adjacent buildings or natural features such as hills or gullies do not serve to trap discharged waste gases close to the ground. Also to be avoided are backwash, a downdraft obtained on the leeward side of any large surface feature, or eddies which may occur in the same plane around the corners of obstructions. Any of these factors, if not allowed for, can cause plume impingement close to the point of discharge and hence cause highly elevated concentrations of the components of the discharged exhaust gases close to the ground.

A primary objective of stack discharge of waste gases is to obtain the minimum or zero elevation of the ground level concentrations of the discharged

gases in the immediate area of the stack. The stack discharge point should be high enough that sufficient diffusion of the discharged gases occurs to make the concentrations of these gases acceptable at the point the diffusion cone intersects the ground. For any given mass emission rate, Equation 2.28

$$\text{ground level concentration} = K(m/uH^2) \quad (2.28)$$

where m = mass rate of emission
u = wind velocity
H = *effective* stack height

may be used as a rough guide to the effect on the ground level concentration of a discharged gas which would be expected for different stack heights and wind velocities at the point of discharge [8, 174]. In this expression the "effective stack height", corresponds to the physical stack height plus the plume rise. The plume rise is obtained from a combination of thermal (buoyant) rise and momentum rise and the height component from this factor is defined as the vertical distance between the top of the stack and the point at which the centre line of the plume levels off to within 10° of the horizontal [7]. In the case of a warm or fan-forced discharge this plume rise can add considerably to the actual stack height, significantly raising the effective height of the discharge [175, 176]. Most real situations, however, are far more complex than can be reasonably approximated using this expression, so that more complex expressions are usually required to accommodate the greater number of variables [7, 8].

The point of impingement for particulates, or the location on the ground where discharge products of the stack are at a maximum, corresponds to about 3 or 4 times the effective stack height. For gases this corresponds to about 10 times the effective stack height. Because of the influence of the "virtual stack effect", further downward diffusional dilution of flue gases beyond the point of impingement is prevented by the presence of the ground (Figure 2.9 b). Therefore, at points further from the stack than this the concentration of any discharged gases at ground level still continues to decrease, but is found to be roughly double what would be predicted from simple diffusion cone calculations [7].

The effective height of a chimney, or the efficiency of effluent dispersion may be improved by any of a number of measures. Addition of a jockey stack, a smaller diameter supplementary stack installed on top of an existing installation, can help. Or im-

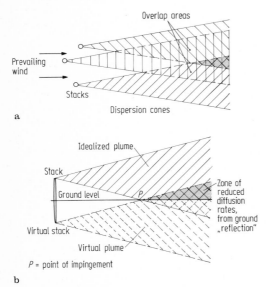

**Figure 2.9.** Horizontal(**a**) and vertical sections(**b**) of stack discharges showing how overlapping plumes, and virtual stack effects increase ambient air concentrations of pollutants at ground level above that to be expected from simple diffusion cone calculations

provement may be gained by raising the flue gas temperature or velocity to obtain greater plume rise. Valley inversions may be avoided by running ductwork up a hillside for discharge via a stack at the top, taking advantage of both the increased height above the inversion layer and the higher windspeeds for more efficient dispersal [177]. Diffusion and dispersal rates under stable air conditions may also be improved by the use of vortex rings [178], or super high stacks [179, 180]. But ultimately, particularly with very large installations discharging in areas where inversions are frequent, containment measures of one sort or another in conjunction with stack discharge have to be considered [181].

## 2.6 Air Pollution Abatement by Containment

With the increasing dilution and dispersal burdens being placed on a finite volume of atmosphere it is becoming more and more clearly evident that containment, or neutralization of air pollutants is required before discharge in order to prevent further deterioration in air quality [115, 182]. Air emission control measures which fall into this category may be precombustion or predispersal measures, which clean up a fuel or combustion process or modify a chemical process in such a way that the pollutant of concern is never dispersed into the exhaust gases. Or they may comprise measures taken to remove or neutralize the particulates, aerosols, and gases to be discharged *after* the combustion or chemical processing step. The choice of which action or combination of actions to be taken in any particular case depends on whether the process is already in operation or in the planning stage for construction, and the type of pre- or post-combustion emission control strategies which are available or may be readily developed. These options are each discussed separately.

### 2.6.1 Pre-combustion Removal Methods

If the technology is accessible and the cost of applying it is not too great, pre-combustion removal

of potential air pollutants has a great deal of merit [183]. At this stage not only is the offending substance (or substances) present in the highest concentration but it has not as yet been dispersed into the combustion or process exhaust gases.

For example the incombustibles in coal, which can contribute significantly to the particulate content of the flue gases on combustion (as well as to the bottom ash volume), are now removed before combustion for more than half of the coal produced in North America [8]. This is accomplished by using one or more of the following cleaning methods with coal that has been previously finely ground [184, 185]. It may be washed with water on riffle boards, during which the less dense coal particles are carried by the water stream while the more dense rock particles tend to sink and are captured in the cavities. In air jigs the powdered mixture is suspended on a bed of air in a fluidized form. In the jig the heavier rock particles tend to sink and may be drawn off the bottom, and the cleaned coal is drawn off the top. Or an actual liquid or fluidized dense solid may be used to obtain a more direct sink-float or froth flotation separation of the coal, of density about $1.5\,g/cm^3$, from the denser rock particles and other impurities.

In the process of the pre-combustion cleaning of powdered coal a significant proportion of the sulfur is also removed. Since about half of the sulfur compounds present in uncleaned coal reside in inorganic pyrite and sulfates, and about half in organic sulfur compounds present in the coal, it is much of the inorganic sulfur that is removed by the particulate cleaning methods [186]. This level of improvement can be further increased to some 80 % or so of the total sulfur present by using various aqueous leaching chemicals such as ferric sulfate [187-189] or sodium hydroxide [190], in the coal cleaning procedure, or by extraction with an organic solvent such as molten anthracene, *after* conventional coal cleaning [191-193] (Equations 2.29, 2.30).

Microbial methods of coal desulfurization have also been tested [194]. By precombustion cleaning measures such as these the emission characteristics of coalburning installations may be dramatically

Leaching:
$$FeS_2 + 8\,H_2O + 14\,Fe^{3+} \rightarrow 15\,Fe^{2+} + 2\,HSO_4^- + 14\,H^+ \tag{2.29}$$
Regeneration of leach solution:
$$2\,FeSO_4 + 1/2\,O_2 + H_2SO_4 \rightarrow Fe_2(SO_4)_3 + H_2O \tag{2.30}$$

improved. Coal conversion to a liquid or gaseous fuel permits much greater improvement in the residual pre-combustion impurities present [195-198], but also at greater cost [199].

The same kinds of considerations apply in the pre-combustion emission control measures used for the non-distilled fractions of petroleum, the residual fuel oils, which are used as utility fuels. The residual fuel oils contain all of the involatiles of the original crude oil, concentrated by a factor of 6 to 10 times since the volatile fractions have been removed. As a result, these fuels possess a significant potential for particulate and sulfur oxides emission on combustion. Normal refinery crude oil desalting practice is capable of removing 95 % or more of the particulate discharge problem before the crude is ever distilled [8]. But to lower the sulfur dioxide emission requires either use of only low sulfur crude oils for residual fuel formulation [200, 201], or more complicated and expensive catalytic de-sulfurization processing of the residual feedstocks prior to formulation as a fuel [201-204]. Even the volatile petroleum fractions present in natural gas, principally methane and ethane, are not without their sulfur gas polluting potential if burned directly in the state in which it is obtained from the well head. But this is normally thoroughly de-sulfurized using an amine scrub before it is piped to industrial and domestic consumers as another pre-combustion cleaned fuel [205, 206]. More details of this process are given later.

## 2.7 Post-combustion Emission Control

### 2.7.1 Particulate and Aerosol Collection Theory

Of the two non-gaseous air pollutant classifications, particulates and aerosols, the aerosols are much more difficult to control because they do not readily settle out of a gas stream of their own accord. The separation of aerosols from a waste gas stream is however greatly improved if the very small particles or droplets are agglomerated (coagulated or coalesced) to larger particles or drops, before removal is attempted. There are four primary means by which the agglomeration of aerosols can occur. Being aware of these contributing factors can provide a useful background to enable measures to be taken to accelerate the process.

Brownian agglomeration, an important contribut-

ing process to the natural removal of aerosols from the atmosphere, is the first of these. The microscopic particles of fumes and fogs are so small that the frequent collision of gas molecules with them causes them to move in an erratic path, commonly called Brownian motion. When this occurs with reasonably high concentrations of aerosol particles, i.e. high particle *numbers,* it results in frequent particle particle collisions. Particle collisions usually result in agglomeration by poorly understood mechanisms, forming loose clusters or chains with solid particles and larger drops with liquid droplets. The coagulation rate for any particular aerosol system is predictable to a significant extent [8] (Equation 2.31).

$$\text{Coagulation rate} \to K_b c^2, \text{ where} \qquad (2.31)$$

$K_b$ is the rate constant for Brownian coagulation of the particular type of particle, and

$c$ is the number of particles per $cm^3$.

The coagulation rate is seen to be proportional to the square of the particle concentration, making the number of particles per unit volume the single most important factor affecting the rate of Brownian coagulation. Table 2.10 not only illustrates this aspect, but also gives the indication that concentrations of aerosols above about $10^6/cm^3$ are unstable, and concentrations above this will rapidly drop to near this concentration range by agglomeration processes. As particles larger than ca. $5\text{-}10\,\mu m$ form, i.e. in the particulate rather than the aerosol size range, they are removed by sedimentation. It is a combination of processes such as these which contribute to keeping the aerosol concentration in

**Table 2.10** The relationship of coagulation time to the original particle concentration [a]

| Initial concentration of particles per $cm^3$ | Time for the number of particles to be decreased to $\frac{1}{10}$ initial conc. |
|---|---|
| $10^{12}$ | 0.03 sec |
| $10^{11}$ | 0.3 sec |
| $10^{10}$ | 3.0 sec |
| $10^9$ | 0.5 min |
| $10^8$ | 5.0 min |
| $10^7$ | 50 min |
| $10^6$ | 8.3 hours |
| $10^5$ | 17 hours |
| $10^4$ | 1.3 days |

[a] Data cited and extended by calculation using the methods given in [8].

the earth's atmosphere in the range below about $10^4$ to $10^5$ particles per cm$^3$ [7].

Turbulent coagulation occurs when particles in multiple flow line intersections collide, a process stimulated by turbulent flow, or eddy conditions. The turbulent coagulation rate is again proportional to the square of the particle concentration [8] (Equation 2.32).

$$\text{Coagulation rate} = K_s c^2, \text{ where} \qquad (2.32)$$

$K_s$ is the rate constant for turbulent coagulation.

The magnitude of $K_s$, the proportionality constant for turbulent coagulation, is in the range of 10 to 4,000 times $K_b$. Hence, turbulent coagulation can be used to greatly accelerate the natural agglomeration processes useful for emission control.

In natural situations, such as when burning wood out-of-doors, a smoke containing $10^6$ particles/cm$^3$ may be generated. By a combination of Brownian and turbulent coagulation processes the average particle size in this smoke can increase from $0.2\,\mu$m to $0.4\,\mu$m before it has travelled 1,000 m [8].

Sonic agglomeration applies to the use of high intensity sound waves to accelerate particle coagulation [207]. This procedure is based on the premise that particles of different sizes will tend to vibrate with different amplitudes, increasing the opportunities for collision and coagulation [8]. Also, the particle concentration in the compression zone of a high intensity sound wave will temporarily be raised, artificially increasing Brownian coagulation rates. Applying this technique involves the exposure of a particle-laden gas stream for a fraction of a second to a high intensity siren or oscillation piston, operating at 300 to 500 Hz and 170 decibels in a sound insulated chamber [208]. After exposure the particulate collection efficiency of downstream containment devices is significantly improved from the greater proportion of large particles produced.

Electrostatic methods, the fourth class, may also be used to obtain efficient coagulation of particles of the aerosol size range and larger. By passing a particle-laden gas stream past the negative side of a high D.C. voltage corona (intense electric field), the particles become negatively charged. They are then efficiently attracted and coagulated against positively charged plates which are in close proximity. Periodic rapping of the plates or a trickle of water dislodges the large particles into collection hoppers. People involved with stack sampling of

gases downstream of an operating electrostatic precipitator should remember that the 20,000 to 30,000 volts normally used is high enough to leave a hazardous residual charge in the very small fraction of particles not retained by the precipitator.

## 2.7.2 Particulate and Aerosol Collection Devices

Gas cleaning devices vary in their removal efficiencies, but almost invariably are more efficient at removing particulate, rather than aerosol size range material. But even the least efficient device for removing small particle sizes, the gravity settling chamber (Figure 2.10), still plays an important role in emission control. When the gas stream to be cleaned has only large particles present which can be effectively removed by the device, then this device alone represents an inexpensive method of control. When the particle size range to be collected is wide, and a dry cyclone is to be used for final particle collection, very large particles (100 to 200 $\mu$m) *should* be removed prior to the gas stream entering the cyclone. Otherwise the interior of the cyclone is subjected to very high abrasion rates.

Since the efficiency of gas cleaning devices is almost inevitably better for larger particles than for small and since the number of particles represented by any given mass goes up exponentially as the particle size goes down, mass removal, rather than particle removal efficiencies have become the standard method of specifying air cleaning capabilities. In one example, a particulate control device with a 91 % mass efficiency worked out to have only 0.09 % particle count efficiency for the same hypothetical dust-laden gas sample [28]. Thus, to be fully aware of the collection characteristics of an air cleaning device one must have both mass removal efficiencies, plus the particle size range over which the mass removal efficiency was measured [201]. Typical characteristics of a number of particle collection devices have been collected in Table 2.11.

The operating characteristics of some common particulate emission control units are illustrated in Figure 2.10. Cyclones, which are extensively used in industry for this purpose, are most efficient in small diameters which forces the entering gas stream through higher tangential velocities. It is high tangential velocities which most efficiently separate particles from the gas. To obtain the same

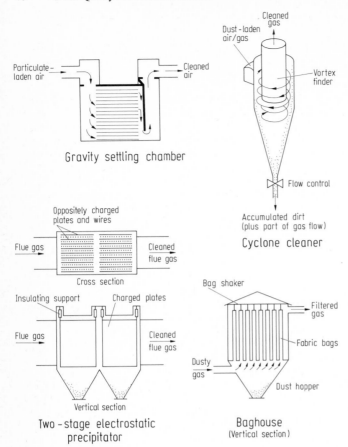

**Figure 2.10.** Simplified diagrams of some of the common types of particulate and aerosol emission control devices. Some of these are also useful for gas or vapour emission control. See text for details

**Table 2.11** Examples of typical collection efficiencies for some common types of particulate and aerosol gas cleaning devices [a]

| Air cleaning device | Particle size range for 90% removal, μm | Remarks |
|---|---|---|
| Gravity settling chamber | 50 – 100 | effective only at low gas velocities, <200m/min |
| Baffle chamber | 20 – 100 | gas velocity ca. 300 m/min |
| Dry cyclone | 10 – 50 | gas velocity 300 to 1500 m/min 2 – 750 cm diameter, smaller are more efficient |
| Spray washer | 5 – 20 | gas velocity 70 – 100 m/min |
| Packed tower scrubber | 1 – 5 | gas velocities to 4000 m/min high resistance to gas flow (pressure drop) |
| Intensive rotary scrubber | 0.1 – 1 | high restistance to flow, energy intensive |
| Cloth filter | 0.2 – 1 | used in bag houses or continuous roll filters |
| Absolute filter | 0.01 – 0.1 | highly efficient but considerable resistance to flow |
| Electrostatic precipitator | 0.01 – 10 | must have low gas flow rates, very low pressure drop, very large units (> 3000 m³/min) |
| Activated charcoal | molecular | normally for control of gases, vapours, and odours; also retains particles, about as cloth filter |

[a] Compiled from data in [7, 8, 28 and 280].

high efficiencies with very large gas volumes a large number of small cyclones are connected in parallel, rather than using a single, very large cyclone.

Absolute filters, developed by the U.S. Atomic Energy Commission for containment of radioactive particles, are extremely efficient for retention of very fine particles but require considerable operating energy because of a large pressure drop across the filter element.

Electrostatic precipitation is also very efficient for retention of very fine particles, as long as these are not highly electrically conductive. But the structural features of these units are such that they are only cost effective for particulate and aerosol emission control for very large volumes of gas, hence their typical use in fossil-fueled power plants.

### 2.7.3 Hydrocarbon Emission Control

Adoption of automotive control devices has had the most significant effect on the gross hydrocarbon emissions for regulated areas, because of the large contribution to the total mass of hydrocarbons originating from this source (Tables 2.6, 2.7). But this measure on its own will not necessarily be effective for photochemical smog abatement since it is the reactive hydrocarbons, rather than total hydrocarbons which are of greatest significance [210, 211].

Measures which can reduce the ventilation losses of hydrocarbons from the tank farms and other points of an operating oil refinery are discussed with the technology related to this industry, Plate 2.1.

In the drycleaning business or in vapour degreasing plants, losses of hydrocarbon, or chlorinated hydrocarbon vapours occur from the use and re-

covery of solvents. These losses can be prevented by the use of activated adsorbent systems to capture the organic vapours from the vents or hoods of these devices [212]. Activated carbon, which is available in versions providing up to $1,000\,m^2/g$ (ca. 8 acres/oz) of adsorption area is one of the best adsorbents for this purpose because of its selectivity for organic vapours. Solvent present in high concentrations in air may also be recovered by cooling the air to liquid nitrogen temperatures to condense the solvent [213].

Air contaminated with low concentrations of organic vapours, which can also arise from a number of processes, may be economically controlled by using this air stream as boiler feed air. Any combustibles present are burned to provide fuel value, in the process destroying any hydrocarbons present before discharge [214, 215]. This method is particularly appropriate when the concentration of organic vapours in air is too small for economic containment using adsorption [216, 217].

The paint industry, with its high dependence on solvents and active, filmforming components some of which come into the regulatory "reactive hydrocarbon" class, has also been hard pressed to meet air quality tests [218-220]. But by the further development of water-based coatings, even for many metal-finishing applications [221-223], and by innovations into such areas as high solids, or even dry powder coating technologies the industry is decreasing its dependence on organic solvents for many finishing applications [224-226]. In turn these measures have decreased the emission of hydrocarbons when the coating is applied, and when combined with electrostatic application technology to better direct the coating particles or droplets onto the surface being coated these techniques simultaneously result in much less wastage as well.

**Plate 2.1.** Floating roof petroleum storage tank for the control of hydrocarbon vapour loss. Roof is sunk below the rim of this partially empty tank, shown by inclined service walkway.

### 2.7.4 Control of Sulfur Dioxide Emissions

Total sulfur dioxide discharges for the U.S.A. for 1981 amounted to 22.5 million tonnes [227]. Combustion sources concerned with power generation produced about 60 % of this, and a further 37 % was evenly split between other stationary combustion sources (mainly space heating) and industrial processes [228]. More than half of that originating from industrial processes, or about 11 % of the total, arose from the smelting of sulfide minerals and metal refining activities. Only about 3 % of the total originated with the transportation sector.

The extent of sulfur dioxide emission is of importance for the direct effect it has on the ambient air levels, for which guidelines have been laid down to protect public health (Table 2.12). However, the atmospheric half life of discharged sulfur dioxide is estimated to be short, of the order of 3 days [229]. This rapid return of sulfur dioxide to the earth's surface, both as sulfur dioxide itself [230] and as its oxidized and hydrated products is the reason for its significance as a contributor to the acidity of rainfall, about which comment has already been made [109-113]. "Sulfurous acid", from hydrated sulfur dioxide, and sulfuric acid, formed from the oxidation of sulfurous acid or the hydration of sulfur trioxide, both contribute to low pH rain. Sulfur trioxide forms through the gas phase and heterogeneous oxidation of sulfur dioxide on particulate matter, which occurs to a significant extent both in plumes and in the ambient air [231-233].

The industrial origins of two of the major sulfur dioxide discharges, combustion sources, and sour gas plants (in Canada) generally produce waste gas streams containing 0.15 to 0.50 % sulfur dioxide, too low to be economically attractive for containment. Only with the smelter sources, which commonly produce roaster exhaust gas streams containing 2 to 5 % sulfur dioxide and with modern equipment up to about 15 % sulfur dioxide, is it anywhere near an economic proposition to contain the sulfur dioxide. This will either be sold as liquid sulfur dioxide, or converted to sulfuric acid for sale. The smelter acid product, being less pure than sulfuric acid from sulfur burning sources, commands a much lower market price but is still a useful grade for such applications as metal pickling and the manufacture of fertilizers.

A number of solutions exist for decreasing the sulfur dioxide emissions from sources where the concentrations present are too low for economic containment. The choice in any particular situation depends on the existing ambient air quality (particularly with regard to public health) in the operating area, and the time frame within which abatement action has to be taken.

Temporary shutdown or a reduced level of operation may be required of a fossil-fueled power sta-

**Table 2.12** Ambient air quality standards for Canada and the U.S.A., given in µg/m$^3$ and parenthetically in ppm [a]

| Pollutant | Averaging time | Primary standards (Health related, or maximum acceptable) | | Secondary standards (Welfare related, or maximum desirable) | |
|---|---|---|---|---|---|
| | | Canada | U.S.A. | Canada | U.S.A. |
| Total suspended | annual[b] | 70 | 75 | 60 | 60 |
| particulate | 24 hour | 120 | 260 | – | 150 |
| Sulfur dioxide | annual[c] | 60 (0.02) | 80 (0.03) | 30 (0.01) | – |
| | 24 hour | 300 (0.11) | 365 (0.14) | 150 (0.06) | – |
| | 3 hour | – | – | – | 1300 (0.50) |
| | 1 hour | – | – | 450 (0.17) | – |
| Carbon monoxide | 8 hour | 15 (13) | 10 (9) | 6 (5) | 10 (9) |
| Nitrogen dioxide | annual[c] | 100 (0.05) | 100 (0.053) | 60 (0.03) | 100 (0.053) |
| | 24 hour | 200 (0.11) | – | – | – |
| Ozone | 24 hour | 50 (0.025) | – | 30 (0.015) | – |
| | 1 hour | 160 (0.08) | 235 (0.12) | 100 (0.05) | 235 (0.12) |
| Lead | 3 month | – | 1.5 | – | 1.5 |

[a] Compiled from [227 and 281].
[b] Annual geometric mean.
[c] Annual arithmetic mean.

tion or smelter to avoid producing local, dangerously high ambient concentrations of sulfur dioxide during a severe, inversion—promoted smog situation. Over a longer time frame temporary curtailment requirements of this kind may be anticipated, by a power station for instance, by having separate stockpiles of low-cost, relatively high sulfur content coal for periods of power generation during normal atmospheric conditions, and alternative more expensive lower sulfur coal (or even coke, or natural gas, if suitably equipped) which may be burned during severe inversion episodes. With this latter option available emission curtailment by reduction of power production or temporary shutdown, which can be quite costly and inconvenient alternatives, will be required less often.

Emission reduction measures which have greater lead times to institute involve process modifications to avoid generating the sulfur dioxide-containing waste gases. For smelters this might mean adoption of hydrometallurgical technology such as is now available for copper, nickel, and zinc, in order to bypass the older roasting methods of sulfur removal from ores [234-237]. For power stations this option might mean the installation of more rigorous pre-combustion sulfur removal techniques for their coal or residual fuel oil, or increased emphasis on natural gas firing, or selection of hydro-electric, nuclear, or solar-related technologies in the utility's long term plans for power generation.

A third choice to a company faced with a need to practise emission reduction is to remain with the existing front end process, which continues to

**Table 2.13** The chemical details of some examples of stack gas desulfurization processes [a]

| Reaction phases | Process examples | Regeneration method | Product(s) |
|---|---|---|---|
| Gas-gas | Homogeneous: | | |
| | — gaseous ammonia injection plus particle collection | none | $(NH_4)_2SO_3$, $(NH_4)_2SO_4$ |
| | Heterogeneous: | | |
| | — $SO_2$ reduction with $H_2S$ + catalyst (Claus) | none | S |
| | — $SO_2$ reduction with $CH_4$ + catalyst to $H_2S$, then amine scrub (e.g. Girbotol), or | none | conc. $H_2S$ |
| | aqueous AQ (Stretford) recovery[b] | air | S |
| | — $SO_2$ catalytic oxidation with air (chamber, or contact processes), plus hydration | none | $H_2SO_4$ |
| Gas-liquid | Absorption, plus chemical reaction: | | |
| | — by aqueous ammonia solution | none | $(NH_4)_2SO_3$, $(NH_4)_2SO_4$ |
| | — dimethylaniline solution (ASARCO) | heat | conc. $SO_2$ |
| | — aqueous sodium sulfite (Wellman-Lord) | heat | conc. $SO_2$, $Na_2SO_4$ |
| | — citrate process, U.S. Bureau of Mines | $H_2S$ | S |
| | — citrate process, Flakt-Boliden version | heat | conc. $SO_2$ |
| | — eutectic melt($Na_2CO_3/K_2CO_3$), gives dry plume | water gas | conc. $H_2S$ |
| | — limestone (or lime) slurry scrubbing, inexpensive, throw-away slurry product | none | $CaSO_3 \cdot nH_2O$, $CaSO_4 \cdot 2H_2O$ |
| Gas-solid | Physical adsorption: | | |
| | — $SO_2$ onto activated charcoal or alkalized alumina | heat | conc. $SO_2$ |
| | Adsorption, plus chemical reaction: | | |
| | — powdered limestone injection, 20–60% efficient capture with particle collection (precipitator), waste product | none | dry $CaSO_3$, and $CaSO_4$ |
| | — CuO (or other metal oxides) | methane | conc. $SO_2$ |
| | Chemical reaction: | | |
| | — $SO_2$ reduction in heated coal bed, Resox process | none | S |

[a] Compiled from sources cited in the text, where further details are available, plus the reviews of [228, 242, 282 – 285].

[b] AQ, short for anthraquinone disulfonic acids, see text for details.

produce a sulfur dioxide-containing waste gas stream, and move to some system which can effectively remove the sulfur dioxide from this waste gas before it is discharged. Many methods are available, each with features which may make one of them more attractive or suitable than the others for the particular sulfur dioxide removal requirements of the process (Table 2.13). Some of the factors to be considered in arriving at an appropriate process choice are the waste gas volumes and sulfur dioxide concentrations which have to be treated and the degree of sulfur dioxide removal required, remembering that the trend is towards a continuing decrease in allowable discharges. Further considerations are the type of sulfur dioxide capture product which is produced by the process and the overall cost of it, considered, with any product credit which may help to decrease process costs. Finally the type of controlled gas discharge required for the operating situation, i.e. warm or ambient temperature, moist or dry etc., also has to be taken into account. Chemical details of the processes of Table 2.13 are outlined below.

Only one control procedure, ammonia injection, relies solely on gas phase reactions. The moist flue gas containing hydrated sulfur dioxide undergoes a gas phase, acid-base reaction to produce particulate solids (Equations 2.33, 2.34).

$$SO_2 + H_2O + NH_3 \rightarrow NH_4HSO_3 \qquad | \quad (2.33)$$

$$SO_2 + H_2O + 2\,NH_3 + 1/2\,O_2$$
$$\overset{\text{particulate}}{\underset{\text{catalysis}}{\rightarrow}} (NH_4)_2SO_4 \quad 2 \;(2.34)$$

The solid product, now a separate phase, can be readily captured by electrostatic precipitation or any other particulate collection device which is already normally in place for fly-ash control.

Heterogeneous reduction processes still involve the reaction of gases, but in these cases the reaction occurs in the presence of a suitable solid phase catalyst. Sulfur dioxide may be reduced to sulfur with hydrogen sulfide, if this is available, and the sulfur vapour condensed out of the gas stream by cooling, the second half of the Claus process [238, 239] (Equation 2.35).

$$SO_2 + 2\,H_2S \overset{Fe_2O_3}{\rightarrow} 3\,S + 2\,H_2O \qquad 3 \;(2.35)$$

Or the sulfur dioxide may by reduced catalytically with methane or other hydrocarbons to hydrogen sulfide [240]. The hydrogen sulfide produced by this method is captured by amine scrubbing of the reduced gas stream (Equations 2.36-2.38).

The high concentration of hydrogen sulfide obtained by heating the amine salt may then be easily and economically converted to elemental sulfur via the Claus process. The Stretford process, which may also be used for hydrogen sulfide capture, accomplishes both capture, using an aqueous mixture of 1,5- and 1,8-disulfonic acid of 9,10-anthraquinone (AQ), and conversion of the hydrogen sulfide to sulfur in a single step [241] (Equation 2.39).

Regeneration of the quinone from the quinol is accomplished with air (Equation 2.40).

Heterogeneous catalytic oxidation of sulfur dioxide with air, via the contact or the chamber process, so serves to improve the collection efficiency of the sulfur oxides. Sulfur trioxide has a very strong affinity for water, unlike sulfur dioxide [49, 227], so that collection of sulfur trioxide by direct absorption into water is extremely efficient, and the product sulfuric acid is a saleable commodity.

Sulfur dioxide containment by gas-liquid interactions can be as simple as flue, or process gas scrubbing with dilute ammonium hydroxide [242] (Equations 2.41, 2.42).

The ammonium salt products crystallized from the concentrated spent scrubber liquors may be used as

$$3\,SO_2 + 2\,CH_4 \overset{Al_2O_3}{\rightarrow} 2\,H_2S + 2\,CO_2 + 2\,H_2O + S \qquad\qquad 4 \;(2.36)$$

Absorption:
$$H_2S + H_2NCH_2CH_2OH \rightarrow HS^- + H_3\overset{+}{N}CH_2CH_2OH \qquad\qquad 5 \;(2.37)$$

Regeneration:
$$HS^- + H_3\overset{+}{N}CH_2CH_2OH \underset{\text{heat}}{\rightarrow} H_2S + H_2NCH_2CH_2OH \qquad\qquad 6 \;(2.38)$$

Absorption − chemical conversion:
$$AQ{=}C{=}O + H_2S \rightarrow AQ{=}C(SH)(OH) \rightarrow AQ{=}C(H)(OH) + S \qquad\qquad 7 \;(2.39)$$

Regeneration:

$$2\,AQ{=}C(H)(OH) + O_2 \rightarrow AQ{=}C{=}O + 2\,H_2O$$
$$\text{8} \quad (2.40)$$

$$2\,NH_4OH + SO_2 \rightarrow (NH_4)_2SO_3 \qquad \text{9} \quad (2.41)$$

$$2\,NH_4OH + SO_2 + 1/2\,O_2 \rightarrow (NH_4)_2SO_4 \quad (2.42)$$

valuable constituents of fertilizer formulations. Or they may be first reduced to ammonia and hydrogen sulfide with natural gas, followed by conversion of the hydrogen sulfide to sulfur by a Claus-type sequence [243, 244]. In this French designed modification, the ammonia is recycled to the scrubbing circuit.

Similar acid-base chemistry is involved in the American Smelting and Refining Company's (ASARCO's) sulfur dioxide capture process using aqueous dimethylaniline (Equation 2.43).

However, in this case the sulfurous acid-amine salt is heated to regenerate the much more expensive dimethylaniline solution for re-use, and obtain a concentrated sulfur dioxide gas stream. The more concentrated sulfur dioxide is now economic for acid production.

The Wellman-Lord process effectively uses the sodium sulfite/sodium bisulfite equilibrium to capture sulfur dioxide from flue gases [240, 245-247] (Equations 2.44, 2.45).

Most of the sodium bisulfite produced is converted back to sodium sulfite on regeneration, but some sodium sulfate inevitably forms from irreversible oxidation. This is crystallized out, dried, and sold as pulping chemical ("salt cake").

The citrate process, in which much development work has been invested by the U.S. Bureau of Mines and by Pfizer, Inc., uses an aqueous solution of citric acid to capture sulfur dioxide [248, 249] (Equations 2.46, 2.47).

Regeneration of the uncomplexed citric acid at the same time as formation of an elemental sulfur product from the bisulfite anion is obtained by treating the absorption solution with hydrogen sulfide [250] (Equation 2.48).

If hydrogen sulfide is not conveniently available from local process sources it may be produced on site for this requirement by reducing a part of the sulfur product of the process with methane (Equation 2.49).

The Flakt-Boliden version of the citrate process uses thermal regeneration of the citrate absorbing solution to obtain stripped citrate solution and a stream of up to 90 % sulfur dioxide at this stage [251, 252]. Recovery of a sulfur dioxide product gives flexibility to the final stage of processing as to whether liquid sulfur dioxide, sulfuric acid, or sulfur are obtained as the final product.

A further gas-liquid interaction process to a useful product is the sodium carbonate-potassium carbonate eutectic melt process, which operates at a temperature of about 425 °C, above the melting point of the eutectic [253]. Sulfur dioxide absorption takes place with loss of carbon dioxide (Equation 2.50).

The original eutectic is regenerated plus a hydrogen sulfide product obtained in a separate unit by treating the sodium sulfite with "water gas" (Equation 2.51).

$$PhN(CH_3)_2 + SO_2 + H_2O \rightarrow Ph\overset{+}{N}(H)(CH_3)_2 + HSO_3^- \qquad \text{11} \quad (2.43)$$

Absorption:
$$SO_2 + H_2O + Na_2SO_3 \rightarrow 2\,NaHSO_3 \qquad \text{12} \quad (2.44)$$

Regeneration:
$$2\,NaHSO_3 + heat \rightarrow SO_2 + H_2O + Na_2SO_3 \qquad \text{13} \quad (2.45)$$

$$\text{citric acid} \rightarrow HO\text{-}C(CH_2CO_2H)_2CO_2H \rightarrow H_2Cit^- + H^+ \qquad \text{14} \quad (2.46)$$

Absorption:
$$H_2 \cdot Cit^- + HSO_3^- \rightarrow (HSO_3 \cdot H_2Cit)^{2-} \qquad \text{15} \quad (2.47)$$

Regeneration:
$$(HSO_3 \cdot H_2Cit)^{2-} + H^+ + 2\,H_2S \rightarrow 3\,S + H_2Cit^- + 3\,H_2O \qquad \text{16} \quad (2.48)$$

$$4\,S + CH_4 + 2\,H_2O \xrightarrow{\;Al_2O_3\;} 4\,H_2S + CO_2 \qquad \text{17} \quad (2.49)$$

Absorption:
$$Na_2CO_3 + SO_2 \rightarrow Na_2SO_3 + CO_2 \qquad \text{18} \quad (2.50)$$

Regeneration:

$$Na_2SO_3 + CO + 2\,H_2 \rightarrow Na_2CO_3 + H_2S + H_2O \qquad (2.51)$$

$$CaCO_3 + SO_2 \rightarrow CaSO_3 + CO_2 \qquad (2.52)$$

$$CaCO_3 + SO_2 + 1/2\,O_2 \rightarrow CaSO_4 + CO_2 \qquad (2.53)$$

$$Ca(OH)_2 + H_2O + SO_2 \rightarrow CaSO_3 \cdot 2\,H_2O \qquad (2.54)$$

$$Ca(OH)_2 + H_2O + SO_2 + 1/2\,O_2 \rightarrow CaSO_4 \cdot 2\,H_2O \qquad (2.55)$$

The hydrogen sulfide initial product can be subsequently converted to more useful sulfur via the Claus process. In return for the high operating temperatures required for this process, it gives a substantially dry plume, unlike the other gas-liquid interaction processes mentioned, which may be an important consideration if the process is to be operated in a fog susceptible area.

A finely powdered limestone or lime slurry in water used in a suitably designed scrubber is an effective and relatively low cost sulfur dioxide containment method [228, 245] (Equations 2.52-2.55).

With either reagent, however, a throw-away product is obtained which requires the allocation of land for lagoon disposal of spent scrubber slurries, or other systems set up to handle the waste solid [254]. A recent variation of this approach is to employ the alkalinity of flyash itself, in a water slurry, as a means of capturing sulfur dioxide [255].

The simplest gas-solid containment systems conceptually are the direct adsorption ones. These accomplish adsorption on solids such as activated carbon, or alkalized alumina at relatively low temperatures and ordinary pressures [256-259]. In a separate unit, when the absorbent is regenerated by heating, a more concentrated sulfur dioxide stream is produced. The more concentrated stream is now attractive as feed to an acid plant, or for liquefaction or sulfur generation.

Among the simplest systems, at least for power stations already burning powdered coal, is the use of powdered limestone injection [260]. Essentially the same burner assembly is required, and the heat of combustion of the coal is sufficient to form calcium oxide (lime) from the limestone (Equation 2.56).

$$CaCO_3 \rightarrow CaO + CO_2 \qquad (2.56)$$

The alkaline lime, in a gas-solid phase reaction, reacts with sulfur dioxide in the combustion gases to form solid particles of calcium sulfite and cal-

cium sulfate which are captured in electrostatic precipitators (Equations 2.57, 2.58).

$$CaO + SO_2 \rightarrow CaSO_3 \qquad (2.57)$$

$$CaO + SO_2 + 1/2\,O_2 \rightarrow CaSO_4 \qquad (2.58)$$

The process is simple in concept, and to operate, but the gas-solid reaction is only about 20 to 60 % stoichiometrically efficient so that an excess of limestone is required for moderately efficient collection [261]. This procedure also imposes a heavier solids handling load on the precipitators, a factor which has to be considered, and it yields a throw-away product. It does, however, produce a dry plume, which may be an advantage in some situations.

Other gas-solid containment systems, in which a significant amount of the original development work was invested by Shell, are based on copper oxide on an alumina support [262]. The absorption step of this system both oxidizes and traps the sulfur dioxide on the solid support as copper sulfate (Equation 2.59).

Uptake:

$$CuO + SO_2 + 1/2\,O_2 \rightarrow CuSO_4 \qquad (2.59)$$

Regeneration by methane reduction is conducted in a separate unit returning sulfur dioxide, but now in more useful high concentrations (Equation 2.60).

Regeneration:

$$2\,CuSO_4 + CH_4 \rightarrow 2\,Cu + CO + 2\,H_2O + 2\,SO_2 \uparrow \qquad (2.60)$$

Finely divided copper on the solid support is rapidly re-oxidized to cupric oxide when it comes into contact with the hot flue gases on its return to the absorption unit (Equation 2.61).

$$Cu + 1/2\,O_2 \text{ (hot flue gas)} \rightarrow CuO \qquad (2.61)$$

Ferric oxide on alumina has also been tested in a very similar process [263].

A final gas-solid interaction process involves primarily chemical reaction for sulfur dioxide containment. A bed of crushed coal, kept at 650 to 815 °C, is used to reduce a stream of concentrated sulfur dioxide passed through it, to elemental sulfur and carbon dioxide [264] (Equation 2.62).

$$SO_2 + C \rightarrow CO_2 + S \qquad \qquad (2.62)$$

## 2.7.5 Control of Nitrogen Oxide Emissions

Discharges of nitrogen oxides in the U.S.A. in 1981 totalled 19.5 million metric tonnes, only slightly less than the total discharges of sulfur dioxide for the same year [227]. But the transportation sector was much more significant for this classification, contributing about 8.4 million tonnes of this total. Stationary combustion sources, however, were still the most significant contributor to this classification with a total of 10.2 million tonnes, followed by industrial sources which totalled about 950 thousand tonnes.

While both the primary and secondary ambient air quality standards for nitrogen oxides are somewhat higher than for sulfur dioxide the rationales behind the need for control measures are very similar, namely health effects and their influence on the pH of precipitation (Table 2.12). In addition, however, the significance of the role of nitrogen oxides in photochemical smog episodes is a further factor in the matrix of information used to set up the ambient air quality standards [265].

The mechanism of formation of nitrogen oxides ($NO_x$) in flue gases or combustion processes generally, is a composite of the relative concentrations of nitrogen and oxygen present and the combustion intensity (peak temperatures). Thus, small space heaters operating at lower temperatures typically produce combustion gases containing about 50 ppm $NO_x$, while large power plant boilers can produce flue gas concentrations of as high as 1,500 ppm [141, 266]. The formation reactions are relatively simple, involving a combination of atoms of oxygen with nitrogen and atoms of nitrogen with oxygen, both elements having finite concentrations present in atomic form at high combustion temperatures (Equations 2.20, 2.21). Because of the higher concentration of atomic species present at high temperatures rather than low, and because of the very rapid reaction rates at high temperatures, the equilibrium of these reactions is established rapidly. A small further contribution to the nitrogen oxides in exhaust gases can come from the oxidation of nitrogen-containing organics (combined nitrogen) in the fuel.

To decrease nitrogen oxide emissions any one or a combination of measures may be used. In utility applications one of the simplest measures, which also serves to reduce average fuel combustion and hence the costs of power generation, is to take some care over the matching of combustion rate to load requirements [267]. Fuel combustion at any rate greater than the heat removal (or load requirement) tends to produce excessive boiler and flue gas temperatures and higher nitrogen dioxide concentrations.

More fundamental equipment modifications are necessary to adopt staged combustion, or larger flame volume changes as means of decreasing $NO_x$ formation [268]. Measures using a fuel-rich combustion core the gases from which partially reduce nitrogen oxides formed near the fuel-lean walls to nitrogen are also being tested [269] (e.g. Equations 2.24-2.27, 2.63).

$$2\,NO_2 + CH_4 \rightarrow 2\,N_2 + 2\,H_2O + CO_2 \qquad (2.63)$$

A second significant factor which can assist in decreasing nitrogen oxide formation is a reduction of the excess air used in combustion processes. This option requires the installation of oxygen analyzers for the continuous monitoring of the oxygen content of flue gases coupled to automatic controls or manual reset alarms, to ensure that the excess air used for combustion stays within pre-determined limits. Changing from a practice of operating with a 15 % excess of theoretical combustion air to a 2 to 3 % excess can alone serve to decrease nitrogen oxide emissions by some 60 %, and at the same time increases energy recovery efficiency [270].

Other kinds of process changes can also serve to decrease $NO_x$ emissions by avoiding formation to a significant extent. Among these are recirculation of a fraction of the flue gas to decrease peak combustion temperatures [271], a measure which has also had some success in decreasing automotive $NO_x$ emissions [142]. Use of pure, or enriched oxygen instead of air for combustion can also help when used with flue gas recirculation, by decreasing the concentration of nitrogen available for oxidation. Use of fuel cells for power generation could eliminate utility-originated $NO_x$ emission, but the pure gas requirement for present fuel cell technology makes this option prohibitively expensive except in special circumstances [272].

Post-formation nitrogen oxide emission control measures include adsorption in packed beds, with which some decrease is observed [272]. But this method is not particularly effective. It is also not a very practical method either for utilities or for transportation sources. Ammonia or methane reduction of $NO_x$ to elemental nitrogen is also an effective method, which is cost-effective for high concentration sources such as nitric acid plants [273]. Details of this procedure are discussed with nitric acid technology. Two stage scrubbing, the first stage using water alone, and the second using aqueous urea, has also been proposed as an effective post-formation $NO_x$ control measure [274].

## Relevant Bibliography

1. CRC Handbook of Environmental Control, R.G. Bond and C.P. Straub, editors, CRC Press, Cleveland, 1972, Volume 1, Air Pollution
2. Patty's Industrial Hygiene and Toxicology, G.D. Clayton and F.E. Clayton, editors, 3rd edition, Wiley, New York, 1978, Volume 1, General Principles
3. Air Pollution, 2nd edition, A.C. Stern, editor, Academic Press, New York, 1968
   Volume 1. Air Pollution and Its Effects
   Volume 2. Analysis, Monitoring, Surveying
   Volume 3. Sources of Air Pollution and Their Control
4. Air Pollution, 3rd edition, A.C. Stern, editor, Academic Press, New York. Volume 1. Air Pollutants, Their Transformation and Transport, 1976. Also Volumes 2, and 5, 1977
5. W.L. Faith and A.A. Atkisson, Jr., Air Pollution, 2nd edition, Wiley-Interscience, New York, 1972
6. Air Pollution Abstracts, U.S. Environmental Protection Agency, Research Triangle Park, North Carolina, from volume 1, 1969
7. L.B. Lave, E.G. Seskin, and M.J. Chappie, Air Pollution and Human Health, Johns Hopkins University Press, Baltimore, 1977
8. Air Pollution and Lichens, B.W. Ferry, M.S. Baddeley and D.L. Hawkesworth, editors, Athlone Press, London, 1973
9. W.H. Smith, Air Pollution and Forests: Interactions Between Air Contaminants and Forest Ecosystems, Springer-Verlag, New York, 1981
10. Cleaning Our Environment, A Chemical Perspective, 2nd edition, American Chemical Society, Washington, 1978
11. H. Brauer and Y.B.G. Varma, Air Pollution Control Equipment, Springer-Verlag, New York, 1981
12. D.E. Painter, Air Pollution Technology, Reston Publishing Co., Reston, Virginia, 1974
13. J.M. Marchello, Control of Air Pollution Sources, Marcel Dekker, Inc., New York, 1976

## References

1. E. Cook, Scientific American 225 (3), 134, Sept. 1971
2. B. Bolin, Scientific American 224 (3), 124, Sept. 1970
3. R.E. Newell, Scientific American 225 (1), 32, Jan. 1971
4. C.S. Wong, Science 200, 197, April 14, 1978
5. C.F. Cullis, Chem. In Brit. 14 (8), 384, Aug. 1978
6. C.C. Delwiche, Scientific American 224 (3), 136, Sept. 1970
7. S.J. Williamson, Fundamentals of Air Pollution, Addison-Wesley, Reading, Massachusetts, 1973
8. Air Pollution Handbook, P.L. Magill, F.R. Holden, and C. Ackley, editors, McGraw-Hill, New York, 1956
9. R. Findley, National Geographic 159 (1), 3, Jan. 1981
10. Mount St. Helens: Researchers Sort Data, Chem. Eng. News 58 (23), 19, June 9, 1980
11. Symposium to Mark 1883 Eruption of Krakatoa, Chem. Eng. News 61 (28), 138, July 11, 1983
12. A. Nohl and F. Le Guern, Chromat. Rev. 8 (2), 8, Oct. 1982
13. Mount St. Helens Stirs Chemical Interest, Chem. Eng. News 58 (34), 28, Aug. 25, 1980
14. Effects of Mercury in the Canadian Environment, M.B. Hocking and J.A. Jaworski, editors, National Research Council, Ottawa, 1979
15. W.W. Vaughn and J.H. McCarthy, Jr., U.S. Geol. Surv. Prof. Paper No. 501-D, D 128 (1964)
16. S.H. Williston, J. Geophys. Res. 73 (22), 7051, (1968)
17. F.A. Fischback, J. Envir. Health 38 (6), 382, May/June 1976
18. V.V. Kingsley, Bacteriology Primer in Air Contamination Control, University of Toronto Press, Toronto, 1967
19. T.J. Wright, V.W. Greene, and H.J. Paulus, J. Air Polln Control Assoc. 19 (5), 337, May 1969
20. E.D. King, R.A. Mill, and C.H. Lawrence, J. Envir. Health 36 (1), 50, July/Aug. 1973
21. D.C. Blanchard, Saturday Review, 60, Jan. 1, 1972
22. S.A. Berry and B.G. Notion, Water Research 10, 323 (1976)
23. Atmospheric Aerosol: Source Air/Quality Relationships, ACS Symposium Series No. 167, American Chemical Society, Washington, 1981

24. J.A. Dorsey and J.O. Burckle, Chem. Eng. Prog. *67* (8), 92, Aug. 1971

25. Research Appliance Co., Allison Park, Pennsylvania, U.S.A.

26. Five Nights News Roundup, Canadian Broadcasting Corp., Toronto, Nov. 9, 1972

27. L. Svarovsky, Chem. Ind. (London), 626, Aug. 7, 1976

28. Air Pollution, Volume III, Sources of Air Pollution and Their Control, 2nd edition, A.C. Stern, editor, Academic Press, New York, 1968

29. M.H. Chatterton and I.E. Lund, Chem. Ind. (London) 636, Aug. 7, 1976

30. A.W. Gnyp, S.J.W. Price, C.C. St. Pierre, and J. Steiner, Water and Polln Control, *111* (7), 40, July 1973

31. M.J. Pilat, D.S. Ensor and J.C. Bosch, Atmos. Envir. *4*, 671 (1970)

32. Introductory Analytical Techniques for Airborne Particulate Matter, Application Procedure AP 308, Millipore Corp., Bedford, Massachusetts, 1972

33. W.G. Hyzer, Research/Development *26* (4), 50, April 1975

34. New Sampler Separates Large, Small Samples, Chem. Eng. News *56* (44), 23, Oct. 30, 1978

35. Detection and Analysis of Particulate Contamination, Millipore Corp., Bedford, Massachusetts, 1966

36. W.C. McCrone, R.G. Draftz, and J.G. Delly, The Particle Atlas, Ann Arbor Science Publishers, Ann Arbor, Michigan, 1971

37. M. Beer, R.W. Carpenter, L. Eyring, C.E. Lyman, and J.M. Thomas, Chem. Eng. News *59* (33), 40, Aug. 17, 1981

38. K. Anlauf, L.A. Barrie, H.A. Wiebe, and P. Fellin, Can. Research *13* (2), 49, Feb. 1980

39. R. Dams, J. Billiet, C. Block, M. Demuynck, and M. Janssens, Atmos. Envir. *9*, 1099 (1975)

40. M.A. Kay, D.M. McKown, D.H. Gray, M.E. Eichor and J.R. Vogt, Amer. Lab. *5* (7), 39, July 1973

41. Toxic Metals in Air, Chem. In Can. *25* (9), 6, Oct. 1973

42. M.B. Jacobs, The Chemical Analysis of Air Pollutants, Interscience, New York, 1960

43. Air Pollution Vol. II, Analysis, Monitoring, and Surveying, 2nd edition, A.C. Stern, editor, Academic Press, New York, 1968

44. Analysis of Industrial Air Pollutants, Papers by P.R. Harrison, H.W. Georgii et al, MSS Information Corp., New York, 1974

45. J. Walinga and A.H.C. Hendriks, Can. Res. Devel. 7, 36, March/April 1974

46. Air Pollution Control Part III, Measuring and Monitoring Air Pollutants, W. Strauss, editor, Wiley-Interscience, New York, 1978

47. T.A. Rooney, Applic. Note AN 228-6, Hewlett Packard, Washington, 1977

48. Tenax for Trapping, Gas-Chrom Newsletter *21* (2), 8, June 1980

49. M.B. Hocking and G.W. Lee, Water, Air and Soil Polln *8*, 255, (1977)

50. W.J. Blaedel and V.W. Meloche, Elementary Quantitive Analysis, Theory and Practice, Harper and Row, New York, 1963

51. R.B. Fischer, Quantiative Chemical Analysis, 2nd edition, W.B. Saunders Co., Philadelphia, Pennsylvania, 1961

52. M.B. Jacobs, The Chemical Analysis of Air Pollutants, Interscience, New York, 1960

53. K. Leichnitz, Detector Tube Handbook, 3rd edition, Draegerwerk AG, Lubeck, W. Germany, 1976

54. R.M. Ash and J.R. Lynch, J. Amer. Ind. Hygiene Assoc. *32* (8), 552, Aug. 1971

55. J. Rowland, Varian Instr. Applic. *15* (1), 15, 1981

56. J.J. Havlena, D. Hocking, and R.D. Rowe, The Photography of Sulfur Recovery Plant Plumes, Proc. Alta Sulfur Gas Workshop II, Kananaskis, Alta., Jan. 16-17, 1975, page 30

57. Photographing $SO_2$ Emissions, Chem. in Can. *26* (11), 16, Dec. 1974

58. Quantitative Analysis with the Miran Gas Analyzer, Wilks Scientific, South Norwalk, Connecticut, Seminar, May 1973

59. P.L. Hanst, A.S. Lefohn, and B.W. Gay, Jr., Appl. Spectr. *27* (3), 188 (1973)

60. J. Fleming, Chem. In Brit. *13* (9), 328, Sept. 1977

61. J. Fleming, Chem. In Brit. *13* (10), 377, Oct. 1977

62. J.N. Pitts, Jr., B.J. Finlayson-Pitts, and A.M. Winer, Envir. Science and Techn. *11* (6), 568, June 1977

63. E.D. Hinkley and P.L. Kelly, Science *171* (3,972), 635, Feb. 19, 1971

64. B.L. Sharp, Chem. In Brit. *18* (5), 342, May 1982

65. J.R. Alkins, Anal. Chem. *47* (8), 752 A, July 1975

66. J.A. Hodgeson, W.A. McClenny, and P.L. Hanst, Science *182*, 248, Oct. 19, 1973

67. E.D. Hinkley, Env. Science and Tech. *11* (6), 564, June 1977

68. W.F. Herget and W.D. Connor, Env. Science and Tech. *11* (10), 962, Oct. 1977

69. Instrumental Analysis, H.H. Bauer, G.D. Christian, and J.E. O'Reilly, editors, Allyn and Bacon, Boston, 1978

70. F.W. Karasek, Research/Development *27* (11), 42, Nov. 1976

71. M.J.E. Hewlins, Chem. In Brit. *12* (11), 341, Nov. 1976

72. D.A. Lane and B.A. Thomson, J. Air Polln Control Assoc. *31*, 122, Feb. 1981

73. F.W. Karasek, Can. Research *13* (4), 30, April 1980

74. Advances in Mass Spectrometry, Volume 8 B, Proc. of the 8th International Mass Spectrometry

Conf., Oslo, 12-18 Aug., 1979, Heyden and Son, pages 1,422-1,428, and 1,480-1,489

75. Instrumentation in Analytical Chemistry, articles collected by A.J. Senzel, American Chemical Society, Washington, 1973

76. D. Betteridge and H.E. Hallam, Modern Analytical Methods, The Chemical Society, London, 1972

77. R. Perry and R.M. Harrison, Chem. In Brit. *12* (6), 185, June 1976

78. ASTM Validates Air Pollution Test Methods *51*, 13, March 5, 1973

79. M.R. Midgett, Envir. Science and Tech. *11* (7), 655, July 1977

80. Industrial Hygiene and Toxicology, 2nd edition, F.A. Patty, editor, Interscience, New York, 1963, volume II, 1963

81. Cleaning Our Environment, A Chemical Perspective, 2nd edition, Committee on Environmental Improvement, American Chemical Society, Washington, 1978

82. California Air Environment *4* (3), 12, spring 1974

83. Patty's Industrial Hygiene and Toxicology, 3rd edition, G.D. Clayton and F.E. Clayton, editors, John Wiley and Sons, New York, 1981, volume 2 A, Toxicology

84. Air Pollution Damage to Vegetation, J.A. Naegle, editor, American Chemical Society, Washington, 1973

85. D. Hocking and M.B. Hocking, Envir. Polln (London) *13*, 57 (1977)

86. A.A. Loman, Alberta Ambient Air Quality Standards For Sulfur Dioxide, Foliar Sulfur Dioxide Symptoms on Forest Vegetation and Impact on Growth and Vigour, in Proceedings of Alberta Sulphur Gas Research Workshop II, Kananaskis, Alta., Jan. 16, 1975, Queen's Printer, Edmonton, Alberta

87. Environmental Detectives Study Clues, Chemecology 6, June 1983

88. W. Summer, Odour Pollution of Air, CRC Press, Cleveland, Ohio, 1971

89. U. Kirschbaum, R. Klee, and L. Stenbing, Staub-Reinhalt. Luft *31* (1), 21, Jan. 1971, cited by Air Polln Abstr. 10586, June 1971

90. W.C. Denison and S.M. Carpenter, A Guide to Air Quality Monitoring with Lichens, Lichen Technology Inc., Corvallis, Oregon, 1973

91. O.L. Gilbert, The Effect of $SO_2$ on Lichens and Bryophytes around Newcastle upon Tyne, in Proc. 1st European Congress on the Influence of Air Pollution on Plants and Animals, Centre for Agric. Publishing and Documentation, Wageningen, 1969, page 223

92. J.J. Barkman, The Influence of Air Pollution of Bryophytes and Lichens, in Proc. 1st European Congress on the Influence of Air Pollution on Plants and Animals, Centre for Agric. Publishing

and Documentation, Wageningen, 1969, page 197

93. A.C. Skorepa and D.H. Vitt, A Quantitative Analysis of Lichen Vegetation in Relation to $SO_2$ Pollution at Rocky Mountain House, in Proceedings of Alberta Sulfur Gas Research Workshop II, Kananaskis, Alta., Jan. 16, 17, 1975, Queen's Printer, Edmonton, Alberta

94. A.C. Skorepa and D.H. Vitt, A Quantitative Study of Epiphytic Lichen Vegetation in Relation to $SO_2$ Pollution in Western Alberta, Information Report NOR-X-161, Northern Forest Research Centre, Edmonton, Alberta, October 1976

95. Lichen Chart, Things of Science, Cambridge, 1972

96. D.V. Bates, A Citizen's Guide to Air Pollution, McGill-Queen's University Press, Montreal, 1972

97. Man, Health, and Environment, B.Q. Hafen, editor, Burgess Publishing Co., Minneapolis, Minnesota, 1972

98. Air Pollution, Vol. I, Air Pollution and Its Effects, 2nd edition, A.C. Stern, editor, Academic Press, New York, 1968

99. L. Hodges, Environmental Pollution, Holt, Rinehart, and Winston, New York, 1973

100. J. Mage and G. Batta, Chemie und Industrie, *27*, 1961 (1932), cited by reference 8

101. J.A. Kerr, J.G. Calvert, and K.L. Demerjian, Chem. In Brit. *8* (6), 252, June 1972

102. J.G. Calvert, Envir. Science and Tech. *10* (3), 248, March 1976

103. J.N. Pitts, Jr., and B.J. Finlayson, Angew. Chem. (Internat. Edition) *14* (1), 1, Jan. 1975

104. P.L. Hanst, J. Air Polln Control Assoc. *21* (5), 269, May 1971

105. E. Hessvedt, Geophys. Norvegica *31* (2), 1 (1975)

106. Power Plants May Be Major Ozone Source, Chem. Eng. News *52*, 22, Oct. 28, 1974

107. R.M. Harrison and C.D. Holman, Chem. In Brit. *18* (8), 563, Aug. 1982

108. R.M. Harrison and C.D. Holman, Atmos. Envir. *13*, 1,535 (1979)

109. E.B. Cowling, Envir. Science and Tech. *16* (2), 110 A, Feb. 1982

110. G.E. Likens, Chem. Eng. News *54* (48), 29, Nov. 22, 1976

111. Research Summary, Acid Rain, M. Schaefer, editor, U.S. Environmental Protection Agency, RD-674, Washington, ca. 1980

112. G.E. Likens, R.F. Wright, J.N. Galloway and T.J. Butler, Scientific American *241* (4), 43, Oct. 1979

113. L.R. Ember, Chem. Eng. News *59* (37), 20, Sept. 14, 1981

114. B. Hileman, Envir. Science and Tech. *15* (10), 1,119, Oct. 1981

115. L.R. Ember, Chem. Eng. News *60* (47), 25, Nov. 22, 1982

116. B. Hileman, Envir. Science and Tech. *17* (6), 232 A, June 1983

117. R.E. Newell, Scientific American *224* (1), 32, Jan. 1971

118. S.A. Changnon, Jr., Science *205*, 402, July 27, 1979

119. T.F. Hatch, in The Air We Breathe, S.M. Farber, editor, C.C. Thomas, Springfield, Illinois, 1961, Cited by reference 7

120. S.K. Friedlander, Envir. Science and Tech. *7* (13), 1,115, Dec. 1973

121. R.J. Seltzer, Chem. Eng. News *54* (44), 21, Oct. 25, 1976

122. W. Lepkowski, Chem. Eng. News *55* (42), 26, Oct. 17, 1977

123. B. Bolin, Scientific American *225* (9), 124, Sept. 1970

124. C.S. Wong, Science *200*, 197, April 14, 1978

125. E. Aynsley, New Scientist *44*, 66, Oct. 9, 1969

126. T.L. Brown, Energy and the Environment, C.E. Merrill Publishing Co., Columbus, Ohio, 1971

127. A.H. Oort, Scientific American *225* (9), 54, Sept. 1970

128. B.J. Palmer, Envir. Letters *5* (4), 249 (1973)

129. $CO_2$ Buildup's Effect on Climate Explored, Chem. Eng. News *55* (31), 18, Aug. 1, 1977

130. H. Wexler, Scientific American *186* (4), 74, April 1952

131. K.S. Groves, S.R. Mattingly and A.F. Tuck, Nature *273*, 711, June 29, 1978

132. M.J. Molina and F.S. Rowland, Nature *249*, 810 (1974)

133. Nitrogen Fertilizers May Endanger Ozone, Chem. Eng. News *53* (47), 6, Nov. 24, 1975

134. R.P. Turco and R.C. Whitten, Atmos. Envir. *9*, 1,045 (1975)

135. L.C. Glasgow, P.S. Gumerman, P. Meakin, and J.P. Jesson, Atmos. Envir. *12*, 2,159, (1978)

136. R.M. Baum, Chem. Eng. News *60* (37), 21, Sept. 13, 1982

137. Chlorofluorocarbons in the Environment: The Aerosol Controversy, T.M. Sudgen and T.F. West, editors, Society of Chemical Industry, E. Horwood Publishing, Chichester, 1980

138. R.A. Rasmussen, M.A.K. Khalil, S.A. Penkett, and N.J.D. Prosser, Geophys. Res. Letters *7* (10), 809, Oct. 1980

139. R.A. Rasmussen, M.A.K. Khalil, and R.W. Dalluge, Science *211*, 285, Jan. 16, 1981

140. H.S. Sandhu, Meteorological Factors and Photochemical Air Pollutants over Edmonton, Alberta, VIIIth International Photochemistry Conference, Edmonton, Alberta, Aug 7-13, 1975. J. Photochem. *5* (2), 88, March 1976

141. G.S. Parkinson, Chem. In Brit. *7* (6), 239, June 1971

142. G.J.K. Acres, Chem. Ind. (London), 905, Nov. 16, 1974

143. J.J. Mikita and E.N. Cantwell (Dupont), Exhaust Manifold Thermal Reactors—A Solution to the Automotive Emissions Problem, 68th Ann. Meeting, National Petrol. Refiners Assoc., San Antonio, Texas, April 5-8, 1970

144. E.N. Cantwell, Hydroc. Proc. *53* (7), 94, July 1974

145. Why So Many New Cars Run So Poorly, Consumer Reports *37* (4), 196, April 1972

146. M.K. McAbee, Chem. Eng. News *53* (47), 14, Nov. 24, 1975

147. J.D. Butler, Chem. In Brit. *8* (6), 258, June 1972

148. Three Way Catalyst Cleans Auto Exhaust, Chem. Eng. News *55* (11), 26, March 14, 1977

149. H.K. Newhall, California Air Envir. *5* (1), 1, Fall 1974

150. Auto Emissions Control Faces New Challenges, Chem. Eng. News *58* (11), 36, March 17, 1980

151. G.J.K. Acres and B.J. Cooper, Platinum Metals Review *16* (3), 74, July 1972

152. M.E. Hilburn, Chem. Soc. Revs. *8* (1), 63 (1979)

153. L.R. Ember, Chem. Eng. News *58* (25), 28, June 23, 1980

154. P.L. Dartnell, Chem. In Brit. *16* (6), 308, June 1980

155. Airborne Lead in Perspective, National Academy of Sciences, Washington, 1972

156. Lead in the Canadian Environment, National Research Council, Ottawa, 1973

157. T.D. Leah, The Production, Use, and Distribution of Lead in Canada, Environmental Contaminants Inventory Study No. 3, Inland Waters Directorate, Burlington, Ontario, 1976

158. A.N. Manson and H.W. de Koning, Water and Polln Control *111* (9), 70, Sept. 1973

159. D. Turner, Chem. In Brit. *16* (6), 312, June 1980

160. T.J. Chow, Nature *225*, 295 (1970)

161. N.I. Ward and R.R. Brooks, Envir. Polln *17*, 7 (1978)

162. R.M. Harrison and D.P.H. Laxen, Chem. In Brit. *16* (6), 316, June 1980

163. N.I. Ward, R.R. Brooks, E. Roberts, and C.R. Boswell, Envir. Science and Tech. *11* (9), 917, Sept. 1977

164. New Systems Provide Useful Measurements of Auto Particulate Emissions, Hydroc. Proc. *49* (12), 23, Dec. 1970

165. New Particulate Trap Lowers Lead Emissions 90 %, Hydroc. Proc. *51* (12), 24, Dec. 1972

166. K. Habibi, Envir. Science and Tech. *7* (3), 223, March 1973

167. H.K. Newhall, Calif. Air Envir. *5* (1), 5, Fall 1974

168. J. Dunn, Popular Science *207* (5), 102, Nov. 1975

169. J. Daniels, La Recherche *11* (114), 938, Sept. 1980

170. L.R. Babcock, Jr., J. Air Polln Control Assoc. *20* (10), 653, Oct. 1970

171. E.G. Walther, J. Air Polln Control Assoc. *22* (5), 353, May 1972

172. D. Carleton-Jones and H.B. Schneider, Chem. Eng. *75* (10), 166, Oct. 14, 1968

173. Increased Heavy Metal Around Sudbury's Smelters, Water and Polln Control *111* (6), 48, June 1973

174. C.F. Barrett, Roy. Inst. Chem. Revs. *3* (2), 119, Oct. 1970

175. B. Bringfelt, Atmos. Envir. *2* (11), 575, Nov. 1968

176. E.C. Halliday, Atmos. Envir. *2* (9), 509, Sept. 1968

177. M. Overend, Water and Polln Controll *110* (4), 76, April 1972

178. C. Simpson, Eng. *209*, 578, June 5, 1970

179. E. Jeffs and A. Conway, Eng. *207*, 796, May 30, 1969

180. Tallest Chimney in the World, Eng. *209*, 178, Feb. 20, 1970

181. Very Tall Stacks Not Too Useful, Chem. Eng. News *56* (8), 23, Feb. 20, 1978

182. L.R. Ember, Chem. Eng. News *61* (25), 27, June 20, 1983

183. H.F. Elkin and R.A. Constable, Hydroc. Proc. *51* (10), 113, Oct. 1972

184. Chemical Engineers' Handbook, 4th edition, J.H. Perry, editor, McGraw-Hill, New York, 1963

185. A.F. Taggart, Handbook of Mineral Dressing, Ores, and Industrial Minerals, John Wiley and Sons, New York, 1945, page 6-1

186. Coal Agglomeration, Chem. In Can. *23* (11), 10, Nov. 1971

187. R.A. Meyers, Hydroc. Proc. *54* (6), 93, June 1975

188. H.A. Hancock and T. Betancourt, The Effect of the Addition of Chemical Oxidants on the Desulphurization of Cape Breton Coal by Ferric Sulphate Leaching, Proc. 28th Chemical Engineering Conference, Halifax, Nova Scotia, Oct. 22-28, 1978, plus other reports by the same authors

189. Coal and Coke Technology, Chem. In Can. *31* (2), 26, Feb. 1979

190. W. Worthy, Chem. Eng. News *53* (27),24, July 7, 1975

191. New Way to Cope with Sulfur Oxides, Hydroc. Proc. *51* (9), 19, Sept. 1972

192. Clean Fuel from Coal, Hydroc. Proc. *51* (10), 42, Oct. 1972

193. F.K. Schweighardt and B.M. Thames, Anal. Chem. *50* (9), 1,381, Aug. 1978

194. Microbial Desulfurization of Coal, Mech. Eng. *100*, 46, Dec. 1978

195. Coal Conversion: New Processes from Old, Chem. Eng. News *55* (36), 24, Sept. 5, 1977

196. R.L. Rawls, Chem. Eng. News *55* (4), 26, Jan. 24, 1977

197. R. Dagani, Chem. Eng. News *58* (1), 27, Jan. 7, 1980

198. Dow Details Coal Liquefaction Process, Chem. Eng. News *58* (27), 32, July 7, 1980

199. Coal Conversion at a Stalemate, Chem. Eng. News *57* (10), 18, March 5, 1979

200. C.M. McKinney, Hydroc. Proc. *51* (10), 117, Oct. 1972

201. C.W. Siegmund, Hydroc. Proc. *49* (2), 89, Feb. 1970

202. The Technology and Costs to Remove Sulfur from Oil Based Fuels, Chem. In Can. *23* (9), 25, Oct. 1971

203. Desulfurization Refinery Capacities, Env. Science and Tech. *7* (6), 494, June 1973

204. J. Yulish, Chem. Eng. 58, June 14, 1971

205. P. Grancher, Hydroc. Proc. *57* (9), 155, Sept. 1978

206. P. Grancher, Hydroc. Proc. *57* (9), 257, Sept. 1978

207. E.P. Mednikov, Acoustic Coagulation and Precipitation of Aerosols, translated by C.V. Larrick, Consultants Bureau, New York, 1965

208. Sound Pulses Clean Foul Air, Chem. Eng., 52, Aug. 24, 1970

209. M.B. Hocking, J. Envir. Systems *5* (3), 163 (1976)

210. Cleaning Our Environment, The Chemical Basis for Action, American Chemical Society, Washington, 1969, page 36

211. J.N. Pitts, Jr. and B.J. Finlayson, Angew. Chem. (Internat. Edition) *14* (1), Jan. 1975

212. R.R. Manzone and D.W. Oakes, Polln Eng. *5* (10), 23, Oct. 1973

213. Airco Develops Solvent Recovery System, Chem. Eng. News *58* (4), 7, Jan. 28, 1980

214. G.D. Arnold, Chem. Ind. (London), 902, Nov. 16, 1974

215. Fume-incineration Design Saves Fuel, Can. Chem. Proc. *61* (2), 31, Feb. 1977

216. D.E. Waid, Chem. Eng. Prog. *68* (8), 57, Aug. 1972

217. R.D. Ross, Chem. Eng. Prog. *68* (8), 59, Aug. 1972

218. Paint, Energy and the Environment, Chem. In Can. *26* (4), 24, April 1974

219. L.A. O'Neill, Chem. Ind. (London), 464, June 5, 1976

220. Industry Now Leads in Paint Usage, Can. Chem. Proc. *63* (6), 37, June 6, 1979

221. C.M. Hansen, Ind. Eng. Chem., Prod. Res. Dev. *16* (3), 266, (1977)

222. New Mixes Meet Air Quality Tests, Can. Chem. Proc. *63* (6), June 6, 1979

223. C.A. Law, Chem. In Can. *30* (8), 24, Sept. 1978

224. Air Quality Rules Boost High Solids Coatings, Chem. Eng. News *60* (11), 5, March 15, 1982

225. J. Cross, Chem. In Brit. *17* (1), 24, Jan. 1981

226. Powder Coating Group Starts Up Programs, Chem. Eng. News *59* (36), 16, Sept. 7, 1981

227. National Air Quality and Emissions Trends Report, 1981, U.S. Environmental Protection Agency, 450/4-83-011, Research Triangle Park, North Carolina, 1983

228. N.Kaplan and M.A. Maxwell, Chem. Eng. *84*, 127, Oct. 17, 1977

229. W.W. Kellogg, R.D. Cadle, E.R. Allen, A.L. Lazarus, and E.A. Martell, Science *175* (4,402), 587, Feb. 11, 1972

230. J.A. Garland, D.H.F. Atkins, C.J. Readings, and S.J. Caughey, Atmos. Envir. *8*, 75 (1974)

231. D.D. Davis, The Photochemistry and Kinetics of Power Plant Plumes, VIIIth Photochem. Conf., Edmonton, Alberta, Aug. 7-13, 1975. J. Photochem. *5*(2), 88, March 1976

232. M.A. Lusis and C.R. Phillips, Atmos. Envir. *11*, 239 (1977)

233. W.H. White, Envir. Science and Tech. *11*(10), 995, Oct. 1977

234. M.A. Hughes, Chem. Ind. (London), 1,042, Dec. 20, 1975

235. P.J. Bailes, C. Hanson, and M.A. Hughes, Chem. Eng. *83*, 86, Aug. 30, 1976

236. M.C. Kuhn, Mining Eng. *29*(2), 79, Feb. 1977

237. R. Derry, Chem. Ind. (London), 222, March 19, 1977

238. P. Grancher, Hydroc. Proc. *57*(9), 155, Sept. 1978

239. P. Grancher, Hydroc. Proc. *57*(9), 257, Sept. 1978

240. New Units Aimed at Fluegas Clean-up, Chem. In Can. *62*(7), 26, July 1978

241. Hydrogen Sulphide Removal by the Stretford Liquid Purification Process, Chem. Ind. (London), 883, May 19, 1962

242. A.A. Siddiqi and J.W. Tenini, Hydroc, Proc. *56*(10), 104, Oct. 1977

243. Stop Air Pollution and Recover Sulphur, Proc. Eng., 9, Sept. 1976

244. P. Bonnifay, R. Dutriau, S. Frankowiak, and A. Deschamps, Chem. Eng. Prog. *68*(8), 51, Aug. 1972

245. New Scrubbers Tackle SO₂ Emissions Problem, Chem. Eng. News *56*(45), 24, 1978

246. System Removes SO₂ From Stack Gas, Chem. Eng. News *55*(45), 35, Nov. 7, 1977

247. B.H. Potter and T.L. Craig, Chem. Eng. Prog. *68*(8), 53, Aug. 1972

248. Citric Acid Used in SO₂ Recovery, Chem. Eng. News *49*(24), 31, June 14, 1971

249. Pfizer to Test Citric Acid For Converting SO₂ into Sulfur, Chemecology, 7, Dec. 1972

250. F.S. Chalmers, Hydroc. Proc. *53*(4), 75, April 1974

251. S. Bengston, Chem. In Can. *33*(1), 24 (+ cover), Jan. 1981

252. Flue Gas Desulfurization to be Tested, Chem. Eng. News *57*(14), 20, April 2, 1979

253. SO₂ Removal, Chem. Eng. News *49*(6), 52, Feb. 8, 1971

254. Flue Gas Scrubbing With a Limestone Slurry, Can. Chem. Proc. *59*(10), 8, Oct. 1975

255. Flyash Replaces Limestone in Sulfur Dioxide Removal, Chem. Eng. News *57*(32), 26, Aug. 6, 1979

256. G.N. Brown, S.L. Torrence, A.J. Repik, J.L. Stryker, and F.J. Ball, Chem. Eng. Prog. *68*(8), 55, Aug. 1972

257. S.L. Torrence, U.S. Patent 3,667,908, June 6, 1972, to Westvaco Corp. Cited by Air Polln Abstr. *5*(1), 69, Jan. 1974

258. J.E. Newell, Chem. Eng. Prog. *65*(8), 62, Aug. 1969

259. J.H. Russell, J.I. Paige, and D.L. Paulson, Evaluation of Some Solid Oxides as Sorbents of Sulfur Dioxide, U.S. Bureau of Mines, RI 7,582, Washington, 1971

260. H.L. Falkenberry and A.V. Slack, Chem. Eng. Prog. *65*(12), 61, Dec. 1969

261. T.H. Chilton, Chem. Eng. Prog. *67*(5), 69, May 1971

262. F.M. Dautzenberg, J.E. Nader, and A.J.J. Ginneken, Chem. Eng. Prog. *67*(8), 86, Aug. 1971

263. Reduction of Low Concentrations of SO₂ by CO, Can. Chem. Proc. *55*(7), 49, July 1971

264. Coal Converts Sulfur Dioxide to Sulfur, Design News *34*, 18, Nov. 20, 1978

265. J.N. Pitts, Jr. and B.J. Finlayson, Angew. Chem. (Internat. Edition) *14*(1), 1, Jan. 1975

266. D.R. Bartz, Calif. Air Envir. *5*(1), 10, Fall 1974

267. R.B. Rosenburg, Eng. Digest *17*(7), 27, Aug. 20, 1971

268. R. Rawls, Chem. Eng. News *59*(13), 19, March 30, 1981

269. Low NO_x Burner to be Demonstrated in Utah, Chem. Eng. News *59*(17), 16, April 27, 1981

270. A.M. Woodward, Hydroc. Proc, *53*(7), 106, July 1974

271. NO_x Yield Reduced by New Design, Can. Chem. Proc. *63*(2), 30, Feb. 5, 1979

272. Zeroing in on Flue Gas NO_x, Envir. Science and Tech. *3*(9), 808, March 1969

273. R.A. Searles, Platinum Metals Rev. *17*(2), 57, April 1973

274. Two Processes to Control Emissions of Nitrogen Oxides, Can. Chem. Proc. *59*(11), 4, Nov. 1975

275. D.J. Spedding, Air Pollution, Clarendon Press, Oxford, 1974

276. Joint Air Pollution Study of St. Clair-Detroit River Areas For International Joint Commission, Canada and the United States, International Joint Commission, Ottawa and Washington, 1971

277. Matheson-Kitagawa Toxic Gas Detector System, Matheson, Lyndhurst, N.J., July 1980

278. Gastec Precision Gas Detector System, Gastec Corp., Tokyo, ca. 1974

279. The Clean Air Act, Annual Report 1977-78, Ottawa, cited by M. Webb, The Canadian Environment, W.B. Saunders Co., Can. Ltd., Toronto, 1980

280. Industrial Hygiene and Toxicology, 2nd edition, Volume I, General Principles, F.A. Patty, editor, Interscience, New York, 1958

281. Ottawa Sets Timetable for Cleanup, Can. Chem. Proc. *58*(3), 19, March 1974

282. C.G. Cortelyou, Chem. Eng. Prog. *65* (9), 69, Sept. 1969
283. A.V. Slack, Envir. Science and Tech. *7* (2), 110, Feb. 1973
284. R.L. Andrews, Combustion (New York) *49* (19), Oct. 20, 1977
285. R.B. Engdahl and H.S. Rosenberg, Chemtech. *8* (2), 118, Feb. 1978
286. Air Pollution, L.C. McCabe, editor, McGraw-Hill Book Co., New York, 1952. Cited by reference 8

# 3 Water Quality and Emission Control

## 3.1 Water Quality, Supply, and Waste Water Treatment

Water is a vital commodity to industry for process feedstock (reacting raw material), solvent component and cooling purposes, as it is to us as individuals for all our personal water requirements. The global supply appears to be extensive, so much so that many people take it for granted. But when one considers that only 3 % of this total resource exists as fresh water at any one time the concept of this as a globally limited and potentially exhaustible resource becomes more real (Table 3.1). When one further takes into account that roughly three quarters of this 3 % is frozen from immediate use

by the ice caps and glaciers of the world, then an appreciation of the limited extent of the available global fresh water becomes more apparent.

If the total available surface freshwater supply of some 126,000 km$^2$, a quantity which excludes the ice caps and glaciers, was evenly distributed over the total non-frozen land area of the earth, it would amount to only some 1.1 m in water depth. Addition of the net annual land-based precipitation, after evaporation, would add only a further 0.8 m depth. The combined total certainly does not represent a limitless resource to serve the agricultural, industrial, and personal needs of a world population now approaching five billion.

Another aspect of the world's freshwater resource

**Table 3.1** Estimated distribution of the global water resource [a]

| Location | Volume, 10$^3$ km$^3$ | Fraction, percent | |
| --- | --- | --- | --- |
| | | of total | of fresh |
| Overall: | | | |
| Oceans[b]: | 1 319 000 | 97.2 | — |
| Icecaps, glaciers | 29 200 | 2.15 | 76.8 |
| Fresh water on land or air | 8 500 | 0.65 | 23.2 |
| Total | 1 358 000 | 100.00 | 100.00 |
| Freshwater[c]: | | | |
| Antarctic ice cap | 26 900 | 1.99 | 71.33 |
| Greenland ice cap | 2 100 | 0.15 | 5.57 |
| Glaciers | 210 | 0.02 | 0.56 |
| Groundwater, to 4 km depth | 8 360 | 0.62 | 22.17 |
| Freshwater lakes | 125 | 0.009 | 0.33 |
| Rivers, average | 1.25 | 0.0001 | 0.003 |
| Atmosphere, average | 13.0 | 0.001 | 0.03 |
| Total | 37 709 | 2.8 | 100.0 |
| Precipitation, annual: | | | |
| Over oceans | 320 | | |
| Over land area | 100 | | |
| (less evaporation) | 70 | | |
| Net, to land | 30 | | |

[a] Compiled and calculated from data in [1 and 104]

[b] Includes saline lakes, principally the Caspian Sea, of gross volume 10$^4$ km$^3$.

[c] Instantaneous values. Does not consider precipitation, which balances out evaporation etc. on a long term basis.

is that it is not at all uniformly distributed over the land surface, even in the form of lakes and water courses. This occurs partly from the uneven distribution of glaciers and their meltwaters, and partly from the wide disparities in rainfall over the earth's surface. Total rainfalls of above 11 m are experienced in some years in the Mt. Waialeale area of Hawaii and at Cherrapunji, India, while less than 0.2 cm falls in the same period in Arica, Chile, or at Wadi Halfa in the Sudan [1]. This uneven distribution of source waters has already contributed to the need to re-use much of the available natural supply [2]. It has also stimulated the development of economical desalination techniques and the construction of large scale desalination plants to enable brackish or marine sources of water to be used [3].

Multiple usage and re-use of available supplies, and sometimes poor liquid and general waste disposal practices have combined to severely degrade the quality of the surface waters of many parts of the world. As recent examples, in 1969 the Cuyahoga River, which flows into Lake Erie at Cleveland, Ohio, was declared a fire hazard from the accumulated loading of combustible organics floating on its surface [4]. Shortly thereafter it actually did catch fire and the resulting heat seriously damaged two steel railway bridges. News reports from Mexico City in 1976 announced that the city had found it necessary to post "No Smoking" signs beside a promenade which ran alongside the Tlalnepantla River, motivated by similar risks [5]. The Rhine River, in Europe, which flows through several different federal jurisdictions with sometimes confused responsibilities for water quality regulation, has also had its share of degradation [6]. Surface water quality trends such as these have led to a growing move from piped to bottled water for drinking purposes [7].

Stationary bodies of water such as the Great Lakes, and the Caspian and Black Seas, are also not without their quality problems even though they represent very large volumes of water. This is because of the more limited or non-existent water exchange rates possible in these surface water features, as compared to rivers. Limited water exchange, in turn, limits the contribution of dilution to the recovery of water quality from waste discharges. The Mediterranean and Baltic Seas also have little external water exchange. The high level of industrial activity on their shores coupled with the sometimes poor emission control and waste disposal practices have combined to severely affect their water quality [8-11].

The size of the oceans themselves provides a very large buffer capacity for waste disposal before water quality is noticeably affected. But even these very large ultimate waste sinks are showing signs of degradation, particularly around busy ports and coastal industrial cities as well as in the region of major shipping lanes in the open ocean [12].

However, it is possible, with cooperation and a determined effort, to reverse the trends towards a deterioration in water quality in any of these areas. As an example, the Thames River through London, England, as recently as 1958 showed no fish at all in the 60 km stretch of estuary between Fulham and Gravesend. Only tubifex worms and eels, both quite resistant creatures, were found to be living in this region. This low species diversity was mainly a result of the high temperatures, a high biochemical oxygen demand (BOD) and a deficiency of dissolved oxygen. But by 1982, the sweeping powers of the Thames Water Authority enabled control of most of the causes of this poor water quality, returning dissolved oxygen concentrations to an average of 50 % of saturation. Many species of fish had by this time returned to this habitat, and plans were being made to restock the river with salmon. These aspects, together with the return of abundant waterfowl, are all signs of a dramatic improvement [13, 14].

Contamination of the Great Lakes in North America was also clearly recognized in the late 1960's, as was evidenced by increased nutrient and pesticide levels, and decreased catches of commercial species of fish coinciding with increased populations of coarser (more pollutant tolerant) species [15]. But fortunately this was recognized early enough by both the U.S.A. and Canada to enable a joint effort to already result in noticeable improvement [16]. Full restoration of former water quality in these lakes is likely to take longer, however, than it took with the Thames River because of the very much slower water exchange rates. Some 70 years would be required for the St. Lawrence River to drain the water of the Great Lakes system once.

## 3.2 Water Quality Criteria and Their Measurement

A surface water resource of very poor quality is easily recognized as such by sight by anyone, and frequently also by smell. The need for improvement is also self evident. But though valid, these assessments are subjective. They are not quantified or placed into categories appropriate to determine the steps necessary to obtain improvement. Also a body of water may appear to be "all right" to the senses but still be of poor quality for some kinds of uses. As examples of natural waters which on the face of it would appear pristine, ground water supplies from springs and wells are known which are high in toxic arsenic, fluorides, or sulfides [17, 18], and recreational lakes in southern Norway and the Adirondacks (U.S.A.) are known which have pH's of 4 and less [19]. For these reasons it is useful to have a set of quantifiable criteria which may be used to measure water quality. The values obtained for each of these criteria allow a valid and quantitative relative assessment of the water quality of different sources to be made. They also permit identification of the appropriate action required to improve the quality of the surface water. If the surface water source is to be used as a municipal or industrial supply, the values obtained for these criteria also establish the complexity of the treatment measures necessary.

## 3.2.1 Suspended Solids

Insoluble matter in surface waters may be partly settleable, consisting of fairly large particles, or may be non-settleable, falling into the colloidal size range of silts and clays. The dividing line by diameter between these two classifications differs with the density of the particle. But for spheres of density $2 \text{ g/cm}^2$, diameters larger than about $100 \mu m$ (0.10 mm) settle out or less quickly, and smaller than this are relatively more slow to settle (Table 3.2).

The suspended solids content of a water sample may be determined by filtering an appropriate measured volume through a tared, fine glass fiber filter mat, and then rinsing any dissolved salts from the mat with a small portion of distilled water. The solids collected are dried at a standard temperature, usually 100-105 °C, and the dry weight obtained is then related to the original volume filtered for the result [20, 21]. Very fine sintered glass filters, or well prepared asbestos fiber pads (Gooch filters) may also be used, but paper filters perform poorly in this application [20, 21]. Depending on the pore size of the filter medium used, this method gives a result which includes the settleable and

**Table 3.2** Terminal settling velocities versus diameter of sphere of specific gravity 2, in water at 25 °C [a]

| Particle diameter, $\mu m$ | Terminal velocity, mm/sec | Characteristic description[b] |
|---|---|---|
| 1000 | 100 | coarse sand |
| 500 | 60 | coarse sand |
| 200 | 30 | coarse/fine sand |
| 100 | 6 | fine sand |
| 50 | 1.5 | fine sand |
| 20 | 0.2 | fine sand/silt |
| 10 | 0.06 | silt |
| 5 | $1.5 \times 10^{-3}$ | silt |
| 2 | $2.2 \times 10^{-4}$ | silt/clay |
| 1 | $6.0 \times 10^{-5}$ | clay |

[a] Settling velocities estimated from a linear graphical representation of [109].

[b] Approximate dividing lines indicated by joint descriptions.

much of the non-settleable suspended solids present in the sample. The size distribution of the collected particles may be determined using a Colter Counter [22].

Centrifugation may also be used for suspended solids determination using the proper conical end centrifuge tubes [20]. After centrifuging the sample the supernatant water is decanted from the precipitate, and the precipitate is rinsed with a small portion of distilled water to get rid of any dissolved salts. Centrifuging again, decantation, and drying the residue as before gives the suspended solids result. A smaller proportion of the non-settleable suspended solid is retained by this method than by filtration, a factor which has to be considered when comparing results obtained by the two systems.

An Imhoff cone is used to determine the settleable solids content of treated waste waters. This cone-shaped measuring device, of 1 L capacity and made of transparent glass or plastic, is graduated down the side to units of tenths of a mL at its lower apex. The sample is placed in the cone, and allowed to stand for a period of one hour, when the volume of the solids layer settled is read directly from the graduations on the side.

*In situ* turbidity determinations, which are related to, but do not strictly correspond to suspended solids determinations, are carried out using a Secchi disk [23]. The disk, a circular plate of about 20 cm in diameter made of sheet steel or other metal, is painted with alternating black and white quadrants on its upper face. A rope or chain calibrated in meters is attached to the disk via an eye bolt at the center of the disk, so that the disk hangs horizontally. For a reading, the disk is lowered into the water until the black and white painted quadrants appear to be uniformly gray from the turbidity of the water, and the depth at which this occurs is recorded. It is then lowered a few meters more and then gradually raised again, noting the depth at which the black and white quadrants just become discernible again. The average of these two readings gives the turbidity result, which can range from 0.2 m or less for a silt-laden river in spring flood, to 25 m or more for an oligotrophic (geologically young) mountain lake.

## 3.2.2 Dissolved Solids

Water quality and monitoring programs also have an interest in the dissolved solids or salts content of water systems. Among the monovalent cations, there is an interest in the concentrations of sodium, potassium, and ammonium ions (and neutral ammonia, which is in equilibrium). Among the polyvalent cations, calcium and magnesium are the main ones of concern because of their strong tendency to precipitate and form useless curds with natural soaps, and to form adhering deposits or scale on the walls of all kinds of water heating appliances, from domestic kettles to industrial boilers. Occasionally this category might also include iron, aluminium, or other polyvalent cations, since, if these are present, they also contribute to this tendency. This collective property of polyvalent cations is referred to as hardness. Thus, hard waters, with a high polyvalent cation content do not launder well with ordinary soaps, whereas soft waters do (Table 3.3). Soft waters also heat more cleanly in water heating appliances.

Among the principal anions of interest in water treatment programs are chloride, sulfate, carbonate, and fluoride. High chloride concentrations are of concern because of the tendency of chloride to accelerate the corrosion rates of pipelines and local water distribution systems. Sulfate at concentrations above 150 ppm can cause severe digestive upset, essentially the symptoms of diarrhea, especially in non-acclimatized people. The concentration of carbonate present has important consequences in relation to hardness, about which more is said later. Concentrations of

**Table 3.3** A rough guide to the scale of hardness of natural waters according to their content of calcium carbonate (or the equivalent) [a]

| Description | Hardness, as ppm (mg/L) of $CaCO_3$ |
|---|---|
| Soft | 0 – 50 |
| Moderately soft | 50 – 100 |
| Slightly hard | 100 – 150 |
| Hard | 200 – 300 |
| Very hard | > 300 |

[a] Data from Klein [20].

fluoride of ca. 1 ppm in water supplies are beneficial in reducing dental caries, but natural concentrations much above this which do occur occasionally can give rise to toxic symptoms [24-26].

The concentrations of the nutrient anions comprising nitrogen, primarily as nitrate, $NO_3^-$, and nitrite, $NO_2^-$, and phosphorus, primarily as phosphate, $PO_4^{3-}$, are also of interest in water monitoring programs. Nitrite is also of independent interest because of its high toxicity, particularly to infants [27]. Since nitrate is also reduced to nitrite in the digestive system, the presence of either ion at concentrations above 10 mg/L in a water supply is cause for concern since 328 cases of methaemoglobinemia (blue baby syndrome) were reported for infants from this cause, in the U.S. during the 1945-69 period, and 39 fatalities [28]. Both groups of ions, but particularly phosphate, are limiting nutrients for the growth of algae and weeds in water systems [27]. Hence, if the concentration of phosphate is allowed to rise above 0.015 mg/L (calculated as P) and nitrate above 0.3 mg/L (as N) in surface waters, then algal blooms (prolific algal growth) are a likely outcome [28-30].

Specific conductance is a useful rough guide to the total dissolved ionic solids present in a water sample, and is a technique which also readily lends itself to the continuous monitoring of a river or waste stream for the total ion content (Figure 3.1). It is also a simple method which can easily be used to check the accuracy of other types of analyses conducted for specific ions. Specific conductance is measured via a pair of carefully spaced platinum electrodes which are placed either directly in the water to be measured, or in a sample withdrawn from the stream. The water temperature should be 25 °C, or the result corrected to this temperature.

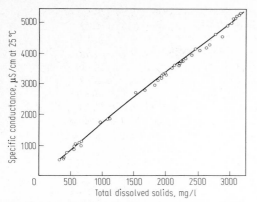

**Figure 3.1.** Total dissolved solids concentration (by analysis) versus specific conductance, shown by Gila River (Bylas, Arizona) water samples taken over a one year period [302]

Voltages in the 12 to 14 range, 60 to 1000 Hz AC are used, plus a Wheatstone bridge circuit to obtain a conductivity reading in $\mu$mhos/cm or $\mu$m/cm (micro Siemens/cm) [31]. The response obtained is linear with total ion content over a wide range of concentrations (Figure 3.1). Examples of the conductance ranges and seasonal variation of some typical Canadian rivers are presented in Figure 3.2. Rivers of both low and high dissolved solids, and with little or wide seasonal variations are clearly evident from these plots.

More specific and detailed information about the dissolved ionic solids present may be obtained through the application of conventional specific analytical techniques for the ions of interest. Complexation titration, for example, using a stable

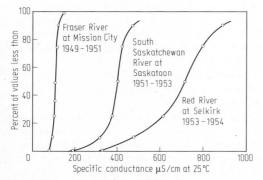

**Figure 3.2.** Distribution of the specific conductance readings for some Canadian rivers during two-year data collection periods by Thomas [303]

complexer such as EDTA (ethylenediaminetetraacetic acid) is useful for the determination of dissolved calcium, as well as other ions [32]. EDTA, being a hexacoordinate complexer, forms very stable 1:1 complexes with many metal ions, including calcium. An 0.01 M standardized EDTA solution, usually as disodium EDTA, is first made up, 1.00 mL of which is exactly equivalent to 400.8 $\mu$g Ca, or 1,000 $\mu$g $CaCO_3$. Then titration of a sample containing calcium ion in the presence of a weakly complexing dye indicator gives displacement of the calcium from the indicator plus a corresponding change in colour of the solution, which is used as the end point of the determination (Equations 3.1, 3.2).

$$M^+ + dye \rightarrow M^+ \cdot dye \text{ (weak complex)} \quad (3.1)$$

$$M^+ \cdot dye + EDTA \rightarrow dye + M^+ \cdot EDTA \quad (3.2)$$
$$\text{(strong complex)}$$
$$\text{(colourless)}$$

Murexide, or Eriochrome blue-black R are suitable indicators for this analysis [32]. A number of other metal ions may also be determined by related complexometric titrations [31].

Ion selective electrodes provide a simple and accurate method for the determination of many ions in solution, not only hydrogen ion, which has been known for many years [33]. Thus concentrations of $Na^+$, $K^+$, $Ca^{2+}$, and $Pb^{2+}$ as examples of cations, and $F^-$, $Cl^-$, $Br^-$, $I^-$, $S^{2-}$ and $CN^-$ as anions may all be measured using this method. With the appropriate measuring precautions and attention to possible interferences most of these ions can be determined at concentrations as low as $10^{-5}$ to $10^{-6}$ M, and lead for example down to $10^{-7}$ M [33, 34].

Of the nutrient ions of current concern, "nitrogen" refers to "combined nitrogen", that is nitrite, nitrate, ammonia or ammonium salts, and occasionally amines and the like. The dissolved element itself is not of concern.

Nitrite concentrations are determined by first, diazotization of sulfanilamide under acid conditions using the acidified waste sample to be analyzed as the source of the nitrous acid (Equation 3.3) When the diazotization is conducted in the presence of N-(1-naphthyl)ethylenediamine, the reactive diazonium salt couples with the amine to form an intensely-coloured, reddish purple azo compound (Equation 3.4). The concentration of azo compound obtained can be determined either colourimetrically or spectrophotometrically. The nitrite

$$H_2N-\!\!\bigcirc\!\!-SO_2NH_2 + HNO_2 + H^+ \longrightarrow 2\,H_2O + N_2^+\!\!-\!\!\bigcirc\!\!-SO_2NH_2 \qquad (3.3)$$

sulfanilamide          nitrous                    diazonium salt $(= I)$
                        acid

$$1 + H_2NCH_2CH_2N(H)\!\!-\!\!\bigcirc\!\!\bigcirc \longrightarrow$$

amine

$$H_2NCH_2CH_2N(H)\!\!-\!\!\bigcirc\!\!\bigcirc\!\!-N\!\!=\!\!N-\!\!\bigcirc\!\!-SO_2NH_2 \qquad (3.4)$$

reddish purple

detection limit of this method is 0.02 mg/L of nitrite, as N [32] (equivalent to 0.02 ppm or 20 ppb N in fresh waters).

Nitrate is determined on a separate water sample by first reducing nitrate to nitrite with either hydrazine sulfate or with an easily oxidized metal such as cadmium [32] (Equation 3.5)

$$NO_3^- + Cd \rightarrow CdO + NO_2^- \qquad (3.5)$$

The sensitive nitrite method is then used to determine total nitrite, which will be equivalent to nitrite plus the original nitrate. Subtraction of the original nitrite from this total nitrite result gives the value for the nitrate concentration.

Reduced nitrogen, as ammonia or ammonium salts, may be analyzed by spectrophotometric determination of the absorbance of the yellow-brown colloidal suspension formed on the addition of Nessler's reagent, which is potassium mercuriiodide (Equation 3.6)

$$2\,K_2HgI_4 + 2\,NH_3 \rightarrow NH_2Hg_2I_3 + 4\,KI + NH_4I$$
$$\text{yellow-brown} \qquad (3.6)$$

If amines are present, then the determination requires prior Kjeldahl decomposition of the amine to an ammonium hydrogen sulfate using boiling sulfuric acid, before the quantitative analysis. Following this, the ammonia content is measured using Nessler's reagent. Again subtraction yields an amine concentration independent of the original ammonia content for the water sample analyzed.

Phosphate concentrations may be measured gravimetrically, by weighing the yellow precipitate formed on the addition of a solution of ammonium molybdate to the water sample acidified with nitric acid (Equation 3.7, product stoichiometry certain, although structure not) [35].

$$PO_4^{3-} + 12\,(NH_4)_2MoO_4 + 48\,HNO_3 \qquad (3.7)$$
$$\rightarrow (NH_4)_3PO_4 \cdot 12\,MoO_3 \cdot 3\,H_2O +$$
$$21\,H_2O + 21\,NH_4NO_3 + 3\,NO_3^-$$

If higher sensitivities are desired then the ammonium phosphomolybdate product may be reduced with stannous chloride or ascorbic acid to give soluble molybdenum blue [32]. Spectrophotometric determination of the blue absorbance at 885 m$\mu$ gives a detection limit of 0.003 mg/L phosphate, as phosphorus (3 ppb in fresh water) or 0.0092 mg/L as phosphate [32].

### 3.2.3 Total Solids or Residue Analysis

Analysis of a water sample for total solids content (suspended and dissolved) requires evaporation of the water from a measured volume of sample, usually 1 L. For water samples the residue is usually dried at 180 °C, to put calcium sulfate in the anhydrous state, and magnesium sulfate as the monohydrate, $MgSO_4$ x $H_2O$, in the residue [20]. In any case, because of the variable dependence of the level of hydration of many salts on temperature, and different standard drying temperatures used [32], the temperature used should be stated with the results.

To determine the total solids content of river muds or sewage sludges, either of which may have a high organic content, lower drying temperatures of 100-105 °C are normally used to reduce the risk of thermal decomposition of the organic content [20, 32]. For this reason, azeotropic drying is sometimes used for samples of this kind [20]. Using perchloroethylene as the azeotroping solvent, for instance, on a weighed amount of mud or sludge, drops the temperature required for water removal to 88.5 °C, the boiling point of the perchloroethylene-water azeotrope. When azeotropic methods are employed the water removed from the sample is read from a graduated solvent/water separator, from which the solvent from the condensate of the process is returned to the sample flask.

## 3.2.4 Dissolved Oxygen Content

Oxygen, nitrogen (as $N_2$), carbon dioxide and the other trace gases of the atmosphere are in a dynamic equilibrium between surface waters of the earth, and the air. Gases in the surface layer of water move through (e.g. during river flow), or are largely restricted from moving through (e.g. in a thermally-stratified lake) the water column below it, depending on the degree of mixing of the water phase. The solubility of oxygen in water is low, but its presence is extraordinarily significant to the existence of all kinds of water organisms and to the perceived and actual water quality.

The solubility of any gas in a liquid with which it does not react is proportional to the pressure of that gas (i.e. the partial pressure) above the liquid (e.g. Equation 3.8 for oxygen).

$$P_{O_2} = k_{O_2} X_{O_2}, \text{ where} \tag{3.8}$$

$P_{O_2}$ is the partial pressure of oxygen above water, in atmospheres,
$k_{O_2}$ is the Henry's Law constant, and
$X_{O_2}$ is the mole fraction of oxygen in the water phase.

This solubility relation, also known as Henry's Law, is quite closely followed for pressures not a large multiple removed from normal atmospheric pressure, and for any given temperature. Oxygen, comprising about 21 % (actual value 20.95 %) of

the atmosphere by volume, therefore contributes 0.21 atmosphere partial pressure above surface waters at sea level, and lower pressures at higher altitudes.

Gas solubility in a liquid is also inversely proportional to temperature (Equation 3.9).

$$T = k_t / X_{O_2}, \text{ where} \tag{3.9}$$

$k_t$ is the proportionality constant at absolute temperatures.

This factor means that discharge of warmed waste waters into a river or lake decreases the solubility of oxygen in that water, and thus tends to decrease the actual concentration of dissolved oxygen present.

The combined influence of temperature and pressure on the solubility of some common gases important in water quality studies is given in Table 3.4. A couple of interesting features to note are that oxygen, with a Henry's Law constant of $1.91 \times 10^7$ moles/(mole mmHg), is more than twice as soluble in water as nitrogen. Oxygen also has a very similar molar solubility as argon, a reflection of the diatomic nature of oxygen gas which has a similar molar mass as that of monatomic argon, 39.95. Also evident here is the rationale for the use of oxygen-helium mixtures for deep sea diving [36]. The much lower solubility of helium in water, and hence in the blood, reduces the risk of gas bubble formation in the bloodstream on the return of a deep sea diver to ordinary atmospheric pressure, such as can readily occur with nitrogen (the "bends") if ordinary air is used for diving under these circumstances.

A third factor influencing the solubility of oxygen or other gases in water is the presence of other solutes in the water. A high concentration of dis-

**Table 3.4** Henry's law constants for the solubility of some common dry gases in water [a, b]

| Temperature °C | Henry's law constants/$10^7$ | | | | |
| | Argon | Carbon dioxide | Helium | Nitrogen | Oxygen |
|---|---|---|---|---|---|
| 0 | 1.65 | 0.0555 | 10.0 | 4.09 | 1.91 |
| 10 | 2.18 | 0.0788 | 10.5 | 4.87 | 2.48 |
| 20 | 2.58 | 0.108 | 10.9 | 5.75 | 2.95 |
| 30 | 3.02 | 0.139 | 11.1 | 6.68 | 3.52 |
| 40 | 3.49 | 0.173 | 11.0 | 7.60 | 4.14 |
| 50 | 3.76 | 0.217 | 10.9 | 8.20 | 4.50 |

[a] For the equation p = kX, where p = partial pressure of the gas in mm of Hg,
and X = mole fraction of the gas in water,
and k = the Henry's law constant (e.g. $1.91 \times 10^7$ for oxygen at $0\,°C$).
[b] Values selected from those given in [293].

solved solids in the water phase decreases the solubility of oxygen [37]. Other things being equal, therefore, marine waters tend to have a lower dissolved oxygen content than fresh waters.

The combined influence of all of these solubility factors on the actual oxygen content of ordinary surface waters is presented in Table 3.5. From this data it can be seen that the air-saturated oxygen content of sea water is normally about 20 % less than the oxygen content of fresh water, at the same temperature. It can also be seen that the effect of elevated temperatures on solubility is such that at 30 °C only about half the oxygen content is obtained in the water phase at 0 °C, for either fresh or salt waters. Lakes at higher altitudes, exposed to lower atmospheric pressures and hence lower oxygen partial pressures will tend to have lower concentrations of dissolved oxygen, other things being equal. Similarly the salt lakes which lie below sea level, such as the Dead Sea, Salton Sea etc., will tend to have higher concentrations of dissolved oxygen than would be predicted for similarly saline waters at sea level.

The Winkler test, or variations of it, is the standard wet chemical procedure for measuring the concentration of dissolved oxygen in water samples [32, 38, 39]. This test uses a standard 300 mL BOD sample bottle closed with a glass stopper which has a polished cone-shaped end to it to ensure that it is completely filled during the test. This avoids inadvertent aeration of the sample during the test. Initially a white precipitate of manganous hydroxide is prepared in the BOD bottle (Equation 3.10).

$$MnSO_4 + 2\,KOH \rightarrow Mn(OH)_2 \downarrow + K_2SO_4$$
$$\text{white} \tag{3.10}$$

Dissolved oxygen in the water sample being tested oxidizes the manganous hydroxide to a brown suspension of manganic hydroxide in the test solution (Equation 3.11).

$$2\,Mn(OH)_2 + O_2 \rightarrow 2\,MnO(OH)_2 \downarrow \tag{3.11}$$
$$\text{brown ppte}$$

If no brown colouration is observed at this stage of the test, there is no dissolved oxygen in the water sample being tested. If there is a brown colouration the amount of manganic hydroxide is accurately determined by first putting it into solution with sulfuric acid (Equation 3.12).

$$MnO(OH)_2 + 2\,H_2SO_4 \rightarrow Mn(SO_4)_2 + 3\,H_2O \tag{3.12}$$

**Table 3.5** Oxygen content of water in equilibrium with air saturated with water vapour at 760 torr (mmHg) [a]

| Temperature °C | Oxygen content, mg/L | | Water vapour pressure, mmHg |
|---|---|---|---|
| | Fresh water | Sea water, 3.5% salinity | |
| 0 | 14.6 | 11.22 | 4.58 |
| 10 | 11.3 | 8.75 | 9.21 |
| 20 | 9.2 | 7.17 | 17.54 |
| 30 | 7.6 | 6.10 | 31.82 |
| 40 | 6.6 | 5.14 | 55.32 |
| 50 | 5.6 | – | 92.51 |

[a] Compiled from [20, 32, 109].

To determine the original levels of oxygen present, excess aqueous potassium iodide is added, releasing iodine equivalent to the manganic sulfate present in the most rapid reaction of this sequence (Equation 3.13).

$$Mn(SO_4)_2 + 2\,KI \rightarrow MnSO_4 + K_2SO_4 + I_2$$
$$\text{brown solution} \tag{3.13}$$

The iodine generated is prevented from being lost as vapour by ensuring that an excess of potassium iodide is present, which forms the stable ionic complex, $KI_3^-$. The liberated iodine is measured by the familiar sodium thiosulfate titration, using starch as an indicator near the end point of the titration to sharpen its observation (Equation 3.14).

$$I_2 + 2\,S_2O_3^{2-} \rightarrow 2\,I^- + S_4O_6^{2-} \tag{3.14}$$
$$\text{colourless solution}$$

Each mole of liberated iodine is equivalent to a half mole of dissolved oxygen in the original water sample, so that each mL of 0.0250 M thiosulfate is equivalent to 200 $\mu$g of original dissolved oxygen content. The Winkler test is reliable and accurate, to of the order of 20 $\mu$g/L for a 200 mL sample of ordinary waters and somewhat less accurate for sewage and industrial effluents where interfering substances may be present [32]. It is also a primary determination method, which means that it may be used to calibrate or check other kinds of dissolved oxygen instrumentation such as dissolved oxygen meters based on the oxygen electrode. Improved designs of these instruments, such as those produced by the Yellowstone Instrument Company [40], provide rapid and convenient dissolved oxygen readings as long as care is taken to operate at or near calibration temperatures, and periodic performance checks are carried out [41]. Water samples of a range of dissolved oxygen levels may be readily prepared for this purpose. A zero mg/L (ppm) test may be obtained by sparging a quantity of boiled distilled water with nitrogen for 3 or 4 hours and in a thermostat bath of the required test temperature. Sparging air through at 20 °C for the same period would give 9.2 mg/L, and sparging pure oxygen through should give 44.0 mg/L at this temperature.

For meaningful short-term results, particularly with waters having prolific weed or algae growth, care must be taken to sample or measure the dissolved oxygen *in situ* at the same time of day, and note the ambient temperature. This is because photosynthetic production of oxygen towards the end of a sunny day can almost double the dissolved oxygen content of the water, even to well above saturation levels, as compared to the values in the very early morning after a cloudy day [42]. Metabolic consumption of oxygen under these circumstances can reduce the dissolved oxygen concentrations to near zero.

Dissolved oxygen concentrations in surface waters below about 5 mg/L or 50 % of saturation are generally unsatisfactory for a diverse biota [6]. In fact game (or sport) fish require more like 60 to 70 % of saturation to do well. Lower dissolved oxygen concentrations than these tend to limit the habitat to the growth and reproduction of coarser species of fish. No dissolved oxygen eliminates the survival of gill breathers entirely, and discourages many other species so that only organisms capable of air breathing, such as eels and certain Tubifex worms, can survive in this situation.

The dissolved air, or more particularly dissolved nitrogen, in water can also be too high for fish well-being, 110 % of saturation already putting them at risk. If gill breathers are exposed to waters supersaturated in air, their blood becomes supersaturated too, through gill action. Thus at 120 % of saturation the excess nitrogen in the bloodstream, which is not metabolically consumed, comes out of solution behind the membranes of the eye sockets causing "pop-eye", and also behind the membranes of the fins, tail, and mouth causing gas blisters [43-45]. These symptoms, collectively referred to as nitrogen narcosis, acutely stress and when severe can kill the fish [43]. Exposure to 120 to 140 % of saturation for any length of time causes symptoms to appear, and some fish mortality. Exposure to concentrations above 140 % of saturation causes high mortalities.

Air supersaturation of natural waters can occur from the warming of a cold, air-saturated stream as it flows into a shallow reservoir, or from being used as a source of cooling water, which amount to thermal causes [46]. It is also a normal occurrence while releasing reservoir water into a plunge basin, at the bottom of which the pressure of air bubbles entrained in the water may be raised nomentarily to 2 or more atmospheres from the pressure of the head of water over the bubbles. This increased pressure is sufficient to supersaturate the water, causing problems here and downstream of the plunge basin [46]. Plunge basin supersaturation can be avoided by the use of a "flip-lip" or "flip-bucket" spillway designs, for example, which spread spilled water widely and as a fine spray to avoid carrying bubbles to depth.

From the point of view of disposal of oxidizable wastes to water, low concentrations of dissolved oxygen in a receiving body of water greatly exaggerate the impact of the waste discharge on water quality. Too low a waste dilution ratio on discharge can cause the dissolved oxygen content of the water course to drop to zero at any time. Discharge of any oxygen-consuming waste causes a decrease in the dissolved oxygen content of the receiving lake or stream. But if this discharge coincides with low dissolved oxygen in the receiving waters the resultant drop in oxygen concentration will be greater and the recovery of the stream to normal oxygen concentrations will be less rapid than at higher initial oxygen concentrations. Also it is known that fish and other aquatic animals that are stressed by low dissolved oxygen concentrations are more susceptible on exposure to any additional stresses such as high temperatures, toxic substances, pH deviations and the like.

### 3.2.5 Relative Acidity and Alkalinity

The pH of a water supply, in conjunction with its dissolved solids content, is important because it relates to the corrosivity of the supply. Corrosivity is of concern in municipal water distribution systems, as well as for the feed water of power boilers and process cooling water. pH's in the neutral to slightly alkaline range are generally preferred for these applications. Most aquatic creatures require a pH in the 5 to 8.5 range for optimal growth and reproduction, although they may survive for a time at pH's somewhat outside this range [6, 27]. Natural waters in the acid range, which can arise by absorption of acidic atmospheric gases [47] (see Chapter 2) or from the accumulation of humic acids on percolation through peaty soils, have the potential to mobilize elements of the rocks and soils through which they flow. In limestone, dolomites and like rock formations, calcium and magnesium (and other elements) may be dissolved out, in the process increasing the pH but raising the hardness of the water (e.g. Equations 3.15, 3.16).

$$CaCO_3 + H_2CO_3 \rightarrow Ca(HCO_3)_2 \quad (3.15)$$
insoluble                    soluble

$$MgCO_3 + 2\,HNO_3 \rightarrow Mg(NO_3)_2 + H_2O + CO_2$$
insoluble                    soluble                    (3.16)

Industrial liquid waste streams, in general, have the greatest potential of any of the ordinary range of aqueous wastes to influence the pH of receiving waters, because of their sometimes significant deviation from neutral conditions and the large volumes commonly involved. Pulp mills, particularly those producing fully bleached kraft pulp, have some very acidic and some strongly alkaline waste streams generated by their processes. Drainage waters from coal mines and from metal mines working sulfide ores can be quite acidic from acid generated by the bacterial oxidation of sulfides (e.g. Equations 3.17-3.19).

$$2\,FeS_2 + 7\,O_2 + 2\,H_2O \rightarrow 2\,FeSO_4 + 2\,H_2SO_4$$
$$(3.17)$$

$$2\,FeS_2 + 7\tfrac{1}{2}O_2 + H_2O \rightarrow Fe_2(SO_4)_3 + H_2SO_4$$
$$(3.18)$$

$$Fe_2(SO_4)_3 + 6\,H_2O \rightarrow 2\,Fe(OH)_3 + 3\,H_2SO_4$$
$$(3.19)$$

Chemical plants producing phosphatic fertilizers or phosphoric acid may also have quite acidic discharges, and ammonia plants or coal-coking plants quite alkaline streams, as further examples.

Monitoring the pH of waste streams or surface waters is most frequently accomplished using a pH meter. An important preliminary for accurate readings is that the meter be calibrated using buffers at slightly higher and lower pH's than those of the samples being measured [32]. Also the temperature calibration should be properly adjusted and the electrode given sufficient time in the sample being measured to come to thermal equilibrium. Using a meter allows determination of pH free from any interferences due to colour, colloidal matter, coarse turbidity, free chlorine or the presence of other oxidants or reductants. However, pH measurement of distilled water, or the purer natural waters may give inaccurate readings from too low conductivities unless small amounts of a neutral salt, such as 1 g/L potassium chloride, are added to raise the total ionic strength to about 0.1 M [48]. If the pH measurement is not being conducted *in situ*, it should be measured promptly after sampling to avoid errors due to gas exchange or biochemical processes altering the pH after collection. In this connection it should be noted that the pH of eutrophic waters can be highly time dependent from the influence of variable rates of metabolic activity on carbonate-bicarbonate buffer [20]. Marine waters, and any other streams with a relatively high sodium ion content will require a sodium ion correction, especially if the measurement is being conducted at high pH [32].

For low cost occasional pH measurement and for measurements in situations unfavourable to the placing of a fragile glass pH electrode, narrow range single indicator and wide range multiple indicator pH papers are convenient and sufficiently accurate for most purposes. Many of these read to 0.5 pH unit and some to 0.3 pH unit [49-51].

### 3.2.6 Toxic Substances

Here are considered a number of potential water contaminants which, for want of space for details, are grouped together. Included are such candidates as toxic heavy metals, pesticides, water and weed treatment chemicals, radioactive particles and the like.

The relative toxicities to fish of some appropriate examples of toxic substances are given in Table 3.6. It should be remembered however that aquatic toxicity, like toxicity measurements in general, are difficult properties to pin down to an exact value since they depend not only on the particular species and the ages and state of health of the exposed individuals, but also on the time exposed, water temperature, simultaneous presence of other contaminants, oxygen content, prior acclimatization, hardness and other associated factors. As a rule though the high sensitivities of fish to the substances tabulated is the consequence of the good blood/water exchange processes necessary in gill breathing animals for their respiration. A parallel in man's case is our high sensitivity to many types of atmospheric contaminants.

**Table 3.6** Approximate acute toxicities of some common potential water contaminants to freshwater fish [a]

| Substance | Approximate lethal concentration mg/L (ppm) |
|---|---|
| chlorine | 0.05 – 0.2 |
| chloride | ca. 6000 |
| copper | 0.05 |
| cyanide | 0.04 – 0.10 |
| DDT | 0.02 – 0.10 |
| fluorides | 2.6 – 6.0 |
| malathion | 13.0 |
| mercury, $Hg^{2+}$ | ca. 0.01 |
| phenols | 1 – 10 |
| pentachlorophenol | 5 |
| natural soap [b] | 10 – 12 |
| (in hard water) | (900 – 1000) [b] |
| synthetic detergent [c] | 6 – 7 |
| toxaphene | 0.01 |
| zinc, $Zn^{2+}$ | 0.15 – 0.60 |

[a] Exposure times and toxicity criteria vary. Compiled from selected data of [27, 84 and 293].
[b] Sodium oleate ($NaO_2C(CH)_{2)16}CH_3$), which is much less toxic in hard water, parenthesized value.
[a] Sodium dodecylbenzenesulfonate ($NaO_3S(CH_2)_{12}C_6H_5$).

More than 20 metals or metalloids, Al, Sb, As, Ba, Be, Bi, Cd, Co, Cu, Ce, In, Pb, Hg, Mo, Ag, Te, Tl, Sn, Ti, W, U and Zn are significant in industrial hygiene programs and therefore could potentially be involved in water (or other) hazards [52, 53].

Some wet chemical sample preparation, such as pH or oxidation state adjustment, is normally required for most metal ion determinations. Then complexometric titration using EDTA, as already mentioned, or diphenylthiocarbazone ("dithazone"; Equation 3.20)

$$Ph-N=N-C(S)-NH-NH-Ph \qquad (3.20)$$

may be used for cadmium, copper, lead, mercury, or zinc determinations. With the latter reagent, the concentration of the coloured complex obtained may be measured colourimetrically, by comparison with solutions made up from complexes of known concentrations of the metals of interest. This method is easily applied in the field. Or it may be determined somewhat more accurately, but at greater expense, using a spectrophotometer in the laboratory. Other versatile and sensitive techniques for determination of metal ions in water samples are atomic absorption, which uses the attenuation of a beam of light of the appropriate wavelength by the atomized metal as the measure of concentration [54-56] and anodic stripping voltammetry, which is an electrochemical technique [57]. Further details of all these procedures are available in standard texts [20, 21, 32, 58].

Determination of pesticide and herbicide content, or the presence of other organics in surface waters, usually involves a preliminary concentration step by extraction with an organic solvent such as hexane or heptane. But the concentrations of these substances present are often so low that special highly purified grades of solvent, so-called "pesticide grades", are required to avoid analytical problems from contamination of the sample by traces of pesticide already present in the solvent. After extraction, the small volume of organic phase obtained is then dried and carefully concentrated for analysis by injection into a high sensitivity gas chromatograph [59-62]. A series of peaks of differing retention times is obtained on a strip chart from which probable identities of the compounds represented by each peak may frequently be established by comparision with the retention times for a solution of standard reference materials separately injected into the chromatograph. More unequivocal and rapid peak identification may be obtained from a mass spectrometer coupled to the outlet of the gas chromatograph [62, 63]. The mass spectrometer can be used to establish accurate masses and fragmentation patterns of the constituents of each of the peaks obtained on the chromatograph trace.

Radioactive particles are generally present in surface waters at such low levels that a preliminary concentration step is necessary for determination. Then gross counting of the concentrated sample is carried out to give an overall measure of the radionuclide content [32, 64]. The result obtained can then be related back to the original sample volume. Related to this it should be mentioned that deliberate radiolabelling provides a safe and very sensitive tracer for the tracking of water flow patterns and the dispersion of waste discharges [65, 66].

### 3.2.7 Micro-organisms

The types and populations of micro-organisms present in water are an important part of any water quality considerations [67]. For this purpose they can be conveniently grouped into two main classes.

The indigenous, or naturally-occurring organisms, which includes many ambient types of bacteria, phytoplankton and zooplankton, algae and diatoms, comprise one of these. This class is relatively harmless in water supplies. There are also the bacteria, viruses, and parasites arising from the excreta of warm-blooded animals or from human sewage contamination. This group includes examples which are harmless, i.e. they form a component of the normal human intestinal flora, and some examples which are pathogenic (disease-causing). Detection of the presence of any of this second group of organisms in water supplies should be taken as a warning of the risk of contamination by pathogens [68].

A survey of the micro-organism status of a water sample can be taken by means of a standard plate count. To do this, the sample plus several dilutions are inoculated onto a nutrient agar medium in separate petri dishes, shallow plates with loosely-fitted lids. The spotted dishes are then incubated for a period of time, after which the spots of growth in each are counted. Very high micro-organism numbers in the water sampled tend to cause a merging of spots, making accurate counting impossible, and agglomerations or clumping of organisms tend to give lower counts. Both problems are minimized by the sample dilutions chosen before inoculation of aliquots of each dilution onto the plates.

Since it is the degree of contamination by the bacteria of warm-blooded animals that is most significant from a water quality standpoint, this is measured by means of the differential ratio test [67] (Equation 3.21).

Differential ratio test $= (20\,^{\circ}\text{C count})/$
    $(37\,^{\circ}\text{C count})$ (3.21)
    If $> 10$, reasonably good supply
    If $< 10$, indicative of contamination

The count at $20\,^{\circ}\text{C}$ is taken for a 48 hour incubation of the plates at this temperature. The $37\,^{\circ}$ count is taken for a 24 hour incubation at this temperature, to approximate the propagation conditions provided by warm-blooded animals. If the ratio obtained by the differential ratio test is above 10, it is taken as an indication of a reasonably good water supply [67].

Use of different nutrients in the petri dish, as well as different incubation conditions and various staining and slide-making techniques permit positive identification of the particular micro-

organisms collected [69-71]. Tests giving results comparable to the differential ratio test may also be carried out by using different plating out media for inoculation, or by membrane filtration techniques [72]. There is also a simple dipstick method which may be used for more qualitative bacterial monitoring, the Coli-Count water tester, which uses a filter paper matrix already factory-impregnated with dry nutrient and squared to facilitate counting after incubation [72]. The prepared paddle is dipped into the water to be sampled, which both inoculates and activates the medium. When followed by incubation for the required time in the sterile container provided, a convenient indication of microbiological water quality is obtained.

Viral analysis of waters is more difficult because of their very much smaller size than bacteria, about 10 to 300 nm, and the need for susceptible living cells such as chick embryos or tissue cultures and the like for cultivation and identification in the laboratory [67, 73]. Nevertheless, viruses represent an important microbiological class for water and wastewater monitoring programs since serious diseases such as polio and hepatitis are transmitted in this way.

## 3.2.8 Temperature

Unnaturally affected surface water temperatures are most often caused by the warmed discharges from industrial process or thermal power station cooling requirements. This may seem to be a relatively unimportant parameter to be concerned with, but on the contrary. Increasing the temperature of the surface waters of a lake, for example, increases the rate of evaporation of water from the lake. The net water consumption caused by this increased evaporation rate from the cooling load of a 100 megawatt thermal power station has been estimated to equate to the consumptive load of a city of 100,000 people [74]. This increased evaporation also has the effect of increasing the dissolved solids and nutrients in the water system [75]. These effects in turn tend to increase turbidities due to algal blooms and nuisance weed growth, as a result of the increased nutrient concentrations. The higher average year round temperatures also tend to decrease the average concentrations of dissolved oxygen, from the decreased solubility of oxygen at higher temperatures [76].

High thermal loadings also affect the distribution of fish populations, since coarse fish such as suckers, carp, and catfish can better tolerate higher temperatures and lower dissolved oxygen concentrations than can game fish such as salmon, trout or other commercially valuable species [6, 77]. But even for coarse fish, exposure to temperatures above 40 °C for a brief period is usually fatal. Warm water temperatures can also favour year-round spawning of the coarser species of fish, which tends to further improve the competitive position of coarse fish in this habitat and gradually displaces any pre-existing population of sport fish. Temperature measurement of surface waters using ordinary thermometers is relatively straightforward. Measurement at depth to delineate thermal gradients is more difficult. Flip thermometers may be used, which are lowered to the depth of interest in the normal, vertical temperature-sensing position. At the depth of interest the thermometer is flipped upside down, which breaks a slender portion of the column of mercury in the expansion capillary, preserving the temperature reading for examining at the surface. Or rather more robust thermocouples may be used for continuous temperature measurement at the surface or at depth, a system which also lends itself to continuous automatic recording. A thermocouple, which uses the electrical potential generated by a pair of dissimilar metal junctions, commonly iron/constantan or chromel/alumel, gives either a millivolt reading which can be translated to temperature using a table, or a direct temperature readout via solid state circuitry. The fast response of this system facilitates the taking of a series of readings while gradually lowering the sensor through a single depth profile, or for monitoring the spread of warmed waters with distance.

Thermistors (thermal resistors), which are semiconductor devices which have a very large drop in resistance with temperature, of the order of 20,000 ohms per degree over a 500 K temperature range, may also be used for temperature studies requiring quick response times [78]. But for field overviews of a river or a lake receiving warmed discharges aerial infra-red photography [79] (an aerial thermogram) or ERTS (Earth Resources Technology Satellites, Landsat and Seasat) studies can reveal more details more quickly than the majority of surface methods [80]. Satellite studies can also give an instantaneous thermal picture of very large areas of water, combined with the distribution of algae and the like, in a manner not possible by any other means.

### 3.2.9 Oxygen Demand

Any dissolved oxidizable material present in water places a demand on the dissolved oxygen content of the water, as biochemical processes consume oxygen to utilize it. Therefore the oxygen demand level of surface waters or waste streams discharged into them has an important bearing on water quality and the concentration of dissolved oxygen that can be maintained in surface waters. The biochemical oxygen demand (B.O.D. or BOD), the chemical oxygen demand (C.O.D. or COD), and the total organic carbon content (T.O.C. or TOC) tests are used to measure this parameter, so named because of the respective measurement methods used. Each of these tests is complementary to the others in defining the oxygen demand profile of a water sample, and each has special situations in which it is a more useful or practical measure of oxygen demand, for one reason or another, than the other tests.

The biochemical oxygen demand, as the name suggests, is a measure of the biochemically oxidizable material present in the water sample expressed in terms of the oxygen required to consume it. More precisely BOD is defined as the number of milligrams of oxygen taken up by a one liter sample of water on incubation in the dark for 5 days at 20 °C [20]. For fresh water samples 1 mg/L equates to 1 ppm (weight for weight) so both sets of units have been in common use. But to avoid potential ambiguity, since about 1970 the use of mg/L (or multiples of this) has been growing. In marine waters, because of density differences, it is more accurate to stick to mg/L units. The 5 subscript sometimes used with the BOD label refers to the 5 day test period, which is normally assumed. If the test period differs from this, it should be stated by the subscript used. The BOD test is the most lengthy of the oxygen demand tests to perform and the answers obtained may be difficult to interpret, but it is also a useful test in that the result comes the closest to reproducing the natural oxidation and recovery conditions in a river or lake [81]. This rationale is also the reason for the selection of the standard test temperature of 20 °C. The test can also be performed with relatively simple, inexpensive equipment requiring little operator time, even through there is a lengthy delay before the result is obtained.

There are three possible approaches to measurement of biochemical oxygen demands. The oxygen required to oxidize the organic matter in the water sample can come solely from the dissolved oxygen content of the sample itself or from added aerated dilution water, procedures referred to as direct and dilution techniques, respectively. Or the oxygen can come from a closed air space above the water sample to be analyzed, in which case the procedure is referred to as a manometric, or respirometer, method. Because it is difficult to measure the small volume of gas phase oxygen consumption obtained from a low BOD water sample, the manometric method is normally reserved for use with sewages and high oxygen demand industrial waste streams where the oxygen consumption will be sufficient to obtain a valid reading.

Using the direct technique, which is appropriate for ordinary clean river waters, requires complete filling of two 300 mL BOD bottles with the sampled water. The oxygen content of one is measured within 15 to 20 minutes of collection. The other bottle is incubated in the dark, to avoid photosynthetic contribution to the oxygen content of the water, for 5 days at 20 °C. The extent of biochemical oxidation of any organics taking place is then determined by measuring the oxygen content of the second bottle. The difference between the measured oxygen contents of the first and second bottles is a direct measure of the BOD of the river sampled (Equation 3.22)

Direct biochemical oxygen demand:
$$BOD_5, mg/L = DO_1 - DO_5, \text{where} \qquad (3.22)$$
$DO_1$ = the dissolved oxygen content (mg/L) after 15 minutes, and
$DO_5$ = the dissolved oxygen content after 5 days

Since the oxygen content of fresh waters is ordinarily limited to 9.2 mg/L (Table 3.5) and will usually be somewhat lower than this, BOD measurement by *filling* both BOD bottles with the sampled waters is limited to measurement of BOD's which are less than the dissolved oxygen content of the water. For sewage and some types of industrial effluents, where BOD's of from 100 up to 20,000 are not uncommon, the waste water stream must be diluted with distilled aerated dilution water containing added nutrient salts before the BOD test is carried out. Two BOD bottles will be filled for each of several dilution ratios for a sample with an unknown BOD, so as to ensure that a test with at least one of the dilutions gives a consumption of about 50 % of the initial dissolved oxygen content. Usually the right types of bacteria required will naturally be present. But if not, a small amount of treated municipal sewage (stored frozen) will also be seeded into each bottle to inoculate it. Again the dissolved oxygen of one bottle of each dilution will be measured immediately and of the other after 5 days in the dark, at 20 °C. The BOD of the sample from the test will be calculated using Equation 3.23, or 3.24 if a sewage seed was used.

Biochemical oxygen demand with dilution:
$$BOD_5 = \frac{(DO_1 - DO_5) \times 300 \, mL}{(\text{volume of sample per bottle, mL})} \qquad (3.23)$$

Biochemical oxygen demand with dilution plus seed: $\qquad (3.24)$

$$BOD_5 =$$
$$\frac{(DO_1 - DO_5 - \text{seed correction}) \times 300 \, mL}{(\text{volume of sample per bottle, mL})}$$

where the seed correction =

$$\frac{(DO_1 - DO_5)}{(\text{mL of seed per bottle})}$$

which is determined in a separate test using only seed plus dilution water in BOD bottle

Various refinements to these techniques have been proposed [82]. Substances such as glucose or potassium hydrogen phthalate, which are completely oxidized during the normal 5 day period of the BOD test, may be used as test substances to check the experimental technique of the BOD method used [32]. (Equations 3.25, 3.26).

$$C_6H_{12}O_6 + 6 O_2 \rightarrow 6 CO_2 + 6 H_2O \qquad (3.25)$$

$$C_8H_5O_4K \, (= Ph(CO_2H)(CO_2K)) + {}^{15}\!/_2 O_2 \rightarrow \\ 8 CO_2 + 2 H_2O + KOH \qquad (3.26)$$

Parallel classifications of river water quality based on their BOD loadings, and independently based on their dissolved oxygen content have been proposed (Table 3.7). As a rough rule of thumb intended to maintain river quality a guideline has been suggested that no discharge to a river should bring the BOD of the river to less than 4 mg/L.

The manometric method of BOD determination measures the volume of oxygen uptake by a measured volume of the undiluted sample, placed in a brown glass bottle to prevent any photosynthetic influence (Figure 3.3). After filling, the system is closed and the mercury manometer is set to a zero reading. Stirring is continued during the test to fa-

**Table 3.7** Classification of river quality based on dissolved oxygen content, and independently based on biochemical oxygen demand (BOD) [a]

| Dissolved oxygen % of saturation | River pollution status [b] | BOD loading, mg/L [c] |
|---|---|---|
| 90 or more | very clean | 1 or less |
| ca. 90 | clean | 2 |
| 75 to 90 | fairly clean | 3 |
| 50 to 75 | moderately polluted | 5 |
| 25 to 50 | heavily polluted | 10 to 20 |
| < 25 | severely polluted | 20 or more |

[a] Compiled from separate tables of [20] plus the ranges of data given for the Trent River system [98].
[b] Oxygen content of the river water is a result independent from the BOD result, even though it may be the first part of the two steps required for BOD measurement. While the two readings are not necessarily connected, frequently there will be a low oxygen saturation level coinciding with high BOD occurrences.
[c] Whole river BOD loadings of at least 70 mg/L have been reported for the Trent River systen in the U.K. [98], and exceeding 450 mg/L for the River Irwell (Radcliffe) [27].

cilitate gas transfer across the air-water interface while bacterial attack of the organics present consumes oxygen and produces carbon dioxide (Equation 3.27).

$$C \text{ (organics)} + O_2 \rightarrow CO_2 \tag{3.27}$$

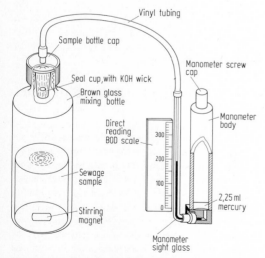

**Figure 3.3.** Diagram of a manometric apparatus for determination of biochemical oxygen demand [304] (Hach Chemical Company)

However, there would be little or no gas volume change without some means of taking up carbon dioxide as it is formed. This is accomplished by a wick, moistened with aqueous potassium hydroxide and placed in a stainless steel cup above the surface of the liquid (Equation 3.28).

$$CO_2 + 2\,KOH \rightarrow K_2CO_3 + H_2O \tag{3.28}$$

The net oxygen uptake is read directly from a mercury manometer, which makes it easy to take intermediate readings during the test to obtain a profile of the oxygen consumption during the five days, or longer if desired. The manometer reading obtained after the elapse of five days' stirring is the BOD for the sample tested. Samples of a higher BOD than the range of the manometer would require dilution before loading the apparatus, to get a valid result within the capacity of the instrument [83].

In contrast to the BOD test, which is designed to measure the oxygen demand of only the biochemically oxidizable carbon compounds present in the sample, the chemical oxygen demand (COD) test gives a close measure of the total oxygen demand of the sample. It is also a practical method to obtain a valid oxygen demand result when the sample contains toxic substances which cannot be easily neutralized. A standard BOD test conducted under these conditions would give a low or zero result from toxicant action on the micro-organisms.

Using strong chemical oxidants such as potassium permanganate [20, 84], or potassium dichromate [32] under strongly acidic conditions, both easily oxidized and more resistant organic materials are converted to carbon dioxide and water (Equation 3.29) [85].

$$C_nH_aO_b + c\,Cr_2O_7^{2-} + 8c\,H^+ = $$
$$n\,CO_2 + [(a + 8c)/2]H_2O + 2c\,Cr^{3+}$$
$$\text{where } c = \tfrac{2}{3}n + \tfrac{a}{6} - \tfrac{b}{3} \tag{3.29}$$

During the process any inorganic salts present are also converted to the oxidized forms. The result is specified in terms of the mg/L of oxygen which would be consumed equivalent to the amount of chemical oxidant required (e.g. Equations 3.30, 3.31)

$$3\,CH_2O + 2\,Cr_2O_7^{2-} + 16\,H^+ \rightarrow$$
$$3\,CO_2 + 11\,H_2O + 4\,Cr^{3+} \tag{3.30}$$

$$3\,CH_2O + 3\,O_2 \rightarrow 3\,CO_2 + 3\,H_2O \tag{3.31}$$

Under the strong oxidizing conditions of the test most organic compounds give 95 to 100 % of the theoretical oxygen consumption although some of the more stable aromatics, such as benzene and to-luene, give lower oxidant consumptions than this [32]. A micro semi-automated method of COD determination of surface and wastewaters has been developed which can help process large numbers of samples through this test [86].

Oxygen demand by total organic carbon determination is an instrumental method originally developed by the Dow Chemical Company [87, 88] and now sold commercially by Beckman under license from Dow [89]. Either normal surface water samples or waste streams high in organics may be analyzed [90, 91]. The bulk sample is first acidified to pH 2 with hydrochloric acid, and then sparged with nitrogen or helium to remove any carbon dioxide and other inorganic carbon present (Equations 3.32, 3.33).

$$Na_2CO_3 + HCl \rightarrow 2\,NaCl + H_2CO_3 \qquad (3.32)$$

$$H_2CO_3 \rightarrow H_2O + CO_2 \uparrow \qquad (3.33)$$

A 10 to 30 $\mu$L sample is then injected into the instrument. Initially the water and any volatile organic carbon compounds present are vaporized at 200 °C. The volatilized carbon compounds are first all completely oxidized to carbon dioxide, and then the carbon dioxide is all catalytically converted to methane, in a hydrogen atmosphere (Equation 3.34)

$$CO_2 + 4\,H_2 \overset{350^\circ}{\underset{Ni}{\rightarrow}} CH_4 + 2\,H_2O \qquad (3.34)$$

The final step to obtain a volatile organic carbon reading, is a quantitative analysis of the methane produced in a gas chromatograph using a sensitive flame ionization detector, which is unaffected by water. This instrumental configuration gives a very good linear response to the quantity of methane produced, which is the rationale behind the preliminary conversion of organic compounds to methane.

The TOC instrument also has provision for the analysis of refractory organic carbon (more oxidation/volatilizaton resistant carbon) by a vaporization and oxidation mode at 850 °C, which gives a separate reading. The total organic carbon reading for the sample is the sum of the two component readings (Equation 3.35)

$$TOC = VOC + ROC, \text{ where} \qquad (3.35)$$
$$\quad\text{VOC is volatile organic carbon, and}$$
$$\quad\text{ROC is refractory organic carbon}$$

The result is specified in mg/L carbon, but may be readily translated into oxygen demand for the carbon content *only,* by calculating the oxygen consumption required to convert this to carbon dioxide.

Each of these measures of oxygen demand will give different results because of the differing analytical methods used [92, 93]. But each will generally be relatable to the others by a factor, depending on the particular kind of waste analyzed [94]. The COD for ordinary sewages, for example, is frequently about twice the BOD. The COD reading minus the BOD reading is sometimes referred to as the refractory organics, i.e. referring to the organics which are more resistant to oxidation.

### 3.2.10 Biological Indicators

By the number of different species represented (diversity) and the relative sensitivity to pollutants of the species found, much can be learned about the condition of fresh or marine waters. Sessile (more or less stationary) organisms are preferred for biological assessments, since they are unable to avoid discharged wastes. These can either be sampled by scooping up portions of the bottom material for examination and counting, or artificial substrates may be placed in the stream bed for colonization, and later examined [95].

Using solely biological methods to assess water quality, if properly quantified against appropriate control stations or against surveys conducted prior to discharge, is said to give a superior indication than either chemical or physical data alone [96]. Several numerical methods for indexing pollution levels have been published [97-99].

## 3.3 Water Quality Related to End Uses

The surface water quality requirements differ according to the application for which the water is required. While the highest quality surface waters can generally be used for all freshwater needs, many applications do not have as stringent quality demands which can be met with a supply with less than the optimum properties [100]. A detailed consideration of water quality requirements, use by use and parameter by parameter, is a difficult and extensive exercise beyond the scope of this treatment of the subject, and about which whole books have been written [101, 102]. But the anecdotal

Table 3.8 has been put together to give some appreciation of the considerations involved. This illustrates, in a general way, how the required levels of the various water quality parameters may change with the anticipated end use so that several different qualities of surface or ground water may all have appropriate end uses. It should also be remembered that the exact requirements for any particular end use will vary with local conditions and with the objectives of the governing jurisdiction of the area of concern. Setting of standards is a complicated exercise, requiring consideration of not only the obvious factors such as toxicity and aesthetics (e.g. for recreational use) but also consideration of the implications of long term consumption and use, the effectiveness of raw water processing on the constituents present in the supply, the seasonal variations in supply characteristics, alternative supplies which may be available, local (particularly upstream) waste water disposal practices and the like, which involve the skills of many disciplines. Guidelines are available to assist in interpreting the rationales for the regulatory standards to be set [100, 103].

**Table 3.8** The water quality requirements for various primary classes of end use [a]

*Agricultural Uses*
Livestock — low bacteria, $< 40/100$ mL, and low concentrations of toxic substances (e.g. $F^-$)
Irrigation — low dissolved solids, $< 500$ mg/L, (to avoid increased soil salinity)
— total bacteria;
  allowable 100 000/100 mL
  desirable $< 10\,000/100$ mL
— low heavy metals

*Fish, Aquatic Life, Wildlife Requirements*
— concentrations of toxic substances low
— pH near neutral, 6.5 – 8.5
— low BOD, 1 – 2 mg/L or less
— high dissolved oxygen:
  cold, 6 – 7 mg/L at 15 °C or less
  warm, 4 – 5 mg/L at ca. 20 °C
— low temperature, turbidity, etc.

*Industrial Uses*
Cooling — low hardness,
  $< 50$ ppm of $(Mg^{2+} + Ca^{2+})$,
    (usually as $SO_4^{2-}$ and $CO_3^{2-}$)
— low corrosivity
Food processing, brewing and soft drinks
— as public drinking water,
  but $F^- < 1$ ppm
Thermal power
— total dissolved solids $< 0.1$ ppm
  (and lower for boiler pressures above 135 bars)

*Public Recreational Requirements*
— free of colour, odour, taste, and turbidity
— total bacteria $< 1000/100$ mL, coliforms $< 100/100$ mL
— low nutrients, to avoid nuisance algal growths

*Public Drinking Water (treated)*
— no bacteria
— low nitrates, nitrites $< 10$ ppm
— very low pesticides, none $> 0.05$ ppm
— fluoride allowable to 2.4 ppm
— toxic substances (metals etc.) below criteria levels
— total dissolved solids $< 500$ ppm

[a] Drastically condensed from [100, 101, 103 and 294], from which further details may be obtained.

# 3.4 Treatment of Municipal Water Supplies

The primary physiological water requirement of an adult human not subjected to heat stress is only 2.5 to 3 liters per day [104]. Yet in households equipped with facilities for running water the actual daily per capita consumption for such uses as human waste transport, bathing, washing/cleaning, and garden irrigation has been reported to be 95-380 L/day (25-100 U.S. gal/day) [105]. When one wraps into these domestic requirements the very large water demand of most industries which often purchase at least part of their requirements from a municipality, it can easily be seen why the facilities for the production of a safe and convenient water supply rank among the largest scale materials processing operations.

## 3.4.1 Simple Municipal Water Treatment

Depending on the raw water quality and the seasonal stability of its characteristics many municipalities are able to use a simple two-stage supply treatment involving preliminary filtration, followed by a disinfection step [106]. Fast filtration methods use a pressure differential to force the raw water through a simple bed of finely granulated clean sand, crushed anthracite coal or sometimes a mixture of media, to remove suspended solids present in the supply (e.g. see Figure 3.4). As the filtration proceeds, accumulated solids in the filter bed cause a gradual slowing of the water flow rate through the filter. To periodically clean this accumulation the filter is back-flushed using filtered water at a sufficiently high flow rate to lift the filter medium and free the accumulated algae, diatoms, silt etc. from the medium. The flushate is discarded to the sewer. This need for periodic filter cleaning requires a water treatment plant to have either two filters, one of which can continue operation while the other is in the back-flush mode, or a large enough holding basin for filtered water to provide for both back-flushing and normal filtered water requirements while a single filter is being cleaned. There is also a "slow filtration" variant of filtration used by some water treatment plants, which combines both a physical separation of suspended solids plus some biological consumption or adsorption of undesirable dissolved substances in the water supply in a single, two, or three stage unit [107, 108]. Bacteria, algae, and diatoms accumulated on the coarse sand layer of the filter bed metabolize nutrients etc. from the supply, thereby removing them. This slow filtration system has been reported to be capable of removing up to 50 % of the chlorinated pesticide content of the influent water [107]. Slow filtration may require a prior aeration step to ensure that the biochemical processes employed remain aerobic, since anaerobic operation can contribute bad odours or tastes to the supply in the form of sulfides and amines.

After suspended matter has been removed a disinfection step, usually with chlorine, is necessary to ensure that the supply is free of any viable pathogenic organisms. The active bactericide, undissociated hypochlorous acid, is formed immediately on contact of the chlorine gas with water (Equations 3.36, 3.37).

$$Cl_2 + H_2O \rightarrow H^+ + Cl^- + HOCl \qquad (3.36)$$

$$HOCl \rightarrow H^+ + ClO^- \qquad (3.37)$$

active     relatively inactive

Sufficient gaseous chlorine is added to the water to leave a "residual chlorine" of 0.1 to 0.2 mg/L, after the normal consumption of a part of the added chlorine in reactions with any dissolved or residual suspended matter in the supply has taken place [107]. A preferred contact time of 1 to 2 hours is recommended, but in any event 20 to 30 minutes' contact time should be ensured before use. This residual chlorine content is necessary to maintain safe transmission of the treated water supply through local piping system.

Since the dissociation constant for hypochlorous acid is very small, in neutral solutions undissociated hypochlorous acid is the dominant species of this equilibrium [109] (Equation 3.38).

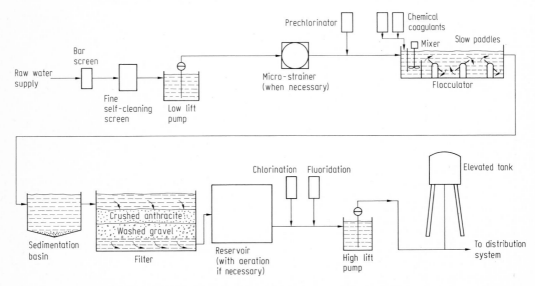

**Figure 3.4.** Flow diagram for an intermediate complexity municipal water treatment plant

$$k = \frac{[H^+][OCl^-]}{[HOCl]} = 2.95 \times 10^{-8} \text{ at } 18\,°C. \quad (3.38)$$

Under acid conditions the higher concentration of hydrogen ion present tends to depress the hypochlorous acid dissociation reaction (Equation 3.34) which consequently tends to raise the concentration of hypochlorous acid above that present in near neutral solutions (100 % at pH 5 or lower). Thus, the effectiveness of chlorine disinfection is maintained at low pH.

At a pH of 7.5, near neutral conditions, the hypochlorous acid will be roughly 50 % dissociated and disinfection will still be quite effective.

However, in disinfections of somewhat alkaline water supplies, the decreased hydrogen ion concentration present depresses the concentration of undissociated hypochlorous acid, i.e. at pH's of 10 or higher, hypochlorous acid is present almost entirely (ca. 99.7 %) as hypochlorite anion. This has the effect of decreasing the effectiveness of chlorine disinfection under these conditions [107], which can be remedied by pH adjustment of the supply before disinfection, by a higher chlorine dose rate, or by a longer contact time before water use.

When the time between disinfection and delivery at the householder's tap is long, such as when treatment is carried out at a reservoir some distance away from the consuming urban center, ammonia may be added at the same time as the chlorine. The more stable chloramines which form confer a longer term residual chlorine to the supply than is possible with chlorine alone [107, 108] (Equations 3.39-3.41).

$$HOCl + NH_3 \rightarrow NH_2Cl + H_2O \quad (3.39)$$

$$HOCl + NH_2Cl \rightarrow NHCl_2 + H_2O \quad (3.40)$$

$$HOCl + NHCl_2 \rightarrow NCl_3 + H_2O \quad (3.41)$$

But chloramines also require a much longer disinfection period than chlorine, to obtain the same effectiveness, and so are only used in situations like this. Hypochlorites (e.g. household bleach) are also effective for disinfection of water supplies, and are an economic and convenient choice for smaller volume requirements such as for a single household, a hamlet, or for small campsites [110, 111]. With hypochlorite salts similar equilibria operate to rapidly establish a concentration of hypochlorous acid which corresponds to the amount of hypochlorite salt added (e.g. Equation 3.42, Table 3.9).

$$Ca(OCl)_2 + 2\,H_2O \rightarrow Ca(OH)_2 + 2\,HOCl \quad (3.42)$$

| | $Ca(OCl)_2$ | $2\,H_2O$ | $Ca(OH)_2$ | $2\,HOCl$ |
|---|---|---|---|---|
| mol. wgts | 143 | 18 | 74 | 52.5 |
| moles | 1 | 2 | 1 | 2 |
| mass ratios | 143 | 36 | 74 | 105 |

Table 3.9 "Available chlorine" content of various chlorine-containing water disinfectants [a]

| Disinfectant | Molecular weight | Moles of equivalent chlorine [b] | Percent by weight | |
|---|---|---|---|---|
| | | | Actual chlorine | Available chlorine [c] |
| chlorine | 70.9 | 1 | 100.0 | 100.0 |
| hypochlorous acid | 52.5 | 1 | 67.7 | 135.0 |
| sodium hypochlorite | 74.5 | 1 | 47.7 | 95.2 |
| calcium hypochlorite | 143.0 | 2 | 49.6 | 99.2 |
| chloramine ($NCl_3$) | 120.4 | 3 | 88.5 | 265.5[d] |

[a] Data recalculated from that of [112].
[b] The number of moles of hypochlorite (oxidizing chlorine) which could form on dissolving one mole of the disinfectant in water.
[c] This equates to the calculated weight of elemental chlorine required to produce the same amount of hypochlorous acid in water as the given disinfectant.
e.g. for calcium hypochlorite

$$\frac{(70.9 \text{ g/mol} \times 2 \text{ mol/mol} \times 100)}{143.0 \text{ g/mol}} = 99.2\%$$

[d] Action is very slow.

Thus, "% available chlorine" from calcium hypochlorite, which by definition includes $Cl_2$, HClO, and OClˉ (but not Clˉ), is 99.2 %, or nearly the same in effectiveness, on a weight for weight basis, as treating with elemental chlorine [112]. As long as the residual chlorine level used with hypochlorite salt disinfections is the same as with chlorine gas, there is no difference in the relative effectiveness of the two methods.

Chlorine is the most common disinfectant for public water supplies, and there is no doubt that its effectiveness, simplicity of application, and low cost have been important factors in the widespread adoption of chlorination as a significant public health measure. However, questions have been raised about the presence of chloroform and other chlorinated organics, formed from chlorinations of substances in public drinking water supplies [113-120]. Other methods have also been used for water disinfection including the hypochlorites already mentioned, which would behave much the same as chlorine in aqueous chlorinations. However chlorine dioxide ($ClO_2$) has a much lower chlorination tendency than chlorine or hypochlorites, and is also an effective disinfectant at high pH's, when chlorine and hypochlorites are relatively ineffective [108]. Other water disinfectants such as ozone, iodine, ultraviolet irradiation, and gamma irradiation [121], avoid chlorination entirely. Chlorine dioxide, favoured in the Federal Republic of Germany, has an essentially similar mode of action to chlorine [122, 123]. Ozone, which is favoured in France, not only functions as a disinfectant but also serves to remove odour, taste and colour from its more general oxidizing properties, which is an important consideration with a poor quality raw water supply [124, 125]. However, it is not possible to maintain a disinfection residual with ozone, making it necessary in some instances to back up the ozone treatment with a low chlorine dosage to provide this transmission safeguard [107, 122]. Iodine, as crystals, or more conveniently as a tincture or iodine-releasing tablets, is also an effective field disinfection method but is generally too expensive for municipal water treatment [108, 111]. The remaining methods mentioned, while they are effective, have cost, maintenance, or lack of familiarity problems attached to their use.

## 3.4.2 More Elaborate Municipal Water Treatment Methods

When there is little control over raw water quality, such as when the supply is near the mouth of a river flowing through a populous watershed, or when the natural supply from a groundwater source is poor, more complicated treatment methods are required (Figure 3.4, or more complex then this). Following the preliminary treatment steps outlined below, filtration plus chlorination, or chlorination alone can be used to produce the finished supply for the area served.

It is generally desirable, if not always economic, to remove excessive hardness (> 150 ppm) from water supplies before distribution. Synthetic detergents have circumvented much of the poor performance of carboxylate soaps in hard waters. This does not, however, help the aesthetics of personal bathing for which ordinary soaps are almost universally preferred. Hardness removal is also not particularly necessary from a health standpoint since, if anything, a degree of hardness in the water appears to reduce the risk of heart attack [126-129]. However, excessive hardness is still a serious problem in the scale build-up on water heating devices of all kinds, from kettles and domestic water heaters all the way up to commercial power boilers.

It is the carbon dioxide content of natural waters which helps to contribute much of the temporary hardness, also referred to as bicarbonate hardness, into water supplies (Equations 3.43, 3.44).

$$CO_2 + H_2O \rightarrow H_2CO_3 \qquad (3.43)$$

$$H_2CO_3 + 2\,CaCO_3 \rightarrow 2\,Ca(HCO_3)_2 \qquad (3.44)$$
$$\text{insoluble} \quad \text{soluble}$$

Temporary hardness may be removed by heating the water, which reverses the reaction which originally put the calcium and magnesium into solution (Equations 3.45, 3.46).

$$Ca(HCO_3)_2 \xrightarrow{\text{heat}} CaCO_3 \downarrow + H_2O + CO_2 \uparrow \qquad (3.45)$$

$$Mg(HCO_3)_2 \rightarrow MgCO_3 \downarrow + H_2O + CO_2 \uparrow \qquad (3.46)$$

These also represent the processes which cause scale build-up from temporary hardness, and therefore are not of very great practical use. Oddly enough however, addition of calcium in the form of slaked lime ($Ca(OH)_2$) can efficiently remove much of the temporary hardness of natural water supplies (Equations 3.47-3.49).

$$Ca(HCO_3)_2 + Ca(OH)_2 \rightarrow 2\,CaCO_3 \downarrow + 2\,H_2O \tag{3.47}$$

$$Mg(HCO_3)_2 + Ca(OH)_2 \rightarrow \\ MgCO_3 + CaCO_3 \downarrow + 2\,H_2O \tag{3.48}$$

$$MgCO_3 + Ca(OH)_2 \rightarrow \\ Mg(OH)_2 \downarrow + CaCO_3 \downarrow \tag{3.49}$$

The second equivalent of calcium hydroxide is necessary with magnesium temporary hardness because magnesium carbonate is still somewhat soluble, unlike calcium carbonate.

Neither calcium hydroxide nor heat is effective for removal of permanent hardness, that is the hardness caused by the presence of the sulfates (or other anions), rather than the bicarbonates of calcium and magnesium [130]. However, treatment with sodium carbonate (soda ash) is effective for this purpose (Equation 3.50).

$$2\,Na_2CO_3 + CaSO_4 \rightarrow CaCO_3 \downarrow + Na_2SO_4 \tag{3.50}$$

For effective removal of both temporary and permanent hardness, sodium hydroxide addition is required (Equation 3.51)

$$Ca(HCO_3)_2 + 2\,NaOH \rightarrow \\ CaCO_3 \downarrow + Na_2CO_3 + 2\,H_2O \tag{3.51}$$

The sodium carbonate formed from the removal of temporary hardness is available to precipitate any calcium or magnesium salts present as permanent hardness. Thus, the usual water treatment practice for removal of hardness involves adding a mixture of quicklime (CaO) and sodium carbonate in accordance with the composition of the raw water, as determined by analysis. Frequently a coagulant such as alum $(Al_2(SO_4)_3 \cdot 14\,H_2O)$ or ferrous sulfate will be added to help avoid the carryover of precipitated but still finely divided hardness. Effectively the same steps may be conducted at the household level to decrease the undesirable effects of a hard water supply by the addition of washing soda $(Na_2CO_3 \cdot 10\,H_2O)$ or household ammonia to the laundry supply. Or a water softener containing beads of a sulfonated polystyrene resin in the sodium form may be installed (Figure 3.5). As hard water passes through, sodium ions are exchanged on a charge equivalent basis with any calcium, magnesium, or other polyvalent cations in the supply. In the process the household water is softened, and the resin content of the water softener gradually accumulates polyvalent cations until its capacity is exhausted. Regeneration of the resin requires passage of a saturated sodium chloride brine solution through the softener, to displace the polyvalent ions by sodium again. During the regeneration step the outlet of the softener is connected to

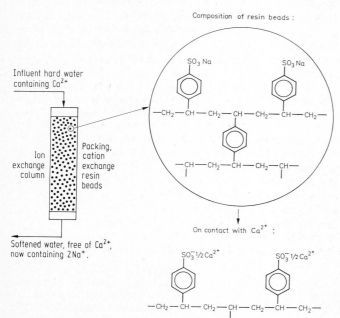

Figure 3.5. Details of operation of a household water softener

the sewer. After a short period of flushing with fresh water to remove the residual brine the hardness removal operating cycle begins again.

Special techniques are required for the removal of other metal ions from raw water supplies, although lime treatment followed by flocculation serves to remove some heavy metal content by adsorption [107]. Excessive fluoride (i.e. > 2.4 ppm) too, is decreased by lime treatment, fluoride precipitation being greatly assisted by the presence of 50 mg/L or more magnesium ion in the water at the time of treatment (Equation 3.52).

$$2\,F^- + Ca(OH)_2 \rightarrow CaF_2 \downarrow + 2\,OH^- \qquad (3.52)$$
$$\text{adsorbed on } Mg(OH)_2$$

Bone ash, synthetic apatite, both consisting of essentially calcium phosphate $(Ca_3(PO_4)_2)$, or a synthetic apatite, calcium hydroxide mixture, are also effective methods for fluoride removal because of the affinity of fluoride for these phosphate salts [107].

A taste or smell in the water supply may be removed by single or double aeration, if the causative agent is volatile [120]. Otherwise adsorption on activated carbon, or chemical oxidation with ozone or chlorine, or both methods used in series may be required to control the problem. Fortunately, many organic toxic substances which may be present, such as pesticides, are also efficiently adsorbed by activated carbon and may be removed in this way. Lime treatment, if not already practised for other reasons, is also efficient at removing dissolved coloured substances, such as humic acids, when these are present in the supply. However, sometimes liming may have to be accompanied by activated carbon adsorption or coagulation plus filtration for effective colour removal [108].

### 3.4.3 Municipal Water by Desalination

Coastal communities in arid areas may need to undertake more extensive and usually somewhat more costly methods to produce a potable supply from seawater or brackish supplies. The dissolved solids content of sea water, roughly 3.5 %, 35 ‰ (i.e. 35 "parts per thousand") or 35,000 mg/L, needs to be reduced to about 1 % of this figure to meet the dissolved solids requirement of a potable water supply. Brackish waters, with a lower salt concentration than seawater, are a preferred raw water source when available since the energy costs for salt removal are less.

Distillation can accomplish salt removal with reasonable thermal efficiencies if it is carried out under reduced pressure using several stages in series [131] (Figure 3.6). Up to six stages have been found to be economic for maximum energy utilization at current energy costs. More than 95 % of the world's existing desalination plants in 1969, representing a total capacity of nearly a million $m^3$/day (240 million U.S. gal/day), utilized some form of distillation for saline water conversion, the greatest capacity being installed in Kuwait and Oman, the coastal arid states of the U.S.A., the U.S.S.R., and the Netherlands. [105].

Membrane processes of the reverse osmosis (hyperfiltration) or electrodialysis types are also used to a smaller extent by some large scale facilities [105] (Figure 3.7). Reverse osmosis units operate by using high brackish or sea water charging pressures on one side of a semipermeable membrane, sufficient to exceed the osmotic pressure of the de-ionized water on the product side of the membrane [132]. Thus, water free of ions permeates through the membrane at a rate proportional to the applied pressure (Table 3.10). Desa-

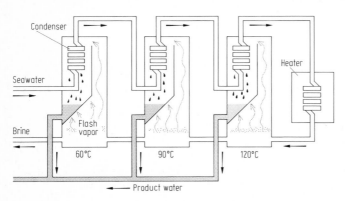

**Figure 3.6.** Schematic of the mode of operation of a multistage flash distillation unit [305]. Reprinted courtesy of Cummings Publishing Co

lination by electrodialysis again uses membranes, but this time of a type permeable to ions. The driving force for ion removal in this instance is derived from the opposing electrochemical potentials on either side of the membrane. Electrochemical removal of ions is continued in staged electrodialysis cells until the salts concentration is brought down to the required level.

Differential freezing methods may prove particularly useful in water-short polar regions, where the energy costs for freezing would be low, but may also be utilized in more common desalination sites in hot climates [132]. Reducing the brine temperature to below freezing generates substantially salt-free ice crystals which are separated from the brine mother liquor [108] (Figure 3.8). For efficient energy utilization the heat removal capability of the thawing ice may be used to prechill incoming brine, prior to freezing.

### 3.4.4 Water Quality Requirements of Industry

Most industrial cooling water requirements may be met by the usual characteristics of the surface waters available, as long as the hardness and the concentrations of chloride or other corrosible substances are low. Bacteriological, colour, dissolved oxygen etc. requirements for this use are minimal. The food processing industry is one exception to this generalization, however, in that for the cooling of processed foods, either in bulk or in the final packages of bottles or cans, potable water quality is required to avoid any risk of contamination in event of leakage. This quality is required even for the cooling of packaged products after pasteurization since a pinhole leak plus the reduced interior pressure produced on cooling could cause coolant, and hence organism leakage into the package.

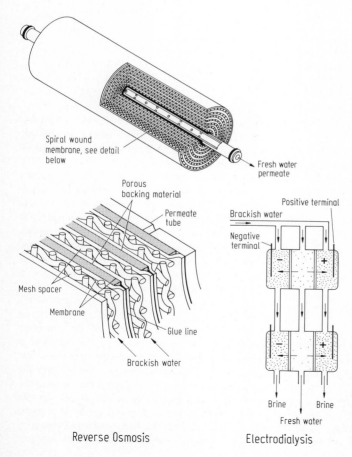

Spiral wound membrane, see detail below

Fresh water permeate

Porous backing material

Permeate tube

Positive terminal

Brackish water

Negative terminal

Mesh spacer

Membrane

Glue line

Brackish water

Brine    Brine

Fresh water

Reverse Osmosis

Electrodialysis

**Figure 3.7.** Principles of operation of two types of membrane water desalination units, reverse osmosis and electrodialysis [305]. Reprinted courtesy of Cummings Publishing Co

**Table 3.10** A comparison of the performance characteristics of three commercial reverse osmosis membrane configurations [a]

| Characteristic | Hollow fiber cellulose acetate | Hollow fiber polyamide | Spiral wound sheets of cellulose acetate |
| --- | --- | --- | --- |
| Mandatory pre-treatment | pH control | softener or pH | pH control |
| pH range (ideal) | 4 – 7.5 (5.5) | 4 – 11 (4 – 11) | 3 – 6 (5.5) |
| Prefiltration, μm | 5 | 5 | 25 |
| Maximum $Cl_2$, ppm | 1.0 | 0.1 | 1.0 |
| Maximum temperature, °C | 35 | 35 | 35 |
| Normal operating pressure | 2700 kPa. (400 psig) | 2700 kPa (400 psig) | 2700 – 4100 kPa (400 – 600 psig) |
| Max. product back pressure | 550 kPa (80 psig) | 275 – 340 kPa (40 – 50 psig) | 275 – 340 kPa (40 – 50 psig) |
| Rated TDS level [b] | 2000 ppm | 2000 ppm | 2000 ppm |
| TDS rejection rates [b] | 90 – 95% | 90 – 95% | 90 – 98% |
| Water recovery | 75% | 50 – 75% | 22 – 75% |
| Max. suspended solids, | 1.0 JTU [c] | 1.0 JTU [c] | 1.0 JTU [c] |
| Ease of cleaning | good | poor | very good |
| Surface water performance | good | poor | excellent |
| Well water performance | excellent | good | excellent |
| Packing density, $m^2$ membrane/$m^3$ module | 29 500 | – | 660 |
| Flux, $L/m^2$/day (pressure) | 7.3 (4100 kPa) | – | 730 (5500 kPa) |
| Flux density, $L/m^3$ module/day | 220 000 | – | 480 000 |

[a] Properties selected, calculated and compiled from those given in [295].
[b] TDS = total dissolved solids. Newer membranes can handle up to 35 000 ppm, i.e. sea water.
[c] Jackson turbidity units on an arbitrary scale [32].

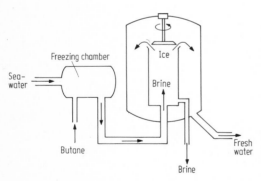

**Figure 3.8.** Schematic of differential freezing as a method for salt removal from sea water [305]. Reprinted courtesy of Cummings Publishing Co

A potable quality water supply or better is also required by the soft drinks and alcoholic beverage industries as well as for medical and pharmaceutical applications and for the water needs of many types of research institutions. The beverage industry has a fluoride requirement of < 1 mg/L, lower than that required for ordinary potable water supplies, probably to accommodate possible higher individual consumption rates. De-ionized or distilled grades of water quality are required by, and are sufficient for the majority of the other applications areas mentioned.

But among the most stringent of all water quality standards are those set for the feedwater requirements of modern high pressure steam boilers [133]. Even by 1970 high pressure boiler operation required direct conversion of feedwater to superheated (dry) steam at 160 bars (2,400 psi) in the boiler tubes, so that any dissolved solids present would be deposited and gradually accumulated in the tubes, contributing to eventual failure. Consequently, this type of boiler requires water with a total impurity level of 0.03 mg/L (0.03 ppm, or 0.000003 %), or less. Water of this standard amounts to the highest purity of any commercial chemical, more pure than the best grades of analytical reagents. But an objective has just recently been set to raise even this high boiler water standard up to a requirement of 0.01 mg/L (10 ppb) total impurities [133]. This equates to or exceeds the present requirements of the electronics industry which currently has among the most stringent

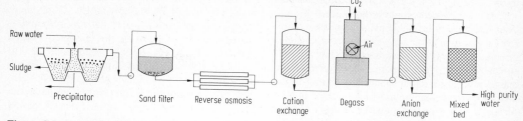

**Figure 3.9.** Schematic flow sheet of the water treatment system used by the electronics industry to obtain the required level of finished water quality (see text) [133]

water quality requirements of all: neutral salts, 0.02 mg/L; total dissolved solids, 0.02 mg/L; free base or mineral or organic acid, nil; and a conductivity of $0.05\,\mu$S/cm or better [133, 134]. The detailed processing steps necessary to meet these standards on a continuous basis requires more or less a composite of all the best existing water purification technology wrapped up in a single processing train (Figure 3.9)

## 3.5 Treatment of Municipal Waste Waters

To be able to deal most efficiently with the various available domestic or sanitary sewage disposal options one has to establish the ranges of the principal chemical and physical characteristics to be expected (Table 3.11). While the tabulated list of impurities appears long it should be remembered that

**Table 3.11** Approximate values for the principal physical and chemical characteristics of municipal waste waters [a]

|  | Domestic sewage | | | Urban storm waters |
|---|---|---|---|---|
| Volume: | 70 – 200% of supply (from infiltration) | | | 0 – 5000 + % of supply |
| Temperature: | 1 – 2 °C above supply | | | ambient |
| B.O.D.: | ca. 100 – 300 mg/L | | | 1 – > 700 mg/L |
| C.O.D.: | ca. 2 × B.O.D. | | | 5 – 3100 mg/L |
| Micro-organisms, | ca. 100 – 3000 × $10^6$ | | | 200, to 146 × $10^6$ |
| (as bacteria, viruses etc. no. per 100 mL) | (mostly non-pathogenic) | | | |
| Total Solids: mg/L | 655 | | | 450 – 14 600 |
| Solids Distribution, | Inorganic | Organic | B.O.D.[b] | |
| Suspended * | 65 | 170 | 110 | 2 – 11 300 |
| Dissolved ** | 210 | 210 | 30 | 450 – 3 300 [c] |
| Totals | 275 | 380 | 140 | |

\* Suspended solids: $\frac{1}{3}$ non-settleable ($< 50\,\mu$m)
  (by volume)      $\frac{2}{3}$ settleable ($> 100\,\mu$m diam.)

\*\* Dissolved solids:  Approx. twice municipal
                  supply (dry weather)
                  N 50 mg/L (mostly $NH_3$/urea)
                  P 30 mg/L (mostly as $PO_4^{3-}$)
                  plus NaCl etc.

[a] Compiled from the data of [27, 28, 135, 153 and 296].

[b] Distribution of biochemical oxygen demand between the suspended and dissolved matter, of sewage with total BOD 140 mg/L.

[c] At high end of range found particularly in areas using highway de-icing chemicals ($CaCl_2$, NaCl etc.).

the total impurities present still constitute less than 1 % of the sewage volume to be treated. From a chemical standpoint municipal sewage thus represents a dilute solution and suspension of a heterogeneous mixture of constituents — not an easy stream to treat. Nevertheless, since untreated domestic sewage represents about one quarter of the total BOD waste loading to surface waters, at least for the U.S.A., treatment and/or disposal selected can have a significant influence on surface water quality [135].

### 3.5.1 Discharge Requirements, and Remedies to Post-Discharge Degradation

If permitted at all, discharge of untreated sewage of the properties outlined could only be condoned if two composite requirements are met. These are that there be sufficient dilution by the receiving stream, and that there is good mixing with this stream on discharge. The dilution requirement means that the gross BOD leading of the river should not be raised to more than 2 mg/L, and the dissolved oxygen should not be depressed to below about 4 mg/L, (5 mg/L Alberta [135], 3 mg/L, New York [137]), in order to meet the criteria of a clean river. To discharge raw sewage of an average BOD of about 200 mg/L demands a river flow of more than 100 times the sewage volume at low river flow periods, plus no other discharges in close proximity, i.e. a predischarge river BOD of 1 mg/L or less, to meet the dilution requirement alone. The second requirement, good mixing, is also difficult to achieve and maintain [138-140]. So there are very few situations where the requirements for raw sewage discharge to rivers could be compatible with these criteria, even if it were aesthetically acceptable.

Similar considerations apply to the discharge of untreated domestic effluent to the sea [141, 142]. Raw sewage discharge into shallow, poorly mixed salt waters can give the same problems with odorous anaerobic decomposition and microbial contamination on the shoreline and in the surrounding air of the discharge area, as would be experienced with a river or a lake [143-146].

It is possible via some special procedures to ameliorate temporarily depressed oxygen levels or high BOD's in a river or lake, such as may occur for short periods during times of low flow or process upsets, even with normally adequate waste treatment. Electrically-driven aerators can be placed on

floats across the stream, such as has been used for the Lippe River, West Germany [6]. Using air of oxygen partial pressure 0.21 bar this system was found capable of boosting the dissolved oxygen content by about 1 kg/900 watts/hour (2 1b/ horsepower/hour) in the Upper Passiac River, New Jersey [147]. A similar arrangement has also been used temporarily in the North Saskatchewan River just below Prince Albert, Saskatchewan [148].

Alternatively, to take advantage of the increased exchange rates provided by increased air pressure, the whole of the low flow river water may be passed through a U-tube bored 14 m or more beneath the river bed. Passage of air bubbles into the downflow side boosts the oxygen partial pressure in the bubbles about 2 1/2 times to about 0.50 bar, greatly accelerating oxygen dissolution rates (Figure 3.10). For the Red Deer River in Alberta this procedure more than doubles the dissolved oxygen content when used at times of low flow [136], which is particularly useful in winter when ice cover prevents reaeration. When the river flow is high, and aeration therefore unnecessary, the bulk of the water flows over the weir by-passing the U-tube.

Aeration using pure oxygen, i.e. providing an oxygen partial pressure of 1 bar (plus) has been proposed for the rejuvenation of highly eutrophic lakes by Union Carbide, a producer of industrial gases [149]. Certainly this system would provide excellent oxygen transfer conditions, but it would

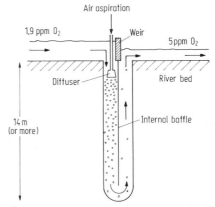

**Figure 3.10.** Automated U-tube river aeration, at times of low flow. A 1.3 meter head drives the whole flow through the U-tube, which in practice can consist of a single or several parallel boreholes with an internal baffle [136]

be likely to be unduly expensive except for experimental or other special cases.

## 3.5.2 Stream Assimilatory Capacities

Adequate sewage treatment is necessary to minimize the frequency of these less efficient, post-discharge solutions to improve river water quality. The degree of treatment required is significantly dependent on the capacity of the receiving body of water to accommodate oxidizable wastes. In general this will be higher for a river which has continuous water turnover than for a lake. For the same gross water flow a shallow, fast flowing river will have a higher assimilatory capacity than a deep, slow flowing river. This is because the initial content of dissolved oxygen is likely to be higher, and the oxygen exchange rate with the air will also be more rapid for the shallow river. By the same token riffles, or the shallow fast-flowing stretches of a river, have 2 or 3 times the reaeration capacity of deeper, slow moving pools [150]. These kinds of considerations have led to the tabulation of stream assimilatory capacities as related to their volume of flow, depth, and oxygen exchange rates [150] (Table 3.12).

Knowing the approximate assimilatory capacity of a stream, which equates to its recovery or reaeration rate, allows an estimate of the acceptable BOD loadings which could be discharged to the river without severely degrading river quality. Assuming a discharge of $2.4 \times 10^6$ L/day of untreated sewage of 200 mg/L average BOD, a volume which could reasonably arise from a small town of 24,000 inhabitants, this would equate to a daily BOD loading of about 480 kg/day (Equation 3.53).

$$200 \text{ mg/L} \times 2.4 \times 10^6 \text{ L/day} =$$
$$4.8 \times 10^8 \text{ mg/day} \qquad (3.53)$$

Discharge of this volume of oxygen demand into a Class I river, if evenly distributed through the day, would result in a river BOD loading of 14 mg/L (Equation 3.54).

$$\frac{4.8 \text{ m}^3/\text{sec} \times 10^8 \text{ mg/day}}{0.4 \text{ m}^3/\text{sec} \times 86\,400 \text{ sec/day} \times 1000 \text{ L/m}^3}$$
$$= 13.9 \text{ mg/L BOD} \qquad (3.54)$$

i.e. a very heavy loading sufficient to severely degrade river quality, even if good mixing is achieved. However, if the BOD of this sewage was decreased to 20 mg/L by suitable treatment before discharge, i.e. to one tenth of the raw BOD values, then discharge could occur without severe river degradation.

Another approach to this problem is to use a figure for the oxygen demand loading per capita, estimates of which range from 54 g to 115 g per day [106, 135, 151]. For the community of 24,000 this method gives a range for the gross BOD discharge of 1,296 to 2,760 kg/day for untreated sewage, much higher than the estimate based on the previous method. If discharges on this scale were actually practised to a Class I river this would correspond to severe river degradation from gross BOD loadings of 37.5 to 80 mg/L.

From Table 3.12, the background calculations allow a quick estimate for a Class I river that 23 kg of BOD discharge would cause a drop in river BOD of 1 mg/L, one kilometer below the point of discharge. It can also be seen that although the flow volume of a Class III river is of the order of 50,000 times greater than a Class I river, the assimilatory capacity is only 130 times more. This fact is a reflection of the much slower reaeration rate of a deep, slowflowing river with vertical stratification as compared to a shallow, faster moving river. It

**Table 3.12** Waste assimilatory capacity related to river size [a]

| Stream class | Common examples [b] | Assimilatory capacity [c] | Average depth, m | Average flow, m³/sec |
|---|---|---|---|---|
| I | Numerous local | 0.023 | 0.17 | 0.4 |
| II | Allegheny, Kansas, N. & S. Saskatchewan, Rhone, Po | 0.56 | 1.5 | 300 – 1500 |
| III | Mississippi, St. Lawrence, Orinoco, Ganges, Yenisei | 2.90 | 13.7 | 20 000 |

[a] Compiled and calculated from the data of [105, 297 – 299].
[b] Based on flow rates of first example of each class.
[c] Round number estimates in tonnes BOD per km per unit oxygen deficiency for the physical data as given. Originally stated in units of tons BOD/mile/unit oxygen demand: Class I, 0.04; Class II 1.0; Class III, 5.2.

also is a reminder that just because one river has ten times the average flow volume of another does *not* necessarily mean that it can accept ten times the oxidizable waste volume without degradation: hence the need for sewage treatment.

### 3.5.3 Primary and Secondary Sewage Treatment

Methods for municipal sewage treatment may be conveniently grouped into three categories according to the stage of waste water processing and the types of treatment used since these two aspects generally coincide quite closely. The primary treatment step applies basically mechanical processes, amounting to physical separation procedures to sewage clean-up. Secondary sewage treatment uses mainly biological methods to remove and consume wastes by a variety of configurations, and provides a further significant degree of improvement of effluent quality over that obtained by primary treatment alone. Tertiary treatment, sometimes called advanced treatment, involves one or more of biological, chemically-based, or physical separation processes for a final refinement of effluent quality. The level of treatment required in any particular situation is decided on the basis of the population loadings to be served by the system and the assimilatory capacity of the receiving body of water. If this is a river, the decision re level of treatment required has to reflect the times of lowest flows [152], periods of ice cover, and whether it is used for recreational purposes etc. A plot of the interrelationships between the dissolved oxygen demand helps to make the influence of these considerations clearer (Figure 3.11).

Primary treatment involves the use of a combina-

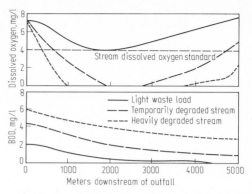

**Figure 3.11.** Plots of plausible degrees of degradation and recovery rates for oxygen-consuming discharges to a stream which heavily degrades, temporarily degrades, or stays within reasonable guidelines to maintain stream quality

tion of one or more of the following types of processing units [153] (Figure 3.12). First the effluent is screened by self-cleaning bar screens of 2 to 5 cm spacing to remove any coarse material, followed by a set of finer screens which are also usually self-cleaning. These preliminaries serve to protect the downstream pumps and sludge handling equipment from fine gravel and any other larger suspended matter which often enters combined sewerage systems used to collect both sanitary and storm run-off waters in the same piping. After screening, the waste stream passes through a comminutor, an electrically-driven fragmenter, which chops any residual coarse material ready for further processing. Following comminution the wastewaters are passed into quiescent channels with underflow discharge which remove floatables,

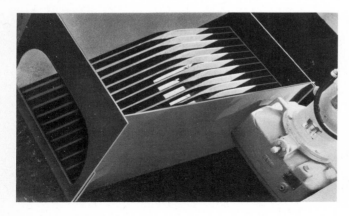

**Plate 3.1.** Bar screen and comminutor drive system for the influent sewage to a small scale municipal sewage treatment plant at Central Saanich, British Columbia

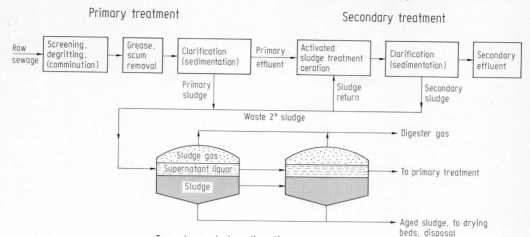

Figure 3.12. Schematic flow diagram for the principal steps involved in municipal sewage treatment to the secondary level [135, 306]

grease, hair, plastics, scum, etc. The underflow discharge from these channels feeds into a large sedimentation tank (clarifier), which has a retention time of a few hours. The supernatant overflow from this tank comprises primary treated sewage effluent. These processing steps achieve about a 60 % reduction in suspended solids content and a 30 % reduction in oxygen demand over the influent sewage, plus a smaller improvement in other parameters, as summarized in Table 3.13. Sludge col-

lected from the bottom of the primary settling tank, which at this stage contains only 2.5-5 % solids, has to be dewatered and further treated for disposal. Details of disposal options are discussed later.

Biological consumption and immobilization, and further coagulation plus settling of residual wastes in the sewage stream are the objectives of secondary treatment. Greatly enhanced concentrations of micro-organisms are employed to bring about ac-

Table 3.13 Approximate cumulative waste removal efficiencies of various sewage treatment procedures, in percent [a]

| | Primary | Primary + secondary | Primary + Secondary + Tertiary | | |
| --- | --- | --- | --- | --- | --- |
| | | | Shallow lagoons | Chemical coagualtion | R. Osmosis, electrodialysis [c] |
| B.O.D. | 35% | 90% | 95% | 95% | 95% |
| C.O.D. | 30 | 80 | 90+ | 85 | 95 |
| Refractory organics | 20 | 60 | 85 | 80 | 90 |
| Suspended solids | 60 | 90 | 95 | 95 | 95+ |
| Total N | 20 | 50 | 85 | 60 | 90 – 95 |
| Total P | 10 | 30 | 85 | 85 | 90 – 95 |
| Dissolved minerals | 1 – 2 | 5 | 10 | 10 | 50 – 90 |
| Incremental cost, $ per 100 m$^3$ [c] | 0.80 – 1.05 | 1.85 – 4.00 | 1.60 – 3.20 | 3.20 – 6.60 | 33 – 100 |
| Bacteria | 35% | 90% | 99% | | |
| Disinfected [b] | 90% | 99% | 99.9+ | | |

[a] Data tabulated from [135, 174, 300 and 301]. Costs approximate, varies greatly with scale.
[b] Approximate values post-disinfection, usually by chlorination [166, 178].
[c] 100 m$^3$ equates to roughly 22 000 Imperial gallons or 26 400 U.S. gallons.

celerated biochemical and biological consumption and destruction of residual wastes, aided by a vigorous supply of air (oxygen). By these measures the BOD and suspended solids are both decreased to about 1/10 of influent values and other qualities of the effluent are also improved somewhat over that provided by only primary treatment (Table 3.13). With the augmented micro-organism populations and artificial aeration the treatment time required is reduced to a matter of hours, instead of the several days that the same processes would take in a natural setting. Thus the effluent improvement may be conducted in much more compact equipment than would otherwise be possible. The detailed methods used to achieve this rapid waste utilization by micro-organisms vary widely with country and the local conditions which have to be met [154].

In North America, and to a lesser extent in Europe, activated sludge plants or variations of this process are common for secondary treatment, as shown schematically in Figure 3.12. High bacterial populations are maintained by the return of settled sludge and artificial aeration is maintained by compressed air or "brush" aerators to stimulate a high level of aerobic microbial activity in this tankage. By this means both dissolved and suspended matter is rapidly converted to settleable bacterial biomass. Biomass, plus absorbed and adsorbed impurities are removed in a secondary sedimentation tank to produce a supernatant sewage effluent treated to the secondary level, of BOD 20-30 mg/L, plus a residual secondary sludge (Table 3.13). The Unox and Oxitron systems are secondary treatment variants promoted by Union Carbide, which use oxygen instead of air for aeration to obtain accelerated rates of waste treatment [155-158]. By the use of oxygen instead of air the capacity of a sewage treatment plant may be increased without an increase in size.

Other variants of this procedure to obtain efficient microbiological action include the trickling filter, common in Europe, and the biodisk, or rotating biological contactor ("RBC"), which are particularly suitable for smaller installations [159, 160]. Both of these systems use efficient air exchange on a microbial support medium, rather than using compressed air, to obtain rapid aerobic consumption of wastes in the effluent [106]. Where land is available and inexpensive, and the volumes of sewage to be treated are not unduly large, the Dutch Oxidation Ditch also provides a technologic-

ally simple method of obtaining roughly equivalent effluent quality [106, 161]. Where land is scarce, on the other hand, there is also the deep shaft treatment method, originally developed by I.C.I. (U.K.) which accomplishes the same type of improvement in a bore hole of 0.3 to 10 m diameter and 150-300 m deep [162-165]. The pressure of the head of water is said to boost aeration rates to a factor of 10 higher than a regular activated sludge plant, and in the process halves the volume of sludge produced. Apparently the working bacteria, with their very high surface to volume ratio and hence high intracellular-extracellular gas exchange rates, are not afflicted with the "bends" under these conditions.

Bacterial counts in the supernatant liquor from a secondary treatment plant are greatly decreased, to about 10 % of the influent raw sewage count, but are still high. Therefore, if secondary treatment produces adequate improvement in effluent quality it is normally disinfected by a fairly heavy dose of chlorine (ca. 10 ppm) before discharge, causing a further reduction to about 1 % of the influent bacterial counts [166, 167]. Each 1 ppm of post-chlorination also tends to decrease the BOD's of the effluent by about 2 ppm from the oxidizing action of the chlorine [166], which is estimated to consume 50 to 80 % of the chlorine added [119]. Because of the higher chlorine dose rates used for sewage disinfection, together with the higher concentrations of a variety of organics also present, sewage disinfection processes probably represent a more significant source of chlorinated organics to surface waters than that contributed from the disinfection of municipal water supplies [168, 169]. Gamma irradiation and bromine chloride are other effective measures which have been proposed for disinfection of treated sewage effluent [170-172]. At least the first of these disinfection methods does not contribute any halogenated organics, or additional toxicity to the effluent which may require neutralization before discharge.

If the final chlorine-disinfected effluent is to be discharged to a relatively sensitive watercourse, dechlorination may be necessary to avoid problems from the toxicity of the residual chlorine. After a 20 to 30 minute holding time for completion of the disinfection to take place, dechlorination is accomplished by converting chlorine to chloride with a reducing agent such as sulfur dioxide followed by pH neutralization with lime (Equations 3.55, 3.56).

$$HOCl + SO_2 + H_2O \rightarrow HCl + H_2SO_4 \qquad (3.55)$$

$$2\,Ca(OH)_2 + 2\,HCl + H_2SO_4 \rightarrow$$
$$CaCl_2 + CaSO_4 + 4\,H_2O \qquad (3.56)$$

It should be realized that this dechlorination step, while it does decrease chlorine-induced toxicity, does nothing to remove any chlorinated organics which may also have formed during the disinfection step.

One of the outstanding problems of this sequence of sewage treatment is the wide variability of influent volumes resulting from the use of combined sewerage systems. Combined systems, in which domestic sewage and storm runoff are collected in the same piping as an installation economy, may at times have outflows of 50 or more times normal domestic flows, far exceeding the throughput capacity of the treatment plant. To deal with this situation overflow points are provided at critical junctions in the collection system to release combined sewage flows directly to surface waters at the time of a storm. Towards the end of a prolonged storm these outflows become relatively dilute, low toxicity wastes for which direct discharge does not constitute a serious problem. But the initial storm flows, where accumulated street wastes such as soil, vehicle residues, animal droppings, pesticides and fertilizers etc. during the preceding dry spell, as well as material sedimented out of domestic sewage in quiescent flow periods of the sewerage system itself, are all flushed out of the system, can have characteristics significantly worse than those of ordinary untreated domestic sewage, in terms of its effect on discharge to surface waters [27, 28] (Table 3.11).

Solutions to these problems of combined sewerage systems include gradual separation of the two systems, as the construction costs become budgetarily feasible, plus provision of separate storm water treatment facilities. Initial storm flows may then be retained for treatment while allowing discharge of later, cleaner flows [28]. This treatment philosophy can also be applied to combined sewerage systems by provision of initial storm water storage using the excess storage capacity within the ducting itself, simply by raising the regulation levels of overflows. Or system storage plus holding basins at the treatment facility can be combined to retain most of the highly contaminated initial storm flows for later gradual treatment, combined with the discharge of later flows from a prolonged storm, if necessary, when holding capacity is exceeded [27].

### 3.5.4 Tertiary, or Advanced Sewage Treatment

If the effluent quality produced by secondary treatment is still not adequate, as may well be the case for discharges to a protected watershed or to a small stream, then one or more of the tertiary treatment options may have to be employed to follow the primary and secondary steps. Tertiary treatment, sometimes referred to as advanced treatment, covers a variety of methods which may be used either singly or in combination for the upgrading of secondary effluent quality [173, 174].

One or more shallow lagoons in series having a total secondary effluent holding time of the order of days can alone accomplish significant further improvement [175]. The shallowness requirement ensures good sunlight penetration and oxygen exchange to the bed of the lagoon, assuring the continued action of aerobic bacterial as well as photosynthetic waste utilization [176]. Element uptake by the rooted and free-floating plants and bacterial action in the lagoons continues to remove nutrient ions plus C, H, O, and S compounds in the secondary effluent, albeit at a slower rate (Table 3.13). This is why a longer holding time is required with this step than for primary or secondary treatment. It also provides sufficient time to further decrease the suspended solids content by a combination of biological coagulation processes and settling. Bacterial counts are also reduced by a factor of about 10 by a combination of soil adsorption and coagulation/settling [177, 178]. Tertiary lagoons, however, while simple in concept, do have a significant land requirement for their use.

Chemical methods of effluent improvement, being generally fast acting, have a smaller treatment volume and land requirement but have chemical costs attached to their use. Coagulation of suspended solids may be accomplished using an inorganic coagulant, such as alum. Chemically, alum is $K_2SO_4$ x $Al_2(SO_4)_3$ x $24\,H_2O$, but since it is the aluminum ion of this double salt which is the active coagulant, "alum" for sewage treatment normally refers to aluminum sulfate itself, $Al_2(SO_4)_3$. Alum treatment removes much of the residual colloidal matter by promoting agglomeration to larger particles which then settle. It also can remove 90-95 % of the dissolved phosphate, dropping this from 35-55 mg/L to 0.5 mg/L or less [179-181] (Equations 3.57, 3.58).

$$Al^{3+} + \text{negatively charged colloids} \rightarrow$$
$$\text{settleable floc} \qquad (3.57)$$

$$Al_2(SO_4)_3 + 2\,Na_3PO_4 \rightarrow$$
$$\text{soluble}$$
$$2\,AlPO_4 \downarrow + 3\,Na_2SO_4 \qquad (3.58)$$
$$\text{insoluble}$$

However, alum dosages of 200-300 mg/L are required to achieve this, which contributes a significant inorganic content to the sludge. This sludge mineral loading may be a nuisance in some disposal systems.

Lime, employed at similar dosages as alum, has about the same phosphate removal efficiency (Equation 3.59).

$$3\,Ca(OH)_2 + 2\,PO_4^{3-} \rightarrow Ca_3(PO_4)_2 \downarrow + 6\,OH^-$$
$$(3.59)$$

Lime treatment is generally lower in cost than alum treatment, but it still has the attendant disadvantage of adding significant inorganic (and hence incombustible) mass to the sludge disposal system.

Some commercial organic polyelectrolytes are also efficient for colloid flocculation. These, being combustible, avoid adding to the sludge disposal problem [182]. They are also active at dosages of only 1-2 ppm which also helps in this respect. Thus, even though they are more expensive than inorganic flocculants on a weight for weight basis, they are competitive on a treated sewage volume basis. If used in conjunction with ferrous chloride or sulfate, which are byproducts of the steel pickling industry, they too are capable of decreasing the phosphate concentration by some 80 % [183, 184] (e.g. Equation 3.60).

$$3\,FeCl_2 + 2\,Na_3PO_4 \rightarrow$$
$$Fe_3(PO_4)_2 \cdot 8\,H_2O \downarrow + 6\,NaCl \qquad (3.60)$$
$$\text{vivianite}$$
$$\text{(finely divided)}$$

Subsequent settling removes the flocculated insoluble vivianite which forms, as the second step of this two stage chemical treatment.

Phosphate may also be removed from treated effluent by passage over aluminum electrodes carrying an AC current [185]. With high effluent flow rates and low treating currents about 1.4 g of aluminum is consumed per gram of phosphate removed.

If nitrogen as well as phosphate removal is required, combinations of chemical and biological action are required. Secondary effluent treated with lime to decrease the pH and then subjected to anaerobic digestion effectively converts up to 90 % of the combined nitrogen compounds present to ammonia (Equation 3.61).

$$NO_3^-,\ NO_2^- \text{ (anaerobic digestion)} \rightarrow$$
$$NH_3 \text{ in solution} \qquad (3.61)$$

The ammonia, still in solution, is air-stripped from the effluent before discharge in a procedure that was adopted for the treatment plant in the sensitive Lake Tahoe, California watershed [28].

Alternatively, removal of combined nitrogen may be accomplished by denitrification of nitrate and nitrite using appropriate bacteria in the presence of a carbon source such as methanol [174] (Equation 3.62).

$$6\,NO_3^- + 5\,CH_3OH + 6\,H^+ \rightarrow$$
$$3\,N_2 + 5\,CO_2 + 13\,H_2O \qquad (3.62)$$

With this option the nitrogen end product altogether gets rid of the combined nitrogen originally present in the effluent from the watershed, unlike the possible return in subsequent precipitation of a fraction of the ammonia produced by anaerobic digestion.

Finally, removal of other dissolved salts from the effluent, which may be necessary for water re-use in some water conservation programs, may best be accomplished using reverse osmosis or electrodialysis [186, 187]. Ion exchange may also be used [3, 188, 189]. More information concerning the operating details of these processes is given with municipal water treatment.

For some water re-use applications activated carbon may be employed to adsorb 90 to 98 % of any residual degradation-resistant organics when necessary [174]. The resultant effluent quality is certainly adequate for many industrial coolant or irrigation applications, and could even be re-used for potable purposes in emergency situations after minimal further treatment such as re-aeration and chlorination [190, 191]. However, not enough is known about the potential for accumulation of trace toxins to recommend this procedure for long term potable water use.

## 3.5.5 Sludge Handling and Disposal

The large volumes of sludge produced during sewage treatment and its low solids content pose a secondary environmental problem, its disposal. Frequently the costs of sludge treatment and ultimate disposal or destruction equal all the other costs of sewage treatment [135, 192]. Primary sludges, of 2.5-5 % solids, and the sludges from trickling filters and activated sludge plants of 0.5-5 % and 0.5-1 % solids, respectively, are particularly difficult to dewater because much of the water is tied up intracellularly by micro-organisms and in the sedimented flocs, both of which retain water tenaciously. In terms of disposal weights, roughly 1 ton dry weight of sludge is produced daily for each 10,000 population (domestic wastes or industrial equivalent) connected to the sewerage system [193].

Anaerobic sludge digestion at 30 to 35 °C reduces sludge volume by about 90 % while producing a medium grade fuel gas of methane and carbon dioxide, in the proportions of roughly $2\,CH_4 : CO_2$ [73, 194] (e.g. Equations 3.63-3.66, Figure 3.12).

$$CO_2 + 4\,H_2 \rightarrow CH_4 + 2\,H_2O \qquad (3.63)$$

$$4\,HCOOH \rightarrow CH_4 + 3\,CO_2 + 2\,H_2O \qquad (3.64)$$

$$CH_3COOH \rightarrow CH_4 + CO_2 \qquad (3.65)$$

$$2\,CH_3CH_2OH \rightarrow 3\,CH_4 + CO_2 \qquad (3.66)$$

The gas produced may be usefully used for energy requirements in the sewage treatment plant itself, or it may be relatively easily upgraded by carbon dioxide removal (water washing) and added to a town gas distribution system [195]. Small volumes of digested sludge may be disposed of by plowing into land, though there remain safety questions relating to residual pathogens and possible toxic heavy metal accumulations, particularly in the sludges from heavy industrial areas, which may be contributed to the land in this way [196-200].

Sludge may be disinfected by chlorination, effected by the direct addition of 2-4 mg/L chlorine [166], or by sludge electrolysis producing chlorine *in situ* from the residual sodium chloride present in the wet sludge [201-204]. Gamma irradiation has also been used for sludge disinfection [135, 205, 206], and various heat treatments are also effective for this purpose [207-209]. Any of these disinfection alternatives improves the safety and appropriateness of land disposal of sludges, provided that this is to a level area removed from any surface water courses, that the heavy metal content of the sludge is low and that the sludge application to land is not too frequent or heavy [210-212]. For the year following application to agricultural land the site should be clearly marked, and the land used only for pasture, fallow, or forage crops, not for produce or dairy cattle. This sludge disposal option has, however, been usefully applied to assist in the reclamation of strip-mined areas [213, 214]. The sludge may also be composted with the organic content of municipal waste or wood chips to produce a useful soil conditioner [215, 216]. Or dried, alone, it can be more economically shipped greater distances and marketed as a low analysis fertilizer, as is promoted by the city of Milwaukee, with their Milorganite [217].

If none of the disposal options above or their variations can be practised, then wet or dry incineration are the only alternatives. Wet oxidation with air at 300-350 °C, the Zimmerman process, yields liquid effluent containing a much reduced volume of easily settled ash [135, 218, 219]. Dry incineration in multiple hearth units uses much of the heat generated by combustion to complete the drying of the entering sludge [220, 221]. The final product, a sterile ash of 3-5 % of the original dewatered volume, comprises a much smaller volume for discarding to landfill or the like than the original sludge. But one should be aware of the possibility of metal elution from the ash in the landfill, and take appropriate precautions [222]. Or the sludge can be more thoroughly dewatered by filtration or centrifugation techniques, combined with milled coal to form briquets as developed by BASF [223], and then burned in a coal-fired power station for energy recovery from the composite fuel.

## 3.6 Industrial Liquid Waste Disposal

Industrial liquid wastes can only be considered for discharge into municipal sewerage systems if they meet certain strict criteria, because of frequent incompatibility problems. They may have very high BOD's, in the tens of thousands instead of 100 to 300 mg/L, corresponding to carbon loadings of 1 % or more. Oxygen demands that are this high in liquid wastes fed to a municipal sewage treatment plant can seriously overload the BOD removal capabilities. Industrial waste streams may contain toxic constituents such as cyanide ion, heavy metals, or toxic organics which could kill the active micro-organisms employed for secondary sewage treatment. They may also contain dissolved or immiscible solvents which are not only hazardous to the sewage treatment plant but also to the sewage collection system itself. An accident with released hexane, for instance, recently left the main street of Louisville, Kentucky pock-marked with 7 m craters as a result of a sewer explosion [224]. Therefore, many kinds of industrial liquid wastes require at least some kind of preliminary treatment before acceptance by a municipal system and many are not acceptable to a municipal system under any circumstances.

With any byproduct or waste stream, utilization as far as possible, is a far more attractive proposition than paying for ultimate disposal [225]. For instance, diphenyl ether is obtained as a byproduct of the chlorobenzene to phenol process roughly according to the stoichiometry of Equation 3.67 and is not a direct objective of the process [226].

$$20\,C_6H_5Cl + 20\,NaOH \rightarrow$$
$$18\,C_6H_5OH + (C_6H_5)_2O + 20\,NaCl + H_2O$$
$$(3.67)$$

Diphenyl ether production may be decreased by recycling a part of it back into the process, but markets have also been developed for this byproduct as a component of high temperature heat transfer fluids, brake fluid formulations and the like. In these applications diphenyl ether earns a profit rather than imposing a disposal cost burden on the process.

As other examples of this philosophy, the anhydrous hydrogen chloride in excess of market requirements produced during chlorinations of hydrocarbons to halocarbon solvents, or during the manufacture of toluene di-isocyanate, may be usefully employed in the production of phosphoric acid from phosphate rock. Fluorides associated with the production of phosphoric acid and phosphatic fertilizers may be used to manufacture synthetic cryolite ($Na_3AlF_6$), useful in aluminum smelting, or fluorspar, valuable in iron and steel metallurgy. In turn the iron-containing red muds from alumina purification prior to smelting can be used to prepare chemicals useful in water and sewage treatment. Aluminum itself can also be economically recovered from flyash [227].

The number of examples of these kinds of utility being found for process byproducts are almost endless, but each new process developed and started up poses new byproduct problems to test the ingenuity of the developers to come up with a use for it, rather than generate a waste disposal problem [225]. Thoughts of this kind are behind the "Product Stewardship" ideas put forward within the Dow Chemical Company some time ago to promote this concept early in the formative stage of new process ideas [228, 229], and have recently been adopted and promoted by the Chemical Specialties Manufacturers Association [230].

### 3.6.1 Aqueous Wastes With High Suspended Solids

Liquid waste streams with a high suspended solids content can be cleaned up by solids removal in clarifiers, thickeners, liquid cyclones, accelerated settling by inclined "Chevron" settlers or the like [231-234]. For waste streams with very finely-divided solids in suspension (i.e. < ca. 100 $\mu m$) dissolved air flotation techniques have been shown to be more efficient than methods employing sedimentation [235, 236]. Final dewatering of the sludges obtained may be carried out on a continuous filter or a centrifuge [235, 237]. Water clarified in these ways can be accepted for more potential options of re-use or final disposal than untreated water, and the separated solids may be burned or discarded to landfill, as appropriate [238].

### 3.6.2 Aqueous Wastes Containing an Immiscible Liquid

Liquid waste streams containing an insoluble liquid, such as can arise from extraction processes, from steam ejectors operating on solvent distillation systems, or from the loss of heat exchange fluid from a heat exchanger, should be phase-separated before final disposal measures are undertaken. A simple settler, or a unit such as an API separator can be used to accomplish this step. When the initial separator is coupled to an entrained or dissolved air flotation unit, the concentration of residual organics can be further reduced [236, 239]. The recovered organics can be returned to their processing origins, via a further clean-up if required, and the water phase more safely discarded.

### 3.6.3 Heated Effluent Discharges

Various options exist to deal with high thermal loads to water [240]. If the cooling water supply is plentiful the impact of the heated discharge can be decreased by reducing the design temperature rise of the cooling water flow. This can be accomplished within limits by increasing the cooling water flow rate to existing heat exchangers, or by combining this step with the installation of larger heat exchangers. If limited water availability precludes this option, the cooling water may be reused by passage through a cooling pond or through convective or forced air circulation evaporative cooling towers [241, 242] (Figure 3.13). When heated effluent is available near an urban center it can be used for community heating requirements, with mutual benefit. Otherwise it can be employed for the warming of greenhouses or soil heating, or for accelerated fish rearing [243]. If water supplies for this purpose are very limited, indirect cooling towers, not employing evaporation, or air cooling options may have to be adopted [244, 245]. These options may also be necessary in the event of severe fogging in the vicinity of evaporative cooling towers [246], caused by the very large water vapour discharges from such installations.

### 3.6.4 Aqueous Waste Streams with a High Oxygen Demand

Aqueous waste streams with BOD's of 5,000 to 20,000 but which are essentially non-toxic, such as those produced by the food processing industry or by a distillery, may be efficiently treated by extended aeration in ponds or lagoons, or in deep shaft aeration units [162, 247-249]. With added nutrients in the form of ammonium phosphate to promote bacterial waste utilization, BOD's may be decreased by 90-95 % [220]. Spray irrigation of this type of waste stream with or without prior BOD reduction has also been advocated, in which case infiltration to the soil and the activity of soil bacteria are relied upon to deal with the oxygen demand [220]. If this disposal method is used it is important not to overload the land disposal area to avoid placing the soil microbial system under anaerobic conditions.

### 3.6.5 Highly Coloured Waste Waters

Activated carbon has been tested for the clean-up of highly coloured waste streams such as result

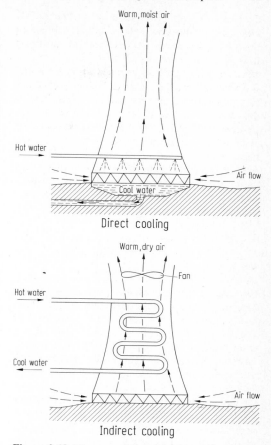

**Figure 3.13** Diagrams of direct (evaporative) and indirect cooling towers designed to shed heat loads to air rather than surface waters

from dyeing operations [236, 250]. Coal carbons were 99 % effective for dye removal and could accept dye loadings as high as 0.40 kg dye per kg of carbon. Lignite carbon was also useful, but less effective.

### 3.6.6 Fluid and Solid Combustible Wastes

Combustible wastes which are fairly fluid at ordinary temperatures may be simply atomized and burned for disposal. But the combustion units to handle organics containing a heteroatom such as nitrogen or chlorine must be designed to handle more corrosive than normal combustion gases to avoid corrosion problems in the combustion chamber, and to have a water scrubber or other types of exhaust gas emission controls integral with the unit. Very viscous or solid combustible wastes are normally disposed of by direct burning in fluidized bed [251] or multiple hearth combustors [238] (Figure 3.14).

Aqueous solutions of combustible organics such as phenols, alcohols, lower ketones and the like can also be burned for disposal, but proper combustion may have to be assisted by co-combustion with some external fuel [252]. If, however, fairly fluid combustible wastes are burned at the same time as these aqueous wastes containing combustibles in special burners designed to accommodate dual fuels, both may be disposed of simultaneously without having to use purchased fuel to assist the process.

### 3.6.7 Neutralization and Volume Reduction of Intractible Waste Streams

Non-combustible or toxic liquid wastes that do not lend themselves to disposal by any of the means just outlined are the most difficult to dispose of safely. With these kinds of wastes in particular, it is in the producer's best interests to minimize ultimate disposal costs. Thus, acidic pickling plant wastes, which will contain iron plus unused sulfuric or hydrochloric acids, may be used to precipitate phosphate from secondary sewage effluent [184] (Equation 3.55). Or they may be blended with aqueous alkaline wastes containing phenolate

and unexpended sodium hydroxide to neutralize the pH extremes of both streams.

Wastes containing cyanide ion and various heavy metals may be at least partially de-toxified (cyanide neutralized) by combining these wastes in proper proportions with the toxic waste streams from chlorohydrin-based ethylene oxide or propylene oxide manufacture, which will usually contain unexpended hypochlorite [253] (Equation 3.68). At higher ratios of oxidant to cyanide, complete conversion to carbon dioxide and nitrogen can be achieved [254-256] (Equation 3.9).

$$2\,NaCN + Ca(OCl)_2 \rightarrow 2\,NaCNO + CaCl_2 \tag{3.68}$$

$$4\,NaCNO + 3\,Ca(OCl)_2 + 2\,H_2O \rightarrow 2\,N_2 + 4\,NaCl + CaCl_2 + 2\,Ca(HCO_3)_2 \tag{3.69}$$

Since a slight excess of chlorine is usually employed to obtain cyanide destruction, the waste stream may still be quite toxic from the residual chlorine present. Dechlorination using sulfur dioxide effectively reduces any residual chlorine to chloride, removing the toxicity from this source before discharge (Equation 3.55).

The metals content of an aqueous waste stream can be substantially decreased by complexation or adsorption methods and the recovered metal used to offset a part of the treatment costs [257-259]. Unexpended hypochlorite can also be neutralized by

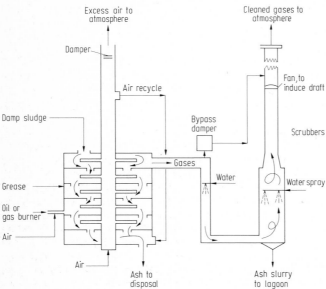

**Figure 3.14** Multiple hearth combustion unit with emission controls, for the disposal of solid wastes [238]

contacting this stream, in the correct proportions, with spent scrubber liquor from sulfur dioxide emission control systems (Equation 3.70).

$$Ca(OCl)_2 + 2\,CaSO_3 \rightarrow CaCl_2 + 2\,CaSO_4$$
(3.70)

Decreasing the volume of waste streams containing a high concentration of dissolved salts by such means as ion exchange, or reverse osmosis can also help reduce the ultimate disposal costs of the solutions of salts produced by these neutralization reactions [189, 260].

Since ingenuity in dealing with these kinds of wastes frequently requires good communication between companies in widely differing business areas, participation in a waste trading exchange [261-269] or the setting up of an independent waste disposal company [270-275] can be useful options. These clearing houses can then contract workable price schedules for proper disposal of various classes of wastes, and by so doing are set up to maximize the benefit of neutralization interactions or co-disposal options between the various kinds of wastes handled.

### 3.6.8 Ultimate Destruction or Disposal of Hazardous Wastes

Incineration at the high continuous operating temperatures of a cement kiln [276] or by large ocean-going incinerators such as the Chemical Waste Management, Inc. and Bayer vessels, Vulcanus and Vesta, operating at sea [277, 278], are two ultimate disposal options for more than usually hazardous wastes. Chemical fixation into absorbent solids by proprietary processes such as those trade-named Sealosafe, Chemfix, or Poz-o-tec, and then use of the fixed product as landfill or as foundation material is a further alternative [275, 279, 280]. Direct landfilling and acceptance of long term sterilization of the land with proper management, recording, and site posting has also been used for hazardous waste disposal [275, 281]. A clay lining to the disposal site does decrease the transmission of larger molecules through the soil, but is still quite permeable to smaller organics [282]. More rigorous sheet or cement linings plus a drainage system under the lining may be called for in some circumstances.

Deep well disposal, generally to a brine aquifer at depth, has also been used to a significant extent, particularly for waste brines (primarily $CaCl_2$, NaCl, etc.) [283]. For this to be effective without

serious damage to important groundwater resources requires careful attention to the basic requirements. These are that the wells be located in a seismically stable region, and that the discharge take place via a two casing (double pipe lined) well to a porous stratum capped by one or more impermeable strata to ensure retention of the waste at the level discarded. Drilling the well itself is expensive, US$ 30-100 per meter, and there are also a number of important precautions to be taken in the well installation, such as prior removal of suspended solids and low injection pressures, to ensure safe operation [283, 284] (Figure 3.15). So deep well incineration is not an inexpensive disposal method, although it does have some virtues, e.g. for brine disposal into a brine aquifer. The long term costs of poorly-operated deep well disposal installations if groundwater contamination should occur [285-287], are so great that licensing of these operations should be required by more jurisdictions, as it is by Ontario. The potential for contamination of the groundwater resource is by no means limited to deep well disposal accidents, however, but can also occur from bacteria and viruses in groundwater recharge operations using treated sewage [288], or from toxins leached from poorly placed or operated municipal [289, 290] and industrial waste dumps [291].

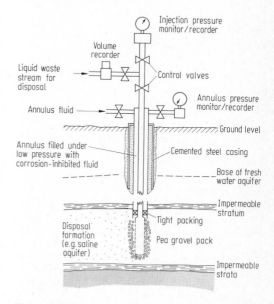

**Figure 3.15** Details of an injection installation for deep well disposal of liquid wastes

# Relevant Bibliography

1. L.O. Benefield, J.F. Judkins, and B.L. Weand, Process Chemistry for Water and Wastewater Treatment, Prentice-Hall, Englewood Cliffs, N.J., 1982
2. L.J. Thibodeaux, Chemodynamics, Environmental Movement of Chemicals in Air, Water, and Soil, John Wiley, New York, 1979
3. N.L. Nemerov, Industrial Water Pollution, Origins, Characteristics, Treatment, Addison Wesley, Reading, Mass., 1978
4. M.L. Hitchman, Measurement of Dissolved Oxygen, Wiley-Interscience, New York, 1978
5. Water Quality and Treatment: A Handbook of Public Water Supplies, 3rd edition, American Water Works Assoc., McGraw-Hill, New York, 1971
6. D.W. Sundstrom and H.E. Klei, Wastewater Treatment, Prentice-Hall, Englewood Cliffs, N.J., 1979
7. P.A. Vesilind, Treatment and Disposal of Wastewater Sludges, Ann Arbor Science Publishers, Ann Arbor, Mich., 1975
8. W. Leithe, Analysis of Organic Pollutants in Water and Wastewater, translated by STS, Inc., Ann Arbor Science Publ., Ann Arbor, Mich., 1973
9. Toxic and Hazardous Waste Disposal, Volume 2, R.B. Pojasek, editor, Ann Arbor Science Publishers, Ann Arbor, Mich., 1979
10. I.J. Kugelman, Water Reclamation and Re-use, J. Water Polln Control Fed. 46 (6), 1,195, June 1974
11. T.J. Tofflemire and F.E. Van Alstyne, Land Disposal of Wastewater, J. Water Polln Control Fed. 46 (6), 1,201, June 1974
12. A.V. Bridgewater and C.J. Mumford, Waste Recycling and Pollution Control Handbook, Van Nostrand Reinhold, New York, 1980
13. Characterization, Treatment and Use of Sewage Sludge, Proc. 2nd European Symposium, Vienna, 21-23, 1980, edited by P. L'Hermite, and H. Ott, D. Riedel, Dordrecht, 1981
14. W.E. Isman and G.P. Carlson, Hazardous Materials, Collier Macmillan for Glencoe, West Drayton, U.K, 1980. (Re managing hazardous material incidents.)

# References

1. Water Resources of the World, F. van der Leiden, compiler-editor, Water Information Center, Port Washington, N.Y., 1975
2. Water Reuse, L.K. Cecil, editor, Chem. Eng. Prog. Symp. Series No. 78, 63, 261 pp., 1967
3. D.G. Downing, R. Kunin, and F.X. Pollio, Chem. Eng. Prog. Symp. Series 64 (90), 126, (1968)
4. G. Young and J.P. Blair, National Geographic 138 (6), 743, Dec. 1970
5. River Could Catch Fire, Victoria Times, page 1, April 20, 1976
6. C.G. Wilber, The Biological Aspects of Water Pollution, C.C. Thomas, Springfield, Illinois, 1969
7. Hitting the Bottle, Newsweek 78 (5), 52, Aug. 2, 1971
8. Mediterranean Conference, Chem. Eng. News 54 (8), 19, Feb. 23, 1976
9. J. Bugler, New Scientist, 4, Jan. 5, 1978
10. D.A. O'Sullivan, Chem. Eng. News 55 (9), 17, Feb. 28, 1977
11. D.A. O'Sullivan, Chem. Eng. News 58 (22), 10, June 2, 1980
12. Wesley Marx, The Frail Ocean, Coward McCann, New York, 1967
13. A. Wheeler, Science J. 6 (11), 28, Nov. 1970
14. R. Gore and J. Blair, National Geographic 162 (6), 694, Dec. 1982
15. Pollution of Lake Erie, Lake Ontario and the International Section of the St. Lawrence River, Report to the International Joint Commission by the Internat. L. Erie Water Pollution Board and the Internat. L. Ontario-St. Lawrence River Water Pollution Board, Windsor, Ontario, 1969, 3 volumes
16. Great Lakes Water Quality, 1978 Annual Report to the International Joint Commission, Great Lakes Water Quality Board, Windsor, July 1979
17. S.K. Gupta and K.Y. Chen, J. Water Polln Control Fed. 50 (3), 493, March 1978
18. J. Cholak, Archives of Ind. Health 21, 312 (1960)
19. D. O'Sullivan, Chem. Eng. News 54 (25), 15, June 14, 1976
20. L. Klein, River Pollution, Volume 1, Chemical Analysis, Butterworths, London, 1959
21. J. Rodier, Analysis of Water, translated from the French by Israel Program for Scientific Translations, John Wiley, New York, 1975
22. G.P. Treweek and J.J. Morgan, Envir. Science and Tech. 11 (7), 707, July 1977
23. The Aerial Photo — Water Quality Link, Envir. Science and Tech. 10 (3), 229, March 1976
24. U.K. Study Vindicates Fluoridation, Chem. Eng. News 54 (4), 13, Jan. 26, 1976
25. D. Rose and J.R. Marier, Environmental Fluoride, National Research Council No. 16,081, Ottawa, 1977
26. H.J. Sanders, Chem. Eng. News 58 (8), 30, Feb. 25, 1980
27. L. Klein, River Pollution, Volume II, Causes and Effects, Butterworths, London, 1962
28. Cleaning Our Environment, A Chemical Perspective, 2nd edition, American Chemical Society, Washington, 1978
29. C.N. Sawyer, J. New Engl. Water Works Assoc. 61, 109 (1947)
30. C.N. Sawyer, Sewage Industrial Wastes 26, 317 (1954). Cited by reference Klein, vol II, III, pages 125, 159

31. W.J. Blaedel and V.W. Meloche, Elementary Quantitative Analysis, Theory and Practice, 2nd edition, Harper and Row, New York, 1963

32. Standard Methods for the Examination of Water and Wastewater, 13th edition, American Public Health Assoc., Washington, 1971

33. M.S. Frant, J.W. Ross, Jr., and J.M. Riseman, American Laboratory, Jan. 1969

34. Analytical Methods Guide, Orion Research, Cambridge, Mass., May 1975

35. Lange's Handbook of Chemistry, 10th edition, N.A. Lange, editor, McGraw-Hill, New York, 1969

36. E.A. Hemmingsem, Science 167, 1,493 (1970)

37. E.J. Green and D.E. Carritt, Science 157, 191 (1967)

38. L.W. Winkler, Berichte Dtsch. Chem. Ges. 21, 2,843 (1888)

39. W.R. Stagg, J. Chem. Educ. 49 (6), 427, June 1972

40. Science Specialities, Cole-Parmer Instrument Co., Chicago, Summer 1977, page 26

41. M.L. Hitchman, Measurement of Dissolved Oxygen, Wiley-Interscience, New York, 1978

42. G.J. Shroepfer, Sewage Works J. 14, 1,030 (1942). Cited by reference 20

43. M.J.R. Clark, Annotated Extracts of Some Papers Dealing With Various Aspects of Dissolved Atmospheric Gases, Water Resources Service, Victoria, B.C., 1977, plus Appendix 2, reference list for the above, 209 pp

44. W.J. Ebel, Fisheries Bulletin (U.S. Fish and Wildlife Service) 68, 1 (1969), cited by reference 43

45. H.H. Harvey and S.B. Smith, Can. Fish Cult. 30, 39 (1961), cited by reference 43

46. B.G. D'Aoust and M.J.R. Clark, Trans. Amer. Fisheries Soc. 109, 708 (1980)

47. L. Granat, Tellus 24 (6), 550, June 1972

48. H.L. Youmans, J. Chem. Educ. 49 (6), 429, June 1972

49. Indicator Paper, eight indicators, Carl Schleicher and Schuell Co., Keene, N.H./Hannover, W. Germany

50. Baker Dual Tint Paper, J.T. Baker Chem. Co. Phillipsburg, N.J.

51. Alkacid pH Paper, Fisher Scientific Ltd., Toronto

52. G.S. Fell, Chem. In Brit. 16 (6), 323, June 1980

53. Industrial Hygiene and Toxicology, 2nd edition, F.A. Patty, editor, Interscience, New York, 1963, volume II

54. J.F. Cosgrove and D.J. Bracco in Water and Water Pollution, L.L. Ciaccio, editor, Marcel Dekker, New York, 1971, volume 4, page 1,315

55. J.C. Van Loon, Chem. In Can. 28 (5), 15, May 1976

56. A.R. Knott, Atomic Absorption Newsletter 14 (5), 126, Sept.-Oct. 1975

57. E.J. Maienthal and J.K. Taylor in Water and Water Pollution, L.L. Ciaccio, editor, Marcel Dekker, New York, 1971, volume 4, page 1,751

58. J.W. McCoy, Chemical Analysis of Industrial Water, Chemical Publishing Company, New York, 1969

59. F.W. Karasek, Research/Development 26 (11), 40, Nov. 1975

60. F.W. Kawahara, Envir. Science and Tech. 5 (3), 235, March 1971

61. K.W. Kawahara, Anal. Chem. 40 (11), 2,073, Nov. 1968

62. R. Farrell, Can. Research 12 (1), 29, Feb. 1979

63. J. Krieger, Chem. Eng. News 59 (27), 28, July 6, 1981

64. B. Kahn in Water and Water Pollution, L.L. Ciaccio, editor, Marcel Dekker, New York, 1971, volume 4, page 1,357

65. N.E. Armstrong and E.F. Gloyna, Chem. Eng. Prog. Symp. Series No. 90, 64, 47 (1968)

66. K.E. White, Chem. In Brit. 12 (12), 375, Dec. 1975

67. P.V. Scarpino, in Water and Water Pollution Handbook, L.L. Ciaccio, editor, Marcel Dekker, New York, 1971, volume 2, page 639

68. R.C. Cooper, J. Envir. Health 37 (3), 217, Nov./Dec. 1974

69. Current Laboratory Practice, Gelman Sciences Inc., Montreal, May 1980

70. J.P. Sherry, S.R. Kuchman and B.J. Dutka, Can. Research 12 (1), 33, Feb. 1979

71. R. Hare, An Outline of Bacteriology and Immunity, 3rd edition, Longmans, London, 1967

72. Biological Analysis of Water and Wastewater, Millipore Corp., Bedford, Mass., 1973

73. J.E. Zajic, Water Pollution, Disposal and Reuse, Volume 1, Marcel Dekker, New York, 1971

74. T.L. Brown, Energy and the Environment, C.E. Merrill Publishing Co., Columbus, Ohio, 1971

75. R.J. Allan and D.J. Richards, Effect of Thermal Generating Station on Dissolved Solids and Heavy Metals in a Prairie Reservoir, Inland Waters Directorate, Regina, Sask., 1978

76. R.V. Thomann, D.J. O'Connor, and D.M. Di-Toro, Chem. Eng. Prog. Symp. Series No. 90, 64, 21 (1968)

77. J.R. Clark, Scientific American 220 (3), 18, March 1969

78. As produced by Omega Engineering Inc., Stamford, Conn., for example

79. J.R. Clark, Scientific American 220 (3), 3, March 1969

80. T.M. Lillesand and R.W. Kiefer, Remote Sensing and Image Interpretation, John Wiley and Sons, New York, 1979, pages 393, 583

81. T.J. McGhee, R.L. Torrens, and R.J. Smaus, Water and Sewage Works *119*, 58, June 1972
82. V.T. Stack, Jr., in Water and Water Pollution, L.L. Ciaccio, editor, Marcel Dekker, New York, 1971, Volume 3, page 801
83. D. Lui, Can. Research and Devel. *6*, 35, May-June 1973
84. D.W. Ryckman, A.V.S. Prabhakara Rao, and J.C. Buzzell, Jr., Behavior of Organic Chemicals in the Aquatic Environment, Manufacturing Chemists Assoc., Washington, 1966
85. Reaction Site — Chemical Oxygen Demand, Can. News and Notes *1* (1), 2, Jan. 1983, Hach Chemical Co., Loveland, Colorado
86. A.M. Jirka and M.J. Carter, Anal. Chem. *47*, 1,397, July 1975
87. C.E. Van Hall, J. Safranko, and V.A. Stenger, Anal. Chem. *35* (3), 315, March 1963
88. C.E. Van Hall and V.A. Stenger, Anal. Chem. *39* (4), 503, April 1967
89. R.H. Jones and A.F. Dageforde, ISA Transactions *7* (4), 267, April 1968
90. Organic Carbon Pollution in Surface Waters Measured with TOC Analyzer, Beckman Application Data, Data No. 4,374, ca. 1975
91. R.H. Jones, Water and Sewage Works *119* (3), 72, March 1972
92. J.C. Buzzell, Jr., R.H.F. Young, and D.W. Ryckman, Behavior of Organic Chemicals in the Aquatic Environment, Part II. Behavior in Dilute Systems, Manufacturing Chemists Assoc., Washington, April 1968
93. J.W. Helms, Rapid Measurement of Organic Pollution by Total Organic Carbon and Comparison with Other Techniques, U.S. Geological Survey, Open-file Report, Menlo Park, California, May 21, 1970
94. C.P. Hwang, G. Goos, and E. Davis, Water and Polln Control *112* (3), 28, March 1974
95. Biological Method Tests River Pollution, Chem. Eng. News *49* (23), 31, June 7, 1971
96. H.B.N. Hynes, Chem. Ind. (London), 435, March 14, 1964
97. T.W. Beak, in Advances in Water Pollution Research, Proc. 2nd International Conf., Tokyo, Pergamon Press, Oxford, 1964, Volume I, page 191
98. F.S. Woodiwiss, Chem. Ind. (London) No. 11, 443, March 14, 1964
99. J. Cairns, Jr. and K.L. Dickson, J. Water Polln Control Fed. *43*, 755 (1971)
100. R.N. McNeeley, V.P. Neimanis and L. Dwyer, Water Quality Source Book, A Guide to Water Quality Parameters, Inland Waters Directorate, Water Quality Branch, Ottawa, 1979
101. J.E. McKee and H.W. Wolf, Water Quality Criteria, 2nd edition, California State Water Quality Control Board, Publication No. 3-A, Sacramento, 1971
102. Water Quality Interpretive Report, Ontario, 1967-1977, R.N. McNeely, editor, Inland Waters Directorate, Ottawa, 1979
103. Guidelines and Criteria for Water Quality Management in Ontario, Ontario Water Resources Commission, Toronto, June 1970
104. F. Franks, Chem. In Brit. *12* (9), 278, Sept. 1976
105. D.K. Todd, The Water Encyclopedia, Water Information Center, Port Washington, N.Y., 1970
106. Chemical Technology: An Encyclopedic Treatment, J.F. Van Oss, editor, Barnes and Noble, Inc., New York, 1968, volume I, page 169
107. Water Treatment Handbook, Degremont Co., translated from the French by D.F. Long and Co., London, 1973
108. J.E. Singley in Water and Water Pollution, L.L. Ciaccio, editor, Marcel Dekker, New York, 1971, volume 1, page 365
109. Handbook of Chemistry and Physics, 56th edition, R.C. Weast, editor, CRC Press, Cleveland, Ohio, 1975
110. A Refresher on Sodium Hypochlorite, Water and Polln Control *115* (1), 11, Jan. 1977
111. M.B. Hocking, Chem. Eng. News *61* (1), 50, Jan. 3, 1983
112. G.C. White, Handbook of Chlorination, Van Nostrand Reinhold, New York, 1972, page 189
113. J.G. Smith, J. Chem. Educ. *52* (10), 656, Oct. 1975
114. J.G. Smith, Civic *28* (3), 38, March 1976
115. J.G. Smith, S-F. Lee, and A. Netzer, Water Research *10*, 985 (1976)
116. B.G. Oliver and J.H. Carey, Envir. Science and Tech. *11* (9), 893, Sept. 1977
117. R.D. Smillie, Water and Polln Control *115* (12), 8, Dec. 1977
118. Evidence that Chlorination May Form Mutagens, Chem. Eng. News *56* (3), 33, March 27, 1978
119. B. Hileman, Envir. Science and Tech. *16* (1), 15 A, Jan. 1982
120. R.J. Bull, Envir. Science and Tech. *16* (10), 554 A, Oct. 1982
121. Chlorine Looks Secure as Water Reagent, Can. Chem. Proc. *60* (3), 28, March 1976
122. M. Hyde, Chem. In Brit. *12* (9), 293, Sept. 1976
123. E.J. Calabrese, G.S. Moore, R.W. Tuthill, J. Envir. Health *41* (1), 26, July/Aug. 1978
124. M. Peleg, Water Research *10*, 361 (1976)
125. H. Mitchell, Water and Polln Control *115*, 22, Jan. 1977
126. M.D. Crawford, M.J. Gardner, and J.N. Morris, The Lancet *1*, 827, April 1968
127. E.F. Winton and L.J. McCabe, J. Amer. Water Works Assoc. *62*, 26 (1970)
128. M.L. Bierbaum, Chemtech *6* (5), 314, May 1976
129. Calcium May be Key to Health, Chem. Eng. News *55* (16), 17, April 18, 1977

130. R.A. Horne, The Chemistry of Our Environment, Wiley-Interscience, New York, 1978

131. A. Porteous, Saline Water Distillation Processes, Longman, London, 1975

132. K.S. Spiegler, Saltwater Purification, 2nd edition, Plenum, New York, 1977

133. T.V. Arden, Chem. In Brit. *12* (9), 285, Sept. 1976

134. A.W. Michalson, Chem. Eng. Prog., Symp. Series No. 90, L.K. Cecil, editor, *64,* 134 (1968)

135. Cleaning Our Environment, The Chemical Basis for Action, American Chemical Society, Washington, 1969

136. A First for Alberta, Aeration by U-tube, Water and Polln Control *110* (4), 42, April 1972

137. C.C.K. Lui, Can. Research *10* (5), 23, Sept.-Oct. 1977

138. K.M. Mackenthum, The Practice of Water Pollution Biology, Federal Water Pollution Control Administration, U.S. Dept. of the Interior, 1969

139. Water Pollution in Edmonton: The North Saskatchewan River, D. Hocking, editor, Edmonton Anti-pollution Group, Edmonton, 1971

140. C.G. Patterson and J.R. Nursall, Water Research *9,* 425 (1975)

141. D.V. Ellis, Sewage Disposal to the Sea, in Western Geographical Series Volume 12, University of Victoria, Victoria, B.C., 1977, page 289

142. W. Bascon, Scientific American *231* (2), 16, Aug. 1974

143. P.C. Loh, R.S. Fujioka, and L.S. Lau, Water, Air, and Soil Polln *12,* 197 (1979)

144. P. Burdyl and F.J. Post, Water, Air, and Soil Polln *12,* 237 (1979)

145. S.A. Berry and B.G. Notion, Water Research *10,* 323 (1976)

146. H.P. Savage and N.B. Hanes, J. Water Polln Control Fed. *43* (5), 854, May 1971

147. W. Whipple, Jr., in Water, 1969, L.K. Cecil, editor, Chem. Eng. Prog. Symp. Series, No. 97, *65,* 75 (1969)

148. P. O'Dwyer, Toronto Globe and Mail, March 6, 1971

149. When a Lake Bottom is Starved for Oxygen, Can. Chem. Proc. *58* (2), 10, Feb. 1974

150. R.J. Benoit, in Water and Water Pollution Handbook, L.L. Ciaccio, editor, Marcel Dekker, New York, 1971, volume 1, page 243

151. K. Mellanby, Pesticides and Pollution, Collins, London, 1967

152. C.C.K. Liu, Can. Research *10* (5), 23, Sept.-Oct. 1977

153. H.A. Painter, in Water and Water Pollution Handbook, L.L. Ciaccio, editor, Marcel Dekker, New York, 1971, volume 1, page 329

154. D.L. Smith and R.V. Daigh, J. Water Polln Control Fed. *53* (8), 1,272, Aug. 1981

155. A New Secondary Wastewater Treatment Process, Can. Chem. Proc. *55* (11), 12, Nov. 1971

156. J.F. Hanahan, Chem. Eng. News *49* (17), 31, April 26, 1971

157. Waste Treatment is Undergoing Rapid Change, Can. Chem. Proc. *61* (11), 4, Nov. 1977

158. Prospects Strong for Wastewater Oxygenation, Chem. Eng. News *55* (13), 17, March 28, 1977

159. J.A. Chittenden and W.J. Wells, Jr., J. Water Polln Control Fed. *43* (5), 746, May 1971

160. Biodisc Treats Small Community Wastes, Water and Polln Control *111* (2), 20, Feb. 1973

161. Incredible, But it Works, Water and Polln Control, 1973-74 Directory Issue, 69, June 1974

162. CIL to Test New Sewage Unit, Can. Chem. Proc. *59* (10), 32, Oct. 1975

163. B. Appleton, New Civil Engineer, April 17, 1975. See also Water Resources Abstracts *8* (22), Report W 75-10,981, page 55, April 1975

164. New Units Groomed for Sewage, Garbage, Can. Chem. Proc. *63* (2), 18, Feb. 2, 1979

165. O.C. Collins and M.D. Elder, J. of Inst. of Water Polln Control *79* (2), 272, Feb. 1980

166. H.E. Babbit and E.R. Baumann, Sewerage and Sewage Treatment, 8th edition, John Wiley and Sons, New York, 1958

167. K.L. Murphy, Water and Polln Control *115* (1), 13, Jan. 1977

168. J.G. Smith, S.-F. Lee, and A. Netzer, Envir. Letters *10* (1), 47, (1975)

169. B.G. Oliver, Can. Research *11* (6), 21, Oct. 1978

170. K.L. Murphy, Water and Polln Control *112* (4), 24, April 1974

171. Bromine Chloride Looks Good as Disinfectant, Chem. Eng. News *55* (41), 8, Oct. 10, 1977

172. Field Trials Confirm Efficacy of BrCl, Chem. Eng. News *56* (41), 6, Oct. 9, 1978

173. F.B. DeWalle, W.G. Light, and E.S.K. Chian, Envir. Science and Tech. *16,* 741 (1982)

174. R. Eliassen and G. Tchobalaglous, Chem. Eng. *75,* 95, Oct. 14, 1968

175. C. Murray, Chem. Eng. News *54* (12), 23, March 22, 1976

176. Process Treats Waste Water, Produces Algae, Chem. Eng. News *48,* 47, Aug. 3, 1970

177. D.W. Hendriks, F.J. Post, and D.R. Khairnar, Water, Air and Soil Pollution *12,* 219 (1979)

178. R.C. Cooper, J. Envir. Health *37* (4), 342, Jan./Feb. 1975

179. A. Shindala and J.W. Stewart, Water and Sewage Works *118* (4), 100, April 1971

180. S.G. Grigoropoulos, R.C. Vedder, and D.W. Max, J. Water Polln Control Fed. *43* (12), 2,366, Dec. 1971

181. R.E. Finger, J. Water Polln Control Fed. *45* (8), 1,654, Aug. 1973

182. W.J. Storck, Chem. Eng. News *56* (4), 9, Jan. 23, 1978

183. The Dow Process For Phosphate Removal, Dow Chemical of Can., Ltd., Sarnia, Ontario, 1972

184. Fe Wastes Help Remove Phosphates, Can. Chem. Proc. *58* (6), 44, May 1974

185. L.A. Campbell and A.J. Horton, Water and Polln Control *110* (3), 28, March 1972

186. Reverse Osmosis Studied for Sewage Treatment, Water and Polln Control *112* (4), 15, April 1974

187. Reverse Osmosis Can Handle Phosphates, Can. Chem. Proc. *55* (3), 50, March 1971

188. F.X. Pollio and R. Kunin, Envir. Science and Tech. *2* (1), 54, Jan. 1968

189. R. Kunin and D.G. Downing, Chem. Eng., 67, June 28, 1971

190. Canadian Researchers Say, Can. Chem. Proc. *57* (6), 41, June 1973

191. Water Reuse, E.J. Middlebrooks, editor, Ann Arbor Science Publ., Ann Arbor, Mich., 1982

192. J.L. Jones, D.C. Baumberger, Jr., F.M. Lewis, and J. Jacknow, Envir. Science and Tech. *11* (8), 968, Oct. 1977

193. Calculated from Cleaning Our Environment, The Chemical Basis for Action, reference 135

194. Anaerobic Methane Digestion Has Come Of Age, B.C. Guidelines, B.C. Research (Vancouver), (11), 1, Nov. 1980

195. D.W. Osborne, J. Inst. Sew. Purif. *3*, 195 (1962). Cited by L. Klein, River Pollution, volume III, Control, Butterworths, London, 1966

196. G.S. Stoewsand, Chem. Eng. News *60* (18), 4, May 3, 1982

197. Ontario Collects Some of Its Own Data, Water and Polln Control *110* (8), 25, Aug. 1972

198. Why There's Concern for Cadmium, Water and Polln Control *110* (8), 26, Aug. 1972

199. D. Lui, J. Stephenson, and K. Kwasniewska, Can. Research *8* (6), 21, Nov.-Dec. 1975

200. And Where Have All The Sludges Gone, Water and Polln Control *110* (8), 20, Aug. 1972

201. N.S. Wei, and G.W. Heinke, Water and Polln Control *112* (5), 31, May 1974

202. Spinoff Process Aids Sewage Treatment, Envir. Science and Tech. *5* (9), 756, Sept. 1971

203. H.W. Marson, The Engineer, 591, April 1965

204. R.F. Milton and J.L. Hoskins, Nature *158* (4,019), 673, Nov. 9, 1946

205. Irradiation of Sewage Gains Adherents, Chem. Eng. News *53* (34), 30, Aug. 25, 1975

206. Cesium Irradiation to Treat Sewage, Chem. Eng. News *61* (8), 19, Feb. 21, 1983

207. C. Limb, J. Proc. Inst. Sewage Purif., Part I, 5, (1951)

208. Porteous Unit Ready For Start-up, Envir. Science and Tech. *2* (11), 1,068, Dec. 1968

209. First for Toronto Treatment Plant, Can. Chem. Proc. *63* (7), 19, Oct. 10, 1979

210. Ontario's Guidelines for Sludge Disposal, Water and Polln Control *110* (8), 30, Aug. 1972

211. Treatment and Utilization of Sewage Sludge on Land, Compost Science *13* (3), 26, May-June 1972

212. The Land Can Be Returned, Water and Polln Control *112* (4), 18, April 1974

213. Sewage Sludge For Reclamation, Water and Polln Control *112* (1), 12, Jan. 1974

214. Philadelphia Sludge Used to Reclaim Land, Chem. Eng. News *59* (13), 16, March 30, 1981

215. G.D. Zarnett, Water and Polln Control *110* (5), 37, May 1972

216. D. Brown, Focus on Great Lakes Water Quality, *8* (2), 12, April 1982

217. Personal Communication, Sewerage Commission, City of Milwaukee, Wisconsin

218. C.K. Brown, Water and Polln Control *112* (12), 36, Dec. 1974

219. Wet Air Oxidation Processes, Chem. In Can. *29* (5), 15, May 1977

220. Industrial Pollution Control Handbook, H.F. Lund, editor, McGraw-Hill, New York, 1971

221. J. McCarthy, Water and Sewage Works *119* (4), 66, April 1972

222. Incineration, Another Imponderable, Water and Polln Control *110* (8), 27, Aug. 1972

223. Energy From Sludge, Chem. In Brit. *16* (2), 98, Feb. 1983

224. Ralston Purina Suspect in Louisville Blast, Chem. Eng. News *59* (8), 11, Feb. 23, 1981

225. M. Campbell and W. Glenn, Profit from Pollution Prevention, Pollution Probe, Toronto, 1982

226. F.A. Lowenheim and M.K. Moran, Faith, Keyes and Clark's Industrial Chemicals, John Wiley and Sons, New York, 1975, page 616

227. Process Extracts Aluminum From Flyash, Chem. Eng. News *60* (16), 28, April 19, 1983

228. Quarterly Report, Oct. 1971, page 3; and Annual Report, The Dow Chemical Company, Midland, Mich., 1975

229. Dow Cleans Up Pollution At No Net Cost, Business Week, 32, Jan. 1, 1972

230. Product Stewardship Seminar Set for Chicago, Chem. Eng. News *58* (35), 17, Sept. 1, 1980

231. I.E. Puddington, U.S. Patent 2,868,384, Jan. 13, 1959

232. W. Graham and R. Lama, Can. J. Chem. Eng. *41*, 162, Aug. 1963

233. Chevron Tube Settler, Permutit Sybron Corp. Tech. Bull. *9* (2), Feb. 1971

234. M.B. Hocking and G.W. Lee, Fuel *56* (7), 325, July 1977

235. Water Pollution Control, Chem. Eng. *78* (14), 65, June 21, 1971

236. G.L. Shell, J.L. Boyd, and D.A. Dahlstrom, Chem. Eng. *78* (14), 97, June 21, 1971

237. D. Dahlstrom, Chem. Eng. *75*, 103, Oct. 14, 1968

238. F.P. Sebastian and P.J. Cardinal, Jr., Chem. Eng. *75*, 112, Oct. 14, 1968

239. A.E. Franzen, V.G. Skogan, and J.F. Grutsch, Chem. Eng. Prog. *68* (8), 65, Aug. 1972

240. Engineering Aspects of Thermal Pollution, F.L.

Parker and P.A. Krenkel, Vanderbilt University Press, Nashville, Tenn., 1969

241. A.R. Thompson, Chem. Eng. *75*, 100, Oct. 14, 1968

242. D.B. Leason, Atmos. Envir. (8), 307 (1974)

243. P. Bienfang, New Scientist and Science J., 456, Aug. 26, 1971

244. P.T. Doyle and G.J. Benkly, Hydroc. Proc. *52* (7), 81, July 1973

245. K.V. Shipes, Hydroc. Proc. *53* (5), 147, May 1974

246. Pollution Solutions, Chemical Week *108* (13), 25, March 31, 1971

247. E.L. Barnhart, Chem. Eng. Prog., Symp. Series No. 90, *64*, 111 (1968)

248. Treatment of Wet Milling Wastes, Chem. In Can. *28* (6), 8, June 1976

249. New Sarnia Bio-ox Plant Treats 8 % Salt Stream, Civic *28* (3), 28, March 1976

250. P.B. Dejohn and R.A. Hutchins, Textile Chemist and Colorist *8* (4), 34, April 1976

251. T.J. Chessell, Can. Chem. Proc. *62* (6), 42, June 1978

252. T.W. Wett, Chemical Processing *33* (10), 58, Oct. 1970

253. R.F. Curran, Chem. Eng. Prog., Symp. Series No. 90, *64*, 162 (1968)

254. G.E. Eden, B.L. Hampson and A.B. Wheatland, J. Soc. Chem. Ind. (London) *69*, 244 (1950)

255. Treatment of Cyanide Waste Liquors by Alkaline Chlorination, Technical Service Report 14.55 R, Allied Chemical, Syracuse, New York, Aug. 1961

256. G.R. Mapstone and B.R. Thorne, J. Appl. Chem. Biotechnol. *28* (2), 135, Feb. 1978

257. A.B. Wheatland, C. Gledhill, and J.V. O'Gorman, Chem. Ind. (London), 632, Aug. 2, 1975

258. Acid Complexes Recover Effluent Metals, Chem. Eng. News *48* (24), 84, June 8, 1970

259. Peatmoss Traps Heavy Metals, Can. Chem. Proc. *60* (2), 12, Feb. 1976

260. R.W. Newkirk and P.J. Schroeder, Hydroc. Proc. *51* (10), 103, Oct. 1972

261. Pulp Mill Uses Refinery Waste, Chem. In Can. *25* (4), 13, April 1973

262. Denmark Opens First Nationwide System to Recover, Dispose of Chemical Wastes, Chemecology, 7, June 1975

263. L.J. Ricci, Chem. Eng. *83* (14), 44, July 5, 1976

264. Waste Exchange Achieves Initial Target, Can. Chem. Proc. *63* (2), 12, Feb. 1979

265. Waste Exchange Receives Ottawa Support, Can. Chem. Proc. *62* (1), 22, Jan. 1978

266. Waste Exchangers, Focus on Great Lakes Water Quality (Windsor) *7* (3), 11, Fall 1981

267. Canadian Recycling Market, biweekly published by Corpus, Don Mills, Ontario

268. More on Waste Exchanges, Focus on Great Lakes Water Quality (Windsor), *8* (2), 16, April 1982

269. R. Anliker and E.A. Clarke, Chem. In Brit. *18* (11), 796, Nov. 1982

270. A.C. Faatz, Water and Sewage Works *116*, R 197, Nov. 28, 1969

271. Treatment Plant Near Montreal, Water and Polln Control *110* (12), 48, Dec. 1972

272. A.K. Coleman, Chem. and Ind. (London), 534, July 5, 1975

273. New Jersey Plant to Treat Industrial Wastes, Chem. Eng. News *56* (10), 10, March 6, 1978

274. New Twist for Tricil System, Can. Chem. Proc. *66* (2), 10, March 1982

275. D.C. Wilson, Chem. In Brit. *18* (10), 720, Oct. 1982

276. Cement Kilns Can Destroy Toxic Chlorinated Compounds, Can. Chem. Proc. *61* (5), 6, May 1977

277. W. Worthy, Chem. Eng. News *60* (10), 10, March 8, 1982

278. Shipboard PCB Destruction is Successful, Chem. Eng. News *61* (22), 24, May 30, 1983

279. Waste Solidification, Chem. In Brit. *10* (3), 105, March 1974

280. IU Conversion Systems, Inc., Philadelphia, Pa.

281. B. Orchard, Can. Chem. Proc. *65* (1), 29, Feb. 1981

282. W.J. Green, G.F. Lee, R.A. Jones, J. Water Polln Control Fed. *53* (8), 1,348, Aug. 1981

283. J.S. Talbot, Chem. Eng. *75*, 108, Oct. 14, 1968

284. R.O. Van Everdingen, and R.A. Freeze, Subsurface Disposal of Waste in Canada, Department of the Environment, Technical Bulletin No. 49, Ottawa, 1971

285. R.K. Ballentine, S.R. Reznek, C.W. Hall, Subsurface Pollution Problems in the U.S., U.S Environmental Protection Agency, Report No. TS-00-72-02, Washington, May 1972

286. D.M. Evans and A. Bradford, Environment *11* (8), 3, Oct. 1969

287. Quality Assurance for Groundwater, Envir. Science and Tech. *10* (3), 226, March 1976

288. B.H. Keswick and C.P. Gerba, Envir. Science and Tech. *14* (11), 1,290, Nov. 1980

289. A.E. Zanoni, CRC Critical Reviews in Envir. Control *3* (3), 225, May 1973

290. J.R. McBride, E.M. Donaldson, and G. Derkson, Bull. Envir. Contam. Toxiol. *23*, 806 (1979)

291. Waste Sites Pose Risk to Water Supply, Chem. Eng. News *58* (40), 6, Oct. 6, 1980

292. Handbook of Chemistry and Physics, 39th edition, C.D. Hodgman, editor, Chemical Rubber Publishing Company, Cleveland, Ohio, 1957, page 1,608

293. C.F. Mason, Biology of Freshwater Pollution, Longman, London, 1981

294. Canadian Drinking Water Standards and Objectives 1968, The Joint Committee on Drinking Water Standards, The Advisory Committee on Public

Health Engineering, and The Canadian Public Health Assoc., Ottawa, 1969

295. D. McBain, Chem. In Brit. *12* (9), 281, Sept. 1976

296. R. Field and C.-Y. Fan, J. Envir. Eng. Div., Amer. Soc. Civ. Engineers, *107*, 171, Feb. 1981

297. W.B. Langbein and H.W. Durum, U.S. Geological Survey, Circular No. 542, U.S. Dept. of the Interior, Washington, 1967

298. R.K. Linsley, Jr., M.A. Kohler, and J.L.H. Paulhus, Applied Hydrology, McGraw-Hill, New York, 1949

299. Discharge of Selected Rivers of Canada, The Secretariat, Canadian National Committee for the International Hydrological Decade, Ottawa, 1972

300. L.W. Weinberger, D.G. Stephan, and F.M. Middleton, Annals of the New York Acad. of Sciences *136*, 131, July 8, 1966

301. R. Davies, Water and Polln Control *109* (8), 32, Aug. 1971

302. YJ.D. Hem, Study and Interpretation of the Chemical Characteristics of Natural Water, U.S. Geological Survey Water Supply Paper 1,473, Washington, 1959

303. Surface Water Quality in Canada — An Overview, Inland Waters Directorate, Water Quality Branch, Ottawa, 1977

304. 6 Bottle Manometric Apparatus, Model 2,173 B, 3rd edition, Hach Company, Ames, Iowa, 1982

305. L.T. Pryde, Environmental Chemistry, An Introduction, Cummings Publishing Co., Menlo Park, California, 1973

306. Introduction to Popular Treatment Methods for Municipal Wastes and Water Supplies, Ontario Water Resources Commission, Toronto, ca. 1972

# 4 Natural and Derived Sodium and Potassium Salts

## 4.1 Sodium Chloride

Sodium chloride, or common salt, is among the earliest chemical commodities produced by man, prompted by its essential requirement in the diet and the scattered accessibility of available land-based supplies. The word salary itself is derived from the Roman "salarium", which was a monetary payment given to soldiers for salt purchase to replace the original salt issue. While the initial production and harvesting of sodium chloride was from dietary interests, today this application represents less than 5 % of the consumption, and uses as a chemical intermediate far exceed this (Table 4.1). The wide availability of sodium chloride has contributed to the derivation of nearly all compounds containing sodium or chlorine from this salt, and to the establishment of many large industrial chemical operations adjacent to major salt deposits. Three general methods are in common use for the recovery of sodium chloride, which in combination were employed for the world-wide production of 167 million tonnes of this commodity in 1976 and slightly less in 1981 (Table 4.2).

**Table 4.2** Major world salt producers [a]

| | Thousands of tonnes | | |
|---|---|---|---|
| | 1976 | 1980 [b] | 1981 [b] |
| U.S.A. | 40 114 | 36 607 | 35 800 |
| China | 30 000 [b] | 17 280 | 18 100 |
| U.S.S.R. | 14 000 [b] | 14 500 | 14 500 |
| United Kingdom | 7 900 | 6 586 | 6 600 |
| West Germany | 7 495 | 12 970 | 13 200 |
| France | 6 416 | 7 100 | 7 284 |
| Canada | 6 398 | 7 029 | 7 100 |
| Poland | 5 470 | 3 357 | 3 300 |
| Australia | 5 350 | 5 315 | 5 300 |
| Mexico | 4 591 | 5 990 | 6 900 |
| India | 4 480 | 7 262 | 7 300 |
| Italy | 4 012 | 5 270 | 5 400 |
| Other Countries | 31 199 | 35 488 | 34 200 |
| Total | 167 425 | 164 754 | 163 984 |

[a] From Canadian and U.S. Minerals Yearbooks [2, 49].
[b] Estimates, or preliminary figures.

**Table 4.1** Sodium chloride use profile for Canada and U.S.A., 1976 [a]

| Application | U.S.A. | | Canada | |
|---|---|---|---|---|
| | thousand tonnes | % of total | thousand tonnes | % of total |
| Highway snow and ice control | 8 101 | 20.1 | 2 224 | 49.3 |
| Chlorine and sodium hydroxide production | 19 003 | 47.2 | 1 348 | 39.9 |
| Sodium carbonate (soda ash) | 3 684 | 9.1 | 616 | 13.6 |
| All other chemicals | 1 742 | 4.3 | | |
| Fishing industry | — | — | 97 | 2.1 |
| Food processing | 1 080 | 2.7 | 91 | 2.0 |
| livestock feeds | 1 746 | 4.3 | 47 | 1.0 |
| Meatpacking, tanning | 524 | 1.3 | 49 | 1.1 |
| Pulp and paper | 193 | 0.5 | 38 | 0.8 |
| Table salt | 905 | 2.2 | | |
| Textile and dyeing | 185 | 0.5 | 3 | 0.1 |
| Miscellaneous uses | 3 146 | 7.8 | 3 | 0.1 |
| Total | 40 311 | 100.0 | 4 516 | 100.0 |

[a] Calculated from data in Minerals Yearbooks [1, 2], and the Statistical Yearbook [51].

## 4.1.1 Solar Salt

Outdoor recovery of sodium chloride by evaporation of sea water or natural brines is carried out in areas with a high evaporation rate and low rainfall. Because of the net evaporative rate restriction on this process solar salt represents only a small fraction of the production of temperate countries but the low capital, energy, and labour costs still makes this a dominant and attractive method for world salt recovery. For instance the fraction of the total world salt which is produced by solar evaporation is about 45 %. Some examples of the range of respective contributions to this total by various countries are: Canada, 0 %, U.S.A 4 %; West Germany, 11 %; Italy, 15 %; France, 22 %; Spain, 41 %; Portugal, 42 %; Ethiopia, 88 %; and Namibia (South-West Africa), 100 % [3].

Feed brines for solar salt recovery may be derived from sea water of salinity about 3.5 % or from one of the enclosed seas of natural salt lakes of higher salinities (Table 4.3). The Caspian Sea, salts of which are about 25 % lower in chloride and about 3 times higher in sulfate than ordinary sea salt [3], is not used for solar salt production because of its much lower salinity (1.3 %) and its temperate location. The brine is moved, with tidal assistance if feasible, into large shallow initial evaporation ponds where solar evaporation is allowed to proceed until it reaches a specific gravity of about 1.21 g/cm. It is then pumped to lime ponds where further evaporation, sometimes with the absorptive assistance of added dyes, causes crystallization and precipitation of most of the calcium sulfate (phase diagrams, reference 8). When the brine density reaches about 1.24, gates are opened to allow gravity flow of the brine to the harvesting or crystallizing ponds where sodium chloride crystallization is allowed to proceed until the brine density reaches 1.25 to 1.29, when some 75 % of the sodium chloride will have crystallized. The mother liquor, or bittern, is then run off, and the residual crystallized salt scraped into long piles or windrows to drain. This product is frequently marketed in moist form (8 to 10 % water) [10]. Simple drying, either by natural of artificial means, produces an industrial grade solar salt consisting of about 95 % sodium chloride. Yields range from about 50 tonnes per hectare per year (or about 20 tonnes per acre) in the San Francisco Bay area [11], 150 to 175 tonnes per hectare per year for the May to September season from the Great Salt Lake, and 12 to 185 or more from seawater for areas with higher evaporation rates [13].

To obtain a food or dairy grade of product the solar salt is redissolved and the solution evaporated to induce crystallization, which greatly reduces the calcium sulfate concentration. Pure salt is hygroscopic and will tend to cake with changes in relative humidity. To avoid this, salt crystals are coated with 0.5 to 2 % of finely powered hydrated calcium silicate, magnesium carbonate, or tricalcium phosphate to give the crystals free-flowing characteristics. Iodized salt for table use would have, in addition, 0.01 % potassium iodide plus stabilizers added to the product. Cattle blocks comprise highly densified dyed aggregations of salt, to which various micronutrients are added according to local soil deficiencies of the marketing area.

The bittern from solar salt production containing 300-400 g/L dissolved solids, and now relatively enriched in the less concentrated salts, may be either discarded or further worked to recover other elements of value. Brine from the Great Salt Lake, for instance, is processed for magnesium chloride hexahydrate recovery [9] for later conversion to metallic magnesium [10], and Dead Sea brines, which are processed primarily for potassium chloride (potash) are also worked for sodium and magnesium chlorides and derived products such as bromine and hydrochloric acid [7] (Sections 4.2.2 and 6.6).

Table 4.4 Properties of sodium chloride and
          solutions

| | |
|---|---|
| Density | 2.165 g/cm$^3$ |
| Melting point | 800.8 °C |
| Boiling point | 1465 °C |
| Solubility: | 35.7 g per 100 g water at 0 °C |
| | 39.8 g per 100 g water at 100 °C |

Saturated sodium chlorine brine has a boiling point of 108.7 °C, and contains 28.41% NaCl. At 25 °C it has a specific gravity of 1.1978, and contains 26.48% NaCl.

## 4.1.2 Sodium Chloride by Conventional Mining

By the evaporation in geologic history of large inland seas or land-locked lakes of significant drainage basins extensive subterranean deposits of sodium chloride and other salts in layers of cumulative thickness as great as 400 metres, have

**Table 4.3** Salinities and compositions, in percent by weight of dissolved cells for some brine sources of solar salt.

| | Sea Water[a] | Red Sea[b] | | Searles Lake, Calif.[c] | | Dead Sea[a,d] | | Great Salt[e] Lake, Utah |
|---|---|---|---|---|---|---|---|---|
| | | Surface | Below 2000 m | Upper | Below ~40 m | Upper | Below 100 m | |
| **Salinity: %** | | | | | | | | |
| average | 3.0 | 3.8 | 25.6 | 34.83 | 34.60 | – | 26.7 | 25.0 |
| variation | 1 to 5 | 3.6–4.1 | – | – | – | 19.5–25 | – | 13.8–27.7 |
| | place, depth | depth | | depth | | time, depth | | |
| Density g/mL | 1.025 | – | – | 1.30 | 1.31 | 1.17 | 1.24 | 1.10–1.22 |
| NaCl | 77.8 | – | 89.1 | 46.66 | 46.97 | 31.3 | 40.2 | 84.4 |
| MgCl$_2$ | 10.9 | – | – | – | – | 48.4 | 51.5 | 3.8 |
| MgSO$_4$ | 4.7 | – | trace | – | – | – | – | 8.4 |
| CaSO$_4$ | 3.6 | – | trace | – | – | 0.56 | 0.02 | – |
| KCl | 2.5 | – | – | 13.92 | 8.67 | 4.35 | 4.69 | 3.1 |
| CaCO$_3$ | 0.34 | – | 1.6 | – | – | – | – | 0.3 |
| MgBr$_2$ | 0.22 | – | 2.4 | – | – | 1.94 | 2.29 | – |
| CaCl$_2$ | – | – | 5.3 | – | – | 13.50 | 14.33 | – |
| Na$_2$SO$_4$ | – | – | – | 20.67 | 19.51 | – | – | – |
| Na$_2$CO$_3$ | – | – | – | 13.35 | 18.35 | – | – | – |
| Other | – | – | trace[f] | 1.48[g] | 1.52[g] | – | – | traces[h] |
| Total salts | $8.44 \times 10^{16}$[i] | – | – | – | – | $1.15 \times 10^{10}$[j] | – | $5.44 \times 10^9$ |

[a] Calculated from data in Encyclopaedia Brittannica [3].

[b] Calculated from data of Degens and Ross [4].

[c] Data from Shreve and Brink [5]. Composition of the nearly dry Searles Lake salts is very similar to that of nearby Owens Lake, from which it was separated by evaporation [6].

[d] Calculated data [3] agree quite closely with the analyses reported by Epstein.

[e] Calculated from data [6, 8 and 9]. Traces of magnesium carbonate and other more minor constituents are also present.

[f] Predominantly of other sulfates.

[g] Includes predominantly Na$_2$B$_4$O$_7$, Na$_3$PO$_4$, and NaBr.

[h] Includes ionic strontium and borate.

[i] Corresponds to a volume of $1.87 \times 10^7$ km$^3$.

[j] It estimated that a total of some 910 000 tonnes of salts per year is entering via the Jordan River.

been laid down in many parts of the world. Deposits less than 500 to 600 metres below the surface, in well consolidated strata, are usually economic to work by conventional mining techniques, in Canada amounting to about 74 % of the total [2], and in the U.S. about 35 % of the total [1].

The extent and quality of the deposit is determined from a pattern of core drillings (samples of the subterranean rock brought to the surface) obtained for the specific area of interest. Access to the horizontal, or near horizontally bedded salt is obtained by sinking a vertical shaft of about 5 metres diameter and then by undercutting, drilling, blasting and underground haulage the rock salt is brought to the shaft and lifted to the surface. Because deposits of this type are quite dense, underground cavities can range up to 15 metres in each dimension but they require stout supporting pillars of 20 to 60 metres square at regular intervals because of the plastic nature of sodium chloride (flows under pressure). In this way salt recoveries from the conventional mining operation may range from 25 to 40 % and rarely as high as 60 % of that present in the deposit. Crushing and screening to size may either be carried out underground or at the surface. Mechanical and electronic sorting then gives an unrefined but beneficiated product of better than 95 % NaCl (occasionally better than 99 %). Most of the Canadian and American mined rock salt is employed as mined in highway de-icing, although a further significant fraction of the American rock salt is consumed in the production of chlorine, sodium hydroxide, and other chemicals (Table 4.1). For the small quantities of higher purity sodium chloride made from this raw material the beneficiated material is dissolved in water, concentrated and crystallized. The purified sodium chloride crystals are filtered and dried. Further finishing steps for marketing are as described in Section 4.1.1.

### 4.1.3 Solution Mining of Sodium Chloride

To recover salt from underground deposits located more than about 600 metres below the surface, or with less well consolidated intervening strata, solution mining is usually the method of choice, and provides the dominant source of brine for Canadian chloralkali operations [2]. This may be accomplished, for low production rates, by sinking a single drilled well and fitting this with a casing and

a smaller bore tube, usually of fibre-glass reinforced plastic to avoid corrosion problems, and pumping water down one to displace brine up the other. For larger rates of production two (or more) bore holes will normally be sunk into the salt formation, and water will be pumped at high rate and pressure into one to introduce an expansion fracture connecting the two wells (= "fracturing") and allowing water passage and hence brine recovery through this crack.

Fields of wells in Canada recover brine from deposits as deep as 2,000 metres, and may have as many as 20 wells for saturated brine production from a single deposit [2]. At intervals, new wells will be bored some distance removed from both the initial solution mining site and surface producing facilities to leave intervening support columns underground and reduce the risk of subsidence.

Saturated brine emerging from the underground cavity will contain 26 % by weight sodium chloride, small amounts (0.01 to 0.1 %) of calcium and magnesium salts and possibly traces of ferrous and sulfide ions. Insoluble matter will have largely remained underground. This brine will be suitable as obtained or after some specific chemical treatment (see Section 6.1) for many process uses directly requiring a sodium chloride solution.

To obtain a high purity crystalline salt from this solution it is aerated to remove most of the sulfide as hydrogen sulfide and then chlorinated to oxidize any residual sulfide and ferrous ions.

$$Cl_2 + H_2O \rightarrow HCl + HClO \qquad (4.1)$$

$$HCl + HClO + 2\,FeCl_2 \rightarrow H_2O + 2\,FeCl_3 \quad (4.2)$$

Calcium, magnesium, and ferric ions are then removed as their hydroxides by treating the brine with small amounts of sodium carbonate (soda ash) and sodium hydroxide (caustic soda) and filtering or allowing the mixture to settle [11].

$$CaCl_2 + Na_2CO_3 \rightarrow CaCO_3 \downarrow + 2\,NaCl \quad (4.3)$$

$$MgCO_3 + 2\,NaOH \rightarrow Mg(OH)_2 \downarrow + Na_2CO_3 \qquad (4.4)$$

$$FeCl_3 + 3\,NaOH \rightarrow Fe(OH)_3 \downarrow + 3\,NaCl \quad (4.5)$$

Water is then removed from the treated brine by multiple-effect vacuum evaporation with low pressure steam providing the heat for the first effect (Figure 4.1).

To evaporate water from brine by heating with steam one can only obtain somewhat less water evaporation than the mass of steam consumed (a heat transfer efficiency limitation). However, by applying a partial vacuum to the heated brine in the first stage, and using the hot water vapour from this stage to heat another stage and so on, allows far more efficient use of steam and the evaporation of progressively more water for each tonne of primary steam used as the number of stages is increased (Figure 4.2). Further thermal efficiencies are introduced by countercurrent movement of

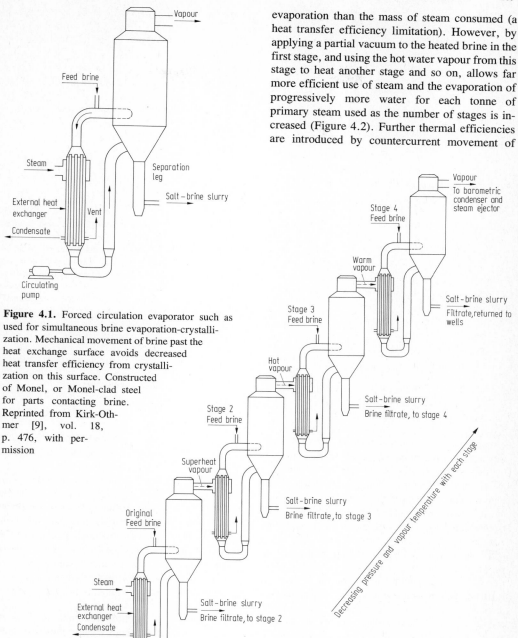

**Figure 4.1.** Forced circulation evaporator such as used for simultaneous brine evaporation-crystallization. Mechanical movement of brine past the heat exchange surface avoids decreased heat transfer efficiency from crystallization on this surface. Constructed of Monel, or Monel-clad steel for parts contacting brine. Reprinted from Kirk-Othmer [9], vol. 18, p. 476, with permission

**Figure 4.2.** Quadruple effect evaporator-crystallizer. Pressure and boiling temperature decrease from left to right, temperature about 20 to 30 °C across the four effects. Feed brines to each effect may be directly from brine wells, or may be fed cascade fashion from the filtrate to the salt brine slurry of the previous effect, or a combination of these measures. Adapted from Kirk-Othmer [9]

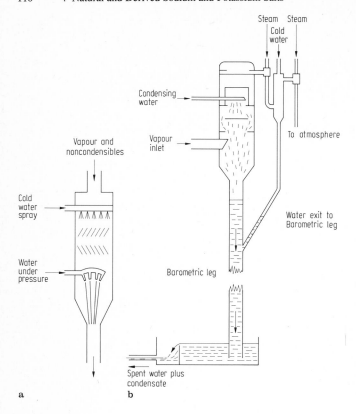

**Figure 4.3. a** Detail of water ejector using a supply presure of 1.4-1.7 40 $10^2$ pascals (gauge pressure; 20 to 25 psig) or higher. **b** Diagram of barometric condenser plus steam driven ejector to remove and condense vapour and provide the reduced pressures required for the last stage of evaporation. With steam pressures of 2.8-3.5 × $10^2$ pascals (gauge pressure; 40 to 50 psig) or higher and six stages of ejection absolute pressures in the range 0.005 to 0.50 mm mercury are available [50]

brine and steam (Chapter 1) and by use of the still warm exit brine to prewarm incoming brine. In combination, these measures can give evaporation efficiencies of the order of 1.75, 2.5, 3.2, and 4.0 tonnes of water per tonne of steam from 2, 3, 4, and 5 stages of evaporation, respectively [14]. Capital costs increase with the number of stages, and the percent improvement obtained with each additional effect decreases so that the maximum practical number [14] of effects is 12. Partial vacuum for the last effect (stage of evaporation) is economically provided without moving mechanical devices by the use of cooling water with a barometric condenser, and a steam or water ejector for removal of non-condensible gases (Figure 4.3).

A stream of a saturated brine suspension of crystals is continuously withdrawn from each evaporator-crystallizer, and the salt crystals separated on a continuous rotary filter with return of the brine to the evaporator. Much of the salt may be marketed in moist condition, or it may be passed through a

drier countercurrently to heated air to give "vacuum salt" of typically 99.8 to 99.9 % purity [11, 15].

There must be some net throughput of brine to waste through the evaporator circuit to avoid gradual build-up of impurities. In some operations, in fact, when the feed brine is of consistently high purity the bleed rate of brine mother liquor can be controlled at an appropriate level by continuously keeping track of calcium and magnesium concentrations, to give a finished sodium chloride purity of 99.5 % or better, without requiring brine pretreatment [11].

The crystal size of vacuum salt is consistently quite small because of the high evaporation rates and turbulence in the evaporator crystallizers, which makes this product less suitable for some uses. For this reason there is still a small segment (ca. 1 %) of the brine evaporation business conducted at atmospheric pressure in open pans, the "Grainer" process. Here, after special treatment to remove calcium sulfate, the brine evaporation takes place in a quiescent manner at 95° giving large crystals of high surface area and purity suitable for the dairy industry [11]. This process is, however, more extravagant of heat than the vacuum process. Alternatively, for the large dense crystals required for the regeneration of ion-exchange water softeners, quantities of vacuum salt may simply be heated to about 815° and the melt cooled and broken up for this purpose.

### 4.1.4 New Developments in Sodium Chloride Recovery

As pressure on existing freshwater supplies tightens, sea water desalination plants, by multi-effect vacuum distillation or reverse osmosis, are going to be required in increasing numbers. The residual evaporated brines from these plants contain much higher salt concentrations than ordinary sea water and are also obtained near potential salt markets. These features are likely to encourage further development of vacuum or solar evaporation salt recovery operations to work these brines, in close proximity to the desalination plants. In this way various salts may be more profitably recovered from these artificially enriched salinity sea waters for reasons similar to the present incentives to use rich natural brine sources for sodium chloride production, netting similar energy savings.

## 4.2 Potassium Chloride

Potash, as potassium chloride and as indeed all potassium salts are collectively referred to in industry, is the name derived from the methods used for original recovery of potassium carbonate from wood ashes. About one tonne of crude potassium carbonate was recovered from the water leachate of the ash generated from the burning of some 400 tonnes (more than 200 cords) of hardwood [16]. Perhaps in part because of the large wood requirement, Canada was the world's largest potash exporter during much of the first half of the 19th century.

At this time purified potassium carbonate was primarily valued as an ingredient in glass-making. The discovery of mineral potassium chloride at Stassfurt, Germany in 1852, rapidly provided competition to the wood ash leachate industry since this mineral could be readily converted to potassium carbonate by the Leblanc process which was al-

ready in commercial scale operation (see Chapter 5). The mineral potash is also more specifically referred to as "muriate of potash" (muriatic acid = hydrochloric acid) or "sulfate of potash" when it is desired to limit the term to these particular salts.

### 4.2.1 Potassium Chloride Production and Use Pattern

Prior to 1960, world production of potassium chloride was dominated by the U.S.S.R., U.S.A., East and West Germany and France (Table 4.5). But since the incidental discovery of potash mineralization in Saskatchewan during oil prospecting in 1943, and the first commercial production there in 1962, Canada became the world's leading producer in 1968 and 1969. Canada is at present the leading exporter and stands second after the U.S.S.R. in potash production [17]. World production volume is presently doubling in something just over 10 years, from 9.82 million tonnes of $K_2O$ equivalent in 1961 to 19.1 million tonnes in 1971, and about 25.8 million tonnes in 1976 [2, 17]. At 1976 levels of production, present Canadian reserves of 107 x $10^9$ tonnes of potassium chloride (67 x $10^9$ tonnes, $K_2O$ equivalent) primarily in sylvinite and carnallite minerals, would be sufficient to supply the world demand for some 2600 years [17]. West German and Russion reserves too appear to be of a similar order of magnitude [13, 18].

Table 4.5 Major world producers of potash, 1976 [a]

|  | Potash produced, as $K_2O$ equivalent [b] | |
|  | thousand tonnes | % of total |
| --- | --- | --- |
| Canada | 5 000 | 19.4 |
| East Germany | 3 175 | 12.3 |
| France | 1 606 | 6.2 |
| Israel | 680 | 2.6 |
| China | 490 | 1.9 |
| Spain | 530 | 2.1 |
| U.S.A. | 4 010 | 15.6 |
| U.S.S.R. | 8 260 | 32.0 |
| West Germany | 2 032 | 7.9 |
| World Total | 25 783 | 100.0 |

[a] Data from 1976 Minerals Yearbook [1].
[b] Dominant product is potassium chloride, e.g. for Canada 99%, but small amounts of potassium sulfate and potassium nitrate are also produced and are included in these figures.

Some 95 % of the potassium chloride is consumed directly in fertilizer applications. Most of the rest is used to make potassium hydroxide (see Chapter 6) which is ultimately employed to manufacture some types of glass, and also for the saponification of fats to produce liquid soaps, among other small scale uses.

## 4.2.2 Potassium Chloride Recovery from Natural Brines

While it would seem feasible to use sea water bitterns from solar salt facilities (Section 4.1.1) as the raw material for potassium chloride recovery, there are as yet no commercial processes operating on this basis [19]. Virtually all present production facilities use natural brines from sources already relatively richer in potassium than raw seawater (Table 4.3), and all have to cope somehow with the solubility differentials with temperature (Table 4.6) In general it is expected that mixtures of salts, such as in seawater, will tend to come out of solution in the reverse order of their solubilities, though this will also be strongly affected by their initial relative concentrations. Thus, a precipitation sequence something like calcium carbonate > gypsum ($CaSO_4$ x $2 H_2O$) > sodium chloride > magnesium sulfate ($MgSO_4$ x $6 H_2O$) > magnesium chloride ($MgCl_2$ x $6 H_2O$) > potassium chloride > sodium bromide would normally be expected. For this reason recovery of the less concentrated minerals, particularly if they are more soluble than sodium chloride, requires progressively more complex processing than required for sodium chloride recovery.

Searles Lake, California, is a remnant of a former salt lake which is now worked for minerals and consists primarily of beds of salts containing brine in the interstices, the brine lying from 15 cm below

to 15 cm above the beds of salts depending on season and extent of flooding [20]. By comparison with seawater these brines contain far higher concentrations of sodium and potassium ions, but much of it as sulfate not chloride, and negligible calcium and magnesium (Table 4.3). Initial triple-effect evaporation allows separation of mainly crystals of sodium chloride and Burkeite ($Na_2CO_3$ x $2 Na_2SO_4$) with a small amount of dilithium sodium phosphate. With a complex series of subsequent steps, sodium sulfate (Glauber's salt) and sodium carbonate (soda ash) are recovered, and the lithium salt is converted to lithium carbonate and phosphoric acid for recovery of both components [4, 21]. Brines from the initial evaporation are then rapidly chilled (by cooling water and vacuum) to bring down potassium chloride crystals on a scale of some 675 tonnes per day, and leaving a supercooled solution of sodium borate. Seed crystals and more prolonged crystallization time produces crystalline sodium borate as a further product. Full details have been described [4, 13].

Solar evaporation, from primary and secondary ponds of $100 km^2$ and $30 km^2$ in extent, are the initial stages for potassium chloride recovery from Dead Sea brines [19]. The smaller number of constituent ions present in these waters significantly simplifies salts recovery, and the fact that they contain nearly twice the relative potassium chloride concentration of sea water also improves profitability. Developed from a process which was first operated in 1931, evaporation in the first pond reduces the volume to about one half of the initial volume and brings down much of the sodium chloride together with a small amount of calcium sulfate (Figure 4.4). The concentrated brines are then transferred to the secondary pond where evaporation of a further 20 % of the water causes carnallite ($KCl \cdot MgCl_2 \cdot 6 H_2O$) and some further sodium chloride to crystallize out. With care a 95 % potassium chloride product on a scale of some 910,000 tonnes per year is obtained either by counter-current extraction of the carnallite with brines, or by hot extraction of potassium chloride from the sylvinite matrix followed by fractional crystallization for its eventual recovery [22].

The "end brines" from the secondary evaporation pond, consisting largely of concentrated calcium and magnesium chlorides, are mostly returned to the sea. However, some 73,000 tonnes of sodium chloride, 900 tonnes of magnesium chloride and with some chemical conversion 23,000 tonnes of

**Table 4.6** Aqueous solubility of potassium sulfate with temperature [a]

| Temperature °C | Solubility g/100 mL |
|---|---|
| 0 | 7.35 |
| 20 | 11.11 |
| 40 | 14.76 |
| 60 | 18.17 |
| 80 | 21.40 |
| 100 | 24.10 |

[a] Data from Chemical Engineers' Handbook [43].

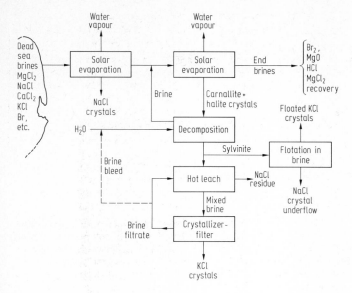

**Figure 4.4.** Flow sheet for potassium chloride recovery from Dead Sea brines [19]

magnesia (MgO), 91,000 tonnes of hydrochloric acid and 18,000 tonnes of bromine are also obtained annually [21] from these brines. Magnesium chloride is obtained from further processing of the sodium chloride residual bittern [9].

Potassium is also recovered from the Great Salt Lake, but mainly as the sulfate [10].

### 4.2.3 Potassium Chloride by Conventional Mining and Froth Flotation

Dominant minerals which occur in potash-bearing deposits are halite (NaCl), sylvite (KCl), and carnallite ($KCl \cdot MgCl_2 \cdot 6H_2O$), in which the two salts appear as separate crystals ranging in size from a few $\mu$m (microns) to 25 mm. Halite and sylvite minerals also occur together as physical mixtures known as sylvinite ($nNaCl \cdot KCl$; n normally about 2). Sylvite and carnallite occur both separately and in closely associated mixtures with one another. All of these potash bearing combinations also contain varying amounts of insolubles such as clays, anhydrite ($CaSO_4$), dolomite ($CaCO_3 \cdot MgCO_3$) and calcite ($CaCO_3$) which variously affect potash benefication steps.

Recovery of sylvinite from the shallower deposits of about 1,000 m depth at the northern margin of the sloping 30 m thick deposits in Saskatchewan, Canada, is entirely by the shaft and room method. One natural obstacle to overcome was an under-

ground 60 m thick bed of poorly consolidated water saturated sand (Blairmore, and other formations) lying below the 380 m level which introduced considerable accessing difficulty [24]. This was eventually solved by prolonged chilling of these formations with lithium chloride brine at −46 °C after which the frozen formation was then safely penetrated by rapid excavation and placement of water tight caissons through the unstable formation. Apart from the accessing difficulties for this particular deposit, conventional mining of sylvinite is conducted in a parallel manner to the mining of salt (Section 4.1.2).

The high grade of sylvinite obtained from deposits in Saskatchewan (about 40 % KCl; n about 2) favoured the application of the differential surface activity between the finely ground potassium and sodium chlorides present to effect a separation of the two minerals, rather than the considerably more energy intensive solution fractional crystallization technology [11]. This process marks an unusual application of froth flotation technology, commonly applied to the concentration of metal ores (Chapter 11), and here employed in the beneficiation of a water soluble mineral. Mined sylvinite lumps are first dry crushed in gyratory or roll crushers and then ground to a fine pulp in saturated brine in a ball mill (Figure 4.5). After classification in a cyclone, the coarse material is returned for regrinding and the fine pulp is deslimed

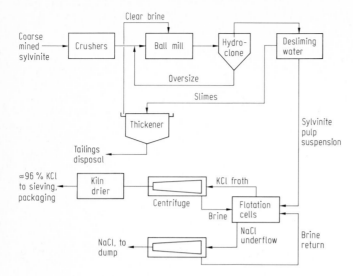

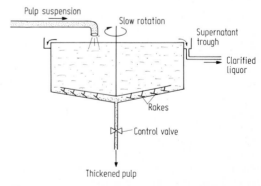

**Figure 4.5.** Flow diagram of the separation of fertilizer grade potassium chloride (sylvite) from sylvinite by froth flotation in saturated brine

(fine clays removed) by rigorous agitation with brine and removal of the clay laden brine fraction. Brine is recycled after clay removal in thickeners (Figure 4.6). The cleaned pulp is then treated with guar gum, a suppressant which coats any remaining clay to prevent adsorption of the amine collector by clays. For each tonne of ore processed, 100 g of a tallow amine (long alkyl chain, primary amine) is added to induce attachment of the potassium chloride particles to air bubbles (does not affect sodium chloride), plus 110 g of a polyethylene glycol to encourage formation of a stable froth [25]. An anionic (fatty acid salt) collector may be used if it is desired to float sodium chloride and give a potassium chloride underflow [18].

Froth flotation of the surface-sensitized pulp by vigorous aeration and agitation in saturated brine (density ca. 1.18 g/mL), first in a series of "rougher" cells with further refinement in "cleaner" units produces a stable froth consisting almost entirely of beneficiated potassium chloride and an underflow of sodium chloride (particle densities of 1.984 and 2.165 g/mL respectively). Potassium chloride recoveries from sylvinite by this procedure are 90 to 95 % [18].

Flotation streams are individually centrifuged (Figure 4.7) for separate recovery of the potassium chloride and sodium chloride and the brine is returned, usually via a thickener, to the flotation circuit for re-use. At present the sodium chloride is simply dumped. Potassium chloride is dried in gas-fired rotary kilns, and then sized by sieving into one of the four commercial grades: Granular, Coarse, Standard, and Soluble, all of which fall into the range of 60.2 to 60.7 % $K_2O$ equivalent purities. Potash dust from the drying operation is captured, dissolved in hot brine, and chilled under vacuum to cause purified potassium chloride crystals to form. After filtration and drying, this produces a further Soluble grade of 62.5 % $K_2O$ purity. Re-solution of this material, and crystallization from water again gives Refined, 99.9 % pure

**Figure 4.6.** Diagram of a single deck mechanical thickener for producing a clear supernatant and a solids enriched slurry from a pulp suspension in water. Slow moving rakes transport sedimented solids towards thickened pulp outlet and an exit valve for this stream allows control of the solid/liquid ratio

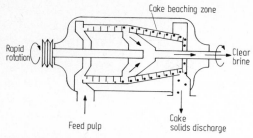

Cake beaching zone

Rapid rotation

Clear brine

Feed pulp

Cake solids discharge

**Figure 4.7.** Solid bowl continuous centrifuge for separation of a solid suspension from a liquid. With sodium chloride or potassium chloride slurries in brine it is capable of producing a dry cake of 92 to 99 % solids

potassium chloride, mostly used in chemical (not fertilizer) applications.

The bulk of the fertilizer market demand is for the two coarse grades, necessitating some compaction of the finer grades to produce these. Passage of the finer Standard and Soluble grades of material through closely-spaced steel rollers under pressure, forms a fused sheet of potassium chloride, because of the plastic nature of this salt. This fused sheet is then broken up and re-screened to raise the proportion of Granular and Coarse products produced [17].

Potassium chloride as a component in fertilizers is specified in terms of its "$K_2O$ equivalent". On this basis 100 % pure potassium chloride is equated to 63.18 % $K_2O$ equivalent. The approximately 96 % pure potassium chloride product of the flotation separation thus is equivalent to 60.7 % $K_2O$ (see Chapter 9).

Chemical grades correspond to the once-crystallized, soluble product of about 99.5 % KCl, and 99.95 % KCl which is the Refined, twice crystallized material. The price differential, 60 to 62.4 % $K_2O$ equivalent at 44 to 51 US$ per tonne, and 99.95 % KCl at 44 to 72 US$ per tonne (1977 prices [25], range depending on quantities and contracts) is sufficient to cover the cost of the additional processing for the small amount of chemical grade potash produced, and, together with the decreased environmental impact resulting from dust containment which is the source of the crystallization feed [25], are normally considered to be worthwhile steps by most operators.

## 4.2.4 Solution Mining of Potassium Chloride

Recent innovative technology has been successfully applied to the preferential solution mining of potassium chloride from the sylvinite deposits 215 m thick and lying some 1,600 m below the surface situated towards the southern extremity of the bed underlying much of southern Saskatchewan, which would be only marginally if at all profitable to mine by conventional means. This unique operation takes advantage of the relative differential which exists between the hot and cold water solubilities of sodium chloride and potassium chloride (Table 4.7), to recover a far larger proportion of the potassium chloride than the sodium chloride from the deposit [28]. Present capacity of this particular operation is 1.35 million tonnes per year [29].

The feature of this process which distinguishes it from conventional solution mining for sodium chloride lies in its application of a heated, mixed sodium chloride/potassium chloride brine, rather than fresh water at ambient temperature for salt dissolution. The warm extracting solution is pumped into the ore body via a concentric pair of pipes placed in a drill hole such that the reservoir brine temperature is maintained at 45 °C or higher [29]. Natural formation temperatures aid in maintaining these temperatures in a large reservoir. By this means, some sodium chloride but significantly more potassium chloride is dissolved from the formation and brought to the surface via the "up-brine" piping for potassium chloride recovery. Any suspended sediment is removed from the mixed brine in thickeners (plus a filtration circuit) following which it is concentrated to 99 % sat-

**Table 4.7** Aqueous hot and cold solubility differential between potassium chloride and sodium chloride [a]

| Temperature °C | Solubility (g/100 mL $H_2O$) | |
| --- | --- | --- |
| | Potassium chloride | Sodium chloride |
| 0 | 27.6 | 35.7 |
| 20 | 34.7 | 36.0 |
| 40 | 40.0 | 36.6 |
| 60 | 45.5 | 37.3 |
| 80 | 51.1 | 38.4 |
| 100 | 56.7 | 39.8 |

[a] Data from Chemical Engineers' Handbook [43].

uration at $100°$ in two sets of quadruple-effect vacuum evaporators operated in a series—parallel arrangement [30] (Figure 4.8). Sodium chloride and any precipitated calcium sulfate is removed, hot, in a Bird solid bowl centrifuge (Figure 2.8). Chilling the saturated, mixed brine from the centrifuges by a combination of vacuum and cooling water then produces a suspension of potassium chloride crystals in a predominantly sodium chloride brine. The crystals are separated on horizontal table filters and dried by counter-current application of the clean flue gases produced from gas-fired steam generation, in a fluid bed dryer (Figure 4.9). This product, $99+\%$ potassium chloride ($62.5\%$ $K_2O$ equivalent), is somewhat higher purity than the crude froth flotation product.

Part of the centrifuged sodium chloride is processed for table and cattle block salt production. Unused sodium chloride is slurried to a 162 hectare ($1.62\,km^2$) impounding basin, where wet storage minimizes windage losses. The bulk of the mother liquor, together with a small amount of make-up water, is reheated and returned to the formation for saturation with potassium chloride. Any waste water has been disposed of underground to a water bearing formation of the 1,200-1,500 m level. Dust from drying operations and the Tyler-Hummer screens, which are used for grade classification, is captured in water scrubbers and the brines produced by dust capture are returned to process streams.

### 4.2.5 Environmental Aspects of Sodium and Potassium Chloride Recovery

Processes which recover sodium or potassium chlorides from natural brines originating from the

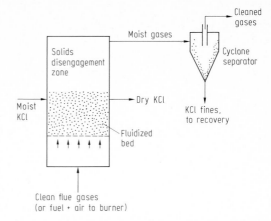

**Figure 4.9.** Fluidized bed drier. Indirectly heated air rather than the combustion gases from burning natural gas may also be used for bed fluidization, for cleaner drying action

ocean or salt lakes probably have the least salination impact on surrounding lands or water, and since many of these operations use solar evaporation they also have a low external energy requirement. The ratio of reserves to annual production rates, even for the salt lakes, is so large that there is not likely to be any noticeable salinity decrease for many years.

Estimates for the Dead Sea indicate that the present production rate exceeds the rate of influx by only some 23 %. However, these types of operations do not have proximity to all major markets, hence the economic incentive to recover both salts from inland deposits as well as from the obvious saline water sources [1].

Purification of mined inland sodium chloride has

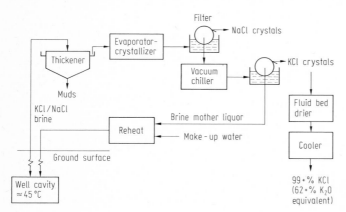

**Figure 4.8.** Potassium chloride from subterranean sylvinite by solution mining

had problems with the disposal of the waste mother liquors obtained from the purification process. Deep well disposal has been practised, frequently into a deep brine aquifer to minimize subterranean impact [31]. River discharge has also been employed, but with growing demands on this fresh water resource and because of the significant negative impact of a salt discharge on many other fresh water uses, this method is now being generally more strictly regulated [1].

For the potash recovery operations in Saskatchewan significant concerns, particularly for those in the neighborhood of conventional mining operations, are the sodium chloride stockpiles and waste disposal basins [32] presently accumulating sodium chloride at somewhere near the annual four million tonnes of potassium chloride production. Precautions are being taken to avoid significant contamination of the limited surface and aquifer fresh water courses [33] of the area by such methods as surrounding the waste disposal basins with impermeable clay dikes. Aeolian salt contamination of the surrounding soils is, however, being felt to some degree despite windage control measures such as stockpile wetting and polymeric spray coatings, and thus may be affecting plant life in the vicinity [34].

"No particular health hazards have been demonstrated ..." to workers, from potassium chloride dust in mines and mills, though a comprehensive baseline study of these operations has been initiated [35].

## 4.2.6 New Developments in Potassium Chloride Recovery

Recovery of potassium salts from natural brines using solar evaporation ponds has been made less dependent on the weather in one operation by a chloride to sulfate conversion process followed by recovery of the sulfate, as is being practised by Great Lakes Minerals and Chemicals Corporation [1, 10]. Highly absorbing dyes, particularly when added to the brine at later stages of production, may also allow significant increases in production from the same evaporation area by improved efficiency of solar energy absorption [1, 36].

Artifical sylvinite, formed by evaporation of marine waters has been found to be less subject to potassium chloride and sodium chloride crystal agglomeration if a small amount of a crystal habit modifier such as sodium ferrocyanide is added du-

ring carnallite decomposition stages [19]. In this way, flotation recoveries of potassium chloride are markedly enhanced. Improvements in the processing of mined potash ores have also been achieved, principally in the methods by which clay, brought up from intrusions in the deposit are dealt with. Clay in the pulped ores causes both higher consumption of collector reagents (by adsorption) and, simultaneously, reduced sylvite recoveries. The clay may be advantageously separated from the pulped ore prior to amine collector addition by a desliming decantation or a gentle pre-flotation collection of the clays in a clay-laden froth. The brine from the clay/brine froth is readily recycled to the flotation circuit after flocculation of the clay with organic reagents followed by removal of the floc in a thickener [18].

Collector adsorption on any residual clays present has been found to be markedly decreased by employing various triethanolamine derivatives (molecular weight < 5,000) as blocking agents for the clay active sites, in addition to the usual starches or guar gums which are also used traditionally for this purpose. Combining this treatment with the employment of straight chain (not branched) fatty amines (and not amine salts) as the potassium chloride collector assures improved recoveries [18]. To provide a closer particle size match between the crude product obtained and market demands, it has been proposed that more emphasis be placed on straight pneumatic flotation assisted by the presence of more viscous reinforcing collector oils, ideally of the butylpolyoxypropylene type $C_4H_9-(O-CH_2-CH(CH_3))_n-OH$, n = 2 to 5, to allow the flotation separation of potash of larger average particle size. Adoption of this procedure might permit abandonment of roller fusion of potash fines and enable direct production of a size distribution more closely matching market demands. Modified fracturing techniques and multiple well sitings have been suggested as means of boosting ultimate potassium chloride recoveries from solution mining operations to the 60-70 % range of in-place potash (versus ca. 40 % experience for conventional shaft mining)[36]. Improvements may also be achieved by employing an oil blanket, with or without a gas head, and solvent density stratification to allow greater control of zones of dissolution [37]. Controlling brine inlet and exit pressures at the same time enables achievement of calculated overburden support from the cavity fluids which thus may also allow solution mining of deeper deposits than previously considered possible. Cavity interconnections may be logged by dye injections into down-brine wells to assist in avoiding or controlling subsidence. Occasionally it is found that further solution of sylvite is prevented by a coating of sodium chloride crystallized out on the walls of the cavern. However it has been found that addition of about 100 ppm of manganese ion (or a number of other transition metal ions) prevents sodium chloride recrystallization and deposition in the well, and hence avoids the blocking of further dissolution of sylvite [28].

Relevant to the potash recovery part of solution mining, suggestions to employ high ammonia concentrations in water to achieve greater sodium chloride/potassium chloride selectivity in solubility [38] do not appear to have offered sufficiently attractive improvements to be commercially adopted. Overall however, solution mining of potash has matured sufficiently from the pioneering work of Kalium Chemicals at Belle Plaine, Saskatchewan [39], to encourage adoption of this technology by at least two operations which were formerly based on conventional mining [40-42]. These recently modified processes are both advantageously sited to favour solar evaporation rather than employing multiple effect evaporator-crystallizers for potash recovery. One of these makes use of 162 hectares $(1.62 \text{ km}^2)$ of PVC-lined evaporation ponds, in this way achieving significant energy economies in the design production rate of 236,000 tonnes of potassium chloride per year [40]. Good pond solution containment is vital to the economic success of this method of salts purification and recovery [9]. Experiments with solar absorbing, non-emitting dyes at these operations, and application of mechanical spray systems such as has been proposed to reduce brine disposal problems [44], may lead to further boosts in production from the existing evaporation area.

## 4.3 Sodium Sulfate

Synthetic or manufactured sodium sulfate was originally produced (in Canada in the late 1800's)[45] as a by-product of the Mannheim furnace method for the production of hydrochloric acid [46]. This process (Equations 4.6, 4.7), consisting of

Acidulation:
$$2\,NaCl + H_2SO_4 \xrightarrow[600-800\,°C]{} 2\,HCl + Na_2SO_4 \quad (4.6)$$

crude salt cake (ca. 98% on NaCl)

Purification (by crystallization):
$$Na_2SO_4 + 10\,H_2O \rightarrow Na_2SO_4 \cdot 10\,H_2O \quad (4.7)$$
Glauber's salt
(ca. 98% recovery)

is the basis of one operating plant in the U.S. and is still a significant source of synthetic sodium sulfate in Europe, provided the origin of the term "salt cake" which is used synonymously with sodium sulfate to specify industrial grade material.

### 4.3.1 Production and Use Pattern for Sodium Sulfate

World production of sodium sulfate has averaged about four million tonnes per year for 1976 and the two preceding years, approximately 1/6 the scale of potassium chloride production (Table 4.8). Very nearly half of this is from sodium sulfate recovered from natural sources, generally from captive lake basins in areas with high evaporation rates or from aquifers with a high dissolved sodium sulfate content, and the remainder is recovered as by-product material from other industrial processes [47]. The major use for sodium sulfate is as make-up for the chemical losses incurred in the kraft process for the production of pulp and paper, some 79 % of the

Table 4.8 Major world producers of sodium sulfate, including saltcake and detergent grades [a].

|  | Thousands of tonnes | | |
|---|---|---|---|
|  | 1972 | 1976 | 1980 [b] |
| Belgium | – | 308 | 249 |
| Canada | 460 | 490 | 454 |
| Japan | 410 | 315 | 290 |
| Mexico | 128 | 250 | 400 |
| Spain | 210 | 290 | 308 |
| West Germany | 300 | 260 | 202 |
| U.S.S.R. | 420 | 535 | 603 |
| U.S.A. | 1 205 | 1 120 | 1 142 |
| Other countries | 627 | 352 | 854 |
| World Total | 3 760 | 3 920 | 4 502 |

[a] Calculated from Mineral Yearbooks [47 – 49].
[b] Estimates.

total supply in Canada and 70 % in the U.S.A. in 1976 [47, 49]. Further significant quantities, 14 %, and 20 % in these two countries respectively, are employed as builders in the formulation of synthetic detergents followed by about 4 % and 10 % respectively, as ingredients in glass making.

### 4.3.2 Recovery from Natural Brines

In recent years, only about one half the sodium sulfate in the U.S.A. has been produced from natural sources (e.g. see Table 4.3), while almost all the Canadian product is obtained in this way. Some plants use a floating dredge to mine lake bottom crystal beds of the minerals mirabilite ($Na_2SO_4 \cdot 10 H_2O$), laid down by successive natural seasons of evaporation and chilling, or thenardite ($Na_2SO_4$), which tends to crystallize out in the presence of significant concentrations of sodium chloride [49]. Processing of natural salts recovered in this way may simply involve dehydration by melting the decahydrate (melting point 32.4°) and evaporation of the water of hydration which comprises more than one half the weight of the crude product, in a submerged combustion unit. Kiln drying of the wet crystal mass at higher temperatures then produces a "salt cake" grade of product, with a minimum 97 % $Na_2SO_4$ content, suitable for kraft process use and in 1977 selling at $50 to $60 per tonne [49].

For a "detergent grade" product brines may be ponded for natural evaporation or may be put through multiple effect evaporators for concentration, and then chilled to 0°C or below to induce crystallization of the decahydrate, Glauber's salt [46, 49]. More frequently, however, by-product sodium sulfate, which is frequently recovered from process liquors by a crystallization step, is produced on a sufficient scale and adequate purity (up to 99.77 % $Na_2SO_4$) to supply the small detergent grade market at prices ranging from $60 to $121 per tonne.

### 4.3.3 By-product Sodium Sulfate

By-product sources include that material available from the Mannheim furnace process for the production of hydrogen chloride which directly yields by-product sodium sulfate. This may be sold in the form obtained as a "salt cake" product, or it may be recrystallized after neutralization and removal of insoluble matter to give a detergent grade product. Careful recovery of the sodium sulfate solution generated from rayon spinning (1.1 kg of sodium sulfate is obtained for each kg of rayon spun) [46] gives Glauber's salt on a scale second only to natural brines, and a mother liquor still containing some sodium sulfate which may be returned to the spinning process. Dehydration of the Glauber's salt from this source yields a detergent grade sodium sulfate.

At present, the Hargreaves-Robinson process to produce hydrogen chloride is also providing a significant proportion of the by-product sodium sulfate in Europe, and at least one plant in the U.S.A. still uses this process [46]. Air, steam and sulfur dioxide are passed over a heated bed or porous salt granules causing a heterogeneous reaction which yields hydrogen chloride and sodium sulfate (Equation 4.8).

$$4 NaCl + 2 SO_2 + O_2 + 2 H_2O \rightarrow \\ 2 Na_2SO_4 + 4 HCl \\ 93 \text{ to } 98\% \text{ yields} \tag{4.8}$$

The crude salt cake by-product may be purified for detergent markets in a manner similar to that used for the crude Mannheim furnace product.

## Relevant Bibliography

1. R.B. Tippin. Potash flotation handles variable feed, Chem. Eng. 84, 73, July 18, 1977
2. Potash production moves to lower grade ores, Eng. and Mining J. 176, 166, June 1975
3. Solution mining symposium, Dallas, Texas, Feb. 25-27, Abstracts of papers, Eng. and Mining J. 175, 62, July 1974
4. S. Serata and W.G. Schultz. Application of stress control in deep potash mines, Min. Congr. J. 58, 36, November 1972
5. Vinyl-lined ponds, key to solution mining of potash, Civil Eng. 42, 98, June 1972
6. J.B. Mitchell. Three ways to process potash, Min. Eng. 22, 60, March 1970
7. J.B. Davis and D.A. Shock. Solution mining of thin bedded potash, Min. Eng. 22, 106, July 1970
8. B.S. Crocker, J.D. Dew and R.J. Roach. Contemporary potash plant engineering, Can. Min. and Metal. Bull. 62, 729, July 1969

## References

1. 1976 Minerals Yearbook, Vol. I, Metals, Minerals and Fuels, U.S. Dept. of the Interior, Bureau of Mines, Washington, D.C., 1978, pages 1136, 1156, 1184, 1190

2. Canadian Minerals Yearbook 1977, Energy, Mines and Resources Canada, Minister of Supply and Services, Hull, 1979, pages 382 to 388
3. Encyclopaedia Brittannica, 15th edition, Macropaedia, London, 1974, Volume 16, page 192
4. E.T. Degens and D.A. Ross, Sci. Amer. *222* (4), 32, April 1970
5. R.N. Shreve and J.A. Brink, Jr., Chemical Process Industries, 4th edition, McGraw-Hill Book Co. Inc., Toronto, 1977, page 266
6. K. Rankama and T.G. Sahama, Geochemistry, Univ. of Chicago Press, Chicago, 1950, pages 283-287
7. J.A. Epstein, Hydrometallurgy, *2*, 1, 1976
8. G.E. Hutchinson, A Treatise on Limnology, Vol. 1, Geography, Physics and Chemistry, John Wiley and Sons, New York, 1957, page 569
9. Kirk-Othmer's Encyclopedia of Chemical Technology, 2nd edition, Interscience Publishers, Toronto, 1968, supplement, pages 438, 444
10. Chemicals Output Rises, Chem. Eng. News *55* (11), 10, March 14, 1977
11. F.A. Lowenheim and M.K. Moran, Faith, Keyes and Clark's Industrial Chemicals, 4th edition, Wiley-Interscience, Toronto, 1975, page 723
12. Salt, Morton Salt Company, Chicago, 1962(?), page 9
13. R.M. Stephenson, Introduction to the Chemical Process Industries, Reinhold, New York, 1966, page 38
14. B.G. Reuben and M.L. Burstall, The Chemical Economy, Longman, London 1973, page 149
15. R.W. Thomas, and P.J. Farago, Industrial Chemistry, Heinemann, Toronto, 1973, page 79
16. C.J. Warrington and B.T. Newbold, Chemical Canada, Chemical Institute of Canada, Ottawa, 1970, page 5
17. Potash in Saskatchewan, Department of Mineral Resources, Province of Saskatchewan, Regina, April 1973, pages 12, 23
18. V.A. Arsentiev and J. Leja, Can. Min. Metal. Bull. *70A*, 154, March 1977
19. J.A. Epstein, Chem and Ind. (London) *55*B, 572, July 16, 1977
20. R.W. Mumford, Ind. Eng. Chem. *30*, 872, 1938
21. W.A. Gale, Ind. Eng. Chem. *30*, 867, 1938
22. J.A. Epstein, Hydrometallurgy, *2*, 1, 1976
23. D. Ashboren, J. Appl. Chem. Biotechn. *23*, 77, 1973
24. B. Cross, Chem. Eng. *69* (23), 176, Nov. 12, 1962
25. Potash Corporation of Saskatchewan, Saskatoon, 1979. Booklet, 24 pp.
26. Prices calculated in U.S. dollars per tonne of KCl, based on "unit $K_2O$ equivalent" prices of 65 to 75c unit ton (1 short ton unit $K_2O$ equivalent = 20 lbs. $K_2O$ equivalent[27]) quoted in Chemical Marketing Reporter, Dec. 12, 1977, page 71

27. Canadian Minerals Yearbook 174, Energy, Mines and Resources Canada, Supply and Servives, Ottawa, 1977, page 401
28. J.B. Dahms and B.P. Edmonds, Can. Pat. 672, 308, (to Pittsburgh Plate Glass Co.), Oct. 15, 1963
29. Potash Product Profile, Can. Chem. Proc. *60* (3), 58, March 1976
30. J.A. Paquette, personal communication, June 4, 1974
31. Industrial Pollution Control Handbook, H.F. Lund, editor, McGraw-Hill, Toronto, 1971, page 7-34
32. J.A. Vonhof, A Hydrogeological and Hydrochemical Investigation of the Waste Disposal Basin at IMCC (Canada) Ltd. K2 Potash Plant, Esterhazy, Saskatchewan, National Hydrology Research Institute, Ottawa, 1980. Manuscript, 167 pp.
33. A. Vandenberg, An Unusual Pump Test Near Esterhazy, Saskatchewan, Technical Bulletin No. 102, Inland Waters Directorate, Water Resources Branch, Ottawa, Canada, 1978. 25 pages
34. R.S. Edwards, The effects of air-borne sodium chloride and other salts of marine origin on plants in Wales, In: Proc. of the 1st European Congress on Air Pollution Effects on Plants and Animals, Centre for Agric. Publ. and Documentation, Wageningen, 1969, page 99
35. John Markham, cited in Chem. in Can. *27* (9), 14, Oct. 1975
36. New process may aid solution mining, Can. Chem. Proc. *54*, 44, Dec. 1970
37. H.H. Werner, Can. Pat. 838, 477, April 7, 1970
38. R.A. Gaska, R.D. Goodenough and G.A. Stuart, Sr., Chem. Eng. Prog. *61* (1), 139, Jan. 1965
39. First solution-mined potash, Chem. Eng. *71* (22), 84, Oct. 26, 1964
40. R.L. Curfman, Mining Congr. J. *60*, 32, March 1974
41. Solution mining/solar evaporation of potash nears capacity at Cane Creek, Eng. Mining J. *173,* 30, Sept. 1972
42. D. Jackson, Jr., Eng. Mining J. *174,* 59, July 1973
43. Chemical Engineers Handbook, 4th edition, R.H. Perry, editor, McGraw-Hill, Toronto, 1969, pages 3-92, 3-93
44. W.G. Smoak, C.M. Wong, W.F. Savage, W.W. Rinne, and C. Gransee. Spray systems—a method of increasing evaporation rates to facilitate brine disposal from desalting plants. Report No. 480, U.S. Dept. of the Interior, Bur. Reclamation, Res. and Dev. Prog., Washington, 1969. Cited by reference 32
45. C.J.S. Warrington and R.V.V. Nichols, A History of Chemistry in Canada, Pitman, Toronto, 1949, page 213
46. F.A. Lowenheim and M.K. Moran, Faith, Keyes and Clark's Industrial Chemicals, 4th edition, Wiley-Interscience, Toronto, 1975, page 762

47. 1976 Minerals Yearkbook, Vol. I Metals, Minerals and Fuels, U.S. Dept. of the Interior, Bureau of Mines, Washington, D.C., 1978. page 1251

48. 1974 Minerals Yearbook, Vol. I. Metals, Minerals and Fuels, U.S. Dept. of the Interior, Bureau of Mines, Washington, D.C., 1976, page 1214

49. Minerals Yearbook 1980, Vol. I. Metals and Minerals, U.S. Dept. of the Interior, Bureau of Mines, Washington, D.C., 1981, page 763

50. The Encyclopedia of Chemical Process Equipment, W.J. Meade, editor, Reinhold, New York, 1964, page 329

51. Statistical Yearbook, United Nations, New York, 1977

# 5 Industrial Bases by Chemical Routes

## 5.1 Calcium Carbonate

The dominant source of calcium carbonate is limestone, the most widely used of all rocks, but this occurs in nature with at least traces of clay, silica, and other minerals which may interfere with some applications. However, high calcium limestone consists of about 95 % or better calcium carbonate. White marble, a metamorphosed form of pure limestone with a closely packed crystal structure, is chemically suitable. But marble is usually of higher value for other applications than as a chemical feedstock. Dolomitic limestones consist of calcium and magnesium carbonates present in a near one to one molar basis, though this ratio can vary widely in ordinary dolomites. For some applications at least, the presence of the magnesium carbonate is not a handicap to the use of this calcium base component of the dolomite. The remaining principal natural sources of calcium carbonate more closely reflect their biogenetic origin in the forms of chalk, which comprises the shells of microscopic marine organisms, bivalve shells, which for example are accessible in sufficient quantities on shores of the Gulf of Mexico to be employed as an industrial feedstock, and coral, which consists of the massive, sub-marine fused skeletons of multiple stationary organisms.

Both to obtain higher purity and higher specific bulk density grades of calcium carbonate, and as an expedient to a saleable product from the waste streams of some processes, synthetic calcium carbonate is also produced on a large scale [1]. As early as 1850 J. & E. Sturge Ltd. of Birmingham, England were treating a calcium chloride Solvay waste stream with sodium or ammonium carbonates to produce a high grade of calcium carbonate. In the U.S., certainly by 1900, not only Solvay byproduct calcium chloride, but also causticizer

$$CaCl_2 + Na_2CO_3 \rightarrow 2\,NaCl + CaCO_3 \downarrow \quad (5.1)$$

$$CaCl_2 + (NH_4)_2CO_3 \rightarrow 2\,NH_4Cl + CaCO_3 \downarrow \quad (5.2)$$

sludges (Equation 5.3) and direct carbonation of

**Table 5.1** Calcium Carbonate Dissociation Pressures [a]

| Temperature °C | Pressure mm Hg [b] | Temperature °C | Pressure mm Hg |
|---|---|---|---|
| 500 | 0.1 | 852 | 381 |
| 550 | 0.41 | 898 | 760 |
| 605 | 2.3 | 950 | 1 490 |
| 701 | 23.0 | 1 082.5 | 6 758 |
| 749 | 72 | 1 157.7 | 14 202 |
| 800 | 183 | 1 241 | 29 711 |

[a] Compiled from data of Kirk Othmer [1], and the Handbook of Chemistry and Physics [2].
[b] 1 mm Hg = 1.33322 mbar.

slaked lime (Equation 5.4) were used to produce fine grades of calcium carbonate.

$$Ca(OH)_2 + Na_2CO_3 \rightarrow 2\,NaOH + CaCO_3 \downarrow \quad (5.3)$$

$$Ca(OH)_2 + CO_2 \rightarrow H_2O + CaCO_3 \quad (5.4)$$

Industrial uses of natural calcium carbonate range from the direct, very large consumption in the cement, iron and steel, and other metal refining industries as well as the Solvay ammonia soda process and directly or indirectly in the manufacture of at least 150 other chemicals. In recent years, direct or indirect use in emission control measures for combustion processes has taken a growing fraction of limestone production. The higher purities and more tailored properties of the synthetic grades of calcium carbonate both command a higher price and for this reason limit the appropriate applications somewhat. More than 75 % of U.S. synthetic calcium carbonate is applied as a surface coating to paper to impart a smoother, whiter surface. Smaller amounts are used in paints, to give flatness, and low gloss, in rubber, as a reinforcing agent, in plastics, as a filler, and in many other consumer items such as foods, cosmetics, toiletries, pharmaceuticals and the like.

**Plate 5.1.** Operating lime kiln plus associated emission control equipment at the B.C. Forest Products pulp mill at Crofton, British Columbia. Kiln is the large, slightly inclined cylinder across center of picture.

## 5.2 Calcium Oxide

Calcium oxide, burned lime, or simply lime as it is known in industry, is obtained by heating calcium carbonate to a temperature of 900 to 1,000 °C to cause loss of carbon dioxide (Equation 5.5). Temperatures below this range slow the rate of carbon dioxide loss sufficiently to leave some unburned limestone in the product from the usual calcination, and fail to confer any significant energy saving if the limestone retention time is prolonged as a result to ensure complete reaction, because of the greater heat loss experienced for the longer calcination period. Also, when a moderate size range of crushed limestone, rather than finely ground material, is used to reduce dust carryover in the kiln, a certain amount of additional excess carbon dioxide partial pressure must be developed in the larger granules to ensure carbon dioxide diffusion to the outside of the granule during the hot period. In Canada, and probably elsewhere in the early nineteenth century, burned lime manufacture was carried out as a batch process using pot kilns heated with wood or coal [3]. These produced 6 to 27 tonnes per cycle, as did the draw kilns of the late nineteenth century, which were designed to use fuel more efficiently. Today a variety of vertical flow kilns are in common use, these mainly designed to operate with still lower energy requirements than the early models [4]. One further feature of

some of these newer designs is the separation of the function of gas flow required for combustion from the limestone heat requirement, thus obtaining a gas stream from above the limestone as high in carbon dioxide concentration as possible. This is a desirable feature, for example as a component of the Solvay ammonia-soda process (Section 5.2).

Probably the most broadly useful design of modern lime kiln, allowing as it does for feed of lumps or fines of limestone or marble, and wet or dry calcium carbonate sludges, some of which would plug vertical kilns, is the rotary horizontal kiln (Figure 5.1). The significant component of this unit for chemical conversion is the 2.5 to 3.5 m diameter by 45 to 130 m long firebricklined inclined steel tube to which heat is applied via oil, gas, or coal [5] burners at the lower end. The feed to be calcined is fed in at the top end and slow rotation of the tube on its axis gradually moves the feed down the tube, tumbling counter-current to the hot combustion gases. In this way, wet feed is dried in the first few meters of travel. Further down the tube, for wet or dry feeds, carbon dioxide loss from the feed begins as the temperature of the feed is raised. By the time the solid charge reaches the lower, fired end of the kiln it reaches temperatures of 900 to 1,000 °C and carbon dioxide evolution is virtually complete. Normally the temperature of the lower end of the kiln is not allowed to go much above this range since this results in reduced kiln lining life. It also adversely affects the crystal structure of the lime

$$CaCO_{3(S)} \overset{900-1000°}{\rightleftarrows} CaO_{(S)} + CO_{2(g)} \overset{\Delta H}{+177.8 \text{ kJ} (+42.5 \text{ kcal})}$$ (5.5)

100.09    56.08    44.01  molecular weights

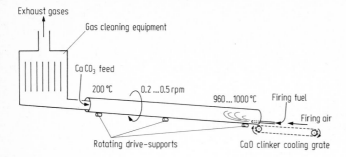

**Figure 5.1.** Details of operation of a rotary horizontal lime kiln. Temperature of calcium carbonate charge rises as it moves down the kiln. The hot calcium oxide, as it cools, is used to simultaneously preheat the firing air. Fuel may be natural gas or oil

product producing a so-called "dead-burned" or "overburned" lime, which introduces slaking difficulties on conversion to calcium hydroxide, and raises fuel costs proportionally. The calcined product is normally packed in air-tight sheet steel barrels or bulk containers for shipment.

With a specific heat of about 0.90 J/g/K (0.214 cal/g/°C, at 100 °C; varies with temperature [6]) calcium carbonate requires some 7.92 x $10^5$ kJ/tonne of sensible heat simply to bring the salt up to the final kiln temperature. The endotherm of the reaction consumes a further 1.78 x $10^6$ kJ/tonne, which, when combined with the sensible heat requirement and an allowance of a further 2 x $10^5$ kJ/tonne for imperfect heat transfer efficiencies gives a total heat requirement for this process of about 2.8 x $10^6$ kJ/tonne. Production estimates for different fuels range from 2.5 to 3.4 kg of lime per kg of coke [5] or coal [7], approximately equivalent to 4.5 kg per kg of Bunker C fuel oil [8].

$$0.90 \text{ J/g/K} \times 10^6 \text{ g/tonne} \times (900-20)\,°C$$
$$= 7.92 \times 10^8 \text{ J/tonne}$$

$$177.8 \text{ kJ/mol} \times \frac{10^6 \text{ g/tonne}}{100.09 \text{ g/mol}}$$

$$= 1.776 \times 10^6 \text{ kJ/tonne}$$

Improved energy efficiencies are obtained by the practice of employing "recuperator chains" loosely slung inside the kiln to aid in heat transfer from hot gases to drying and reacting solids, and by using the heated air obtained from lime cooling as the firing air for the kiln. Some other kiln designs provide better heat transfer and recovery than the rotary kiln but have certain other restrictions in the form of suitable feeds etc. which do not affect the rotary horizontal kiln [4]. There is also a potential energy gain available from the use of oxygen-enriched air, accessible through one of the new low cost enrichment techniques [9], since this would avoid the necessity of heating the larger volumes of associated nitrogen in normal air. Apart from the much greater additional expense, there would be little point in utilizing pure oxygen since one of the functions of gas throughput is to displace the dissociated carbon dioxide.

### 5.2.1  Lime Kiln Emission Control

Environmental concerns of lime kiln operations are limited primarily to exhaust gas dust control, usually solved by water scrubbing. The slaked lime (calcium hydroxide) produced by the scrubber can be otherwise employed in the operations of a chemical complex, or sold. Electrostatic precipitation of the at least partially cooled gases is also used. Occasionally both control measures are employed in series. The dissociated carbon dioxide discharged is not ordinarily regarded as a pollutant and the amount discharged from this source is far exceeded by the volumes of carbon dioxide discharged from fossil fuel combustion. Lime kilns associated with Solvay ammonia-soda plants may be able to recycle a part of the dissociated carbon dioxide, particularly if the concentrations are raised by the use of oxygen-enriched combustion air. Recently there has been an interest shown in the collection of carbon dioxide from industrial sources, such as lime kilns, for use in enhanced oil recovery operations.

### 5.2.2  Uses of Calcium Oxide

The extensive industrial uses of calcium oxide (lime) are a reflection of both its low cost and its chemical properties. It is consumed on a very large scale by both the cement and the iron and steel industry. In both cases *in situ* lime preparation from limestone by heat is an integral part of the process.

The metals smelting industry also employs lime in a general way as a basic refractory lining material. This basicity is also employed in processes for the heterogeneous absorption of acid gases produced from combustion or metallurgical roasting processes. The strong affinity of lime for water is applied to the industrial drying of alcohols as, for example, for water removal from the glycerin obtained from the saponification of fats, or from the 5 % water azeotrope of ethanol. Most of the remaining uses are indirect since they employ slaked lime or calcium hydroxide, the product of the reaction of lime with water.

# 5.3 Calcium Hydroxide

Calcium oxide, or burned lime can be slaked with the theoretical amount of water to form calcium hydroxide (slaked lime) as a white powder, but this must be conducted with care because the highly exothermic reaction may produce sufficiently high temperatures that the rate of hydration may actually be decreased [5] (Equation 5.6). This process is sufficiently exothermic that it is

$$CaO + H_2O \rightarrow Ca(OH)_2$$
$$\Delta H = -64.9 \text{ kJ } (-15.5 \text{ kcal}) \qquad (5.6)$$

dangerous to store large quantities of calcium oxide in combustible bags or even in non-fireproof buildings if there is any risk of contact of the burned lime with water or air, since even moist air will slake burned lime. A simple calculation demonstrates that the high heat of hydration coupled with the relatively low specific heats ($CaO = 0.946$ J/g; $Ca(OH)_2 = 1.197$ J/g) [10] is theoretically capable of generating a temperature rise of 730 °C above ambient, more than enough to ignite combustible materials. This is apart from the explosive risks connected with possible sudden steam generation in an enclosed space.

Other factors in the preparation of slaked lime powder by the addition of the theoretical amount of water to lime are the original calcining temperature used and the magnesia (MgO) content. Very high calcining temperatures, which produce a "deadburned" lime, can have a retarding effect on the hydration reaction [5]. The presence of any significant concentrations of magnesia, such as may be obtained from the calcination of a dolomitic limestone (a "dolime") also shows this effect.

However, the hydration retardation shown by burned dolimes is actually regarded as a working advantage in its application as a plaster [4].

Both continuous rotary and batch hydrators are used commercially to produce calcium hydroxide (slaked lime) powder. The continuous version uses a gently inclined, slowly rotating steel cylinder of about 1 by 6-7 meters in length. The calcium hydroxide is fed into the upper end and a water spray added at the same point correctly proportioned to this, plus follow-up tumbling in the cylinder produces an acceptably uniform product. Open trough-type batch hydrators, in which the reacting components are mechanically combined, provide a greater degree of control of hydration rate and temperature. By this method there are some improvements in the quality of the product for certain uses, but there is also some variability in properties. A relatively recent announcement of an "explosion process", in which the hydration is conducted in a pressure vessel, produces a slaked lime with better flow characteristics [11]. The smaller mean particle size obtained by this method yields a product more suitable than conventionally slaked limes for uses where small particle size is important, such as for filters.

The same quantity of heat is still produced when lime is slaked with a liberal amount of water to produce a slurry or suspension of calcium hydroxide in water. However, coping with this heat evolution is less of a problem because of the much higher heat capacity of the excess water present. In the preparation of cement mixes, morters, and plasters the low solubility of slaked lime in water (1.84 g/L at 0°; 0,77 g/L at 100°) [12] does not pose any problems. However, this factor does have to be considered in the applications of calcium hydroxide when used as an industrial base.

## 5.3.1 Uses of Calcium Hydroxide

Some of the uses of slaked lime are direct, in this form, and some are indirect via calcium oxide addition to water or in an aqueous medium. These uses range all the way from its employment as a component of insecticides, to its value as a component of water treatment programs. The venerable but still viable use as an active pesticide component in lime-sulfur sprays, as well as its substrate utility in many insecticidal dusts are also important. The chemical basicity of this commodity is used in its medicinal applications as an antacid. This property also prompts its use for pH adjustment in animal food formulations, and for neutralization of overly acid soils. Large quantities are employed as a slurry in water to absorb acid contaminants from flue gases and from roaster off-gases of small smelters. It is also used in paper making. In the sulfite process it is used directly to provide the active cation of the pulping liquor. In the kraft process it is used indirectly for the regeneration of the sodium hydroxide component of the pulping liquor from sodium carbonate recovered from the combustion of spent pulping liquor.

Calcium hydroxide frequently performs a dual role in water treatment. Firstly, it aids the removal of colloidal suspended solids present by encouraging coagulation and settling. Secondly, it reacts with any dissolved calcium, magnesium or other cations which are present as their bicarbonates to precipitate these ions and in this way softening the water supply for domestic and industrial uses (Equations 5.7-5.9).

$$Ca(OH)_2 + Ca(HCO_3)_2 \rightarrow 2\,CaCO_3 \downarrow\, + 2\,H_2O \quad (5.7)$$

$$Ca(OH)_2 + Mg(HCO_3)_2 \rightarrow MgCO_3 \quad + CaCO_3 \\ + 2\,H_2O \quad (5.8)$$

$$Ca(OH)_2 + MgCO_3 \quad \rightarrow Mg(OH)_2 \downarrow + CaCO_3 \quad (5.9)$$

## 5.4 Natural and Synthetic Sodium Carbonate

Up until the end of the eighteenth century the only commercially available sodium carbonate was a low grade product containing only 5-20 % of this salt. It was obtained from the burning of kelp or barilla (a stout, berry-bearing Spanish shrub), followed by extraction of the ashes with water [13]. Early in the 19th century the availability of sodium carbonate was enhanced by the introduction of the Le Blanc process, which both increased the potential supply and produced a higher analysis product (Equations 5.10, 5.11).

$$2\,NaCl + H_2SO_4 \rightarrow Na_2SO_4 + 2\,HCl \quad (5.10)$$

$$Na_2SO_4 + 2\,C + CaCO_3 \xrightarrow{\Delta}$$
$$CaS + Na_2CO_3 + 2\,CO_2 \quad (5.11)$$

The early versions of this process were significant contributors to localized air pollution since about 3/4 of a tonne of hydrogen chloride was discharged for each tonne of sodium carbonate produced. In 1836 it was discovered that the acid discharge could be controlled by absorption in water, but it was 1863 before this control measure was made compulsory [13]. The other negative feature of LeBlanc process operations at that time was the large piles of waste calcium sulfide and calcium oxysulfide which were generated and disposed of nearby, reclamation of which was very difficult.

The ammonia-soda process to sodium carbonate, which is still practised on a large scale today, was known in 1822 but this early version had no ammonia recovery phase [13] (Equations 5.12-5.15).

$$NH_3 + H_2O \rightarrow NH_4OH \quad (5.12)$$

$$NH_4OH + CO_2 \rightarrow NH_4HCO_3 \quad (5.13)$$

$$NH_4HCO_3 + NaCl \rightleftarrows NaHCO_3 \downarrow\, + NH_4Cl \quad (5.14)$$

$$2\,NaHCO_3 \xrightarrow[\Delta]{175^\circ} Na_2CO_3 + CO_2 + H_2O \quad (5.15)$$

The cost of the ammonia required to work this process was prohibitive and hence the process was not commercialized until 1863 when E. Solvay devised a method employing a suspension of slaked lime in water to accomplish ammonia recovery (Equation 5.16).

$$2\,NH_4Cl + Ca(OH)_2 \rightarrow \\ 2\,NH_3 \uparrow\, + CaCl_2 + 2\,H_2O \quad (5.16)$$

Modern practice of the Solvay ammonia-soda process involves initial successive saturation of a purified saturated solution of sodium chloride in water, first with ammonia gas and then with carbon dioxide. The carbonation step is used to drive reaction 5.14 to about 73 % to the right [14]. Sodium bicarbonate, being the least soluble salt component of this mixed solution, is precipitated and filtered off. It is then dried and mildly calcined in a rotary kiln to yield the product, sodium carbonate (Table 5.2). Conducting the calcination in a sealed rotary kiln enables recovery of the liberated carbon dioxide (Equation 5.15) which provides some of that required for the primary carbonation (Figure 5.2). The remaining carbon dioxide is obtained from the operation of a separate lime kiln which produces calcium oxide. The calcium oxide (burned lime) is required to produce calcium hydroxide for ammonia recovery, and also for carbon dioxide recovery from boiler flue gases as well, if necessary. The ammonia recovery system of modern configurations of Solvay plants are quite efficient. Only about 3 kg of ammonia make-up is required to

**Table 5.2** Quality of typical sodium carbonate from the ammonia-soda (Solvay) process [14]

| ANALYSIS | |
| --- | --- |
| $Na_2CO_3$ | 99.50 (58% $Na_2O$) |
| NaCl | 0.20 to 0.25 |
| $Na_2SO_4$ | 0.02 |
| insolubles | 0.02 |
| water | balance |

maintain the 300 kg or so of absorbed ammonia which is needed for each tonne of sodium carbonate produced [15]. The development of this high ammonia recovery efficiency was the feature that gradually spelled the end of the commercial Le Blanc sodium carbonate plants, the last of which closed down in the 1920 to 1930 period [13].

The crude sodium carbonate from the Solvay process, while of quite good purity, is a low density fluffy material of bulk density of about 0.5 g/cm³, and referred to as "light soda ash". The density may be increased to a bulk density of about 1.0 g/cm³ required for some uses by conversion to the monohydrate, $Na_2CO_3 \cdot H_2O$. This is effected by tumbling the anhydrous light ash with the appropriate quantity of water and then drying at < 100 °C. Sal soda or washing soda, which is the decahydrate, $Na_2CO_3 \times 10\,H_2O$, is generally obtained by crystallization of the light ash from water and drying at ambient temperatures (< 35.5 °C).

Oddly enough the sodium bicarbonate ($NaHCO_3$) initial product of the Solvay process is not the article of commerce principally because of impurities, in particular ammonia, that are present. Instead the relatively much smaller market for the bicarbonate is satisfied by the carbonation of a saturated solution of sodium carbonate, the final Solvay product, and recovery of the precipitated, much less soluble sodium bicarbonate by filtration [16] (Equation 5.17). This product is the household "baking soda". The same source is also

$$Na_2CO_3 + CO_2 + H_2O \rightarrow 2\,NaHCO_3 \qquad (5.17)$$

used to supply the sodium bicarbonate used for the manufacture of baking powder, for the preparation of carbonated mineral waters, and also used as the filling for one type of dry powder fire extinguisher.

### 5.4.1 Environmental and Related Concerns of Sodium Carbonate Production

An area of difficulty of present day Solvay operations is related not so much to the technology of the process but to the basic chemistry, which determines the co-production of more than a tonne of calcium chloride for each tonne of the product sodium carbonate obtained. For producers operating in areas with well developed dissipative markets for calcium chloride in applications such as high early strength cement, winter ice and snow removal for highways, or dust control on unimproved roads, the co-product can be sold at reasonable prices. Non-dissipative uses of calcium chloride such as employment as the dissolved salt of an aqueous heat transfer fluid in artificial ice rinks and the like, or as a component of solar energy storage systems [17], do not at the moment consume large amounts of the salt. Because of this, some producers are faced with high brine disposal costs since the waste, low pH calcium chloride

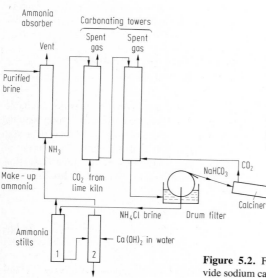

**Figure 5.2.** Flowsheet for the ammonia-soda (Solvay) process to provide sodium carbonate. Units numbered 1 and 2 are ammonia stills, the first being used to recover free ammonia and the second to recover that produced by the addition of slaked lime (Equation 5.16)

brines cannot be discharged to surface fresh water courses without creating significant pollution problems. Some processors have adopted calcium chloride brine discharge into underground brine aquifers as a disposal method.

Brine disposal problems can be avoided by modifications to the Solvay process which have been proposed, which employ magnesium hydroxide for ammonia recovery [15]. The magnesium chloride formed in the process could be recycled and hydrogen chloride obtained as a co-product by contacting the magnesium chloride with superheated steam (Equations 5.18, 5.19). However, this process has not as yet been commercially proven.

$$MgCl_2 + H_2O \xrightarrow{\Delta} MgO + 2\,HCl \qquad (5.18)$$

$$MgO + H_2O \rightarrow Mg(OH)_2 \qquad (5.19)$$

A dominant alternate source of sodium carbonate, at least in the U.S.A. where large reserves of natural brines containing sodium carbonate are accessible, has been to process these for sodium carbonate recovery. In part because of calcium chloride disposal problems of the Solvay process, and also because of the lower capital cost per tonne of capacity for sodium carbonate recovery from natural brines, the capital cost of a natural sodium carbonate plant is only about one half the cost of a similarly sized Solvay plant [18]. Because of this, there has been a strong trend in the U.S.A. towards natural sources in recent years (Table 5.3). Only two other countries, Chad and Kenya, are reported to be recovering natural sodium carbonate, and these were operating on the scale of about 10,800 and 119,000 metric tonnes annually, in 1977 (calculated from reference 19). All other world producers still rely heavily on the ammonia-soda process [18-20] (Table 5.4).

An expedient but not large scale additional source of sodium carbonate is via the in situ or separated carbonation of sodium hydroxide, as it is formed, in an electrolytic cell. The precipitated sodium hydrogen carbonate is calcined to obtain sodium carbonate from the initial product in a manner very similar to the last step of the Solvay process (Equation 5.15). This is not much practised because of high cost.

### 5.4.2 Uses of Sodium Carbonate

Nearly half of the sodium carbonate produced in the U.S.A., whether from synthetic or natural sources, is consumed by the glass industry. A fur-

**Table 5.3** Growth of natural sodium carbonate production at the expense of synthetic material in the United States [a]

| Year | Sodium carbonate produced, thousands of tonnes | | Natural, as percentage of total |
|------|------------|---------|---|
| | Manufactured | Natural | |
| 1940 | 2 744 [b] | 118 | 4.1 |
| 1945 | 3 969 | 176 | 4.2 |
| 1950 | 3 621 | 318 | 8.1 |
| 1955 | 4 452 | 557 | 11 |
| 1960 | 4 134 | 734 | 15 |
| 1965 | 4 473 | 1 355 | 23 |
| 1970 | 4 004 | 2 439 | 38 |
| 1975 | 2 542 | 3 926 | 61 |
| 1976 | 2 126 | 4 730 | 69 |
| 1977 | 1 644 | 5 650 | 78 |
| 1978 | 1 360 [c] | 6 160 | 82 |

[a] Calculated from data in Mineral Yearbooks, U.S. Bureau of Mines [19, 20].
[b] A further 22,680 tonnes was made in this year by the carbonation of electrolytic sodium hydroxide.
[c] Estimated.

ther quarter of the product goes into general chemicals production. The detergent industry, pulp and paper, and water treatment processes each consume somewhat less than 10 % of the total [19], although there are also industries such as kraft paper production, which produce and consume large quantities of sodium carbonate internally. Therefore, the figures for this captive production by the pulp industry do not appear in the consumption data reported.

## 5.5. Sodium Hydroxide by Causticization

Prior to 1850 sodium hydroxide was not available commercially. Processes such as soap-making which required sodium hydroxide (caustic soda) had to obtain this by causticization of a solution of purchased sodium carbonate with lime (Equation 5.20). Sodium hydroxide, which first became commercially available in

$$Na_2CO_{3(aq)} + Ca(OH)_{2(aq)} \rightleftarrows$$
$$2\,NaOH_{(aq)} + CaCO_{3(s)} \qquad (5.20)$$
$$\Delta H = -8.4\ kJ\ (-2.0\ kcal)$$

the early 1850s, was also made in this way [13] until about 1890 when the first electrolytic product started to be produced. Despite the advantages of

**Table 5.4** Major world producers of sodium carbonate (thousands of metric tonnes) [a]

|  | 1960 | 1965 | 1970 | 1975 | 1977 | 1978 | 1979 |
|---|---|---|---|---|---|---|---|
| Bulgaria | 128 | 223 | 300 | 989 | 1 194 | 1 294 | 1 498 |
| China, Mainland | – | – | – | – | 1 090 | 1 330 | 1 785 |
| France | 848 | 1 089 | 1 419 | 1 279 | 1 365 | 1 353 | 1 355 |
| German Dem. Republic | 594 | 682 | 676 | 818 | 840 | 852 | 853 |
| German Fed. Republic | 1 117 | 1 165 | 1 334 | 1 249 | 1 351 | 1 230 | 1 180 |
| India | 145 | 322 | 446 | 541 | 568 | 590 | 610 |
| Italy | – | – | 662 | 647 | 710 | 725 | 725 |
| Japan | 516 | 762 | 1 230 | 1 124 | 1 179 | 1 162 | 1 180 |
| Poland | 522 | 602 | 644 | 716 | 671 | 663 | 662 |
| Romania | 180 | 350 | 582 | 693 | 861 | 900 | NA |
| U.S.S.R. | 1793 | 2 728 | 3 485 | 4 692 | 4 876 | 5 350 | 5 350 |
| United Kingdom | – | – | – | 1 397 | 1 497 | 1 600 | 1 630 |
| United States [b] | 4 135 | 4 469 | 3 859 | 2 534 | 1 643 | 1 360 | NA |
| Total World [c] | 11 480 | 14 160 | 16 610 | 17 580 | 18 610 | 18 409 | |

[a] Does not include natural sodium carbonate. Data derived from U.N. Statistical Yearbooks [21], and calculated from Mineral Yearbooks [19, 20].
[b] Natural sodium carbonate represents a major proportion of the U.S. total. See Table 5.3.
[c] World market growth is greater than would be suggested by these totals for synthetic $Na_2CO_3$. For 1960 the gross figure, including natural $Na_2CO_3$, would be about $12.2 \times 10^6$ and for 1979, $24.6 \times 10^6$ metric tonnes.

electrolytic production of caustic soda, the causticization or "lime-soda" process dominated the production of trade sodium hydroxide in the U.S.A. until at least 1940 [22] and is still of commericial significance in the regeneration of waste pulping liquors in the manufacture of pulp and paper by the kraft process. It is also used for the production of sodium hydroxide without chlorine at times when the market demand ratio of sodium hydroxide to chlorine exceeds the stoichiometric ratio available from electrolysis of sodium chloride solutions.

The chemical change involved in this process depends for its success on the low solubility of calcium carbonate, which is also the feature which lends suitability of this process for small scale low capital production of sodium hydroxide by batch processes. For batch operation, several of the functions such as slaking, mixing, and settling, may be carried out in the same wooden (or steel) vessel unlike the separate units required by the continuous process. A further chemical feature important to the recycle of the spent lime of this process is the relatively easier thermal loss of carbon dioxide from calcium carbonate than from sodium carbonate. It might be expected that, because sodium bicarbonate ($NaHCO_3$) may be calcined at 175 °C to obtain carbon dioxide loss, sodium hydroxide

may be most simply obtained by merely calcining sodium carbonate at a somewhat higher temperature followed by hydration of the resulting sodium oxide (Equations 5.21, 5.22). Whilst this series of reactions is possible, very

$$Na_2CO_3 \underset{\Delta}{\rightarrow} Na_2O + CO_2 \qquad (5.21)$$

$$Na_2O + H_2O \rightarrow 2\,NaOH \qquad (5.22)$$

much higher temperatures are required for sodium carbonate dissociation to sodium oxide and carbon dioxide, than for calcium carbonate dissociation. hot, concentrated sodium hydroxide is also very corrosive. Therefore it is not nearly as easy to use this sequence of reactions as it is to go through the same sequence with calcium carbonate, which has a much higher decomposition pressure at all temperatures (Table 5.5). This dissociation can then be followed by the slaking and the causticization reactions (Equations 5.6, 5.20).

A hot (80-90 °C) 20 % solution of sodium carbonate is stirred with a slight stoichiometric excess of a slurry of calcium hydroxide in water ("milk-of-lime"). Since this reaction is only slightly exothermic (Equation 5.20) the equilibrium is not significantly affected by operating the reaction hot. The reaction is also both faster and produces larger, more readily precipitated particles of calcium

**Table 5.5** Carbon dioxide dissociation pressures of sodium
hydrogen carbonate and sodium carbonates [a]

| NaHCO₃ | | CaCO₃ | | Na₂CO₃ |
|---|---|---|---|---|
| Temperature °C | Pressure mm Hg [b] | Temperature °C | Pressure mm Hg [b] | Pressure mm Hg [b] |
| 30 | 6.2 | 700 | 22.6 | 0.99 |
| 50 | 30.0 | 820 | 237 | 3.0 |
| 70 | 120.4 | 880 | 595 | 9.9 |
| 90 | 414.3 | 990 | 3 324 | 12.0 |
| 100 | 731.1 | 1 100 | 8 490 | 21 |
| 110 | 1 252.6 | 1 150 | 13 440 | 28 |
| 115.5 | 1 654.6 | 1 180 | 18 068 | 38 |
| | | 1 200 | 21 534 | 41 |

[a] Dissociation pressures for sodium salts calculated from International Critical Tables [23], and for calcium carbonate, by interpolation from data in the Handbook of Chemistry and Physics [24].
[b] 1 mm Hg = 1.33322 mbar.

carbonate at high, than at ambient temperatures. An additional advantage gained by operation at temperatures near the boiling point of water is obtained from the higher concentrations of sodium carbonate which may be employed (Table 5.6). Starting with high concentrations of sodium carbonate raises the concentration of sodium hydroxide obtained directly from the reaction to about 12 % [26], sufficient for direct utilization in pulping liquor preparation and certainly decreasing evaporation costs for applications which demand higher sodium hydroxide concentrations.

Operation of a causticizer at 80 °C or higher is not without an attendant drawback, however. While the solubility product for calcium carbonate (Equation 5.23) is low in pure water, it is higher in water containing sodium hydroxide which, in effect, displaces the equilibrium of the formation reaction

$$K_{sp\,(CaCO_3)} : [Ca^{2+}]\,[CO_3^{2-}]$$
$$= 0.87 \times 10^{-8} \text{ at } 25\,°C \qquad (5.23)$$
$$K_{sp\,[Ca(OH)_2]} : [Ca^{2+}]\,[OH^-]^2$$
$$= 1.3 \times 10^{-6} \text{ at } 25\,°C \qquad (5.24)$$

somewhat to the left. In practical terms this means that at inital 2, 5, 10, 15, and 20 % sodium carbonate concentrations one obtains approximately 99.4, 98.9, 96.5, 91.5, and 83.5 percent conversions of the sodium carbonate to sodium hydroxide, respectively [27]. Industrial yields on sodium carbonate for a lime recycle process on the above bases amount to about 90 % [28] giving a yield per batch,

or yield per pass for a continuous process of about 75 % (0.835 x 0.90) when using a 20 % initial sodium carbonate concentration.

The situation described above, showing gradually diminishing sodium carbonate conversions for increasing concentrations, is the concern that limits the practical sodium carbonate concentrations used in this process to well below what would be technically feasible. An empirical expression which may permit estimation of sodium carbonate conversions from the known solubility products and the initial sodium carbonate concentrations has been derived [29], but has been found to be un-

**Table 5.6** Solubilities of sodium carbonate and
sodium hydrogen carbonate in water [25].

| Temperature °C | Solubility, weight % | |
|---|---|---|
| | Sodium carbonate | Sodium hydrogen carbonate |
| 0 | 6.5 | 6.5 |
| 10 | 11.1 | 7.5 |
| 20 | 17.7 | 8.8 |
| 25 | 27.8 | 9.4 |
| 30 | 31.4 | 10.0 |
| 40 | – | 11.3 |
| 42 | 32.7 | – |
| 50 | – | 12.6 |
| 60 | – | 13.8 |
| 70 | 31.7 | – |
| 80 | 31.4 | – |
| 90 | – | – |
| 100 | 31.1 | 19.1 |

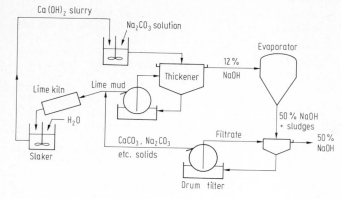

**Figure 5.3.** Flowsheet giving the details of a continuous causticization process for the production of sodium hydroxide. Slaker water is normally derived from lime mud washing. The process may also be operated in a batch mode requiring a smaller number of process vessels, and with, or without lime recycle

reliable [30]. This lack of rigour is partly because the solubility of calcium hydroxide is too high to obtain a solubility product constant, but also so low that only partial ionization precludes the application of ionic equilibria to the determination of ion concentrations. Thus, this expression would predict higher sodium carbonate conversions for higher excess ratios of slaked lime to sodium carbonate whereas in practice this has been found to make very little difference [27].

The insoluble calcium carbonate is separated from the sodium hydroxide solution by simple settling in a batch process, or by settling in a thickener followed by filtration of the more concentrated slurry from the thickener in a continuous process (Figure 5.3). Continuous filtration, as used in this process, involves aqueous sodium hydroxide removal from a sludge by drawing this solution through a filter element fastened to the periphery of a slowly rotating drum. The driving force for fluid flow through the filter is the reduced pressure maintained in the interior of the drum (Figure 5.4). Normally in the larger scale causticization processes the calcium carbonate filter cake, or "lime muds", will be calcined to obtain calcium oxide. After slaking, this regains the calcium hydroxide required for causticization (see Section 5.1.2).

The crude sodium hydroxide solution is obtained at a concentration of about 12 %. This concentration may be suitable for a commercial product if it is to be used at near this concentration at the same location. This might be the case, for example, to prepare the white liquor for the chemical pulping of wood or for soap making. At this stage the crude sodium hydroxide will contain impurities to the extent of about 2-3 % sodium carbonate and traces of sodium chloride.

To deliver this product any significant distance, however, shipping costs are greatly reduced if most (or all) of the water is removed first. Water evaporation is carried out either batchwise, in corrosion resistant close-grained cast iron pots over an open flame, or in continuous caustic evaporators heated via an industrial heat exchange fluid (Section 6.1). At a concentration of 50 % sodium hydroxide, when cooled, sodium carbonate and sodium chloride residues are less soluble and crystallize out. These salts are removed from the 50 % product by settling or filtration. For distant shipping or small scale customers, all the water may be removed from the solution. Because of the strong

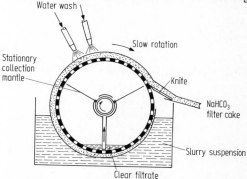

**Figure 5.4.** Cross section through the drum and slurry trough of a drum filter, used for the continuous separation of a solid from a liquid suspension. This version includes a water wash for scavenging the last traces of a solution of value. Sodium hydroxide washings are retained in the process by collecting them separately in the filter and using this to dissolve fresh sodium carbonate, and to slake burned burned lime. (Redrawn from Chemical Technology: An Encyclopaedic Treatment. L.W. Codd, et al., editors, Barnes and Noble, New York, 1968. Vol. 1, p. 23.)

affinity of sodium hydroxide for water this requires temperatures of 500 to 600 °C, leaving molten sodium hydroxide containing less than 1 % water [31]. If this sodium hydroxide is manufactured from Solvay sodium carbonate it would give roughly the following analysis [26]:
NaOH, 98,62 %; $Na_2CO_3$, 0.90 %;
NaCl, 0.30 %; $Na_2SO_4$, 0.18 %.
Detailed use profiles for sodium hydroxide are given in chapter six.

## 5.5.1 Emission Controls for the Causticization Process

Emission control problems in a causticization operation arise chiefly in dust control of the exit gases in the lime kiln since the water circuit is virtually self-contained. Effective containment is obtained by the use of scrubbers which achieve some 99 % dust mass removal from the exhaust gases [32]. Spent scrubber liquor may be returned to the causticization circuit for recycle. Using spent scrubber liquor either for the slaking of lime, or to prepare fresh sodium carbonate solutions for causticization, avoids creating a water emission problem from this aqueous waste stream. It also improves the raw material balance of the process. Some facilities use electrostatic precipitation for lime kiln dust control, which avoids generation of a waste water stream. The captured dust can be recycled to the slaker, avoiding dust disposal costs and assisting the material balance of the operation.

## Relevant Bibliography

1. S.R. Dennison, Fine Natural Calcium Carbonate in Paper Coatings, Tappi 62, 65, Jan. 1979
2. M. Hartman, J. Pata, and R.W. Coughlin, Influence of Porosity of Calcium Carbonates on their Reactivity with Sulfur Dioxide, Indust. and Eng. Chem., Proc. Design and Development 17, 411, Oct. 1978
3. C. Wen and C. Chang, Absorption of Sulfur Dioxide in Lime and Limestone Slurry: Pressure Drop Effect on Turbulent Contact Absorber Performance, Env. Sci. and Tech. 12, 703, June 1978
4. T.R. Lawall, Flash Calcining Applied to Limestone, Rock Prod. 81, 82, Aug. 1978
5. Ceramic Paper Lining in Rotary Kilns Saves Fuel and Bricks, Fiberfrax 970-J paper, Eng. and Mining J. 178, 171, Sept. 1977
6. New Source Performance Standards Set for Refineries, Pulp Mills and Lime Plants, J. Air Polln Control Assoc. 28, 610, June 1978
7. B.B. Broms and P. Boman, Lime Softening Helps Cooling Tower Operators, Chem. Eng. 86, 60, Feb. 12, 1979
8. L.M. Bingham and P.A. Angevine, Modern Recausticizing and Lime Calcining System, Tappi 60, 111, Nov. 1977
9. Lime, Chem. Eng. News 55 (20), 12, May 16, 1977
10. Product Profile Soda Ash, Can. Chem. Proc. 60 (5), 66, May 1976; and 56 (5), 65, May 1972
11. Industrial Mineral Production; Soda Ash, Min. Eng. 30, 541, May 1978
12. W. Worthy, Natural Product Leads in Soda Ash, Chem. Eng. News 53 (47), 10, Nov. 24, 1975
13. G. Parkinson, Kerr-McGee Expands Soda Ash Output Nine-fold from Searles Lake Brines, Eng. and Mining J. 178, 71, Oct. 1977
14. C.C. Templeton, Pressure-temperature Relationship for Decomposition of Sodium Bicarbonate from 200 to 600 °F, J. Chem. Eng. Data 23, 7, Jan. 1978
15. D.R. Larrymore, Enhancing the Ice Melting Action of Rock Salt, Public Works 109, 54, Aug. 1978

## References

1. Kirk-Othmer Encyclopedia of Chemical Technology, 2nd edition, John Wiley and Sons, New York, 1964, volume 4, pages 2,7
2. Handbook of Chemistry and Physics, 51st edition, Chemical Rubber Publishing Company, Cleveland, Ohio, 1970, page F-64
3. C.J.S. Warrington and R.V.V. Nichols, A History of Chemistry in Canada, Sir Isaac Pitman and Sons, Toronto, 1949, page 96
4. R.N. Shreve and J.A. Brink, Jr., Chemical Process Industries, 4th edition, McGraw-Hill, Toronto, 1977, page 166
5. R.W. Thomas and P. Farago, Industrial Chemistry, Heinemann, Toronto, 1973, page 63
6. C.D. Hodgman and N.A. Lange, Handbook of Chemistry and Physics, 15th edition, Chemical Rubber Publishing Company, Cleveland, Ohio, 1930, page 866
7. W.L. Faith, D.B. Keyes, and R.L. Clark, Industrial Chemicals, 3rd edition, John Wiley and Sons, New York, 1965, page 483
8. R.H. Perry, C.H. Chilton, and S.D. Kirkpatrick, Chemical Engineers' Handbook, McGraw-Hill, Toronto, 1963, page 9-6
9. R.N. O'Brien and W.F. Hyslop, Oxygen Enhancement of the Air, British Patent Application 8,003,884, Jan. 28, 1980
10. Chemical Engineers' Handbook, reference 8, page 3-116
11. Corson (of Lime Hydration Process) Honoured, Chem. Trade J. 156, 414, April 1, 1965
12. Handbook of Chemistry and Physics, reference 2, page B-78

13. F.S. Taylor, A History of Industrial Chemistry, Reprint Edition, Arno Press, New York, 1972, page 182

14. Riegel's Industrial Chemistry, J.A. Kent, Reinhold Book Corp., New York, 1962, page 129

15. F.A. Lowenheim and M.K. Moran, Faith, Keyes and Clark's Industrial Chemicals, 4th edition, John Wiley and Sons, Toronto, 1975, page 706

16. R.N. Shreve and J.A. Brink, Jr., reference 4, page 211

17. G.A. Lane, J.S. Best, E.C. Clarke, D.N. Glew, G.C. Karris, S.W. Quigley, and H.E. Rossow, Solar Energy Subsystems Employing Isothermal Heat Sink Materials, The Dow Chemical Company, ERDA Contract No. NSF-C 906, Midland, Michigan, March 1976

18. F.A. Lowenheim and M.K. Moran, reference 15, page 714

19. Minerals Yearbook 1978-79, Vol. I, Metals and Minerals, U.S. Department of the Interior, Washington, 1980. Page 836

20. Minerals Yearbook 1977, Vol. I, Metals, Minerals and Fuels, U.S. Department of the Interior, Washington, 1979, pages 857, 862, and equivalent pages earlier years

21. Statistical Yearbook 1978, Department of International Economic and Social Affairs, United Nations, New York, 1979, page 301, and equivalent pages earlier years

22. R.N. Shreve, The Chemical Process Industries, McGraw-Hill, New York, 1945, page 274

23. International Critical Tables, E.W. Washburn, Editor, McGraw-Hill, New York, 1930, volume VII, page 305

24. Handbook of Chemistry and Physics, 56th edition, R.C. Weast, editor, Chemical Rubber Publishing Co., Cleveland, Ohio, 1975, page F-86

25. Solubilities of Inorganic and Organic Compounds, H. Stephen and T. Stephen, editors, Pergamon Press, New York, 1963, volume I, page 115

26. Riegel's Industrial Chemistry, J.A. Kent, editor, Reinhold Book Corp., New York, 1962, page 158

27. J.C. Olsen and O.G. Direnga, Indust. Eng. Chem. *33*, 204 (1941)

28. W.L. Faith, D.B. Keyes and R.L. Clark, Industrial Chemicals, 3rd edition, John Wiley and Sons, New York, 1965, page 694

29. L.F. Goodwin, J. Soc. Chem. Ind. *45*, 360-361 T (1926)

30. R.M. Stephenson, Introduction to the Chemical Process Industries, Reinhold, New York, 1966, page 56

31. F.A. Lowenheim and M.K. Moran, Faith, Keyes and Clark's Industrial Chemicals, Wiley-Interscience, Toronto, 1975, page 741

32. H.F. Lund, editor, Industrial Pollution Control Handbook, McGraw-Hill, Toronto, 1971, page 18

# 6 Electrolytic Sodium Hydroxide and Chlorine and Related Commodities

## 6.1 Electrochemical Background and Brine Pretreatment

All electrolytic routes to sodium hydroxide and chlorine from sodium chloride brine have to contend with keeping the highly reactive products separated. Sodium and chlorine would react explosively together to return.

$$Na^+ + e^- \rightarrow Na \ (or \ NaOH + \frac{1}{2} H_2) \quad (6.1)$$

$$Cl^- - e^- \rightarrow \frac{1}{2} Cl_2 \quad (6.2)$$

starting salt, and sodium hydroxide reacts vigorously with chlorine to give sodium chloride and sodium hypochlorite (Equations 6.3, 6.4), so that similar precautions are required with either sodium product. Two main procedures have

$$2 Na + Cl_2 \rightarrow 2 NaCl \quad (6.3)$$

$$2 NaOH + Cl_2 \rightarrow NaCl + NaClO + H_2O \quad (6.4)$$

dominated the methods used to keep these products apart. One involves the separation of the electrochemical cell into two compartments by a porous vertical diaphragm which permits the needed passage of ions, but keeps the products separated. The other employs a flowing mercury cathode to continuously carry away sodium, as an amalgam, from the brine and chlorine components of the cell. Diaphragm cells, with which efficient production can be maintained using natural brines (brines solution-mined from underground deposits) account for some 3/4 of North American chloralkali production. Mercury cells, which generally require a higher degree of brine purity for efficient operation, are used by the majority of European chloralkali producers who mostly use vacuum pan or recrystallized salt to make up the electrolyte feed. Two other types of solution cells have been employed in the past, the Billiter, or Bell-jar type cell in which chlorine was drawn off by a bell-shaped housing suspended in the brine and surrounding the anode [1, 2], the horizontal diaphragm cell [3] which is not presently in use in North America but is still employed in many European installations.

The initial stages of brine pretreatment are quite similar for both the diaphragm and the mercury cell. If calcium sulfate is present, removal requires the addition of a barium carbonate suspension (Equation 6.5) [2] or if present as the chloride it is removed by the addition of sodium carbonate (Equation 6.6).

$$CaSO_4 + BaCO_3 \rightarrow CaCO_3 \downarrow + BaSO_4 \downarrow \quad (6.5)$$

$$CaCl_2 + Na_2CO_3 \rightarrow CaCO_3 \downarrow + 2 NaCl \quad (6.6)$$

Magnesium ion requires a higher pH for effective removal, accomplished by the addition of small amounts of sodium hydroxide (Equation 6.7). The bulk of the precipitated salts is removed by settling, either in the treating unit itself

$$MgCl_2 + 2 NaOH \rightarrow Mg(OH)_2 \downarrow + 2 NaCl \quad (6.7)$$

or in a separate clarifier. This is followed by a polishing filtration to give a brine containing 5 ppm or less of $Ca^{2+}$ and $Mg^{2+}$. Acidification with hydrochloric acid to about pH 2 produces a brine which is then suitable as feed for a diaphragm cell. For mercury cell operation brine purity requirements are more stringent, particularly with regard to traces of transition metal ions which can cause severe process upset by lowering the hydrogen overvoltage on mercury. Normally much of the control required will be achieved by simple adsorption of interfering ions onto the precipitated calcium and magnesium flocs. Iron is simultaneously precipitated as the hydroxide (Equation 6.8). If this is not sufficient, the

$$FeCl_3 + 3 NaOH \rightarrow Fe(OH)_3 \downarrow + 3 NaCl \quad (6.8)$$

general practice is to further increase the volume of precipitate by deliberate addition of calcium and/or iron salts until control of vanadium, chromium, manganese, and iron to 0.5 ppm or less is achieved. To further increase the hydrogen overvoltage on mercury the feed to mercury cells is also acidified with hydrochloric acid to pH 2 to 3 prior to use. From this point on details of the two processes differ and will be treated separately.

## 6.1.1 Brine Electrolysis in Diaphragm Cells

The first electrolytic production of chlorine was by the electrolysis of a potassium chloride brine with the coproduction of potassium hydroxide, by the Griesheim Company, in Germany, in 1888 [2, 4]. Sodium hydroxide and chlorine were first commercially produced electrolytically in 1891 in Frankfurt, Germany using technology advanced in the patents of Mathes and Weber and in 1892, in the U.S. at Rumford Falls, Maine by the Electrochemical Company [2]. Both of the early cells used a mode of operation which amounted to a scaled-up version of a laboratory electrolyzer (Figure 6.1a) and were operated in batch modes.

The cell was loaded with saturated potassium chloride brine, indirectly heated to 80-90 °C with steam pipes, and then a current of about 3,500 amperes was passed through the cell for a period of about three days. During this time the chlorine produced was collected, and the potassium hydroxide concentration in solution rose to about 7 % (Equations 6.9, 6.10). The electrolyte was then pumped out and processed to recover the potassium hydroxide, and the cell was recharged with fresh potassium chloride and water to repeat the whole process on a batch basis.

Anode:  $2\,Cl^- - 2\,e^- \rightarrow Cl_2$        (6.9)

Cathode: $2\,H^+ + 2\,e^- \rightarrow H_2$        (6.10)
    Electrolyte accumulation of $K^+$, $OH^-$.

The Griesheim method was both successful and a significant advance in simplicity on the earlier causticization procedures used to produce potassium hydroxide and chlorine. At the same time, it was slow, and it made for inefficient material handling with the frequent necessity for emptying and filling of cells. It also had a low electrical efficiency, partly because solution concentrations were not optimum and partly because of side reactions caused by diffusion of the products through the diaphragm. These side reactions consumed both electrolytic products, in the process producing potassium chlorate (Equations 6.11-6.15).

$2\,OH^- + Cl_2 \rightarrow 2\,HClO$        (6.11)

$HClO \rightleftarrows H^+ + ClO^-$        (6.12)

$2\,ClO^- \rightarrow ClO_2^- + Cl^-$        (6.13)

$ClO_2^- + ClO^- \rightarrow ClO_3^- + Cl^-$        (6.14)

$K^+ + ClO_3^- \rightleftarrows KClO_3$        (6.15)

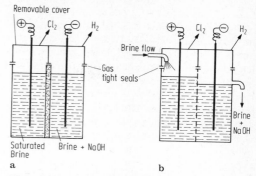

**Figure 6.1. a** Schematic diagram of a Griesheim-type cell which operated on a batch basis. The product was pumped out after a given electrolysis time. **b** Schematic diagram showing the LeSueur percolating diaphragm which changed the mode of cell operation to a continuous basis and decreased the concentration of impurities in the sodium hydroxide product

Many of these difficulties were overcome by the Rumford Falls, Maine, chloralkali operation in 1892, by incorporating the improvements developed by E.A. LeSueur of Ottawa, Canada just prior to this. These innovations centered on a percolating diaphragm which allowed the continuous passage of both ions and brine solution from the anode to the cathode compartments [5] (Figure 6.1b). To obtain brine flow in the desired direction a tube fed saturated brine continuously into the anode compartment. An outlet was placed at the desired cathode compartment solution level, lower than the anode compartment level, allowed for the continuous removal of non-electrolyzed brine and the caustic product. These measures jointly contributed to a positive pressure in the anode compartment and a continuous brine flow through the diaphragm to the cathode compartment. In this way back-diffusion of hydroxide ions through the diaphragm was largely prevented, which both avoided consumption of products and formation of impurities, as well as achieving continuous, rather than batch operation.

These features, the percolating diaphragm together with continuous brine entry and brine and caustic take-off are incorporated into virtually all modern diaphragm cell designs. This includes the North American Hooker, Diamond Alkali, and Dow filter press designs, which all use vertical diaphragms, as well as the European horizontal diaphragm designs such as the Billiter cell and the

I.G. Farben-Industrie modification, which introduced large flutes to the diaphragm and cathode for increased capacity [2, 7, 8]. A particular feature of interest with the horizontal diaphragm is the manner in which hydroxide ion back-diffusion was further minimized by maintaining a hydrogen gas head between the catholyte (cathode compartment electrolyte) and the diaphragm.

The percolating diaphragm of most modern cells consists of a thick asbestos fiber pad, supported on a crimped heavy iron wire mesh or punched steel plate located on the cathode side of the diaphragm. In this way the wire diaphragm support also serves as the cathodic electrode which serves to place it as close as possible to the anode for electrical efficiency while remaining on the cathode side of the diaphragm (Figure 6.2). The iron does not corrode under these severe conditions (contact with the hot saturated brine) because it is made electrochemically cathodic from its function in the cell. In this

way it is continuously supplying electrons to the solution such that the metal itself is not scavenged ($Fe - 3 e^- \rightarrow Fe^{3+}$) to do so.

In this severe service, however, even the durable asbestos fiber pad has to be replaced at intervals ranging from 1 to 3 years. During this time, graphite particles from the anode and insoluble impurities in the brine gradually accumulate on the anode side and into the pad of the diaphragm, reducing its permeability. Also, the passage of electrical power through the diaphragm causes the asbestos fibers of the pad to swell. To maintain the chlorine and caustic production rate under these conditions, the brine level in the anode compartment has to rise, which raises the differential pressure across the diaphragm and in this way maintains the brine flow rate through it. During this time the operating voltage is also raised to offset the decreased ion mobility. Diaphragm permeability can be improved several times, while in service,

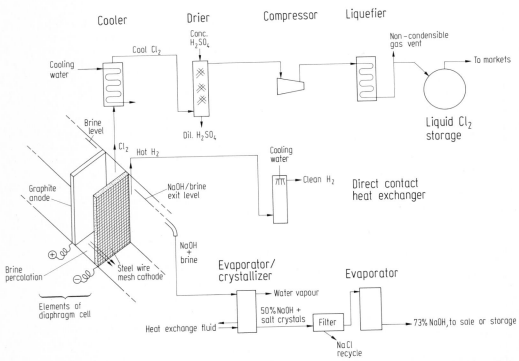

**Figure 6.2.** Diagram of the components of a commercial asbestos diaphragm cell, and flow sheet for the electrolytic production of chlorine and sodium hydroxide. Many commercial cells employ a multiple series of anodes and diaphragms in a single brine container, only one of which is shown here. The asbestos layer is located on the anode side of the steel wire grid cathode, and located relatively much closer than shown here to minimize operating voltage

by a weak acid wash which serves to remove "hardness". But eventually, when there is no further adjustment possible, the cell is shut down and dismantled for overhauling. The spent asbestos is removed from the iron grid with high pressure water jets, and the water slurry is collected in a disposal pit. New asbestos is then slurried in brine, placed in the anode compartment, and by means of a vacuum on the hydrogen exit of the cell it is applied in an even coating to the cathode grids. After fitting new anodes, if necessary, the cell is then reassembled. It may then be placed back in service. The saturated (about 300 g/L NaCl) pretreated brine is preheated to about 90 °C and fed to the anode compartment as a *broken* stream, broken to avoid electrical losses. Preheating the brine both increases cell conductivity by increasing ion mobility from the lower solution viscosities resulting, and decreases chlorine solubility in the brine. In this way, less dissolved chlorine is carried through the diaphragm in the percolating brine which contributes to higher cell current efficiencies and lower concentrations of impurities in the cathode exit brine (Equations 6.11-6.15). It also tends, however, to result in increased graphite wear rates. Anode electrode reactions produce chlorine gas which is removed from the vapour space above the anode compartment by ducting made of polyvinyl chloride or other chemically resistant material, under slight negative pressure (Equations 6.16-6.18). High brine temperatures assist removal of chlorine from the brine

$$Cl^-_{(aq)} \rightarrow Cl^\circ_{(aq)} + e^- \qquad (6.16)$$

$$2\,Cl^\circ_{(aq)} \rightarrow Cl_{2(aq)} \qquad (6.17)$$

$$Cl_{2(aq)} \rightarrow Cl_{2(g)} \qquad (6.18)$$

and the slight negative pressure both aids in chlorine removal and provides a safeguard in the event of any minor leaks in the chlorine collection system.

At the cathode, which is the crimped iron wire grid supporting the diaphragm, electrons are added to water to form hydrogen gas and hydroxide ions (Equations 6.19-6.21). The hydrogen gas is removed under slight positive pressure. It might be thought reasonable to expect that it would be sodium ions

$$H_2O + e^- \rightarrow H^\circ_{(aq)} + OH^- \qquad (6.19)$$

$$2\,H^\circ_{(aq)} \rightarrow H_{2(aq)} \qquad (6.20)$$

$$H_{2(aq)} \rightarrow H_{2(g)} \qquad (6.21)$$

which are electrolyzed, not water, since these are also present at high concentrations [9]. Sodium, if formed, would react with water under these conditions to yield sodium hydroxide and hydrogen gas, the observed products. But examination of the relevant electrochemical data shows that this is not so. The deposition potential for hydrogen is −0.828 volts, which together with the −0.30 volts hydrogen overpotential on iron gives a total of −1.128 volts required for hydrogen generation. The single electrochemical potential for sodium deposition is −2.71 volts [10], more than twice that required for hydrogen deposition. Therefore, hydrogen and hydroxide ion are the direct, not indirect cathode electrochemical products.

In practice, brine and current flows to a diaphragm cell are synchronized to only electrolyze about 30-40 % of the sodium chloride present in the feed brine solution. To electrolyze a larger fraction of the sodium chloride present than this would appear to be advantageous in that less salt would have to be removed from the cell effluent to produce a grade of sodium hydroxide suitable for sale. However, to do this would require an increased current flow for the same rate of brine flow through the diaphragm which would, in turn, necessitate a higher operating voltage. The higher voltage requirement would result in higher electrical power cost for each tonne of product made. Or, to do this would require that the current flow be maintained at its original value and the brine flow rate decreased. Decreasing the brine flow would increase hydroxide ion back-diffusion through the diaphragm with two undesirable consequences. Loss of both hydroxide ion and chlorine cell products would occur, resulting in an impurity in the sodium hydroxide (Equations 6.11-6.15). Also the graphite consumption would increase from the increased rate of reaction of the anode material with oxygen from the electrolyzed hydroxide ion (Equations 6.22-6.23). Even with only about 30-40 % electrolysis of the brine, anode life is limited

$$2\,OH^- \rightarrow H_2O + \tfrac{1}{2}O_2 + 2\,e^- \qquad (6.22)$$

$$2\,C + O_2 \rightarrow 2\,CO \uparrow, C + O \rightarrow CO \uparrow \qquad (6.23)$$
$$CO + O \rightarrow CO_2 \uparrow, 2\,CO + O_2 \rightarrow 2\,CO_2 \uparrow$$

to about two to three years from graphite consumption at the rate of 2-3 kg of carbon per tonne of chlorine [2].

## 6.1.2 Brine Electrolysis in Chlorate Cells

A cell similar in principle to the diaphragm cell described above but operated without a diaphragm, allows the initial electrolytic products to react. These ultimately form sodium chlorate as the final cell product from sodium chloride brine electrolysis [11-13], (Equation 6.24). The component reactions

$$NaCl_{(aq)} + 3\,H_2O \underset{6\ Faradays}{\overset{electrolysis}{\rightarrow}} NaClO_{3(aq)} + 3\,H_2$$

$$(6.24)$$

involved in this process carried out in what is known as a chlorate cell include both chemical disproportionation reactions (Equation 6.11-6.15) and electrochemical reactions (Equation 6.25). Normally chlorate cells are operated

$$6\,ClO^- + 3\,H_2O - 6\,e^- \rightarrow$$
$$2\,ClO_3^- + 6\,H^+ + 4\,Cl^- + {}^3\!/_2\,O_2 \qquad (6.25)$$

with a small concentration of sodium dichromate inhibitor present in the brine to decrease the extent of corrosion caused by hypochlorite ion.

## 6.1.3 Purification of Crude Diaphragm Cell Products

Chlorine and hydrogen as obtained directly from the diaphragm cell are both hot and wet, and the 8 to 12 % sodium hydroxide liquid cell effluent still contains 12 to 19 % dissolved (and "undecomposed") sodium chloride. A few applications of these products can use them in the form obtained directly from the cell. For example, propylene chlorohydrin may be ring-closed to propylene oxide using the crude sodium hydroxide solution still containing sodium chloride (Equation 6.26). However, the majority of uses require at least a simple clean up procedure of the crude products prior to use.

$$\overset{Cl}{\underset{|}{}}\ \overset{OH}{\underset{|}{}}$$
$$CH_3-CH-CH_2 + NaOH/NaCl \rightarrow$$

$$\overset{O}{\overset{/\backslash}{}}$$
$$CH_3-CH-CH_2 + 2\,NaCl + H_2O \qquad (6.26)$$

Elemental chlorine under ordinary conditions is a heavy (dense) greenish-yellow gas, which may be maintained in the liquid form at temperatures below $-35\,^\circ C$, its atmospheric pressure boiling point. Important considerations for its collection

and purification are that it is highly toxic (hence the slight negative pressure for collection), it is highly reactive with many ordinary metals, especially when wet, and with organic substances generally. It is also slightly soluble in water (partially reacting). The first step in clean-up is direct or indirect cooling in process water-cooled heat exchangers constructed of polyvinyl chloride, titanium, or fiber-glass reinforced plastic which removes much of the water vapour and entrained brine droplets from the chlorine. It is then passed countercurrently to concentrated sulfuric acid, which has a strong dehydrating action, in one or more fiber-glass reinforced plastic towers packed with chemical stoneware and placed in series. This packing, which can consist of Berl saddles, for example, provides a sufficiently large surface area for contact that the sulfuric acid consumed is only of the order of 2 to 5 kg per tonne of chlorine. Once dried, steel piping may be safely used to move chlorine but its reactivity must still be kept in mind during compression and liquefaction of the gas since chlorine will spontaneously react with iron at temperatures above $150\,^\circ C$. Care must also be taken to ensure maintenance of anhydrous conditions.

**Plate 6.1.** Chlorine direct contact cooling tower (L.H.S.) and drying towers. Sea water of bottom section oif the cooling tower is heated to prevent loss of dissolved chlorine

Much of the chlorine produced will be simply compressed and piped if it is to be consumed locally, either by the same company producing it ("captive consumption") or by another. But if it has to be shipped any distance chlorine will normally be liquefied by compression and cooling to minimize storage and shipping volumes. Sulfuric acid-filled centrifugal eccentric compressors (Nash pumps), or reciprocating compressors operating in a similar fashion to a piston engine but with graphite piston rings and using no cylinder lubricant were commonly used prior to World War II. For larger facilities today several stage centrifugal turbine compressors are employed [14]. The compressed hot gases are first air or water cooled and then further chilled either by Joule-Thomson effect cooling obtained by expansion of a portion of the cooled compressed gas, or occasionally by using liquid ammonia or other refrigerant in a separate circuit. Liquefaction may be achieved at 700 to 1200 kPa (about 7-12 atmospheres) pressure at 18 °C or at 100 kPa at −40 °C, just below the atmospheric pressure boiling point. Most producers settle for intermediate liquefaction conditions of about 200 kPa and 4 °C [8]. Small amounts of non-condensible gases, "blow gas" or "sniff gas", consisting of traces of hydrogen, oxygen, carbon dioxide, and air accumulate during liquefaction and are vented, usually under aqueous sodium hydroxide stream or the like for safety. In mercury cell plants, the aqueous sodium hypochlorite stream produced in this way may be used for mercury recovery. Hydrogen gas, which tends to get more concentrated in the chlorine gas stream as lique faction proceeds, must not be allowed to rise to 5 % or higher in the liquefaction or venting systems or there is risk of an explosion [8].

The primary hydrogen stream from the cell is also hot and wet, but because of the low solubility in water it is normally simultaneously cooled and cleaned of chlorine, salt, etc., by a direct, counter-current contact with a cool spray of dilute sodium hydroxide in water. Clean hydrogen exits from the top of the cooler through baffles placed so as to remove entrained water droplets from the gas stream. This is occasionally followed by a plastic mesh to assist in removing any water "mists".

Crude 10 % sodium hydroxide containing sodium chloride is upgraded in a similar manner to the processing of the product of the causticization process. Initial evaporation of the water takes place in E-Brite (a high chrome stainless steel), nickel, or nickel-clad steel (to reduce corrosion and iron contamination) multiple-effect evaporators to about 50 % sodium hydroxide concentration. At this concentration sodium chloride is only about 1 % soluble (2 %, dry basis) in the more concentrated caustic so that the bulk of it crystallizes out and is filtered off. This quite pure sodium chloride is recycled to the cells. For many purposes, such as for pulp and paper production, this purity of 50 % sodium hydroxide is quite acceptable. If higher purities are required sodium hydroxide may be separated from residual water and salt, either as minimum melting point (m.p.) double hydrate crystals $NaOH \times 2 H_2O$, m.p. about 6 °C, or as $NaOH \times 3.5 H_2O$, with a m.p. of about 3 °C, sometimes chilled by the direct addition of ammonia [12]. Other proposals advocate the use of counter-current extraction [12, 15]. In either case the bulk of residual sodium chloride and traces of sodium chlorate are removed in the process, producing sodium hydroxide containing only two or three ppm sodium chloride [16]. This purity is equivalent to the purity of the mercury cell product ("rayon grade"). Concentrations of 73 % and 100 % sodium hydroxide (see details, Section 5.1) are also marketed.

## 6.2 Electrochemical Aspects of Brine Electrolysis

The overall chemical changes involved in the electrolysis of aqueous sodium chloride can be expressed as given in Equation 6.27.

$$NaCl_{(aq)} + H_2O_{(l)} \xrightarrow{\text{electrol.}}$$
$$NaOH_{(aq)} + \tfrac{1}{2} H_{2(g)} + \tfrac{1}{2} Cl_{2(g)} \qquad (6.27)$$

To enable thermodynamic insight into the overall energetics of this process, it may be broken down to its component enthalpy changes as follows (Equations 6.28-6.30). To obtain a net enthalpy change for the process of Equation 6.27, the signs of the first two of the component processes have to be changed to correspond to these being exercised in the opposite direction to that given. Then summing the three components (+ 407 + 286 − 469) kJ gives a net enthalpy change of + 224 kJ, that is it is an endothermic process.

The net enthalpy change of a process may be used to obtain a theoretical voltage required for a true reversible electrochemical equilibrium under spec-

$$\Delta^\circ H$$

$$Na_{(s)} + \tfrac{1}{2}Cl_{2(g)} \rightarrow NaCl_{(aq)} \qquad -407 \text{ kJ } (-97.0 \text{ kcal}) \qquad (6.28)$$

$$H_{2(g)} + \tfrac{1}{2}O_{2(g)} \rightarrow H_2O_{(l)} \qquad -286 \text{ kJ } (-68.3 \text{ kcal}) \qquad (6.29)$$

$$Na_{(s)} + \tfrac{1}{2}O_{2(g)} + \tfrac{1}{2}H_{2(g)} \rightarrow NaOH_{(aq)} \qquad -469 \text{ kJ } (-112 \text{ kcal}) \qquad (6.30)$$

ified conditions by employing this enthalpy value in the Gibbs-Helmholtz equation [16]. It also can be summed for the two component electrochemical changes taking place (Equations 6.31, 6.32) [8]. In both cases a value of close to 2.20 volts is obtained, which is defined as the "theoretical cell voltage" or "theoretical decomposition voltage".

$$2 Cl^- \rightarrow Cl_2 + 2 e^- \qquad -1.360 \text{ V} \quad (6.31)$$
$$2 H_2O + 2 e^- \rightarrow H_2 + 2 OH^- \quad -0.828 \text{ V} \quad (6.32)$$

Factors such as the finite spacing required between electrodes, the cathode overpotential requirement for deposition of hydrogen, the good (but not perfect) conductivity of the hot brine, and the resistance to desirable ion diffusion imposed by the presence of the diaphragm (and other minor components [8]) all contribute to the higher actual operating voltages required of about 3.7 to 4.0 volts for a diaphragm cell. The ratio of the theoretical cell voltage to the actual operating voltage is defined as the "voltage efficiency" of the cell. Expressed as a percentage (x 100) the voltage efficiency normally lies in the 55 to 60 % range. Voltage efficiencies usually drop somewhat with increased production rates, as the anodes wear (increasing electrode separation) or as diaphragm permeability decreases.

The current consumption, or electrical flow rate in amperes required to operate the cell is the quantity that directly relates to the amount of electrochemical product collected. One Faraday (one mole of electrons, after the original discoverer) of electricity is required to deposit one gram equivalent weight of a substance in an electrochemical reaction (Equation 6.33).

$$1 \text{ Faraday} = 96\,494 \text{ coulombs (g equiv. wgt.)} \qquad (6.33)$$

$$1 \text{ Faraday} = 96\,494 \text{ ampere seconds/}$$
(g equiv. wgt.)
since 1 coulomb (C) = 1 ampere second (A ·s)

So chlorine, being a monovalent ion, would theoretically require $3.15 \times 10^4$ amperes to be produced at the rate of 1 tonne per day (Equation 6.34).

$$\frac{96\,494 \text{ A} \cdot \text{s/(equiv. wgt.)}}{(24 \times 60 \times 60) \text{ s/day}} \times \frac{10^6 \text{ g/tonne}}{35.453 \text{ g/(equiv. wgt.)}}$$
$$= 3.1502 \times 10^4 \text{ ampere days/tonne chlorine} \qquad (6.34)$$

Side reactions from hydroxide ion back-diffusion across the diaphragm and from the presence of some dissolved chlorine in the brine percolating from the anode to the cathode compartment, as well as the production of small amounts of oxygen at the anode, all consume electrical power without contributing to the primary products of the cell. These factors reduce the cell efficiency to somewhat less than theory, so that about 32.5 kA are actually required for a daily tonne of chlorine. The ratio of the theoretical current requirement to the actual current consumed is called the "current efficiency", and (x 100) is thus about 92-96 %. By convention, the current efficiency is normally understood to refer to the cathode current efficiency, which for a diaphragm chloralkali cell is nearly the same, though marginally higher than the anode current efficiency. Ordinarily, cell operators expect current efficiencies to fall slightly with increasing decomposition efficiencies of the cell (decreasing salt to sodium hydroxide ratios in effluent brine) for the reasons outlined earlier.

Current density refers to the total current flow in kiloamperes divided by the anode electrode area in square meters. It is desirable for the result, $kA/m^2$, to be large particularly for electrochemical processes which yield unstable products. With high current densities electrolytic products are rapidly moved away from the site of formation which decreases side reactions and maximizes current efficiencies [12, 17]. High current densities, however, do increase heat generation, anode wear, and the operating voltage so that lower current densities (and more cells) are better if the cells can be made cheaply.

The overall "energy efficiency" of the cell is the voltage efficiency times the current efficiency and relates to the gross *power* consumption as opposed to the current flow required to produce any given quantity of product, e.g. a tonne of chlorine. It is here that an increased operating voltage can be seen to directly increase the power cost per tonne of product. For example, the mean direct current power requirement for chlorine production in a diaphragm cell amounts to some 2,700-2,900 kWh/tonne [2, 4]. But if the cell is operated below capacity, at 3.2 volts, it would only require some 2,496 kWh/tonne (3.2 x 32.5 x 24 h). Operating the cell at near the end of its diaphragm life and/or pushing the cell to produce somewhat beyond its rated capacity by running at 4.2 volts would require *3,276* kWh/tonne, an additional 31 % more power (and power cost) for the same mass of chlorine production as at 3.2 volts. Since the power costs comprise some 60 to 70 % of the total chlorine production costs [18], these voltage increments represent significant operating cost components. Nevertheless, for temporarily slack or tight markets operating cells at 5 to 10 % below or above rated capacities (or actually taking cells out of service) are ways by which production rate may be most economically adjusted to match demand.

Finally, the electrochemical "decomposition efficiency" is defined as the equivalents produced divided by the equivalents charged (or entering) the electrochemical cell (Equation 6.35). As such, this quantity relates to the

$$\text{Decomposition efficiency} = \frac{(\text{equivalents produced})}{(\text{equivalents charged})}$$
$$= \frac{(\text{moles electrolyzed})}{(\text{moles fed})} \quad (6.35)$$

proportion of the salt present in the incoming solution which is electrolyzed on its passage through the cell. For the reasons previously discussed, for a diaphragm cell this electrolyzed fraction is about 0.30 to 0.40, or (x 100) 30 to 40 %. Since this represents a mole (not weight) fraction electrolyzed, the crude 10-12 % sodium hydroxide in water produced also contains about 12-15 % dissolved sodium chloride.

The interrelationships between these various operating and electrolytic characteristics for diaphragm cells are given in Table 6.1.

## 6.3 Brine Electrolysis in Mercury Cells

Use of the knowledge that an electrolytic cell employing mercury as a cathode would cause sodium deposition in the mercury, rather than hydrogen generation, was made simultaneously in

**Table 6.1** Typical operating characteristics of relatively small and large diaphragm chloralkali cells

| | Rated current flow, amperes | |
| --- | --- | --- |
| | 30,000 [a] | 150,000 [b] |
| Current efficiency | 96.5 | 96 |
| Current density [c], kA/m$^2$ | 1.29 | 2.7 |
| Operating voltage [c] (nominal) | 3.82 | 3.48 |
| Operating temperature, °C | 90 | 85 – 90 |
| Cell effluent: | | |
| Temperature | 95° | — |
| % NaOH | 10.5 | 10.5 |
| NaClO$_3$ g/L | 0.07 | 0.05 – 0.25 |
| Cell output, tonnes per day: | | |
| Chlorine | 0.92 | 4.60 |
| NaOH | 1.03 | 5.19 |
| Power [c], kWh/tonne Cl$_2$ | 3 000 | 2 740 |

[a] Data calculated from Chlorine, its Manufacture, Properties and Uses [2].
[b] Data adapted from The Modern Inorganic Chemicals Industry [4], and S. Puschaver [18].
[c] The improved performance of the larger cell is at least partly the consequence of the dimensionally stable anodes used (see Section 6.5).

1892 by H.Y. Castner in the USA, and by Karl Kellner in Austria, each independently of the other [2]. Their design employed a rocking cell to move mercury from the sodium amalgam (sodium-/mercury alloy or solution of sodium in mercury) forming side of the cell to the sodium amalgam hydrolysis side (Figure 6.3). This design became a commercial success, on further development by Solvay et Cie, unlike the less practical designs such as that put forward by Nolf [19] sometime earlier. The main advantage of this method of brine electrolysis was that there was no direct contact between the brine and the sodium hydroxide streams. Because of this, it was possible to produce a high purity (and high concentration) sodium hydroxide solution directly, without the need for concentration, sodium chloride crystallization etc. as required for the sodium hydroxide product of the diaphragm cell. This provided such clear advantages for the high purity sodium hydroxide market that Castner-Kellner rocking cells were used at an Olin-Mathieson chloralkali operation at Niagara Falls until as recently as 1960 when they converted to a new mercury cell design [2]. These newer designs now universally involve variations on the theme of a long shallow sloping flat-bottomed troughs in which the electrolysis is conducted. Pumps are used to circulate mercury on a continuous basis, as a thin film flowing on the bottom of the trough.

Pretreatment of brine to be used as feed for a mercury cell involves all the same steps as for diaphragm cell feed plus additional precautions to ensure that certain critical transition metals are absent. Molybdenum, chromium, and vanadium ions, in particular, must be maintained at levels below $1 \mu g/L$ (1 ppb) and the concentration of (magnesium + iron) salts combined below $1 mg/L$ (1 ppm) to avoid reducing the hydrogen overvoltage on mercury relative to the sodium deposition voltage [2]. If one or more of these ions exceed these limits, then hydrogen gas evolution in the electrolyzer (rather than sodium deposition in mercury) begins to become a significant process. Thus hydrogen starts to seriously contaminate the normal chlorine product of the electrolyzer. This contamination in turn introduces an explosion hazard at the time the chlorine is compressed, especially if compression is accompanied by liquefaction (Equation 6.36). Generally these interfering ions are removed down to sufficiently low

$$H_{2(g)} + Cl_{2(g)} \rightarrow 2\,HCl_{(g)}\ \Delta H = -92.3\ \text{kJ/mol}$$
$$(-22.1\ \text{kcal/mol})$$
$$(6.36)$$

concentrations by simple adsorption onto the colloidal precipitates of calcium carbonate, magnesium carbonate, and barium sulfate produced by conventional brine pretreatment. If not, more involved measures may be necessary [2].

The only additional brine pretreatment step required is acidification of the brine to about pH 4 or less with hydrochloric acid, a further measure designed to maintain a high hydrogen overvoltage on mercury to minimize the formation of hydrogen in the electrolyzer (Table 6.2). The brine is then preheated to about 65 °C before passage to the electrolyzer, for the same reasons as diaphragm cell feed.

Modern mercury cells comprise two key parts, the electrolyzer and the decomposer, each of which

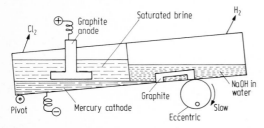

**Figure 6.3.** Vertical section of a Castner-Kellner rocking mercury cell demonstrating the principle of operation. The cell had provision for brine entry and removal of depleted brine from the left compartment, and for purified water entry and removal of sodium hydroxide from the right compartment. The slowly revolving eccentric moved the mercury on a regular basis between the two compartments. A cell divider extended close enough to the bottom of the cell to maintain a mercury seal which prevented any mixture of the water phases of the two compartments

**Table 6.2** A comparison of mercury cell gas composition for graphite versus dimensionally stable anodes [a]

|  | Graphite anode, % | Dimensionally stable anode[b], % |
|---|---|---|
| Chlorine | 97.5 – 98.5 | 97.5 – 98.5 |
| Hydrogen | 0.4 – 0.6 | 0.4 – 0.6 |
| Oxygen | 0.1 – 0.3 | 0.2 – 0.5 |
| Nitrogen | 0.3 – 0.7 | 0.3 – 0.7 |
| Carbon dioxide | 0.4 – 0.9 | 0.3 – 0.7 |

[a] From S. Puschaver [18].
[b] For details, see Section 6.5.

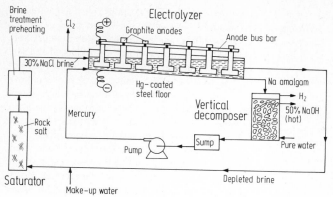

**Figure 6.4.** Mercury cell process for the electrolytic production of chlorine and sodium hydroxide. Packing in the vertical decomposer is graphite lumps. For treatment of cell products, see Figure 6.2

function independently of the other (Figure 6.4). The electrolyzer consists of a slightly inclined (about 1 in 100) ebonite (a hard rubber) lined steel trough with an unlined, smoothly machined steel floor connected to the negative side of a DC potential, and hence the mercury which covers it serves as the cathode. The anode elements of the electrolyzer consist of graphite blocks with spaces between them to allow the escape of chlorine. They are suspended with a 3 to 4 millimeter gap between the mercury film and the graphite. Both saturated brine and mercury enter at the top of the electrolyzer and move co-currently through the electrolyzer while a high current flow is maintained. In its passage through the electrolyzer the sodium chloride concentration in the brine is decreased by electrolysis from an initial 300 g/L or so, to 260-280 g/L. In the process chlorine gas is generated and collected (Equations 6.16-6.18), and a solution of 0.3 to 3 % sodium metal in the mercury stream is produced (Equation 6.37). Since the electrolyzed fraction of the

$$Na^+ + Hg + e^- \rightarrow Na/Hg \qquad (6.37)$$

sodium chloride feed concentration is much lower than for the diaphragm cell, the decomposition efficiencies are correspondingly less, about 7 to 14 %. The mercury cell is operated in this way primarily to assist in keeping the electrolyzer operating voltage down.

The equilibrium electrochemical potential, or theoretical voltage requirement for the overall electrolyzer reaction (Equation 6.38) is 3.13 volts, significantly higher than the 2.20 volts required by the diaphragm cell.

$$NaCl_{(aq)} + Hg_{(l)} \xrightarrow{\text{electrolysis}} Na/Hg_{(l)} + \tfrac{1}{2}Cl_{2(g)} \qquad (6.38)$$

However, because of better internal cell conductivities the operating potential is only about 4.3 volts, giving a voltage efficiency of about 70 %.

The sodium amalgam cathode product of the electrolyzer of the mercury cell is the significant technological difference between this and the diaphragm cell. The production of sodium hydroxide from this intermediate stream is carried out using a separate set of electrochemical reactions conducted in a decomposer or denuder. This unit is usually located below the electrolyzer to allow gravity feed of the sodium amalgam from the electrolyzer to the decomposer (Figure 6.4). De-ionized water is fed into the bottom of the decomposer to provide counter-current flows of sodium amalgam and water. Mercury, stripped (or denuded) of sodium is continuously drawn off the bottom of the decomposer, and a hot solution of 50 to 70 % sodium hydroxide in water plus hydrogen gas from the top. The decomposer is packed with graphite lumps or plates which are in direct contact with each other and with the metal shell of the decomposer. These provide an electrochemical short circuit for the potential generated by the sodium solution reaction. Thus, in this unit the sodium amalgam becomes the anode, electrons being given up by the sodium to the mercury matrix (Equation 6.39)

$$Na(Hg) \rightarrow Na^+ + Hg + e^- \qquad (6.39)$$

in the process of going into solution in the water phase. At the same time hydrogen evolution occurs at the graphite surfaces which provide the electrons for this process by the direct contact with the mercury phase, simultaneously generating hydroxide ions in solution (Equation 6.40). There have been many

$$H_2O + e^- \rightarrow OH^- + \tfrac{1}{2} H_2 \qquad (6.40)$$

attempts to utilize the approximately 1.8 volts [2] generated by these exothermic electrochemical reactions occuring in the decomposer. However, it has not been found possible to both do this *and* maintain high sodium hydroxide concentrations and low residual sodium in the stripped mercury. The sodium hydroxide product obtained from the decomposer of a mercury cell is very pure, containing 0.001 % or less sodium chloride. This product is frequently referred to as "rayon grade caustic" because the high purity and low sodium chloride content makes it particularly suitable for rayon manufacture. This high purity is achieved without any special purification steps as are required to obtain a similar degree of purity from the diaphragm cell product.

Typical operating characteristics of mercury cells, which are generally capable of much higher production rates per cell than the largest commercially available diaphragm cells, are given in Table 6.3. Current efficiencies of the two types of cells are approximately the same, so that the approximately 20 % additional power requirement per tonne of chlorine of the mercury cell results almost entirely from the higher operating voltage. This is not regarded as a significant penalty, since there is considerable energy saving in the areas of caustic evaporation and purification with the mercury cell. However, if the impure direct product of the diaphragm cell can be used for captive consumption of the caustic right at the plant site, there is a significant power saving to be gained by so doing. This is one of the reasons why some chloralkali producers operate both types of cells, so that the lower cost lower grade product may be economically used locally, where feasible. Mercury loss in the products and waste streams of the mercury cell has also become a significant negative factor now in the choice of the chloralkali process used in many cases, as has the wild swings in mercury prices.

## 6.4 Emission Control Aspects of Brine Electrolysis

Any of the products of brine electrolysis, chlorine, sodium hydroxide, and hydrogen can in themselves be hazardous if released. However, releases of these materials largely result from process upsets or accidental breakdowns, which may be minimized by the design and construction of fail-safe plants, as far as feasible, and by safe transport and storage practices [20]. Probably of greater long term public and industrial concern than the occasional inadvertent release of chloralkali products is the mercury loss experienced through the products and other process streams of a mercury cell chlor-

**Table 6.3** Typical operating characteristics of small and large mercury chloralkali cells

| | Rated current flow, amperes | |
| --- | --- | --- |
| | 30,000 [a] | 400,000 [b] |
| Current efficiency, % | 95 | 96 |
| Current density [c], kA/m² | 4.8 | 12 |
| Operating volatage [c] (nominal) | 4.5 | 4.32 |
| Cell output, tonnes per day: | | |
| Chlorine | 0.95 | 12.2 |
| NaOH | 1.08 | 13.8 |
| Power, kWh/tonne Cl₂ | 3 580 | 3 400 |
| Mercury: inventory, kg | 1 410 | – |
| loss rate, g/tonne Cl₂ | 250 | – |
| Graphite consumption, kg/tonne Cl₂ | 2.6 | |
| Electrode gap | – | 3 – 4 mm |

[a] Data calculated from J. S. Sconce [2, page 182].
[b] Data adapted from R. Thompson [4].
[c] Improved operating characteristics of the larger cell achieved by the use of dimensional stable anodes (see Section 6.5).

alkali operation. These losses can also carry over to the products of the diaphragm cell even though this does not use mercury, if a common brine well or common dissolver is used for both sets of cells. Precautions have always been practiced to contain the considerable mercury inventory (Table 6.3) of operating mercury cells, primarily because of the value of the metal itself. However, it was not until about 1970, when it was convincingly demonstrated that not only mercury ions but also the free metal could be converted by natural processes [21, 22] to the far more toxic forms of mercury, methylmercury salts and dimethyl mercury. That this could occur even under water, which has been and still is used as a means of reducing mercury vapour loss from stored mercury inventory, was what alarmed industrialists, toxicologists, and legislators alike and led to the rapid installation of control measures to drastically reduce mercury loss rates in Europe, Japan and North America (Table 6.4) [23].

Electrolyzer mercury losses occur in the spent brine as very finely divided mercury droplets and as dissolved mercuric chloride and also as the very stable tetrachloromercury complex ($HgCl_4^{2-}$). Droplet entrainment in the brine stream occurs

both directly, from the use of a high speed centrifugal pump to return sodium-stripped mercury from the lower end of the decomposer to the upper end of the electrolyzer, and in part just from the interaction of the two flowing liquids. Because of the presence of elemental chlorine in the electrolyzing brine, oxidizing conditions prevail in this phase. Any contact of this hot, low pH layer with suspended and finely divided mercury droplets oxidizes the mercury, to mercuric chloride and tetrachloromercury dianion, which ultimately go into solution in the brine (Equations 6.41, 6.42). Total mercury concentrations in

$$Hg^\circ + NaClO + H_2O + NaCl \rightarrow$$
$$HgCl_2 + 2NaOH \tag{6.41}$$
$$Hg^\circ + NaClO + H_2O + 3NaCl \rightarrow$$
$$HgCl_4^{2-} + 2Na^+ + 2NaOH \tag{6.42}$$

the spent brine may range from 3-5 $\mu g/g$ (ppm) amounting to a potential mercury loss rate of 2-15 g/tonne chlorine produced. However, since the depleted brine normally flows in a closed circuit either to underground wells or to a dissolver, this generally does not result in a direct mercury loss to the biosphere. Brine losses which occur during

**Table 6.4** Grams of mercury lost per tonne of chlorine produced by mercury cell electrolysis plants [a]

| Source of loss | Canadian practice prior to 1970 [a] | Canada 1971 | Sweden 1969 | Stenungsund | Britain 1976 | Canada 1974 | Japan 1974 |
|---|---|---|---|---|---|---|---|
| In Products | | | | | | | |
| Chlorine | 0.05 | 0.05 | <0.5 | slight | 1.4 | 0.6 | 0.024 |
| Caustic soda | 1– 11.5 | 1.75 | 1.0 | 0.4–1.0 | | | |
| To Sewer | | | | | | | |
| Hydrogen coolers | 0–100 | – | – | – | – | – | – |
| Basements drains | 15–150 | 0.15–1.25 | <5.0 | 0.5 | 18 | 0.43 | 0.00 |
| Brine sludges | 1– 25 | – | <2.0 | 0.11 | | | |
| To Atmosphere | | | | | | | |
| End box vents | 1– 10 | 1.25 | not given | – | 10–25 | | |
| Hydrogen | 10–200 | 1.25 | 1.0 | <0.5 | 3.6 | 5.58 | 0.76 |
| Ventilation | 1– 10 | 2.5–5.0 | variable | ≈1.0 | 1.5–2.5 | | |
| Losses not Accounted for | 15 | – | – | – | 0.2 | 6.50 | 8.68 |
| Solid Wastes | | | | | | | |
| From brine system | – | 1.25–5.0 | – | – | | | |
| Caustic filter, cell room muds, effluent treatment | 1.25 | – | – | 0.2–0.6 | 0.82 | 3.00 |
| Totals, g/Tonne $Cl_2$ | | | | | | | |
| Range | 32–510 | 9.45–16.8 | ~9.5 | 2.51–5.62 | 34.6–51.3 | ~14.0 | 12.5–116 |
| Average experience | 196 | – | – | – | – | – | 50.1 (5 y) |
| Prior experience | – | – | 150–200 | new facility | 250 | – | new facility |
| Other Comparisons | 300, U.S.; 50–250, World | 0.77–3.58 [b] (1972) | – | – | – | – | – |

[a] From M. B. Hocking and J. F. Jaworski [23], with permission.
[b] In liquid effluents only.

process upsets, and brine pretreatment sludges (discussed later), however, must both be contained and treated before discharge or disposal to avoid mercury losses.

If diaphragm and mercury cells were being run on the same dissolver or brine wells, operators still had to cope with possible mercury losses in the diaphragm cell products since the resaturated brine from either source still contained from 0.2-1 $\mu g/g$ (0.2-1 ppm) mercury.

Some mercury settling occurs during resaturation, reducing the concentration slightly from that leaving the cell. One way of solving this problem was to add an additional sulfide treatment step [24-27] to the brine pretreatment for the diaphragm cells only (Equations 6.43, 6.44).

$$S_n^{2-} + Hg^\circ \rightarrow HgS \downarrow + S_{n-1}^{2-} \qquad (6.43)$$

$$Na_2S + Hg^{2+} \rightarrow HgS \downarrow + 2\,Na^+ \qquad (6.44)$$

Care was required however, to avoid treating the brine with excess sulfide, a natural tendency, or resolution of mercuric sulfide would occur (Equation 6.45)

$$HgS_{(s)} + S^{2-} \rightarrow HgS_2^{2-}{}_{(aq)} \qquad (6.45)$$

Thorough prior dechlorination of the brine by sparging with air was also necessary, sometimes also followed by the additional precaution of an addition of a mild reductant such as hydrazine prior to sulfide treatment. Otherwise precipitate reoxidation and resolution would occur (Equation 6.46). A

$$HgS + 4\,ClO^- \rightarrow HgCl_4^{2-} + SO_4^{2-} \qquad (6.46)$$

more permanent and less troublesome alternative was simply to separate the diaphragm and mercury cell resaturation systems by installing an additional dissolver or by drilling additional brine wells.

There is some mercury droplet and mercuric chloride as well as some mercury vapour present in the hot moist chlorine gas produced by the mercury cell electrolyzer. Carryover of entrained mercury in the chlorine is decreased by passage through a "Demister", a labyrinth filter of titanium ribbon. The low concentration of mercury vapour present is condensed along with water vapour in an indirect, process water-cooled tube and shell heat exchanger. The condensate, containing 0.1-0.2 $\mu g/g$ mercury [28], may be recycled to the brine circuit or charged to a waste brine treatment system prior to discharge.

Mercury loss in the decomposer section of a mercury cell occurs as finely divided metal droplets in the sodium hydroxide stream. This entrainment happens as a result of the vigorous interaction between the de-ionized water and sodium amalgam streams which is encouraged by decomposer internal geometry to ensure as complete sodium removal from the mercury as feasible. With mercury concentrations in the range of 0.5-10 $\mu g/g$ (0.5-10 ppm) in the sodium hydroxide as it leaves the decomposer this would represent a mercury loss rate of 1-20 g/tonne of chlorine produced. This loss generally occurs indirectly through secondary processes, such as pulp and paper production, and not directly to the environment. Efficient centrifuging or filtration on porous carbon or other caustic resistant media (occasionally both) are procedures capable of reducing the mercury concentrations in 50 % sodium hydroxide to the 0.1 $\mu g/g$ range.

A far more significant mercury loss used to occur from the hydrogen stream of the decomposer. Hydrogen, having a relatively low water solubility, used to be efficiently cooled and washed by direct counter-current contact with a water shower. Since it left the decomposer hot and nearly saturated in mercury vapour, i.e. it contained about 297 mg/m$^3$ Hg in hydrogen at 60 °C, this loss represented about 92 g Hg per tonne of chlorine [23]. When this hydrogen came into direct contact with the cooling water most of the mercury was condensed into the water phase producing a water stream high in suspended mercury. The hydrogen stream too, still contained sufficient mercury to represent a potential loss of 2-11 g/tonne chlorine produced. Current practice dictates that hydrogen be cooled indirectly, in two stages, sometimes including prior compression to 2-3 bars to improve mercury recovery. The first cooling stage normally uses process water to cool economically to about 20 °C. This is followed by a stage using a refrigerant such as vaporizing ammonia, liquid propane or a Freon to chill the hydrogen to the −10 to −20 °C range. Coupled with good mercury mist control as the hydrogen is chilled, these combined measures bring the mercury content down to < 1 mg/m$^3$ and the loss rate to any subsequent applications of the hydrogen down to < 1 g/tonne chlorine.

Sludges result from the pretreatment of re-saturated brine for removal of impurities. They also result from brine which has been treated with sulfide for mercury removal prior to discharge, occasionally necessary since water build-up in the brine circuit is a far more common problem than water depletion. These sludges contain 8 to

15 mg/g (dry basis) or higher mercury concentrations, as a complex mixture of compounds. To recover the mercury from these sludges the bulk of the water is first removed. Following this, the sludge is resuspended in aqueous sodium hypochlorite, such as is available from the absorption of "blow gas" in aqueous sodium hydroxide. The hypochlorite oxidizes the sulfide and any elemental mercury present (Equations 6.41, 6.46) producing a concentrated aqueous stream of dissolved mercury salts. Insoluble sludge components are then removed by filtration, and the solution is then returned to the brine circuit. When this reaches the electrolyzer, electrolytic reduction simply and effectively recovers the dissolved mercury present (Equation 6.47).

$$Hg^{2+} + 2\,e^- \rightarrow Hg^o \qquad (6.47)$$

Direct recovery of mercury in this way provides a neat example of the use of ingenuity and simplicity to recycle mercury salts present in aqueous streams.

Disposal of the residual solid filtered from the hypochlorite dissolving solution, and now containing less then $50\,\mu g/g$ mercury, is usually accomplished by blending this with some inert diluent material such as sand and then burial. The land disposal areas used for this are especially well sealed using plastic film or layers of impermeable clays for containment, and have a ring of drainage tile placed around the perimeter with provision for treatment of any collected eluate. Other measures, like the addition of a chemical reductant such as zinc followed by vacuum distillation of the sludges for recovery of the mercury, have also been used [28, 29].

General mercury vaporization losses to the cell room air amount to 1-5 g/tonne chlorine [28, 30]. Therefore, the cell room requires careful attention to ventilation design to ensure safe working conditions. The general hygiene requirement is that the mercury vapour concentration be kept below $0.05\,mg/m^3$. This is achieved by a combination of tight cell construction and localized hooding and venting of critical cell areas, and by using general cell room ventilation rates of 6-8 air changes per hour. Mercury cell chloralkali plants in moderate climates are able to operate their cells without a building enclosure, which avoids these ventilating problems.

The provisions outlined above, together with good housekeeping practices in the cell room area, particularly as regards mercury handling and containment, do provide a safe working environment. But at the same time the ventilation requirements cause a noticeable elevation of the mercury concentration in the air [31] and in the precipitation [32] in the immediate area of an operating mercury cell chloralkali plant. It would be possible to employ the chemistry of Equations 6.41 and 6.42 in massively-sized scrubbers to treat the large volumes of ventilation air and achieve some reduction of discharge. But the very high installation costs and attendant difficulties have meant that this measure has never been adopted. So the best overall compromise for this area of loss has been in the area of cell design, particularly of the upper and lower end boxes and other cell modifications which maintain as tight vapour control as feasible at the sources of the loss. It may also be economically feasible to treat the lower volume, high mercury vapour concentration streams obtained from the localized hooding ducts.

Complications of mercury containment measures have combined to convince the North American chlorine and caustic producers to build new facilities, or at the time of a major overhaul to convert existing facilities to diaphragm cell technology [33-36]. Nearly 60 % of U.S. chloralkali production was by mercury cells in 1970, whereas by 1977 it had dropped almost half of this value. In Japan guidelines are in place to phase out mercury cell chloralkali production entirely [4]. In Europe these complications have not had such a dramatic effect on the cell choice since a large proportion of the total chlorine is still produced by mercury cells.

With growing environmental awareness conventional diaphragm cells too are not without their hazards, principally from fiber dispersion into the cell products from the asbestos diaphragm. The industry is aware of the hazard potential here, however, and has taken steps to be able to monitor the degree of hazard [37]. While there is a risk of asbestos fiber dispersal into all of the cell streams these are all generally contained either in closed circuits, or are fed as components to following processes so that the risk of outside asbestos dispersal from these sources during normal operation is low. During diaphragm changes, however, when it is required that the fiber be dispersed in water or brine for removal and placement, precautions to avoid both personnel exposure and undue dispersal of waste fiber on final disposal should be taken. While inhalation of large numbers of fibers is known to be injurious, little is yet known about the hazards of oral ingestion [38].

The other emission control aspect of diaphragm cell operation which should be mentioned concerns the use of the crude cell product, still containing sodium chloride, to carry out base catalyzed reactions such as ring closure of propylene chlorohydrin (Equation 6.26) or hydrolysis or chlorobenzene (Equation 6.48). It is desirable that the water phase from these reactions, now high in

$$C_6H_5Cl + NaOH + H_2O \rightarrow$$
$$C_6H_5OH + NaCl + H_2O \qquad (6.48)$$

sodium chloride, be re-used in chloralkali cells, but the traces of organics present after use interfere with efficient cell operation. By providing the right metabolic conditions in ponds and seeding this stream with a heterogeneous bacterial population using sewage sludge it has been found possible in some cases to cleanly remove the organic constituents, in what is referred to as "Bio-ox" units [39]. After clarification, the brine may be resaturated and fed to electrolysis cells without causing problems. The discharge alternative, rather than bio-oxidation and recycle, would result in contributing a waste stream, which has both a high oxygen demand and a heavy ionic loading, on any receiving water body.

## 6.5  New Developments in Brine Electrolysis

Adoption of dimensionally stable anodes (DSA) made of a corrosion resistant titanium screening has been significant by both mercury cell and diaphragm cell operators. A baked-on rutile titanium dioxide paste is used with these electrodes to decrease the electrochemical overvoltage and to improve practical operating current densities. While significantly more expensive than graphite initially, this type of anode has thirty times the conductivity of graphite [4] and allows the maintenance of closer tolerances on the anode, mercury film spacing in the operation of mercury cells, resulting in savings of both power and operator time [18]. For diaphragm cells these electrodes mean that the cell overhaul time is dictated more by the asbestos diaphragm life rather than by the 3 to 4

year graphite anode life that used to be experienced. Not only is anode life lengthened, but because dimensionally stable anodes do not continually shed particles as graphite anodes do, even the percolation lifetime of the asbestos diaphragm is extended somewhat. A further advantage, particularly with diaphragm cells, is a decrease in the carbon dioxide content of the chlorine [18] (Tables 6.2, 6.5) and a significant (2 to 5 %) increase in chlorine yields.

A further experimental improvement suggests the use of tert-butyl hydroperoxide, as a diaphragm cell cathode depolarizer [40]. This cell would operate at only about 3 volts, and would produce tert-butanol instead of hydrogen in a much more compact cell design.

Even asbestos fiber, which is highly chemically resitant, is degraded and the fiber length shortened somewhat with time during normal diaphragm cell operation. Replacing the asbestos with a more inert synthetic polymer fiber allows longer periods of operation without cell overhaul. Teflon, polytetrafluoroethylene fiber has been successfully used in this capacity, and has been found to maintain diaphragm porosity and percolation capacity longer than is possible with asbestos [4].

Conventional diaphragm cell technology still has the significant disadvantage over mercury cell technology in that the former produces a relatively dilute sodium hydroxide product still mixed with sodium chloride, as opposed to the pure sodium hydroxide solution obtained directly from the mercury cell at high concentrations. But the new ion-

Table 6.5 A comparison of diaphragm cell gas composition for graphite versus dimensionally stable anodes [c]

|  | Graphite anode, % | Dimensionally stable anode, % |
|---|---|---|
| Chlorine | 96.5 – 98.0 | 96.5 – 98.0 |
| Hydrogen | 0.1 – 0.4 | 0.1 – 0.4 |
| Oxygen | 0.5 – 1.5[a] | 1.0 – 3.0[a] |
| Nitrogen | 0.1 – 0.5[a] | 0.1 – 0.5[a] |
| Carbon dioxide | 1 – 2 | 0.1 – 0.3[b] |
| Non-volatiles [d] | 0.02 – 0.24 | 0.008 – 0.016 |

[a] From air.
[b] From brine.
[c] From S. Puschaver [18].
[d] Includes iron, asbestos, sodium chloride, hexachloroethane, chloroform, carbon tetrachloride, etc. Personal communication, R. F. Wilson.

selective membrane cells already in use in the industry to some extent, promise to substantially remove this handicap [17, 36, 41]. These cells use a polyetrafluoroethylene, ion permeable membrane through which no percolation occurs, to replace the usual non-discriminating asbestos (or synthetic polymer) fiber brine percolation diaphragm (Figure 6.5). By using carboxyl or sulfonic acid pendant groups on the membrane it becomes selectively permeable to sodium ions and rejects chloride ions [42-47]. In this way it becomes possible to produce a sodium hydroxide stream low in sodium chloride.

As the saturated brine passes through the anode compartment of a membrane cell it becomes depleted in sodium chloride through the formation of chlorine gas (Equations 6.16-6.18) and through diffusion of sodium ions through the microporous membrane. The pendant negatively-charged groups on the membrane sheet prevent both the forward diffusion of chloride ion and the backward diffusion of hydroxyl ion. Purified water which is added to the cathode compartment is partially electrolyzed to hydrogen gas and hydroxide ions which, when combined with the diffused sodium ions, gives the sodium hydroxide solution cathode product. Since there is neither chloride diffusion nor brine percolation through the diaphragm the product is a nearly pure sodium hydroxide solution in water. Depending on current and water flow rates concentrations of sodium hydroxide of around

15 % are common, and up to 28 % occasionally have been obtained from continuously operating membrane cells. This caustic contains only about $50\,\mu g/g$ sodium chloride [4], as compared to about $30\,\mu g/g$ in 50 % sodium hydroxide from mercury cells. Higher sodium hydroxide ion concentrations from membrane cells lead to decreased current efficiencies from hydroxide ion back-diffusion, the reason for the present sodium hydroxide concentration limitation of these cells. Other remaining technical concerns with the further development of membrane cells are related to both the generally somewhat lower current efficiencies and to membrane lifetimes. At present this is limited to 2 to 3 years operation when this is coupled to much more careful brine pretreatment than is required for conventional asbestos diaphragm cells.

A more drastic realignment of diaphragm cell technology which accomplishes production of sodium hydroxide free of chloride from a sodium chloride/zinc chloride melt has been successfully tested in experimental prototypes [48-50]. In this system a salt mixture is used to obtain an eutectic depression of melting point to about 330 °C rather than the 801 ° required to melt pure sodium chloride. Electrolytic ion mobility is thus obtained from the melt rather than from a solution of the salt. This anolyte melt is separated from the cathode compartment by a sodium ion permeable $\beta$-alumina ceramic sheet (Figure 6.6). Passage of an electric current through the cell, which requires a potential of about 5 volts, produces dry chlorine from the graphite anode and a sodium hydroxide melt (containing a small amount of water) plus hydrogen at the pure nickel

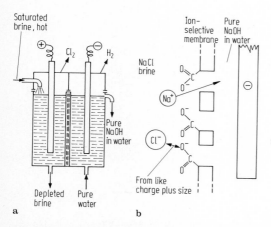

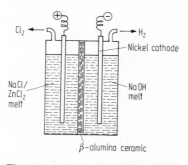

**Figure 6.5. a** Schematic diagram showing a vertical section of a membrane cell for the electrolysis of brine, and **b** a details of a section of membrane illustrating the ion selection mechanism

**Figure 6.6.** Operating details of a $\beta$-alumina ceramic cell for the production of chlorine, hydrogen, and nearly anhydrous sodium hydroxide. The $\beta$-alumina ceramic divider serves an ion selective membrane function to the sodium chloride/zinc chloride anolyte melt

cathode. The much higher temperatures required for operation of this cell are expected to be maintained directly from internal cell resistance. When the smaller surface to volume ratio is considered, together with the larger electrical current flow of full size, commercial cells, this seems to be a reasonable prospect. Commercial success of this cell rests on developments with the $\beta$-alumina ceramic to improve lifetime and ion flow capabilities, and on the completion of larger scale tests where voltage and current efficiencies may be determined more accurately under conditions more appropriate to industrial rates of production.

## 6.6 Chlorine and Sodium Hydroxide Production, Use, and Balance

Production figures for the major producers of sodium hydroxide together with the world totals are given in Table 6.6. From the growth of sodium hydroxide production on a country by country and a world-wide basis, it can be seen that this commodity chemical has shown a slightly more than doubling increase every ten years from 1950 to 1970, an indication that it has reached a relatively "mature" status as a product, and has relatively stable markets. From the trend since 1970 it appears that even this steady growth has slowed signi-

ficantly, so that the production for 1980 will only show about a 50 % increase. This has probably been the result of two worldwide business recessions complicated by increased energy costs during this period.

Electrolytic production of sodium hydroxide and chlorine from sodium chloride solutions so heavily dominates the supply of these chemicals that the production of chlorine can be quite closely approximated by multiplying the sodium hydroxide figures by the fraction calculated from the theoretical ratio of chlorine to sodium hydroxide (70.906:79.996) of 0.886:1.000. For the U.S. production and marketing area for example, the actual production ratio of chlorine to sodium hydroxide is 0.953 to 1.000 [8], very close to this theoretical ratio.

The use pattern for both sodium hydroxide and chlorine is wide, reflecting the broad importance of these commodities to the chemical and other industries (Table 6.7). Most important in the consumption of both commodities is the manufacture of organic chemicals which alone consumes about one third of the sodium hydroxide and almost two thirds of the chlorine produced. The pulping of woods for paper, and the bleaching steps for pulp brightening together consume about 1/7 of the sodium hydroxide and chlorine. The production of inorganic chemicals altogether accounts for only about 1/10 of each product. Water treatment and

**Table 6.6** Major World Producers of Sodium Hydroxide [a]

Thousands of Metric Tonnes

|  | 1950 | 1960 | 1970 | 1978 | 1979 | 1980 | 1981 | 1982 |
|---|---|---|---|---|---|---|---|---|
| Canada | 134 | 339 | 860 | 1 142 | 1 138 | 1 495 | 1 460 | 1 367 |
| China | – | 330 [b] | – | 1 640 | 1 826 | 1 923 | 1 923 | 2 073 |
| France | 242 | 263 | 1 094 | 1 338 | 1 410 | 1 336 | 1 318 | 1 278 |
| Italy | 159 | 426 | 999 | 987 | 984 | 951 | 842 | 841 |
| Japan | 195 | 843 | 2 606 | 2 777 | 3 021 | 3 159 | 2 871 | 2 792 |
| U. Kingdom | – | – | – | 1 033 [d] | 1 095 [d] | 970 [d] | 922 [d] | 995 [d] |
| U.S.A. | 2 278 | 4 510 | 9 200 | 9 719 | 1 124 | 10 549 | 9 663 | 8 293 |
| U.S.S.R. | – | – | 1 783 | 2 763 | 2 680 | 2 755 | 2 800 | 2 867 |
| W. Germany | 336 | 776 | 1 682 | 3 062 | 3 219 | 3 191 [d] | 3 210 [d] | – |
| Other [c] | 269 | 986 | 5 616 | 5 149 | – | – | – | – |
| World | 3 613 [c] | 8 473 [c] | 23 880 | 29 610 | | | | |

[a] Compiled primarily from U.N. Statistical Yearbooks [51], Verband der Chemischen Industrie e.V. (VCI), and [52] and [53].
[b] For 1959, closest year for which data available.
[c] Does not include the U.S.S.R.
[d] Estimates, from reported chlorine production for these years.

**Table 6.7** Recent use profile for sodium hydroxide and chlorine in the U.S.A. [a]

| Sodium Hydroxide | | Chlorine | |
|---|---|---|---|
| Application | % of Total | Application | % of Total |
| Organic chemicals | 34 | Vinyl chloride | 17 |
| Inorganic chemicals | 9 | Propylene oxide | 10 |
| Petroleum refining | 5 | Methylene chloride | 3 |
| Food processing | 1 | $Cl_xC_2H_{6-x}$ | 14 |
| Pulp and paper | 15 | Fluorocarbons | 8 |
| Soap and detergents | 6 | Other organic chemicals | 10 |
| Alumina from bauxite | 5 | Inorganic chemicals | 11 |
| Rayon and cellulose | 3 | Pulp and paper | 13 |
| Textiles | 3 | Water and wastewater | |
| Other, miscellaneous | 7 | treatment | 5 |
| Exports | 12 | Other miscellaneous | 9 |
| | 100 | | 100 |

[a] Based on data from Chem. Engin. News [54] and F. A. Lowenheim and M. K. Moran, Faith, Keyes and Clark's Industrial Chemicals [12].

sewage disinfection, traditional uses of chlorine, only account for about 5 % of the total produced. Chlorine and sodium hydroxide production by the electrolysis of brine solutions necessarily locks the ratio of the two products to the theoretical ratio of the process. When the market for sodium hydroxide exceeds the market for chlorine, the causticization of sodium carbonate to sodium hydroxide (Section 5.3) may be used by some suppliers and consumers to supplement the available sodium hydroxide without producing large amounts of excess chlorine. Another expedient for large scale chloralkali producers faced with this situation is to stimulate the chlorinated solvent, or hydrochloric acid markets in an attempt to increase the consumption of chlorine to restore the balance. But this type of measure is not usually sufficiently rapid to respond over the short term, unless the solvent plant too is operated by the chloralkali producer.

Occasionally the market will swing the other way leaving chlorine in short supply. Fused sodium chloride is commercially electrolyzed in Down's cells [16] to give chlorine and metallic sodium as the cell products. Sodium production in the U.S.A. has averaged 135,000-150,000 metric tonnes annually since 1968 [55], which represents only a 2-3 % contribution to the chlorine supply from this source. In the U.K. it is estimated that as much as 10 % of the available chlorine arises from Down's cell technology [4]. Potassium chloride solutions are also electrolyzed for relatively small volumes of commercial potassium hydroxide, but the contribution to the chlorine supply from this source is

even less than from fused sodium chloride electrolysis.

Chlorine has also been made commercially until very recent years by the chemical oxidation of sodium chloride with nitric acid, the Salt process [2]. Nitrate is used to oxidize chloride to chlorine and in the process is converted to nitric oxide (Equation 6.49). The co-produced sodium nitrate could be utilized and sold as a fertilizer constituent to somewhat affect the costs of chlorine production by this route.

$$6\,NaCl + 12\,HNO_3 + 2\,Na_2CO_3 \rightarrow$$
$$3\,Cl_2 + 10\,NaNO_3 + 2\,CO_2 + 2\,NO + 6\,H_2O$$
$$(6.49)$$

Routes to chlorine from hydrogen chloride oxidation have been refined from the original commercial ventures of Weldon [2] which employed manganese dioxide (Equation 6.50). The Weldon process permitted a maximum of 50 % chlorine recovery from the hydrogen chloride consumed.

$$4\,HCl + MnO_2 \rightarrow MnCl_2 + Cl_2 + 2\,H_2O \quad (6.50)$$

Measures of varying success and complexity were taken to recycle the manganese (II) chloride [56], until the adoption of the Deacon process in which manganese dioxide was conveniently replaced by air as the oxidizer (Equation 6.51).

$$4\,HCl + O_2 \underset{cool}{\overset{Cu/heat}{\rightleftarrows}} 2\,Cl_2 + 2\,H_2O \quad (6.51)$$

Further tuning of the Deacon process resulted in the relatively recent development of the Kel Chlor variant which employs concentrated sulfuric acid as a dehydrating agent. Dehydration greatly assists the success of this process by driving the equilibrium of Equation 6.51 to the right [57-59]. The chlorine product of Equation 6.51 is also efficiently removed by olefin, which has the same net effect on the equilibrium as water removal by sulfuric acid (Equation 6.52). This variant is what is known as the oxychlorination process [60, 61].

$$2\,HCl + CH_2{=}CH_2 + \tfrac{1}{2}\,O_2 \xrightarrow{\text{catalyst}}$$
$$CH_2Cl{-}CH_2Cl + H_2O \qquad (6.52)$$

The steps outlined above may also be carried out separately by electrolysis of a hydrochloric acid solution to obtain hydrogen and chlorine, and the chlorine then used for chlorination (e.g. Equations 6.53-6.55).

$$CH_4 + Cl_2 \rightarrow CH_2Cl_2 + 2\,HCl \qquad (6.53)$$

$$CH_2{=}CH_2 + Cl_2 \rightarrow CH_2Cl{-}CH_2Cl \qquad (6.54)$$

$$CH_2Cl{-}CH_2Cl \rightarrow CH_2{=}CHCl + HCl \qquad (6.55)$$
$$\text{vinyl chloride}$$

Another expedient to improve the balance in the supply of chlorine and caustic is to increase the consumption of sodium hydroxide by diverting part or all of this product to production of sodium carbonate (soda ash) [62]. This process is straightforward, requiring simply that sodium hydroxide be contacted with clean flue gas (Equation 6.56). It also simplifies shipment of the product in

$$2\,NaOH + CO_2 + H_2O \rightarrow Na_2CO_3 + 2\,H_2O \qquad (6.56)$$

dry form over having to evaporate to 100 % sodium hydroxide, the alternative. As a general rule, however, recovery of sodium carbonate from naturally occurring trona is less expensive.
In one or more of these ways, these variants of old processes may be used to produce additional chlorine to ease the demand on electrochemical chlorine or to increase the consumption of sodium hydroxide at times when chlorine is in short supply, and so achieve a balance in the supply and demand of these commodities.
The hydrogen co-product from the cells of large chloralkali facilities may be used to contribute to that required for ammonia (Equation 6.57) or

$$N_2 + 3\,H_2 \rightarrow 2\,NH_3 \qquad (6.57)$$

methanol production (Equation 6.58). If the chloralkali plant is operating in

$$CO + 2\,H_2 \rightarrow CH_3OH \qquad (6.58)$$

the vicinity of an oil refinery it can provide a part of the gas requirements of hydrocracking or various hydrotreating processes used in refining. Hydrogen may also be burned in chlorine to profitably produce hydrochloric acid [6] (Equation 6.59), either for their own use or for sale. However, since the quantities of

$$\tfrac{1}{2}\,H_{2(g)} + \tfrac{1}{2}\,Cl_{2(g)} \rightarrow HCl_{(g)}$$
$$\Delta H = -92.3\ kJ\ (-22.1\ kcal) \qquad (6.59)$$

hydrogen produced in all except the largest facilities are relatively minor, many of the smaller plants simply burn the hydrogen in air to produce energy to assist in steam-raising. For a diaphragm cell plant producing 50 % sodium hydroxide this option can reduce the external fuel requirement by as much as one third.

# Relevant Bibliography

1. System Saves Mercury in Chloralkali Electrolysis, Proc. Eng. 7, Feb. 1977
2. L.C. Mitchell and M.M. Modan, Catalytic Purification of Diaphragm Cell Caustic, Chem. Eng. *86*, 88, February 26, 1979
3. R.N. Chakrabarty, A.Q. Khan, and S.N. Chattopadhya, Treatment and Disposal of Caustic Chlorine Waste, Envir. Health *11* (3), 239 (1969)
4. M. Seko, Ion Exchange Membrane Chloralkali Process, Ind. and Eng. Chem., Prod. Res. and Dev. *15*, 286, Dec. 1976
5 .F. Hine, M. Yasuda, and T. Yoshida, Studies on the Oxide-Coated Metal Anodes for Chloralkali Cells, J. Electrochem. Soc. *124*, 500, April 1977
6 .S. von Winbush and A.F. Sammells, Zinc Chloride Molten Salt Cell, J. Electrochem. Soc. *123*, 650, May 3, 1976
7. R.G. Smith, Chlorine: An Annotated Bibliography, The Chlorine Institute, Inc., New York, 1971
8. Environmental Mercury and Man, Pollution Paper No. 10, Department of the Environment, Her Majesty's Stationery Office, London, 1976
9. L. Buffa, Review of Environmental Control of Mercury in Japan, Report No. EPS-WP-76, Environment Canada, Ottawa, 1976
10. Mercury-bearing Wastes, Waste Management Paper No. 12, Department of the Environment, Her Majesty's Stationery Office, London, 1977

11. The Decline in Mercury Concentration in Fish from Lake St. Clair, 1970-1976, Ministry of the Environment Report No. AQS77-3, Toronto, Ontario, 1977
12. Modern Chlor-alkali Technology, M.O. Coulter, editor, The Society of Chemical Industry, Ellis Horwood Ltd., Chichester, U.K., 1980

# References

1. Riegel's Industrial Chemistry, J.A. Kent, editor, Reinhold Book Corp., New York, 1962, page 172
2. Chlorine, Its Manufacture, Properties, and Uses, J.S. Sconce, editor, Reinhold, New York, 1962, pages 85-89
3. Encyclopedia of Chemical Process Equipment, W.J. Mead, editor, Reinhold, New York, 1964, page 81
4. The Modern Inorganic Chemicals Industry, R. Thompson, editor, The Chemical Society, London, 1977, page 106
5. C.J. Warrington & B.T. Newbold, Chemical Canada, Chemical Institute of Canada, Ottawa, 1970, page 38
6. E.A. LeSueur, British Patent No. 5983, April 7, 1891. Also cited by reference 5
7. H.A. Sommers, Chem. Eng. Prog. 61 (3), 94, March 1965
8. Kirk Othmer Encyclopedia of Chemical Technology, 3rd Edition, John Wiley and Sons, Toronto, 1978, volume 1, page 709
9. C.J.S. Warrington and R.V.V. Nichols, A History of Chemistry in Canada, Chemical Institute of Canada/Pitman and Sons, Toronto, 1949, page 207
10. Handbook of Chemistry and Physics, 56th Edition, R.C. Weast, editor, Chemical Rubber Publishing Company, Cleveland, Ohio, 1976, page D-146
11. Kirk Othmer, Encyclopedia of Chemical Technology, 2nd edition, John Wiley and Sons, Toronto, 1964, volume 5, page 54
12. F.A. Lowenheim and M.K. Moran, Faith, Keyes, and Clark's Industrial Chemicals, Wiley-Interscience, Toronto, 1975, page 716
13. The Encyclopedia of Chemical Process Equipment, W.J. Mead, editor, Reinhold, New York, 1964, page 95
14. E.J. Dickert, Multistage Centrifugal Compressors, Reprint 116, Compressed Air and Gas Institute 4 (1), Cleveland, Ohio (1977)
15. H.C. Twiehaus and N.J. Ehlers, Chem. Ind. (New York) 63, 230 (1948). Chem. Abstr. 45, 4415 (1951)
16. R.N. Shreve and J.A. Brink, Jr., Chemical Process Industries, 4th edition, McGraw-Hill, Toronto, 1977, page 214
17. R.L. Dotson, Chem. Eng., July 17, 1978, p. 106
18. S. Puschaver, Chem. Ind. (London) 236, March 15, 1981
19. A.L. Nolf, British Patent No. 4349, Sept. 12, 1882. Also cited by reference 2
20. R. Papp, Chem. Ind. (London) 243, March 15, 1981
21. S. Jensen and A. Jernelov, Nature 223, 753 (1969)
22. J.M. Wood, F.S. Kennedy, and C.G. Rosen, Nature 220, 173 (1968)
23. Effects of Mercury in the Canadian Environment, M.B. Hocking and J.F. Jaworski, editors, National Research Council of Canada, Ottawa, 1979
24. G.L. Bergeron and C.K. Bon, U.S. Patent 2,860, 952, November 18, 1958. To the Dow Chemical Company
25. L.H. Dunsmoor, U.S. Patent 3,835,217, September 10, 1974. To Pittsburgh Plate Glass Industries Inc.
26. J.H. Entiwsle and R.W. Griffiths, U.S. Patent 3, 718, 457, February 27, 1973. To Imperial Chemical Industries Limited, England
27. D.M. Findlay and R.A.N. McLean, Can. Patent, 1, 083,272, 1979. To Domtar Inc., Montreal
28. E.J. Laubusch, Mercury Emissions Control in Scandinavia, Pamphlet No. R-105 (unpublished), The Chlorine Institute, New York, 1970
29. R.A. Perry, Chem. Eng. Prog. 70 (3), 73-80, March 1974
30. F.L. Flewelling, Chem. in Can. 23 (5), 14 (1971)
31. M.J. Bumbaco, J.H. Shelton, and D.A. Williams, Ambient Air Levels of Mercury in the Vicinity of Selected Chloralkali Plants, Report EPS 5-AP-73-12 Ottawa, July 1973, 43 pages
32. A. Jernelov and T. Wallin, Atmos. Envir. 7, 209-214 (1973)
33. A Chloralkali Plant is converted for Weyerhauser, Hydroc. Proc. 54 (6), 107, June 1975
34. Canada Will House, Can. Chem. Proc. 58 (7), 6, July 1974
35. Industrial Chemicals, Chem. in Can. 28 (6), 13, June 1976
36. C. Law, Can. Chem. Proc. 63 (8), 37, Nov. 14, 1979. M.B. Hocking and M. Gellender, Chem. International, 7, April 1982
37. D.R. Beaman and D.M. File, Anal. Chem. 48 (1), 101-110, January 1976
38. R.W. Durham and T. Pang, Asbestiform Fibre Levels in Lakes Superior and Huron, Scientific Series No. 67, Canada Centre for Inland Waters, Burlington, Ontario, 1976
39. Dow Bioclarification Test Looking Good, Chem. Eng. News 54 (32), 24, Aug. 2, 1976
40. H.B. Johnson, Cited in "Brine Electrolysis: A More Efficient Cathode", Chem. Eng. News 58 (14), 38, April 7, 1980
41. E.G. Grot, U.S. Patent 3,718,627, (1973). D.J. Vaughan, Dupont Innovation 4 (3), 10, Spring, 1973
42. Permionic Membrane Use, Can. Chem. Proc. 58 (7), 6, July 1974
43. E.H. Cook et al, U.S. Patent 3,948,737, (1976).

Cited by J.J. Leddy, J. Chem. Educ. *57* (9), 640, Sept. 1980

44. Membrane Cell, Chem. in Can. *28* (6), 13, June 1976
45. Chloralkali Membrane Cell Set for Market, Chem. Eng. News *56* (12), 20, March 20, 1978
46. Chloralkali Membrane Data, Chem. Eng. News *58* (30), 23, July 28, 1980
47. S.C. Stinson, Chem. Eng. News *60* (11), 22, March 15, 1982
48. Y. Ito, S. Yoshizawa, S. Nakamatsu, J. Appl. Electrochm. *6*, 361 (1976)
49. S. Yoshizawa and Y. Ito, 30th Meeting, International Society of Electrochemistry, Trondheim, Norway, August 26-31, 1979. Extended Abstracts, page 38
50. I. Fukuura, Development of $\beta$-Alumina Ceramics for a Separator for Molten NaCl Electrolysis, in: Applications of Solid Electrolytes, T. Takahashi and A. Kozawa, editors, JEC Press Inc., Cleveland, Ohio, 1980
51. Statistical Yearbook 1978, 30th edition, United Nations, New York, 1979, page 300; plus earlier editions
52. Facts and Figures for the Chemical Industry, Chem. Eng. News *58* (23), 33, June 9, 1980
53. Facts and Figures for the Chemical Industry, Chem. Eng. News *60* (24), 31, June 14, 1982
54. Steady Growth Ahead for Chlorine-Caustic, Chem. Eng. News *54* (21), 11, May 17, 1976
55. Minerals Yearbook, 1978-79, Volume 1, Metals and Minerals, U.S. Bureau of Mines, Washington, 1980, page 843
56. F.S. Taylor, A History of Industrial Chemistry, Arno Press Inc., New York, 1972, page 192
57. New Process May Reshape Chlorine Industry, Chem. Eng. News *47* (19), 14, May 5, 1969
58. Chlorine Recovery (Kel-Chlor Process) Pullman Kellogg, Hydroc. Proc. *56*, 139, Nov. 1979
59. L.E. Bostwick, Chem. Eng. *83* (21), 46, Oct. 11, 1979
60. P.J. Thomas, Chem. Ind. (London), 249, March 15, 1975
61. F.A. Cotton and G. Wilkinson, Advanced Inorganic Chemistry, 3rd edition, Interscience, Toronto, 1972, page 461
62. Chlorine Production is Higher, Can. Chem. Proc. *59* (5), 8, May 1975
63. More HCl Producers Turn to Chlorine Burning, Chem. Eng. News *57* (27), 9, July 2, 1979

# 7 Sulfur and Sulfuric Acid

## 7.1 Commercial Production of Sulfur

Sulfur is widely distributed in the earth's crust, but only to the extent of about 0.1 % by weight. It is found here chiefly as the element, as sulfides, and as sulfates. The annual industrial mass of global sulfur production is about twice that of sodium hydroxide, another important commodity chemical, which is an indicator of the significance of sulfur in the chemical marketplace. However, in keeping with the varied geologic forms in which sulfur occurs and its non-uniform distribution in the crust, as well as the differences in the degree of industrialization of a country, there is considerable variation in the level of production, country by country (Table 7.1). The influence of these diverse factors is reflected in the per capita level of production of say Poland and Canada, which have market circumstances or natural factors which tend to make sulfur production advantageous and produce around 150 and 300 kg per capita per year. These annual per capita production figures are much higher than the 25 to 50 kg per capita per year ex-

perienced by Japan, the U.S.A., and the U.S.S.R., other large scale producers. Thus, the level of sulfur production of a country is markedly influenced by natural or market factors. However, the per capita sulfur consumption is generally a good indicator of the level of industrial activity of a country.

A diversity of methods have been used to commercially recover this element to cope with the varied forms in which sulfur occurs (Table 7.2). For example, about 85 % of Canada's sulfur production is from sulfides removed from natural gas to "sweeten" it. That is it results from, or is an involuntary byproduct of natural gas production and is not a product sought for its own sake. Poland, on the other hand, obtains about 85 % of its annual sulfur by employing Frasch recovery of natural sulfur, a process which is more discretionary in its volume of production.

For the world's two largest producers of sulfur, the U.S.A. and the U.S.S.R., the picture is much more diverse. The Frasch process still dominates the American sulfur industry, for reasons which will become evident later. The working of pyrites for

**Table 7.1** Major world producers of sulfur [a]

| | Thousands of metric tonnes | | | | | |
|---|---|---|---|---|---|---|
| | 1960 | 1965 | 1970 | 1975 | 1980 | 1981 [b] |
| Canada | 249 | 1 877 | 4 440 | 7 538 | 7 405 | 6 842 |
| China | 244 | 254 | 250 | 1 070 | 2 300 | 2 300 |
| France | 791 | 1 521 | 1 736 | 1 994 | 2 213 | 2 156 |
| W. Germany | 84 | 76 | 176 | 1 283 | 1 775 | 1 695 |
| Italy | 120 | 98 | 143 | 703 | 604 | 567 |
| Japan | 256 | 250 | 343 | 2 490 | 2 784 | 2 493 |
| Mexico | 1 349 | 1 586 | 1 381 | 2 217 | 2 252 | 2 202 |
| Poland | 25 | 431 | 2 711 | 5 083 | 5 535 | 5 072 |
| Spain | 42 | 44 | 1 278 | 1 371 | 1 236 | 1 250 |
| S. Africa | n.a. | 7 | 16 | 366 | 618 | 629 |
| U.S.A. | 5 898 | 7 586 | 8 678 | 11 443 | 11 866 | 12 145 |
| U.S.S.R. | 863 | 1 453 | 1 600 | 7 982 | 11 000 | 11 215 |
| Other | 8 769 | 10 091 | 9 728 | 6 664 | 7 047 | 7 103 |
| World | 18 690 | 25 274 | 32 480 | 50 204 | 56 635 | 55 669 |

[a] Calculated and compiled from data in Minerals Yearbooks [98].
[b] Preliminary estimates.

**Table 7.2** Breakdown of the sulfur production of major producing countries in 1980, by methods of recovery [a], given as percentages of total.

| | Canada | Poland [b] | U.S.A. | U.S.S.R. [b] | World |
|---|---|---|---|---|---|
| **Primary** | | | | | |
| Elemental: | | | | | |
| Frasch | — | 84.3 | 53.8 | 8.2 | 25.3 |
| Native | — | 9.4 | — | 25.4 | 6.7 |
| Pyrite | 0.2 | — | 2.7 | 32.3 | 18.2 |
| **Byproduct:** | | | | | |
| Coal | — | — | — | 0.4 | 0.1 |
| Metallurgy | 12.2 | 5.4 | 8.5 | 21.0 | 13.7 |
| Natural gas | 81.0 | — | 14.8 | 10.9 | 21.5 |
| Petroleum | 2.6 | 0.5 | 19.5 | 1.8 | 8.8 |
| Tar Sands | 4.0 | — | — | — | 0.5 |
| Miscellaneous | — | 0.4 [c] | 0.7 | — | 6.2 |
| Totals, percent | 100.0 | 100.0 | 100.0 | 100.0 | 100.0 |
| (Thousands tonnes) | (7 405) | (5 535) | (11 866) | (11 000) | (56 635) |

[a] Table compiled from data in Minerals Yearbooks, 1981 [98].
[b] Estimates throughout.
[c] Produced from gypsum.

sulfur recovery makes the largest contribution to the Russian sulfur industry, and is also of historical interest.

Apart from distorting an orderly production pattern based on market needs, the large scale production of sulfur as a process secondary to natural gas production, such as has developed in Canada since 1951 when the first plant was built [1], also temporarily dislocates any logical local pricing structure. It is evident from Figure 7.1 that, within a transportation cost differential, Canadian ancillary sulfur production tended to hold American sulfur prices down. However, growing energy restrictions, which add significantly to the cost of Frasch sulfur, and the development of a more orderly marketing structure [2] have served to bring North American sulfur prices to a more realistic level.

Traditionally almost 90 % of all sulfur produced has been converted to sulfuric acid [3, 4], which while a broadly based market, is not a volume-flexible one. The only other significant uses of sulfur are by the pulp and paper industry, about 5 %, and in the manufacture of carbon disulfide, 2.5 %, neither significant enough to have a major impact on sulfur markets. Hence, a search for new viable uses for sulfur has been stimulated by the low prices and the widespread surplus of production of sulfur during the late 1960's and early 1970's. The sulfur (or sulfur-oxide) production associated with other industries such as natural gas processing or

the smelting of sulfide ores have been particularly important contributors to the sulfur industry as sulfur gas containment, rather than discharge and dispersal, has been increasingly imposed on this sector with the growing recognition of environmental damage from this source.

Closed-foam sulfur slabs poured *in situ* from a melt have been successfully tested for highway subgrade insulation in arctic regions for prevention of frost-

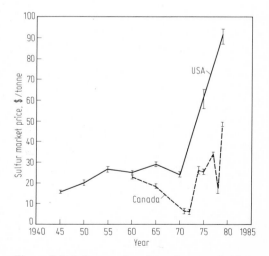

**Figure 7.1.** Influence of market and non-market factors on the price of sulfur, U.S.A. and Canada

induced damage [5, 6]. As little as 10 cm of foam was found to replace the equivalent of 120 cm thickness of gravel in this application [7]. It has also been found useful to incorporate up to about 40 % sulfur by weight into asphalt formulations. Finished paving made from this asphalt plus stone aggregate contains about 2.5 % sulfur, sufficient to contribute to better pavement flexibility, particularly under very cold conditions. It also retains sufficient hardness and durability to be useful under high summer temperatues [1, 8-12]. There has even been limited testing of a paving material where the sulfur plus an "additive" entirely replaces petroleum-based asphalt as an aggregate binder [13, 14]. Also associated with highway uses of sulfur is an application as a traffic paint, applied as a melt. This produces a pale yellow, durable highway marker which can accept traffic within a few seconds after its placement [1, 15].

Construction applications for sulfur have also been tested. Interlocking bricks made predominantly of sulfur have been found to speed construction, but have an inherent fire hazard which restricts their use [16]. Concrete blocks impregnated with molten sulfur have shown a more than ten fold increase in compressive strength, from 5.9 to 83.9 MPa (Megapascals; $1 MPa = 10^6 N/m^2$), at the same time as an improvement in tensile strength from 1.3 MPa to 8.5 MPa for a 13-15 % sulfur content [17, 18]. It also improves its chemical resistance. Sulfur-coated bamboo has been found to be an economical yet effective concrete reinforcing agent as a replacement for steel, in areas where bamboo occurs naturally [19].

Adding sulfur to a sulfur deficient soil can boost crop yields by 1,000 % or more, particularly of seed crops having a high sulfur content [20]. A deficiency of sulfur in the soil has traditionally been corrected by the application of sulfate-containing fertilizers such as ammonium sulfate and superphosphate (Section 8.5). Today, however, methods of sulfur application have been extended to include finely divided sulfur itself in weakly granulated form, or as a 10-12 % solution of sulfur in anhydrous ammonia as ways of raising the sulfur analysis of the fertilizer formulation [21]. Other variations along this theme include the coating of highly soluble nitrogenous fertilizers, such as urea, with sulfur to produce a slow release, and hence greater long term effectiveness [21, 22].

Of these uses some are evidently developing into viable markets for sulfur, although a levelling off of the increases in the sulfur production from recovery processes coupled with increased energy costs and increased sulfur demand for traditional uses has, to a significant extent, offset this pressing need for new markets.

## 7.2  Properties of Elemental Sulfur

Sulfur is an intriguing element in the multitude of forms in which it can be obtained. Under ordinary conditions it occurs as a solid which exists in the well-known bright yellow form as either rhombic or monoclinic crystals, or as a dark, amorphous, moldable mass referred to as plastic sulfur (Figure 7.2). The rhombic form is most stable at room temperature. The transitional equilibrium between the rhombic and the monoclinic forms occurs at 95.5°C. Just above this temperature, at 114.6°C, it melts to a transparent, pale yellow mobile liquid. In all the above forms sulfur occurs as molecules of eight-membered, crown-shaped rings.

Somewhat below 160° the rings start to break up into chains. At about 160° the ring-bonded form of sulfur is largely lost to a much darker, reddish-brown, viscous linear polymeric form, where n, the number of bonded sulfur atoms, is significantly more than eight. Above 187°C depolymerization of this linear form occurs, decreasing the viscosity again. Rapid chilling of the viscous liquid gives the dark plastic sulfur at room temperature, thought to be a linear molecular sulfur occurring in helices. Slower cooling of liquid sulfur allows the transitions through the eight-membered ring, monoclinic and rhombic crystal forms to occur, and yields the more familiar yellow powder modification of sulfur.

## 7.3  Sulfur Recovery by Mining and Retorting

Of both historical interest and as a general method still used to produce nearly a quarter of the world's sulfur (Table 7.2), is the application of conventional mining techniques to bring to the surface lumps of either a volcanic, or one of the many pyritic forms of sulfur. Some of the pyritic forms are pyrite ($FeS_2$) itself, chalcocite ($Cu_2S$), and chalcopyrite ($CuFeS_2$). The sulfur content of the

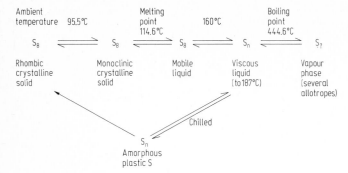

Figure 7.2. Allotropic forms and physical properties of sulfur

raw mineral is usually 25-35 %, but may run as high as 50 %. To obtain the sulfur in separated form the original procedure was to pile the lumps of ore outside and seal these with clay or earth. Burning a part of the contained sulfur sealed into these mounds, with careful control of the air, generated sufficient heat to melt any elemental sulfur present and thermally decompose the pyrite (Equations 7.1-7.3) [23].

$$3\,FeS_2 + 8\,O_2 \rightarrow Fe_3O_4 + 6\,SO_2 + heat \quad (7.1.)$$

$$S + O_2 \rightarrow SO_2 \quad (7.2)$$

$$8\,FeS_2 \rightarrow S_8 + 8\,FeS \quad (7.3)$$

This primitive method gave sulfur recoveries of about 50 % of that contained in the ore, the rest being burned to provide the heat requirement of the process. More advanced sulfur recovery techniques using dual chamber furnaces are capable of recovering as much as 80 % of the contained sulfur [23].

Sulfur dioxide co-produced by this method was generally converted to sulfuric acid, at least in the early operations, by using the chamber process. Using this source for the sulfur dioxide for acid-making put all of the volatilized arsenic vapour from the pyrites directly into the sulfuric acid product. But this aspect was not considered to be of serious concern at the time since most of the sulfuric acid was consumed in the Le Blanc process to produce sodium carbonate (Equations 7.4, 7.5) causing arsenic volatilization and loss

$$2\,NaCl + H_2SO_4 \rightarrow 2\,HCl + Na_2SO_4 \quad (7.4)$$

$$Na_2SO_4 + 2\,C + CaCO_3 \rightarrow$$
$$CaS + Na_2CO_3 + 2\,CO_2 \quad (7.5)$$

with the hydrogen chloride of the first step [24]. If sulfuric acid production was the primary goal of the operation, the air supply to the piles of ore was not restricted and the whole of the contained sulfur was converted to sulfur dioxide (Equations 7.1, 7.2). On the other hand, if the sulfur dioxide could not be used on site it was also possible to practice sulfur dioxide reduction back to elemental sulfur by using incandescent coke [23-27] (Equation 7.6).

$$SO_{2(g)} + C_{(s)} \rightarrow CO_{2(g)} + S_{(g)} \quad (7.6)$$

## 7.4 Frasch Sulfur

Native sulfur deposits were discovered in 1865 while prospecting for oil in a dome-like formation in Louisiana. These beds lay under a thick bed of unstable quicksand, eliminating conventional mining as an economic recovery method. But from initial successful experiments in 1891, H. Frasch succeeded in developing a method using superheated water and air to produce sulfur from this deposit on a commercial scale by 1902. Since this first large scale production by the Frasch method the procedure has been the chief source of the success of the U.S.A. as a dominant sulfur producer. The state of Texas alone supplied 70 % of the world's sulfur by this means in 1935 [28]. Essentially the same method has also been used for most of the sulfur produced by Poland in recent years, although it is not referred to as the Frasch process there [29, 30].

A primary geologic requirement for the success of the Frasch process lies in the rather unique salt dome structures which occur in the coast area of the Gulf of Mexico, in the states of Louisiana and Texas, plus parts of Mexico (Figure 7.3). Only about 28 of the some 400 or so known dome formations in the area have produced commercial quantities of sulfur [23, 31], because only a relatively few possess all the correct geological features to be exploited in this way. Nevertheless, enormous reserves are accessible from a single dome with the right features; one dome in Texas, for example, has produced 12 million tonnes of sulfur during its useful life [32].

Frasch sulfur recovery today is still practised much as originally developed and initially involves conventional oil well drilling equipment to reach the sulfur-bearing calcite zone. After placing a string of three concentric pipes into the well (Figure 7.4), superheated water at 140 to 165 °C is pumped down to the formation in the outermost, 15-20 cm diameter pipe to melt the native sulfur in place. After passage of hot water for some time, heated compressed air is forced down, inducing vigorous agitation of the superheated water and molten sulfur mixture present. Slugs of water, sulfur, and air are forced to enter the intervening sulfur return pipe. The outermost pipe carrying superheated water serves to keep the sulfur (melting point 114.6 °C) molten in the return pipe, which is jacketed by it. At the same time, the heated compressed air both agitates the mixture in the formation, and by means of the air lift principle (Figure

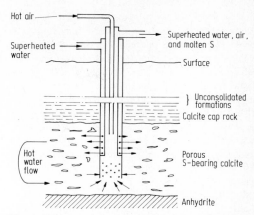

**Figure 7.4.** Details of a conventional Frasch sulfur string. Superheated water (140-160 °C, 17 bar) enters outermost pipe, and air at 150-200 °C and 34 bar the innermost pipe

7.5) and some air incorporation into the sulfur, assists in bringing slugs of molten sulfur (density 2.0 g/cm$^3$) with water to the surface.

At the surface, tanks which are kept hot receive this mixture, still under pressure, and allow a lower liquid sulfur phase to separate from the water. Much of this product is simply filtered, while still molten, to remove any excess carbonaceous material, before it is shipped in this state via insulated tank trucks, rail tank cars, or barge. Traces of heavy petroleum which also occasionally occurs with the sulfur from dome sources are removed by sublimation of the sulfur from the oil, if necessary.

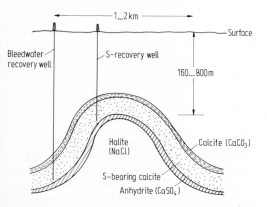

**Figure 7.3.** Vertical cross section of a typical salt dome formation occurring in the area around the Gulf of Mexico. Approximate placements of S-recovery and bleed water wells are shown

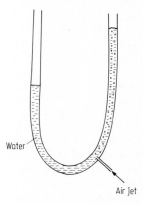

**Figure 7.5.** The air lift principle shown applied to a U-tube in which it is apparent that the tube leg with a mixture of bubbles and water will rise to a higher level to counterbalance the leg having water alone

For a solid sulfur product the sulfur layer is pumped to large outside vats where it is allowed to cool. The simplest containment measure for this method of storage consists of a sheet metal rim which is used to hold the molten sulfur until it solidifies. Once the product is solid, the metal rim is moved upward ready to receive the next layer of molten sulfur. The large, compact dense masses of product thus formed represent one of the least expensive and best environmentally controlled forms of storage of excess sulfur inventory. For delivery, the large mass is broken up, piecemeal, using small explosive charges or mechanical means [33]. In other sulfur recovery operations better control of product particle size is obtained. The molten sulfur is chilled on a steel belt to produce a roughly half-inch thick flake sulfur, or sprayed into water [34] or tumbled in drums to obtain granulated sulfur [35]. Pelletized sulfur, similar in particle size range to the granulated variety, is also obtained when a melt is sprayed into the top of a tower to form droplets of molten sulfur which harden into shot-sized beads as they fall through a current of air. In this form it is referred to as prills [36]. The narrow size range of particles of prilled sulfur, as well as the negligible dust content in this form, make it more convenient to use and as such it normally commands a price premium.

## 7.4.1 Environmental Aspects of Frasch Operations

A major cost item in Frasch processing is fuel for water heating, usually heavy oil or natural gas, since some 4 to 50 tonnes of superheated water are required for each tonne of sulfur recovered at the surface [37]. Some hot water is lost from the actual producing zone via underground fractures and by slumping into the cavities from which sulfur (and some soluble salts) has been withdrawn. But much of this "leaked" water is recovered via "bleed wells" located downslope on the dome from the producing zone (Figure 7.3). The residual heat content of bleed water and production waters which return sulfur to the surface may be salvaged by reheating the once-used water (with some withdrawal to allow a partial purge of impurities) to the required 140-165° production temperatures. This is usually carried out indirectly using steam rather than by a direct firing boiler, to decrease the corrosion and fouling problems which can occur from the high total solids content of the recycle water

[33, 38]. The scale of water heating capacities of these plants ranges from 4 to 40 million liters per day (1-10 million U.S. gallons/day) [33, 39].

Direct risk to boiler equipment and sulfur storage vats from over-burden slumping is avoided by locating these surface facilities a kilometer or so away from producing wells. Risk to other parties not involved in the Frasch operations are minimized by purchase of the surface rights overlying the areal extent of the mined dome. Production upsets from slumping or subsidence are minimized by providing double sets of connecting pipe and much of the associated equipment to minimize down-time (lost production) in the event of a rupture. The extended period of formation heating required before the sulfur production rate becomes steady, and the problems associated with pipe blockage by solidified sulfur make it important that the injection of superheated water be maintained as much as possible without interruption, once production from a particular well has been started. Bleed-water or process water in excess of injection requirements, which contains dissolved salts at similar concentrations to sea water [40], may be safely discarded to the sea after heat exchange energy recovery [32] and treatment to remove toxic sulfides. Initial heat exchange to fresh process water not only conserves fuel but decreases calefaction (thermal pollution) into the receiving body of water. Prolonged aeration of these streams for both stripping and oxidation of sulfide has been used to decrease the sulfidic toxicity of this stream before discharge. When this is carried out in the presence of nickel salts, or by the addition of acid [33] which also increases stripping through the loss of hydrogen sulfide, it has been found possible to reduce the sulfide content down to acceptable levels in 5 to 10 hours. Alternatively, hypochlorite (Equation 7.8) or permanganate have both been found to be effective [41]. These

$$Na_2S + 4NaOCl \rightarrow Na_2SO_4 + 4NaCl \qquad (7.8)$$

chemical oxidants are used particularly for treating small volumes of sulfidic aqueous wastes because of the higher reagent costs. Flue gas scrubbing of the waste water for sulfide removal has also been proposed, but with the associated disadvantage that this method is only workable when it is acceptable to vent hydrogen sulfide with the spent flue gases at the site of operations [33].

# 7.5 Sulfur from Sour Natural Gas

In the same way as Frasch sulfur recovery rapidly became the dominant source of the world's sulfur in the early 1900's, sulfur recovered from the hydrogen sulfide contained in sour natural gas as well as other petroleum streams began to exceed that produced by the Frasch process in 1970 [42]. This present dominance of sulfur from petroleum sources has been given added impetus by the dramatically increased energy costs of the 70's which has had a negative impact, worldwide, on Frasch production. It has also been stimulated by the increased demand for natural gas, which has given a boost to the secondary, sulfur-recovery aspects of natural gas processing. Canada, a major contributor to sulfur from petroleum sources (Table 7.2), started processing of sour natural gas in 1951, and Canadian production from this source probably peaked in 1976 [1]. However, as tar sands development begins to supply a larger fraction of Canadian petroleum, sulfur recovery from these bituminous sources may begin to dominate the Canadian petroleum-recovered sulfur supply. At Lacq, in France, the large amounts of sulfur recovered from natural gas is leading to the development of a sulfur-based chemicals industry at this site [43].

Why should sulfur recovery from natural gas be so important to the natural gas industry? Firstly, hydrogen sulfide, particularly in the presence of moisture with which it almost always occurs, is extremely corrosive to steel pipe lines. Hence, the desirability of removing it as near as possible to the producing well-head. And secondly, while some natural gas streams are virtually free of hydrogen sulfide as they emerge from the deposit, many gas wells yield a product containing 5 to 25 % hydrogen sulfide by volume, and occasional wells as much as 85 % $H_2S$ [44, 45]. The latter wells were usually capped because the expense of sulfur recovery was so great as to make the production of natural gas unprofitable. While natural gas has a reputation for being a clean, and an easily controlled fuel, if burned with even 5 to 25 % $H_2S$ present, as commonly obtained from the well-head, it would generate flue gases with a very high sulfur dioxide content. Also the hydrogen sulfide itself, and its sulfur dioxide combustion product are both very toxic so that there is a safety aspect to its removal as well. Hence, it is only after sour natural gas is treated to remove sulfur compounds to obtain a "sweetened" natural gas that this fuel can truly be regarded as clean.

## 7.5.1 Amine Absorption Process for Hydrogen Sulfide Removal

Sulfur recovery from sour natural gas is conducted in two stages, the first involving removal of the reduced sulfide gases from the natural gas stream. This is usually achieved by scrubbing the gases with an amine solution, for example by countercurrent contacting of the raw gases with a solution of monoethanolamine (MEA) in diethylene glycol (about 1 MEA : 2 DEG by volume) [46]. Monoethanolamine is a condensation product of ethylene oxide with ammonia (Equation 7.9), and is a weakly basic organic liquid. When contacted with

$$NH_3 + H_2C \overset{O}{\overbrace{\qquad}} CH_2 \rightarrow H_2N\text{-}CH_2CH_2\text{-}OH$$
$$\text{boiling point } 170\,^{\circ}C$$
$$(7.9)$$

hydrogen sulfide in natural gas under high pressures and at near ambient temperatures in a tray type absorber (Figure 7.6), this solvent mixture forms a monoethanolammonium salt which remains dissolved in the absorbing fluid to collect as a solution at the bottom of the absorbing unit (Equation 7.10). This

$$H_2N\text{-}CH_2CH_2\text{-}OH + H_2S \rightleftarrows$$
$$\overset{+}{H_3N}\text{-}CH_2CH_2\text{-}OH$$
$$HS^-$$
$$(7.10)$$

scubbing mixture also serves to free the incoming

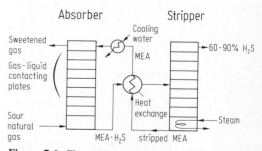

**Figure 7.6.** Flow sheet giving the schematic details of Girbotol and related sour gas sweetening processes. Twenty to twentyfour plates are used in the absorbers and strippers [51] for efficient contacting and desorption with monoethanolamine (MEA)

natural gas of any water vapour. The light hydrocarbon component of the gas stream is only very slightly soluble in the absorbing fluid and passes through the absorber unchanged. If used on a one pass basis the volumes of monoethanolamine required would be prohibitively expensive. However, since the hydrogen sulfide is only weakly associated to the monoethanolamine, it is readily driven off and the MEA recovered by heating this solution indirectly with steam in a regenerator or stripper (Figure 7.6; Equation 7.11). In this way the volatile hydrogen

$$\begin{matrix} + \\ H_3N-CH_2CH_2-OH \\ HS^- \end{matrix} \xrightarrow{100-140\,^\circ C}$$

$$H_2NCH_2CH_2OH + H_2S \qquad (7.11)$$

sulfide (b.p. $-$ 60.7 °C) is separated as an overhead gas stream from the monoethanolamine (b.p. 170 °C) and diethyleneglycol (b.p. 245 °C) which emerge as a regenerated solution, hot, from the bottom of the stripping column. As a heat conservation measure the hot, $H_2S$-lean monoethanolamine stream from the stripper is normally heat exchanged with the relatively cool $H_2S$-rich stream from the base of the absorber, sometimes with additional cooling with process water before it enters the absorber.

Most natural gas as obtained from the well contains almost no carbon dioxide although some wells yield gas containing up to 10 % carbon dioxide and a few contain more than this [44]. Carbon dioxide in the presence of moisture or other proton source is also acidic in nature. If present in the gas it is also collected by the monoethanolamine about as efficiently as hydrogen sulfide although it reacts more slowly [47] (Equations 7.12-7.15; Table 7.3). While

Moist conditions:
$$CO_2 + H_2O \rightarrow H_2CO_3 \qquad (7.12)$$
$$H_2CO_3 + H_2O \rightarrow H_3O^+ + HCO_3^- \qquad (7.13)$$
$$H_2CO_3 + H_2N-CH_2CH_2-OH \rightarrow$$
$$\overset{+}{H_3N}-CH_2CH_2OH\ HCO_3^- \qquad (7.14)$$
Dry conditions:
$$CO_2 + 2\,H_2N-CH_2CH_2-OH \rightarrow$$
$$HO-CH_2CH_2-NH\cdot COO^- +$$
$$\overset{+}{HOCH_2CH_2-NH_3} \qquad (7.15)$$

carbon dioxide removal does raise the fuel value of the natural gas, it also has the undesirable effect of raising the acid gas loading of the absorbing stream in the natural gas sweetening plant, decreasing the possible gas throughputs.

Tertiary amines, dry, do not react with carbon dioxide and yet still rapidly form a salt with hydrogen sulfide. This selectivity feature allows preferential absorption of hydrogen sulfide over carbon dioxide from a natural gas stream. Realization of this has led to methyldiethanolamine ($CH_3N(CH_2CH_2OH)_2$) being proposed as an absorbing fluid in situations where both acid gases occur in the natural gas stream [47]. Even for wet gas streams, where the formation of carbonic acid (Equation 7.12) could cause interaction and absorption on the tertiary amine, it was found that the salt-forming reaction with carbon dioxide was relatively much slower than with hydrogen sulfide. Hence, a short natural gas contact time for the absorbing fluid with moist natural gas streams was found to effectively minimize carbon dioxide absorption and yet still efficiently remove hydrogen sulfide [47].

A further variable of potential concern in gas pro-

**Table 7.3** Solubilities of carbon dioxide and hydrogen sulfide in 15.3% by weight monoethanolamine in water [a]

| Gas partial pressure mmHg | Moles carbon dioxide per mole amine | | Moles hydrogen sulfide per mole amine | |
|---|---|---|---|---|
| | 40 °C | 100 °C | 40 °C | 100 °C |
| 1 | 0.383 | 0.096 | 0.128 | 0.029 |
| 10 | 0.471 | 0.194 | 0.374 | 0.091 |
| 50 | 0.542 | 0.299 | 0.683 | 0.203 |
| 100 | 0.576 | 0.347 | 0.802 | 0.279 |
| 500 | 0.672 | 0.458 | 0.959 | 0.536 |
| 1000 | 0.727 | 0.509 | — | — |

[a] Data selected from Chemical Engineers' Handbook [100].

cessing is the presence of carbon disulfide ($CS_2$) or carbonyl sulfide (COS) in the natural gas stream, whether from natural sources or, for the former component at least, occasionally arising from the use of carbon disulfide to remove sulfur blockages from sour gas wells or components of the Girbotol plant [48]. These particular sulfur compounds react with monoethanolamine to yield complex thiazolidine and oxazolidine heterocycles and polymerization products of these, which are not dissociated in the regeneration step of the gas cleaning plant [49, 50] (Equation 7.16 below).

Diethanolamine ($HN(CH_2CH_2OH)_2$) and di-isopropanolamine are far less susceptible to this irreversible reaction than monoethanolamine and hence have largely replaced MEA at locations where carbon disulfide and carbonyl sulfide occur to a significant extent in the natural gas [48, 51].

A minor variant to the amine scrubbing process described above is the Sulfinol process, which still uses an alkanolamine base, diisopropanolamine (35 %), but in a solvent consisting of a mixture of sulfolane (40 %, tetramethylene sulfone, $(CH_2)_4SO_2$, which is a good hydrogen sulfide solvent) and water [52]. Other processes are based on hydrogen sulfide absorption in aqueous alkaline carbonate solutions, such as used with the Catacarb and Benfield systems [52] (Equations 7.17, 7.18). Still further process variations use physical absorption of

Absorption:
$$Na_2CO_3 + H_2S \quad \rightarrow \quad NaHCO_3 + NaSH$$
$$(7.17)$$

Regeneration:
$$NaHCO_3 + NaSH \underset{heat}{\rightarrow} Na_2CO_3 + H_2S \uparrow$$
$$(7.18)$$

hydrogen sulfide by the solvent, rather than chemical reaction with it, to effect sweetening of natural gas.

## 7.5.2 Claus Process Conversion of Hydrogen Sulfide to Sulfur

Hydrogen sulfide separated from natural gas by amine scrubbing is a highly odourous, toxic, low boiling gas and as such is difficult to store or ship in large quantities. Hence, there is normally a Claus unit closely associated with each amine scrubber which is designed to convert the hydrogen sulfide gas primary product of the scrubber to elemental sulfur, a commodity which is much easier and safer to store and ship in large quantities.

The technology of Claus conversion of hydrogen sulfide to sulfur was worked out in Germany in about 1880, but it was not until 1940 that this process was commercially adopted in the U.S.A. By 1967 annual sulfur production in the U.S. from this process had already reached 4.8 million tonnes, and by 1970 exceeded U.S. Frasch production of sulfur for the first time [23].

Two reactions are employed in Claus units, the first a simple combustion of one third of the hydrogen sulfide stream in air, carried out in a waste heat boiler to capture the excess heat evolved, as steam (Equation 7.19). Enough heat

$$H_2S_{(g)} + \tfrac{3}{2} O_{2(g)} \rightarrow SO_{2(g)} + H_2O_{(g)}$$
$$\Delta H = -519 \text{ kJ} (-124 \text{ kcal}) \qquad (7.19)$$

is retained in the boiler combustion gas stream for the catalytic stage of the process to proceed using this (Figure 7.7). The sulfur dioxide/steam output is blended with the remaining two thirds of the hydrogen sulfide. The resulting gas mixture is then fed to heated iron oxide catalyst beds where sulfur dioxide is reduced, and the hydrogen sulfide is oxidized, both to a sulfur product, with the further evolution of heat (Equation 7.20). A portion of the heat obtained from

$$SO_{2(g)} + 2 H_2S_{(g)} \underset{300-320\,°C}{\rightleftarrows} 3 S_{(1)} + 2 H_2O_{(1)}$$
$$\Delta H = -143 \text{ kJ} (-34 \text{ kcal}) \qquad (7.20)$$

thiazolidine-2-thione    oxazolidine-2-thione    2-oxazolidine

$$(7.16)$$

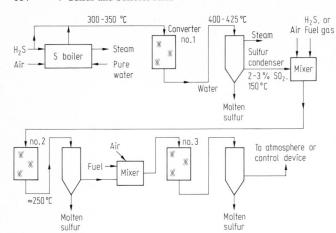

**Figure 7.7.** Schematic diagram of a three stage Claus reactor unit showing steam generation both by direct combustion of one third of the Girbotol output in a hydrogen sulfide fueled boiler, plus further steam recovery from the sulfur condensers. These are designed to remain hot enough to keep sulfur liquid

this reaction is also recovered as steam which, together with that produced by the sulfur-burning boiler, is frequently sufficient to heat the stripper of the associated Girbotol (amine scrubber) plant. The net overall Claus chemistry thus involves oxidation of hydrogen sulfide with air to yield sulfur and water (Equation 7.21). In fact, in most Claus reactors this same overall reaction

$$H_2S + \tfrac{1}{2}O_2 \rightarrow S + H_2O \qquad (7.21)$$

occurs in the first hydrogen sulfide combustion unit, so that a part of the final sulfur product is already present in the combustion gases of the first unit.

The hydrogen sulfide to sulfur conversion part of this process (Equation 7.20) is a reaction in which the equilibrium lies very close to 100 % on the sulfur side at temperatures below 125°, but drops to barely 50 % at 560°C [48, 53] (Figure 7.8). Above this temperature, the equilibrium climbs again, to reach almost 75 % at 1,300°C. With suitable catalysts the rate can be made acceptably fast at temperatures of 300°C or so, sufficient to keep the sulfur in the vapour state, and obtain about 90 % conversion of hydrogen sulfide to sulfur (theoretical equilibrium conversion at 300°C would be about 93 %). In ordinary terms this would seem to give excellent sulfur recoveries. But when it is considered that many Claus plants, such as those operating at Lacq [53] in France and at Waterton and Aquitaine [54] in Canada, are producing sulfur on the scale of 2,000-3,000 tonnes per day, a 10 % sulfur discharge would still represent hundreds of tons of sulfur dioxide loss

per day. For the province of Alberta alone losses from this source could amount to up to 5,000 tonnes per day [54]. However, if the sulfur vapour from one Claus stage is condensed out, and the residual gases are blended with further hydrogen sulfide and passed over catalyst, the equilibrium of Equation 7.20 is displaced further to the right to obtain up to 94 %-95 % conversion to sulfur and process containment is greatly improved (Figure 7.7). A third Claus reactor stage can similarly boost recoveries to the 96 %-98 % range [53, 58]. But this is still not deemed adequate in Japan and the U.S.A., or for large gas processing plants in

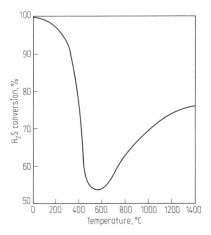

**Figure 7.8.** Equilibrium conversion of hydrogen sulfide to sulfur vapour based on thermodynamic calculations, one atmosphere total system pressure and with no sulfur removal [53, 63]

Alberta where recoveries of about 99.5 % are required [53]. This can theoretically be achieved with a fourth Claus stage but the order of magnitude diminished sulfur return for the large increased investment required for the additional stage makes this alternative economically unattractive relative to other types of control devices [56].

# 7.6 New Developments and Emission Controls, Claus Technology

Primary Claus plant effluent gas from a single stage may still contain 2 to 3 % sulfur dioxide, which would represent both a significant loss of feedstock and an emission problem if not further processed. Both problems may be alleviated by addition of extra stages of Claus reactors but only to approach the limit imposed by the equilibrium relation of Equation 7.20 (Figure 7.8) at Claus operating temperatures, or about 1 % sulfur dioxide. The Sulfreen process takes advantage of Claus chemistry at temperatures of 125-150 °C to use the higher equilibrium sulfur conversion accessible at these temperatures. Low reactor operating temperatures are combined with the practice of sulfur removal as a liquid from the gas phase as it forms, which further displaces the equilibrium to the right. This gives the process a theoretical 100 % conversion efficiency of sulfur dioxide to sulfur [55-57]. As a result of the condensed sulfur, which gradually accumulates on the surface of the catalyst, the reactor has to be purged with a periodic high temperature sulfur vaporization cycle. Exit gas sulfur dioxide concentrations from the process are decreased to the 2,000-2,500 ppm range, as compared to the 1 % or so experienced by the normal Claus technology. Actual sulfur containments of 99.9 % have been claimed for the Sulfreen process [58, 59].

A related procedure is used in the Westvaco Process, except that sulfur dioxide is catalytically oxidized to sulfur trioxide using activated carbon at 75-150 °C. The sulfur trioxide is then hydrated to sulfuric acid which is absorbed onto the active carbon [60]. Sulfur recovery from the sulfuric acid is as sulfur dioxide, which is formed in a regenerator by raising the temperature of the carbon and adding hydrogen sulfide.

Operation of the initial hydrogen sulfide burner in a mode to favour initiation of the high temperature Claus reaction in this unit rather than to optimize on percent sulfur dioxide produced has also been found to raise sulfur recoveries [61]. So has the installation of a continuous ultraviolet spectrophotometer to monitor the relative ratios of sulfur dioxide and hydrogen sulfide in the exit gases from a Claus, first stage reactor [62]. In this way a firmer, appropriate control basis is provided for oxygen or reducing gas addition prior to the subsequent stages.

Other integral measures to improve sulfur recoveries which have a sound theoretical basis but which are more or less speculative at the moment include the use of oxygen-enriched air for primary hydrogen sulfide combustion [48]. This would usefully serve to raise partial pressure of the sulfur gases throughout the process, for the same total system pressure. The integral inclusion of water removal as a part of the sulfur-forming reaction should also serve to raise conversion efficiencies by displacing the equilibrium of Equation 7.20 further to the right [63]. Thus far, neither of these alternatives appear to have been tested under process conditions.

A further alternative to consider in this category is the suggestion that hydrogen sulfide be simply thermally decomposed to sulfur and hydrogen (Equation 7.22) [64]. Thermodynamically it appears that burning only a part of the hydrogen

$$2\,H_2S \rightarrow 2\,H_2 + \tfrac{2}{x}\,S_x$$
$$\Delta H = +159 \text{ kJ} \ (+38 \text{ kcal}) \qquad (7.22)$$

product should be sufficient to provide the energy required for thermal decomposition.

Add-on tail end emission control options include the use of polyethylene glycol scrubbing (the Clauspol process [57, 61), and the use of methane or hydrogen as a sulfur dioxide reducing agent in a follow-up stage, rather than hydrogen sulfide (the SCOT process; Shell Claus Off-gas Treating [57, 59, 61]). Hydrogen sulfide formed in the SCOT process is scrubbed and recycled to the front end of the Claus sequence. Another control option is to react the dilute sulfur dioxide stream with oxygen and water vapour to form sulfuric acid (Equation 7.23), and absorb this onto activated carbon at 75 to 150 °C [65]. A 90 % sulfur dioxide stream, a high enough concentration for recycling, is then regained by a combination of heating in the absence of air and partial reduction with hydrogen sulfide, as already described in detail. There are also a number of other variations of these themes.

$$SO_2 + H_2O + \tfrac{1}{2}O_2 \rightarrow H_2SO_4 \qquad (7.23)$$

## 7.7 Sulfuric Acid

Sulfuric acid, though not often evident in the final product, plays some chemical or refining function in the preparation of a very large number of chemicals (Table 7.4). This wide utilization has occurred because it is a strong, inorganic acid that is also low in cost. Prices over the period 1954 to 1974 have ranged from $25 gradually rising to $36 per tonne in the U.S.A. [66]. This relatively low cost has meant that its chemical properties of high acidity, dehydrating action, ability to sulfonate (as oleum), oxidize, and react with unsaturated hydrocarbons have all been employed commercially on a large scale.

The large volume and wide application of sulfuric acid in the chemical and petroleum refining industries has for a long time meant that the per capita production of sulfuric acid by a country is one of the better chemical activity indicators of industrial development. Less circumstantial anomalies occur in a country-by-country listing of annual sulfuric acid production (Tabel 7.5), a chemical product, than for the per capita sulfur production, an extractive product and hence a commodity which is frequently more strongly influenced by local availability factors than by market requirements. Thus, developed countries average sulfuric acid production levels of from 50 up to 200 kg per capita per year as compared to values of generally less than 5 kg per capita for third world countries.

World sulfuric acid production grew by a factor of 1.4 to 1.6 every ten years since 1930, with an accelerated rate of near doubling during the 1950 to 1960 period, a rate which has slowed again since. A growth in sulfuric acid production exceeding this rate is thus evidence that a country is in the process of development of a technological base, whereas those countries with a parallel or only slightly slower growth rate in their sulfuric acid production generally possessed well developed technologies prior to the period considered (Table 7.5).

Two processes are used commercially to produce sulfuric acid, the contact process, where the important sulfur dioxide to sulfur trioxide oxidation step

**Table 7.4** Use pattern for sulfuric acid in industry, percent of total consumption [a]

| | U.S.A. 1975/78 | United Kingdom 1976 | W. Germany 1976 |
|---|---|---|---|
| Phosphate fertilizers | 38 | 32 | 15 [c] |
| Ammonium sulfate | 7 | | 6 |
| Detergents | — | 11 | c |
| Fibres and cellose film | — | 9 | 46 |
| Petroleum refining | 5 – 8 | — | — |
| Alcohols | 7 | — | d |
| Titanium dioxide | 5 | 15 [b] | 22 |
| Iron and steel pickling | 3 | — | 11 |
| Explosives | 2.5 | — | — |
| Other chemicals | 10 | 16 | d |
| Battery acid | 0.5 | — | — |
| Miscellaneous | 19 – 22 | 17 | d |
| | 100 | 100 | 100 |

[a] Prepared from data of [66, 67, 101, 102]. Due to a different breakdown in the listings used, a hyphen does not necessarily indicate zero consumption of a category.
[b] Includes other paint and pigment uses as well as titanium dioxide.
[c] Detergent consumption included in phosphate fertilizer category.
[d] Plastics, petrochemicals, and miscellaneous uses included with fibres and cellulose film.

**Table 7.5** Major world producers of sulfuric acid [a]

| | Thousands of metric tonnes | | | | | | |
|---|---|---|---|---|---|---|---|
| | 1950 | 1960 | 1970 | 1975 | 1978 | 1980 | 1982 |
| Australia | 617 | 1 776 | 1 762 | 1 770 | 1 799 | 2 175 | 2 027 |
| Belgium | 880 | 1 403 | 1 794 | 1 844 | 2 112 | – | – |
| Canada | 686 | 1 517 | 2 475 | 2 723 | 3 261 | 4 295 | 3 131 |
| China | | | | – | – | 7 640 | 8 170 |
| France | 1 215 | 1 983 | 3 682 | 3 758 | 4 584 | 4 943 | 4 019 |
| W. Germany | 1 446 | 3 170 | 4 435 | 4 157 | 4 671 | 4 108 | 3 781 |
| India | 104 | 354 | 1 189 | 1 333 | 2 107 | – | – |
| Italy | 1 276 | 2 299 | 3 327 | 3 006 | 2 945 | 2 822 | 2 338 |
| Japan | 2 030 | 4 452 | 6 925 | 6 000 | 6 437 | 6 777 | 6 530 |
| Mexico | 43 | 249 | 1 235 | 2 047 | 2 372 | – | – |
| Poland | 285 | 685 | 1 901 | 3 413 | 3 172 | 3 019 | – |
| Spain | 456 | 1 132 | 2 021 | 3 624 | – | – | – |
| U. Kingdom | 1 832 | 2 745 | 3 352 | 3 166 | 3 453 | 3 376 | 2 581 |
| U.S.A. | 11 820 | 16 223 | 26 784 | 29 376 | 34 854 | 40 071 | 29 314 |
| U.S.S.R. | n.a. | 5 398 | 12 059 | 18 645 | 22 411 | 23 033 | 23 800 |
| Other | 2 310 | 4 834 | 13 239 | 16 188 | 28 547 | | |
| World | 25 000 | 48 220 | 86 180 | 101 050 | 121 725 | | |

[a] Derived from data in Statistical Yearbooks for 1957, 1969, and 1978 and Chemical and Engineering News [103, 104]. Those countries that have had an annual production in excess of two million metric tonnes at any time are listed. In addition the following countries had 1978 production in the range of 1 to 2 million metric tonnes per year: Czechoslovakia, Greece, Korea, Netherlands, Roumania, and Tunisia.

is accomplished heterogeneously over a solid catalyst with air, and the chamber process where this transfer of oxygen from air to sulfur dioxide is accomplished with a gaseous catalyst. Of the two, the contact process is used to produce more than 95 % of the supply of sulfuric acid at present, both in Europe [67] and in North America [68].

## 7.7.1 Contact Process Sulfuric Acid

The original conception of the essential features of the contact process is credited to a patent issued to P. Phillips in 1831 [69], but the practice of the principal components taught by this patent took nearly 50 years to bring to commercial success. The key step, the reaction of sulfur dioxide and air over a yellow-hot platinum surface to obtain sulfur trioxide, took extensive development work to obtain reasonable conversions. Coupling this initial catalytic oxidation to the hydration of the sulfur trioxide product was eventually achieved on the scale of 17,000 tonnes per year by 1880, rising to 105,000 tonnes per year by 1890, by Badische Anilin-und-Soda-Fabrik, in Germany [69].

The basic chemical reactions involved in the operation of modern contact acid plants are the same as

outlined in the patent which laid the basis for this industry. However, many improvements in practice, occasioned by the present more detailed knowledge of the catalytic gas phase kinetics involved in the sulfur dioxide oxidation step, as well as a better present day appreciation of the gas-liquid equilibria associated with the hydration step, have now been instituted.

Initially sulfur is burned in air to produce sulfur dioxide and heat (Equation 7.24). Many operations also obtain sulfur dioxide for sulfuric acid

$$S_{(g)} + O_{2(g)} \rightarrow SO_{2(g)}$$
$$\Delta H = -297 \text{ kJ } (-70.9 \text{ kcal}) \tag{7.24}$$

production via the oxidation of pyrites or from the roasting of other sulfidic ores. But because of the additional capital cost involved in dust removal equipment to clean sulfur dioxide from these sources, they are of significant though secondary importance. Roaster sources of sulfur dioxide are more often prompted by emission control incentives on metallurgical operations than by sulfuric acid production incentives.

The second reaction, oxidation of sulfur dioxide to sulfur trioxide with air, is a somewhat less but still highly exothermic, equilibrium reaction (Equation 7.25). It is catalyzed in most modern plants by a 6-10 % vanadium

$$SO_{2(g)} + \tfrac{1}{2}O_{2(g)} \overset{V_2O_5}{\leftrightarrows} SO_{3(g)}$$

$$\Delta H = -98.2 \text{ kJ } (-23.5 \text{ kcal}) \qquad (7.25)$$

pentoxide coating on a support such as powdered pumice or kieselgur. Supported platinum is a more active catalyst than vanadium pentoxide. It gives the same sulfur dioxide for the same contact time at lower initial gas temperatures, and also was used in many of the early contact acid plants [69]. But because of its significantly higher cost and its susceptibility to poisoning, particularly by arsenic, vanadium pentoxide has virtually taken over this function [67]. A modern, potassium hydroxide promoted vanadium pentoxide catalyst bed is certainly quite efficient, being capable of virtually 100 % sulfur dioxide conversion at 380-400 °C with the correct initial gas proportions and sufficient contact time. It also has the durability (freedom from poisoning etc.) to maintain near this level of activity over a 20 year life [69]. Normally, however, a fraction of the catalyst is replaced annually to maintain high activity [66].

A catalyst is a substance which speeds up a chemical reaction but is not itself consumed in the reaction. However, a catalyst cannot alter the equilibrium position of a chemical reaction, i.e. in this case the relative proportions of sulfur dioxide and sulfur trioxide present after reaction. Thus, the reaction rate for an equilibrium reaction, such represented by Equation 7.25, is the speed with which equilibrium is reached (*not* the speed to complete conversion).

The equilibrium for a gas phase reaction may be written in a parallel manner to the operation of the law of mass action in solution equilibria (Equation 7.26), in this case the component concentrations being expressed in

$$k_P = \frac{P_{SO_3}}{P_{SO_2} \times (P_{O_2})^{1/2}} \qquad (7.26)$$

terms of partial pressures. The value of the equilibrium constant, for the correct ratio and concentrations of sulfur dioxide and oxygen (Equation 7.25) has been determined experimentally at a number of temperatures (Table 7.6). The Le Chatelier principle states in essence that if a system in equilibrium is disturbed by a change in conditions, the position of the equilibrium will shift in the direc-

**Table 7.6** Partial pressure equilibrium constants for the oxidation of sulfur dioxide [a]

| Temperature °C | $K_p$ |
|---|---|
| 400 | 397 |
| 500 | 48.1 |
| 600 | 9.53 |
| 700 | 2.63 |
| 800 | 0.915 |
| 900 | 0.384 |

[a] Conditions, one atmosphere total system pressure, partial pressures of components given in atmospheres. Data from M. Bodenstein and Pohl, Z. Electrochem, *11*, 373 (1905). Cited by [68].

tion that will minimize the effect of the change on the system. Since this reaction is exothermic in the direction towards the sulfur trioxide product, from the Le Chatelier principle it would be expected that the equilibrium would lie more towards sulfur trioxide at low temperatures than at high temperatures.

Considering the equilibrium data, at 400 °C, with a value of $K_p$ of 397, the equilibrium of Equation 7.25 is about 96 % on the side of sulfur trioxide. But at this temperature the time required for sulfur dioxide to react with oxygen is relatively long, requiring a large reactor and large catalyst volume (and thus higher capital costs to use these conditions) to obtain significant sulfuric acid production rates.

At 500 °C the rate of reaction is about 100 times as fast as at 400 °C, requiring a much smaller reactor volume for the same sulfuric acid throughput, but the equilibrium constant, $K_p$, drops to about 50. Hence, at this temperature only about 85 % of the sulfur is present as sulfur trioxide.

At 600 °C, the rate of reaction is some 30 to 50 times faster again, requiring an even smaller reactor for the same throughput, but the rate of dissociation of sulfur trioxide to sulfur dioxide becomes appreciable. $K_p$ drops to about 10, giving only about 60-65 % of the sulfur as sulfur trioxide at this temperature, and the remainder as sulfur dioxide. For process purposes there is no point in considering the sulfur equilibrium situation for any higher temperatures than this since with a promoted vanadium pentoxide catalyst bed at 600 °C a 2 to 4 second contact time is already sufficient to obtain essentially equilibrium concentrations at this temperature [66].

For process optimization therefore, advantage is taken of the very fast reaction rates at 550 to 600 °C to operate at these temperatures to about a 60 to 65 % conversion [67]. The gas mixture is then cooled to 400-450 °C, generating further steam, to take advantage of the more favourable equilibrium at this temperature, before being passed over three (or more) additional catalyst beds to reach about 97 to 98 % sulfur dioxide conversion to sulfur trioxide. Occasionally additional air is added at this stage to assist in displacing the equilibrium further to the right [68].

Since one and one-half moles (volumes) of gas on reaction are converted to one mole (volume) of gas in the oxidation of sulfur dioxide it would be expected that carrying out this process under pressure, again by the principle of Le Chatelier, would tend to drive the reaction more to the right. This has been confirmed in practice, but the improvement obtained has not yet been deemed to be worth the additional capital costs required to operate the whole process under pressure by the majority of operators [68, 70]. The same sort of effect may also be achieved by raising the sulfur dioxide and oxygen concentrations, still in a total system pressure of $1.013 \times 10^5$ Pa (equivalent to 1 atmosphere; $1\,Pa = 1\,N/m^2$), as could be obtained by carrying out the sulfur combustion in oxygen-enriched air but again, so far, economics have dictated against this option. New oxygen enrichment developments may change this [71, 72].

These chemical principles are put into practice in a contact acid plant by first burning an atomized jet of filtered molten sulfur in a stream of dry air (Figure 7.9). The air required is dried, prior to combustion, by upward countercurrent passage through a tower containing a stream of near-concentrated sulfuric acid, which is trickled down an acid-resistant chemical stoneware packing. Preliminary air drying is necessary to avoid corrosion problems from the moist gases which would otherwise result after combustion, and to decrease the problems from sulfuric acid mist formation on eventual disposal of the spent waste gases. With a stoichiometric excess of air for this reaction a gas stream containing 9 to 12 % sulfur dioxide in air at a temperature of nearly 1,000 °C is obtained [73].

A fire tube boiler is used to reduce gas temperatures to the 400 to 240 °C range, simultaneously generating steam, before passage of these hot gases over the first catalyst bed of the four-pass converter. Sulfur dioxide is 60-65 % converted to sulfur trioxide at this stage, simultaneously raising gas temperatures to about 600 °C from the exotherm of the reaction and in so doing taking advantage of the very rapid reaction rates at these temperatures. Then, to take advantage of the higher equilibrium proportion of sulfur trioxide accessible at lower temperatures, the gases are again cooled to about 400 °C, generating further steam, before passage over the second catalyst bed. Conversion to 80-85 % sulfur trioxide occurs in the second catalyst bed accompanied by a smaller temperature rise. Cooling by gas-gas exchange and/or by addition of small amounts of dry air at ambient temperatures is used to bring the reacting gases to the 400 °C range for each of the third and fourth catalyst stages. Under ideal conditions these last two stages, in combination, give an overall 98 %

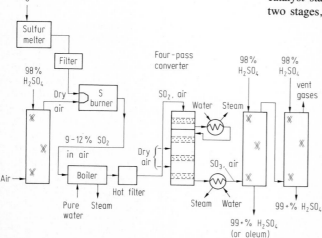

**Figure 7.9.** Contact process for making sulfuric acid and oleum from sulfur. Additional heat control measures are usually present for the last three stages of conversion

conversion of sulfur dioxide (or sulfur) to sulfur trioxide.

After cooling to near ambient temperatures sulfur trioxide concentration is about 10 % by volume. This product is absorbed in concentrated or near-concentrated sulfuric acid, where both absorption and hydration (Equation 7.27)

$$SO_3 + H_2O \rightarrow H_2SO_4$$
$$\Delta H = -130 \text{ kJ} \ (-31.1 \text{ kcal}) \qquad (7.27)$$

occur via counter-current contact in a chemical stoneware packed tower. Using nearly concentrated sulfuric acid for hydration reduces the vigour (the exotherm) of this reaction. Because 98 % sulfuric acid is also the concentration having the lowest vapour pressure (boiling point 338 °C), using this concentration for hydration decreases the tendency to form sulfuric acid mists on absorption. After cooling the stream leaving the bottom of the absorber, product acid is drawn off, and water is added to the remaining acid. Water addition to the process occurs from both the diluted acid produced by the air drier and by the metered addition of pure water, to produce the concentration required for absorption. Commercial grades of sulfuric acid analyze about 96 % $H_2SO_4$ (freezing point, about $-7°C$ [69]) to reduce risks of freezing during storage and transport [76], since pure 100 % sulfuric acid freezes at 10.3 °C [74]. The contact acid plant product is also quite pure, normally containing less than 20 $\mu$g/g (ppm) sulfur dioxide and less than 10 $\mu$g/g iron or nitrogen oxides.

If oleum, a solution of sulfur trioxide in sulfuric acid, is the desired product from the contact plant, this is normally obtained from the first sulfur trioxide absorber. Absorption here, where the highest sulfur trioxide concentration is available in the entering gases, yields up to 20 % oleum directly. This is equivalent to 20 kg of sulfur trioxide dissolved in 80 kg of 100 % $H_2SO_4$. At this concentration the oleum vapour pressure becomes so high that sulfur trioxide loss occurs almost as readily as absorption so that this is also the limit of oleum concentrations available by direct absorption. Higher concentration oleums may be obtained by distillation of the sulfur trioxide (boiling point 45 °C) from 20 % oleum, and condensing this now 100 % sulfur trioxide vapour stream into 20 % oleum until the desired concentration of oleum is obtained. Several commercial oleum grades are marketed, including 60 % and 100 %. At 60 % and higher concentrations it is thought that a significant fraction of the

mixture actually exists as pyrosulfuric acid, $H_2S_2O_7$, a different compound [75]. Pure sulfur trioxide, 100 % oleum, is marketed with about 0.25 % of stabilizers such as sulfur, tellurium, carbon tetrachloride or phosphorous oxychloride added to prevent crystallization or polymerization [69]. Sulfur trioxide is easier to handle in the liquid state. Also, if either crystallization or polymerization were to occur during storage or shipping these processes are sufficiently exothermic that they have the potential to raise the temperature of the bulk of stored sulfur trioxide to above its boiling point. If this should happen it risks sulfur trioxide release or vessel rupture from overpressure.

Operating a contact sulfuric acid plant essentially as described above thus allows straightforward production of concentrated sulfuric acid, oleums, and pure sulfur trioxide which may be used in sulfonations or to chemically rejuvenate process-diluted sulfuric acid by simple addition. Typical raw material and utility requirements are given in Table 7.7, from which it can be seen that the contact process is actually a net *producer* of high pressure steam, sometimes a useful feature in a chemical complex.

Provision of the sulfur dioxide feed gas for a contact plant is dominated by direct sulfur combustion, but many other sources may also be tapped. Pyrite, $FeS_2$, is burned (Equation 7.28) and other sulfidic minerals are roasted (Equations 7.29, 7.30), the

$$FeS_2 + {}^5\!/_2 O_2 \rightarrow FeO + 2 SO_2 \qquad (7.28)$$

$$CuS + {}^3\!/_2 O_2 \rightarrow CuO + SO_2 \qquad (7.29)$$

$$ZnS + {}^3\!/_2 O_2 \rightarrow ZnO + SO_2 \qquad (7.30)$$

Table 7.7 Typical raw material requirements and utilities consumed in the production of one tonne of 100% sulfuric acid [a]

| | Contact Plant 180 tonne/day | Chamber Process, 50 tonne/day |
|---|---|---|
| Sulfur, kg | 337 to 344 | 337 |
| Nitrogen oxides, kg (or ammonia burned) | — | 2 to 2.5 |
| Water, L (process, plus cooling) | 16,700 | 10,000 |
| Electricity, kWh | 5 – 10 | 15 – 16.5 |
| Steam, kg | 1,000 (credit) | — |
| Air, m³ | 7,800 | 8,600 |

[a] Calculated and compiled from [28, 66 and 68].

latter primarily for their metal values rather than for the sulfur dioxide. Extensive dust removal facilities are required to clean up the sulfur dioxide stream from these sources, which adds significantly to the capital cost of the plant and serves to negate the raw material cost advantage conferred by these sulfur sources. Sometimes hydrogen sulfide is simply burned to produce sulfur dioxide when the source of the hydrogen sulfide is near the producing sulfuric acid plant. If this method is used then elemental sulfur production by the Claus process is by-passed.

If, however, none of these more straightforward sulfur sources is available and anhydrite ($CaSO_4$) is available on-site, or close by, then it is possible to practise thermal reduction of anhydrite with coke to obtain sulfur dioxide [76] (Equations 7.31, 7.32). Again the sulfur dioxide gas stream requires

$$CaSO_4 + 2\,C \;\rightarrow\; CaS + 2\,CO_2 \qquad (7.31)$$

$$CaS + 3\,CaSO_4 \overset{\Delta}{\rightarrow} 4\,CaO + 4\,SO_2 \qquad (7.32)$$

efficient dust removal prior to conversion. By using the lime clinker produced from this process as raw material for cement production (about equal amounts of sulfuric acid and clinker are produced) the economics have been sufficiently favourable to have had roughly 20-25 % of United Kingdom (ICI) sulfuric acid produced via this means at one time [77].

# 7.8 Chamber Process Sulfuric Acid

The chamber process for the production of sulfuric acid is by far the older of the two large scale commercial processes. Records exist at least back to 1746, when Dr. Roebuck, in Birmingham was burning sulfur and nitre ($KNO_3$) in the presence of steam in lead-lined rooms (chambers, hence the name given to the process) to produce sulfuric acid [69, 78]. Less detailed references exist to a similar practice by B. Valentine as far back as the late 1400's. In addition there are even less well documented reports of the use of processes of this type dating back to 1,000 A.D. [69]

The early chamber process facilities were relatively small scale operations practised from the experience that the desired product was obtained by carrying out certain steps, but with little appreciation of the detailed chemistry involved. The science of chemistry itself had not developed sufficiently to be able to determine the chemical details when this process was first practised. Even if it had, the number and complexity of the reactions involved were sufficient to forestall complete understanding until about 1910 [77]. This was, however, the only existing process for sulfuric acid production until the late nineteenth century [69] and until as late as 1945 was used to produce close to one half of the sulfuric acid produced in the United Kingdom [77] and the United States [69]. Since then, however, the proportion of the total sulfuric acid produced via the chamber process has rapidly declined, in the U.S.A. to about 10 % of the total in 1958 [79], and further to only about 0.5 % of the total in 1973 [66], mainly because of the lack of flexibility in the concentrations of acid produced by this process.

In the Roebuck version, as in the present version of this process, nitrous acid is the active oxidant used to convert sulfurous acid to sulfuric acid (Equation 7.33). At the normal operating temperature (about 110 °C) the

$$H_2SO_3 + 2\,HNO_2 \rightarrow H_2SO_4 + 2\,NO + H_2O$$
$$\text{nitric oxide} \quad (7.33)$$

nitrous acid is in the vapour state, hence oxygen transfer is via a gas phase catalyst, i.e. here vapourized nitrous acid, unlike the heterogeneous solid phase catalyst employed in the contact process. Sulfurous acid is obtained from the steam hydration of sulfur dioxide produced by the burning of sulfur, and the nitrous acid via the hydration of nitrogen oxides generated by the heating of potassium nitrate (Equations 7.34, 7.35). The thermal decomposition of potassium

$$2\,KNO_3 \rightarrow K_2O + NO_2 + NO + O_2 \qquad (7.34)$$

$$NO + NO_2 + H_2O \rightarrow 2\,HNO_2 \qquad (7.35)$$

nitrate is not well understood, and at best may be non-stoichiometric [80, 81]. As operated in the 1750's, the 50-60 % (50 ° Bé) sulfuric acid product was collected from the floor of the lead-lined chamber, and the associated gaseous nitric oxide was vented to the atmosphere.

In the early days of the process, it had been thought that the proportion of potassium nitrate heated, to the amount of sulfur burned was crucial. The importance of the presence of air for re-oxidation of nitric oxide to nitrogen dioxide (Equation 7.36) was only discovered in 1806, and has been profitably made

$$2 NO + O_2 \rightarrow 2 NO_2 \tag{7.36}$$

use of by operators since that date [78]. Since the potassium nitrate (nitre) was an expensive ingredient this discovery alone significantly improved the economics.

It was not until 1827 that Gay-Lussac developed the absorption tower that then made it possible to capture nitrogen oxides from the sulfuric acid chamber(s) to produce nitrosyl sulfuric acid, so-called "nitrous vitriol" (Equation 7.37) [69, 78]. This would have improved not only the economic but also

$$NO + NO_2 + 2 H_2SO_4 \rightarrow 2 ONOSO_3H + H_2O$$
nitric  nitrogen           nitrosyl      (7.37)
oxide  dioxide            sulfuric
       (nitrogen          acid
       IV oxide)

the environmental aspects of this early process, except that is was not easy to return the active nitrogen oxides to the working chambers without diluting the solution of nitrosyl sulfuric acid in sulfuric acid, with water. This, most producers did not want to do and, therefore, did not put into practice because of the high cost of reconcentrating the diluted acid to commercial strength after the release of nitrogen oxides.

The development of the Glover tower, in 1860, allowed both nitrogen oxide release by water dilution (Equation 7.38) and, in the same unit reconcentration

$$ONOSO_3H + H_2O \rightarrow H_2SO_4 + HNO_2 \tag{7.38}$$

of the acid via the hot gases generated from sulfur combustion [69, 78]. This additional innovation now made the combination of the Glover tower front end unit, as a generator-concentrator, and the Gay-Lussac tower tail gas recovery unit more attractive to sulfuric acid producers, but still led to only slow adoption. Even by 1890 only about half of the sulfuric acid plants in the U.S.A. used the nitrogen oxide conserving towers as an integral part of their processes [78]. But gradually the ability to recycle the nitrogen oxides plus the trend towards an increase in the scale of chamber operations caused a marked drop in price of acid, from £ 33-38 per tonne in 1800, £ 3.75 in 1820 and down to £ 1.30 per tonne in 1885 [78]. Thus, with the

lesser chemical background possessed by the early chamber process operators, it took some 110 years, from the early initial appreciation of the requirements for sulfuric acid formation in 1746, to the time that the knowledge of the nitrogen oxide recycle and sulfuric acid concentration steps of the Glover tower were worked out in 1859, for the full technology of this process to be developed. This is more than twice as long as the time required for the development of the more recent contact process, which occurred when the science of chemistry was further advanced.

The modern format of a chamber plant parallels these early developments quite closely (Figure 7.10). Significant differences exist only in the Glover tower accessories for the make-up of any small amounts of lost nitrogen oxides which now occurs either via the heating of nitric acid (Equation 7.39), or by the combustion of ammonia in air (Equation 7.40, 7.41).

$$2 HNO_3 + NO \rightarrow 3 NO_2 + H_2O \tag{7.39}$$

$$2 NH_3 + {}^5\!/_2 O_2 \rightarrow 2 NO + 3 H_2O \tag{7.40}$$

$$2 NO + O_2 \rightarrow 2 NO_2 \tag{7.41}$$

The nitrogen oxides produced by either method supplement the bulk of the nitrogen oxides which are supplied by recycle of nitrosyl sulfuric acid at the Glover tower by Equation 7.38. In this way, the total amount present is adequate to provide sufficient nitrous acid (Equation 7.35) for the oxidation of sulfurous to sulfuric acid. This oxidation occurs in both the Glover tower, and the chambers. So the lead-lined, quartz-packed Glover tower performs multiple functions in the process. It hydrates the sulfur dioxide, which enters at the bottom, and the nitrogen oxides, and it concentrates the incoming 50 °Bé (62 %) sulfuric acid entering the top of the tower, to about 60 °Bé (78 %) sulfuric acid, which emerges from the bottom. Sulfuric acid of about 78 % concentration is ordinarily the most concentrated product of the chamber process plant. At best, under normal operating conditions, only about half of the direct product of the chamber plant could be of 78 % concentration and the remainder would be the product collected directly from the chambers which would be about 60 % concentration.

The Baumé density scale for specifying sulfuric acid concentrations, which is still in use today, is a hold-over from the early days of acid manufacture

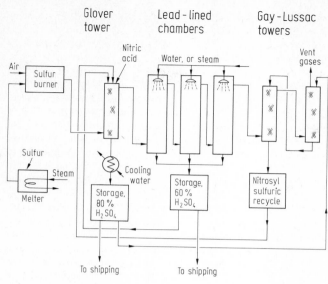

Figure 7.10. Simplified flowsheet of the chamber process for the production of sulfuric acid. The Glover tower, used for concentrating the sulfuric acid and recycling the nitrogen oxides recovered by the Gay-Lussac tower, actually comprises a volume of only 1-1.5 % of the whole chamber set. The three to six chambers are made of wood or steel, lined with lead, size about 7-10 m roughly square cross section and 15-35 m long. Water addition is by steam or liquid water as necessary for temperature control

when there was no clear understanding of either acid composition, or the exact correlation between density and concentration. As such, it is an arbitrary density scale based on the formula of Equation 7.42

$$\text{Specific gravity} = \frac{m}{(m - d)} \qquad (7.42)$$

where m = 145, and d = degrees Baumé reading [82], which was originally devised for specifying sugar concentrations in water [83]. The relationship between some common process and trade acid concentrations and degrees Baumé and density readings is given in Table 7.8.

Most of the formation of sulfuric acid in the chamber process occurs in the liquid layer on the

Table 7.8 Relationship between degrees Baumé and density for some common sulfuric acid concentrations

| Concentration % $H_2SO_4$ (wgt) | °Baumé | Density at 15.6 °C | Remarks |
|---|---|---|---|
| 0.0 | 0.0 | 1.00 | pure water |
| 10.9 | 10.0 | 1.075 | "dilute" acid |
| 26.0 | 23.0 | 1.190 | |
| 29.8 | 26.0 | 1.219 | |
| 33.5 | 29.0 | 1.250 | conc. of battery acid |
| 37.5 | 31.80 | 1.281 | |
| 62.5 | 50.0 | 1.526 | "chamber", or "fertilizer" acid |
| 73.1 | 57.0 | 1.647 | $H_2SO_4 \cdot 2 H_2O$ |
| 77.7 | 60.0 | 1.706 | "Glover tower" acid |
| 84.5 | 63.2 | 1.773 | $H_2SO_4 \cdot H_2O$ |
| 90.0 | 65.1 | 1.8144 | |
| 93.2 | 66.0 | 1.8354 | "oil of vitriol" |
| 95.0 | b | 1.8337 | ordinary concentrated acid |
| 100.0 | b | 1.8305 | "monhydrate" sulfuric acid |

a Data calculated, and compiled from [3 and 105].
b °Baumé readings become unreliable indicators of sulfuric acid concentration because of a decrease in density at these high concentrations [105]. Lange's Handbook of Chemistry, 10th Edition, McGraw-Hill Book Company, New York, 1969, page 1156.

walls of the lead chambers and in the enclosed vapour space. While the sulfurous acid oxidation reaction, Equation 7.33, is relatively rapid, at the moderate 110-120°C temperature of the chambers sufficient volume has to be provided for virtually all of the sulfurous acid to be oxidized before the residual gases and vapours move to the Gay-Lussac tower. Thus these units must be large. Spent chamber gases move into the bottom of the quartz packed Gay-Lussac tower while 60°Bé (about 78 %) sulfuric acid (part of the Glover tower product) is trickled in at the top. In this way the nitrogen oxides present are captured, forming nitrosyl sulfuric acid for recycle (Equation 7.37), and the residual gases, mainly nitrogen and water vapour, are discharged.

Therefore, the two conveniently available concentrations of sulfuric acid from a chamber plant are about 60 % and 80 %. Higher concentrations require additional capital investment in a separate concentrator unit which uses distillation to remove water. This lack of flexibility for higher acid concentrations is probably the main reason for the small commercial operating capacity of this process today. Traces of oxides of arsenic, nitrogen and selenium, and sulfates of iron, copper, mercury, zinc, and lead may also be present in the product acid depending on the presence of these components in the sulfur or sulfide minerals used to provide the sulfur dioxide feed gas. For applications such as the production of fertilizers these impurities can be tolerated, but for applications requiring high purity acid they need to be removed.

# 7.9 Emission Containment for Sulfuric Acid Plants

Form the stoichiometry of the balanced combined equation representing the raw material requirements (Equation 7.43), the theoretical sulfur, oxygen, and

$$S + \tfrac{3}{2}O_2 + H_2O \rightarrow H_2SO_4 \quad (7.43)$$

| molecular weights | 32.06 | 32.00 | 18.02 | 98.08 |
| mass ratios | 32.06 | $\dfrac{\times 1.5}{48.00}$ | 18.02 | 98.08 |

water requirement for any given quantity of sulfuric acid production may be calculated. The theoretical sulfur requirement works out to

326.9 kg (Equation 7.44),

$$\frac{32.06 \text{ g/mol}}{98.08 \text{ g/mol}} \times 1 \text{ tonne} \times 1000 \text{ kg/tonne}$$

$$(7.44)$$

noticeably less than 337 to 344 kg of sulfur per tonne experienced by many sulfuric acid producers (Table 7.7). Material balance and conservation of matter considerations allow us to say that this 10 to 14 kg per tonne difference between the theoretical and actual sulfur consumption, less a proportion to account for any impurities which may be present in the feed sulfur, represents the sulfur loss from the process.

The experience reported above thus represents about 95 to 97 % sulfur to sulfuric acid conversion efficiencies. With sulfuric acid plants of 200 tonnes per day capacity becoming common and the occasional plant operating on the scale of 1,800 tonnes per day [66], even these relatively high conversion efficiencies are not enough to avoid localized emission problems. Thus, the United Kingdom has regulated a requirement of 99.5 % containment of the sulfur burned as a feedstock, while the U.S.A. requirement is for 99.7 % sulfur dioxide to sulfur trioxide conversion efficiency [67, 73, 84]. With modern process modifications and/or emission control devices in place these particular requirements, which in fact work out overall to be very similar for the two countries, are being met.

## 7.9.1 Contact Process Sulfuric Acid Emission Control

To enable improvements to be made to operations which do not meet these guidelines requires, first of all, determination of the types of emissions and their points of loss. The main pollutant losses from a contact acid plant occur through the absorber exit gases, but the origins of these losses are a composite of several primary process variables. While the gas composition of this stream is mainly nitrogen and oxygen, prior to abatement measures being adopted by the majority of sulfuric acid producers it also contained 0.13 to 0.54 % with an average of 0.26 % by volume sulfur dioxide, and 39 mg/m$^3$ to 1,730 mg/m$^3$ (1.1-48.8 mg/ft$^3$) with an average of 457 mg/m$^3$ (12.9 mg/ft$^3$) sulfuric acid mists [85]. The harmful effects of sulfuric acid mists to plant life have been considered by Wedding [86] and to human populations by Patty [87], among others. Environmental chemistry and other effects of the gaseous discharges have been considered in Chapter 2. Since there is contact of nitrogen and oxygen with heated metal surfaces in a contact acid plant there is also a potential for the formation of nitrogen oxides ($NO_x$), and their loss in these exit gases. But the concentration of $NO_x$ present in contact acid plant vent gases would be expected to be less than that present in fossil-fueled power stacks, and certainly less than obtained from an uncontrolled chamber acid plant.

One method of decreasing the discharge of unconverted sulfur dioxide is to optimize the sulfur dioxide to oxygen ratio entering the converter [88]. If, simultaneously, the thickness of the catalyst beds in the four-pass converter is increased, or the gas velocity through the converter is reduced, the increased gas to catalyst contact time obtained will also increase the percent conversion of sulfur dioxide to sulfur trioxide. The first catalyst bed is the one most subject to dust accumulation and partial loss of activity. A regular maintenance schedule for clean-up and partial replacement of the catalyst of the first catalyst bed has been found to help maintain conversion efficiencies [66]. Addition of a water, or ammonia scrubber for tail gas clean-up from the last absorber (Figure 7.11) coupled with the above steps has also been found to significantly improve the gross emission level. Sale of the recovered ammonium sulfate from the ammonia scrubber spent liquor can help to offset installation and operating costs of this unit.

But more than any other single measure, the interpass absorption (IPA) system has introduced a process integral measure which has the effect of re-

ducing sulfur dioxide emissions to an order of magnitude below that obtained from a normal 4-pass converter, that is to the range of 0.01 to 0.03 % of the sulfur feed [85]. This system employs intermediate cooling and sulfur trioxide absorption after the gas mixture has gone through only three stages of conversion. By removing much of the sulfur trioxide product of the conversion *before* the last catalyst stage, the equilibrium of the conversion reaction of Equation 7.25 is strongly shifted to the right. Any residual sulfur dioxide still present after absorption is then much more completely converted to sulfur trioxide as it passes through the last stage of the converter at a moderate 400-420 °C [67, 73] (Figure 7.12). The net effect of these modifications is to bring the overall sulfur conversion to acid to the 99.7 % range. Combining interpass absorption with operation under a pressure of about $5 \times 10^5$ Pa (about 5 atmospheres) for both conversion and absorption has been found to yield further improvement in conversion efficiencies to the 99.80 to 99.85 % range, as practised by the Ugine Kuhlmann Process [89]. Sulfur consumption for this system was reported to be 330 kg per tonne, very close to the theoretical 326.9 kg.

The acid mist emission problem of contact acid plants may arise from one or more of several factors. Water vapour in the air feed to the sulfur burner may cause this since, as the water vapour plus sulfur trioxide stream drops below the dew-point temperature, aerosol (suspension of fine droplets) formation inevitably occurs, even if this level of cooling happens in the absorbers. Mist formation has been routinely minimized for years by drying the air fed to the sulfur burner with concentrated sulfuric acid. Problems may arise with this drying system if the feed acid, for some reason, becomes too dilute, or if the air/acid contact in the drying tower(s) is inadequate, either of which can increase mist formation [90].

Water may also get into the converter gases from hydrocarbons present in the sulfur, a consideration particularly with the dark grades of Frasch sulfur

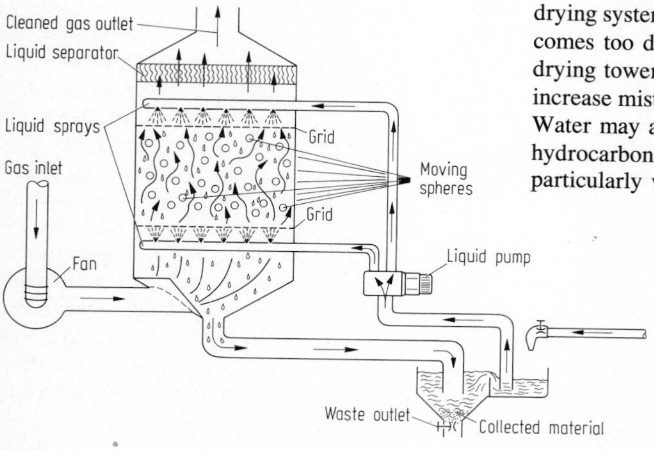

Cleaned gas outlet

Liquid separator

Liquid sprays

Gas inlet

Grid

Moving spheres

Grid

Fan

Liquid pump

Waste outlet

Collected material

**Figure 7.11.** Floating bed scrubber, such as might be used to control sulfur dioxide and sulfuric acid mists. (European Plastic Machinery Mfg., A/S, Copenhagen)

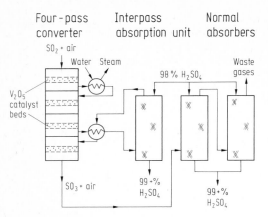

**Figure 7.12.** Schematic diagram of the converter section of a contact sulfuric acid plant employing an interpass absorption (IPA) system for both better sulfur conversion and emission abatement. Product acid (or oleum) is cooled indirectly with process water prior to storage for sale

which may contain up to 0.3 % bitumen [73, 91]. On burning the bitumen content of this sulfur the hydrogen component of the bitumen inevitably gives water vapour, and at a point in the process beyond where the air drying tower can help. The best that can be accomplished in this situation is to keep the gas mixture above the dewpoint throughout, until it enters the absorbers, to minimize corrosion problems in the intervening process equipment [92], and to rely on ancillary devices to minimize mist discharge after absorption.

Occasional mist problems occur during start-up when the concentration of acid used for air drying may not be adequate, and while oleum is being produced in the first absorber, both situations which seem to promote aerosol formation [67, 73, 85, 92]. While these situations prevail the cause is normally intermittent and discharge prevention relies heavily on mist control equipment.

The mist control device most frequently employed is the Brink mist eliminator, consisting of an impaction type of filter made of fiber glass packed between stainless steel screens [73, 90]. Installation of these units is not only useful for the exit gas stream from the last absorber, but also for the intermediate dried air stream leaving the sulfuric acid drier and for the sulfur dioxide air stream after an interpass absorption unit, if present. Mist eliminator mass efficiencies (as opposed to particle efficiencies) of 97 to 98 % have been reported, effectively reducing sulfuric acid aerosol concentrations

from the 850 to $1,275 \, \text{mg/m}^3$ (24 to 36 mg per standard cubic foot (scf)) range to 18 to $21 \, \text{mg/m}^3$ (0.5 to 0.6 mg/scf) [91].

Scrubbers may be used operating either on water [90] or on a closed-cycle solvent [93], and sequenced either as devices employed to follow after a mist eliminator or to replace a mist eliminator at least in its function at the last absorber. Mass efficiencies of 90 to 99 % have been reported for water-wash scrubbers with acid mist loadings of 3.5 to $7 \, \text{g/m}^3$, even for particle size distributions determined to be 60 % less than $3 \, \mu\text{m}$ [91]. Electrostatic precipitators have also occasionally been used for mist elimination [85].

## 7.9.2 Emission Control for Chamber Process Acid Plants

For the chamber process the main point of loss of both sulfur and nitrogen oxides is via the exit gas stream of the Gay-Lussac towers, which are the nitrogen oxide absorbers. Both sulfur dioxide and nitrogen oxide concentrations are regularly found to be in the 0.1-0.2 volume percent range, with nitrogen dioxide comprising about 50 to 60 % of the total nitrogen oxides [85]. There is also sulfur loss as sulfuric acid mist which happens at more variable concentrations in the range of 180-$1,000 \, \text{mg/m}^3$ (5-30 mg/ft³), but which usually comprises less than 0.1 % of the acid produced [85].

To decrease the concentrations of these components requires attention to the operation of the Gay-Lussac tower itself [85]. Low gas velocities, low temperatures and high impaction efficiencies in this unit are all measures which will assist in minimizing losses. But these are all improvements which are best instituted at the design stage. Many recent plants operate with two Gay-Lussac towers, partly for this reason [37, 94]. If the concentration of nitrogen oxides in the chambers is increased, it has the effect of increasing the rate of oxidation of sulfurous acid and thus may help reduce exit gas sulfur dioxide concentrations. For maximum effectiveness this measure also requires efficient control of the nitrogen oxides at the tail end of the process [95]. A final water scrubbing of Gay-Lussac tower exit gases has been reported to decrease sulfur dioxide concentrations by 40 % and nitrogen oxide concentrations by 25 % [85]. By routing the scrubber effluent as feed water to the chambers, both captured feedstock components are recycled back into the process.

Combined sulfur dioxide and nitrogen oxides in power station flue gases have been controlled by adsorption in alumina [96], an expedient which may also lend itself to adoption for chamber plant operation. Regeneration by heating the adsorbent on swing cycles would produce a gas stream containing a high concentration of sulfur dioxide and nitrogen oxides which could beneficially be fed back to the chambers. A similar suggestion, using a solution of sodium hydroxide and sodium sulfite in water could be workable, but probably at higher cost [97].

## 7.10  Recycling of Sulfuric Acid

Motivated partly by decreasing supplies and increasing costs of elemental sulfur, but more from the high costs of environmentally acceptable disposal methods for spent sulfuric acid, an increasing number of acid consumers are recycling, or are considering recycling of used acid. To minimize the effects of land-based disposal of waste acid the costs of disposal options such as barge transport of waste acid 180 km out to sea [41, 106] have become comparable to the costs of recycling options, stimulating the adoption of these alternatives by major acid users. For example, in West Germany by 1976 some 13 % of the total sulfuric acid used by each of two largest use categories, organic chemicals and inorganic pigments (Table 7.4), was already being recycled [102]. The percentage of acid recycle by these sectors is likely to increase further while at the same time adoption of acid recycle by smaller sulfuric acid users also grows. For the latter group, even if the high cost of small scale recycle options is unattractive on an individual basis, collective recycle arrangements may still prove to be practical.

The properties of the spent sulfuric acid to be dealt with vary widely from acid which is simply diluted, but otherwise virtually uncontaminated, to acid which is still quite concentrated but contaminated with metal or other inorganic or organic impurities, to acid which is both diluted and contaminated. Measures for recovery and recycle therefore vary in complexity from simple reconcentration, to water removal accompanied by various chemical purification steps. Occasionally thermal destruction of impurities accompanied by dissociation of the acid and recycle of the sulfur content for acid remanufacture may be required.

For these reasons the appropriate acid recycle method very much depends on the use to which the original acid was put, and the condition of the spent acid obtained. Specific examples follow.

Regeneration of high acid concentrations from sulfuric acid which has been diluted by water but is otherwise uncontaminated, such as is obtained from air or other gas drying functions, can be accomplished by simply boiling off the water in either a batch (pot) or continuous (heat exchanger) mode [107, 108, 109]. Temperatures of the order of 300 °C are required for product acid concentrations of 95 % or better when distillation is conducted at normal atmospheric pressure. However, of a reduced pressure of 20 mm Hg or so is used, which is easily accessible by water ejectors, water removal may be accomplished at about 200 °C, the temperature of low superheat steam [100]. In both cases lead, or lead-lined equipment is necessary for the dilute acid stages to avoid corrosion problems although steel may be used for containment of concentrations above 95 % (65 °Bé) [90, 110].

Regeneration of high concentrations of sulfuric acid may also be achieved by oleum or sulfur trioxide addition to diluted acid. In the process the inventory of acid in circulation is increased by a corresponding amount (Equation 7.27). A third method of reconcentration, useful when the acid consumption occurs as a part of, or adjacent to a producing contact sulfuric acid plant, is to pass the diluted acid through the acid plant absorption tower. This option amounts to on-site addition of sulfur trioxide, and the increased acid inventory obtained can be sold to markets through the normal producer channels.

Large scale nitrations, which use a mixture of high concentrations of nitric plus sulfuric acids, yield a spent acid which mainly consists of diluted sulfuric acid but also contains traces of residual nitric acid and organics [111]. Acid recovery from this spent acid can be accomplished by initially removing the organics by extraction with a solvent such as toluene. The aqueous acid raffinate is then steam-stripped to remove most of the residual nitric acid. Any residual nitric acid is destroyed by treatment with hydrogen peroxide and hydroxylamine. Subsequent reconcentration regains a useful purity of high strength sulfuric acid [112]. A simpler but less rigorous procedure for the recovery of sulfuric acid used in nitrations involves stirring of the spent acid with activated carbon (1-3 %) and diatomaceous earth (2-5 %) to yield directly a chamber grade acid suitable for reconcentration [113]. Simple de-nitration of spent nitration acid by addition of ammonium ion or urea has also been tested [114, 115]. Methods have also been developed for destruction of the organic content of spent acid using hypochlorite [116] or hydrogen peroxide [117]. If the concentration of organics in the waste acid is not high even the residual nitric acid content may be adequate to destroy these, in the process consuming much of the residual nitric [108] (Equations 7.45, 7.46).

$$5\,C + 4\,HNO_3 \rightarrow 5\,CO_2 + 2\,N_2 + 2\,H_2O \quad (7.45)$$

$$C + 2\,HNOSO_4 \rightarrow CO_2 + 2\,H_2SO_4 + N_2 \quad (7.46)$$

Metal ion impurities in spent acid impose different recovery problems. For instance, iron residues in the spent acid from titanium dioxide manufacture may be removed by crystallization of first $FeSO_4 \cdot 7\,H_2O$ followed by crops of various iron (II) ammonium salts [118] or by electrodialysis [119]. Incidentally electrodialysis has also been used for sulfuric acid recovery from waste water, either directly [120, 121] or after preconcentration on cation exchange resin [122, 123]. Iron removal methods for acid recovery from the waste liquor of steel pickling baths have recently been reviewed [124].

Removal of copper ion from plating bath liquors has been accomplished electrochemically, by plating it out [125], and also by extraction with tributyl phosphate [126]. The tributyl phosphate extract brings nickel, antimony, arsenic, and copper all into the organic phase, leaving a significantly cleaner acid raffinate. Phosphate extraction has also been used for iron removal from pickling waste liquors [127].

Sometimes, however, the spent acid is so severely contaminated with inorganics or intractable tars that high temperature oxidation of the impurities plus thermal dissociation of the sulfuric acid for sulfur dioxide recovery becomes the only feasible option to recycle the acid [128, 129] (Equation 7.47).

$$2\,H_2SO_4 \rightarrow 2\,SO_2 + 2\,H_2O + O_2 \quad (7.47)$$

Temperatures of the order of $1,100\,^\circ C$ are necessary in the presence of added air to promote complete oxidation of any organic impurities and thermolysis of the acid. Subsequently, the sulfur dioxide is recovered and passed through a conventional contact plant for regeneration to sulfur trioxide and

thence via rehydration to concentrated sulfuric acid (Equations 7.26, 7.27).

If disposal appears to be the only practical alternative despite this range of recycle options, then pH neutralization prior to discharge is to be preferred and is required by many jurisdictions. One inexpensive method has been tested for effectiveness and operating details which employs the waste product from dry lime-based sulfur dioxide emission control systems [130] (Equation 7.48).

$$CaO + \tfrac{1}{2}O_2 + SO_2 \rightarrow CaSO_4 \quad (7.48)$$

Only a fraction of lime employed in such processes for sulfur dioxide neutralization is consumed in this application. The residual calcium oxide, comprising up to two thirds of the weight of the precipitator catch from these control systems, provides the inexpensive and convenient source of base for neutralization of waste but unrecoverable sulfuric acid. This measure, by utilizing the residual chemical activity of both wastes, decreases the disposal problems and costs of dealing with them.

It is also possible to use ammonia to neutralize waste acid, which consumes a more costly base. But in the process ammonium sulfate is obtained which can be sold as a valuable fertilizer constituent, helping to offset the cost of the base [131]. The purity of the ammonium sulfate obtained is improved by extraction with an organic solvent such as methyl ethyl ketone before product crystallization [132].

## Relevant Bibliography

1. P.R. Cote, Canadian Elemental Sulphur from Sour Natural Gas, Dept. of Energy, Mines, and Resources, Ottawa, 1972
2. F. Tuinstra, Structural Aspects of the Allotropy of Sulfur and the Other Divalent Elements, Waltman, Delft, The Netherlands, 1967
3. Analytical Chemistry of Sulfur and its Compounds, J.H. Karchmer, editor, Wiley — Interscience, New York, 3 volumes, 1970-1972
4. Sulfur Removal and Recovery from Industrial Processes, J.B. Pfeiffer, editor, American Chemical Society, Washington, D.C., 1975
5. Symposium on Sulfur Utilization; a Progress Report, D.J. Bourne, editor, American Chemical Society, Washington, 1978
6. National Inventory of Natural Sources and Emissions of Sulphur Compounds, Environmental Applications Group Ltd., for Environment Canada, Ottawa, 1980, 91 pages

7. G.H.K. Pearse, A National Strategy for By-product Sulfur, Energy, Mines, and Resources Canada, Ottawa, 1980

8. Effects of Airborne Sulphur Compounds on Forests and Freshwaters, Department of the Environment and the Natural Environment Research Council, London, 1976

9. A.G. Maadah and R.N. Maddox, Predict Claus Products, Hydroc. Proc. 57, 143, Aug. 1978

10. L. Milica, Reaction Mechanisms in Sulphuric Acid and Other Strong Acid Solutions, Academic Press, London, 1971

11. J.R. Shafer et al, Pollution Control Practices; Sulfuric Acid Plants for Handling H$_2$S Gases, Chem. Eng. Prog. 74, 62, Dec. 1978

12. W.H. Megonnell, Efficiency and Reliability of Sulfur Dioxide Scrubbers, J. Air Polln Control Assoc. 28, 725, July 1978

13. K.S. Gaur, Pollution Control with SO$_2$ Recovery; Wellmann-Lord Sulfur Dioxide Recovery System, Polln Eng. 10, 51, May 1978

14. J.L. Gras and G.P. Ayers, Sizing Impacted Sulfuric Acid Aerosol Particles, J. Appl. Meteorology 18, 634, May 1979

15. Continuous Process for Sulfuric Acid Concentration, Chem. Eng. 86, 109, Aug. 13, 1979

16. Sulphur: New Sources and Uses, M.E.D. Raymont, editor, ACS Symposium Series No. 183, American Chemical Society, Washington, 1983

11. Sulfur-Asphalt Process, Chem. Eng. News 56 (21), 7, May 22, 1978

12. Sulfur-Asphalt Blend, Chem. Eng. News 57 (42), 8, Oct. 15, 1979

13. Sulfur Paving Material, Chem. Eng. News 57 (9), 30, Feb. 26, 1979

14. Test Road Paved, Chem. Eng. News 58 (35), 25, Sept. 1, 1980

15. Sulfur Surfeit, Chem. In Can. 24 (1), 10, Jan. 1972

16. Sulfur Block House, Montreal Gazette, Aug. 17, 1972

17. M. Gellender, Can. Chem. Proc. 62 (1), 14, Jan. 1978

18. Sulfur Concrete Offers Corrosion Resistance, Chem. Eng. News 58 (39), 45, Sept. 29, 1980

19. Uses of Low Cost Construction Materials, Chem. Eng. News 54 (48), 20, Nov. 22, 1976

20. Sulfur Response, Chem. In Can. 24 (7), 5, Summer 1972

21. J. Platou, Hydroc. Proc. 51 (7), 86, July 1972

22. Sulphur-coating Urea in a Spouted Bed, Sulphur (London) No. 134, 35, 40, January-February, 1978. Cited by Minerals Yearbook 1978-79, U.S. Bureau of Mines, Washington, 1980, page 897

23. Kirk-Othmer Encyclopedia of Chemical Technology, 2nd edition, John Wiley & Sons, Toronto, 1969. Volume 19, page 337

24. F.S. Taylor, A History of Industrial Chemistry, reprint edition, Arno Press, New York, 1972, pages 184, 189

25. S.D. Kirkpatrick, Chem. Metal Eng. 45, 492, Sept. 1938

26. Riegel's Industrial Chemistry, J.A. Kent, editor, Reinhold Book Corp., New York, 1962, page 84

27. Coal Converts Sulfur Dioxide to Sulfur, Design News, 34, 18, Nov. 20, 1978

28. R.N. Shreve, Chemical Process Industries, McGraw-Hill, New York, 1945, page 353

29. Sulfur Faces Delicate Supply-Demand Balance, Chem. Eng. News 59 (20), 26, May 18, 1981

30. Polish Sulphur Industry Revisited, Sulphur (London) No. 144, 17, Sept.-Oct. 1979

31. Thorpe's Dictionary of Applied Chemistry, Longmans, Green and Co., Toronto, 1954, Volume IX, page 218

32. F.A. Lowenheim and M.K. Moran, Faith, Keyes, and Clark's Industrial Chemicals, 4th edition, Wiley-Interscience, New York, 1975, page 786

33. C.E. Butterworth and J.W. Schwab, Indust. Eng. Chem. 30 (7), 746, July 1938

34. Melted Sulfur Yields New Form, Can. Chem. Proc. 60 (11), 7, Nov. 1976

35. Sulfur Shippers Favour Procor Unit, Can. Chem. Proc. 63 (3), 20, May 2, 1979

36. Sulfur Prilling Tower Starts Operating, Chem. Eng. News 58 (13), 37, March 31, 1981

37. R.N. Shreve, Chemical Process Industries, 3rd edition, McGraw-Hill, Toronto, 1967, page 322

# References

1. A.H. Vroom, Hydroc. Proc. 51 (7), 79, July 1972

2. Currently, Alberta Producers are Building, Can. Chem. Proc. 58 (7), 4, July 1974

3. W.L. Faith, D.B. Keyes, and R.I. Clark, Industrial Chemicals, 3rd edition, Wiley-Interscience, New York, 1965, page 744

4. Canadian Minerals Yearbook 1978, Department of Energy, Mines and Resources, Ottawa, 1980, page 437

5. H. Mock, U.S. Pat. 1,297,583, March 18, 1919. Cited by A.H. Vroom, Hydroc. Proc. 57 (7), 79, July, 1972

6. Sulfur-based Foam Protects, Chem. Eng. News 55 (14), 24, April 4, 1977

7. New Technology May Make Roads, St. John's Edmonton Report, 5 (6), 21, Jan. 16, 1978

8. Sulfur in Roadways, Can. Chem. Proc. 60 (3), 20, March 1975

9. Sulphur, Chem. In Can. 25 (11), 17, Dec. 1973

10. Paved with Gold, Can. Research 10 (5), 7, Sept.-Oct. 1977

38. D.T. McIver, J.B. Chatelain, and B.A. Axelrad, Indust. Eng. Chem. *30* (7), 752, July 1938

39. R.N. Shreve and J.A. Brink, Jr., Chemical Process Industries, 4th edition, McGraw-Hill, New York, 1977, page 293

40. F.G. Deiler, Freeport Sulfur Company's Pollution Control Program, In Water — 1969, L.K. Cecil, editor, Chemical Engineering Symposium Series *65* (97), 53 (1969)

41. Industrial Pollution Control Handbook, H.F. Lund, editor, McGraw-Hill, New York, 1971, page 14-10

42. L.F. Hatch, Hydroc. Proc. *51* (7), 75, July 1972

43. Lacq Targeted as Sulfur Chemicals Site, Chem. Eng. News. *53* (28), 11, July 19, 1976

44. Acid Gas Content of Alberta Natural Gas, listing to April 30, 1974, Energy Resources Conservation Board, Calgary, Alberta, 1974

45. Canadian Minerals Yearbook, 1978, Energy Mines and Resources Canada, Queen's Printer, Ottawa, 1980, page 437

46. R.N. Shreve, Chemical Process Industries, 3rd edition, McGraw-Hill, New York 1967, page 87

47. F.C. Vidaurri and L.C. Kahre, Hydroc. Proc. *56*, 333, Nov. 1977

48. J.B. Hyne, Chem. In Can. *30* (5), 26, May 1978

49. J.B. Osenton and A.R. Knight, Reaction of Carbon Disulphide with Alkanolamines used in the Sweetening of Natural Gas, Can. Nat. Gas Proc. Assoc. Meeting, Calgary, Alta., Nov. 20, 1970

50. A.R. Knight, personal communication, Nov. 6, 1972. Polderman, Dillon, and Steel, Oil and Gas. J. *54* (2), 180 (1955)

51. R.F. Smith and A.H. Younger, Hydroc. Proc. *51* (7), 98, July 1972

52. Chemical and Process Technology Encyclopedia, D.M. Considine, editor, McGraw-Hill, New York, 1974, pages 12-15

53. Pollution control in Claus Sulphur Recovery Plants, Sulphur, No. 109, 36, Nov/Dec. 1973

54. Design Capacity of Gas Processing Plants in Alberta, Energy Resources Conservation Board, Calgary, Nov. 1974

55. Z.M. George, Phosphorus and Sulfur *1*, 315 (1976)

56. J.E. Martin and G. Guyot, Can. Gas. J. *62* (2), 26 (1971). Can. Chem. Proc. *57* (8), 40, Aug. 1973

57. P. Grancher, Hydroc. Proc. *57*, 257, Sept. 1978

58. H. Krill and K. Storp, Chem. Eng. *80* (17), 84, July 23, 1973

59. Sulfur Recovery Routinely Hits 99 + %, Can. Chem. Proc. *62* (5), 21, May 1978

60. F.G. Ball, G.N. Brown, J.E. Davis, et al, Hydroc. Proc. *51* (10), 125, Oct. 1972

61. P. Grancher, Hydroc. Proc. *57*, 155, July 1978

62. Photometer keeps Claus Unit on Track, Can. Chem. Proc. *62* (5), 29, May 1978

63. M.J. Pearson, Hydroc. Proc. *52* (2), 81, Feb. 1973

64. M.E.D. Raymont, Hydroc, Proc. *54* (7), 139, July 1975

65. F.J. Ball, G.N. Brown, J.E. Davis, A.J. Repik, and S.L. Torrence, Hydroc. Proc. *51* (10), 125, Oct. 1972

66. F.A. Lowenheim and M.K. Moran, Faith, Keyes and Clark's Industrial Chemicals, Wiley-Interscience, New York, 1975, page 795

67. A. Phillips in The Modern Inorganic Chemicals Industry, R. Thompson, editor, The Chemical Society, London, 1977, page 183

68. R.N. Shreve and J.A. Brink, Jr., Chemical Process Industries, 4th edition, McGraw-Hill, New York, 1977, page 296

69. Kirk-Othmer Encyclopedia of Chemical Technology, 2nd edition, John Wiley, New York, 1969, Vol. 19, page 441

70. R.A. Bauer and B.P. Vidon, Chem. Eng. Prog. *74*, 68, Sept. 1978

71. Cheaper Oxygen? Engineering *219*, 99, Feb. 1979

72. R.N. O'Brien and W.F. Hyslop, Oxygen Enhancement of the Air, British Patent Application 8,003,884, Jan. 28, 1980

73. A. Phillips, Chem. In Brit. *13* (12), 471, December 1977

74. Handbook of Chemistry and Physics, 51st edition, Chemical Rubber Company, Cleveland, Ohio, 1970, page B-144

75. F.A. Cotton and G. Wilkinson, Advanced Inorganic Chemistry, 3rd edition, Interscience, New York, 1972, page 180

76. D.M. Samuel, Industrial Chemistry-Inorganic, 2nd edition, Royal Institute of Chemistry, London, 1970, page 80

77. Thorpe's Dictionary of Applied Chemistry, Longman's, Green and Co., Toronto, 1954, Volume IX, page 290

78. F.S. Taylor, A History of Industrial Chemistry, reprint edition, Arno Press, New York, 1972, page 187

79. Riegel's Industrial Chemistry, J.A. Kent, editor, Reinhold, New York, 1962, page 66

80. G.L. Billington, Chem. In Brit. *16*, 452 (1980)

81. G.W. Rayner-Canham, M. Thompson, F.R. Paulsen, R.A.H. Hillman, J.L. Jorris, Chem. In Brit. *17* (1), 13, Jan. 1981

82. Handbook of Chemistry and Physics, 51st edition, R.C. Weast, editor, Chemical Rubber Publishing Co., Cleveland, Ohio, 1970, page F-3

83. H.G. Jerrard and D.B. McNeill, A Dictionary of Scientific Units, Chapman and Hall, London, 1964, page 42

84. M.F. Tunnicliffe, Chem. In Brit. *14*, 2 (1978)

85. E.F. Spencer, Jr. in Industrial Pollution Control Handbook, H.F. Lund, editor, McGraw-Hill, New York, 1971, page 14-3

86. J.B. Wedding, M. Ligothe and F.D. Hess, Envir. Sci. Tech. *13,* 875 (1979)

87. Industrial Hygiene and Toxicology, 2nd edition, Volume II, F.A. Patty, editor, Interscience, New York, 1963, page 895

88. J.R. Donovan, J.S. Palermo, and R.M. Smith, Chem. Eng. Prog. *74,* 51, Sept. 1978

89. R.A. Bauer and B.P. Vidon, Chem. Eng. Prog. *74,* 68, Sept. 1978

90. P.D. Nolan, Can. Chem. Proc. *61* (4), 35, April 1977

91. O.H. York and E.W. Poppele, Chem. Eng. Prog. *66* (11), 67, Nov. 1970

92. D.R. Duros and E.D. Kennedy, Chem. Eng. Prog. *74,* 70, Sept. 1978

93. A Regenerable Solvent, Can. Chem. Proc. *56* (12), 34, Dec. 1972

94. Riegel's Handbook of Industrial Chemistry, 7th edition, J.A. Kent, editor, Van Nostrand Reinhold, New York, 1974, page 66

95. R.N. Shreve, The Chemical Process Industries, McGraw-Hill, New York, 1945, page 375

96. P.M. Medellin, E. Weger and M.P. Dudukovic, Ind. Eng. Chem., Proc. Design and Devel. *17,* 528, Oct. 1978

97. H. Takeuche and Y. Yamanaka, Ind. Eng. Chem., Proc. Design and Devel. *17,* 389, Oct. 1978

98. Minerals Yearbooks, 1981, 1978-79, 1975, 1970, and 1965, U.S. Bureau of Mines, Washington, D.C., for these years

99. Minerals Yearbook, 1978-79, Vol. I, U.S. Bureau of Mines, Washington, 1980, pp. 890-895

100. Chemical Engineer's Handbook, 4th edition, J.H. Perry, editor, McGraw-Hill, New York, 1969, page 14-10

101. Sulfuric Acid, Chem. Eng. News *56* (13), 11, March 27, 1978

102. U. Sander and G. Daradimos, Chem. Eng. Prog. *74,* 57, Sept. 1978

103. Statistical Yearbook, United Nations, New York, 1957, 1969, 1978, and 1979/80

104. Facts and Figures, Chem. Eng. News *61* (24), 26, June 13, 1983

105. Industrial Pollution Control Handbook, H.F. Lund, editor, reference 41, page 14-30

106. R.F. Vaccaro, G.D. Grice, G.T. Rowe, and P.H. Wiebe, Water Research *6,* 231 (1972)

107. A Continuous Process for Acid Concentration, Chem. Eng. *86* (17), 109, Aug. 13, 1979

108. H.R. Kueng and P. Reimann, Chem. Eng. *89,* 72, April 19, 1982

109. G.M. Smith and E. Mantius, Chem. Eng. Prog. *74,* 78, Sept. 1978

110. W.L. Faith, D.B. Keyes, and R.L. Clark, Industrial Chemicals, 3rd edition, John Wiley and Sons, New York, 1965

111. G.P. Edwards and W.T. Ingram, Sewage Ind. Wastes *26,* 1484 (1954)

112. K. Blanck, B. Leutner, and W.D. Back, German Offen. 2,831,941 to BASF, A.G. Chem. Abstr. *93,* P 10,344 z (1980)

113. C. Gloria, and N. Rainaldi, Braz. Pedido PI 79 07,380. Chem. Abstr. *95,* P 135,147 K (1981)

114. Japan Kokai Tokkyo Koho 79 46,198, to Bayer A.-G. Chem. Abstr. *91,* P 59,581 x (1979)

115. M. Mistrorico, Braz. Pedido PI 79 07,415. Chem. Abstr. *94,* P 68,090 a (1981)

116. R.B. Valitov, G.G. Garifzyanov, R.M. Sadrislamov, and A.M. Tukhvatullin, U.S.S.R. 567,664. Chem. Abstr. *87,* P 186,615 d (1977)

117. E.I. Elbert, N.E. Tsveklinskaya, R.A. Bovkun, V.N. Stepanov, N.V. Grazhdan, and A.S. Zlobina, U.S.S.R. 601,222. Chem. Abstr. *89,* P 8,414 w (1978)

118. P. Kadlas, J. Michalek, and K. Haas, Chem. Prum. *26* (11), 577 (1976)

119. J.J. Barney and J.L. Hendrix, Ind. Eng. Chem., Proc. Res. and Dev. *17,* 148, June 1978

120. A. Koehling, H. Behret, and R. Eggersdorfer, Dechema-Monogr. *86,* 1743 (1979). Chem. Abstr. *93,* 74,926 (1980)

121. T. Utsonomiya, T. Kawahara, and H. Shibata, Japan Kokai 77,101,690, to Asahi Glass Co. Chem. Abstr. *88,* P 64,606 b (1978)

122. F. Perschak, Ernaehrung (Vienna) *2* (8), 341 (1978). Chem. Abstr. *92,* 78,508 r (1980)

123. B.R. Nott, Ind. Eng. Chem., Proc. Res. and Dev. *20,* 170 (1981)

124. G. Hitzemann, Wire *26* (2), 45 (1977)

125. E. Korngold, K. Kock, and H. Strathman, Desalination *24* (1, 2, 3), 129 (1978)

126. A. De Schepper, A. Van Peteghem, Can. Pat. 1,044,825. Chem. Abstr. *90,* P 139,717 r (1979)

127. M. Watanabe and S. Nishimura, Ger. Offen. 2,610,434 to Solex Research Corp. Chem. Abstr. *86,* P 159,402 e (1977)

128. F.P. Albul, I.V. Kelman, A.S. Kostenko, S.S. Kudryashov, A.I. Manoila, and Y.V. Shevelev, Neftepererab. Neftekhim. (Moscow) (6) 38, (1981). Chem. Abstr. *95,* 172,238 m (1981)

129. U. Sander and G. Daradimos, Chem. Eng. Prog., *74,* 57, Spet. 1978

130. P.-W. Lin, Envir. Science and Tech. *12* (9), 1,081, Sept. 1981

131. G.V. Annenkova, V.K. Beiden, A.F. Kozhevnikov, Mater. Vses. Konf. 94, (1976). Chem. Abstr. *89,* 196,163 k (1978)

132. A.S. Romashev, V.A. Korin'ko, B.F. Filimonov, et al, Otkrytiya, Izobret., Prom. Obraztsy, Tovarnyl Znaki (3), 73 (1979). Chem. Abstr. *90,* P 170,948 r (1979)

# 8 Phosphorus and Phosphoric Acid

## 8.1 Phosphate Rock Deposits and Beneficiation

Phosphorus occurrence in the lithosphere is predominantly as phosphates, $PO_4^{3-}$ with variations, although a rare iron-nickel phosphide, schreibersite $((Fe, Ni)_3P)_8$ is also known in nature [1]. For this reason, phosphates are the primary source of elemental phosphorus for chemical process requirements. Only 0.20 to 0.27 % phosphate (0.15 to 0.20 % as $P_2O_5$; 0.07 to 0.09 % as P) is present in ordinary crustal rocks, but nevertheless this is the location of the bulk of the phosphorus present in the lithosphere [2].

Fortunately, however, a significant fraction of phosphate deposits have a much higher $P_2O_5$ equivalent concentration than the average crustal concentrations. About 80 % of current world production is derived from the sedimentary phosphorites of marine origin, containing phosphate concentrations of 29 to 30 % (as $P_2O_5$ equivalent) in unweathered deposits and 32 to 35 % in leached deposits [2]. These frequently occur as a variety of apatite in beds of loosely consolidated granules. Second in commercial importance are the igneous apatites (general formula: $Ca_5 (Cl, F, OH) (PO_4)_3$) which average about 27 % $P_2O_5$, and comprise some 15 % of current phosphate production mainly from deposits in the U.S.S.R., South Africa, Uganda and Brazil. Guano sources, resulting mostly from deposition from sea fowl, can range from 10 to 32 % $P_2O_5$ depending on whether the deposits are relatively modern, or leached. As examples the source of most of the material produced from Nauru in the Western Pacific, and from Christmas Island in the Indian Ocean (Table 8.1), is from large reserves of this origin.

The U.S.A. and U.S.S.R. are dominant producers (Table 8.1) and users of phosphate, and the state of Florida is by far the largest single producing area. From phosphorite deposits about 10 m thick covering an area of about 5,000 km$^2$, Florida alone supplied one third of the world total in 1979 [5]. Phosphorite deposits may require beneficiation by froth

**Table 8.1** Major world producers of phosphate rock [a]

| Producer | Thousands of metric tonnes | | | | | |
| | 1960 | 1965 | 1970 | 1975 | 1980 | 1981[b] |
|---|---|---|---|---|---|---|
| China | 300 | 910 | 1 700 | 3 400 | 5 500 | 5 500 |
| Christmas Island | 580 | 859 | 989 | 1 487 | 1 638 | 1 336 |
| Israel | 224 | 392 | 880 | 882 | 2 307 | 2 290 |
| Jordan | 362 | 828 | 913 | 1 353 | 3 911 | 3 523 |
| Moroco | 7 492 | 9 824 | 11 424 | 14 119 | 18 824 | 19 696 |
| Nauru | 1 248 | 1 720 | 2 200 | 1 534 | 2 087 | 2 000 |
| Senegal | 198 | 1 002 | 998 | 1 682 | 1 632 | 2 017 |
| South Africa | 268 | 610 | 1 685 | 1 866 | 3 282 | 2 910 |
| Togo | | 974 | 1 508 | 1 660 | 2 933 | 2 244 |
| Tunisia | 2 101 | 3 040 | 2 969 | 3 512 | 4 582 | 4 596 |
| U.S.S.R. | 7 000 | 15 500 | 17 800 | 24 150 | 29 450 | 30 950 |
| U.S.A. | 17 797 | 23 986 | 35 143 | 44 276 | 54 415 | 53 624 |
| Vietnam | 543 | 550 | 455 | 1 400 | 500 | 550 |
| Others | 1 332 | 180 | 2 956 | 17 019 | 7 272 | 7 394 |
| World | 39 445 | 60 375 | 81 620 | 118 340 | 138 333 | 138 630 |

[a] Producers of over one million tonnes of phosphate rock in 1977. Average phosphate content range from 29 to 36% (as $P_2O_5$). Data assembled from [3 to 6].
[b] Estimated.

flotation for removal of slimes (clays and other finely-divided material) to raise the phosphate content of the ore for market. Igneous apatite deposits require only simple ore selection and crushing to market size requirements. Guano deposits merely require loading to ocean freighters for transport. Thus, phosphate rock marketing is very little dependent on mineral technology and more dependent on solids material handling and shipping economics for competitive export. Prices in 1974 ranged from about $ 33 per tonne in Tampa (U.S.A.) to $ 52 per tonne in Casablanca [4] (Morocco) at 77 % TPL (triphosphate of lime, $Ca_3(PO_4)_2$).

### 8.1.1 End Use Areas for Phosphate Rock

About 87 % or about 32 million tonnes of the phosphate rock consumed in the U.S.A. in 1978 went to agricultural uses, the remainder going as feedstock for elemental phosphorus production [6]. Phosphate rock of fluorapatite stoichiometry $(CaF_2 \cdot 3 Ca_3 (PO_4)_2)$ which is destined as a mineral supplement for animal feeds, or as a bulking agent in fertilizers is normally defluorinated first, by heating in a rotary kiln with silica and steam (Equation 8.1). The volatile

$$3 CaF_2 \cdot 3 Ca_3(PO_4)_2 + SiO_2 + H_2O \rightarrow$$
fluorapatite ca. 1500 °C
$$SiF_4 \uparrow + 2 HF + 3 CaO + 9 Ca_3(PO_4)_2 \quad (8.1)$$

fluorides formed by this process are controlled by capture in water scrubbers (Section 8.4.3). However, about 58 % of agricultural rock, or 50 % of the total, ultimately goes into wet process phosphoric acid production (Section 8.4). The 13 % of the total destined for non-agricultural uses is a feedstock for elemental phosphorus production, from which both high purity phosphoric acid and other phosphorus derivatives are made. Nearly half of this, about 6 % of the total, goes into the manufacture of phosphate builders for detergents, about a tenth into food and beverage additives, and the remainder into a multitude of small scale applications.

### 8.1.2 Environmental Impacts of Phosphate Rock Processing

The principal environmental concerns of phosphorite surface mining operations relate to water consumption, storage of large volumes of waste slimes from beneficiation, and reclamation of the mined out area. Some concern has also been expressed about exposure to elevated radiation levels on reclaimed land from the exposed slightly elevated uranium concentrations [6]. Natural dewatering of the waste slimes produced is a slow process, it being reported to take two years before the solids content reaches 25 to 30 % [7]. This effectively decreases water recycle capability and increases the volume of the stored slimes, which consequently exceeds the volume of the mined out areas. Thus slime impoundment ponds extend above the level of the natural terrain, which is regarded as at least a nuisance [8]. Mining of igneous apatites, however, does not pose any significant environmental problems particular to the recovery of the ore. And guano sources, which typically involve surface mining and no beneficiation, only have surface reclamation to look after as necessary.

## 8.2 Elemental Phosphorus

Phosphorus was first isolated in the seventeenth century, by a procedure which gave a recovery of about one ounce of phosphorus from a hogshead (about 50 Imperial gallons) of urine [9]. In 1769, a more convenient method that also made larger quantities accessible was developed by Scheele. This involved the treating of crushed bones with sulfuric acid, followed by extraction of the phosphoric acid from the gypsum also produced with small amounts of water. Evaporation of most of the water from the phosphoric acid and then mixing of the residue with charcoal and vigorous heating gave elemental phosphorus (Equations 8.2, 8.3). The phosphorus vapour which formed was captured

$$Ca_3(PO_4)_2 + 3 H_2SO_4 + 6 H_2O \rightarrow \quad (8.2)$$
from crushed bone
$$3 CaSO_4 \cdot 2 H_2O \downarrow + 2 H_3PO_4$$
gypsum (insoluble)

$$4 H_3PO_4 + 10 C \xrightarrow[\text{heat}]{\text{white}} P_4 \uparrow + 6 H_2O + 10 CO$$
white (8.3)
(or yellow)
phosphorus

by condensation under water, since yellow phosphorus spontaneously catches fire in air. Produced in this manner, phosphorus was a small scale article of commerce in France and Britain in the eighteenth century, mainly for match production.

**Table 8.2** Relationship of $P_2O_5$ content and $P_4$ content to phosphate rock containing different concentrations of calcium phosphate, $Ca_3(PO_4)_2$ [a]

| Calcium phosphate content, % (Equivalent to % BPL, or % TPL [b]) | Percent $P_2O_5$ equivalent | Percent $P_4$ equivalent |
|---|---|---|
| — | 100.00 | 43.64 |
| 100 | 45.76 | 19.97 |
| 80 | 36.61 | 15.98 |
| 60 | 27.47 | 11.98 |
| 40 | 18.30 | 7.99 |

[a] Pure fluorapatite $CaF_2 \cdot 3\ Ca_3(PO_4)_2$ analyzes 92.26% $Ca_3(PO_4)_2$, 42.26% $P_2O_5$, and 18.44% $P_4$ equivalent.

[b] BPL and TPL are the still commonly used trade designations *b*one *p*hosphate of *l*ime and *t*riphosphate of *l*ime respectively, used synonymously with calcium phosphate ($Ca_3(PO_4)_2$).

Elemental phosphorus, as obtained by this procedure, is a soft solid at ordinary temperatures comprised of tetrahedral molecules of four phosphorus atoms with three bonds to each atom. The six bonds of a white phosphorus molecule are at 60° to each other placing them under great strain, and thought to be the origin

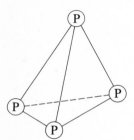

**White phosphorus:**
— tetrahedral
— density 1.82 g/cm$^3$
— melting point 44 °C
— boiling point 280 °C
— virtually insoluble in water
— Luminesces in the dark (phosphorescence)
— very toxic

of the high reactivity [10] (e.g. the spontaneous inflamation in air). In the dark it is seen to luminesce, the property which originally gave this element its name. The other properties important to both its original isolation and its commercial recovery are its boiling point of 280°, low enough to encourage easy volatilization as it is formed in the heated reduction mixture and yet high enough to permit condensation of the vapour by cold water. By being insoluble in, and denser than water, direct contact ("spray", rather than heat exchange) condensation is the most efficient and least expensive option for vapour capture and gives the product an immediate protective covering of water to avoid its contact with air. The moderate melting point permits easy transfer of phosphorus in the liquid state merely by keeping the protective water layer warm.

## 8.2.1 Electric Furnace Phosphorus

The forerunners of the modern electric furnace for phosphorus production were first operated in England by J.B. Readman (1888) and Albright and Wilson (1893) the latter unit with a production capacity of 180 tonnes per year [11]. By allowing control of the heating level independently of the atmosphere above the heated charge, unlike earlier methods which required a combustion source for the white heat necessary, this development revolutionized commercial production of phosphorus.

Today, some 13 to 15 % of American phosphate rock produced is consumed for elemental phosphorus production [6, 12]. Prior to charging to the phosphorus furnace, the phosphate rock of about 65 to 70 B.P.L. level (65 to 70 % *b*one *p*hosphate of *l*ime, $Ca_3(PO_4)_2$, from the early origins) is sintered at 1,200 to 1,250 °C in a rotary nodulizing kiln to obtain marble to golf ball-sized porous lumps. This preliminary step increases charge porosity in the furnace, facilitating the release of phosphorus vapour as it forms. It also decreases the amount of entrained dust that is carried out of the furnace with the product. The nodulized rock is fed in a carefully metered ratio with silica (sand) and coke to the furnace of steel lined with monolithic carbon and fitted with three adjustable carbon electrodes. Two hundred to three hundred volts of three phase AC power is fed to the electrodes which serve to heat the charge to 1,260 to 1,480 °C. These temperatures are sufficient to fuse the mixture and initiate the fluxing and reducing reactions which lead to phosphorus formation. This process may be a composite of phosphate partial reduction by silica followed by reduction of the phosphorus pentoxide ($P_4O_{10}$) on contact with the incandescent coke (Equations 8.4, 8.5). Or it may be that silica merely serves as a fluxing agent or solvent for

$$2\,Ca_3(PO_4)_2 + 6\,SiO_2 \rightarrow P_4O_{10} + 6\,CaSiO_3 \quad (8.4)$$
$$\text{sublimes}$$
$$\begin{array}{c} 300\,^{\circ}C \\ P_4O_{10} + 10\,C \rightarrow P_4 \uparrow + 10\,CO \quad (8.5) \\ \text{vapour} \end{array}$$

the reduction, in which case the overall process is best represented by Equation 8.6. Electrical energy provides the heat required

$$2\,Ca_3(PO_4)_2 + 6\,SiO_2 + 10\,C \xrightarrow[1400\,^{\circ}C]{1300\,-} \quad (8.6)$$
$$\text{BPL} \qquad \qquad \text{flux}$$
$$P_4 \uparrow + 6\,CaSiO_3 + 10\,CO$$
$$\text{vapour} \qquad \text{slag}$$
$$\Delta H = +2840\ \text{kJ}\ (+680\ \text{kcal})$$

for the endotherm of this reaction, which is an indirect electrical contribution to the chemical change. This transformation is therefore called an electrothermal, rather than an electrochemical process. The carbon monoxide, phosphorus vapour mixture produced is first fed to an electrostatic precipitator where any entrained dust is removed while the gases are still hot (Figure 8.1). In this way, non-volatile contaminants are excluded from the water spray condenser which is used to capture most of the liquid white phosphorus. Carbon monoxide, which is non-condensible under these conditions, simply passes through the water condenser. The direct contact condenser may also be followed by an indirect, tubular type condenser to raise phosphorus containment efficiency [13]. The product is pumped from the conserser kept as a liquid by hot water-jacketed pipes, to sumps, where it is also stored in this form. A layer of hot water lying over the phosphorus maintains it in this form for easy transfer, and keeps it out of contact with air. Typical specifications of this product are $P_4$, > 99,85 %; Pb, < 2 ppm; As$_2$O$_3$, < 250 ppm; and F, < 3 ppm. The now clean carbon monoxide gas stream, which comprised about 93 % of the gases leaving the furnace by volume, is used to fuel the phosphate rock nodulizing kiln (Figure 8.1).

The white phosphorus allotrope, often also called yellow phosphorus, is the direct product. It is only occasionally converted to phosphoric acid on site since the shipping of phosphoric acid involves considerably more mass than shipping of the element itself. Thus, molten white phosphorus is commonly shipped by road or rail, in insulated tank cars of some 50 tonnes capacity [14], and by sea in mild steel, beehive-shaped tanks of 1,100 tonne capacity under water and inert gas [15].

Small amounts of red and black phosphorus, two more stable phosphorus allotropes, are also produced for special purposes from the white phosphorus of the electric arc furnace. Red phosphorus is obtained by heating white phosphorus at 400 °C for several hours yielding a complex polymeric material considerably more stable than the white variety [10, 14]. Red phosphorus is not only stable in air, but far less toxic than white phosphorus. Black phosphorus, of a different complex structure, is obtained by heating the white variety at 220 to 370 °C for eight days, and in addition requires either pressures exceeding $10^9$ Pa or a seed crystal of black phosphorus for its formation. This product has a density of 2.70 g/cm$^3$, much higher than either of the other allotropes, has a structure resembling graphite, is a good electrical conductor and can be lit with a match only with difficulty (Table 8.3) [10, 14].

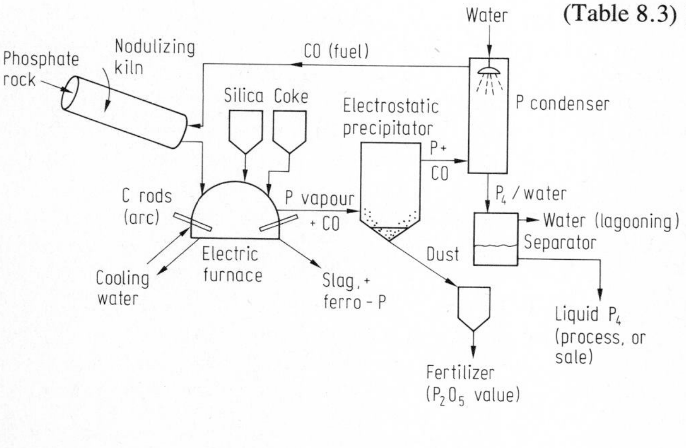

**Figure 8.1.** Flowsheet for the electric furnace process for the production of elemental phosphorus. Phossy water from the phosphorus phase separator (sump) is normally maintained in a closed circuit holding pond system which also provides the feed water for the phosphorus direct contact condenser. Some facilities have a further indirect, heat exchange condenser for the carbon monoxide stream, after the direct contact condenser

**Table 8.3** Properties of allotropic forms of phosphorus

|  | Phosphorus allotrope | | |
|---|---|---|---|
|  | White | Red | Black |
| Density, g/cm$^3$ | 1.82 | 2.34 | 2.70 |
| Melting point, °C | 44 | ca. 400 | – |
| Flammability in air | spontaneous | when heated to 200° | when heated to $> 550$°C |
| Solubility | – in CS$_2$, many organic solvents – about 3 mg/L [a] in water at 15 °C | insoluble | insoluble |
| Toxicity, [b] acute (man) | $< 0.2$ g | relatively harmless | relatively harmless |
|     chronic | 1 mg per day | – | – |

[a] As quoted by the Handbook of Chemistry and Physics [16]. More recent data gives 0.56 mg/L at 20°.

[b] Details see [18].

A single present day phosphorus furnace produces from 60 to 160 tonnes of phosphorus per day, almost up to the *annual* production figures of the early arc furnaces [13] and requiring a power supply of some 90,000 kW for a single furnace at the upper end of this production range. Power consumption per tonne of phosphorus produced varies with the percent calcium phosphate in the rock (% BPL level) and furnace size among other factors but ranges from around 12,000 to 14,000 kWh/tonne (Table 8.4). Hence, power costs are a major component of electric furnace phosphorus production.

Raw material addition and phosphorus vapour removal from the furnace occur more or less on a continuous basis. Removal of the molten calcium silicate slag, more than 7 tonnes of which is formed for each tonne of phosphorus, and ferrophosphorus, which is formed from iron present in the charge, takes place periodically through appropriately placed tap holes on the lower furnace wall. The slag may be made into a rock wool in-

**Table 8.4** Operating experience of an electric furnace phosphorus plant, per tonne of phosphorus produced [a]

| Furnace size, kW | 15 000 | 25 000 | large | range | |
|---|---|---|---|---|---|
| Raw materials: | | | | | |
|   Phosphate rock, kg | 10 000 | 9 340 | 7 250 | 6 | – 10 000 |
|   Silica, kg | 1 500 | 1 560 | 2 660 | 0.5 – | 3 500 |
|   Coke, kg | 1 500 | 1 550 | 1 430 | 1 200 – | 1 800 |
|   Power, kWh | 14 320 | 11 830 | 11 850 | 11 600 – | 16 000 |
|   Electrode C, kg | – | – | 15 | 20 – | 60 |
| Water: | | | | | |
|   Spray condenser | 13 500 L | 5 400 L | 13 000 L | – | |
|   Cooling | – | 145 m$^3$ | – | – | |
| Products: | | | | | |
|   Phosphorus, kg | 1000 [b] | 1 000 | 1 000 | 1 000 [b] | |
|   Ferrophosphorus, kg | 300 | 360 | 90 | 90 – | 380 |
|   Slag, kg | 8 900 | 8 100 | 7 100 | 7 100 – | 8 900 |
|   Gases | 2 800 (as CO) | 2 820 | 3 300 | – | |

[a] Data compiled from [7, 13, 14, 19, 20].

[b] 87 – 90% elemental phosphorus recovery from phosphate rock.

sulating material or may be crushed for use in glass making or liming of soils but is more frequenty employed as an aggregate for highway or railway roadbed construction. Ferrophosphorus is normally sold for steelmaking, providing a convenient source for phosphorus addition to alloy steels.

### 8.2.2 Uses of Elemental Phosphorus

White (or yellow) phosphorus used to be used to a significant extent for the manufacture of matches and fireworks. But the debilitating occupational illness of phossy jaw among workers, and the occasional hot pockets and singed purses among the public led to its discontinuance in these applications by the early 1900's [18]. Today the much safer red phosphorus, or phosphorus sesquisulfide ($P_4S_3$; Equation 8.7)

$$P_4 + 3\,S \xrightarrow[\text{heat}]{CO_2} P_4S_3 \tag{8.7}$$

are used for matches, flares and other kinds of incendiary devices. Both of these more recent ignition components possess a lower vapour pressure, are much less toxic, and are much less likely to ignite inadvertently.

The phosphorus halides were, and still are, an important class of chemical intermediates and are prepared by combination of the elements (Equations 8.8-8.10).

$$P_4 + 6\,Cl_2 \rightarrow 4\,PCl_3 \tag{8.8}$$

$$2\,PCl_3 + O_2 \rightarrow 2\,POCl_3 \tag{8.9}$$

$$P_4 + 10\,Cl_2 \rightarrow 4\,PCl_5 \tag{8.10}$$

The products, phosphorus trichloride, phosphorus oxychloride, and phosphorus pentachloride are each useful in their own right as well as being of value for the direct preparation of phosphite and phosphate esters used as extractants, plasticizers and gasoline and oil additivies, pesticides, etc.

But by far the largest fraction of the elemental phosphorus produced, some 90 %, is used for the preparation of high purity grades of phosphoric acid (see Section 8.3) [12] and various phosphate salts for detergent formulation from this. The decline in detergent phosphate concentrations as a result of a concern about phosphate pollution in water courses has been the main factor responsible for the steady decline in the U.S. production of phosphorus in recent years, from about 540 thousand tonnes in 1970 to about 390 thousand tonnes in 1977 (Table 8.5). Production seems to have levelled off now with detergent consumption of phosphorus stabilized at about 45 % of the total [23]. U.S. prices in the latter years have ranged around $ 1.20 per kilogram, depending on purity [21].

### 8.2.3 Environmental Aspects of Phosphorus Production

Electric furnace phosphorus production is relatively tolerant of the impurities present in the poorer grades of phosphate rock which contain lower concentrations of calcium phosphate. A high silica content in the phosphate rock can be accommodated by appropriate decrease in the proportion of silica fed to the phosphorus furnace, and the pre-

**Table 8.5** Annual production of yellow phosphorus by selected countries [a]

| | Thousands of metric tonnes [b] | | | | | | |
|---|---|---|---|---|---|---|---|
| | 1965 | 1975 | 1976 | 1977 | 1978 | 1979 | 1980 |
| Canada | (27) | (85) | (85) | | | (85) | (82) |
| China | (ca. 18) | | (35) | | | | (18) |
| E. Germany | (18) | | (15) | | | | (18) |
| France | (14) | — | 17 | 17 | 13 | | (20) |
| Italy | (13) | | | | | | (16) |
| Japan | (30) | 16 | 19 | 14 | 16 | | (26) |
| U.K. | (32) | | | | | | |
| U.S.A. [c] | (555) | 408 | 397 | 390 | 400 | 417 | 392, (512) |
| U.S.S.R. | (ca. 100) | | (250) | | | | (435) |
| W. Germany | (73) | | (80) | | | | (80) |

[a] Countries included having a 1965 annual production capacity of more than 10 000 tonnes.
  Sources: see [14, 19, 21, 22, and 79].
[b] Where only capacity data was available, this is given in parentheses.
[c] U.S. production in 1970 was 542 000 tonnes.

sence of carbonaceous matter may be allowed for similarly. The presence of more, or less iron oxide affects the yield of ferrophosphorus byproduct, again without seriously affecting operation. In fact, occasionally lumps of iron will be deliberately added to the furnace charge to boost the yield of ferrophosphorus at times when the market for this alloying material is strong.

Primary areas of environmental concern in the operation of a phosphorus furnace are from potential losses of dissolved and suspended elemental phosphorus in water. Even low concentrations of the element present a potent toxicity hazard to freshwater and marine animal life [24]. Losses of fluoride to air or freshwater which might occur, particularly during process upsets, are also of concern. If severe, or prolonged, this too may cause damage to adjacent plant life or freshwater organisms [25, 26]

The nodulizing of phosphate rock before it is charged to the furnace is a measure designed for good gas release from the heated furnace charge. It also decreases dust entrainment in the large volume of gaseous products produced by the furnace by decreasing the fraction of small particles in the charge. If fluorapatite (theoretical 3.7 % F) comprises a significant fraction of the phosphate rock feed used, the rock will contain about 2.8 % fluoride. In the rotary nodulizing kiln which is used to prepare the furnace charge, dusts are entrained in the hot gases and about 8 percent of the fluoride present is volatilized to hydrogen fluoride and silicon tetrafluoride (Equation 8.1). The dust is collected from the kiln exhaust gas stream following which the fluorides are removed by scrubbing with water [27]. Hydrogen fluoride is extremely soluble in water and is effectively trapped in this form. Silicon tetrafluoride reacts with scrubber water to form soluble fluosilicic acid and colloidal silica (Equation 8.11).

$$3\,SiF_4 + 2\,H_2O \rightarrow 2\,H_2SiF_6 + SiO_2 \qquad (8.11)$$

A safeguard additional to nodulization of the charge to prevent dust entry to the phosphorus condenser from the furnace gas stream is provided by two stage electrostatic precipitation of any involatiles from the carbon monoxide and phosphorus vapour mixture leaving the furnace [13]. This measure not only raises the purity of the condensed phosphorus recovered from the sumps of the phosphorus condenser, but it also assists in decreasing the amount of phosphorus which is condensed in

water in colloidal form rather than as easily settleable droplets. Colloidal phosphorus is difficult to separate from the water.

Condensation of phosphorus vapour by direct contact with a water shower in a mild steel tube is both the most economical method of obtaining the efficient heat transfer necessary, and immediately provides the protective water covering required for the condensed phosphorus to prevent inflammation in air. At the same time, it places dissolved (solubility about 3 mg/L [28]) and colloidal, highly toxic elemental phosphorus into the water layer. For this reason phosphorus condensers are normally operated on a closed circuit, water recycle basis from a lagoon or holding pond reservoir, to avoid a toxic discharge problem.

During electrothermal phosphate reduction, 80 to 90 % of the fluoride contained in the rock remains with the slag as its calcium salt [27, 29] (Equation 8.12).

$$CaF_2 \cdot 3\,Ca_3(PO_4)_2 + 9\,SiO_2 + 15\,C \rightarrow$$
$$\tfrac{3}{2}P_4 \uparrow + CaF_2 + 15\,CO \uparrow + 9CaSiO_3 \quad (8.12)$$

Since calcium fluoride is insoluble in water, fluoride departure from the process in this form is relatively innocuous. This fluoride present in the slag may, however, be mobilized in the water phase by procedures which contact the hot slag with water to cause fracture [27]. The product, a "crushed" slag, is highly appropriate for some of the end uses mentioned.

The fluoride which is not captured in the slag is converted to the volatile substances silicon tetrafluoride and hydrogen fluoride, both of which leave the furnace with the phosphorus vapour and carbon monoxide but are captured in the water stream of the phosphorus condensers (Equation 8.11). Thus, the water from the phosphorus condenser contains not only dissolved and colloidal phosphorus, but also dissolved fluoride as the ion and as the complex fluosilicate anion.

Thus, the water outflow of the sumps fed by the phosphorus condensers ordinarily contains about 1,700 ppm $P_4$, which is joined by further phossy water streams such as that displaced by incoming phosphorus in the water-blanketed liquid phosphorus storage tanks. This stream also contains scrubbed fluorides [28, 30]. Gradual build-up of these contaminants occurs in the recycle pond water to the point where it may contain phosphates and phosphorus at 1.7 g/L (specified as $P_2O_5$), and fluoride and ammonium ion each at 10 g/L (when ammonia is used for pH control). Either from the gradual accumulation of excess dissolved solids, which start to affect process requirements and aggravate corrosion problems, or from an excess of precipitation over evaporation rates in the holding ponds of the area in which the plant operates, a portion of the pond water content must be continuously or periodically purged from the closed system. Percolation of the discharged water through a pile of granulated slag has been found to decrease the fluoride concentration from about 1 % (10,000 ppm) to 30 ppm by reaction of the fluoride with residual calcium in the slag, forming insoluble calcium fluoride [27]. The eluate from this procedure may be further treated with lime in a second lagoon, if necessary, to decrease fluoride concentrations to lower than this (Equation 8.13).

$$2\,NH_4F + Ca(OH)_2 \rightarrow CaF_2 \downarrow + 2\,NH_4OH \tag{8.13}$$

Alternatively the waste waters may be treated with aluminum slags and/or sodium hydroxide in a variety of ways to generate normal, or ammonium cryolite of marketable value for aluminum production (Equations 8.14, 8.15) [27].

$$6\,NH_4F + NaAlO_2 + 2\,NaOH \rightarrow$$
$$Na_3AlF_6 + 6\,NH_3 + 4\,H_2O \tag{8.14}$$
cryolite

$$12\,NH_4F + Al_2(SO_4)_3 \rightarrow$$
$$2\,(NH_4)_3AlF_6 + 3\,(NH_4)_2SO_4 \tag{8.15}$$
ammonium
cryolite

Alternatively, magnesium fluosilicate, for preservation of Portland cement surfaces, or sodium fluosilicate for fluoridation of water supplies may be prepared and marketed (Equations 8.16, 8.17).

$$MgO + H_2SiF_6 \rightarrow MgSiF_6 + H_2O \tag{8.16}$$
$$2\,NaOH + H_2SiF_6 \rightarrow Na_2SiF_6 + 2\,H_2O \tag{8.17}$$

The phosphorus content of the pond recycle water can vary widely depending on recycle rates and process upsets but seldom goes below about 23 ppm $P_4$ at the inlet and averaging 0.3 to 0.5 ppm occasionally rising to 1.0 to 2.0 ppm at the exit. These concentrations are much lower than those direct from the phosphorus condensers, but are still not low enough for safe discharge since phosphorus concentrations of even a few parts per

billion are toxic to fish [24, 28, 30]. Since phosphate ($PO_4^{3-}$) is relatively innocuous at these concentrations it might be expected that aeration of phossy water should rapidly render it safe. But several tests have shown that oxidation of aqueous phosphorus by direct aeration is both relatively slow and ineffective [18, 24, 28, 30]. Ozone, or potassium permanganate have been found to be effective oxidants, but are too expensive for commercial use. Chlorine, too, is an effective reagent for treatment [17] but metering the amount fed relative to the phosphorus content of the waste water was found to be difficult, and an excess of chlorine was found to be not much better than the pollutant it was intended to neutralize [28]. The most effective combination for phossy water treatment appears to be addition of lime and one or more chemical coagulants, coupled with settling and occasionally centrifugation leaving treated waste water containing 50 parts per billion phosphorus (ppb as $P_4$) or less [28]. If these measure are still inadequate for a particular situation then 24 hour or longer holding of the treated waste water in a large settling lagoon provides a further safeguard and an additional decrease in phosphorus concentrations to 2-3 $\mu g/L$ [30].

## 8.3 Phosphoric Acid via Phosphorus Combustion

Three main routes are employed for the commercial production of phosphoric acid. The dry process, also called the combustion process and yielding a furnace acid, obtains phosphoric acid by hydration of phosphorus pentoxide obtained via the combustion of yellow phosphorus. The other two processes, both referred to as wet processes, produce phosphoric acid by acidulation of phosphate rock. The process using sulfuric acid acidulation is also referred to as the Dorr process, while acidulation with hydrochloric acid is commonly referred to as the Haifa process. The wet processes will each be discussed separately later.

For the combustion process, liquid phosphorus is displaced from a closed, steam heated storage tank by metering in hot water (Figure 8.2), to a burner which uses compressed air to finely atomize the phosphorus. Additional combustion air then accomplishes complete oxidation producing phosphorus pentoxide and a great deal of heat (Equation 8.18). Phosphorus pentoxide is so named because

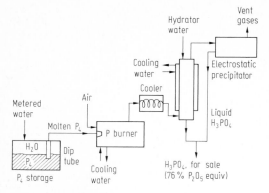

**Figure 8.2.** Diagram of process paths for the production of phosphoric acid by the combustion of elemental phosphorus. Graphite tubes are frequently employed to provide the initial cooling for the extremely corrosive combustion products

$P_2O_5$ was originally

$$4\,P + 5\,O_2 \rightarrow P_4O_{10} \qquad (8.18)$$
$$\text{melts } 580-585\,^{\circ}C$$
$$\text{sublimes } 300\,^{\circ}C$$
$$\Delta H = -3010 \text{ kJ } (-720 \text{ kcal})$$

thought to be the molecular formula of $P^{5+}$-oxide, but is now known to be $P_4O_{10}$. The combustion chamber may be operated as a separate unit and the phosphorus pentoxide vapour ducted via an intermediate cooler to the base of the hydration tower [31]. Or the combustion chamber itself may form the base of the hydration tower. In either case, cooling by water addition for hydration, and indirect cooling of the combustion chamber with water jackets are employed to absorb the heat produced by the reaction.

Hydration is accomplished in a packed, acid-proof steel tower by passing phosphorus pentoxide vapour in at the base, and trickling in phosphoric acid and water at the top, thereby accomplishing both cooling and hydration (Equation 8.19). Hydration, too, releases some additional heat.

$$P_4O_{10} + 6\,H_2O \rightarrow 4\,H_3PO_4 \qquad (8.19)$$
$$\Delta H = -188 \text{ kJ } (-45 \text{ kcal})$$

The nature of phosphorus pentoxide and the absorption process inevitably leaves as much as 25 % of the oxide plus a phosphoric acid mist in the exit gases from the absorption tower [11] which are captured on passage through an electrostatic precipitator. By variations of the process details and the

equipment used, grades of phosphoric acid from 75-105 % $H_3PO_4$ (ortho, or superphosphoric acid) may be made this way [32].

Phosphoric acid production by phosphorus combustion is usually accomplished in a stepwise manner as outlined, with intermediate isolation of the phosphorus. But it may also be conducted by direct contact of phosphorus vapour from the furnace of a phosphorus plant with an airstream and the phosphorus pentoxide produced passed directly into a hydrator without collection of the intermediate phosphorus as a liquid. Direct conversion to phosphoric acid in this way reflects the fact that the bulk of white phosphorus produced proceeds directly to phosphoric acid production.

Furnace phosphoric acid is reasonably pure for most uses as obtained directly from the process. But for food grade applications traces of arsenious oxide must be removed. Arsenic, present in the feed phosphorus to the extent of 50-180 ppm (as $As_2O_3$ equivalent) [11] mainly because of the similarity of its chemical properties to those of phosphorus (same group in the periodic table), ends up in the product acid on oxidation and hydration. For a food grade product this may be removed by the addition of the requisite amount of sodium sulfide or hydrogen sulfide and precipitating as arsenic III and V sulfides (e.g., Equation 8.20). Any excess hydrogen sulfide is removed from the acid by air

$$AS_2O_3 + 3\,H_2S \rightarrow As_2S_3 \downarrow + 3\,H_2O \qquad (8.20)$$

stripping, after which it is clarified by passage through a sand filter. By this means arsenic concentrations (as $As_2O_3$) can be brought to below $0.1\,\mu g/g$ and the only other impurities present at levels above $1\,\mu g/g$ will be $Na_2O$, 0.02 % and iron and chloride at about $2\,\mu g/g$ each [31].

### 8.3.1 Environmental Features of Furnace Phosphoric Acid Production

Losses of phosphoric acid mists or phosphorus pentoxide fumes represent the most commonly experienced emission control problems with furnace acid plants. This can occur either from corrosion induced rupture of ductwork or a vessel, or from a process operational parameter. When this happens, a white plume, or "ghost" remains visible downwind of the point where the steam component of the stack discharge has dissipated. Corrosion problems are gradually becoming better understood now so that more of the exposed components are being constructed of stainless steel or, if high temperatures are also involved, equipment is lined with graphite [32]. Heat exchanger tubes for the cooler situated between the phosphorus burner and the hydrator towers are also made of graphite. The gas path in the electrostatic precipitator consists of carbon tubes with cathode elements of stainless steel wire to ensure reliable operation of this unit [31]. Process parameters of importance for mist and fume control include measures to minimize the moisture content of the combustion air fed to the phosphorus burners [17, 32]. A high moisture content in this gas stream tends to aggravate mist formation. Solid phosphorus pentoxide has been found to be very difficult to dissolve in either water or phosphoric acid. This has led to the suggestion that temperatures in the hydrator be maintained high enough that phosphorus pentoxide absorption takes place from the vapour phase [31]. Further studies have shown that the uptake of phosphorus pentoxide vapour in 70 % phosphoric acid is only about 60 % and climbs with increasing acid concentrations up to about 88 % phosphoric acid. More recent studies have shown that two fiber beds operated in series, the first bed as an agglomerator, and the second as a collector, may also be used to efficiently control mists and fumes from furnace acid plants [33]. Mass containment efficiencies of better than 99.9 % were reported for a median aerosol particle diameter of 1.1 to $1.6\,\mu m$.

# 8.4 Phosphoric Acid using Sulfuric Acid Acidulation

The oldest process route to phosphoric acid (see Section 8.2) and still the lowest cost route to this product, at least for fertilizer grades, is via the addition of quite high concentrations of sulfuric acid to finely ground phosphate rock. This acidulation releases phosphoric acid from the calcium phosphate salts present and ideally produces insoluble gypsum (calcium sulfate dihydrate) which can be removed by filtration (Equation 8.21). The fluoride ion present in the

$$CaF_2 \cdot 3\,Ca_3(PO_4)_2 + 10\,H_2SO_4 + 20\,H_2O \rightarrow$$
$$6\,H_3PO_4 + 10\,CaSO_4 \cdot 2\,H_2O + 2\,HF \quad (8.21)$$

phosphate rock of the majority of sources is of no commercial value, and in fact is more of a nuisance to phosphoric acid production by this method since strong acid acidulation releases virtually all of the

fluoride originally present in the rock. Considering just the phosphoric acid-forming part of the overall acidulation reaction, this is seen to be a moderately exothermic process [34] (Equation 8.22). Overall plant design and operational features, then, emphasize

$$Ca_3(PO_4)_{2(s)} + 3\,H_2SO_{4(l)} + 6\,H_2O_{(l)} \rightarrow$$
BPL, 70–75 %
$$2\,H_3PO_{4(l)} + 3\,CaSO_4 \cdot 2\,H_2O_{(s)} \quad (8.22)$$
$$\Delta H = -321\ kJ\ (-76.4\ kcal)$$

obtainment of as high a concentration of phosphoric acid as possible, the arrangement of acidulation and vapour control equipment to permit safe containment of the incidental fluorides, and accomplishes all of this with adequate cooling measures in place to permit operation of the whole proess at low to not more than moderate acidulation temperatures.

Operation of the process at not more than moderate temperatures is necessary to avoid the precipitation of anhydrite (anhydrous calcium sulfate) or calcium sulfate hemihydrate ($CaSO_4$ x $0.5\,H_2O$) which tend to form at high operating temperatures, both of which precipitate in a finely divided form which is difficult to filter [34]. The less hydrated forms of calcium sulfate, if formed, may also revert to gypsum at cooler downstream portions of the process and cause solids deposition and plugging of equipment at these points. Gypsum, the coarser dihydrate of calcium sulfate which is easily filtered, while it may form at higher temperatures, is only stable at temperatures below about 40 °C and then only for phosphoric acid concentrations below 35 % (by weight). Hence, it is desirable to keep acidulation temperatures low for this process. Severe corrosion conditions which exist throughout this process have led to the use of wood, lead-lined wood, polyester-fiberglass, and rubber-lined steel for construction of much of the equipment, with dense carbon or a nickel alloy being favoured for heat exchangers.

## 8.4.1 Operation of the Acidulation Process

Phosphate rock of 70 to 75 % BPL, or as high as is reasonably obtainable, is finely ground in a ball mill and then mixed with cooled recycled phosphoric acid-gypsum slurry in a digestion tank (Figures 8.3, 8.4). At this stage the only reaction which occurs is between acid and any carbonates

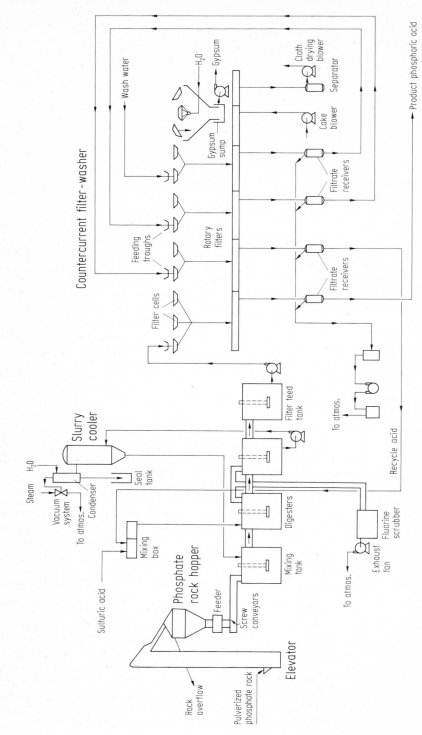

**Figure 8.3.** Process outline for phosphoric acid production by sulfuric acid acidulation of phosphate rock. Fume control and slurry evaporative cooling systems are shown, together with counter-current filtration and washing system which serves to maximize product acid concentrations accessible without evaporation. (Bird Machine Co.)

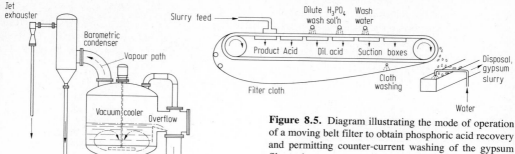

**Figure 8.4.** Details of operation of a slurry vacuum cooler, in which a small water jet exhauster maintains the vacuum, largely produced by direct steam condensation and
the barometric leg of the barometic condenser. Reprodu-
ced from reference 61, page 275, by courtesy of Marcel
Dekker

**Figure 8.5.** Diagram illustrating the mode of operation
of a moving belt filter to obtain phosphoric acid recovery
and permitting counter-current washing of the gypsum
filter cake

present in the rock, and between phosphates and
low concentrations of sulfuric acid which may be
present, in this way minimizing foaming (Equation
8.23)

$$3\,CaCO_3 + 2\,H_3PO_4 \rightarrow$$
$$Ca_3(PO_4)_2 + 3\,CO_2 + 3\,H_2O \qquad (8.23)$$

and the evolution of heat on initial acid contact.
The newly formed mixed slurry is then moved into
the next digestion tank where fresh sulfuric acid
mixed with relatively dilute recycle phosphoric
acid (giving about 50 to 55 % net sulfuric acid) is
added and vigorous mixing continued.
Acidulation is continued through two or more additional digestion tanks like the first, combined with
several passes of the partially reacted slurry
through a slurry cooler. Direct cooling by evaporation of a part of the water present in the slurry,
under reduced pressure, occurs in the cooler (Figure 8.4). This is one of the measures taken to avoid
anhydrite formation. This method of cooling also
helps to keep the final phosphoric acid concentration high. Digestion requires a period of 4 to 8
hours while keeping temperatures to below 75 to
80 °C [12], after which time the slurry proceeds
to the product acid recovery system. Under these
acidulation conditions gypsum is the particular calcium sulfate hydrate which is initially precipitated.
Providing either that the slurry is cooled or that

filtration is accomplished after a reasonably short
time at these temperatures, the gypsum does not revert to troublesome anhydrite or the hemihydrate.
Recovery of phosphoric acid from the fully digested slurry is by continuous counter-current filtration
and washing in moving belt (e.g., Figure 8.5),
moving pan, or horizontal rotary filters [34].
Monofilament polypropylene cloth has been found
to be a durable and fast draining filter element in
this service. Regardless of the particular continuous filtration method adopted, all involve initial
filtration of the cooled, fully digested gypsum
slurry in phosphoric acid to yield a filtrate of 40 to
45 % (29-33 % $P_2O_5$ equivalent) phosphoric acid,
the maximum concentration directly obtainable by
this process. The gypsum filter cake, still damp
with the 40 to 45 % phosphoric acid, is washed
with dilute phosphoric acid collected from a later
wash stage and this filtrate is used to dilute the
fresh incoming sulfuric acid before it contacts the
powdered rock. This wash sequence is repeated
from once, up to three or four times, each time using the filtrate from the subsequent wash stage as
the wash liquor, until the last stage where pure water is used. In this way the gypsum is virtually
freed of adhering residual acid and in the process
the counter-current re-use of the wash water has
also minimized the amount of dilute phosphoric
acid to be incorporated into the product. In this way
the product acid concentration is maintained as
high as is feasible.
The crude acid obtained from this process is not
very concentrated, nor very pure. Concentration
by boiling off water also reduces the solubility of
many of the dissolved impurities such as calcium,
fluoride ions, aluminium and sulfate ions present,
causing precipitation and scaling of these materials
on heat transfer surfaces. As a result, rather spe-

cialized heat transfer equipment is used in order to achieve water evaporation without necessitating frequent equipment cleaning and overhaul. Submerged combustion units having no intervening heat transfer surfaces, or similarly the direct application of hot air [35] or flue gases through carbon nozzles discharging into a shallow stream of acid in an open trough [31], both successfully achieve this. Multiple effect evaporation has more recently been shown to also be feasible, as long as it is accompanied by forced circulation to improve heat transfer efficiencies and decrease deposition of impurities [36].

Raising the concentration of the acid alone serves to remove many of the impurities through precipitation, and about 40 % of the dissolved fluoride by volatilization [34], and produces about 70 % phosphoric acid (50 + % $P_2O_5$) suitable for many uses. But for solution fertilizers, where residual impurities may cause precipitation on long term storage [37] or for the manufacture of a food grade acid, further purification steps are necessary. Fluoride is stripped to less than 50 $\mu g/g$ as volatile hydrogen fluoride, by sparging the crude concentrate itself with superheated steam. Or powdered silica may be added, forming silicon tetrafluoride (Equation 8.24) which is then removed by sparging

$$SiO_2 + 4\,HF \rightarrow SiF_4 + 2\,H_2O \qquad (8.24)$$
$$\text{b.p.} -86\,°C$$

with air or steam [31]. Arsenic, which is also extracted from the phosphate rock by acid, is removed by precipitation as the sulfide on addition of

sodium sulfide [12, 38] (Equation 8.19). Addition of a slight excess of sulfuric acid serves to precipitate the bulk of any dissolved calcium (Equation 8.25) after which filtration of the treated acid through a sand bed produces a

$$Ca(H_2PO_4)_2 + H_2SO_4 + 2\,H_2O \rightarrow$$
$$CaSO_4 \cdot 2\,H_2O \downarrow + 2\,H_3PO_4 \qquad (8.25)$$

clarified, substantially pure product. A few additional purification refinements may be required depending on the origin of the phosphate rock used and the particular impurities common to that deposit, to produce a food grade phosphoric acid (Table 8.6). Overall phosphate recoveries from the rock are about 90 % [12].

### 8.4.2 New Developments and Variations on Sulfuric Acid Acidulation Method

Accessible process choices have been surveyed [39] and the initial separation of calcium sulfate from phosphoric acid as the hemihydrate has been exploited on a commercial scale [34, 40]. In this variation, the initial hemihydrate separation is followed by a second stage where the hemihydrate, with the help of further added sulfuric acid, is recrystallized as conventional gypsum. It is reported that both capital and operating economies are achieved, while the recovery of phosphate from rock is increased to better than 98 % in the process. Conventionally this recovery is about 90 %.

In another variant of the hemihydrate process, the

**Table 8.6** Partial analysis of a typical, food-grade phosphoric acid produced by phosphorus combustion (75.0% $H_3PO_4$; 54.3% $P_2O_5$) [a]

| Component | Concentration ppm | Component | Concentration ppm |
|---|---|---|---|
| $Na_2O$ | 200 | $H_2S$ | 0.1 |
| $O_2$ consumed [b] | 20 | $As_2O_3$ | 0.05 |
| Fe | 2 | Al, $SO_4$ | |
| Cl | 2 | CaO, $K_2O$ } | 0.0 |
| F | 0.4 | $SiO_2$, each | |
| Cr | 0 | Colour | 10 [c] |
| Pb | 0.2 | Turbidity | 1 [c] |
| Cu | 0.1 | Odour | none |

[a] Data from [29].
[b] A measure of phosphorus present in oxidation states lower than phosphate.
[c] In American Public Health Association (APHA) units.

foam layer formed by the reaction of acid with rock carbonates (Equation 8.26) is used to

$$2\,CaCO_3 + 2\,H_2SO_4 \rightarrow$$
$$2\,CaSO_4 \cdot 0.5\,H_2O + 2\,CO_2 \uparrow +H_2O \quad (8.26)$$

distribute a more concentrated sulfuric acid acidulant [41]. The high temperatures employed, 105-120 °C, favour rapid reaction, hemihydrate formation and directly yields a more concentrated phosphoric acid. This results in a net saving on energy costs for phosphoric acid concentration.

A single, isothermal reactor design is said to more reliably yield gypsum crystals because of better temperature control than is possible with multiple reactor designs [42]. This has also been claimed for a two vessel loop acidulation system [43].

Incorporation of methanol at the rate of 2 kg per kg of phosphorus pentoxide in the dried acidulate has been found to significantly improve the purity of the phosphoric acid product in a two stage process. This modification is especially advantageous if low grade phosphate rock, which is increasingly having to be used as raw material [44], provides the rock feed. Acetone has also been found of value in an experimental variant which enables much less expensive sulfurous acid, rather than sulfuric acid, to be used as the acidulant [45]. In this case the solvent serves to form what is probably an α-hydroxy-sulfonic acid with the sulfurous acid

$$\begin{array}{c} H_3C \\ \diagdown \\ \phantom{H_3}C{=}O + SO_2 + 2\,H_2O \\ \diagup \\ H_3C \end{array} \longrightarrow \begin{array}{c} H_3C \quad OH \\ \diagdown\phantom{a}\diagup \\ \phantom{H_3}C \\ \diagup\phantom{a}\diagdown \\ H_3C \quad SO_3^- \end{array}$$
$$+ H_3O^+ \qquad (8.27)$$

(Equation 8.27) which is sufficiently strong to attack phosphate rock at a reasonable rate. Post acidulation applications of solvent either as an extractant [46] or as a precipitant [37] have also been found valuable for some methods of acid purification.

Newer methods of decreasing the fluoride concentrations of phosphoric acid have recently been surveyed [47]. Apart from endorsement of the addition of a reactive form of silica prior (Equation 8.22) to stripping, the procedure of acid dilution prior to re-evaporation has also been recommended, the additional water vapour removal apparently assisting purification by entraining hydrogen fluoride and silicon tetrafluoride.

Uranium recovery from the wet process phosphoric acid filtrate has been of interest for many years [48, 49] and has led to announcements by several U.S. operations which are recovering $U_3O_8$ (yellow cake) on a 200,000 kg per year scale [50]. Prior clean-up of the dissolved organics in the raw phosphoric acid with activated carbon improves the ease of uranium extraction by obtaining cleaner phase separation [51]. Vanadium recovery from wet process acid has also been of interest, particularly from Idaho—Montana—Wyoming ores [34].

### 8.4.3 Emission Control Measures for Wet Process Acid

Contact of the fluoride present in fluorapatite-based phosphate rocks with a strong mineral acid mobilizes much of the fluoride to hydrofluoric acid (HF: boiling point, 19 °C; Equation 8.21) and silicon tetrafluoride ($SiF_4$; boiling point, $-86$ °C; Equation 8.24). Fluorides have an acute oral toxic dose as sodium fluoride of about 200 mg per kg body weight in rabbits [52], and also show severe skeletal effects on excessive chronic exposures [26]. Therefore, it is easily the most important component requiring control in a wet process phosphoric acid plant. Industrial hygiene requirements for workers have been set at 2.5 mg/m$^3$ for dusts and 3 ppm for hydrogen fluoride [52].

From an initial approximately 3.8 % fluoride which may be present in the phosphate rock (e.g. for fluorapatite) the fluoride is distributed to virtually all of the product streams of the acid. But the proportions found in each stream vary widely depending on the origin of the rock and the process details. One study found that, of the original fluoride present in the rock, about 29 % remained in the gypsum filter cake (much less than that captured in the slag of a phosphorus plant), 15 % was evolved to air and about 55 % ended up in the crude phosphoric acid. It is unclear whether the gypsum-bound fluoride is mobile or not since studies of the emissions to air from gypsum settling ponds demonstrate a rate of pond fluoride emission of 0.18 kg per hectare per day [53]. The pond water pH in this study was 1.6 to 1.8 at the time that the fluoride concentration stood at 2,800-5,100 mg/L. This would suggest that separation of the fluoride disposal from the gypsum disposal functions, or liming of the ponds could have significantly decreased these discharge levels. Certainly more recent studies suggest that it may be possible to bind up to 85 % of the total fluoride, and immobilize it in the gypsum [54].

Fluoride losses to air occur in dusts raised in phosphate rock grinding operations, which has been controlled by means of a baghouse [55]. Small amounts of further dust loss occur from the first digester. Fluoride loss, mainly as silicon tetrafluoride vapour, also occurs from the whole of the digestion train $(1,200-3,500 \text{ mg F/m}^3)$ as well as the filtration and vacuum areas. A system of covers for digestion vessels, and hoods over the filtration area, all exhausted to a venturi scrubber system now appears to best meet these control requirements [55, 56]. This system lends itself to low water flow rates on a recycle basis, so that water usage is small. Equilibrium pond water fluoride concentrations of about 0.2 % result from this contribution from the scrubber. Safe disposal of excess pond water volumes may be made by fluoride removal using small separate basins for single, or double liming of the effluent before discharge (Equation 8.13). A recent announcement proposes that the calcium fluoride precipitated from excess pond water be used as a process for the production of calcium fluoride for sale [55].

Crude 28 to 41 % phosphoric acid (20 to 30 % $P_2O_5$ equivalent) may contain as much as 1 to 2 % fluoride, or about 55 % of the fluoride originally present in the rock [57, 58]. This is present mostly as fluosilicic acid ($H_2SiF_6$) or is convertible to fluosilicic acid. Normal evaporative concentration serves to volatilize about two fifths of this [34] which can be increased to a larger proportion of this complex fluoride by variations of this technique (see Section 8.4.1). The vapour control procedures just outlined effectively capture this volatilized fluoride (Figure 8.6). If concentration of the crude acid is not required, however, it may be substantially purified by the addition of sodium carbonate. Sodium fluosilicate precipitates, and may be filtered off, leaving a phosphoric acid containing only about 0.1 % fluoride (Equation 8.28).

$$H_2SiF_6 + Na_2CO_3 \rightarrow Na_2SiF_6 \downarrow + H_2O + CO_2 \quad (8.28)$$

Sodium fluoride is soluble. The sodium fluosilicate may be purified and marketed directly for purposes such as fluoridation of water supplies, or it may be used to prepare other salts, for example synthetic cryolite ($Na_3AlF_6$) [58], of value in aluminum smelting.

Gypsum, while a relatively inert byproduct of wet process acid production, may pose an environ-

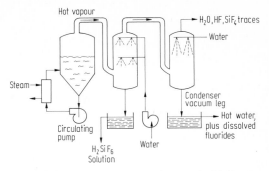

**Figure 8.6.** Emission control devices used with the vapour from a phosphoric acid evaporator to decrease fluoride emissions. The water used to condense steam by direct contact with cold water also serves to quite efficiently control fluoride vapour discharge [61]

mental problem to wet process acid operations merely because of the large mass obtained. Some 5 to 7 tonnes is produced for each tonne of phosphoric acid (100 % basis) [59], depending on the BPL level of the phosphate rock used. In the early days, disposal by some facilities simply involved discharge into the nearest watercourse via a slurry pipeline. But with the growing realization that even relatively inert suspended solids in freshwater streams have a severe effect on both water quality and on the diversity of stream life, from a smothering and siltation action, this practice has now been largely discontinued. Use of the gypsum for wallboard manufacture has been studied with some success [60], and where supplies of natural gypsum are scarce, such as in the U.S.A. and West Germany, this source is used [59, 61]. Application of phosphogypsum as a source of sulfur dioxide for sulfuric acid production (Section 7.7.1) has not attracted much interest. Thus, inland facilities have largely settled on gypsum disposal into extensive holding ponds, or disposal into lagoons laid out with landfill objectives. Many coastal facilities, in particular Europe, dump waste gypsum into the sea. One operation has found that a 90 m outfall is satisfactory for the daily disposal of 400 to 500 tonnes of gypsum, the dispersal of which is assisted by the co-discharge of some 550 $m^3$ per hour of seawater used for process cooling [59].

# 8.5 Phosphoric Acid Using Hydrochloric Acid Acidulation

Other acids than sulfuric have also been used to obtain phosphoric acid from phosphate rock, among them nitric and hydrochloric acids. Process difficulties in obtaining a pure phosphoric acid using nitric acid have resulted in this chemistry only being utilized to prepare granular fertilizers [38, 62] (Equations 8.29, 8.30). The hydrochloric acid acidulation process,

Acidulation:
$$CaF_2 \cdot 3\,Ca_3(PO_4)_2 + 20\,HNO_3 \rightarrow$$
$$6\,H_3PO_4 + 10\,Ca(NO_3)_2 + 2\,HF \qquad (8.29)$$

Ammoniation (full):
$$6\,H_3PO_4 + 10\,Ca(NO_3)_2 + 18\,NH_3 \rightarrow$$
$$6\,(NH_4)_3PO_4 + 10\,Ca(NO_3)_2 \qquad (8.30)$$

however, originally developed by Israel Mining Industries (IMI) at Haifa [63, 64] as a full scale route to phosphoric acid, has been adopted on such a magnitude that the world licensed capacity stood at 50,000 tonnes of $P_2O_5$ equivalent per day in 1965 [38]. Trade hydrogen chloride or hydrochloric acid is normally two to three times the price of sulfuric acid [65]. Hence this process, commonly referred to as the Haifa or IMI process, is primarily of interest to integrated chemical operations having a surplus of hydrogen chloride. Some processes where hydrogen chloride is produced as a byproduct are, for example, operation of a chlorinated solvents plant or from the manufacture of toluene diisocyanate (Equations 8.31-8.33).

$$2\,Cl_2 + CH_4 \rightarrow CH_2Cl_2 + 2\,HCl \qquad (8.31)$$

$$CH_3C_6H_3(NH_2Cl)NHCOCl + COCl_2 \rightarrow$$
$$CH_3C_6H_3(NHCOCl)_2 + 2\,HCl \qquad (8.32)$$

$$CH_3C_6H_3(NHCOCl)_2 \xrightarrow{\text{heat}}$$
$$CH_3C_6H_3(NCO)_2 + 2\,HCl \qquad (8.33)$$

Since markets for anhydrous hydrogen chloride have traditionally been limited the Haifa process provides a useful outlet for this byproduct.

Acidulation of fluorapatite with hydrochloric acid produces phosphoric acid in a very similar manner to sulfuric acid acidulation, except that in this case the coproduct obtained, calcium chloride, is very water soluble (Equation 8.34).

$$CaF_2 \cdot 3\,Ca_3(PO_4)_2 + 20\,HCl \rightarrow$$
$$6\,H_3PO_4 + 10\,CaCl_2 + 2\,HF \qquad (8.34)$$

Therefore, phosphoric acid recovery from the HCl acidulation of phosphate rock is not a relatively simple solid-liquid separation as it is from sulfuric acid acidulation, but requires a solvent extraction step to obtain the product free of dissolved calcium chloride. Suitable solvents must be polar to permit good solution of phosphoric acid, while having relatively low water solubility to minimize solvent losses to this phase. Israel Mining Industries prefer $C_4$ to $C_5$ alcohols [66], though some producers prefer these to be mixed with a water-immiscible solvent such as kerosene. Others prefer trialkylphosphates such as tri-n-butylphosphate or tri-2-ethylhexylphosphate for this function [67, 68], but all have similar physical properties. It is appropriate to mention here that solvent extraction has also been teamed up to sulfuric acid acidulation plants by using solvent discrimination to obtain a purified phosphoric acid extract and leave behind most of the water soluble, solvent insoluble impurities [69, 70].

## 8.5.1 Product Recovery by Solvent Extraction

To establish extraction efficiencies for a substance A one first needs to know, or must experimentally determine, the partition coefficient, $K_p$. The partition coefficient relates the concentration of solute A in the extraction solvent to the concentration of the solute in the extracted phase, at equilibrium (Equation 8.35). This holds true at any given temperature as long as the

$$K_p = \frac{[A]_{\text{solvent}}}{[A]_{\text{water}}} \qquad (8.35)$$

distributing species, solute A, exists in the same form in both phases. Otherwise a more complex "distribution ratio" must be employed [71]. For increased refinement of extraction efficiency determination it may also be necessary to employ activities, rather than concentrations in some cases [71]. To illustrate the partition coefficient aspect of extraction with a simple example, for the fluid system carbon tetrachloride/water and using ammonia or iodine as the solute, the partition coefficient $K_p$ at 25 °C is 0.0042 and 55 respectively [72]. Thus, with ammonia, this solute, being polar, would primarily move into the water phase on extraction. With non-polar iodine, the higher concentration on extraction would be in the non-polar carbon tetrachloride solvent phase. Obviously, using carbon tetrachloride as the extractant, iodine would be much more readily extracted than ammonia, from water. The other features of importance in industrial extractions are whether one, or more than one extraction stage is to be used, which will depend in part on the partition coefficient of the system of interest. Also a further decision has to be made on what flow pattern of solvent and extracted phase (raffinate) is to be used, if more than one extraction stage is employed. Single stage extraction requires simply a mixer, to bring about intimate contact between the two phases, followed by a settler which allows phase separation and a means for independent removal of the two phases (Figure 8.7).

For a solute with a relatively poor partition coefficient one of the two arrangements of multistage extraction, cross-current or counter-current may be selected (Figure 8.8). In order to decide which would be the advantageous choice, assume we have a 1,000 L per hour aqueous waste stream containing 0.02 molar benzoic acid which is to be extracted which 300 L per hour of benzene for benzoic acid recovery. The partition coefficient, $K_p$, for this system is 4.3 at 20 °C. To accomplish this extraction in a single stage system, and as a simplification assuming complete immiscibility of benzene and water, one can set up an equation to calculate the concentration of benzoic acid in each of

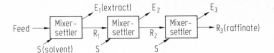

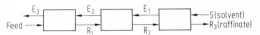

**Figure 8.8.** Possible arrangements of mixer settlers and process streams for multistage extraction processes. **a** Cross-current arrangement **b** Counter-current arrangement

the streams produced. To do this, the value for the concentration of benzoic acid in the first extract $[BzOH]_{E_1}$ obtained from the known equilibrium which must exist between the two streams of Equation 8.36 is substituted into Equation 8.37 which defines the benzoic acid material balance which must exist between the incoming and outgoing stream. In this manner one can derive equation 8.38, which allows

$$[BzOH]_{E_1} = K_p[BzOH]_{R_1},$$
$$\text{from } K_p = \frac{[BzOH]_{E_1}}{[BzOH]_{R_1}} \tag{8.36}$$

going stream. In this manner one can derive equation 8.38, which allows

*Feed Streams*        *Product Streams*
$$S \cdot [BzOH]_S + F[BzOH]_F = E_1[BzOH]_{E_1} + R_1[BzOH]_{R_1} \tag{8.37}$$

$S$, $F$, $E_1$ and $R_1$ represent *volumes* of solvent, aqueous benzoic acid feed stream, the first extract and the first raffinate, respectively. $[BzOH]$ with various subscripts represents the *concentration* of benzoic acid present, in moles per liter, in the subscripted stream.

Calculation of the concentration of benzoic acid in the first raffinate, $[BzOH]_{R_1}$. By substitution of known values into this expression the water phase concentration is found to be $9 \times 10^{-3}$ M. Thus, one extraction stage as just

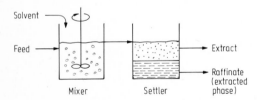

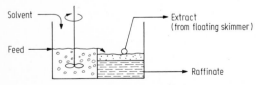

**Figure 8.7.** Arrangement for the physical components of a mixer settler, such as used for solvent extraction processes. Units may be separated or combined, in either case yielding discrete solvent and raffinate (extracted) phases with equilibrium, or near equilibrium concentrations of solute

$$[BzOH]_{R_1} = \frac{[BzOH]_F + \left(\dfrac{S}{F}\right)[BzOH]_S}{1 + \left(\dfrac{S}{F}\right)K_p} \tag{8.38}$$

For single stage extraction, $\dfrac{S}{F} = \dfrac{E_1}{R_1}$

described has dropped the benzoic acid concentration in the effluent to just under $^1/_2$ the original value. Similarly, by substituting this concentration value and $K_p$ into Equation 8.36, the concentration of benzoic acid obtained in the benzene extract is found to be 0.04 M, a ratio corresponding to the benzoic acid partition factor between these two liquids. The same equation may also be used to calculate the single stage extraction parameters for any other solute of interest, providing $K_p$ for the system is known or may first be determined experimentally.

In a similar manner one can derive an equation to calculate the benzoic acid in the raffinate for a three stage, cross-current system as shown in Figure 8.8 (Equation 8.39; cf. [73]). Assuming that 100 L volumes of benzene are

$$[BzOH]_{R_3} = \frac{[BzOH]_F}{\{1 + \left(\dfrac{S}{F}\right)K_p\}^n} \qquad (8.39)$$

Where, for 3 stages, $n = 3$.

If fresh extracting solvent volumes for each stage differ, the denominator requires expanding to individual terms for each solvent volume, thus

$$\{1 + \left(\frac{S_1}{F}\right)K_p\}\{1 + \left(\frac{S_2}{F}\right)K_p\}\{1 + \left(\frac{S_3}{F}\right)K_p\} \ldots$$

$$\ldots \{1 + \left(\frac{S_i}{F}\right)K_p\}.$$

used for each of the three extraction stages in a system of this kind, the benzoic acid concentration in the third stage raffinate is found to be 0.007 M (Equation 8.40), or about one third that of the incoming stream.

$$[BzOH]_{R_3} = \frac{0.02 \text{ mol L}^{-1}}{\{1 + \dfrac{100}{1000}(4.3)\}^3} = 0.007 \text{ mol L}^{-1} \qquad (8.40)$$

Carrying this approach to three stage, counter-current extraction and solving the pairs of equations for each of the three stages of the process one can derive Equation 8.41 which allows the determination of the benzoic acid concentration in raffinate 3, the aqueous effluent stream.

$$[BzOH]_{R_3} = \frac{[BzOH]_F}{\left[1 + \left(\dfrac{S}{F} \cdot K_p\right)\right]\left[1 + \left(\dfrac{S}{F} \cdot K_p\right)^2\right]} \qquad (8.41)$$

Substituting into this expression, one obtain a value of 0.003 mol/L for the benzoic acid concentration in the final raffinate, one seventh that of the incoming stream.

Comparing the calculated results for the three modes of extraction demonstrates the clear superiority of the extraction efficiency of counter-current contacting and provides the rationale for the frequent use of this arrangement for industrial processes using extraction. The additional advantage gained by any of the extraction modes as the total solvent volume employed is increased, and with one additional stage of counter-current contacting is shown by the calculated data of Table 8.7. These data clearly show the choice to be made between increasing the volume of extracting solvent or the mode and number of extracting stages to achieve the level of solute recovery desired.

### 8.5.2 Haifa (or IMI) Phosphoric Acid Process Details

A slight excess of the theoretical requirement of hydrochloric acid, based on the BPL level

**Table 8.7** Calculated molar benzoic acid concentrations in 1000 L of residual aqueous stream after extraction with benzene in various modes at 20 °C. Initial benzoic acid concentration 0.02 mol/L, $K_p = 4.3$

| Volume of benzene, L | Single stage extraction | Three stage cross-current | Counter-current | |
|---|---|---|---|---|
| | | | Three stage | Four stage |
| 100 | 0.0014 | 0.013 | 0.012 | 0.010 |
| 300 | 0.009 | 0.007 | 0.003 | 0.001 |
| 600 | 0.006 | 0.003 | 0.0007 | $4 \times 10^{-5}$ |
| 1 200 | 0.003 | 0.001 | 0.0001 | $8 \times 10^{-7}$ |

$(Ca_3(PO_4)_2$ concentration) in the rock is used for acidulation to avoid the formation of monocalcium phosphate $(CaHPO_4)$ which is insoluble and hence would result in phosphate losses. The acidulation reaction (Equation 8.31) is more rapid with hydrochloric acid than is the equivalent reaction with sulfuric acid, and yields soluble products from both the calcium and phosphate. Thus, there is neither a need to heat the mixture to speed up the reaction, nor any particular crystal form problems requiring temperature stabilization.

About 10 % of the raw rock, comprising silica and the like, does not dissolve but this material is readily removed by settling the quite dilute acidulated slurry in a thickener (Figure 8.9). Occasionally coagulants may have to be used to facilitate this separation of insolubles. The clarified solution from the thickeners is then extracted with a $C_4$, $C_5$ alkohol mixture or trialkylphosphate in a series of three or more mixer settlers, producing a solvent phase rich in phosphoric acid, and a raffinate of aqueous calcium chloride brine freed of phosphate. Presence of the calcium chloride salt in the aqueous

phase undoubtedly assists in driving the phosphoric acid transfer into the organic phase.

To recover the phosphoric acid, the solvent phase is back-extracted with pure water in a further mixer settler. In the absence of dissolved calcium chloride the aqueous phase now accepts the bulk of the phosphoric acid from the solvent, mixed with some residual hydrochloric acid and solvent. Distillation of this stream, usually in three separate evaporators, enables simultaneous recovery of solvent plus any unused hydrochloric acid which are both recycled, and phosphoric acid of appropriate concentrations for sale [74].

The product acid from this process is often purer than that obtained from sulfuric acid acidulation processes [75], and in addition this process is claimed to recover 98 to 99 % of the available $P_2O_5$ in the rock, as opposed to the 94 to 95 % experience of even the most efficient sulfuric acid processes [74]. Refined, to food grades of phosphoric acid may be obtained after chemical treatment, stripping, and if necessary anion and cation exchange to remove impurities.

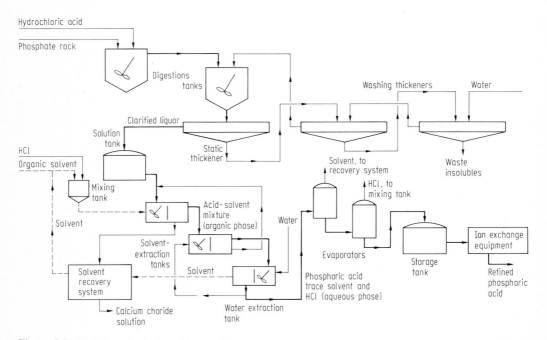

**Figure 8.9.** Flowsheet for the Haifa (Israel Mining Industries) hydrochloric acid acidulation route to phosphoric acid. Digestion for a short time is followed by solvent extraction in polyvinyl-chloride mixer-settlers, which tolerate the corrosive conditions. Reprinted by special permission from Chemical Engineering, Dec. 24. 1962, by McGraw-Hill Inc., New York, N.Y. 10020

An important economic feature of Haifa process operation is the need to maintain good solvent containment, which extends to solvent recovery from the brine raffinate as well as from the washed waste solids and the extracted acid. By these containment precautions it has been found possible to maintain solvent losses at not more than about 5 kg per tonne of $P_2O_5$ equivalent produced [76].

### 8.5.3  Haifa Process Byproducts and Waste Disposal

Fluoride containment still has to be considered in the operation of this process, as it does with sulfuric acid acidulation, since the acid treatment similarly mobilizes the chemically bound fluoride. Problems from fluoride losses are minimized by prior calcination of the feed phosphate rock, as has been practised for the early Japanese experimental operations using this process [77]. In any case, fluoride tends to follow the aqueous phase throughout to end up as a component of the calcium chloride raffinate solution. From this stream, fluoride may either be recovered for sale or removed from the calcium chloride stream and discarded by methods which have already been discussed (Section 8.4.3).

The washed insoluble slurry from acid digestion is generally discarded to landfill, or as a slurry to the sea if the plant operates on ocean frontage.

The calcium chloride brine, containing some two tonnes of the salt for each tonne of phosphoric acid (100 % basis) produced, may be a saleable commodity in its own right for such uses as highway de-icing, a low temperature heat transfer fluid, or as a heat sink as has been outlined in more detail in

connection with the Solvay process (Section 5.2). Otherwise, in the absence of ocean frontage, the brine raffinate may pose a disposal problem. An expedient for inland locations is to employ deep well disposal into a brine aquifer, but geological and water table factors do not always make this locally possible.

## 8.6  Major Producers and Users of Phosphoric Acid

Available production figures for many of the large world producers of phosphoric acid are given in Table 8.8. This product does show a more generally increasing trend with time since fertilizer demands on phosphoric acid output have generally climbed.

End uses of phosphoric acid are highly dependent on the process used. About 550 thousand tonnes (as $P_2O_5$) of furnace phosphoric acid was produced in the U.S.A. in 1980, about 5 % of the total U.S. production [78]. But nearly half of this small component of high purity acid, which has commanded an 80 % or so price premium (at $475 per tonne, $P_2O_5$ equivalent) over fertilizer grade wet process acid, is still dedicated to the production of sodium, potassium, and calcium phosphates for use in detergents. A further 15 %, as food grade product is used, mostly as salts in foods, bakery goods and soft drinks.

But these uses are all relatively small scale compared to the massive consumption for fertilizers. In the U.S. about 85 % of the total phosphoric acid production is used for this purpose, mostly for

**Table 8.8** Phosphoric acid production by selected countries [a]

|  | Thousands of tonnes, $P_2O_5$ equivalent [b] | | | | | | |
|---|---|---|---|---|---|---|---|
|  | 1976 | 1977 | 1978 | 1979 | 1980 | 1981 | 1982 |
| Belgium | 120 | 117 | 118 | | | | |
| Canada | 520 | 613 | 611 | 748 | 808 | 744 | 503 |
| Finland | 104 | 98 | 126 | | | | |
| France | — | 348 | 338 | 202 | 203 | 181 | 208 |
| Italy | 426 [c] | 476 [c] | — | 901 | 474 | 444 | 425 |
| Japan | — | 194 | 198 | | | | |
| U.K. | 482 | 502 | 500 | | | | |
| U.S.A. | 7 210 | 7 940 | 8 750 | 9 360 | 9 840 | 8 997 | 7 736 |

[a] Sources: see [22]. Includes both furnace acid and wet process methods.
[b] To obtain production as 100% $H_3PO_4$, entry shown should be multiplied by 1.38 (i.e., 196 ÷ 142).
[c] Specified as 82% $H_3PO_4$.

ammonium phosphates and triple superphosphate (Chapter 9). Another small component of use of the partially purified grades is a further 5 % in the preparation of animal feeds. This consumption picture outlined for the U.S.A. may be somewhat distorted compared to the consumption pattern of other industrial nations since the U.S. is a major exporter of phosphoric acid, about 10 % of the total U.S. production.

# Relevant Bibliography

1. L.K. Thompson, S.S. Sidhu and B.A. Roberts, Fluoride Accumulations in the Soil and Vegetation in the Vicinity of a Phosphorus Plant, Envir. Polln. *18* (3), 221, March 1979
2. Oldbury Electro-chemical's Phosphorus Plant Solves Problem of Waste Disposal by Installing Unit for Treatment of Effluent with Chlorine, Chem. Eng. News *30*, 466, 1952
3. Y.C. Athanassiaclis, Air Pollution Aspects of Phosphorus and its Compounds, NTIS PB-188-073, Washington, D.C. 17 pp
4. K.L. Elmore, Oxidation of Phosphorus with Steam, U.S. Patent 2613,134, Oct. 7, 1952. Chem. Abstr. *47*, 2,948 c, 1953
5. P. Miller, R.A. Wilson and J.R. Tusson, Continuous Process for Phosphorus Production, Ind. Eng. Chem. *40*, 357, 1948
6. S. Skolnik, G. Tarbutton and W.E. Bergman, Conversion of Liquid White Phosphorus to Red Phosphorus, J. Am. Chem. Soc. *68*, 2310, 1946
7. M.S. Silverstein, G.F. Nordblum, C.W. Dittrich and J.J. Jakabein, Stable Red Phosphorus, Ind. Eng. Chem. *40*, 301, 1948
8. G.L. Bridger *et al.*, New Process for Production of Purified Phosphoric Acid and/or Fertilizer Grade Dicalcium Phosphate from Various Grades of Phosphatic Materials, Ind. Eng. Chem., Process Design and Devel. *20*, 416, 1981
9. J.C. Barker, Air Pollution: the Cost of Pollution Control, Chem. Eng. Prog. *64*, 78, Sept. 1968

# References

1. Phosphorus and its Compounds, J.R. Van Wazer, editor, Interscience, New York, 1961, volume II, page 957
2. Environmental Phosphorus Handbook, E.J. Griffith, A. Beeton, J.M. Spencer and D.T. Mitchell, editors, John Wiley, New York, 1973, page 24
3. Statistical Yearbooks, 1966 and 1978, United Nations, New York, 1967 and 1979
4. Phosphate, Mineral Bulletin MR 160, Energy Mines and Resources Canada, Ottawa, July 1976
5. D.W. Leyshon and R.T. Schneider, The Evaluation of Phosphate Rock, 2nd Chemical Congress of the N. American Continent, Las Vegas, Nevada, Aug. 24-29, 1980. Abstracts, FERT 26
6. Mineral Facts and Problems, 1980 edition, U.S. Bureau of Mines, Bull. 671, Washington, 1980, page 663. Minerals Yearbook, Centennial Edition 1981, U.S. Bureau of Mines, Washington, 1982
7. Riegel's Industrial Chemistry, J.A. Kent, editor, Reinhold, New York, 1962, page 652
8. R.A. Horne, The Chemistry of our Environment, Wiley-Interscience, New York, 1978, page 603
9. F.S. Taylor, A History of Industrial Chemistry, Reprint Edition, Arno Press, New York, 1972, page 203
10. F.A. Cotton and G. Wilkinson, Advanced Inorganic Chemistry, 3rd edition, Interscience, New York, 1972, page 370
11. T.L. Hurst in: Phosphorus and its Compounds, J.R. Van Wazer, editor, Interscience, New York, 1961. Vol. II, page 1158
12. F.A. Lowenheim and M.K. Moran, Faith, Keyes and Clark's Industrial Chemicals, Wiley—Interscience, New York, 1975, page 641
13. P. Ellwood, Chem. Eng. *72*, 54, Feb. 1, 1965
14. Kirk-Othmer Encyclopedia of Chemical Technology, 2nd edition, Interscience, New York, 1968, volume 15, page 276
15. Industrial Chemicals, Chem. In Can. *24*, 12, Oct. 1972
16. Handbook of Chemistry and Physics, 51st edition, R.C. Weast, editor, Chemical Rubber Co., Cleveland, Ohio, 1970. page B-117
17. R.E. Bentley, K.S. Buxton and B.H. Sleight, Acute toxicity of five munitions compounds to aquatic organisms, U.S. Army Medical Research and Development Command, Interim Report, Washington, D.C., 1975. Contract No. DAMO-17-74-R-4755. Cited by reference 18
18. D.R. Idler, G.L. Fletcher and D.F. Addison, Effects of Yellow Phosphorus in the Canadian Environment, National Research Council of Canada, Ottawa, 1981
19. The Modern Inorganic Chemicals Industry, R. Thompson, editor, The Chemical Society, London, 1977, page 357
20. Super Problems Belabor Super Plant, Can. Chem. Proc. *55* (10), 64, Oct. 1971
21. Verband der Chemischen Industrie, E.V., Frankfurt am Main, July 1980 and earlier
22. Facts and Figures for the Chemical Industry, Chem. Eng. News *59* (23), 30, June 8, 1981; Chem. Eng. News *60* (24), 31, June 14, 1982; Chem. Eng. News *61* (24), 26, June 13, 1983
23. Phosphorus, Chem. Eng. News *55* (20), 14, May 6, 1977

24. Effects of Elemental Phosphorus on Marine Life, P.M. Jangaard, editor, Circular No. 2, Fisheries Research Board of Canada, Halifax, 1972

25. S.S. Sidhu and B.A. Roberts, Bi-monthly Research Notes, Forestry Service, Environment Canada *32* (6), 29, Nov.-Dec. 1976

26. D. Rose and J.R. Marier, Environmental Fluoride, National Research Council of Canada, Ottawa, 1977

27. J.C. Barber and T.D. Farr, Chem. Eng. Prog. *66* (11), 56, Nov. 1970

28. J.C. Barber, Chem. Eng. Prog. *65* (6), 70, June 1969

29. R.N. Shreve and J.A. Brink, Jr., Chemical Process Industries, 4th edition, McGraw-Hill, New York, 1977, page 256

30. D.R. Idler, Chem. In Can. *21* (11), 16, Dec. 1969

31. Kirk-Othmer, Encyclopedia of Chemical Technology, 2nd edition, Interscience, New York, 1968, volume 15, page 257

32. J.Q. Hardesty and L.B. Hein, in Reigel's Handbook of Industrial Chemistry, 7th edition, J.A. Kent, editor, Van Nostrand Reinhold, Toronto, 1974, page 537

33. J.F. Coykendall, E.F. Spencer and O.H. York, J. Air. Polln. Control Assoc. *18* (5), 315, May 1968

34. H.M. Stevens in Phosphorus and its Compounds, J.R. Van Wazer, editor, Interscience, New York, 1961, volume II, page 1025

35. Acid's Value Enhanced, Chem. Eng. *71* (1), 26, Jan. 6, 1964

36. G. Kleinman, Chem. Eng. Prog. *74* (11), 37, Nov. 1978

37. J.F. McCullough, Chem. Eng. *83*, 101, Dec. 6, 1976

38. G.S.G. Beveridge and R.G. Hill, Chem. Proc. Eng. *49* (8), 63, Aug. 1968

39. G.S.G. Beveridge and R.G. Hill, Chem. Proc. Eng. *49* (7), 61, July 1968

40. W.E. Blumrich, H.J. Koening and E.W. Schwer, Chem. Eng. Prog. *74* (11), 58, Nov. 1978

41. TVA Displays Energy Saving, Chem. Eng. News *56* (46), 32 Nov. 13, 1978

42. P.L. Olivier, Jr., Chem. Eng. Prog. *74* (11), 55, Nov. 1978

43. L.E. Bostwick, Chem. Eng. *77* (8), 100, April 20, 1970

44. G.L. Bridger and A.H. Roy, Chem. Eng. Prog. *74* (11), 62, Nov. 1978

45. TVA Uses $SO_2$ to Process Phosphate Rock, Chem. Eng. News *54*, 32, Sept. 6, 1976

46. J. Bergdorf and R. Fischer, Chem. Eng. Prog. *75* (11), 41, Nov. 1978

47. W.E. Rushton, Chem. Eng. Prog. *74* (11), 52, Nov. 1978

48. R.E. Stedman, Chem. Ind. (London), 150, Feb. 1957

49. A.P. Kouloheris, Chem. Eng. *87*, 82, Aug. 11, 1980

50. Uranium Recovery from Phosphoric Acid, Chem. Eng. News *55* (42), 17, Oct. 17, 1977. Two More Uranium Recovery Moves, Chem. Eng. News *57* (42), 10, Oct. 15, 1979

51. B.D. Wells, 2nd Chem. Congress of the N. American Continent, Las Vegas, Nevada, Aug. 24-29, 1980. Abstracts FERT, No. 32

52. Industrial Hygiene and Toxicology, 2nd edition, F.A. Patty, Editor, Interscience, New York, 1963, Vol. II, page 832

53. F.L. Cross, Jr., and R.W. Ross, J. Air Polln. Control Assoc. *19* (1), 15, Jan. 1969

54. A.W. Frazier, J.R. Lehr and E.F. Dillard, Envir. Science and Tech. *11* (10), 1007, Oct. 1977

55. H.O. Grant, Chem. Eng. Prog. *60* (1), 53, Jan. 1964

56. C. Djololian and D. Billaud, Chem. Eng. Prog. *74* (11), 117, Nov. 1969

57. P.S. O'Neill, Ind. Eng. Chem. Prod. Res. Dev. *19*, 250 (1980)

58. K.P.A. Nayar, Chem. Proc. Eng. *50* (11), 117. Nov. 1969

59. G. Watson, Chem. Ind. (London), 805, July 1971

60. D. Kitchen and W.J. Skinner, J. Appl. Chem. Biotechnol. *21* (2), 53 and *21* (2), 56, Feb. 1971

61. Phosphoric Acid, A.V. Slack, editor, Marcel Dekker, New York, 1968. Vol. I. Part II, page 503

62. G. Jorquera, Chem. Eng. Prog. *64* (5), 83, May 1968

63. A. Baniel, R. Blumberg and A. Alon, Chem. Eng. Prog. *58*, 104, Nov. 1962

64. R. Blumberg in Solvent Extraction Reviews, Vol. I, Y. Marcus, editor, Dekker, New York, 1971, page 93. R. Blumberg, D. Gonen and D. Meyer in Recent Advances in Liquid-Liquid Extraction, C. Hanson, editor, Pergamon, Oxford, 1971, page 93

65. Chemical Marketing Reporter (New York) *219* (24), 33, June 15, 1981

66. British Patents 805,517, (1958), and 1,051,521, (1967). Cited by reference 64

67. R.S. Long and D.A. Ellis, U.S. Patent 3,072,461, (1963) to the Dow Chemical Company

68. Dow to Test Phosphoric Acid Process, Chem. Eng. News *41* (41), 35, Oct. 14, 1963

69. I. Raz, Chem. Eng. *83*, 52, June 10, 1974

70. T.-T. Huang, Indust. Eng. Chem. *53* (1), 31, Jan. 1961

71. R.M. Diamond and D.G. Tuck in Progress in Inorganic Chemistry, Vol. 2, F.A. Cotton, editor, Interscience, New York, 1960, page 109

72. Chemical Engineers' Handbook, R.H. Perry, editor, McGraw-Hill, New York, 1969, pages 14-45

73. D.F. Rudd, G.J. Powers and J.J. Siirola, Process Synthesis, Prentice-Hall, Englewood Cliffs, 1973, page 141

74. A. Baniel, R. Blumberg, A. Alon, M. El-Roy and D. Goniadski, Chem. Eng. Prog. *58* (11), 100, Nov. 1962

75. HCl Routes to Phosphoric Acid, Chem. Eng. *70,* July 8, 1963

76. Israelis Pioneer New Route, Chem. Eng. *69* (12), 88, June 11, 1962

77. Hydrochloric-Based Route, Chem. Eng. *69* (52), 34, Dec. 24, 1962

78. Key Chemicals, Chem. Eng. News *59* (24), 10, June 15, 1981

79. Kirk-Othmer, Encyclopedia of Chemical Technology, 3rd edition, John Wiley and Sons, New York, 1982, volume 17, page 484

# 9 Ammonia, Nitric Acid and their Derivatives

## 9.1 Ammonia, Historical Background

Early samples of ammonia were obtained by such means as the bacterial action on the urea present in urine (Equation 9.1), or by the dry distillation

$$H_2N-CO-NH_2 + H_2O \xrightarrow{\text{bacteria}} CO_2 + 2\,NH_3 \tag{9.1}$$

of protein-containing substances such as bone, horns, and hides. Small amounts were also obtained from the manufacture of coke or coal gas from coals. Ammonia is still recovered at the rate of 134,000 tonnes per year as an incidental part of U.S. coal-based operations [1, 2], but this amounts to less than 1 % of current U.S. ammonia production (Table 9.1). By far the bulk of ammonia produced today is by the direct combination of the elements. Other synthetic processes have been tested and found to be either impractical for commercial exploitation, or have gradually been supplanted by direct elemental combination for economic reasons.

As an example of what turned out to be a commercially impractical method, the French Serpek process, depended upon the formation of aluminum

nitride from the heating of bauxite in nitrogen (Equation 9.2)[3].

$$Al_2O_3 + N_2 + 3\,C \xrightarrow{1600\,°C} 2\,AlN + 3\,CO \tag{9.2}$$

The aluminum nitride was then hydrolyzed with water to yield ammonia and aluminum hydroxide (Equation 9.3), which could ultimately be converted to alumina in pure form for aluminum production.

$$AlN + 3\,H_2O \rightarrow NH_3 + Al(OH)_3 \tag{9.3}$$

A related route, since again high temperatures were used to fix elemental nitrogen, is the cyanamide route developed by Frank and Caro in Germany in 1898 [4]. In this process an electric arc furnace provides the more practical but still very high temperatures required to obtain calcium cyanamide from calcium carbide (Equation 9.4).

$$CaC_2 + N_2 \xrightarrow{900-1000\,°C} CaNCN + C \tag{9.4}$$

Hydrolysis of the calcium cyanamide then produces ammonia and calcium carbonate (Equation 9.5). Cyanamide production for many years predominantly went into

$$CaCN_2 + 3\,H_2O \rightarrow 2\,NH_3 + CaCO_3 \tag{9.5}$$

**Table 9.1** Production of synthetic ammonia by selected areas, thousands of tonnes [a]

|  | 1970 | 1975 | 1978 | 1979 | 1980 | 1981 | 1982 |
|---|---|---|---|---|---|---|---|
| Africa | (772) | (1 761) | — | — | — | — | — |
| Canada | 1 220 | 1 336 | 2 398 | 2 405 | 2 556 | 2 654 | 2 509 |
| China | — | — | 9 287 | 10 727 | 12 150 | 11 990 | — |
| France | 1 618 | 1 941 | 2 035 | 2 150 | 2 084 | 2 245 | 1 896 |
| W. Germany | — | — | 2 373 | 2 157 | 2 040 | 1 958 | 1 567 |
| Italy | 1 277 | 1 453 | 1 756 | 1 773 | 1 714 | 1 472 | 1 231 |
| Japan | 3 261 | 3 636 | 2 879 | 2 831 | 2 573 | 2 228 | 2 010 |
| Mexico | — | — | 1 579 | 1 653 | 1 883 | 2 183 | 2 469 |
| U.S.A. | 12 541 | 14 871 | 15 530 | 16 803 | 17 262 | 17 275 | 14 061 |
| U.S.S.R. | 7 638 | 11 998 | 13 936 | 14 928 | 16 732 | 17 894 | — |
| World | (62 500) | (84 460) | — | — | — | (99 800) | (105 600) |

[a] Production *capacities* given in parentheses. Data compiled from issues of Canadian Chemical Processing, Chemical and Engineering News, Verband der Chemischen Industrie, e.V. (VCI), and [2].

ammonia synthesis, and even though the direct synthesis of ammonia has now displaced cyanamide hydrolysis for this purpose, cyanamide is still produced on a very large scale for other purposes [4].

Ammonia was probably first synthesized directly from its elements in the laboratory by Döbereiner in 1823, as a secondary result in experiments primarily designed to study the direct combination of gaseous hydrogen and oxygen [5, 6]. But it was not until the early 1900's, as Nernst developed the theory by which equilibrium concentrations could be approximated from thermochemical calculations and Haber proved the direct reaction of nitrogen and hydrogen on a laboratory scale, that any promise of adoption of this process on an industrial scale developed [7, 8, 9]. From an early bench scale reactor which produced about 80 g of ammonia per hour, in collaboration with Badische Anilin- und Soda-Fabrik (BASF) this was rapidly scaled up to plants operating in Germany producing 27 tonnes per day by 1913, and at a rate of 180,000 tonnes per year for the country by 1918. Haber in 1918, and Bosch in 1931, earned Nobel prizes for their respective contributions to this area of chemistry, and jointly lent their names to the process which originated as a result of their work. Today there are many single train ammonia plants worldwide operating at capacities of 900 tonnes per day, and some at 1,650 tonnes per day [2].

### 9.1.1 Principles of Ammonia Synthesis: the Haber, or Haber-Bosch Process

From Equation 9.6, which represents the ammonia synthesis reaction, two moles (volumes) of ammonia are produced for each four moles (volumes) of initial

$$N_{2(g)} + 3H_{2(g)} \leftrightarrows 2NH_{3(g)} \qquad (9.6)$$
$$\Delta H = -92.0 \text{ kJ } (-11.0 \text{ kcal})$$

gases reacting. Therefore, by the principle of Le Chatelier which states that when a change is placed on a system in equilibrium that equilibrium will shift so as to reduce the effect of the change (quantities, pressure, temperature etc.), it would be expected that increasing the pressure on this system would increase the equilibrium ammonia concentration. It does (Table 9.2).

The synthesis reaction is also faster, that is equilibrium conditions are reached more rapidly at higher temperatures. At the same thime the reaction is

**Table 9.2** Equilibrium percent concentrations of ammonia at various temperatures and pressures, from reacting a gas mixture of one mole nitrogen to three moles hydrogen [a]

| Temper- | Absolute pressure, bars | | | | | | |
|---------|------|------|------|------|------|------|------|
| ature, °C | 1 | 10 | 50 | 100 | 300 | 600 | 1000 |
| 200 | 15.3 | 50.7 | 74.4 | 81.5 | 89.9 | 95.4 | 98.3 |
| 300 | 2.2 | 14.7 | 39.4 | 52.0 | 71.0 | 84.2 | 92.6 |
| 400 | 0.4 | 3.9 | 15.3 | 25.1 | 47.0 | 65.2 | 79.8 |
| 500 | – | 1.2 | 5.6 | 10.6 | 26.4 | 42.2 | 57.5 |
| 600 | – | 0.5 | 2.3 | 4.5 | 13.8 | 23.1 | 31.4 |
| 700 | – | – | 1.1 | 2.2 | 7.3 | 11.5 | 12.9 |

[a] Compiled from data of Haber [119] and Larson [120]. Nielson gives more recent and more detailed data for conditions of commercial interest [121].

exothermic to the extent of 46 kJ mol$^{-1}$ (about 54 kJ mol$^{-1}$ at normal conversion temperatures), so that as the synthesis temperature is raised the ammonia equilibrium is shifted to the left, again in accord with Le Chatelier's principle (Table 9.2). Thus, while raising the synthesis temperature does obtain a higher reaction rate allowing a greater volume of production to be obtained in the same size reactor, it also displaces the equilibrium of the synthesis, Equation 9.6, to the left giving a smaller potential nitrogen conversion to ammonia. Fortunately, substantially increasing the pressure on the system materially improves the equilibrium conversion conditions to make the whole exercise practical. For these reasons the majority of synthetic ammonia processes tend to cluster around 100-300 atmospheres pressure and 400-500 °C operating temperatures in an attempt to maximize conversions and conversion rates while moderating compression costs. The Claude, and the Casale ammonia synthesis processes, operating in France at 900 atmospheres and in Italy at 750 atmospheres respectively, correspond to among the highest pressure large scale routine industrial processes. The first of these at least relies on a nitrogen to hydrogen feed gas system connected to multiple, small inside diameter (10 cm) bored nickel chrome ingots to contain the enormous pressures and in return gains conversions to ammonia of 40 % or better [10, 11]. This compares to conventional converter dimensions of 7 to 10 meters high by 3 meters outside diameter requiring alloy steel wall thicknesses of the order of 20 centimeters for pressure containment in this size vessel.

Important to the successful operation of ammonia converters was the discovery of a suitable catalyst which would promote a sufficiently rapid reaction at these operating temperatures to utilize the moderately favourable equilibrium under these conditions. Otherwise higher temperatures would be required to obtain appropriately rapid rates, and the less favourable equilibrium under these conditions would necessitate higher pressures as well, in order to maintain a sufficiently favourable percent conversion. While the original experiments were conducted with osmium catalyst, Haber later discovered that reduced magnetic iron oxide ($Fe_3O_4$) was highly effective, and that its activity was further enhanced by the presence of the promoters alumina ($Al_2O_3$; 3 %) and potassium oxide ($K_2O$; 1 %) [12], probably from the introduction of iron lattice defects. Iron with various proprietary variations still forms the basis of all ammonia catalyst systems today [13, 14].

Space velocity, or the relationship between hourly reacting gas volume under standard conditions to the catalyst volume in the same volume units is a term used to define gas-catalyst contact time for heterogeneous reactions (Equation 9.7).

$$\text{Space velocity} = \frac{m^3 \text{ gas(es)} \cdot \text{hour}^{-1}}{m^3 \text{ catalyst volume}} \quad (9.7)$$

Gas volume to be corrected to 0 °C, 760 mm Hg.

For an initial doubly-promoted magnetic iron oxide catalyst, working on a nitrogen: hydrogen mole ratio of 1:3 at a pressure of 100 bars and 450 °C, and a space velocity of 5,000 hour$^{-1}$ approximately 13-15 % ammonia could be expected [15]. Much lower conversions are obtained from singly promoted or nonpromoted iron.

A space velocity of 5,000 hour$^{-1}$ (corrected to 0 °C, 760 mm) corresponds to a gas change rate in the catalyst space of 132.5 m$^3$ of gas per cubic meter of catalyst space per hour at operating conditions (450 °C, 100 bars; Equation 9.8), or an actual gas-catalyst contact time of about 0.04 seconds.

$$5\,000 \text{ h}^{-1} \times \frac{723 \text{ K}}{273 \text{ K}} \times \frac{1 \text{ bar}}{100 \text{ bars}} \times \frac{1}{3600 \text{ s/hr}}$$
$$(9.8)$$

Increasing the space velocity decreases the gas-catalyst contact time and in so doing decreases the percentage of nitrogen and hydrogen converted to ammonia.

Doubling the space velocity of a converter may well drop the ammonia conversion percentage to say 10 %, which appears to be a change in the wrong direction. But 10 % of twice the original volume of reacting gases passing through the converter still means that the volume of ammonia recovered is roughly one third more or 1,000 m$^3$ h$^{-1}$ $NH_3$(10 % of 10,000 m$^3$) rather than 750 m$^3$ h$^{-1}$ $NH_3$ as obtained at 5,000 space velocity. Hence, high space velocities of 10,000-20,000 hour$^{-1}$ or more are common for commercial ammonia converters.

For this combination of reasons, most large ammonia plants have to recycle a large proportion of the converter exit gases in the form of unreacted nitrogen and hydrogen. Thus, a significant component of the engineering of a modern ammonia synthesis unit is concerned with pressure and heat conservation to enable this recycle to be conducted as economically as possible.

## 9.1.2 Feedstocks for Ammonia Synthesis, Air Distillation

Nitrogen for ammonia synthesis may be either separated from air by liquid air distillation to obtain the nitrogen component, or may be obtained by consumption of the oxygen of air by the burning of a fuel in some air-restricted oxidative process to leave a nitrogen residue.

Cryogenic, low temperature technology is called upon to separate nitrogen from liquid air. Initial steps involve dust removal, usually by filtration, and removal of the normal carbon dioxide present in air (about 330 ppm), either by scrubbing with an organic base such as monoethanolamine (MEA) or with aqueous alkali (Equations 9.9-9.13).

$$CO_2 + H_2O \rightarrow H_2CO_3 \quad (9.9)$$

MEA absorption:

$$H_2CO_3 + H_2NCH_2CH_2OH \xrightarrow[\text{ambient temp.}]{\text{high pressure}}$$
$$HCO_3^- + H_3\overset{+}{N}CH_2CH_2OH \quad (9.10)$$

Regeneration (separate unit):

$$HCO_3^- + H_3\overset{+}{N}CH_2CH_2OH \xrightarrow[\text{heat}]{\text{low pressure}}$$
$$H_2NCH_2CH_2OH + H_2O + CO_2 \quad (9.11)$$

Alkali absorption:

$$H_2CO_3 + NaOH \rightarrow NaHCO_3 + H_2O \quad (9.12)$$

$$H_2CO_3 + 2\,NaOH \rightarrow Na_2CO_3 + 2\,H_2O \quad (9.13)$$

Purification is an important preliminary step, particularly in areas near a refinery or petrochemical complex, since accumulation of combustible dusts, or of condensed vapours such as methane or acetylene in cryogenic lines containing liquid oxygen introduces operating risks. And of course the temperatures in the region of $-200\,°C$ required to liquefy air are sufficiently low to cause carbon dioxide to solidify and plug operating components. After purification the air is normally dried to as low a dewpoint as feasible using towers packed with an adsorbent such as activated alumina, a Molecular Sieve (e.g. Union Carbide type 13 X), or silica gel, to avoid problems similar to those described for carbon dioxide from water vapour present. Two or more drying towers will be operated on swing cycles to provide continuous water adsorption capacity from one tower, while the other is regenerated by heating the adsorbent under reduced pressure.

Liquefaction of the purified air is accomplished using the Joule-Thompson effect, the cooling obtained from a compressed gas when it is allowed to expand. By using this expansion-cooling effect repetitively, and by employing the chilled expanded gas to prechill the compressed gas before expansion, air may be liquefied by employing compression pressures of only about 10 bars ($\sim$ 150 psig, Figure 9.1). To accomplish liquefaction of air by direct compression alone is not possible at temperatures above the critical temperatures of the component gases, $126.1\,K$ ($-147.1\,°C$) for nitrogen and $154.4\,K$ ($-118.8\,°C$) for oxygen, and even at these temperatures, pressures of 33.9 and 50.4 bars respectively would be required for condensation. But by using the technique of repeated compression, external cooling of the compressed gas to well below their critical temperatures followed by expansion-cooling in a well insulated system from relatively moderate pressures is sufficient to liquefy a part of both gases as they approach their atmospheric pressure boiling points, nitrogen, $74.8\,K$ ($-195.8\,°C$) and oxygen, $90.2\,K$ ($-183.0\,°C$).

On fractionation (fractional distillation) of liquid air, nitrogen, the lower boiling constituent, distils off first and the oxygen fraction is frequently vented unless there is a local petrochemical or met-

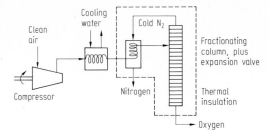

Figure 9.1. Simplified flowsheet for the separation of nitrogen from liquid air

allurgical application to which it can be applied. Liquid air distillations are commonly conducted at a pressure of about 5 bars both to raise the column operating temperatures and to increase the difference in boiling points for the component gases to be separated, improving the separation efficiency. This is particularly true if argon (0.93 % in air) recovery is practised since with a boiling point of $87.5\,K$ ($-185.7\,°C$) it would be lost with the oxygen stream without precautions. Incidentally, the close boiling points of argon and oxygen represent physical evidence of the mono-atomic versus diatomic existence of these gases, Ar atomic and molecular weight 39.95, and $O_2$ atomic weight 16.00 and molecular weight 32.00.

When liquid air distillation is used as the source of nitrogen, the hydrogen also required for ammonia synthesis is obtained from a variety of sources. Some is obtained as the co-product from the electrolytic production of chlorine and caustic soda (which see), some from refinery sources as a by-product of, for instance, cracking processes or olefin synthesis, some from the water gas reaction, and some is produced specifically for the purpose by the electrolysis of alkaline water (e.g. by Cominco, Trail [16], Equation 9.14).

$$2\,H_2O + KOH \xrightarrow[\text{electrolysis}]{} 2\,H_2 + O_2 + KOH \quad (9.14)$$

$\sim$ 5 molar
(26%)

Despite the apparent straightforward nature of these procedures to the feedstocks for ammonia synthesis and the "free" source of nitrogen from the air, this combination of approaches has become feasible only for relatively small ammonia plants of around 100 tonnes per day, or for special process situations where abundant hydrogen is available. When it is realized that several ammonia plants of capacities of 900 tonnes per day are now operating and a few as large as 1,650 tonnes per day it puts this process sequence into proper perspective [2].

### 9.1.3 Ammonia Feedstocks, Reforming and Secondary Reforming

Hydrogen on a sufficient scale for ammonia synthesis is accessible from coal (indirectly), petroleum, natural gas, and water and all of these are used to some extent. In fact, in the original Haber-Bosch process, coke derived from coal was used as the source of hydrogen via the water gas reaction. To do this, initially the coke is burned in air until it reaches red heat (Table 9.3), and then the air feed stream is replaced by steam (Equations 9.15, 9.16).

Combustion phase:

$$C + air \rightarrow CO_2 + 4N_2 + heat \; (air \sim O_2 + 4N_2) \tag{9.15}$$

Water gas reaction:

$$C + H_2O \quad \underset{\sim 1000\,°C}{\rightarrow} \quad CO + H_2 \quad positive \; \Delta H \tag{9.16}$$

The hot coke in an endothermic reaction produces mainly carbon monoxide and hydrogen and cools down, which requires periodic switching of the process to combustion air to maintain the high coke temperatures necessary for the water splitting reaction. Since carbon monoxide is relatively difficult to separate from hydrogen, and also because it is possible to obtain a further mole of hydrogen from each mole of carbon monoxide, the whole water gas stream is normally put through a shift conversion step (Equation 9.17).

Shift conversion:

$$CO_{(g)} + H_2O_{(g)} \underset{\sim 500\,°C}{\overset{FeO + Cr_2O_3}{\rightarrow}} CO_{2(g)} + H_{2(g)}$$

$$\Delta H = -41.2 \; kJ \; (-9.8 \; kcal) \tag{9.17}$$

Here, the carbon monoxide is converted to carbon dioxide at more moderate temperatures in the presence of a chromium-promoted iron catalyst. Carbon dioxide is readily separated from the hydrogen by absorption in water under pressure (Equation 9.18)

$$CO_{2(g)} + H_2O_{(l)} \overset{25 \; bars}{\rightarrow} H_2CO_{3(l)} \tag{9.18}$$

$$\Delta H = -38.5 \; kJ \; (-9.2 \; kcal)$$

leaving a hydrogen stream ready for ammonia synthesis after a final clean up. The nitrogen required is normally taken from a part of the gas stream obtained from the heating, coke combustion phase of the water gas reaction sequence, which is depleted in oxygen. Carbon dioxide removal from the nitrogen feed component is effected in the same manner as from the hydrogen.

Where natural gas supplies are abundant and available at reasonable cost however, a reforming and secondary reforming sequence is generally favoured over the water gas route for ammonia feedstock production, primarily for economic reasons [17]. Coal-based ammonia is estimated to require nearly twice as much capital as a plant based on natural gas, and frequently environmental problems are greater with the former [1] (see also New Developments). In overview terms, reforming and secondary reforming amounts to a burning of methane in the presence of steam and a defi-

**Table 9.3** Relationship between the thermally-induced colour of a substance and its temperature [a]

| Colour | Temperature, °C | Remarks |
|---|---|---|
| dark red | 585 – 870 | Al melts at 660 °, soft glass at ca. 700 °, potters decorating kilns, reformer tubes |
| cherry red | 880 – 980 | Ag melts at 962 °, red clay products |
| orange red | 990 – 1 090 | Au melts at 1 064 °, Cu at 1 083 ° |
| yellow | 1 100 – 1 220 | Lime, cement kiln operation |
| white | 1 230 – 1 280 | Lime, cement kiln operation |
| brilliant white | 1 290 – 1 520 | Ni melts at 1 453 °, firing of porcelain, spark plugs, power line insulators insulators |
| dazzling white | 1 540 – 1 800 | Fe melts at 1535 °, silica bricks fired, fire brick |

[a] Temperature relations from Lange [122].

ciency of air to give a mixture of hydrogen, nitrogen, and carbon dioxide.

In more detail, this option can use any kind of petroleum feedstock, not just natural gas, although the latter is favoured because it has the highest hydrogen content and is the easiest to clean prior to use. If the gas has been thoroughly desulfurized at the well head the only additional treatment required is passage through a bed of activated alumina or bauxite for an absorptive purification. But use of either a gas oil type of distilled petroleum feedstock or a residual fuel oil for synthesis gas generation requires much more stringent precautions to remove sulfur since this is one of a series of elements including, for example, copper, arsenic, and phosphorus, which can permanently decrease the activity (poison) the catalyst system employed for ammonia synthesis [11].

Regardless of the petroleum source employed, the chemistry to obtain an ammonia feedstock is essentially similar. Since methane is of dominant importance this will be used as an example to describe the steps required. Initially methane is mixed with steam and passed into heat resistant nickel-chromium-iron alloy tubes containing a supported nickel catalyst which are heated externally by a further portion of methane consumed as fuel (Figure 9.2, Equation 9.19).

This reaction, which is endothermic and hence requires continued external combustion of methane to maintain the red heat (Table 9.3) of the reformer

Primary steam reforming:

$$CH_{4(g)} + H_2O_{(g)} \xrightarrow[\text{28 bars, 800\,°C}]{\text{Ni catalyst}} CO_{(g)} + 3\,H_{2(g)}$$

$$\Delta H = +208\ \text{kJ}\ (+49.7\ \text{kcal}) \qquad (9.19)$$

tubes, produces most of the hydrogen required for ammonia production.

Addition of air plus methane, in the appropriate proportions, to the exit gas stream from the steam reformer provides all of the eventual nitrogen requirement for ammonia synthesis, plus some further hydrogen (Equation 9.20).

Secondary air reforming:

$$CH_4(g) + air_{(g)} \xrightarrow[\text{950\,°C}]{\text{Ni catalyst}}$$

$$CO_{(g)} + 2\,H_{2(g)} + 2\,N_{2(g)} \qquad (9.20)$$

Air approx. $\frac{1}{2}\,O_2 : 2\,N_2$.

$$\Delta H = +35.6\ \text{kJ}\ (+8.5\ \text{kcal})$$

This reaction is also endothermic and requires external fuel combustion to maintain the high reformer tube temperatures required. About one third of the methane consumed at this stage is used to heat the primary steam, and secondary air reformers.

As with the water gas product of the Haber-Bosch route to hydrogen, the carbon monoxide produced by these two reforming reactions may be used via

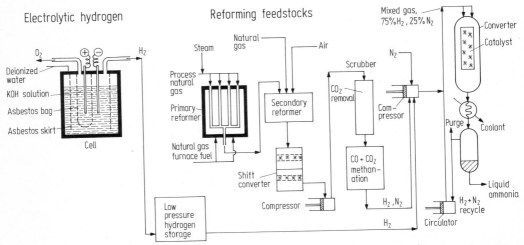

**Figure 9.2.** Overview outline of the main components of an ammonia synthesis plant using reforming and secondary reforming as the principal source of hydrogen. Electrolysis of water is used to supplement this

shift conversion to both provide a further mole of hydrogen per mole of carbon monoxide and, by producing carbon dioxide in the process this may be more readily removed than carbon monoxide from the gases of interest, hydrogen and nitrogen (Equations 9.17, 9.18). To accomplish this, water is sprayed into the very hot gas mixture emerging from the air reformer to bring the temperature down to the 400 °C range for shift conversion which also provides the steam necessary for shift conversion. Since this step is exothermic it is usually carried out in two separate units to permit better temperature control of the process and also, once started, it does not require external heat input. At space velocities of about 100, carbon monoxide conversion to carbon dioxide is achieved to residual monoxide concentrations of less than 1 % [18].

The two reforming reactions and shift conversion are so controlled that the gas mixture resulting contains nitrogen and hydrogen in the mole ratio (close to a volume ratio) of 1:3. But this mixture still contains 20 to 30 % carbon dioxide, resulting from the shift conversion reaction, and traces of unconverted carbon monoxide. Carbon dioxide can yield carbonates and carbamates in the ammonia synthesis cycle, undesirable because they would deposit in the piping, and oxygen, or any of its compounds such as carbon monoxide, water etc. are also ammonia catalyst poisons [11]. Consequently they must be removed.

While carbon dioxide can be removed by scrubbing the pressurized gas mixture with water (Equation 9.19; 21.6 volumes of gas per volume of water at 25 bars, 12 °C) [6] in an ordinary scrubber, energy costs for moving the large volumes of water are significant. Also, even if a low energy contactor is used, such as a U-tube in a deep well [19], it has been found that at sufficiently high pressures for efficient carbon dioxide removal 3 to 5 % hydrogen loss occurs [18].

To avoid these hydrogen losses the more usual techniques for acid gas removal employ gas liquid equilibria which are more favourable than the one for carbon dioxide-water [2]. Several systems are based on employment of monoethanolamine (MEA, e.g. Girbotol) which is capable of decreasing the carbon dioxide content to less than 50 ppm [2, 6, 20] (Equations 9.9-9.12). The potassium carbonate — bicarbonate cycle, in a similar way to MEA allowing recycle of the absorbing base, is also practised but is only sufficient in its normal

configuration to decrease the carbon dioxide content to 500-1,000 ppm (Equations 9.21, 9.22).

Absorption:

$$CO_2 + H_2O + K_2CO_3 \rightarrow 2\,KHCO_3 \qquad (9.21)$$

Regeneration:

$$2\,KHCO_3 \xrightarrow[120\,°C]{} K_2CO_3 + CO_2 + H_2O \qquad (9.22)$$

But employing a two stage contactor with potassium carbonate solutions improves efficiency somewhat, particularly if the second stage is operated at high pressure [21]. Sodium carbonate is a less costly absorbent, but also less efficient. Other acid gas removal techniques use physical absorption of carbon dioxide in organic solvents such as the dimethyl ether of polyethylene glycol (Selexol), or methanol (Rectisol) but for absorption to be effective require high gas pressures [2]. Regeneration of physical absorption solvents is by pressure letdown plus some air stripping (sparging of air through the solvent).

The residual carbon dioxide and any remaining carbon monoxide are then treated catalytically with hydrogen in a step called methanation, which essentially inactivates any catalyst poisoning effect of these oxygen-containing components (Equations 9.23, 9.24).

Methanation:

$$CO + 3\,H_2 \xrightarrow[315\,°C]{Ni\ catalyst} CH_4 + H_2O \qquad (9.23)$$

$$CO_2 + 4\,H_2 \rightarrow CH_4 + 2\,H_2O \qquad (9.24)$$

If, however, the ammonia plant is obtaining nitrogen from an air separation plant, carbon monoxide and carbon dioxide (boiling points $-205$; and $-78.5\,°C$, sublimes) may be condensed and removed by scrubbing with liquid nitrogen [6].

### 9.1.4 Ammonia Synthesis

Whether feedstocks are obtained from air separation plus supplementary hydrogen sources, or from a reforming and secondary reforming sequence the feedstock requirements are the same. Nitrogen and hydrogen gas are required in a mole ratio (at ordinary pressures very close to a volume ratio) of 1:3, and raised to very high pressures to somewhere in the 100-900 bar (10,130-90,120 kPa; 1,500-13,500 psi) range, depending on the synthesis system employed [22, 2, 11]. Choice of synthesis pressure used in any particular instance depends on the balance to be struck between the lower compression energy costs but lower conversions (more recycle cost) at low synthesis pressures, and higher compression costs but higher conversions at high synthesis pressures.

Early in the development of ammonia synthesis technology it became apparent that iron and high carbon steels became brittle and demonstrated intergranular corrosion on exposure to high pressures of hydrogen or ammonia, particularly at high temperatures [23, 24]. This process which originally was little understood caused a modification of the structure of the steel, weakening it, and made vessels so attacked unsafe to use at high pressures after a short time. This process, termed "hydrogen embrittlement", is now understood to involve adsorption of hydrogen onto the steel, followed by ionization (Equation 9.25).

$$2\,H_{2_{(g)}} \xrightarrow{\text{adsorption}} 2\,H_{2_{(ads)}} \xrightarrow{\text{ioniz'n}} 4\,H^+ \xrightarrow{C_{(Fe)}}$$

$$CH_{4_{(Fe)}} \rightarrow CH_{4_{(g)}} \qquad\qquad (9.25)$$

The protons thus formed diffuse into intergranular carbon and react with it, forming methane bubbles which internally stresses the steel, making it much less ductile and causing intergranular cracks [25].

Since the discovery and better understanding of hydrogen embrittlement phenomena [26, 27] the design of the pressure shell of all ammonia converters has incorporated features to avoid weakening by this kind of attack. All use the incoming nitrogen/hydrogen stream to keep the inner wall of the pressure shell cool by passing this entering stream immediately inside the pressure shell, thus eliminating one contributing factor to embrittlement, the heat (e.g. Figure 9.3). By using alloy steels incorporating titanium, vanadium, tungsten, chromium, or molybdenum, methanation of any adsorbed/absorbed hydrogen does not occur, since these metals form very stable carbides. Instrumentation has also been developed to continuously monitor hydrogen penetration rates adding a useful safeguard [28, 29]. Wrapped converter designs are used where only the inner layer of ca. 1.2 cm thick plate is made leak tight, and the outer multi-layers of tightly wrapped 0.6 cm plate are deliberately perforated with small holes to harmlessly release any diffused hydrogen providing another kind of safeguard [10].

Ammonia conversion takes place at 400-500 °C, with most of the input gas heating being provided by indirect heat exchange with already converted

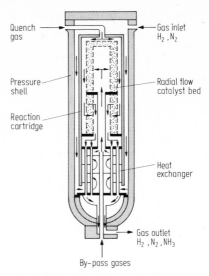

**Figure 9.3.** Diagram showing design features of a large scale ammonia synthesis converter. The quench, and by-pass inlets for feed gas inputs, by foregoing heat exchange before entering the catalyst beds, provide one means of temperature control on reacting gases. The fraction of "inerts" in the feed gas stream is another (see text)

exit gases via an exchanger inside the pressure shell of the converter (Figure 9.3). During the fraction of a second contact time the feed gas stream forms a gas mixture containing 10-20 % ammonia in residual nitrogen and hydrogen, and the pressure drops slightly in the process from both flow resistance through the catalyst beds, and the slight gas volume decrease incurred by the reaction itself. Catalyst life of up to 15 years is known with careful feed and recycle gas treatment.

Ammonia is recovered from this gas mixture by condensation (on cooling the gases) at converter exit pressures. This practice has been adopted for several reasons. Firstly, it conserves compressor capacity and recompression energy requirements once the liquefied ammonia has been separated, by maintaining the unreacted nitrogen and hydrogen at nearly the inlet pressure required for recycle to the converter again. And secondly, ammonia (boiling point − 33 °C) is far easier to condense out of the gas mixture under high pressures, which tend to favour liquefaction, than at atmospheric pressure. In fact, for processes operating at pressures of 400 bars or above, ordinary process cooling water temperatures (10 to 20 °C) are sufficient to cause liquefaction of the bulk of ammonia,

**Table 9.4** Equilibrium ammonia vapour concentrations in a 1:3 nitrogen : hydrogen gas mixture above liquid ammonia at 15 °C [a]

| Pressure, bars | Approximate equilibrium ammonia concentrations, % |
|---|---|
| 50 | 18 |
| 100 | 9 |
| 300 | 6 |
| 600 | 4 |
| 1000 | 3 |

[a] Data from [10].

and it is only at pressures below this that increasingly intensive refrigeration becomes necessary (Table 9.4). The crude liquid ammonia product is then fed to a high pressure separator, and following this to a low pressure separator from which it proceeds to product storage spheres or shipping units (Figure 9.4).

Ammonia recovery from converter exit gases may also be practised by absorption into water (>600 volumes of ammonia per volume of water at 25 °C, 1 bar), when ammonia solutions are to be marketed. Solutions containing 28 to 30 % $NH_3$ by weight are obtained, which is what is ordinarily referred to as concentrated ammonium hydroxide (Equation 9.26).

$$NH_3 + H_2O \rightarrow NH_4OH \qquad (9.26)$$

Before recycle of the uncondensed gases from ammonia recovery the inerts, mainly argon and methane, have to be bled from the system. Otherwise their concentrations build up, diluting the active synthesis gases in the system even to the

point of blowout, a situation obtained when the heat of reaction of low residual concentrations of active gases is inadequate to sustain converter catalyst bed temperatures and the synthesis reaction fails. This dilution is intentionally practiced to control converter temperatures, for example particularly when fresh highly active catalyst has been placed. The purge rate used ultimately determines the concentration of inerts in the system. For instance if the purge rate corresponds to 1 % of the rate of addition of make-up gas, and the make-up gas contains 0.1 % inerts, then the steady state concentration of inerts in the converter feed would be 100 x 0.1 %, or 10 % [30]. At this purge rate 0.9 mole of nitrogen and hydrogen would be discarded for every 100 moles of nitrogen and hydrogen added. At a purge rate of 10 % of the make-up rate the steady state concentration of inerts in the converter drops to 1 % but 9.9 moles of nitrogen and hydrogen are discarded for each 100 moles added. The hydrogen purged is not a complete loss since the hydrogen and methane in the purge gas are at least used to fuel the reformers. But burned hydrogen is not as productive as converted hydrogen so that this conflict of objectives between raising the concentration of inerts fed to the converter versus raising the loss rate of the feed gases as reformer fuel represents another case for process optimization.

Occasionally an ammonia plant practises argon recovery by distillation of liquefied purge gases [31, 32], particularly if air separation is used to provide at least a part of the nitrogen. But when ammonia synthesis is associated with air separation, very often most of the argon and methane are removed in the liquid nitrogen scrub for carbon dioxide clean up, giving very low concentrations of inerts in the make-up gas [6]. Recovered argon is of particular value as an illuminating gas for decorative lighting and as an inert filler for filament light bulbs, and is also used as inert blanketing gas for shielded welding of reactive metals.

Again because of the multiple recycle aspect of ammonia conversion, a very slight deviation from a 1:3, nitrogen to hydrogen mole ratio in the make-up gases fed to a converter can gradually build up to a significant difference from this ratio in the converter feed. The desired ratio is directly stabilized to some extent by the purge rate, and may also be compensated for at regular intervals by adjusting the reformer gas feed rates to affect the proportion of nitrogen to hydrogen in the make-up gas.

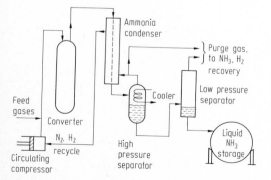

**Figure 9.4.** Simplified flowsheet of the gas recycle and product recovery systems for ammonia synthesis, from the gas mixture leaving the converter

Liquid ammonia from the low pressure separator analyzing 99.5 + % ammonia is led to refrigerated, insulated storage tanks. Ammonia storage capacity needs to be considerable because of the very seasonal nature of its major market, fertilizers. For shipment, it is fed from storage to insulated tank trucks, tank cars, or barges for delivery. Markets and production facilities in the southwestern U.S. are so well developed that here an extensive pipeline network over 4,000 km in extent has been constructed capable of anhydrous ammonia deliveries of up to 4,500 tonnes per day and ensures continuity and efficiency of supply for this area [2]. Smaller quantities are also shipped as ammonia solutions, containing 28 to 30 % ammonia by weight, or in pressurized cylinders.

## 9.1.5 Major Uses of Ammonia

Ammonia is primarily used for fertilizer related applications, which in total represented some 63 % of the U.S. market share in 1968 and climbed to about 80 % of the total by 1973 (Table 9.5). The major proportion of this dramatic increase is in the direct application of ammonia to the soil, either as the gas flowing under pressure to tractor-drawn knives which discharge just under the soil surface or, for smaller agricultural operations, as the 28-30 % solution. The reason for the large increase in this use sector is primarily because of the high nitrogen content, ca. 82 %, and because the applied ammonia by either method is strongly adsorbed by organic matter and clays in the soil to be substantially retained both at the time of application and subsequently. Ammonium nitrate, another popular nitrogen component in fertilizers, consumes ammonia both to make the nitric acid required (Section 9.2) and to form the salt from this. Fertilizer nitrogen from ammonium nitrate, however, is losing favour to urea, both in the U.S. and in Canada. Ammonium phosphate, ammonium sulfate, and other variations are also popular nitrogen components in fertilizers.

Smaller scale uses are mostly indirect, such as the manufacture of explosives via nitric acid or using the new ammonium nitrate/fuel oil or metallized ammonium nitrate slurry explosive systems. Some is consumed in the preparation of mono-, and diamines as monomers for nylons, in the manufacture of acrylonitrile for Acrilan, and melamine (via urea) for the production of melamine-formaldehyde resins. The low cost of ammonia and its appropriate compression-liquefaction properties

**Table 9.5** Trends in the use pattern for ammonia in the U.S.A. since 1968 [a]

| Use | Percent of total | | |
|---|---|---|---|
| | 1968 | 1973 | 1983 |
| Fertilizer | | | |
| ammonia, aqua ammonia | 6.2 | 27.1 | 25 |
| ammonium nitrate | 21.6 [b] | 18.9 | |
| nitric acid | 25.0 | – | 20 |
| urea | 10.1 | 13.9 | 20 |
| other | – | 20.5 | 15 [d] |
| Industrial | | | |
| plastics, fiber monomers [c] | 18.9 | 8.7 | 10 |
| explosives | b | 4.0 | 5 |
| miscellaneous | 17.8 | 6.9 | 5 [e] |
| | 99.6 | 100.0 | 100 |

[a] Compiled from data of [2 and 11], and isues of Chemical and Engineering News, Chemical Marketing Reporter.
[b] Includes fertilizer and explosives applications.
[c] Includes amines produced but not destined for polymer applications.
[d] As ammonium phosphate.
[e] Includes refrigerants, 2%, and water purification, froth flotation and other important applications for small amounts of ammonia.

make it ideal as the working refrigerant in large scale freezing systems such as for frozen food processing and storage, artificial ice rinks and the like. But its significant toxicity and flammability make it of only limited use in household refrigerators in favour of the much more costly but less toxic and non-flammable Freons. Ammonia is also employed as the cation in some sulfite-based wood pulping processes, and as a means of stabilizing residual chlorine in municipal treatment for potable water, among myriads of other small scale applications.

## 9.1.6 New Developments in Ammonia Synthesis Technology

Even with the large price increases experienced in recent years for petroleum-based feedstocks generally it is still held that natural gas is likely to remain a significant, if not a dominant source of the hydrogen required for ammonia synthesis for the foreseeable future [33]. Production facilities that have selected and are committed to this hydrogen source have centered their attention on innovations to boost energy efficiency, or hydrogen production utilization efficiency to obtain operating economies.

Modifications using waste heat recycle and the more extensive use of heat exchange to simultaneously cool product streams and warm feed streams, together with other improvements, are anticipated to reduce the energy cost for ammonia production from 37.1 million kJ per tonne to 30.2 million kJ per tonne, a saving of some 18 % [34, 35]. A new type of ammonia converter utilizing intermediate indirect cooling of the reacting gas stream between a succession of three catalyst beds (rather than quench gas cooling, Figure 9.3), which achieves higher than normal concentrations of ammonia in the converter exit gas has been announced [36]. This feature, which has parallels with contact sulfuric acid technology, is a promising development. More detailed incremental process refinements including adoption of steam turbine drive two stage centrifugal compressors for primary compression in place of reciprocating compressors have been presented [37, 38]. An analysis of extensive computer control of ammonia plant operation has demonstrated that the economies achieved through better regulation of the nitrogen:hydrogen synthesis gas ratio among other operating parameters enables the installation cost of the control system to be written off in as little as one year [39]. Faster startups on commissioning and planned maintenance procedures also contribute favourably to higher actual operating capacities [21]. Shutdowns have been surveyed as to cause and the results usefully interpreted to minimize this source of inefficiency [40], as have recommendations to minimize turnaround time (time to get back in production again) in the event of a shutdown [41, 42].

The efficiency of hydrogen production is being boosted by reducing the steam to carbon ratios in the steam reforming step from 4.5 to about 3, and by improving the carbon dioxide removal system [43, 44, 37]. Skillful control of the concentration of inerts in the recycle gas decreases both feed gas (nitrogen and hydrogen) loss as well as the loss of residual ammonia present in the purge gas when this is burned as reformer fuel [45]. And if efficient control of purge ratios are coupled to recovery of the ammonia and hydrogen from the purge gas prior to combustion, hydrogen utilization efficiencies can be raised to ca. 99.5 % from 92 to 95 % normally experienced [43, 46, 47].

It has been proposed that water be used to a greater extent than at present to provide the hydrogen requirement [48] (Equations 9.27-9.29).

$$2\,CaO + 2\,I_2 \underset{\sim\,100\,^\circ C}{\rightarrow} 2\,CaI_2 + O_2 \qquad (9.27)$$

$$CaI_2 + H_2O \underset{500\,^\circ C}{\rightarrow} CaO + 2\,HI \qquad (9.28)$$

$$2\,HI \underset{300-700\,^\circ C}{\rightarrow} H_2 + I_2 \qquad (9.29)$$

This indirect route to cracking water is claimed to obtain a 20 % improvement in heat efficiency to hydrogen over existing carbon based cracking processes. Taking water as a hydrogen source one stage further, electrolysis of water to provide the dominant or only source of hydrogen has also been proposed with the additional consideration that the electrolyzer used be built to operate at 400 bars pressures [49]. If liquid nitrogen was used as the nitrogen source, this option could be practised without the need for compressors, which in an area providing low cost power for electrolytic hydrogen provides an interesting alternative process sequence to ammonia.

The original Haber-Bosch process relied on coal or coke to produce the hydrogen requirement but has since been largely replaced by natural gas or petroleum-based technologies. However, as these sources have become more expensive and scarcer there has been a resurgence of interest in the coal based systems [50-56]. The dominant sequences which are emerging use coal gasification as the first stage, followed by shift conversion and gas cleanup to obtain ammonia feedstocks (e.g. Equation 9.30).

$$\begin{matrix} \text{Coal} \\ \text{gasification} \to \\ \text{to approx.} \end{matrix} \begin{matrix} CO & 31\% \\ H_2O(g) & 31\% \\ H_2 & 24\% \\ CO_2 & 13\% \\ H_2S & 1\% \end{matrix} \left.\begin{matrix} \text{1. shift conversion} \\ \text{2. clean up} \\ \\ \text{3. } N_2 \text{ wash,} \\ \text{addition} \end{matrix}\right\} \longrightarrow \begin{matrix} \text{feedstocks} \\ N_2 : 3\,H_2 \end{matrix} \qquad (9.30)$$

$$\begin{pmatrix} \text{Coal} \\ 2\,600 \text{ tonnes/day} \\ 4.2\%\ S \end{pmatrix} \to \begin{matrix} 2{,}250 \\ \text{t/day} \end{matrix} \begin{pmatrix} \text{gasified,} \\ \text{cleaned,} \\ N_2 \text{ added} \end{pmatrix} \to \begin{matrix} NH_3,\ 1500 \text{ t/day} \\ \\ S,\ 107 \text{ t/day} \\ \text{wet ash, 550 t/day} \end{matrix} \quad \begin{matrix} (350 \text{ t/day} \\ \text{burned for} \\ \text{steam, flue gas} \\ SO_2 \text{ scrubbed)} \end{matrix} \qquad (9.31)$$

The methanol-based Rectisol process is favoured for gas cleanup because it depends on physical absorption at high pressures, and because it permits the recovery of separated hydrogen sulfide and carbon dioxide streams on solvent regeneration, considerations more important with coal feedstocks because of the higher sulfur gas content in the gasifier product [52, 51]. Mass balance predictions certainly appear to make this alternative worth considering [51] (Equation 9.31), although the increased technological complexity over well-tested natural gas reforming sequences, and greater emission control problems tend to be important negative factors [2, 1].

Solar ammonia production has been demonstrated to occur naturally in desert areas via exposure of nitrogen plus water vapour to catalytic sand grain surfaces and has been confirmed by model experiments in the laboratory [57]. The details of this photochemical water splitting reaction are being further investigated with a view to potential industrial applications (Equation 9.32) [58].

$$2\,N_2 + 6\,H_2O \xrightarrow{h\upsilon} 4\,NH_3 + 3\,O_2 \qquad (9.32)$$

Indirect solar energy utilization via the surface to depth temperature differential existing in warm oceans has also received consideration [2]. This would be exploited using large scale floating facilities. All of these solar ammonia projects remain as longer term experimental prospects at present.

### 9.1.7 Environmental Concerns of Ammonia Production

An analysis of the emissions and their discharge rates was recently completed which allowed gaseous pollutant emission factors per tonne of ammonia produced to be estimated for relatively old and newer production facilities (Table 9.6) [1].

Sulfur dioxide emission rates are generally low using natural gas feedstocks in both types of facilities since the bulk of any sulfur compounds present are removed at the well head (see sulfur). But even the traces of sulfur dioxide present in the reformer product must be removed since sulfur is one of the permanent catalyst poisons and as such its accumulation hastens the need for catalyst renewal. Sulfur dioxide is normally removed from the gas stream by adsorption, and the sulfur dioxide discharge occurs when the adsorptive bed is heated for regeneration. The mass rate of emission with natural gas feedstocks is small, of the order of 60 kg $SO_2$ per day for a 1,000 tonnes per day ammonia plant. Hence, there has not been any perceived need to change the old sulfur dioxide purging practice for the new plants being built. Growing use of fuel oil, and especially coal for hydrogen generation will produce more associated sulfur dioxide on reforming and therefore particularly the latter types of facilities are planned to require integral sulfur dioxide containment [51, 53, 52].

Nitrogen dioxide discharge results from the oxidation of atmospheric nitrogen on combustion of reformer fuel, as occurs to some extent whenever hot nitrogen and oxygen contact heated metal surfaces, with a further small contribution from the combustion of traces of residual ammonia still present in purge gases which are burned. Nitrogen oxidation also occurs, and on a much larger scale from the operation of fossil-fueled power stations. The small improvement noted for modern ammonia plants for this sector results from increased reformer energy efficiency requiring the combustion of less fuel per unit of hydrogen produced and better control of ammonia content of purge gases. These in turn give a lower flue gas discharge volume per tonne of ammonia produced, lowering the mass of $NO_x$ discharged accordingly. A further refinement adopted by some facilities which serves to decrease $NO_x$ emission is to decrease the excess air used for combustion and at the same time improve control of this reformer function, which also improves energy efficiency.

Ammonia plants using aqueous ammoniacal cuprous chloride or formate for final stage scrubbing to remove traces of carbon monoxide [59] experienced significant carbon monoxide discharge on the regeneration cycle of the aqueous scrubber liquor, of the order of 60 tonnes of carbon monoxide per day for 600 tonne per day ammonia plant. Only relatively slight losses occurred on regeneration of the scrubbing liquor of the carbon dioxide scrubbing circuit because of the very high selectivity for carbon dioxide of the solutions used for this step. More recently built ammonia plants employing methanation (Equations 9.24, 9.25) to get rid of the last traces of both carbon monoxide and carbon dioxide avoid the major carbon monoxide discharge source altogether, and only experience losses of the order of 30 kg of carbon monoxide per day for a 1,000 tonne per day facility as traces captured in the carbon dioxide scrubber circuit. Ammonia losses from the synthesis stage primarily

**Table 9.6** Gas emission rates from U.S. ammonia production facilities [a]

| Source of emission | Emission rate, kg/tonne of ammonia produced | | | |
| --- | --- | --- | --- | --- |
| | $SO_2$ | $NO_2$ | CO | $NH_3$ |
| *Old Plants* [b] | | | | |
| Natural gas cleaning | 0.05 – 0.7 | — | — | — |
| Reformer | 0.03 – 0.3 | 0.6 | — | — |
| Carbon dioxide removal | — | — | 0.03 | — |
| Copper liquor scrubbing | — | — | 91.5 | 3.2 |
| Ammonia synthesis | — | — | — | 1.6 |
| Ammonia loading | — | — | — | 0.5 |
| Totals | 0.08 – 1.0 | 0.6 | 91.5 | 5.3 |
| *Modern Plants* [c] | | | | |
| Natural gas cleaning | 0.05 – 0.7 | — | — | — |
| Reformer | 0.03 – 0.3 | 0.5 | — | — |
| Carbon dioxide removal | — | — | 0.03 | — |
| Copper liquor scrubbing | — | — | — | — |
| Ammonia synthesis | — | — | — | 1.6 |
| Ammonia loading | — | — | — | 0.2 |
| Totals | 0.08 – 1.0 | 0.5 | 0.03 | 1.8 |
| Weighted average emission rates, old and modern facilities (kg/tonne $NH_3$ produced) | 0.4 | 0.6 | 6.0 | 1.3 |

[a] Adapted from summary tables of [1].
[b] Plants using ammoniacal cuprous solutions for carbon monoxide removal.
[c] Plants using methanation for residual carbon dioxide and carbon monoxide removal; ammonia-synthesis purge gas is burned as fuel.

arise from purge gas combustion and leakage. Little change is seen in the loss rate from this area with newly adopted technology. But as more facilities adopt purge gas ammonia and hydrogen recovery system [46, 38] the loss rate from this operating area should decrease. Improved loading techniques, including ammonia line vapour recompression provisions and the requirement of continuous control of storage tank vapours [60] have resulted in the decreased loss rate from this sector. Realistic guidelines of daily average ammonia discharge rages of 1 kg per tonne (design) and 1.5 kg per tonne (operating) have recently been set for one jurisdiction compatible with the loss rates outlined above [61].

Water impacts from ammonia producing facilities can occur through thermal loading, from discharge of large volumes of cooling water, or from the discharge of dilute ammonia solutions which may also contain organics from recovered process condensates. Lagooning or forced-air cooling are among the expedients which can decrease the ef-

fect of thermal loadings to water (see chapter 3), and the increasing use of heat exchange of hot exit gases to preheat entering raw materials as an energy conservation measure will also tend to decrease impacts from this source. While steam stripping is effective and may be used to recover ammonia from dilute aqueous waste streams this is a relatively expensive, energy intensive procedure for the small quantities likely to be recovered. A better expedient which has been suggested is to use these streams (containing typically 400-2,000 ppm ammonia) to simultaneously irrigate and fertilize croplands nearby [62], an option of value when the operating location allows. Permissible and desirable criteria for ammonia in public water supplies have been set at 0.5 mg $L^{-1}$, and 0.01 mg $L^{-1}$, respectively [63], limiting the acceptable volumes of such waste streams which can be discharged into surface waters to maintain these criteria.

Noise problems in large ammonia plants, from the movement of large gas volumes, the operation of compressors and the like, are environmental impacts more or less restricted to operating employees but have nevertheless received attention to meet recently placed noise criteria. Procedures to locate and localize or dampen (attenuate) noise sources producing sound levels in excess of 90 dBA have been described [64].

## 9.2 Production of Nitric Acid

### 9.2.1 Nitric Acid Background

Nitric acid *(aqua fortis)* was known and its chemistry practised in the Middle Ages. It was obtained by heating hydrated copper sulfate [3], or sulfuric acid with sodium nitrate (saltpeter, or niter) and cooling the vapours generated to obtain a solution of nitric acid (Equation 9.33).

$$NaNO_3 + H_2SO_4 \xrightarrow{\sim 200\,^\circ C} NaHSO_4 + HNO_3$$
"niter"                      "niter cake"   (9.33)

While it is theoretically possible to obtain 2 moles of nitric acid for each mole of sulfuric acid used, in practice the very high temperatures required to achieve this, about 900 °C, make this stoichiometry impractical. However, it was possible to usefully use the residual acid value of the niter cake to make hydrochloric acid from salt [3] (Equation 9.34).

$$NaHSO_4 + NaCl \rightarrow Na_2SO_4 + HCl \qquad (9.34)$$
"niter cake"        "salt cake"

From both direct nitrogen fertilizer interests and to provide for the nitric acid requirement for the manufacture of explosives in the nineteenth century natural sodium (or potassium) nitrate provided the only source. India produced some 30,000 tonnes per year by 1861 for these purposes and Chile, in 1870, was exporting some 90,000 tonnes annually which gradually climbed to a stabilized figure of about 1.4 million tonnes per year [3], in both cases as the sodium salt.

Present nitric acid production is primarily via oxidation of ammonia and absorption of the oxidation products in water. The chemistry of this process, originally proven experimentally by Kuhlmann [65] had to await the development of efficient and economical routes to the ammonia raw material required before it became commercially significant. Ostwald, working in Germany at around 1900, re-examined Kuhlmann's data and from further more extended experiments established the proper conditions required for the ammonia oxidation step [66]. Very shortly after this, operating processes based on these principles were assembled both in Germany and in the U.S.A. and since then production levels have gradually risen, so that in the U.S.A. since 1976 more than seven million tonnes (100 % basis) of nitric acid have been produced annually (Table 9.7).

**Table 9.7** Production of nitric acid by selected countries [a]

|              | Thousands of metric tonnes | | | | | | |
|--------------|------|------|-------|-------|-------|-------|-------|
|              | 1976 | 1977 | 1978  | 1979  | 1980  | 1981  | 1982  |
| Australia    | —    | —    | 160   | 169   | 180   | 182   | 180   |
| Belgium      | 711  | 985  | 1 048 | 1 055 | 1 167 | 1 316 | —     |
| Canada       | —    | —    | —     | —     | 713   | 1 158 | 977   |
| Finland      | 304  | 291  | 357   | 408   | 422   | —     | —     |
| France       | 3 069 | —   | —     | —     | —     | —     | —     |
| W. Germany   | 633  | 644  | 698   | 3 136 | 3 173 | 2 883 | 2 287 |
| Italy        | 918  | 929  | 1 038 | 1 070 | 1 011 | 1 005 | 1 030 |
| Japan        | 623  | 659  | 655   | 677   | 577   | 518   | 531   |
| Sweden       | 299  | 312  | 348   | 349   | 346   | 354   | —     |
| U.K.         | 2 553 | 2 623 | 2 713 | 2 909 | 2 825 | 2 973 | 3 077 |
| U.S.A.       | 7 162 | 7 215 | 7 306 | 7 771 | 8 102 | 8 203 | 6 884 |

[a] Compiled from [2, 123, 124, 125], plus obtained from data supplied by Verband der Chemischen Industrie e.V. (VCI). U.S. production in thousands of tonnes for earlier yeras was 1960, 3 007; 1965, 4 445; 1970, 6 899.

## 9.2.2 Nitric Acid by Ammonia Oxidation, Chemistry and Process Considerations

Modern nitric acid production, in simple terms, amounts to a catalytic oxidation of ammonia in air followed by absorption of secondary oxidation products in water to yield nitric acid. Unlike the reaction sequence for ammonia feedstock preparation, all the reaction steps in this sequence are exothermic.

Initial ammonia oxidation is conducted using a mixture of ammonia gas (9-11 %) in air which is passed through multiple layers of fine platinum-rhodium alloy gauze (Equation 9.35).

$$4\,NH_{3(g)} + 5\,O_{2(g)} \xrightarrow[\sim\,900\,°C]{Pt/Rh} 4\,NO_{(g)} + 6\,H_2O_{(g)} \quad \text{nitric oxide} \quad (9.35)$$

$$\Delta H = -907 \text{ kJ } (-217 \text{ kcal})$$

While this equation represents a very close approximation of the stoichiometry of the process under these conditions, the detailed chemistry is complex and poorly understood [3]. The exotherm of the oxidation is sufficient to cause a temperature rise of about 70 °C for each one percent ammonia in the mix [67], so that with prewarming of one or both component streams keeps the alloy gauze at close to the optimum 900 °C. Operating at gauze temperatures of 500 °C produces mostly the relatively unreactive nitrous oxide as the ammonia oxidation product [3] (Equation 9.36) which would be lost to acid production. Hence the motivation to provide conditions which maintain gauze temperatures of 900 °C.

$$2\,NH_{3(g)} + 2\,O_{2(g)} \xrightarrow[\text{slow}]{500\,°C} N_2O + 3\,H_2O \quad (9.36)$$

$$\text{nitrous oxide}$$

$$\Delta H = -276 \text{ kJ } (-66 \text{ kcal})$$

It may seem to be a backward step to have gone to the trouble of preparing ammonia from nitrogen and hydrogen, only to turn around and burn the ammonia to obtain nitric oxide. However, if one considers the endothermic nature of the direct combination reaction between nitrogen and oxygen (Equation 9.37)

$$N_{2(g)} + O_{2(g)} \rightarrow 2\,NO_{(g)} \quad (9.37)$$

$$\Delta H = +90.3 \text{ kJ } (+21.6 \text{ kcal})$$

it becomes apparent why nitric oxide yields on direct combination, even at carbon arc temperatures

of around 3,000 °C, are not more than a few volume percent. Using ammonia as the starting material for nitric oxide preparation bypasses the enormous energy requirement for elemental nitrogen bond dissociation, and in so doing achieves the product of interest via a net exothermic (thermodynamically favourable) process.

Nitric oxide obtained from ammonia combustion is then further oxidized to nitrogen dioxide in another, less exothermic step (Equation 9.38).

$$2\,NO_{(g)} + O_{2(g)} \xrightarrow{\text{slow}} 2\,NO_{2(g)} \quad (9.38)$$

$$\Delta H = -113 \text{ kJ } (-27 \text{ kcal})$$

And the nitrogen dioxide obtained immediately participates in a relatively rapid equilibrium to dinitrogen tetroxide (Equation 9.39).

$$2\,NO_{2(g)} \rightleftarrows N_2O_{4(g)} \quad (9.39)$$

$$\text{brown} \qquad \text{colourless}$$
$$\text{m.p. } -11.2\,°C$$
$$\text{b.p. } 21.2\,°C$$

$$\Delta H = -57.4 \text{ kJ } (-13.7 \text{ kcal})$$

At 100 °C the equilibrium composition lies at approximately 90 % $NO_2$, 10 % $N_2O_4$ whereas at 21 °C, the boiling point of $N_2O_4$, only about 0.1 % $NO_2$ is present [68]. This equilibrium may be important for acid production since there is some evidence that dinitrogen tetroxide is the molecular species that reacts with water for acid formation [3].

The final step in acid formation is absorption of nitrogen dioxide in water the net result of which is best described by Equation 9.40.

$$3\,NO_{2(g)} + H_2O_{(l)} \rightarrow 2\,HNO_{3(aq)} + NO_{(g)} \quad (9.40)$$

$$\Delta H = -139 \text{ kJ } (-33.1 \text{ kcal})$$

This is not as straightforward a transformation as it may at first appear since it involves both chemical combination with water and a net redox reaction for the nitrogen (4 +) in nitrogen dioxide to nitrogen (5 +) in nitric acid. One nitrogen dioxide thus takes up two electrons (is reduced) to become nitrogen (2 +) in nitric oxide to balance this oxidation. Probably the actual chemistry involved is better represented as, first of all, a formation of both nitric and nitrous acids (Equation 9.41).

$$2\,NO_2 + H_2O \rightarrow HNO_3 + HNO_2 \quad (9.41)$$

The nitrous acid, being unstable in the presence of any strong mineral acid, disproportionates to yield further nitric acid and nitric oxide (Equation 9.42).

$$3\,HNO_2 \rightarrow HNO_3 + 2\,NO + H_2O \qquad (9.42)$$

Not all of the nitrogen in the ammonia feed ends up as nitric acid. Some of the ammonia reacts with oxygen of the air yielding elemental nitrogen from ammonia nitrogen, and some reacts with nitric oxide causing loss of potential product from both ammonia and oxidant consumption (Equations 9.43a and 9.43b).

$$4\,NH_{3(g)} + 3\,O_{2(g)} \rightarrow 2\,N_{2(g)} + 6\,H_2O_{(g)} \qquad (9.43\,a)$$
$$\Delta H = -1267\text{ kJ } (-303\text{ kcal})$$

$$4\,NH_{3(g)} + 6\,NO_{(g)} \rightarrow 5\,N_{2(g)} + 6\,H_2O_{(g)} \qquad (9.43\,b)$$
$$\Delta H = -1806\text{ kJ } (-432\text{ kcal})$$

Both reactions 9.43a and 9.43b are highly favoured thermodynamically, even relative to the desired reaction, 9.37. The platinum gauze catalytic surface serves to boost the proportion of nitric oxide obtained and decrease nitrogen formation by these pathways, probably by accelerating the rate of the desired reaction without affecting the rates of the others. Use of multiple platinum-rhodium or platinum-palladium alloy gauzes further raises the ammonia conversion to 99 % (for 8.3 % in air; gauze temperature 930 °C), over the 97.5 % accessible with platinum alone [3]. Use of alloy gauzes is also said to decrease the metal erosion rate [69].

The important aspect of the ammonia oxidation reaction, 9.35, from a process design standpoint is that since 9 moles react to produce 10 moles of products, there is little volume change from this step. Therefore, the equilibrium of this reaction will be little affected by a change in operating pressure. But an increase in pressure pushes more material through the same sized equipment in the same time, lowering the capital cost of the plant per unit of product produced. If the pressure increase is moderate, say 5 to 10 bars, the metal wall thickness does not have to be significantly raised to contain it, and this modification provides a further process benefit by slightly increasing the concentration of nitric acid produced from the 50-55 % range when operating at atmospheric pressure, to 57-65 % when under pressure.

Increasing the pressure of the feed gases to the ammonia converter also effectively increases the space velocity through this unit, slightly decreasing the conversion efficiency for ammonia oxidation. But this reaction is so rapid, being essentially complete in $3 \times 10^{-4}$ seconds at 750 °C [4], that addi-

tional layers of catalytic gauze *almost* compensates for this stage of the process.

The nitric oxide oxidation and water absorption reactions, 9.38 and 9.40, are both much slower than the ammonia oxidation reaction, 9.35, and involve a significant volume decrease on reaction, 3 moles (volumes) to 2 and 3 moles to 1, respectively. Thus the absorbers, where the bulk of these reactions occur, must be large to provide sufficient residence time, and cooled to favour the equilibria in the desired direction. In the absorbers, because of this volume decrease on reaction, raising the pressure achieves a very great improvement in performance, in accord with Le Chatelier's principle. For an increase to 8 bars from 1, the rate of the very slow nitric oxide re-oxidation reaction [70] in particular (Equation 9.44) is accelerated by a factor of the cube of this pressure increase, or 512 times.

$$\text{Rate} = k_1(P_{NO})^2(P_{O_2}) - k_2(P_{NO_2})^2 \qquad (9.44)$$

For an operating process absorber, normally not very close to equilibrium, the rate of the reverse reaction (the second term in this rate equation) can be ignored.

And of course the pressure increase also boosts the rate of nitrogen dioxide absorption, once formed, in water. In practice this pressure increase means that the absorber volume required per daily tonne of nitric acid production capacity can be decreased from about 35 m$^3$ to about 0.5 m$^3$ [69].

## 9.2.3 Process Description

To put these principles into practice liquid ammonia is first vaporized by indirect heating with steam, and then filtered to reduce risk of catalyst contamination to obtain an ammonia gas stream at about 8 bars pressure without requiring mechanical compression. An air stream is separately compressed to about the same pressure, preheated to 200-300 °C and filtered prior to mixing with the ammonia (about 10 %) gas stream immediately before conversion. Passage through the red hot platinum-rhodium gauze produces a hot gas mixture of nitric oxide and water vapour plus the unreacted nitrogen and oxygen components of air (Figure 9.5), with a yield efficiency under these conditions of about 95 % [18].

Hot converter gas products are cooled prior to absorption in a waste heat boiler, to give simultaneous production of steam and a product gas stream at more moderate temperatures ready for catalyst recovery. The high pressure nitric acid

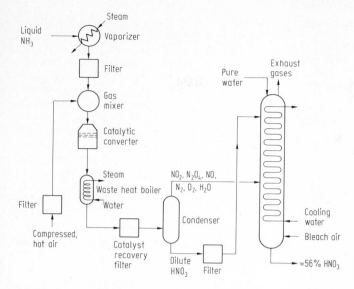

**Figure 9.5.** Production of nitric acid by oxidation of ammonia

process, as described here, experiences a gauze metal loss rate of 250-500 mg per tonne of acid produced which can be kept to the lower end of this range by means of efficient filtration [2]. If operated at atmospheric pressure, catalyst loss rates amount to about 50 mg per tonne. After filtration the gas mixture is sharply chilled by heat exchange with process water to condense some dilute nitric acid from the product gases. The dilute nitric acid condensate is also filtered for further catalyst recovery and then trickled into the top of the absorber to raise the acid concentration. The gas stream, still containing nitric oxide but with the elapse of time and having been cooled now also contains nitrogen dioxide and dinitrogen tetroxide (Equations 9.38, 9.39) plus the residual air, is fed into the absorber at an intermediate level.

In the absorbers nitric oxide is oxidized to nitrogen dioxide at the same time as nitrogen dioxide is absorbed in water to give nitric acid plus further nitric oxide. This is not lost but, like the primary nitric oxide, is being continuously re-oxidized to nitrogen dioxide, while cooling water is used to keep the temperatures down to maintain favourable oxidation and solution equilibria. As the concentration of acid builds up to equilibrium levels in the downward flow of absorption water some dissolved (but unreacted) nitrogen dioxide may also be present giving the acid a brown colour. This nitrogen dioxide is stripped from the product acid by purging the last few plates of the absorber with "bleach air", to give a colourless to pale yellow

product acid containing 60-62 % by weight $HNO_3$ as the bottom product of the absorber. The bleach air, at the same time as stripping excess dissolved nitrogen dioxide from the product, also assists in maintaining an oxygen excess in the middle and upper reaches of the absorber, in this way favouring the formation of nitrogen dioxide from nitric oxide. Venting of residual gases from the top of the absorber is normally via an expander turbine which recovers energy from the vented gas sufficient to provide about 40 % of the power required for driving the inlet air compressor mounted on a common shaft.

### 9.2.4 Nitric Acid Concentrations and Markets

The ammonia oxidation product is sold in several technical and commercial market grades ranging from about 50 % (by weight) $HNO_3$ up to what is referred to by chemists as ordinary concentrated nitric acid (68 %) which is the maximum boiling point, or azeotropic composition (Table 9.8). These are the concentrations obtained either directly from the various types of nitric acid plant or are obtained by raising the concentration of the acid plant product by simple distillation. The concentrated product ranges in colour from water white, through yellow to brown, as the dissolved nitrogen dioxide content gradually rises in the product with time from the operation of the slow equilibrium reaction 9.45.

**Table 9.8** Industrial market grades of nitric acid

| Degrees Baumé [a] | Density, g/mL at 20 °C | Nitric acid conc., % |
|---|---|---|
| 36 | 1.330 | 52.3 |
| 38 | 1.3475 | 56.5 |
| 40 | 1.3743 | 61.4 |
| 42 | 1.4014 | 67.2, ordinary concentrated |
| – | 1.4521 | 80.0 |
| – | 1.4826 | 90.0, or higher, = fuming |
| – | 1.4932 | 95.0 |
| – | 1.5492 | 100.0, anhydrous, unstable [b] |

[a] °Bé, an arbitrary specific gravity of density scale, where specific gravity = 145/(145 – °Bé reading).
[b] Decomposes to $NO_2$, NO and water above the freezing point. Hence. fuming grades are more useful at somewhat lower concentrations.

$$4 HNO_3 \rightleftarrows 2 H_2O + 4 NO_2 + O_2 \qquad (9.45)$$

These grades, or concentrations are what is used in the production of ammonium nitrate, which at present consumes some 65-75 % of the total nitric acid made [18, 71, 72], as well as for many of the other more minor uses. Some of these are for example, adipic acid production (5-7 %), military and industrial explosives (3-5 %), isocyanates for polyurethane manufacture (1-2 %), nitrobenzene (1-2 %), and potassium nitrate preparation from the chloride (~ 1 %). A miscellany of even smaller scale uses consumes the remaining 10 to 15 % of the total acid produced.

For intensive nitrations and other purposes requiring concentrations of nitric acid above 68 % by weight, special techniques have to be used since this is the maximum concentration which may be achieved by simple distillation. Nitric acid concentrations above 86 %, comprising as they do ordinary concentrated nitric acid plus dissolved nitrogen dioxide, are dark brown in colour, and tend to lose the dissolved nitrogen dioxide relatively rapidly if open to the air. For this reason these are collectively referred to as "fuming nitric acid" and require considerable caution in their use because of their significantly greater nitration and oxidation reactivities. A standard commercial fuming nitric is sold containing 94.5-95.5 % $HNO_3$.

One method used commercially to obtain nitric acid concentrations above 68 %, particularly by larger facilities, is to dehydrate the azeotropic composition or slightly less than this, using concentrated sulfuric acid. Countercurrent passage of hot nitric acid vapour against the dehydration acid in a tower packed with chemical stoneware achieves direct dehydration to 90 + % $HNO_3$ and produces a diluted sulfuric acid stream (Figure 9.6). If this concentration process is only practised on a small scale, the sulfuric acid may be reconcentrated by addition of oleum, and a portion of the build-up of sulfuric acid in this circuit may be used to formulate a commercial nitrating mixture with some of the fuming nitric acid made [18]. Operation of the concentrator on a larger scale however, requires the use of a sulfuric acid boiler to reconcentrate the dehydration acid. Thus, the disadvantages of sulfuric acid dehydration are the large heat input required to reconcentrate the high boiling point sulfuric acid, and the noticeable sulfate contamination of the nitric acid product from the direct contact dehydration.

Thus another procedure also used for production of fuming nitric acid, which gets around some of these difficulties, is to employ 72 % aqueous magnesium nitrate ($Mg(NO_3)_2$ x 4 $H_2O$) as a dehydrating agent [73, 74]. As in sulfuric acid dehydration, hot nitric acid vapour is fed countercurrently to hot, molten 72 % $Mg(NO_3)_2$, this time in a multiple plate column of stainless steel, with similar results. But here, with the common anion of the dehydrating agent and nitric acid, there is no sulfate contamination and the diluted magnesium nitrate product (approx. $Mg(NO_3)_2$ x 6 $H_2O$; m.p. 89 °C) from the bottom of the column requires less heat for reconcentration than sulfuric acid.

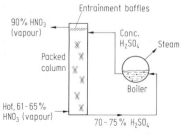

**Figure 9.6.** Dehydration of the nitric acid from an ammonia oxidation plant to fuming grades using concentrated sulfuric acid

## 9.2.5 Nitric Acid Process Variants and New Developments

Early ammonia oxidation processes to nitric acid had absorption trains constructed of chemical stoneware or acid-proof brick, which more or less restricted acid production to near normal atmospheric pressure because of the low strength of the structural materials. The discovery that Duriron [18] (silicon-iron) or high chrome stainless steels [3, 2] could tolerate these corrosive conditions well, allowed adoption of pressure absorption as a means to both markedly decrease the size of the absorbers required and reduce nitrogen oxide stack losses, over what was possible at atmospheric pressure. Pressure operation was easiest to achieve by compression of the feed gases at the front end of the process, and by so doing achieves acid production at very comparable capital costs per unit of product as possible via operation at atmospheric pressure [2].

However, an attendant cost of pressure operation throughout the process is nitric oxide yields of only about 90 % from the ammonia combustion stage [3], by recent improvements raised to about 95 % [2], as compared to yields of 97 to 98 % possible at atmospheric pressure. For this reason atmospheric pressure processes are still viable and are still operating in North America and elsewhere. For the greater ammonia oxidation efficiencies possible at atmospheric or even slightly negative pressures and to retain the much more efficient absorption at elevated pressures many European producers in particular have developed and use split-pressure processes [75, 76]. This combination, which does make more efficient use of ammonia than either of the other technologies discussed, is of particular value when ammonia costs are high. But in general, because of the need to construct the compressor for raising the pressure of oxidized converter gases out of stainless steel or other more exotic metals, the capital costs of such split-pressure plants are $1^1/_2$ to 2 times the cost of the other two alternatives. Nevertheless, the worldwide escalation in ammonia costs has led to announcements of adoption of the split pressure process for some new facilities being constructed in Canada [77].

A relatively recent development in catalyst support systems, either replacing half of the normal 5-10 % rhodium-platinum alloy [74] gauzes by non-noble metal supports [78], or by moving to completely non-noble metal catalysts [79] are said to result in economies in catalyst cost without adversely affecting operating efficiency.

Technologies for concentrating nitric acid have been developed to circumvent the azeotropic restrictions to distillative preparation by obtaining a much higher concentration nitric oxide, nitrogen dioxide gas stream from the acid condenser, of the order of an $NO_x$ vapour pressure of 2-3 bars [74]. Absorption of this gas in cold concentrated (68 %) nitric acid yields a superazeotropic product containing 80 % or more nitric acid. Distillation of this now gives an overhead stream of 96-99 % nitric acid, requiring only six theoretical plates for separation from a column bottom stream of the azeotropic composition (Figure 9.7). Brief operating details of this Espindesa process have recently been announced [80].

### 9.2.6 Emission Control Features

The chief environmental problem of nitric acid plant operation is discharge of residual nitric oxide and nitrogen dioxide from the vent stack of the absorber. For a high pressure process of the 60's, without emission controls such as described here, the total concentration of nitrogen oxides discharged typically amounted to about 0.3 % by volume [63] with a more detailed breakdown given in Table 9.9. Nitric acid plants are not the only source of nitrogen oxides ($NO_x$) but they do correspond to a relatively large point source of emissions, unlike the more diffuse discharge resulting from automobiles, for example. The need to regulate $NO_x$ discharges arises from the implication of nitrogen oxides in photochemical air pollution

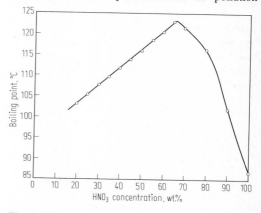

**Figure 9.7.** Plot of the boiling point of nitric acid versus concentration

**Table 9.9** Typical composition of nitric acid plants vent gases, prior to expansion ($\sim$ 30 °C; 7.4 bars) [a]

|  | Percent, Without controls | Percent with methane catalytic reduction |
|---|---|---|
| nitric oxide | 0.05 – 0.20 | 0.01 – 0.20 |
| nitrogen dioxide | 0.05 – 0.20 | |
| oxygen | 3.0 | 0.5 – 2.0 |
| water | 0.6 – 0.7 | 2.5 – 5.0 |
| nitrogen | 96 + | 92 – 93.5 |
| nitric acid | < 0.0 | < 0.0 |
| $CO_2$ | 0.0 | 0 – 2.0 |

[a] Compiled from data of [63, 89, 83, and 85].

problems as well as from the toxic effects of nitrogen dioxide itself, the influence of which may be felt during periods of little air flow when localized build-up may occur.

Nitrogen dioxide, a brown irritating gas existing in equilibrium with dinitrogen tetroxide at ordinary ambient conditions, (Equation 9.39) is dangerous to man at concentrations above 50 ppm [81]. An industrial hygiene standard of 5 ppm for an eight hour workday has been set. Colourless nitric oxide is not itself an irritant but at high concentrations (1,000-2,000 ppm) in air it can cause loss of consciousness and convulsions [81]. But even though nitric oxide *itself* is not frequently a problem, and its discharge is not immediately visible, it does indirectly contribute to atmospheric nitrogen dioxide loadings because the nitric oxide in the vent gases is relatively rapidly oxidized to nitrogen dioxide on exposure to the additional oxygen present in air (Equation 9.38). And nitrogen dioxide, if present in vent gases above about 100 ppm (by volume), gives a visible brown discharge plume from the vent stack.

First generation emission control measures are centered on catalytic reduction of nitrogen oxides using methane (natural gas) or hydrogen (Figure 9.8). For methane reduction, tail gases from the absorber vent are preheated to $\sim$ 400 °C and then blended with the appropriate proportion of methane before passage over platinum or palladium catalytic surfaces for reduction [63]. The concentration of nitrogen oxides is collectively decreased by about 90 % (Table 9.9), from about 0.3 % to 0.01-0.2 % (by volume) depending on conditions, from the reactions of equations 9.46 to 9.48.

$$CH_4 + 2\,O_2 \xrightarrow{\text{cat.}} CO_2 + 2\,H_2O \qquad (9.46)$$

$$\underset{\text{brown}}{CH_4 + 4\,NO_2} \rightarrow 4\,NO \uparrow + CO_2 + H_2O \quad (9.47)$$

$$CH_4 + 4\,NO \rightarrow 2\,N_2 + CO_2 + 2\,H_2O \qquad (9.48)$$

Methane is first rapidly oxidized by the excess oxygen normally present in the vent gases (for reoxidation of nitric oxide to nitrogen dioxide in the absorbers) generating heat but doing nothing to alleviate emissions. Nitrogen dioxide is also rapidly reduced to nitric oxide which at least serves to remove the brown colour from the vent stack discharge, a "decolourization" level of control. And if the reduction unit is large enough to provide adequately low space velocities for the much slower nitric oxide reduction reaction (Equation 9.48), and the added methane is also sufficient to do this then true "abatement" is achieved, at additional in-

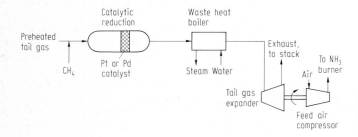

**Figure 9.8.** One emission control system for nitric acid plant tail gas cleanup via catalytic reduction

crements of capital and fuel cost (Figure 9.8) [82, 83]. Fortunately, some of the additional fuel cost of either level of control may be recovered from steam generation and from the energy boost provided to the tail gas expander by the heated, rather than near ambient temperature exhaust gas discharge.

The present regulatory requirements for nitric acid plant discharges in the U.S.A. now stand at 1.5 kg of $NO_x$ (as $NO_2$ equivalent) per tonne of acid, equating to $\sim 230$ ppm $NO_x$, and the discharge must be colourless (opacity of 10 % or less), for new plants [83, 2, 84]. For older plants a maximum 400 ppm (by volume $NO_x$ requirement is likely to be instituted [84]. For European jurisdictions the requirements vary but generally lie near the proposed requirements for older U.S. plants, whereas the U.S.S.R., probably because of either very large single train facilities or because of a concentration of several fertilizer works in close proximity have set a much tighter requirement of 0.55 kg $NO_x$ (calculated as $NO_2$) discharge rate per tonne of acid [84].

Only with very careful operation is the methane catalytic reduction route capable of meeting the more stringent regulatory requirements. Capital cost considerations dictate the maximum economic size of absorbing towers so that early designs, while ammonia and energy costs were low and regulatory requirements were non-existent, were sufficient to only attain 0.1 to 0.3 % $NO_x$ in the tail gases. With the present changed design requirements, new plants with larger absorbers and employing refrigeration chilled water systems particularly on the upper trays of the absorbers are able to reach tail gas $NO_x$ concentrations of < 200 ppm directly [84]. Use of booster compressors to permit absorption at high pressures has also been found to be an economic measure to improve absorption efficiencies, at least for very large plants [2].

From a raw material utilization standpoint as well as for fuel conservation (methane or hydrogen) reasons, discharge controls integral with the process absorbers have obvious advantages, and for new plants these measures are worthy of consideration. But for older facilities having to meet with post-construction regulatory requirements the options are mainly a choice of add-on units. Selective reduction of nitric oxide and nitrogen dioxide by ammonia using alumina-supported platinum catalysts which primarily reduce $NO_x$ compounds

without requiring sufficient ammonia to consume the residual oxygen [85, 2] (Equations 9.49, 9.50), has been shown to be a feasible alternative.

$$6\,NO + 4\,NH_3 \rightarrow 5\,N_2 + 6\,H_2O \qquad (9.49)$$

$$6\,NO_2 + 8\,NH_3 \rightarrow 7\,N_2 + 12\,H_2O \qquad (9.50)$$

Overall this procedure has been shown to confer fuel and capital cost advantages while meeting stringent regulatory requirements [85]. The details of the processing train used are very similar to that given for methane reduction in figure 9.8.

Physical absorption in scrubbers employing water or nitric acid has also been employed [84, 63], as has chemical absorption by alkaline solutions or solutions of urea in water, and all have been shown capable of reducing tail gas $NO_x$ concentrations to below 200 ppm for some combinations [86, 87, 88]. Physical adsorption on molecular sieves (not activated carbon because of explosion hazards) has also been experimentally demonstrated and found to be a feasible control method [89]. In experimental bench scale runs using dried tail gas at 7 bars pressure to maximize sieve $NO_x$ adsorption capacity, about 30 % of the nitrogen oxides were adsorbed. Adsorption, both on molecular sieves and in liquid physical adsorption systems was improved if nitric oxide and nitrogen dioxide were both present in the tail gas rather than just nitric oxide, perhaps from the formation of dinitrogen trioxide ($N_2O_3$). An advantage of the use of all of these physical adsorption systems is that the $NO_x$ collected is recoverable on regeneration of the adsorbent.

Comparative summaries of all of these abatement measures together with brief consideration of variations of these have been recently published [85, 84, 2, 63].

# 9.3 Commercial Ammonium Nitrate

## 9.3.1 Ammonium Nitrate Background

Production of ammonium nitrate is the largest single end use for nitric acid and has ranked 10th to 12th in volume of all inorganic chemicals produced in both Canada and the U.S.A. over the last ten years (Table 9.10). These high levels of production have been developed because of the importance of ammonium nitrate as a fertilizer component and as an ingredient in explosives, combined with its

**Table 9.10** Production of ammonium nitrate and urea in selected countries [a]

|  | | Thousands of tonnes | | | | | | | |
|---|---|---|---|---|---|---|---|---|---|
|  | | 1970 | 1975 | 1976 | 1977 | 1978 | 1979 | 1980 | 1981 | 1982 |
| Canada: | $NH_4NO_3$ |  | 875 | 931 | 1 019 | 960 | 915 | 873 | 1 212 | 1 018 |
|  | Urea |  | 329 | 347 | 887 | 1 342 | 1 212 | 1 274 | 1 303 | 1 231 |
| Japan: | $NH_4NO_3$ | 37 | 115 | 114 | 137 | 139 | 133 | 123 | 116 | 108 |
|  | Urea | 2 463 | 3 162 | 1 572 | 1 718 | 2 090 | 2 083 | 1 770 | 1 483 | 1 187 |
| U.S.A.: | $NH_4NO_3$ | 5 938 | 6 430 | 6 519 | 6 511 | 6 541 | 6 843 | 7 792 | 8 107 | 6 651 |
|  | Urea | 2 829 | 3 446 | 4 080 | 4 601 | 5 691 | 6 374 | 7 103 | 7 350 | 5 380 |

[a] Compiled from data supplied by Verband der Chemischen Industrie e.V. (VCI), calculated from issues of Canadian Chemical Processing, Chemical and Engineering News, and obtained from [126 and 127].

simplicity of production and low cost, $ 100-126 per tonne in the U.S.A. in 1981 [90]. Fertilizer value is derived both from the high nitrogen content ($\sim$ 33 %) split between fast acting nitrate nitrogen and the slower acting ammonium nitrogen, plus the ease of marketing a solid product.

Explosives applications have been varied from blends with nitroglycerin to make this safer for use in mines [8], to admixtures with smokeless powders and with nitro compounds such as trinitrotoluene, to the more recent developments of blends of low density ammonium nitrate prills with fuel oil (so-called A.N.F.O. explosives), or in water slurries with metal powder boosters producing low cost "heaving" power [91]. All explosives applications rely on the large volume of gas released on vigorous detonation of the ammonium nitrate component (Equation 9.51).

$$2\,NH_4NO_{3(s)} \rightarrow 2\,N_{2(g)} + 4\,H_2O_{(g)} + O_{2(g)} \quad (9.51)$$

The oxygen is consumed by the fuel oil or a readily oxidizable metal such as aluminum or magnesium present in the explosive formulation contributing further heat and power to the detonation process. Means to improve the normal handling of ammonium nitrate to decrease the explosion risk [92, 2] under ordinary shipping and storage conditions have been proposed [93, 94].

Under milder conditions a small scale but important application is the preparation of inhalation grade nitrous oxide for use as an anaesthetic by heating pure ammonium nitrate from $\sim$ 200 to 250 °C (Equation 9.52).

$$NH_4NO_3 \rightarrow N_2O + 2\,H_2O \quad (9.52)$$

In this manner the anaesthetic gas is obtained virtually free of nitric oxide or nitrogen dioxide.

### 9.3.2 Production of Ammonium Nitrate

Early methods employed simple neutralization of concentrated aqueous ammonium hydroxide by economic concentrations of nitric acid in the correct proportions, followed by evaporation of much of the water and crystallization of the product [8]. However, because of high capital, labour, and energy costs of this process relative to newer procedures this batch process is being superseded by the Fauser process or variants, which contact ammonia gas with concentrated nitric acid and use the heat of reaction to evaporate a part of the original water [18] or by the Stengel Process, in which the preheat temperatures of ammonia and nitric acid are sufficiently high that complete evaporation of the residual water occurs on contributing the additional heat of reaction to this [18, 22].

The Stengel process, which initially involves blending of ammonia gas preheated to 145 °C with $\sim$ 60 % nitric acid preheated to 165 °C under pressure, in a packed stainless steel reactor (Figure 9.9) is a practised industrial procedure, yet on the face of it unnerving to chemists. By doing this, sufficient additional heat is evolved from the neutralization reaction to nearly completely evaporate the residual water (Equation 9.53) [95].

$$NH_{3(g)} + HNO_{3(aq)} \rightarrow NH_4NO_{3(aq)} \quad (9.53)$$
$$\Delta H = -86.1 \text{ kJ } (-20.6 \text{ kcal})$$

The steam produced [2] is removed from the ammonium nitrate (m.p. 170 °C) via the vortex finder of cyclone separator, and the 99 + % molten salt proceeds to either a cooled stainless steel belt to produce a flaked product (Figure 9.10), or to a 30 m prilling tower where passage of the droplets of melt through a countercurrent dry air stream produces shot-sized prills (beads) of

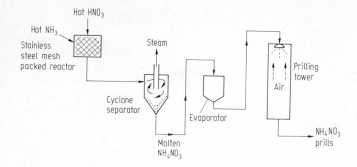

**Figure 9.9.** Ammonium nitrate by the Stengel process. Intermediate evaporation before prilling or granulation would be employed for the production of a high density product

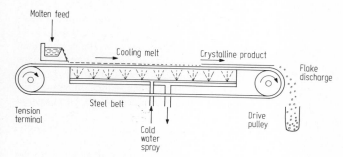

**Figure 9.10.** Steel belt cooler for the preparation of a flaked ammonium nitrate (or other) product

ammonium nitrate [96]. Concentrated hot solutions of ammonium nitrate are explosively sensitized by the presence of traces of acid so that care is taken in these latter stages of working to add sufficient ammonia to the wet melt to keep the pH above about 5. The product is hygroscopic so that it is often coated with an anti-caking compound such as clay or diatomaceous earth before being placed in bulk storage in large moisture-tight silos. Bulk shipment is by covered gondola rail cars, and smaller quantities in vapour tight bags of polyethylene or waxed paper.

## 9.4 Production of Urea

The first synthesis of urea was conducted by Wöhler in 1828 via the heating of ammonium cyanate (Equation 9.54).

$$NH_4OCN \rightarrow H_2NCONH_2 \qquad (9.54)$$

Not only did this preparation mark the first synthesis of urea, but it marked another milestone by being the first time a compound produced by living things was synthesized in the laboratory. As such this event served to open up the study of organic compounds, formerly almost a sacrosanct area, to emerge today with a vast area of products and services based on this area of chemistry.

Production of urea on a commercial scale did not assume much significance until about 1930 and early processes were based on calcium cyanamide hydrolysis in the presence of a base (Equations 9.55, 9.56).

$$CaNCN + 2 H_2O \rightarrow Ca(OH)_2 + H_2NCN \quad (9.55)$$

$$H_2NCN + H_2O \xrightarrow{OH^-} H_2NCONH_2 \qquad (9.56)$$

But since about 1950 most urea producing units have been based on ammonia-carbon dioxide feedstocks put through high pressure equipment of various types. Over the last ten years in North America urea production volumes have grown faster than ammonium nitrate for the supply of fertilizer nitrogen (Table 9.10) [97].

Urea from ammonia and carbon dioxide in the various process sequences operated requires first, reaction of these feedstocks under a pressure of 100 to 200 bars to form ammonium carbamate (Equation 9.57).

$$NH_{3(g)} + CO_{2(g)} \rightarrow$$
$$H_2NCOOH \xrightarrow{NH_3} H_2NCOONH_{4(s)} \qquad (9.57)$$

carbamic       ammonium
acid           carbamate
$\Delta H_{25°} = -159$ kJ ($-38$ kcal)

This thermodynamically favourable step of the process is followed by an endothermic thermal decomposition of ammonium carbamate, in a concentrated solution, to give 50-60 % conversion to urea (Equation 9.58).

$$H_2NCOONH_4 \rightarrow H_2NCONH_2 + H_2O \qquad (9.58)$$
$$\Delta H = + 31.4 \text{ kJ } (+ 7.5 \text{ kcal})$$

A "once-through" modification of a urea process is given in Figure 9.11. Recycle of unconverted ammonium carbamate, and of ammonia and carbon dioxide formed in part from ammonium carbamate decomposition (Equation 9.59) is practised in other variants [98, 99].

$$H_2NCOONH_4 + H_2O \rightarrow (NH_4)_2CO_3 \rightarrow$$
$$2\,NH_3 + CO_2 + H_2O \qquad (9.59)$$

This assists in minimizing the environmental impacts of operation of urea plants [100]. However, despite these recycle measures there is still a small ammonia loss rate, estimated to average 0.6 kg per tonne of urea produced [101, 1]. This loss rate is, however, well below one published guideline for urea plants of 2.7 kg $NH_3$ per tonne [61].

The principal end use of urea is in the provision of the nitrogen values in solid fertilizer formulations although some is also converted to biuret ($H_2NCONHCONH_2$, also called carbamoyl urea) and to a sulfur derivative which are sold as Kedlor [102] and Urasil [103] cattle feed supplements, respectively. Urea is also used to a lesser extent in the manufacture of plastics components, such as melamine, and as a component of urea-formaldehyde resins used as adhesives. It is also used as a component of foamed formulations as a rigid insulating material [97].

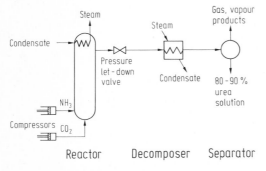

**Figure 9.11.** Production of urea by a simple, one-pass reactor-separator system

# 9.5 Synthetic Fertilizers

Six of the twelve largest volume inorganic chemicals produced in the U.S.A. in 1980 directly or indirectly form part of all the various chemical fertilizer formulations produced [104]. Such is the importance and significance of the fertilizer market in terms of the volume of activity of the chemical process industries.

The value of the chemical fertilizer product, of course, lies in the significantly enhanced agricultural productivity possible with their use [105]. This advantage is more easily afforded by the large highly mechanized farming operations of the western world, so that the per capita consumption in these countries runs from 20 to ~ 100 kg per capita per year as compared to usage rates in the neighbourhood of 1 kg per capita per year for countries such as Bangladesh, mainland China, India and Pakistan. While the per capita fertilizer consumption for these latter countries is small, the effectiveness of chemical utilization to foodstuffs and other agricultural produce is probably greater, and the importance of chemical fertilizer applications in the stabilization of these significantly agricultural economies should not be underestimated.

## 9.5.1 Fertilizer Composition

Plant growth and development requires a total of nine major nutrients or macronutrients to proceed normally [106]. Three of these, carbon, hydrogen, and oxygen are obtained from air (as carbon dioxide) and soil (as water). Another three, calcium, magnesium and sulfur, are present to a sufficient extent in most ordinary soils. And another three, fixed nitrogen, phosphorus, and potassium are present to a small extent in most soils but are rapidly depleted from the soil when the bulk of an agricultural plant is removed from the soil on harvesting. Returning the unused plants to the soil can serve to return some of these elements. But removal from continued cropping [106] as well as from natural soil bacterial processes which causes losses of nitrogen in particular through volatilization as ammonia or nitrogen, requires the regular addition of these elements to the soil to maintain fertility. Compost, manure, and specially treated sewage sludges comprise "organic" methods to return these elements in low analysis forms, but in most situations supplementation by the use of chemical fertilizers can achieve significantly improved yield benefits.

Thus, the major constituents or macronutrients present in most commercial fertilizers are nitrogen, which contributes to early plant development and greening, phosphorus, which assists with early growth and seed or fruit development, and potassium which is of particular value in the production by plants of cellulose and starches. These three principal ingredients are specified in order in the fertilizer analysis: nitrogen, as % N; phosphorus, as % $P_2O_5$ (equivalent) and potassium, as % $K_2O$ (equivalent) as three numbers in sequence separated only by hyphens. Thus, 15-15-15, one of the popular high analysis fertilizers, would contain 15 % N, 15 % $P_2O_5$, and 15 % $K_2O$. For simplicity in shipping and handling, and particularly as machinery for efficient distribution of high analysis fertilizers has been developed for large scale agricultural operations, these more concentrated fertilizers have increasingly occupied an important place in the range of fertilizer products marketed (Tabel 9.11). Some of the other common commercial designations are 0-20-20, 2-12-16, 3-12-12, etc.

In addition to the conventional listing of major constituents a commercial blend may be formulated to contain one or more of the secondary constituents calcium, magnesium, or sulfur, in order to correct gross local soil deficiencies. These are normally specified in a separate listing. The importance of these too cannot be overlooked as, for example, fertilization with soluble sulfate, of a sulfur deficient soil was shown to give over 1,100 % rapeseed (canola) crop improvement in just one example [107].

There are also the trace elements boron, copper, iron, manganese, zinc, molybdenum and occasionally chlorine which are important in some of the plant metabolic pathways [106] and which may be deficient in the soils of some areas [108]. These are

**Table 9.11** Breakdown of commercial mixed fertilizers shipped in Canada in 1969 [a]

| Analysis | Thousand tonnes | Analysis | Thousand tonnes |
|---|---|---|---|
| 0 – 20 – 20 | 29.0 | 5 – 10 – 15 | 22.0 |
| 2 – 12 – 10 | 4.2 | 5 – 20 – 10 | 43.8 |
| 2 – 12 – 12 | 11.6 | 5 – 20 – 20 | 102,6 |
| 2 – 16 – 6 | 1.4 | 6 – 12 – 8 | 5.7 |
| 3 – 15 – 6 | 3.3 | 6 – 12 – 12 | 86.1 |
| 4 – 12 – 10 | 2.2 | 10 – 10 – 10 | 129.4 |

[a] From [128].

added as required to special fertilizer formulations and are also specified separately from the major constituents.

In addition to the above chemically important constituents many fertilizers have fillers and/or conditioners added. Fillers commonly employed include sand, gypsum, or ground phosphate rock (only very slightly water soluble, hence *not* part of $P_2O_5$ specified) which serve to assist in even distribution over small areas such as in home gardens. Sand may also serve a conditioning function by assisting in the break-up of clays, but to be effective usually requires higher application rates than normally used for fertilizers. Some other types of conditioners which serve in different ways to improve soil structure are dried peat moss, specially processed wood waste, some types of divided synthetic polymer particles (e.g. Krylium), and expanded mica (vermiculite).

### 9.5.2 Formulation of Major Active Constituents

Appropriate ingredients for the nitrogen component include ammonia gas itself which, containing 82.5 % N, comprises the highest feasible nitrogen content available in a commercially useful compound. But this is restricted to application methods specifically designed for the purpose. Ammonia solutions, comprising essentially concentrated ammonium hydroxide of about 25 % N require a little less specialized equipment but still is generally restricted to single component applications. Urea, too, (46.6 % N) is occasionally applied to croplands as an aqueous solution.

But for combined nutrient applications, solid forms are normally preferred for ease and lower cost of shipment, and convenient broadcasting. Thus urea and ammonium nitrate (35 % N) serve as nitrogen-containing constituents in solid fertilizer blends. As ammonium sulfate (21.2 % N) nitrogen addition also adds soluble sulfate as a separately specified secondary nutrient. Of course nitrogen may also be added in forms chemically combined with other major constituents such as in diammonium phosphate ($(NH_4)_2 HPO_4$) or potassium nitrate, for example. A few plants, such as peas, clover, and alfalfa have nodules containing nitrogen-fixing bacteria attached to their roots and are able to utilize elemental nitrogen from the air to provide for their nitrogen requirements. Genetic engineering experiments may increase the number of species of plants able to do this [109, 110, 111, 112], and by so doing decrease the agricultural demand for synthetic fertilizer nitrogen.

As already mentioned, finely ground phosphate rock ($Ca_3(PO_4)_2$) is occasionally added to fertilizer formulations as a diluent or filler. However, because phosphate rock has only very limited water solubility its action as a phosphate nutrient is little, and very slow. Hence, this ingredient is not allowed to be included in the % $P_2O_5$ analysis appearing on fertilizer packaging. Before even filler use phosphate rock is normally defluorinated by heating with silica and steam (see phosphate rock) to decrease the risk of soil contamination by fluoride.

A more soluble form of phosphate ($Ca(H_2PO_4)_2$) variously called "superphosphate (of lime)", monocalcium phosphate, or calcium dihydrogen phosphate, is made by the acidulation of phosphate rock with ~ 65 % sulfuric acid (Equation 9.60).

$$CaF_2 \cdot 3\,Ca_3(PO_4)_2 + 7\,H_2SO_4 +$$
$$14\,(\text{or more})\,H_2O \rightarrow$$
$$3\,Ca(H_2PO_4)_2 + 7\,CaSO_4 \cdot 2\,H_2O + 2\,HF \uparrow$$

| calcium dihydrogen phosphate | gypsum | collected in scrubbers |
|---|---|---|

$$(9.60)$$

The thick slurry of calcium dihydrogen phosphate and gypsum obtained is piled outside for 8-10 weeks to complete the reaction ("cure") after which it is milled to produce a granular product, and then bagged for market. Thus the product, superphosphate, contains the inert gypsum component as well as the fertilizer active calcium dihydrogen phosphate, which is some one thousand times more soluble than calcium phosphate. The theoretical content of active material for the composition resulting from the reaction of equation 9.57 is 21.7 % $P_2O_5$ equivalence, just a shade more than the actual 18-21 % $P_2O_5$ range for the commercial product. The presence of silica or iron oxide in the original phosphate rock, plus the formation of the monohydrate of calcium dihydrogen phosphate ($Ca(H_2PO_4)_2$ x $H_2O$) or some dicalcium hydrogen phosphate ($CaHPO_4$) all serve to cause these slightly lower fertilizer values.

As with the nitrogen component, various other alternatives are available for the phosphate rock acidulation step. For instance if acidulation is conducted with nitric, instead of sulfuric acid, calcium nitrate is formed instead of gypsum (Equation 9.61),

$$CaF_2 \cdot 3\,Ca_3(PO_4)_2 + 14\,HNO_3 + 3\,H_2O \rightarrow$$
$$3\,Ca(H_2PO_4)_2 \cdot H_2O + 7\,Ca(NO_3)_2 + 2\,HF \uparrow$$
$$(9.61)$$

producing both $P_2O_5$ and nitrogen value from the single procedure. This mixture may be dried and used as obtained, or it may be partially or fully ammoniated to raise the nitrogen analysis to some higher value. Full ammoniation would correspond to the stoichiometry of equation 9.62.

$$3\,Ca(H_2PO_4)_2 \cdot H_2O + 7\,Ca(NO_3)_2 +$$
$$12\,NH_4OH \rightarrow 3\,Ca(NH_4)_4(PO_4)_2 +$$
$$7\,Ca(NO_3)_2 + 12\,H_2O \qquad (9.62)$$

Of course the ammoniation may also be carried out independently of the use of nitric acid if it is desired that one have only slower acting ammonia nitrogen rather than a mixed nitrogen contribution in the formulation.

Another phosphate contributing alternative is to use phosphoric instead of sulfuric acid for initial acidulation of phosphate rock. In this way virtually the only solid product of acidulation is calcium dihydrogen phosphate, free of gypsum diluent, thus obtaining a product with about three times the "available $P_2O_5$" of superphosphate (Equation 9.63).

$$CaF_2 \cdot 3\,Ca_3(PO_4)_2 + 14\,H_3PO_4 + 10\,H_2O \rightarrow$$
$$10\,Ca(H_2PO_4)_2 \cdot H_2O + 2\,HF \uparrow \qquad (9.63)$$

For this reason, this is the composition referred to commercially as "triple superphosphate". This too may be ammoniated, or mixed with other nitrogen-containing constituents such as urea or ammonium nitrogen to provide both nitrogen and phosphorus for plant growth.

Elemental phosphorus itself has been tested as a fertilizer in its more stable red allotropic form but was not found to be a commercially useful prospect [113, 114].

Potassium chloride, "muriate of potash", is the principal component used as such to supply the potassium macronutrient in fertilizer formulations. However, some crops with a low chloride tolerance such as tobacco may require the chloride of potassium chloride replaced by nitrate, carbonate, or sulfate anion to be useful (e.g. Equation 9.64).

$$KCl + HNO_3 \rightarrow KNO_3 + HCl \uparrow \qquad (9.64)$$

### 9.5.3 Environmental Aspects of Fertilizer Production and Use

Emissions during the production phase of most fertilizer constituents have been discussed with the details of the processes concerned with producing the chemical. However, the production of the phosphatic fertilizers superphosphate and triple superphosphate in particular involve procedures with potential emission problems which differ somewhat from the emission control problems of phosphoric acid production [115]. There is the potential loss of fluoride primarily as calcium fluoride in dusts from the primary grinding of phosphate rock, or from milling of the cured product. Hydrogen fluoride, silicon tetrafluoride and fluosilicic acid vapours are also formed when acid is added to the phosphate rock. Dusts may be controlled in cyclones, for which 80-95 % mass collection efficiencies for a particle size range comprising 40 %, < 10 $\mu$m has been quoted for this application [63]. Vapours may be controlled by dry methods involving adsorption on chalk or limestone followed by capture in a cyclone, for which 95 % control has been claimed [115]. Or, various wet methods employing water scrubbing combinations of various types may be used [116]. The most efficient of the wet systems achieve some 98 % fluoride vapour containment [115]. Of course, while scrubbing of waste gases with water or aqueous solutions can effectively clean the waste gas discharge of contaminants, it also generates an aqueous effluent high in fluoride which must be contained and treated prior to discharge (see phosphoric acid).

Use of chemical fertilizers on agricultural land does introduce the risk of contamination of surface waters from the leaching of soluble constituents from the soil after application. Important to minimize this source of contamination are the use of good farming practices, abiding by the recommended fertilizer application which has the added advantage of providing the best crop return for the fertilizer investment, and maintaining soil structure through measures such as plowing in a fallow crop. The recommendation to maintain soil structure not only serves to minimize erosion losses of the soil itself, but the fiber content provides an insoluble retained matrix on which complexation assists in holding the soluble fertilizer constituents in place. There have also been long term concerns expressed that soil applications of fixed nitrogen in various forms provides more substrate for ammonia, and

nitrous oxide losses to the air via bacterial processes [117, 118, 1]. These sources of fixed nitrogen, contributing about one third of the total atmospheric loading (the remainder arising from natural sources), are in turn speculated to effect depletion of the ozone layer through photochemical reactions.

## Relevant Bibliography

1. K.V. Reddy and A. Husain, Vapour-Liquid Equilibrium Relationship for Ammonia in Presence of Other Gases, Ind. Eng. Chem. Process Res. Dev. *19*, 580 (1980)
2. R.M. Harrison and H.A. McCartney, Some Measurements of Ambient Air Pollution Arising from the Manufacture of Nitric Acid and Ammonium Nitrate Fertilizer, Atmos. Envir. *13*, 1,105 (1979)

## References

1. Ammonia, Subcommittee on Ammonia, Committee on Medical and Biologic Effects of Environmental Pollutants, National Research Council, University Park Press, Baltimore, 1979
2. Kirk-Othmer Encyclopedia of Chemical Technology, 3rd edition, John Wiley and Sons, New York, 1978
3. T.H. Chilton, Strong Water; Nitric Acid: Sources, Methods of Manufacture, and Uses, MIT Press, Cambridge, Mass. 1968
4. R.N. Shreve, Chemical Process Industries, McGraw-Hill, New York, 1945
5. J.W. Dobereiner, Ann. Chim. *24*:2 (2), 91 (1823). Cited by reference 6
6. Kirk-Othmer Encyclopedia of Chemical Technology, 2nd edition, Interscience, New York, 1968. Volume 2, page 268
7. V. Sauchelli, Fertilizer Nitrogen: Its Chemistry and Technology, American Chemical Society Monograph no. 161, Reinhold, New York, 1964
8. F.S. Taylor, A History of Industrial Chemistry, Reprint Edition, Arno Press, New York, 1972
9. B. Timm, Chem. Ind. (London), 274, March 12, 1960
10. Encyclopedia of Chemical Technology, R.E. Kirk and D.F. Othmer, editors, Interscience, New York, 1941
11. C. Matasa and E. Tonca, Basic Nitrogen Compounds, Chemistry, Technology, Applications, Chemical Publishing Co., New York, 1973
12. F. Haber, J. Ind. Eng. Chem. *6*, 325, April 1914
13. J.A. Almquist and E.D. Crittenden, Ind. Eng. Chem. *18*, 1,307, Dec. 1926

14. H. Hinrichs, Brit. Chem. Eng. 12(11), 1,745, Nov. 1867

15. P.H. Emmett and J.T. Kummer, Ind. Eng. Chem. 35, 677 (1943)

16. R.A. King, Nitrogen (London) No. 35, 22, May 1965

17. Methane Reforming to Stay, Chem. Eng. News 59(35), 39, Aug. 31, 1981

18. F.A. Lowenheim and M.K. Moran, Faith, Keyes and Clark's Industrial Chemicals, 4th Edition, Wiley Interscience, New York, 1975

19. A.H. Trotter, U.S. Patent 1,852,763, April 5, 1932 to Atmospheric Nitrogen Corp. Chem. Abstr. 26, 3,077 (1932)

20. R. Wood, Proc. Eng., 7, Sept. 1976

21. J.A. Finneran, L.J. Buividas and N. Walen, Hydroc. Proc. 51(4), 127, April 1972

22. R.N. Shreve and J.A. Brink, Jr., Chemical Process Industries, 4th Edition, McGraw-Hill, New York, 1977

23. J.S. Vanick, Proc. Am. Soc. Testing Materials 24(2), 348. (1924). Chem. Abstr. 19, 1,395 (1925)

24. J.S. Vanick, Trans. Am. Soc. Steel Treating, 4, 62 (1923), Chem. Abstr. 17, 3,313 (1923)

25. K.E. Weale, Chemical Reactions at High Pressures, Spon, London, 1967, page 100

26. M.R. Louthan, Jr., Proc. Ind. Corrosion, 126 (1975)

27. M.R. Louthan, Jr., G.R. Caskey, Jr., J.A. Donovan, and D.E. Rawl, Jr., Mater. Sci. Eng. 10, 357 (1972)

28. C.F. Britton, Chem. in Brit. 17(3), 108, March 1981

29. System Monitors Hydrogen, Can. Chem. Proc. 65(2), 53, March 27, 1981

30. M.D. Wynne, Chemical Processing in Industry, Royal Instit. Chem. Monograph No. 16, London, 1970, page 5

31. Argon Recovery Plant, Chem. Eng. News, 58(4), 32, Jan. 28, 1980

32. W.J. Storck, Chem. Eng. News, 56(34), 11, Aug., 21, 1978

33. Methane Reforming to Stay, Chem. Eng. News 59(35), 39, Aug. 31, 1981

34. Gas Efficiency Key to Fertilizer, Can. Chem. Proc. 64(7), 4, July 1980

35. Ammonia Plants More Efficient, Can. Chem. Proc. 64(10), 10, Oct. 1980

36. F. Forster, Chem. Eng. 87, 62, Sept. 8, 1980

37. T.A. Ring, W.L. Mann, Y.S. Tse, Chem. Eng. Prog. 66(12), 59, Dec. 1970

38. M. Lauzon, Can. Chem. Proc. 65(3), 42, May 1981

39. L.C. Daigre, III and G.R. Nieman, Chem. Eng. Prog. 70(2), 50, Feb. 1974

40. G.P. Williams and J.G. Sawyer, Chem. Eng. Prog. 70(2), 45, Feb. 1974

41. T. Wett, Oil and Gas J. 70(19), May 8, 1972

42. J.G. Sawyer, G.P. Williams, Chem. Eng. Prog. 70(2), 62, Feb. 1974

43. Ammonia Process, Chem. Eng. News 56(48), 19, Nov. 27, 1978

44. Ammonia Unit, Chem. Eng. News 58(33), 24, Aug. 18, 1980

45. R.F. Giles and L.D. Gains, Instr. Techn. 24, 41, Oct. 1977

46. R. Banks, Chem. Eng. 84D, 89, Oct. 10, 1977

47. Monsanto Sells Hydrogen-Recovery, Chem. Eng. News 58(18), 8. May 5, 1980

48. Japanese Technologists, Can. Chem. Proc. 59(5), 8, May 1975

49. A.V. da Rosa, Chemtech 8, 28, Jan. 1978

50. S. Strelzoff, Hydroc. Proc. 53(10), 133, Oct. 1974

51. D. Netzer and J. Moe, Chem. Eng. 84D, 129, Oct. 24, 1977

52. F. Brown, Hydroc. Proc. 56(11), 361, Nov. 1977

53. TVA Ammonia-from-coal Project, Chem. Eng. News 57(33), 27, June 4, 1979

54. D.E. Nichols and P.C. Williamson, 2nd Chem. Congress of N. Amer. Continent, Las Vegas, Aug. 24-29, 1980. Abstr. FERT-14

55. L.J. Buividas, Chem. Eng. Prog. 77, 44, May 1981

56. D. Nichols and G.M. Blouin, Chemtech 9, 512, Aug. 1979

57. Desert Sand Catalyze Ammonia Formation, Chem. Eng. News 56(46), 7, Nov. 13, 1978

58. Prototype Solar Cell, Chem. Eng. News 55(49), 19, Oct. 3, 1977

59. W.L. Faith, D.B. Keyes, and R.L. Clark, Industrial Chemicals, 3rd edition, John Wiley, Toronto, 1965, page 77

60. Guidelines for the Location of Stationary Bulk Ammonia Storage Facilities, Standards and Approvals Division, Alberta Dept. of the Environment, Edmonton, November 1981, 9 pp

61. Guidelines for Limiting Contaminant Emissions to the Atmosphere from Fertilizer Plants and Related Industries in Alberta, Standards and Approvals Division, Alberta Dept. of the Environment, Edmonton, February 1976, 23 pp

62. Ammonia Plant Waste, Chem. Eng. News 58(36), 51, Sept. 8, 1980

63. Industrial Pollution Control Handbook, H.F. Lund, Editor, McGraw-Hill, Toronto, 1971

64. T. Dear, Chem. Eng. Prog. 70(2), 65, Feb. 1974

65. C.F. Kuhlmann, Liebigs Annalen der Chem. 29, 272 (1839), and reference cited therein

66. W. Ostwald, U.S. Patent 858, 904, July 2, 1907, cited by T.H. Chilton, reference 3

67. The Modern Inorganic Chemicals Industry, R. Thompson, editor, The Chemical Society, London, 1977, page 221

68. F.A. Cotton and G. Wilkinson, Advanced Inorganic Chemistry A Comprehensive Text, 3rd Edition, Interscience, 1972, page 357

69. R.M. Stephenson, Introduction to the Chemical

Process Industries, Reinhold, New York, 1966, page 136

70. M. Bodenstein, Zeits. Electrochem. *34*, 183 (1918)

71. Key Chemicals Nitric Acid, Chem. Eng. News *54* (28), 14, July 5, 1976

72. M.C. Manderson, Chem. Eng. Prog. *68* (4), 57, April 1972

73. Maggie Concentrates Nitric, Chem. Eng. News *36B*, 40, June 9, 1958

74. D.J. Newman and L.A. Klein, Chem. Eng. Prog. *68* (4), 62, April 1972

75. New Generation of Nitric Acid Plants, Chem. Eng. News *54* (52), 33, Dec. 20, 1976

76. L. Hellmer, Chem. Eng. Prog. *68* (4), 67, April 1972

77. CIL Plant Uses New $HNO_3$ Process, Chem. In Can. *29* (8), 13, Sept. 1977

78. A New Support System, Chem. Week *109* (4), 39, July 28, 1971

79. Technology Newsletter, Chem. Week *108* (13), April 1, 1970

80. L.M. Marzo and J.M. Marzo, Chem. Eng. *87*, 54, Nov. 3, 1980

81. Industrial Hygiene and Toxicology, 2nd edition, F.A. Patty, editor, Interscience, New York, 1963, Volume II, Toxicology, page 918

82. R.M. Reed and R.L. Harvin, Chem. Eng. Prog. *68* (4), 78, April 1972

83. G.R. Gillespie, A.A. Boyum, and M.F. Collins, Chem. Eng. Prog. *68* (4), 72, April, 1972

84. W. Frietag and M.W. Packbier, Ammonia Plant Safety *20*, 11, 1978

85. M. Yamaguchi, K. Matsushita, and K. Takami, Hydroc. Proc. *55*, 101, Aug. 1976

86. C.G. Swanson, Jr., J.V. Prusa, T.M. Hellman, and D.E. Elliott, Pollution Eng., *10* (10), 52, Oct. 1978

87. New Units For $NO_x$, Can. Chem. Proc. *59* (12), 26, Dec. 1975

88. Two Processes to Control, Can. Chem. Proc. *59* (11), 4, Nov. 1975

89. W. Joithe, A.T. Bell and S. Lynn, Ind. Eng. Chem. Process Res. Develop. *11* (3), 434, 1972

90. Chemical Marketing Reporter *219* (24), 41, June 15, 1981

91. I. Dunstan, Chem. In Brit. *7* (2), 62, Feb. 1971

92. W.J. Storck, Chem. Eng. News *57* (39), 11, Sept. 24, 1979

93. C. Boyars, Ind. Eng. Chem., Prod. Res. Dev. *15* (4), 308, 1976

94. Nitrate Causes Explosion, Science News Letter 118, Aug. 22, 1959

95. Ammonium Nitrate, Hydroc. Proc. *58* (11), 135, Nov. 1979

96. High Demand For Ammonium Nitrate, Can. Chem. Proc. *61* (7), 8, July 1977

97. Product Profile Urea, Can. Chem. Proc. *61* (10), 50, Oct. 1977

98. V. Lagana and G. Schmid, Hydroc. Proc. *54* (7), 102, July 1975

99. E. Otsuka, S. Inove and T. Jojima, Hydroc. Proc. *55* (11), 160, Nov. 1976

100. E.R. Hoffman and R.K. Fidler, Hydroc. Proc. *55* (8), 111, Aug. 1976

101. T. Jojima and T. Sato, Chem. Age (India) *26*, 524 (1975). Cited by Reference 1

102. Dow Feed-Grade Buiret, Chem. Eng. News *49* (25), 23, June 21, 1871

103. A New Liquid Cattle Feed Supplement, Can. Chem. Proc. *56* (10), 4, Oct. 1972

104. W.J. Storck, Chem. Eng. News *59* (18), 35, May 4, 1981

105. Z.I. Sabry, Chem. In Can. *27* (2), 16, Feb. 1975

106. C.J. Pratt, Scient. Am. *212* (6), 62, June 1965

107. Sulphur Response, Chem. In Can. *24* (7), 5, summer 1972

108. R.J.P. Williams, Chem. in Brit. *15* (10), 506, Oct. 1979

109. New Bacteria Fix $N_2$, Chem. Eng. News *56* (27), 5, July 3, 1978

110. Nitrogen Fixation Research Advances, Chem. Eng. News *58* (49), 29, Dec. 8, 1980

111. Bacteria Could Fertilize Soil, Can. Chem. Proc. *61* (10), 6, Oct. 1977

112. New Bacteria Reduce Fertilizer Need, Chem. Eng. News *55* (36), 18, Sept. 5, 1977

113. H.P. Rotbaum, Outlook on Agric. *5*, 123, 1966

114. H.P. Rotbaum and W. Kitt, N.Z.J. Science 7, 67, 1964

115. K. Karbe, The Chem. Eng. *221*, 268, Sept. 1968

116. N.L. Nemerow, Industrial Water Pollution, Origins, Characteristics, and Treatment, Addison-Wesley, Don Mills, Ontario, 1978, page 613

117. Nitrogen Fertilizers May Endanger Ozone, Chem. Eng. News *53* (47), 6, Nov. 24, 1975

118. Fertilizer May Deplete Ozone Layer, Chem. Eng. News *56* (40), 6, Oct. 2, 1978

119. F. Haber, S. Tamaru and C. Ponnaz, Z. Electrochem. *21*, 89, 128, and 191, 1915, and refs. cited therein.

120. A.T. Larson, J. Am. Chem. Soc. *46*, 367, 1924, and earlier refs.

121. A. Nielson, An Investigation on Promoted Iron Catalyst for the Synthesis of Ammonia, 3rd edition, J. Gjellerups Forlag, Denmark 1968. Cited by reference 2

122. Handbook of Chemistry, 10th edition, N.A. Lange, editor McGraw-Hill, Toronto, 1969, page 915

123. Chem. Eng. News *59* (18), 35, May 4, 1981

124. Verband der Chemischen Industrie

125. Statistics Canada, Catalog 46-002, Supply and Services Canada Ottawa, 1982

126. 1974 Annual Fertilizer Review, Food and Agricultural Organization, United Nations, Rome, 1975

127. Japan Economic Yearbook 1981/82, The Oriental Economist, Tokyo, 1981, page 69, and earlier issues

128. Can. Chem. Proc. *54* (9), 55, Sept. 1970

# 10 Aluminum and Compounds

## 10.1 Historical Background

From its relatively small scale utilization of the order of 1/100th that of copper, lead, and zinc prior to 1900, to its present scale of production of three to five times that of these more traditional non-ferrous metals (Table 10.1), aluminum is a metal that has come of age in the twentieth century. Oersted, in Denmark, is credited with first obtaining impure aluminum in 1825, achieved by the reduction of aluminum chloride with potassium amalgam. Wöhler, two years later, obtained higher purity metal and more fully described its properties. Henri St.-C. Deville put aluminum production into commercial practice in France by 1845 using sodium fusion to reduce aluminum chloride (Equation 10.1).

$$AlCl_3 + 3\,Na \rightarrow Al + 3\,NaCl \qquad (10.1)$$
(conducted as a melt)

By 1852 the metal sold commercially for ca. \$545 per pound, but by various process improvements such as cryolite fluxing of the melt and lower sodium costs, the price had come down to 8 US \$ per pound by 1886 [1].

In the early part of 1886, Paul Heroult in France and Charles Hall in the U.S. independently devised electrolytic methods to reduce combined aluminum to the metal, and made this possible on a large scale. In their procedure, the essence of which is still used today, alumina ($Al_2O_3$), the source of the aluminum, was dissolved in a bath of molten cryolite. Passage of a direct current through this bath, via an electrolytic process (not electrothermal, as used for phosphorus), produced a layer of molten aluminum on the bottom of the cell. This procedure both raised the purity of the metal obtained and at the same time dramatically further lowered the price. The advantages in terms of cost and ease of availability that the Hall-Heroult process contributed coupled with particularly the lightness and corrosion resistance of this metal in many applications now led to rapid advances in consumption which stimulated the growth in capacity of producing units (Table 10.1).

Continued growth in the production of aluminum is unlikely to be hampered by a shortage of mineral since it is estimated that the earth's crust consists of about 8 % aluminum, chiefly occurring as aluminosilicates. Even though aluminum thus ranks as the most abundant metallic crustal element, bauxite ores suitable for aluminum recovery only occur in more limited areas where natural processes such as leaching have served to concentrate the aluminum-containing minerals relative to the average crustal content. Since the free metal is quite chemically reactive it is never found in nature in this form.

With the rapid acceptance of the Hall-Heroult electrolytic method of aluminum production, facilities using this process have tended to be constructed in areas with abundant, low cost electric power (Table 10.2). In addition, to minimize shipping costs it is usual practice to process bauxite, the crude ore, at the mine site to produce alumina in purified form for shipment to smelters. In general this factor has meant that a further requirement for

**Table 10.1** Relative growth rates of world production of some non-ferrous metals [a]

|      | Aluminum | | Copper | | Lead | | Zinc | |
|------|-----------------|---------------------|-----------------|---------------------|-----------------|---------------------|-----------------|---------------------|
|      | thous. tonnes | multiple of 1900 | thous. tonnes | multiple of 1900 | thous. tonnes | multiple of 1900 | thous. tonnes | multiple of 1900 |
| 1900 | 5.7 | 1 | 499 | 1 | 877 | 1 | 479 | 1 |
| 1960 | 4 670 | 819 | 4 400 | 8.82 | 2 630 | 3.00 | 3 070 | 6.41 |
| 1970 | 9 666 | 1 696 | 6 227 | 12.48 | 3 299 | 3.76 | 4 905 | 10.24 |
| 1980 | 15 368 | 2 696 | 6 194 | 12.41 | 2 957 | 3.37 | 4 354 | 9.09 |

[a] Data compiled from [5 and 62].

**Table 10.2** Annual production of aluminum by the world's major producers, in thousands of metric tonnes [a]

| | 1945 | | 1965 | | 1970 | 1975 | 1979 | 1980 [b] | |
|---|---|---|---|---|---|---|---|---|---|
| | mass | % | mass | % | mass | mass | mass | mass | % |
| Canada | 196 | 22.1 | 762 | 11.3 | 964 | 913 | 860 | 1 068 | 6.9 |
| France | 37.2 | 4.2 | 340 | 5.1 | 380 | 383 | 395 | 432 | 2.8 |
| Japan | 5.4 | 6.1 | 323 | 4.8 | 733 | 1 013 | 1 011 | 1 091 | 7.1 |
| Norway | 7.0 | 0.8 | 277 | 4.1 | 530 | 591 | 660 | 651 | 4.2 |
| U.S.A. | 449 | 50.6 | 2 498 | 37.1 | 3 607 | 3 519 | 4 557 | 4 654 | 30.3 |
| U.S.S.R. | 86.3 | 9.7 | 1 279 | 19.0 | 1 098 | 1 479 | 1 751 | 1 787 | 11.6 |
| W. Germany | 20.0 [c] | 2.3 | 234 | 3.5 | 308 | 678 | 742 | 731 | 4.8 |
| Other | 37.3 | 4.2 | 1 014 | 15.1 | 2 046 | 3 450 | 4 587 | 4 954 | 32.3 |
| Total | 887.0 | 100.0 | 6 727 | 100.0 | 9 666 | 12 041 | 14 563 | 15 368 | 100.0 |

[a] List includes all countries whose annual primary aluminum production exceeded 400 000 tonnes in 1980. Compiled and calculated from data of Mineral Yearbooks [59].
[b] Estimates.
[c] Totals for Germany before partition.

the siting of aluminum smelters is that they be located in areas providing easy shipping access to permit economical processed alumina delivery from the mine and usually associated ore processing area to the smelter site. Thus, the production of aluminum from bauxite logically can be considered in two steps: the first, production of high purity alumina from the working of natural bauxite deposits; and second, electrolytic reduction of alumina to the metal.

## 10.2 Production of Alumina from Bauxite: the Bayer Process

Bauxite, the principal ore used for aluminum smelting, is named after Les Baux, Provence, the village where the first deposits were discovered. Australia is now the world's largest producer of bauxite, in 1977 accounting for 35 % of the noncommunist production with some 25.2 million tonnes [2]. Bauxite contains hydrated alumina equivalent to as much as 40-60 % $Al_2O_3$ with much of other siliceous materials leached out over time. But it still contains 10-30 % iron oxide, silica, and other impurities making it unsuitable for direct electrolysis. The first commercial scale recovery of alumina from bauxite was practised by Henri Deville, but by 1900 this was largely replaced by the simpler and more economical process devised by Bayer in Austria, based on caustic extraction. Alumina recovery from bauxite by extraction with sodium hydroxide, now frequently referred to as

the Bayer process, relies on the amphoterism of aluminum for its success. Details of the particular alumina extraction procedure required depend on the particular hydrated form of alumina occurring in the bauxite being processed.

Preliminary to any chemical processing the coarse ore is mechanically reduced to finely-divided form and stirred with the requisite concentration of aqueous base prior to pressure leaching. The alumina from trihydrate bauxite ($Al_2O_3$ x $3H_2O$; natural bayerite or gibbsite) is relatively easy to dissolve out using 15-20 % aqueous sodium hydroxide, under pressure, at temperatures of 120-140 °C (Equation 10.2).

$$Al(OH)_3 + NaOH \rightarrow AlO \cdot ONa + 2H_2O \quad (10.2)$$
$$\text{soluble}$$

But if there is any significant proportion of monohydrate alumina ($Al_2O_3$ x $H_2O$; e.g. diaspore) present in the ore both more concentrated sodium hydroxide (20-30 %) and higher temperatures and pressures, 200-250 °C and up to 35 bars are required [3] to effectively leach out the alumina content (Equation 10.3).

$$AlO \cdot OH + NaOH \rightarrow AlO \cdot ONa + H_2O \quad (10.3)$$

Of course the more severe conditions will also serve to extract trihydrate alumina from bauxite so that the deciding factor in the digestion conditions required is the presence or absence of any significant concentration of monohydrate alumina. Fortunately trihydrate alumina is the dominant aluminum species present in major world deposits in Af-

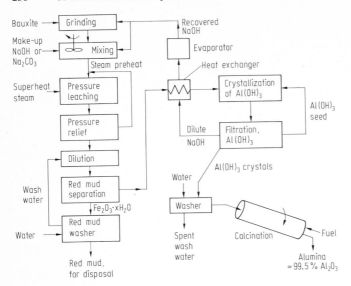

Bauxite → Grinding

Make-up
NaOH or → Mixing
Na₂CO₃

Steam preheat

Recovered
NaOH

Evaporator

Heat exchanger

Superheat
steam → Pressure
leaching

Crystallization
of Al(OH)₃

Al(OH)₃
seed

Pressure
relief

Dilute
NaOH

Filtration,
Al(OH)₃

Dilution

Al(OH)₃ crystals

Wash
water → Red mud
separation

Water

Fe₂O₃·xH₂O

Washer

Water → Red mud
washer

Spent
wash
water

Calcination

Fuel

Red mud,
for disposal

Alumina
≈ 99.5 % Al₂O₃

**Figure 10.1.** Production of alumina from bauxite via the Bayer Process

rica, Australia, the Caribbean, Central and South America and the United States. European bauxites are mainly monohydrate [4].

Iron oxide, clays, and most other impurities do not dissolve under alumina digestion conditions, and are quite finely divided ($1\text{-}10\,\mu\mathrm{m}$). By adding wash water to decrease the viscosity somewhat, it is possible to decant the still hot sodium aluminate solution from the slowly precipitating red muds. These muds, which are red from the high iron content, are then washed with water to minimize losses of alumina and base on ultimate disposal of the muds by lagooning. By using this wash water for dilution of the next digester output, alumina and base are not lost from the Bayer circuit. Some more recently constructed facilities use pressure filtration for both removal and washing of red muds.

Aluminum hydroxide crystals are obtained from the supernatant sodium aluminate solution after a final polishing filtration by ensuring a $Na_2O{:}Al_2O_3$ ratio of 1.5-1.8:1, controlling the solution temperature to about $60\,^{\circ}\mathrm{C}$, and seeding with crystals from an earlier crystallization. Even with seeding crystallization process is slow, requiring 2-3 days, but sufficiently large crystals are obtained by the reverse of the solution process (Equation 10.4)

$$AlO \cdot ONa + 2\,H_2O \rightarrow Al(OH)_3 + NaOH \quad (10.4)$$

to enable recovery of the product by filtration. The aluminum hydroxide product, while still on the fil-

ters, is washed with water and then dried. Evaporation of most of the water from the filtrates allows return of the sodium hydroxide solution to the initial grinding and pressure leaching circuits for reuse at appropriate concentrations for these steps. Calcination of the aluminum hydroxide at about $1{,}200\,^{\circ}\mathrm{C}$ in either fluidized bed or rotary calciners finally yields alumina, of 99.5 % purity (Equation 10.5).

$$2\,Al(OH)_3 \rightarrow Al_2O_3 + 3\,H_2O \quad (10.5)$$

The chief contaminant is 0.3-0.5 % sodium oxide, which fortunately does not affect electrolysis, with $< 0.05\ \%$ calcium oxide, $< 0.025\ \%$ of silica or iron oxide, and $< 0.02\ \%$ of any other metallic oxide [3]. Apart from metal production, some of this high calcination temperature alumina product goes into the manufacture of synthetic abrasives and refractory materials [4]. Activated alumina, destined for adsorptive uses is produced in the same way except that more moderate calcining temperatures of ca. $500\,^{\circ}\mathrm{C}$ are employed, producing a highly porous product with excellent surface activity. The volumes of alumina from the world's major producers is given for several recent years in Table 10.3.

It is also possible to obtain alumina from clays or bauxite via leaching with sulfuric acid, a technique which is especially useful with high silica or high iron oxide bauxites (Equation 10.6).

**Table 10.3** Major world producers of alumina [a]

| | Thousands of tonnes | | | |
|---|---|---|---|---|
| | 1972 | 1975 | 1978 | 1979 |
| Australia | 3 068 | 5 129 | 6 776 | 7 414 |
| Canada | 1 149 | 1 133 | 1 053 | 824 |
| France | 1 274 | 1 250 | 1 221 | 1 238 |
| Jamaica | 2 087 | 2 277 | 2 142 | 2 074 |
| Japan | 1 644 | 1 565 | 1 767 | 1 822 |
| Surinam | 1 378 | 1 129 | 1 316 | 1 250 |
| U.S.A. | 6 114 | 5 304 | 5 960 | 6 450 |
| U.S.S.R. | 2 300 | 2 400 | 2 600 | 2 600 |
| W. Germany | 916 | 1 246 | 1 410 | 1 352 |
| Others | 3 670 | 6 477 | 5 855 | 6 476 |
| World total | 23 600 | 26 500 | 30 100 | 31 500 |

[a] Includes all those countries producing more than a million tonnes of alumina annually in 1978 or 1979. Compiled from data in [63].

$$\text{clays} + H_2SO_4 \rightarrow Al_2(SO)_4 + \text{water}, \quad (10.6)$$
$$\text{soluble} \quad \text{insoluble residue}$$

Crystals of pure aluminum sulfate hydrate $(Al_2(SO_4)_3 \cdot 18 H_2O)$ are obtained from the leach solution, which may then be calcined to yield alumina (Equation 10.7).

$$Al_2(SO_4)_3 \cdot 18 H_2O \rightarrow Al_2O_3 + 3 SO_3 + 18 H_2O \quad (10.7)$$

Cryolite ($AlF_3$ x 3 NaF), the major, though largely re-used component of the electrolytic bath is a rare natural mineral originally found only in Gothaab, Greenland. Most present aluminum smelters function using synthetic cryolite prepared from alumina and hydrogen fluoride with caustic (Equations 10.8, 10.9).

$$CaF_2 + H_2SO_4 \rightarrow 2 HF + CaSO_4 \quad (10.8)$$
fluorspar

$$6 HF + Al(OH)_3 + 3 NaOH \rightarrow$$
briquets
$$Na_3AlF_6 + 6 H_2O \quad (10.9)$$

In this manner a cell electrolyte having characteristics identical to the natural mineral is sufficiently readily available to be employed by the vast majority of smelters.

## 10.3 Aluminum by the Electrolysis of Alumina

Reduction of alumina to aluminum, even though it is carried out at high temperatures, is an electrolytic process like the dominant method for chlorine and caustic production from sodium chloride, not an electrothermal process (e.g. phosphorus). The required ionic mobility could theoretically be provided by melting alumina (melting point 2,050 °C), but the temperatures required are so high (above the useful range of most refractory lining materials) to make this technically unfeasible. Mobility can by provided by dissolving the alumina in an ionizing solution, although not an aqueous one since alumina is insoluble in water. But alumina is sufficiently soluble in molten cryolite ($Na_3AlF_6$, melting point 1,006 °C) to enable electrolysis to be carried out. In practice, addition of a few percent of calcium fluoride and/ or aluminum fluoride depresses the melting point of the electrolyte some 4-5 °C for each 1 % of additive [5] enabling electrolysis at ca 950 °C, temperatures technically a little easier to achieve (Table 10.4). While the aluminum reduction process is primarily electrolytic much of the electrical requirement for aluminum production is consumed through cell resistance to keep the reduction pot temperatures sufficiently high to maintain the electrolyte fluid state.

Virtually all alumina electrolysis cells consist of a rectangular steel shell lined with a 25-35 cm layer of baked and rammed dense carbon, which provides both chemical resistance and the cathode contact with the electrolyte via steel bus bars imbedded in the carbon. Normal lining life is four to six years, after which it is replaced as large preformed slabs. Once a reduction pot has been started the bulk of the cathodic current to the carbon lining

**Table 10.4** Typical melt composition for electrolytic reduction of alumina [5]. Normally alumina reduction pots are operated on the acid side, i.e. a net $AlF_3$ : NaF mole ratio of 1.2 to 1.5 : 1.

| Component | Range, % |
|---|---|
| cryolite, $Na_3AlF_6$ | 80 – 85 |
| fluorspar, $CaF_2$ | 5 – 7 |
| aluminum fluoride | 5 – 7 |
| alumina | 2 – 8 |

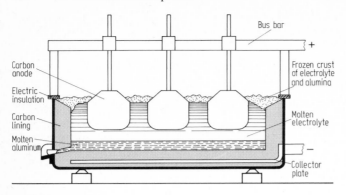

**Figure 10.2.** Cross-sectional diagram of a typical aluminum reduction pot using prebaked carbon anodes. The refractory brick lining and the frozen crust beside and on top of the molten electrolyte provide thermal insulation, raising overall energy efficiencies. (Aluminum Co. of America)

is via the pool of newly-formed molten aluminum in the bottom of the cell (Figure 10.2).

Anodic elements are commonly prebaked low ash carbon blocks, low ash since any ash residue ends up in the electrolyte. These are electrically connected to copper or aluminum bus bars (heavy electrical conductors) suspended over the cell, which also provide mechanical support and a means for vertical adjustment of the anode elements. Another anode variant, the Soderberg paste option, uses a good grade of petroleum coke formed into a paste with hard pitch, to which electrical contact is maintained and mechanical adjustment provided by using specially shaped steel pins [6] (Figure 10.3). As the baked portion of this anode is gradually consumed the paste approaches the molten electrolyte and the volatile components in the paste vaporize leaving a hard baked working anode element. Either type of anode element is consumed at the rate of 1-2 cm a day during normal operation, requiring periodic vertical adjustment to maintain an anode-aluminum metal pool spacing of ca. 5 cm.

To start operation of a pot, anodes are lowered to the lining, the solid electrolyte and alumina components are placed and then the power turned on. Initial melting of the pot charge is by resistance heating, typical cells consuming from 50,000 to 200,000 amperes each, at an operating voltage of about 4.5 to 5.5 V. As the charge melts the anode elements are gradually raised to the normal ca. 5 cm spacing, and electrolysis begins.

At the cathode, aluminum ions pick up electrons to form elemental aluminum (Equation 10.10),

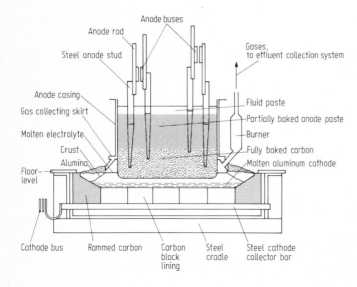

**Figure 10.3.** Diagram of an aluminum reduction pot showing the arrangement of the principal components of a vertical stud Soderberg paste anode [24]. Reprinted courtesy of Alcan Smelters and Chemicals Ltd.

$$Al^{3+} + 3\,e^- \rightarrow Al^0 \qquad (10.10)$$

which, being more dense than the molten cryolite, gradually accumulates as a pool beneath the electrolyte melt layer and rests on the carbon lining. For a 125,000 ampere cell the metal reduction capacity is 810-910 kg per day. Normally, alumina additions are made to each cell at about four hourly intervals. However, if the alumina consumption in any one cell for any reason exceeds this rate of routine addition and the alumina content of the electrolyte drops to about 2 % alumina or less [5], there is a sudden rise in the cell voltage to about 50 volts, primarily caused by polarization of the anode from the near stationary film of carbon monoxide and carbon dioxide on its surface. Simple detection of this situation, called the "anode effect", is provided via a 50-60 volt light wired in parallel to each cell in a multi-cell potline. Normal cell operating voltages are too low to cause bulb incandescence, but an alumina short situation initiates fluorine as well as oxygen deposition at the anodes. The higher anode-cathode voltage differential may also cause arcing to occur across the anode-aluminum pool gap. These severe conditions cause the abnormally present fluorine, like oxygen, to also react with the hot carbon forming carbon tetrafluoride (Equation 10.11)

$$2\,F_2 + hot\,C \rightarrow CF_4 \qquad (10.11)$$

and some hexafluoroethane which, together with carbon monoxide [4], effectively combine to stabilize the polarization of the anode elements by formation of a gas envelope around them. That it is these occurrences which cause the rapid rise of the operating voltage of a cell [3], is confirmed by this effect being temporarily alleviated by stirring the electrolyte, which serves to momentarily clear the gas envelope from the anode surface. But to effectively correct this situation, fresh powdered alumina, already prewarmed on the electrolyte crust of the cell, is broken into the molten electrolyte with a remotely-controlled pneumatic hammer. The anode polarizing gas used to be dispelled by poking wood poles through the crust to flush out the polarizing gases by both the stirring action and the rapid flow of combustion gases. Now, however, air lances are more often used to accomplish this task.

At about daily intervals, as the amount of aluminum produced per cell is sufficient, a stout pipe attached to an evacuable vessel is broken through the crust on the cell and dipped into the pool of molten aluminum. Decreasing the pressure in the receiving vessel pulls the melt into the container, a process which is repeated for each cell down the potline, to collect the molten aluminum in a central holding furnace ready for fabrication.

At the same time as aluminum is being reduced, oxygen ions migrate to the suspended anode elements and in turn are oxidized to atomic and possibly some molecular oxygen (Equation 10.12).

$$O^{2-} - 2\,e^- \rightarrow 2[O]; (2[O] \rightarrow O_2) \qquad (10.12)$$

But because both forms of oxygen are generated in close contact with a hot carbon element, virtually all of it immediately reacts to form carbon monoxide and carbon dioxide (Equations 10.13-10.15).

$$hot\,C + [O] \rightarrow CO \qquad (10.13)$$

$$hot\,CO + [O] \rightarrow CO_2 \qquad (10.14)$$

$$hot\,C + O_2 \rightarrow CO_2 \qquad (10.15)$$

However, as this carbon dioxide moves from the anode region it contacts hot aluminum vapours, and possibly sodium vapours and fine liquid metal droplets (fog) which immediately oxidize to form the corresponding metal oxide and carbon monoxide (Equations 10.16, 10.17).

$$3\,CO_2 + 2\,Al \rightarrow 3\,CO + Al_2O_3 \qquad (10.16)$$

$$CO_2 + 2\,Na \rightarrow CO + Na_2O. \qquad (10.17)$$

Thus, the gas mixture collected from the skirted anode area actually comprises carbon monoxide (from both incomplete initial oxidation, plus metal reduction of carbon dioxide) and carbon dioxide in a weight ratio which varies between 1:2 and 1:3. The toxic carbon monoxide in these hot gas mixtures normally burns spontaneously as it escapes the confining crust. For vertical Soderburg pots, where the gases are captured by the anode skirt, a small burner is used to convert the carbon monoxide content to carbon dioxide at each cell gas exit before the effluent gas mixture proceeds to treatment systems.

The overall electrochemical process being considered is closely represented by Equation 10.18.

$$Al_2O_3 + {}^{3}\!/\!{}_2\,C \rightarrow 2\,Al + {}^{3}\!/\!{}_2\,CO_2 \qquad (10.18)$$
$$\Delta H = +1096\ kJ\ (+262\ kcal)$$

This highly endothermic energy balance in the direction given is what is supplied electrochemically during the electrolysis. Using this figure in the

**Table 10.5** Electrochemical parameters of importance to
Hall-Heroult electrolysis of alumina [a]

| Reduction potentials, volts: | | |
|---|---|---|
| theory | 1.706 | |
| normal operating | 4.5 – 5.5 (to 7, on occasion) | |
| anode effect | 30 – 50 (at ca. 2% $Al_2O_3$) | |
| Operating capacities, kA: | 50 to 260 | |
| Anode characteristics: | | |
| | Resistivity ohm/cm | Current density kA m$^{-2}$ |
| Prebaked | 5 – 6 | 6.5 – 13 |
| Soderberg | 6.5 – 7.8 | 6.5 – 9 |
| Electrode spacing: | ca. 5 cm | |
| Electrolyte bath resistivity: (at 950 °C) | 0.50 ohm/cm | |
| Current efficiency: | ca. 90% (100 kA cells) | |
| Normal energy efficiency: | 34 – 38% | |
| Typical anode gas composition [b]: | 85% $CO_2$, 15% CO | |

[a] Calculated and compiled from [3, 5 and 6].
[b] Gas composition is given as experienced for 90% current efficiencies. At 80% current efficiency the gas composition will be approximately 75% $CO_2$, 25% CO, i.e. the CO content is inversely proportional to current efficiencies.

Gibbs-Helmholtz equation allows determination of the theoretical equilibrium voltage required for deposition of aluminum, which is 1.7 volts. It is generally thought that there is no electrochemical assistance from carbon oxidation by the anode product, hence this electrochemical potential is the same as the standard reduction potential for aluminum, which is $-1.706$ volts [7]. The actual operating potential required is significantly higher than this, 4.5 to 5.5 volts, needed to overcome electrolyte resistivity (Table 10.5) over the 5 cm electrode gap and other factors, as well as to move the process from an equilibrium to a producing situation. Thus, voltage efficiencies range from 24 to about 38 % with the bulk of production taking place at the upper end of this range, significantly lower than either of the common chloralkali electrolysis cell types. As mentioned previously this electrochemical work required for alumina electrolysis is the result of overall cell resistance effects, and appears in the form of heat required to keep the cell contents molten. The smaller surface to volume ratio for larger electrolysis cells (up to 260 kA) [3] tends to confer some voltage efficiency contribution which improves overall energy efficiencies slightly over the values obtained with smaller cells.

Like all electrochemical processes the quantity of product obtained for a given current flow/time period is related through Faraday's Law, i.e. that in an electrochemical reaction one gram equivalent weight of substance is deposited on the passage of 96,494 coulombs (one Faraday) of electricity. Since each aluminum ion, to be deposited, requires three electrons (Equation 10.10), a gram equivalent weight of aluminum is one third of its atomic weight (26.9815 g/mol $\div$ 3 electrons/mol) or 8.994 g. Thus, 8.994 g of aluminum would be expected to be deposited in the cell, for each Faraday of power. But in practice only about 90 % of this amount is collected [3], the lower experience being the result of such processes as the reaction of anode carbon dioxide with already reduced aluminum or sodium, returning these cell "products" on which power has already been expended, to the electrolyte as oxides. It has been found that the addition of aluminum fluoride to the electrolyte tends to decrease the concentration of reduced species in the electrolyte, and in this way raises current efficiencies.

Combining the voltage and current efficiency factors, the overall *energy efficiency* of the cell is usually about 34 % (0.90 x 0.38) although during periods of high voltage operation this can drop as low as 22 %. In production experience terms, this normally means a power consumption of 17,600 kWh per tonne of aluminum for the electrolytic step. Power and other quantitative aspects of aluminum production such as anode consumption rates are given in Table 10.6. For quantitative calculations and reflecting quite closely the actual stoichiometry experienced the molar relationships given in Equation 10.19 have been used.

$$Al_2O_3 + 2\,C \rightarrow 2\,Al + CO + CO_2 \qquad (10.19)$$

Purity of the direct Hall-Heroult product is very good, about 99.5 to 99.7 % aluminum, and adequate for most uses requiring the pure metal, and for alloys. The chief impurities present are iron, 0.06 to 0.10 %, and silicon, 0.04 to 0.30 %, with the total of all other impurities seldom exceeding 0.10 % [5]. Analysis of both the aluminum and the electrolyte of each cell is normally conducted on at least a weekly basis. Among other things this practice provides a warning when the carbon lining of a pot has worn through to the steel shell ahead of schedule, seen as an increase in the iron content of the aluminum. This particular cell may then be taken out of production and repaired before aluminum and electrolyte break through the steel shell. One trace contaminant of chemical, if not commercial interest, is gallium, present to the extent of 60-200 $\mu g/g$. Situated in the same group and immediately below aluminum in the periodic table this element is evidently present from its very similar chemical properties carrying it through the alumina purification and electrolytic steps with the aluminum.

Electrolytic, or fractional crystallization methods are employed to produce 99.99 % pure aluminum.

Electrolytic methods employ a melted aluminum-copper alloy (about 75:25) as the bottom layer, overlain by a high content of a barium salt (either chloride or fluoride) to produce a high density electrolyte, topped by a layer of the electrolytically refined aluminum as it forms [3]. By passage of a direct current through the cell floor as anode and suspended carbon cathodes there is a gradual accumulation of refined aluminum in the top layer. Voltages, and hence voltage efficiencies are similar to those required for primary aluminum production. Zone refining is resorted to when 99.999 % or higher aluminum purities are required [3]. Recycling of scrap aluminum, presently accounting for some 20 % of aluminum produced, requires less than 5 % of the energy required to produce the metal from alumina.

## 10.4 New Developments in Aluminum Production

Despite the world's abundant reserves of bauxite the deposits are not, in many cases, in the major aluminum producing areas. This factor has stimulated the extensive testing of alternative sources of aluminum [8]. High alumina clays, shales, and other alumina-containing minerals have all been used as alumina sources, and some of these such as nepheline syenite $((Na, K)\,(Al, Si)_2O_4)$ in the U.S.S.R., are actually used. Tests in the U.S.A. are being performed on extraction of alumina from kaolin clays (30-35 % $Al_2O_3$) with hydrochloric or nitric acids. Studies being undertaken for alumina recovery from the clays removed from coal (ca. 28 % $Al_2O_3$) during washing [8], and from coal ash, if found to be economically feasible, have the

**Table 10.6** Materials and energy required to produce one tonne of commercial aluminum [a]

| Alumina: | Theory | 1 889.5 kg |
| | Practice | 1 900 – 1 901 kg |
| Carbon anode material: | Carbon, Theory [b] | 445 kg |
| | Practice | 450 – 550 kg |
| | Pitch, Soderberg | 100 – 200 kg |
| Electrolyte: | Cryolite | 30 – 70 kg |
| | Aluminum fluoride | ca. 40 kg |
| Power: | 15 000 – 17 600 kWh | |
| Labour: | 13 – 31 man–hours | |

[a] Calculated from data of [3, 5, 6, 8 and 27].
[b] Assuming this is pure carbon, as is nearly true with prebaked anodes, and the stoichiometry of Equation 10.19.

dual advantage of providing a resource from large scale waste materials. French and Canadian ventures studying alumina extraction from clays and shales using sulfuric and hydrochloric acids have also been announced [9].

Sumitomo Chemical technology, consisting of a number of small changes to conventional Hall-Heroult electrolysis cells combine to cut power consumption to some 14,000 kWh per tonne from the normal experience of 16,000 to 18,000 kWh per tonne [10]. In addition to reduced power consumption, better emission control, extended cell life, and decreased labour requirements are also achieved by these changes. The improvements are obtained through better external cell insulation, changes in their basically Soderberg anode design, and better fume containment by cell skirt construction modifications.

Successful though the above modifications are, and applying as they do to many aluminum producing installations world-wide where an incremental improvement is significant, the most innovative developments involve changes in the actual technology employed. Aluminum can be obtained by metallic, carbon, or electrolytic reduction of appropriate compounds. The earliest processes successfully used metallic reduction via potassium amalgam, or sodium fusion to produce aluminum from aluminum chloride (Equations 10.20, 10.21),

$$AlCl_3 + 3 K(Hg) \rightarrow Al + 3 KCl + Hg \qquad (10.20)$$

$$AlCl_3 + 3 Na \rightarrow Al + 3 NaCl \qquad (10.21)$$

but as Hall-Heroult technology appeared were displaced because of higher costs.

Reduction of alumina with carbon is fraught with all kinds of problems, not the least of which is the very high temperatures required (about 2,000 °C) and the formation of quite stable aluminum carbide (Equation 10.22).

$$2 Al_2O_3 + 9 C \rightarrow Al_4C_3 + 6 CO \qquad (10.22)$$

Metallic aluminum is a sufficiently active metal that it reduces carbon dioxide to carbon monoxide (Equation 10.16) so that the favourable aspects of a solid-gas, alumina-carbon monoxide reduction cannot be utilized. However, good yields have been obtained experimentally by carrying out the reduction in two steps. The first step requires formation of the carbide (Equation 10.22), following which the temperature of the system is raised another 100-200 °C to obtain aluminum by the reaction of aluminum carbide with alumina (Equation 10.23) [11].

$$Al_4C_3 + Al_2O_3 \rightarrow 6 Al + 3 CO \qquad (10.23)$$

At around 2,100 °C, as required for the second step, aluminum (melting point 660 °C, boiling point 2,467 °C) itself has significant volatility causing metal losses and migration problems. Carbothermic reduction in the presence of an alloying element such as copper, iron, or silicon to decrease aluminum vapour pressures decreases volatility problems but requires a second stage to recover aluminum from the alloy product. It may either be selectively dissolved from the alloy with a more volatile metal, such as mercury, lead, or zinc, and then the aluminum recovered by distillation [3]. Or, the tendency for aluminum halides to form more volatile monohalides at high temperatures which also revert to the trihalides at lower temperatures (Equation 10.24) may also be employed.

$$AlCl_3 + 2 Al \underset{\substack{\text{in alloy } 800\,^\circ C}}{\overset{1300\,^\circ C}{\rightleftarrows}} 3 AlCl \qquad (10.24)$$

In this procedure, extensively tested by Alcan at one time [12, 13], dissolution of aluminum from the alloy using aluminum trichloride at ca. 1,300 °C produces the volatile aluminum subhalide. Cooling the subhalide moderately returns a pool of molten product aluminum, and aluminum trichloride (boiling point ca. 183 °C) which serves as a working fluid which can be recycled. Despite problems with the aluminum recovery methods [13] there remain indications of a continuing interest in nonelectrolytic aluminum production such as offered by carbothermic processes [14, 15].

New technology for electrolytic aluminum production employing aluminum is also stimulating fresh interest, because of the about 30 % power savings possible [16]. Since aluminum chloride melts at much lower temperatures and forms a much more fluid melt than the standard Hall-Heroult electrolyte matrix much higher voltage efficiencies are possible, but sublimation and control problems limit the utility of direct, one-component electrolytic methods. But the basis of this idea is employed in the process developed by C. Toth, of Alcoa [17] which has the additional advantage of enabling clay sources of alumina to be tapped [18] (Figure 10.4). While technical details of this proc-

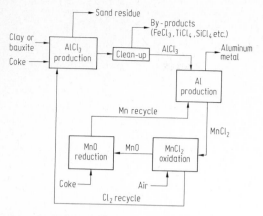

**Figure 10.4.** Simplified flowsheet of new Alcoa aluminum chloride-based route to aluminum. Adapted from [17]

ess have not as yet been released, production probably involves the sequence of reactions given in Equations 10.25-10.28.

$$2\,Al_2O_3 + 6\,Cl_2 + 3\,C \rightarrow 4\,AlCl_3 + 3\,CO_2 \tag{10.25}$$

$$2\,AlCl_3 + 3\,Mn \xrightarrow[\text{anodes}]{\text{electrolysis}} 2\,Al + 3\,MnCl_2 \tag{10.26}$$

$$2\,MnCl_2 + O_2\,(\text{from air}) \rightarrow 2\,MnO + 2\,Cl_2 \\ (\text{Deacon related}) \tag{10.27}$$

$$2\,MnO + C \rightarrow Mn + CO_2 \tag{10.28}$$

Prototype results with this process have been sufficiently attractive for Alcoa to have started production in 1976 using this technology [3] on a scale of about 13,500 tonnes per year [16].

## 10.5 Emission Control Problems and Solutions

### 10.5.1 Environmental Impacts of Bayer Alumina from Bauxite

The mining of bauxite for an alumina product has the same kind and degree of environmental impact as imposed by any type of strip mining operation. These effects may be minimized by spoils replacement and aesthetic contouring, as well as by the replanting of hardy grasses and shrubs, where appropriate.

Processing of bauxite to produce alumina, as outlined, also produces volumes of "red muds"

containing principally water, iron oxide, silica, and the oxides of titanium, chromium, vanadium and aluminum. The solids in this mixture do eventually settle to a relatively high solids content sludge, so that a moderate-sized holding pond may be used for many years [3].

However, as environmental concerns, land costs and raw material costs rise, alumina producing facilities are increasingly seeking ways to decrease or eliminate the volume of this red mud waste. Alcoa has devised the "Alcoa-Bayer" process to enable recovery of much of the formerly discarded alumina fraction, as has been lost in the red muds as combined alumina and precipitated with silica and sodium hydroxide [19]. By adding limestone and sodium carbonate and calcining the mixture to a clinker, it is possible to extract the alumina values out with water and return this extract to the main Bayer circuit. Decreasing the alumina content of the red muds in this way not only directly decreases red mud volumes by removing a formerly significant constituent, particularly from high silica bauxites, but it also indirectly decreases the volume of waste from the decreased volume of bauxite to be processed for a given number of tonnes of alumina product required.

Other tests have shown the residual $Al^{3+}$ and $Fe^{3+}$ ions present in red muds can be effective, and inexpensive flocculants for sewage, particularly efficient at removing phosphate [20]. Processed to a finely-divided dry powder, this might prove to be a generally attractive option yielding a saleable commodity from a waste material.

More recently conversion of dried, pelletized red mud to steel has been tested, via an electric pre-reduction and smelting process [21]. This technology promises to be of value in bauxite processing areas where conventional iron ores are low grade, scarce, or non-existent. Alumina calcination and grinding particulate discharge requirements prior to 1971 were set at 100 and 150 mg/m³ air (standard conditions), respectively, by the VDI (German Engineering Society) [22].

### 10.5.2 Aluminum Smelter Emission Control Problems

Emission problems of conventional aluminum smelters center on fluoride losses, pre-abatement (before ca. 1972) at the rate of some 21 kg per tonne of aluminum produced (Table 10.7). The bulk of this fluoride loss occurred from the operat-

**Table 10.7** Pre-, and post-abatement atmospheric fluoride emission rates in primary aluminum smelters operating in the U.S.A. and Canada [a]

| | Kilograms of fluoride emitted per tonne aluminum | | |
| --- | --- | --- | --- |
| | Uncontrolled potlines [b] | Partial abatement | Efficient abatement |
| Prebaked anodes: | | | |
| Gaseous | 14 | 13 | – |
| Particulate | 9 | 8 | – |
| Total | 23 | 21 | 1.7 |
| Range | (12 – 33) | (5 – 21) | (0.6 – 2.3) |
| Vertical stud Soderberg anodes: | | | |
| Gaseous | 18 | 13 | 3.4 |
| Particulate | 2 – 9 | 2 – 4 | ca. 1 |
| Total | 20 | 21 | 4.4 |
| Range | (18 – 27) | (16 – 22.5) | (1.1 – 4.3) |

[a] Compiled from [23, 24, 27, 36 and 64].
[b] In 1973 represented only 3% of primary aluminum producing facilities in the U.S.A. [23].

ing electrolytic cells, and two-thirds or more as gaseous fluoride [23]. The chief constituents of the fluoride discharge are known to be cryolite ($Na_3AlF_6$), aluminum fluoride, calcium fluoride, chiolite ($Na_5Al_3F_{14}$), silicon tetrafluoride, and hydrogen fluoride. A rough guide as to mass discharge rates may be obtained from consumption rates of these commodities by the industry, in relation to the volume of aluminum produced (Table 10.7). Most of these substances are lost as fumes or vapours, except the hydrogen fluoride gas evolution which results from the reaction of traces of moisture present in alumina added to the cell (Equation 10.29).

$$3\,H_2O + 2\,AlF_3 \rightarrow Al_2O_3 + 6\,HF \qquad (10.29)$$

This mode of fluoride loss is the reason that alumina shipment and transfers are conducted with minimum exposure to air, when moisture adsorption would occur. In addition to these substances the extremely stable fluorocarbons hexafluoroethane and carbon tetrafluoride, which are relatively non-toxic, are known to form during an anode effect [23-25] (e.g. Equation 10.11).

While fluoride emissions have undoubtedly been the aluminum smelting impact area of greatest concern a number of other gases, some arising from the carbon anode baking plant, are also significant. These include carbon monoxide, sulfur dioxide,

hydrogen sulfide, nitrogen oxides (NO and $NO_2$), carbonyl sulfide (COS) and carbon disulfide. In a smelter employing Soderberg anodes, in which the pitch component of the carbon paste is largely vaporized during the *in situ* baking, there will also be measurable generation of hydrocarbon vapours and smokes, including polynuclear aromatic hydrocarbons. The concentration of hydrocarbon vapours collected by the hooding can run to about 3 %, but this is decreased to about 0.1 % on passage through the carbon monoxide burner [26]. Estimates, in kg per tonne of aluminum produced, for sulfur dioxide ranges from 1-30, for particulates (including alumina, soot, etc.) from 5-10, and for hydrocarbon vapours (Soderberg type cells only) from 0.25 to 2 [23, 24, 27].

There are also problems dealing with liquid wastes, such as fluoride contaminated discharge water from wet scrubbers, or drainage of precipitation from areas where spent pot linings are discarded, since during use the carbon becomes impregnated with the fluoride constituents of the electrolyte. Water used for cooling metal castings or transformers is not contaminated. Disposal of spent pot linings or discarded pre-baked anodes requires care to minimize problems of the type mentioned above, but solved by some facilities by fluoride recovery followed by recycle of the carbon content.

Of all the pollutants outlined, fluorides represent

the aluminum smelter discharge component of greatest hazard to plant life, and also indirectly to grazing and predator animals [25, 28-35]. For example, while the injury threshold to plants for sulfur dioxide is about 0.1 ppm (100 ppb) the equivalent value for fluoride, to fluoride-sensitive plants is < 1 ppb [36]. Thus, since fluoride emission has generally represented the largest mass loss to the atmosphere and it may disperse widely by a variety of ways [37-41] it has the potential to produce a significant biotic impact. This, therefore, is the area that has received the most significant control attention [42].

### 10.5.3  Smelter Emission Control Strategies

Aluminum smelting control systems may be considered in two groupings. The primary control system includes the hoods and ductwork for each cell, and the common duct which receives the vent gases from a bank of cells. While the primary system is able to collect and deal with the bulk of the fume and vapour discharge from a cell, particularly when the electrolyte crust between the collection skirt around the anode(s) and the rim of the cell is intact, many normal cell operations require this crust to be broken. Scheduled additions of alumina (several times a day), removal of aluminum (ca. daily), and periodic electrolyte analysis, as well as dealing with the occasional "anode effect" all require breaking the electrolyte crust. For the period of the operation, plus some additional time for a new crust to form, the loss rate to the cell working area, at least for some types of cell, is significantly raised. It is this type of loss that the secondary collection system, generally comprising a series of forced circulation or convective roof exhaust vents together with floor level inlet vents in the sidewalls of the building, is designed to handle [23].

The primary collection system where the bulk of the cell fluoride loss is captured, may simply discharge to a tall stack [27]. This aids in dispersal, which *may* be sufficient for a small smelter. But it does nothing to decrease mass discharge rates, it does not allow recovery of fluoride for re-use, and has been found to be a strategy which even costs more to operate than several true abatement methods employing fluoride capture.

Wet scrubbers comprise the fluoride containment system in longest use for cleanup of primary system gases before discharge, perhaps in part be-

cause of the high affinity for and solubility in water by hydrogen fluoride. The best of these systems, employing a wet scrubber followed by a wet electrostatic precipitator, have been found to be 98-99 % efficient on both gaseous and particulate fluorides [23, 26]. But while most wet scrubber configurations have no difficulty achieving 98 + % mass containment for *gaseous* fluorides some have difficulty trapping as much as 90 % of the particulate mass [23, 27], and achieve efficiencies little better than 10 % on the particle fraction less than 5 $\mu$m in diameter. The other consideration which has stimulated the development of alternate primary control systems is the need, with the wet processes, to have a method of dealing with spent scrubber liquor, preferably to regenerate fluorides of value for electrolyte make-up (Equations 8.14, 8.15). This can be relatively straightforward if a sieve plate column is used for scrubbing, since a spent scrubber liquor containing 4 % hydrogen fluoride by weight may be obtained directly [26] (Figure 10.5).

Various dry scrubbing configurations have been applied to primary control systems for fluoride in-

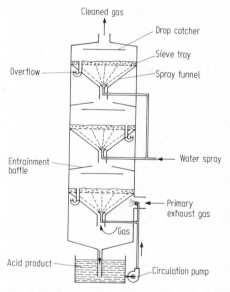

**Figure 10.5.** Hydrogen fluoride removal from smelter primary gas collection system via a sieve tray gas scrubber (only a section shown). Note the recycle of acid product for initial hydrogen fluoride absorption from the exhaust gases to maximize the concentration of recovered acid. Redrawn from reference 26

cluding passage of the gases through coated filters or fluidized beds, or injection of alumina into the waste gas stream followed by recovery of the alumina plus adsorbed fluorides. The dry filter is more efficient than most wet scrubber arrangements for capture of fluoride particulates but the early versions were only about 90 % efficient at controlling gaseous fluoride [23]. But both the fluidized alumina bed [43] (Figure 10.6), and the alumina injection dry scrubbing systems (Figure 10.7) make good use of the high adsorptive power of $\gamma$-alumina to control both gaseous and particulate fluoride as efficiently as good wet scrubbers, 98 + % [23, 44, 45]. $\gamma$-alumina, formed at a lower degree of calcination (lower aluminum hydroxide calcination temperatures) possesses a mosaic crystal structure, probably of a defect spinel nature, with a shortage of cations [46], and has a relatively high specific surface area of 60 m$^2$/g or more, highly suitable for adsorption [47]. Ordinary $\alpha$-alumina which forms a part of most commercial aluminas, is obtained from calcination temperatures above 1,000 °C as hard particles resistant to hydration or adsorption and is consequently less useful in this function. The latter two dry methods have the further advantage that the spent alumina merely has to be heated to re-

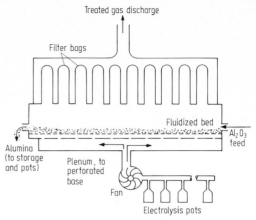

**Figure 10.6.** Typical arrangement of an alumina fluidized bed for primary emission control in electrolytic smelting of aluminum

move adsorbed hydrocarbons which are burned, and at the same time forming aluminum fluoride (Equation 10.30)

$$6\,Al_2O_3 \cdot HF \rightarrow 2\,AlF_3 + 5\,Al_2O_3 + 3\,H_2O \tag{10.30}$$

for direct return of the captured fluoride to the cells

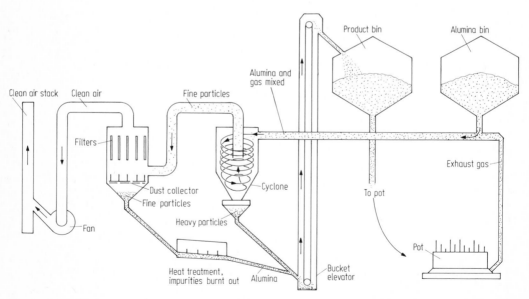

**Figure 10.7.** Typical arrangement of dry scrubber components, as used in aluminum smelter primary emission control. Bucket elevator is used to raise spent, fluoridated alumina to bin used to feed alumina to operating pots. a Captured fluoride is directly recycled [24]. Reprinted courtesy of Alcan Smelters and Chemicals Ltd

at the time alumina is added. These methods form the basis of the primary abatement systems recently adopted by several large North American aluminum smelters [23, 24, 48].

The philosophy of improvement of secondary control systems, that is control of the gaseous and particulate discharges to the air of the buildings in which pot lines are operating has been two-fold. Optimally, by an efficient primary control system, i.e. high containment and collection efficiency via hoods over cells and tight ductwork, the need for secondary control is eliminated. But where this is not the case, such as with older facilities where gas collection efficiencies for primary control systems have been 40-60 % [27], a relatively heavy loading of emissions is left to the wall ventilators and roof exhaust fans of the secondary control systems to disperse. Low primary collection efficiencies also meant that the mass discharge rate of pollutants from the secondary control system was large enough to also require treatment, as abatement became a smelter requirement, to experience any very significant reduction in gross smelter discharge rates. At the secondary treatment stage the concentration of pollutants in air is much lower, and the volume of air to be treated is very much larger than that obtained from the primary collection system, because of the restricted air flow into the hooded cell enclosures of the latter. The large size of treating equipment required to handle the large volumes of air moved through the secondary collection system makes secondary treatment costs very high relative to the overall degree of improvement achieved [26, 27]. This last factor is what has provided a strong stimulus for improvement of the collection efficiency of the primary control systems of the electrolytic cells, even by retrofitting older facilities where this is feasible.

By improvements in the proportion of the cell superstructure which is hooded, as well as by such procedural changes as enclosing the crust breaking/alumina addition operations and increasing the level of sophistication used in timing these, as well as other improvements, have all combined to enable an increase in primary collection efficiencies to about 95 %, at least for new facilities [27, 45]. Some cell operations, such as electrolyte analyses and periodic aluminum removal, require the opening of hooding access doors, thus making 95 % containment about the practical limit for primary control. Some of the incremental improvements for this achievement comprise a part of

the technology package being promoted by Sumitomo [49] and being adopted by some older facilities to upgrade the levels of control achieved [24]. The major advantage to secondary control systems contributed by primary control of 90 + % collection efficiency, apart from much improved cell room working conditions, is that the need to apply any emission abatement procedures to the secondary air discharge is minimized. This has meant that in some cases the pollutant mass discharge rate from the roof vents now exceeds that from the cleaned gas discharge from the primary control system [24], because of the gradual improvement in the degree of pollutant containment possible by either wet scrubber/wet electrostatic precipitator or dry scrubber (injection/cyclone/bag house) combinations on the latter gas stream. As the containment efficiencies of installed primary control systems improve to the technologically feasible levels described, the need for a secondary control system is minimized, although there will always be a ventilating requirement, if for no other reason than to shed the enormous heat load from the operating pots.

Liquid effluents from the operation of wet scrubbers, for those smelters which employ these, and collected precipitation run-off from the operating and waste disposal sites require treatment prior to discharge. Fluoride removal is generally quite straightforward using technology already discussed in connection with the production of phosphorus and phosphoric acid (which see), via the liming of a one or two stage effluent lagoon system. But hydrocarbons and phenols, mainly contributed by scrubber effluents from scrubbers operating on gases from Soderberg cells, have proved to be more of a problem to deal with [26, 27]. This is particularly the case with coastal smelters, which the majority are for ease of alumina delivery, since traces of fluorides and other salts present in *treated* waste water and discharged into seawater of high salts concentration are not an environmental aggravation [35, 41, 50], but a hydrocarbon content can be. The other consideration of importance for discharge of lower density, only slightly contaminated fresh waste into the marine environment, is that for good dispersal this should be conducted via a distributor on the bottom of the receiving waters, and preferably into zones with active tidal exchange. Otherwise, stratification of the fluoride containing waters into top layers will tend to occur and dispersal will be slow [39].

Spent pot linings, discarded after 3-5 years' continuous use and still containing 10-15 % absorbed fluorides also represent a potential environmental hazard unless recycled or disposed of properly. Increased costs for electrolyte fluoride replacement have stimulated more smelters to crush used pot linings and recover the contained fluoride by extraction with aqueous base. The filtered extract is then used for the preparation of synthetic cryolite. If landfill disposal of spent pot linings is used, the drainage of precipitation from the site has to be collected for fluoride removal as necessary. This method of disposal is also subject to occasional fires from spontaneous ignition [27], which is not only troublesome in itself but throws restrictions on any subsequent use of the filled area for some time.

## 10.6 Properties and Uses of Aluminum and its Compounds

Special properties of the metal itself have led to the rapid rise in the applications and in the volume of production of aluminum. Its natural corrosion resistance, due to a firmly-bound chemically inert oxide film contributes to its value in many uses, but in particular to the construction industry for window frames, exterior doors and their fittings, as well as for decorative and insulating panels, siding, roofing, etc. [51]. For some of these applications as well as for some smaller articles, anodizing or an anodic electrical passivation of the alumina surface in chromic or sulfuric acid is used to further enhance corrosion resistance, or to permanently impart colours other than its natural silvery-grey to the surface of the article. Some highly polluted atmospheres will cause pitting or formation of a poorly bound powdery white coating on aluminum articles exposed to these conditions, but otherwise no coating finish is required. Mercury too, should not be allowed to come into contact with an aluminum surface since amalgamation destroys the protective ability of the surface oxidized aluminum. This is the reason for the strict ban on mercury for air freight shipment.

Aluminum, with a density of $2.70 \, g/cm^3$ is the lightest of all metals which are permanent (stable) in air, except for magnesium ($1.74 \, g/cm^3$) and beryllium ($1.85 \, g/cm^3$). This property, plus its relatively low cost not only adds to the advantages in construction-related uses, but also makes it a valued component in the transportation industry.

Many components of aircraft, as well as load carrying components (rather than running gear) of unit rail cars and rapid transit systems, truck cabs and bodies, and some engine components are made of aluminum or alloys. In the automotive industry the applications are more extensive, including not only some of the above, but also wheels, trim, decorative components and transmission housings. Light weight is also an important rationale for the choice of Al/Mg or Al/Mg/Si alloys in many types of small boats for use in fresh or salt waters.

Aluminum is also an excellent electrical and thermal conductor, the latter being a contributing property to the choice of this metal in some of the transportation uses outlined above. The intrinsic electrical conductivity (i.e. for a test piece of the same dimensions) is $2.655 \, \mu\Omega/cm$ or 63 % of that of copper and 60 % of that of silver. But, for an equivalent *weight* of conductor per unit length aluminum is slightly more than twice as good a conductor as copper, and even better than this relative to silver, the reason for its use in long distance high tension transmission lines. Often, the cables used will be a composite of a high tensile strength steel core surrounded by a larger number of aluminum conductors. Great care has to be taken in making proper electrical connections to aluminum because of the non-conductivity of the thin aluminum oxide surface coating always present, complicated by the plasticity and high coefficient of thermal expansion of the metal itself [53]. These properties have led to problems in some aluminum house wiring installations.

Its good thermal conductivity (and non-toxicity) has led to extensive use in domestic cookware as well as in industrial heat exchangers, where its very good chemical resistance (e.g. to hydrogen peroxide, nitric acid) is also valued. And for cryogenic applications such as for construction of components of liquid air distillation plants etc. the conductivity, low temperature toughness, and high reflectivity are also important properties [54]. Related to cryogenic uses, aluminum also has a very low diffusivity (ease of gas transmission through intact metal), less than $^1/_2$ that of steel, a property valuable in the construction of double-walled evacuated vessels. Some steels also become very brittle at low temperatures, another reason for its unsuitability in this type of application relative to aluminum.

Compatibility with most foodstuffs plus its light weight and relatively high strength have led to extensive use of aluminum in the food packaging industry. Aluminum cans are used for soft drinks and beer, an active area for metal recycling, as well as for aerosol, meat, prepared pudding, and fish cans. Heavy foil is used for frozen food packaging, which doubles as a preparation container as well. Lighter foils are used for all types of packaging either on its own or in laminates for foodstuffs, tobacco and confectionery products as well as for other materials where its light weight, conformability, and low gas transmission characteristics prove useful.

## 10.6.1 Preparation and Uses of Aluminum Compounds

Alumina itself, the product of the Bayer process, is of great importance in many industrial applications other than for the production of alumina. The red mud wastes, which comprise the impurities removed from bauxite during alumina preparation, are employed in steel-making, the manufacture of red pigments [55], and for the production of chemicals used for water and sewage treatment.

By heating the washed aluminum hydroxide (alumina hydrate) to relatively moderate temperatures of about 500 °C, most of the water is removed without significantly altering the ionic arrangement leaving a highly porous, surface active product. Activated alumina, obtained in this way, is an invaluable industrial dehydrating agent as it stands and is also extensively employed in this form as a catalyst support in the petrochemical industry. The large surface area, commonly 40-65 $m^2$/g, provides for good distribution of intrinsic activity, or contributes to the effectiveness of other catalytically active agents previously placed on this surface. Alumina-silica composite systems are also of catalytic value.

Calcination of washed aluminum hydroxide at 1,100-1,300 °C causes shrinkage of the particles as they dehydrate and yields a denser, harder product, an $a$-alumina, useful as a constituent of abrasives. The low porosity and high temperature stability (up to ca. 1,800 °C) of this product also make it appropriate for the production of refractories, either as bricks using a little fireclay, or as a loose fill. Ceramics, with an alumina content of 85-95 % and fired for up to two days at 1,450 °C, possess superior strength, hardness, and electrical and thermal insulating properties. Spark plug insulators and textile guides as well as some very demanding ap-

plications in chemical industry are typical uses of these materials. Most recently, $a$-alumina fiber/metal composites are showing great promise in achieving from 4 to 6 times improved stiffness, and 2 to 4 times increased fatigue resistance over unreinforced metal [56].

Aluminum sulfate ($Al_2(SO_4)_3$) or alum is an aluminum salt valuable not only for water treatment (which see) but is also extensively used in papermaking [57] and to a smaller extent in the leather, drug and cosmetic, and dye industries. From a strict chemical standpoint an "alum" is a double salt of a monovalent sulfate and a trivalent metal sulfate crystallized with 24 waters, i.e. $(M^+)_2SO_4$ x $(M^{3+})_2(SO_4)_3$ x 24 $H_2O$. The monovalent cation is usually sodium, potassium, or ammonium, and the trivalent cation, aluminum. Less commonly, chromium or iron is substituted for the aluminum. But since in most traditional applications for alums the trivalent ion is the constituent of value, aluminum sulfate hydrate itself has gradually become known in industry as alum, or "filter alum" [58]. It is made by stirring bauxite with hot sulfuric acid (Equation 10.3),

$$Al_2O_3 \, (+ \, Fe_2O_3, \, SiO_2, \, etc.) + H_2SO_4 \rightarrow Al_2(SO_4) + residue \qquad (10.31)$$

bauxite settling, and then treating with a reducing agent such as sodium sulfide to convert any ferric salts to the less coloured ferrous equivalents. The decanted solution is then concentrated hot, allowed to cool, and crystals of the octadecahydrate, $Al_2(SO_4)_3$ x 18 $H_2O$, filtered off. Partial dehydration of the crystals during drying increases the aluminum analysis of the commercial product to approximately an $Al_2(SO_4)_3$ x 14 $H_2O$ stoichiometry. One variant of this route, used for applications demanding a low iron content such as dyeing, is the use of purified alumina rather than bauxite as the source of the aluminum ion [55]. Another consideration here is that alum production by either method provides a potentially profitable outlet for byproduct sulfuric acid [58] on a reasonable scale, particularly if there is a local market for the product as a solution, since solutions may be more economically shipped short distances without evaporation [59].

Another commercially important aluminum compound is aluminum chloride, valued as a Lewis acid catalyst in many types of petrochemical reactions and for other functions. The usual procedure is to melt clean factory scrap (offcuts from door and window manufacturing operations and the like) and then pass in dry chlorine gas, beneath the surface of the melt (Equation 10.32).

$$2\,Al + 3\,Cl_2 \rightarrow 2\,AlCl_3 \qquad (10.32)$$
m.p. 660 °C    m.p. 190 °C (2.5 bar)

The melt is hot enough to sublime the aluminum chloride out of the melt as it forms (sublimes at > 178 °C), and the product is continuously collected on a water-cooled scraped surface heat exchanger. The largest single use for this anhydrous product is for commercial alkylation reactions, and in particular the alkylation of benzene to ethylbenzene enroute to styrene. Aluminum chloride made by this route is also a classic example of the price role that addition of a large molar quantity of relatively inexpensive chlorine (16 ¢/kg) can play to produce a lower cost product (AlCl$_3$; $ 1.10/kg) than the relatively more expensive aluminum ($ 1.58/kg) [60]. Liquid (solution in water; specific gravity 1.28) aluminum chloride is a byproduct of the manufacture of the anhydrous material, and is also produced by treating alumina with hydrochloric acid (Equation 10.33).

$$Al_2O_3 + 6\,HCl \rightarrow 2\,AlCl_3 + 3\,H_2O \qquad (10.33)$$

This product is employed in the preparation of antiperspirants and some types of pigments.
Virtually equivalent purity anhydrous aluminum chloride (ca. 99.8 %) is also available as a side stream from Alcoa's new aluminum chloride route to the metal [61]. This process is said to require 30 % less energy than either of the procedures described above, and gives a fine grain free flowing product, achieved using a fluid bed reactor.

# Relevant Bibliography

1. Surinam: 62 Years of Bauxite Development, 12 Years of Alumina, Aluminum Production, Eng. and Mining J. *178*, 75, Nov. 1977
2. Aluminum: 1978 Survey of Mine and Plant Expansion, Eng. and Mining J. *179*, 68, Jan. 1978
3. E. Guccione, Steel from Aluminum Waste, Chem. Eng. *78*, 88, Sept. 20, 1971
4. Fluxing Find Cuts Remelt Chlorine Emissions, Iron Age *209*, 25, March 9, 1972
5. T.K. Osag et al., Fluoride Emission Control Costs, Chem. Eng. Prog. *72*, 33, Dec. 1976
6. Singmaster and Breyer, Air Pollution Control in the Primary Aluminum Industry, Singmaster and Breyer Consultants, New York, 1973
7. Aluminum: Properties, Physical Metallurgy and Phase Diagrams, Vol. I, American Society for Metals, Metals Park, Ohio, 1967
8. Handbook of Aluminum, 3rd edition, Aluminum Co. of Canada Ltd., Montreal, 1970
9. Fluorides, Committee on Biologic Effects of Atmospheric Pollutants, National Research Council, Washington, 1971
10. K. Grjotheim, H. Kvande, K. Motzfeldt, and B.J. Welch, Formation and Composition of the Fluoride Emissions from Aluminum Cells, Can. Met. Quart. *11* (4), 585 (1972). Chem. Abs. *78*, 150707 k (1973)
11. J.M. Robinson, G.I. Gruber, W.D. Lusk, and M.J. Santy, Engineering and Cost Effectiveness Study of Fluoride Emissions Control. PB Report (1972), No. 209647, 223 pages. Available NTIS
12. T.G. Pearson, The Chemical Background of the Aluminum Industry, Royal Institute of Chemistry, Lectures, Monographs, and Reports No. 3, London, 1955

# References

1. R.N. Shreve and J.A. Brink Jr., Chemical Process Industries, 4th edition, McGraw-Hill, New York, 1977, page 226
2. Canadian Minerals Yearbook 1977, Energy Mines and Resources Canada, Minister of Supply and Services, Hull, 1979, page 9
3. Kirk-Othmer Encyclopedia of Chemical Technology, 3rd edition, John Wiley, New York, 1978, volume 2, page 129
4. McGraw-Hill Encyclopedia of Science and Technology, McGraw-Hill, New York, 1977, Vol. I, page 326
5. Kirk-Othmer Encyclopedia of Chemical Technology, 2nd edition, John Wiley, New York, 1963, volume 1, page 929
6. The Chemistry of Aluminum, Aluminum Co. of Can. Ltd, Montreal, 1970
7. Handbook of Chemistry and Physics, 51st edition, R.C. Weast, editor, Chemical Rubber Co., Cleveland, Ohio, 1970, page D-111
8. S.H. Patterson, Amer. Scientist 65, 345, May-June 1977
9. Nonbauxite Aluminum Technology, Chem. Eng. News *52*, 19, Dec. 16, 1974
10. M.K. McAbee, Chem. Eng. News 53 (31), 19, Aug. 4, 1975
11. C.N. Cochrane, Calculated Equilibrium for Carbothermic Reduction of Alumina, Electrochemical Soc. Meeting, Toronto, May 1975. Cited by reference 3

12. Alcan Pilots, Can. Chem. Proc. *51* (2), 45 (1967)
13. Recovery of Aluminum, Can. Chem. Proc. *51* (3), 75 (1967)
14. New Refining Processes, Chem. Eng. News *55* (37), 7, Sept. 12, 1977
15. New Aluminum Smelting Process, J. Metals *29*, 6, Nov. 1977
16. Alcan's Work, Can. Chem. Proc. *57* (3), 52, Feb. 1973
17. New Processes Promise Lower Cost Aluminum, Chem. Eng. News *51*, 11, Feb. 26, 1973
18. Aluminum Process, Chem. Eng. News *57*, 23, April 30, 1979
19. N.L. Nemerow, Industrial Water Pollution, Origins, Characteristics and Treatment, Addison-Wesley, New York, 1977, page 502
20. Bauxite Waste Tests OK, Can. Chem. Proc. *60* (3) 26, March 1976
21. Red Mud Converted to Steel, Chem. Eng. News *57* (9) 26, Feb. 26, 1979
22. Industrial Pollution Control, H.F. Lund, editor, McGraw-Hill, New York, 1971, page 4-20
23. R.E. Iverson, J. of Metals, *25*, 19, Jan. 1973
24. R.C. Brewer et al, Environmental Effects of Emissions from the Alcan Smelter at Kitimat, B.C., Ministry of Environment, Province of British Columbia, April 1979, Ms. 151 pp
25. K.T. Semrau, J. Air Polln Control Assoc. 7, 92, Aug. (1957)
26. D.F. Ball and P.R. Dawson, Chem. and Proc. Eng. *52* (6), 49, June 1971
27. Pollution Control costs in the Primary Aluminum Industry, Organization for Economic Cooperation and Development, Paris, 1977
28. P. Macuch, E. Hluchan, J. Mayer, E. Able, Fluoride *2*, 28, (1969)
29. C.E. Carlson, W.E. Bousfield, and M.D. McGregor, Fluoride *10* (1), 14, Jan. 1977
30. G. Balazova and V. Lipkova, Fluoride *7* (2), 88, April 1974
31. E. Groth III, Fluoride *8* (4), 224, Oct. 1975
32. T. Keller, Eur. J. For. Path. *4*, 11 (1974)
33. F. Ender, The Effect of Air Pollution on Animals, in Proc. of the 1st European Congr. on the Influence of Air Pollution on Plants and Animals, Centre for Agric. Publ. and Documentation, Wageningen, 1969
34. F. Leblanc, G. Comeau, and D.N. Rao, Can. J. Bot. *49*, 1,691 (1971)
35. J. Hemens and R.J. Warwick, Water Research *6*, 1,301 (1972)
36. D. Rose and J.R. Marier, Environmental Fluoride, National Research Council of Canada, Ottawa, 1977
37. J.O. Ares, J. Air Polln Control Assoc *28* (4), 344, April 1978
38. C. van Hook, Fluoride *7* (4), 81, Oct. 1974
39. M.B. Hocking, D. Hocking and T.A. Smyth, Water, Air, and Soil Pollution *14*, 133 (1980)
40. E. Kay, Fluoride *7* (1), 7, Jan. 1974
41. R.M. Harbo, F.T. McComas and J.A.J. Thompson, J. Fisheries Res. Board Can. *31*, 1,151 (1974)
42. B. Bohlen, The Chem. Eng. *221*, CE 266, Sept. 1968
43. R. Wood, Proc. Eng. 7, Sept. 1976
44. A.S. Reid, G.J. Gurnon, W.D. Lamb, and K.F. Denning, Effect of Design and Operating Variables on Scrubbing Efficiencies in a Dry Scrubbing System for V.S. Pots, In: 109th A.I.M.E. Annual Meeting, 1980, New York, pages 759-781
45. D. Rush, J.C. Russell, and R.E. Iversen, J. Air Polln Control Assoc. *23* (2), 98, Feb. 1973
46. F.A. Cotton and G. Wilkinson, Advanced Inorganic Chemistry, A Comprehensive Text, 3rd edition, Interscience, New York, 1972, page 262
47. J. Miller and D.F. Nasmith, Operation of Alcan/ ASV Dry Scrubbing Units with Various Aluminas, In: 102nd A.I.M.E. Annual Meeting 1973, A.V. Clack, editor, New York, pages 191-208
48. C.C. Cook, G.R. Swany, and J.W. Colpitts, J. Air Poll'n Control Assoc. *21* (8), 479, Aug. 1971
49. Aluminum Cell Modifications Cut Energy Use, Chem. Eng. News *53* (31) 19, Aug. 4, 1975
50. J. Hemens, R.J. Warwick, and W.D. Oliff, Prog. in Water Technology, *7* (3/4) 579, (1975)
51. Aluminum, Chem. and Eng. News *56*, 14, Sept. 25, 1978
52. A. Jenny and W. Lewis, The Anodic Oxidation of Aluminum and its Alloys, C. Griffin, London, 1950
53. R. Stepler, Pop. Science *207* (5), 150, Nov. 1975
54. R.L. Horst, Ind. and Eng. Chem. *49* (9), 1,578, Sept. 1957
55. D.M. Samuel, Industrial Chemistry — Inorganic, 2nd edition, The Royal Institute of Chemistry, London, 1970
56. Alumina Fibers Used to Strengthen Metals, Chem. Eng. News *58* (26), 24, June 30, 1980
57. Aluminum Sulfate, Can. Chem. Proc. *60* (9), 130, Sept. 1976
58. F.A. Lowenheim and M.K. Moran, Faith, Keyes and Clark's Industrial Chemicals, 4th edition, Wiley-Interscience, New York, 1975
59. Aluminum sulfate, Can. Chem. Proc. *57* (4), 10, April 1973
60. Chemical Marketing Reporter, *219* (24), June 15, 1981, (Schnell Publ. Co., New York)
61. R. Wood, Proc. Eng. 9, Sept. 1976
62. Minerals Yearbook, 1980, Volume 1, Metals and Minerals, U.S. Bureau of Mines, Washington, 1981, and earlier years
63. World Mineral Statistics 1975-79, Institute of Geological Sciences, Her Majesty's Stationery Office, London, 1981, and 1972-76 issue.
64. Air Pollution by Fluorine Compounds from Primary Aluminum Smelting, Organ. Economic Cooperation and Development, Paris, 1973, page 12

# 11 Ore Enrichment and Smelting of Copper

## 11.1 Early Development

Copper rivals gold as one of the oldest metals employed by man. Its first use about 10,000 years ago was stimulated by the natural occurrence of the metal in lumps or leaves in exposed rock formations, so called "native copper". These exposures enabled the fashioning of simple tools directly by hammering and heat working of these fragments, and in so doing signalled the end of the Stone Age [1]. The natural occurrence of elemental copper is a feature of its relatively low standard reduction potential of $+ 0.158$ volts, significantly below hydrogen in the electromotive series and hence relatively easily accumulated in elemental form as a consequence of normal geologic processes. This contrasts with the much more easily oxidized aluminum, which lies above hydrogen in the electromotive series with a reduction potential of $-1.71$ volts, and hence is never found in elemental form in nature.

More purposeful recovery of the metal by mining is known to have occurred since about 3,800 B.C. from early workings discovered in the Sinai. Mines operating in Cyprus around 3,000 B.C. were later taken over by the Roman Empire, and the derived metal product was called *cyprium*, later simplified to *cuprum*, the origin of the Latin name still used for the metal. The incentive for the development of these early metal mines probably derived from the experiments of early man who first relied upon simple hammering of native copper lumps, subsequently elaborated to heat working and hammering, and eventually leading to a need to melt and cast the metal to allow more fabrication scope [2]. It must have first been discovered that coal or charcoal firing was necessary to obtain sufficiently high temperatures to melt the native metal, which is apparently not possible with even a large, hot wood fire. At some time after this must have come the revelation that some highly favourable ores of copper, which did not originally look like copper, could be reduced to the metal under these conditions [3]. The series of steps required for early experimenters to proceed from a knowledge of the requirements to melt metal to the intentional practice of ore reduction by heat in the presence of carbon is still open to speculation [4]. The reddish sheen of the polished metal undoubtedly provided impetus to the early practitioners to develop all forms of fabrication for decorative and artwork reasons as well as for the original weapons, tools and other primarily practical purposes.

From these early beginnings world copper production reached about 8,100 tonnes per year by 1750, 19,000 by 1860, and 50,000 by 1880 [1, 4]. This early production was dominated by smelters in Mansfeld, Germany, and Swansea, Wales [5], the latter center accounting for some three quarters of known world production in 1860, and yet the importance of both centers for the smelting of indigenous ores today is slight to non-existent (Table 11.1). The Rio Tinto Co. in Spain was dominant by 1880 producing in the neighbourhood of 25,000 tonnes of the metal per year [6], and this country still has a significant output though production levels have fluctuated widely in recent years. By 1900, the Anaconda Company, Montana was producing around 50,000 tonnes per year. But all of the production development described above relied on relatively rich (for copper) grades of ore containing from 2-5 % copper [7].

Much larger scale mining and smelting of formerly substandard ores (ores of too low a copper analysis for practical smelting) became feasible and was practised following the development around 1920 of froth flotation methods of ore enrichment [8]. World production, which had reached levels of 1 million metric tonnes by 1912 [7], climbed further to $1^1/2$ million metric tonnes by 1930 with further growth to a volume of 2.3 million tonnes for both 1940 and 1950 [9], stimulated jointly by growing demand and the accessibility to lower grade reserves that the new enrichment methods provided. Froth flotation enrichment of ores has also stimulated a decreased dependence of smelter location on ore location since ore concentrates of 25 to 30 % of a high value metal such as copper selling at 2.2 to 2.85 US \$/kg in 1980 [10], could justify shipping significant distances for eventual metal recovery.

**Table 11.1** Major world producers of copper concentrates and primary smelted copper [a]

| | Thousands of metric tonnes, copper content | | | | | | | |
|---|---|---|---|---|---|---|---|---|
| | 1960 | | 1970 | | 1975 | | 1980 | |
| | ore | metal | ore | metal | ore | metal | ore | metal |
| Australia | 111 | 72 | 142 | 111 | 236 | 189 | 217 | 171 |
| Canada | 399 | 361 | 610 | 453 | 734 | 494 | 710 | 463 |
| Chile | 536 | 502 | 711 | 658 | 831 | 724 | 1 068 | 953 |
| China | n.a. | n.a. | 100 | 100 | 100 | 100 | 200 | 200 |
| Japan | 89 | 188 | 120 | 513 | 85 | 822 | 53 | 861 |
| Peru | 182 | 164 | 218 | 177 | 176 | 157 | 365 | 350 |
| Papua New Guinea | – | – | – | – | 173 | – | 147 | – |
| Philippines | 44 | – | 160 | – | 226 | – | 324 | – |
| Poland | 11 | 22 | 83 | 69 | 230 | 263 | 346 | 364 |
| South Africa | 46 | 46 | 148 | 149 | 179 | 179 | 215 | 181 |
| Spain | 8 | 8 | 10 | 55 | 31 | 110 | 42 | 95 |
| U.S.A. | 980 | 1 119 | 1 560 | 1 489 | 1 282 | 1 313 | 1 168 | 1 008 |
| U.S.S.R. | n.a. | n.a. | 925 | 925 | 1 100 | 1 100 | 900 | 905 |
| W. Germany | 2 | 62 | 1 | 80 | 2 | 168 | 1 | 160 |
| Yugoslavia | 33 | 36 | 91 | 108 | 115 | 162 | 134 | 110 |
| Zaire | 302 | 302 | 387 | 386 | 496 | 463 | 459 | 426 |
| Zambia | 550 | 568 | 836 | 683 | 806 | 640 | 596 | 617 |
| Other | 347 | 220 | 718 | 364 | 598 | 446 | 945 | –37 [c] |
| World Totals | 3 640 [b] | 3 670 [b] | 6 500 | 6 320 | 7 400 | 7 330 | 7 890 | 6 828 |

[a] 1975 and earlier data compiled from [78].
  Most recent data obtained from Minerals Yearbook [10], and hence may not entirely correspond to earlier statistics.
[b] Excludes China, Czechoslovakia, Hungary, North Korea and the U.S.S.R.
[c] Negative value due to rounding and the inclusion of undifferentiated production, as to whether primary or secondary (recycled) e.g. for Poland.
n.a. Not available.

For this reason some countries are large scale ore producers only, e.g. Papua New Guinea and the Philippines, some produce both ore and metal but more of the former, e.g. Australia and Canada, many countries produce roughly equivalent amounts of ore and metal, and a few, e.g. Japan and West Germany, produce metal from mostly imported ore concentrates (Table 11.1). Another feature characteristic of the value of the metal is that copper is extensively re-used. About $^1/_3$ of the current annual consumption is provided by recycled scrap copper, and it is estimated that nearly 60 % of the virgin metal produced is ultimately re-used [1, 10].

# 11.2 Ore Occurrence and Beneficiation of Low Grades

While copper, being a less active metal, does occur naturally in elemental form (e.g. native copper de-posits occur in the "thumb" of Michigan bordering on Lake Superior and at Afton Mine, near Kamloops, British Columbia) the bulk of the commercial metal is produced from sulfide or oxide minerals. Common sulfide minerals are chalcopyrite ($Cu_2S$ x $Fe_2S_3$; 34.5 % Cu), chalcocite ($Cu_2S$; 79.8 % Cu), covellite (CuS; 66.4 % Cu), and hornite ($Cu_2S$ x CuS x FeS; 55.6 % Cu), with many variations possible, both among these minerals and in admixtures with other sulfides. Sulfides of arsenic, antimony, gold, mercury, silver, and zinc in particular, are found with copper ores because of the similarity of some of their chemical properties. The oxide minerals, which are less common than the sulfides, occur as cuprite (red copper oxide, $Cu_2O$; 88.8 % Cu), tenorite (black copper oxide CuO; 79.8 % Cu), malachite (basic, or green copper carbonate, $CuCO_3$ x $Cu(OH)_2$; 57.3 % Cu), azurite (2 $CuCO_3$ x $Cu(OH)_2$; 55.3 % Cu) and other variations.

In the Americas, the minerals containing copper

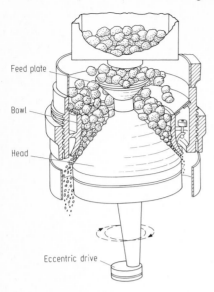

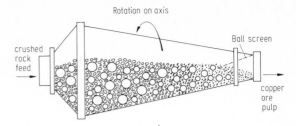

**Figure 11.2.** Cross section of a ball mill used for the wet grinding of copper ore preparatory to froth flotation

**Figure 11.1.** Diagram of a cone crusher. (Inco Ltd.)

occur chiefly in porphyry form, i.e. they are igneous in origin and are widely dispersed in the rock. In Africa and Russia, however, prophyry ores account for less than 20 % of the total. In all areas copper *ores,* or deposits of copper minerals containing a sufficiently high copper concentration to permit economic metal recovery, have a relatively low copper analysis, in the range of 0.3 to 4 % Cu. Many mines operate today on ores of 1 % or lower copper content. Thus, a significant consideration for economical metal production is efficient and low cost removal of igneous rock and recovery of the copper bearing mineral from this.

To minimize ore recovery costs open pit mining is used wherever feasible. For instance in the U.S.A. in 1970, open pit mines accounted for 84 % of mine output (e.g. by the Kennecott Copper Corporation, Salt Lake City) and only 16 % was recovered from underground mining operations [11]. Additionally, by the use of low cost explosives such as ammonium nitrate/fuel oil or aluminum powder supplemented or slurry variants of this, and large scale loading and haulage units, mining costs to withdraw the ore can generally be kept to less than 25 % of the cost of the refined metal [1]. Overburden containing little or no metal value is removed to a spoils disposal area, and the ore with some preselection proceeds to the concentrating circuit.

Concentrating, because of the wide dispersal and hence initial inaccessibility of the valued mineral in the rock, first requires size reduction. Crushing, in two stages in gyratory cone (Figure 11.1) or jaw crushers decreases the initial, randomly-sized masses to first 15-22 cm average diameters, and then in a second unit to 2-5 cm (walnut) sized fragments. The walnut size range is then suitable as feed to rod or ball mills (Figure 11.2) for final wet grinding to 65 to 200 mesh. A "100 mesh" sieve has a nominal 100 openings per lineal inch, equivalent to 10,000 openings per square inch. Thus, a particle passing through a 100 mesh sieve will have an average diameter of 0.01 inch, less the sieve wire thickness of 0.0043 inches or about 0.0057 inches (ca. $0.149 \mu m$). For this particular example then, the 65 to 200 mesh range equates to particle diameters in the 230 to 74 $\mu m$ (0.0090-0.0029 inch) range, or averaging 150 $\mu m$ (0.006 inch). Attainment of a larger particle size range than this decreases grinding costs, but also decreases mineral recovery from the rock by not exposing all copper minerals on the surface of a particle, which is required for flotation. Aiming at a smaller size range than this ensures mineral liberation from the associated rock, thus increasing the selectivity of separation in flotation, but raises grinding costs and may permit losses of non-floated mineral-containing fines from the flotation circuit [12]. For these reasons the ground ore in water produced by the ball or rod mill is sorted in a cyclone (or similar) classifier into a fines stream, which is fed to the flotation circuit, and a coarse stream, which is returned to the grinding circuit for further size reduction.

### 11.2.1 Beneficiation of Low Grade Ores by Froth Flotation

Froth flotation is a procedure used to raise the low initial concentrations of minerals in ores to values that can be economically treated by smelting or hydrometallurgical extraction processes. For copper, the 25-30 % range is suitable for economical smelting. The froth flotation technique was originally developed in about 1910 to raise the copper concentrations of the strip mined ores of Bingham Canyon, near Salt Lake City [13, 14, 15], and was further perfected for the differential separation of lead, zinc, and iron sulfides at Trail, B.C. at about the same time [8]. Now flotation technologies are widely used for such diverse tasks as the beneficiation of lower grade Florida phosphate ores from 20-40 % to 60-70 % concentrations of calcium phosphate (BPL), the separation of about 98 % potassium chloride from sylvinite, a natural mineral mixture of potassium and sodium chlorides, as well as several other mineral separations and differentiations. It is also an important technique for bitumen separation from tar sand, removal of slate from coal, and removal of ink from repulped paper stock preparatory to the manufacture of recycled paper stock.

In general, flotation separations rely on the attachment of properly prepared selected mineral surfaces to air bubbles in water. Surfaces which are easily wetted by water, such as ordinary ground rock, are not attracted to bubbles and thus sink. Elemental sulfur, graphite (and many other forms of elemental carbon), and talc, minerals with layer lattice type structures, have hydrophobic surfaces which are attracted to the surface of an air bubble and, if small enough, may be lifted to the surface of the water and separated from the water as a froth layer.

This distinction in behaviour in aerated water is primarily dependant on the surface activity, or more precisely wettability, natural or induced, of the material under study [16]. Three types of behaviour are observed (Figure 11.3). If a solid surface is hydrophilic, when a bubble is pushed against it the contact angle $\Theta$, is zero (non-existent) or very small and the surface is relatively easily water wetted. If the bubble is released from a holder it will not stick to the surface, nor will small particles of the same material stick to a bubble. Finely ground material of this type, such as rock gangue, will sink.

If the test surface is neither readily wetted nor hydrophobic, i.e., it is indifferent to water, the contact angle is zero or very small and poor differentiation of the unprepared surface can be expected. However, the behaviour of wetted and indifferent surfaces can often be readily modified by treatment with surface active agents to obtain either a sink or a float behaviour, as desired.

If the solid surface is hydrophobic it will tend to develop contact with the air of a bubble pressed against it to give distinct contact angles. For example the contact angle for paraffin wax is 104 °. In this situation, there is a significant force component perpendicular to the solid surface to $\gamma LA$, the liquid-air interface of the force vector diagram (Figure 11.3), and hence there exists a significant force retaining the bubble to the solid surface. Small particles of this type of material will readily cling to small air bubbles and may be raised to a froth in a flotation cell.

The specific reduction in energy, $W_{SA}$, from particle adhesion to, rather than release from a bubble by interaction of the three relevant interfaces is given by Equation 11.1,

$$W_{SA} = \gamma_{SL} + \gamma_{LA} - \gamma_{SA} \tag{11.1}$$

credited to Dupré [15], where S, L, and A repre-

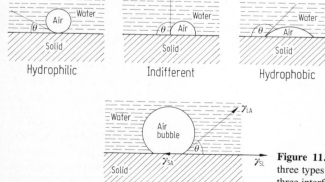

**Figure 11.3.** Behaviour at the air-water interface of three types of solid surfaces and the relationship of the three interfacial tensions involved

sent the solid, liquid and air surfaces, respectively (Figure 11.3). Combining this equation with the Young equation [17], 11.2,

$$\gamma_{SA} = \gamma_{SL} + \gamma_{LA}\cos\theta \qquad (11.2)$$

which relates the equality of the components of the three interfacial tensions involved, allows simplification of the Dupré equation to the form of Equation 11.3.

$$W_{SA} = \gamma_{LA}(1 - \cos\theta) \qquad (11.3)$$

Since $\gamma_{LA}$ for water (ca. 0.07 mbar) is readily determined and for a particular surfactant concentration in water it is lower than for pure water and relatively constant, the contact angle, $\Theta$, is thus also theroretically shown to be a significant variable factor in the work required to obtain particle adhesion. However, it is known that several other factors such as particle and bubble size ranges, the nature and age since grinding of the particle surface, the pH of the aqueous phase, and its influence on particle-bubble interaction via neutralization of electrostatic charges on particles, as well as hydrodynamic effects occurring in agitated mineral suspensions are also variables of importance [18, 19].

Fine particles of many sulfide minerals including those of copper may be, like sulfur, naturally hydrophobic. If they are subjected to flotation soon after grinding they will tend to stick to bubbles and to be raised to a froth. Silica and gangue are water wetted and will tend to sink.

The natural differentiation of valuable mineral from gangue in this way is good, but the "sticking" of sulfide minerals to air bubbles can be improved by the addition of a collector to the water phase. Typical collectors for sulfide minerals are xanthates (ROC(S)S⁻ Na⁺) e.g. potassium ethyl xanthate, $CH_3CH_2OC(S)S^-K^+$ or dithiophosphates ((RO)$_2$P(S)S⁻ Na⁺) e.g. sodium diethyldithiophosphate, $(CH_3CH_2O)_2P(S)S^- Na^+$ which all have in common an oil-attractive (or hydrophobic) group and a sulfide-attractive group [12, 20]. Added to the pulp of ground up ore in water to the extent of 10 to 45 g per tonne of ore processed and just before flotation these collectors confer greater than the natural water repellency and air avidity to the particles of desired mineral, in this case copper sulfide minerals, and in this way improve both the proportion of the total mineral recovered from the gangue and the rejection of gangue in the floated material.

However, if a bubble carrying air-avid mineral particles reaches the surface of the flotation unit and then bursts, the froth raised mineral particles will simply sink again. This problem, one of bubble stability, is minimized by addition of a foam stabilizer or frother which assists in generating a foam layer on the flotation unit sufficiently long-lived to enable skimming of the foam plus associated mineral from the water phase. Usually the frother also serves a function of putting an oily phase on the surface of each bubble as it forms, which serves to aid the mineral-gangue differentiation function of the collector. Typical frothers, used at the rate of 20 to 45 g per tonne of ore, are oily materials of no more than slight water solubility such as pine oil (a mixture of terpenes), or a long chain ($C_5$ or higher) alcohol such as 1-pentanol.

In addition to these two primary flotation agents there are also a number of modifying and conditioning reagents and procedures which are also important to control a variety of ore and flotation circuit variables [18]. For instance, lime is a depressant for pyrite ($FeS_2$) in the flotation separation of copper sulfides [1]. Lime addition would normally take place to the grinding circuit of a copper concentrator to provide sufficient time before flotation is carried out for oxidation of pyrite and the pH regulating effect to take hold. Sodium carbonate and sodium hydroxide are also common pH regulating agents. Molybdenite ($MoS_2$), a valuable by-product of copper ore processing, is at first floated with copper sulfides then depressed, to the underflow of a separate flotation cell, by the addition of colloidal dextrin to the pulp feed. There are also additives, such as sodium cyanide, which are used occasionally to cause depression of some minerals [21], as well as dispersants or deflocculants to avoid non-selective aggregation of particles of mineral value with particles of gangue.

Separation of copper sulfide minerals from the gangue is normally carried on in a series of large rectangular steel tanks of capacities from 2 to 28 m³ equipped with stirrers and compressed air inlets (Figure 11.4) to accomplish the required vigorous agitation and air blowing through a properly conditioned ground ore pulp [22]. The froth collected from a series of rougher cells is then passed to a set of cleaner cells to raise further the copper sulfide concentrations initially obtained (Figure 11.5). The gangue-rich underflow from the rougher cells is passed through a series of scavenger cells to collect any residual copper sulfide

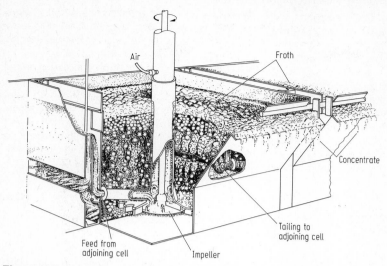

**Figure 11.4.** A cutaway drawing of a froth flotation unit for the concentration of copper ores with a detail photomicrograph showing mineral-loaded bubbles. (Inco Ltd.)

values not picked up by the rougher cells. Dried flotation concentrates obtained in this way may contain from 20 to 30 % copper, and are frequently in the 25-28 % range. In the concentration process 80 to 90 % of the metal content of the ore feed is recovered. The discarded gangue may still contain ca. 0.05 % copper, which may or may not be processed further to recover additional copper values (see Hydrometallurgy).

Copper (or metal) recovery percentages of the flotation circuits of a copper mine/mill are extremely important. For example it is estimated that for a plant processing 50,000 tonnes of ore containing 1 % copper per day, a large but not nearly the largest scale of this kind of operation, increasing the copper ecovery from 86 to 87 % of that contained in the ore could increase revenues by approximately $ 4 million annually. This revenue increase, estimated on the basis of a value of Can $ 1.10/kg allowed for the copper content of the

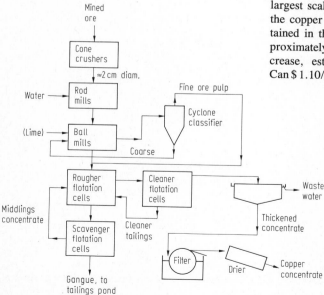

**Figure 11.5.** Flowhsheet of copper sulfide ore pathway, mined ore to dried concentrate ready for smelting. In some operations there may be additional cells and the thickener, such as a molybdenite ($MoS_2$) recovery circuit or the like

concentrate at a time (late 1980) when the metal was selling at Can $ 2.25/kg, would be experienced for basically the same overhead costs. Hence the great importance which is attached to seemingly minor flotation variables in an attempt to gain incremental improvements in metal recovery. Flotation reagents are placed in the wet pulp prior to flotation in the ppm range of concentrations, at a cost of 10-35 cents per tonne of ore. Minor changes in the absolute and relative concentrations of these cause a nominal change in the cost and may exercise a significant influence on differential surface activities.

## 11.2.2  Smelting of Concentrates Derived from Copper Ores

Much of the copper smelted today still uses procedures very similar to those developed by the early

Welsh copper smelting industry [3, 5], although the details of the chemistry involved are better understood today. The procedures required may be divided into four more or less discrete steps.

Roasting involves heating of the concentrate in the presence of air to temperatures of 750-800 °C in a multiple hearth, or fluidized bed roaster, the charge remaining more or less solid throughout. This ensures completion of the drying step as well as serving to remove a part of the sulfur from both the copper and iron present, producing oxides from some of the iron (Equations 11.4-11.7).

$$2\,CuS \rightarrow Cu_2S + S \tag{11.4}$$

$$FeS_2 \rightarrow FeS + S \text{ (pyrite pyrolysis)} \tag{11.5}$$

$$S + O_2 \rightarrow SO_2 \tag{11.6}$$

$$2\,FeS + 3\,O_2 \rightarrow 2\,FeO + 2\,SO_2 \tag{11.7}$$

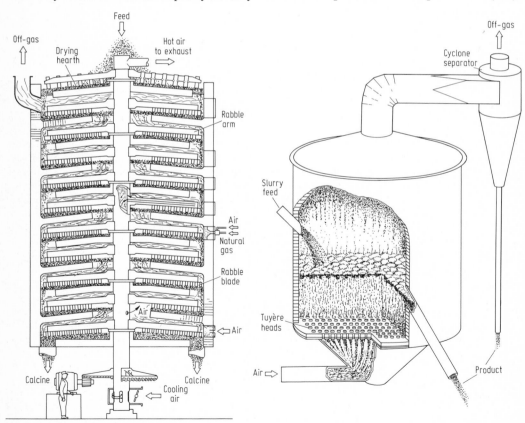

**Figure 11.6.** Diagrams of multiple hearth type and fluidized bed roasters, such as used to decrease the sulfur content of copper sulfide ores. (Inco Ltd.)

Some external fuel, usually oil or gas, is required initially to raise the roaster temperature to the desired range, and smaller amounts subsequently to maintain these temperatures since the sulfur combustion is highly exothermic and assists in providing much of the heat required. In fact, if the ore concentrate analyzes more than about 25 % sulfur and is dry, sufficient heat is evolved from the sulfur burned to the extent desired to maintain the roaster temperature without requiring external fuel sources except at the start. This favourable situation is often called autogenous roasting.

Some sulfur trioxide also forms in the roaster, even though this is not the direct objective of the operation, forming small amounts of ferrous sulfate in the roaster output (Equation 11.8).

$$SO_3 + FeO \rightarrow FeSO_4 \tag{11.8}$$

However, in order to obtain a useful product from formerly discharged waste sulfur dioxide, many modern roasting operations now collect and use the sulfur dioxide present in roaster exit gases to produce sulfuric acid. Fluidized bed roasters, in particular, are favoured for this purpose since the sulfur dioxide concentrations obtained may be as high as 15 %, as opposed to the 3-6 % range normally expected from travelling grate or multiple hearth roaster designs.

Roasting is followed by reverberatory smelting, carried out on the still solid, hot roaster output. Silica, as a fluxing agent, and slag from a previous converter stage are added to the roasted concentrate and then external heat is applied by the combustion of powdered coal, oil, or natural gas, and reflection of the flames and hot gases down from the ceiling of the furnace onto the mixed charge (hence, "reverberatory" furnace). This stage of smelting, where the charge temperature is gradually raised from about 600° to 1,000 °C, accomplishes removal of more of the sulfur, by burning. Much of the iron is also removed as it combines with the fluxing agent to form a lighter ferrous silicate slag which floats on top of the copper melt, as it forms (Equations 11.9-11.13).

$$Cu_2O + FeS \rightarrow Cu_2S + FeO \tag{11.9}$$

$$2\,Cu + FeS \rightarrow Cu_2S + Fe \tag{11.10}$$

$$2\,Fe + O_2 \rightarrow 2\,FeO \tag{11.11}$$

$$Cu_2S + 2\,CuO \rightarrow 4\,Cu + SO_2 \tag{11.12}$$

$$FeO + SiO_2 \rightarrow FeSiO_3 \text{ (slag)} \tag{11.13}$$

The transfer of a part of the iron-bound sulfur to copper, which is an objective of this stage, is the result of the change in the relative heats of formation of cuprous and ferrous sulfides at these high temperatures (e.g. at 1,300 °C, $\Delta H = -174$, and 34.4 kJ respectively) relative to those at 25 °C ($\Delta H = -79.5$, and 95.1 kJ respectively).

The product of the reverberatory smelting stage is matte copper, containing primarily copper metal and cuprous sulfide, and analyzing anywhere from 15 (particularly with non-concentrated ores) to 50 % copper [1]. Ideally, the copper concentration in matte copper is kept at 40-45 % Cu to keep the matte more effective as a collector of precious metals, and to keep the copper loss in the slag to a minimum. Insufficient sulfur (as $Cu_2S$, melting point 1,100 °C) in the matte also makes it difficult to keep the charge molten (melting point of Cu, 1,083 °C; which is eutectic depressed in the matte copper).

Converting, the third stage of copper smelting, comprises blowing air through the mass of molten matte copper with the objective of completing the oxidation of sulfides, since by this stage the bulk of the iron has been removed (Equations 11.14, 11.15).

$$2\,Cu_2S + 3\,O_2 \rightarrow 2\,Cu_2O + 2\,SO_2 \tag{11.14}$$

$$Cu_2S + 2\,Cu_2O \rightarrow 6\,Cu + SO_2 \tag{11.15}$$

Any residual iron which may be present is oxidized and forms a further slag layer with an additonal small portion of added silica (Equation 11.13). This slag layer also serves to capture any traces of oxidized arsenic and antimony, if present, that have not already been volatilized by the blowing. The less volatile metal oxides remaining are what is appropriately called lithophilic, and thus tend to be absorbed by the slag layer. For the same reason, a certain amount of oxidized copper also enters this slag layer, but is recaptured when this slag is used as a part of the charge for subsequent reverberatory smelting. Any precious metals present, silver, platinum, and gold are relatively stable towards oxidation and remain with the copper. The product of converting is termed blister copper, from the pock marks on the surface formed by the escape of sulfur dioxide bubbles from the metal as it cools and solidifies. It is virtually free of sulfur, but contains from 0.6 to 0.9 % oxygen [1] (ca. 5.4-8.1 % $Cu_2O$) at this stage from the more exhaustive oxidation.

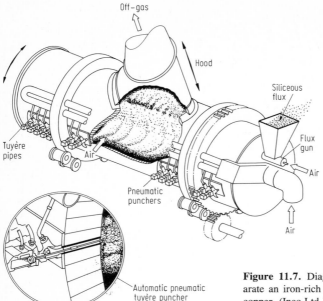

Off-gas

Hood

Siliceous
flux

Flux
gun

Tuyère
pipes

Air

Air

Pneumatic
punchers

Air

Automatic pneumatic
tuyère puncher

**Figure 11.7.** Diagram of a side blown converter used to separate an iron-rich ferro-silicate slag from a sulfur-rich matte copper. (Inco Ltd.)

Fire refining, the final stage of copper smelting, is used to reduce the small percentage of cuprous oxide present in the blister copper back to elemental copper. It is often carried out in the converter itself, once the air flow has been cut off and while the charge is still molten. Originally fire refining was accomplished by the addition of coke and green wood poles, which were firmly pushed under the surface of the melt, to bring about reduction of cuprous oxide by the carbon of both additives. Volatilization of the moisture and partial combustion of the green wood also brought about vigorous agitation of the melt via the subsurface release of steam, carbon dioxide, carbon monoxide, and hydrogen which also combined to add to the reducing action of the carbon (Equations 11.16-11.18).

$$2\,Cu_2O + C \rightarrow 4\,Cu + CO_2 \qquad (11.16)$$

$$Cu_2O + CO \rightarrow 2\,Cu + CO_2 \qquad (11.17)$$

$$Cu_2O + H_2 \rightarrow 2\,Cu + H_2O \qquad (11.18)$$

More frequently now, this "poling" step is carried out in a better controlled manner by the subsurface passage of natural gas or ammonia into the melt by water (or gas) cooled hollow steel lances. Fire refined copper, the product of this smelting stage, normally gives a copper analysis of 99 % or better, particularly if the preliminary smelting stages have

been conducted with care. About 15 % of copper uses, among them the preparation of various alloys whose properties are not seriously affected by small amounts of impurities, directly employ fire refined copper. But the properties of the metal are seriously impaired for some uses by the presence of traces of impurities, making further refining necessary for this reason. Phosphorus, arsenic, aluminum, iron, or antimony in particular cause relatively sharp reductions in electrical conductivity, i.e. some 6 % loss is experienced with the presence of < 50 ppm phosphorus [7]. So for reasons such as this, as well as an interest in the separation and recovery of the precious metal values which will still be present in the fire refined copper if found in the original ore, the bulk of fire-refined copper is further purified by electro-refining.

## 11.2.3 Electro-refining of Smelted Copper

Electrolytic purification of smelted copper is primarily carried out to remove contaminants which adversely affect electrical conductivity, malleability, and other properties of the metal. It also permits recovery of the precious metal content of the fire refined product. Anodes for electro-refining must first be cast from the smelted metal, with care being taken to ensure a uniform thickness across each anode casting as well as from one anode to the next. Sections, or whole anodes which are made too thick give too much scrap anode material for recasting when the spent anodes are changed after a predetermined extent of electrolysis. Newly cast anodes which are too thin or irregular in thickness can cause pieces of metal to drop off in the electrolytic cells as electrolysis proceeds, resulting in electrical shorts and decreased current efficiencies. Anode slabs are connected in parallel to heavy bus bars (high current conductors) and placed in the electrolytic cells, long rectangular vessels of wood or cement lined with lead or plastic (Figure 11.8). Interleaved between the thick anode castings of fire refined metal to be purified are thin starting sheets of electrolytically pure copper which are connected to the cathode (negative) bus bar. The electrolyte consists of ca. 20 % sulfuric acid (ca. 2 molar) which also contains 20 to 50 g/L (0.3 to 0.8 molar) cupric ion. Traces, up to a combined total of 20 to 30 g/L arsenic, antimony, bismuth, and nickel gradually accumulate as a result of the dissolution of these electrolytically more active metals (Table 11.2) from the anodes and which are not subsequently deposited on the cathodes. A low concentration of chloride (ca. 0.02 g/L) from sodium chloride or hydrochloric acid is also maintained in the electrolyte to ensure that silver is precipitated on the bottom of the cell as finely divided silver chloride, rather than being deposited with the copper on the cathodes [1]. As electrolysis proceeds copper is dissolved from the anodes, and is simultaneously deposited from the electrolyte

**Table 11.2** Standard reduction potentials for a series of elements of importance to the electrolytic purification, arranged in order from the most, to the least active

| Ion reduced | Standard reduction potential, volts |
| --- | --- |
| $Ca^{2+}$ | $-2.76$ |
| $Mg^{2+}$ | $-2.38$ |
| $Al^{3+}$ | $-1.71$ |
| $Fe^{2+}$ | $-0.41$ |
| $Ni^{2+}$ | $-0.23$ |
| $Pb^{2+}$ | $-0.126$ |
| $H^+$ | $0.0$ |
| $Sb^{3+}$ | $+0.145$ |
| $As^{3+}$ | $+0.234$ |
| $Bi^{3+}$ | $+0.320$ |
| $Cu^{2+}$ | $+0.158$ |
| $Te^{4+}$ | $+0.63$ |
| $Ag^+$ | $+0.80$ |
| $Pd^{2+}$ | $+0.83$ |
| $Pt^{2+}$ | ca. $+1.2$ |
| $Au^{3+}$ | $+1.42$ |
| $Au^+$ | $+1.68$ |

onto the cathodes causing a movement of metal between the electrodes (Equations 11.19, 11.20),

Anode reaction: $Cu^0 - 2\,e \rightarrow Cu^{2+}_{(aq)}$    (11.19)

Cathode reaction: $Cu^{2+}_{(aq)} + 2\,e \rightarrow Cu^0$  (11.20)

but no overall electrochemical reaction. Hence, the theoretical voltage required is zero, the small applied voltage of about 0.2 volts is required for the process solely to drive the electrolytic purification in the desired direction. Indirect heating of the electrolyte to ca. 60 °C decreases solution viscosity and in this way assists in maintaining a high production rate at these low operating voltages. Current densities of ca. 240 A/m$^2$ are normal, although higher densities of up to 350 A/m$^2$ are successfully employed with periodic current reversals (PCR) [15]. Current efficiencies for copper deposition are of the order of 90-95 %, tending toward the lower end of this range with PCR, so that the overall power requirement for copper purification by electrolysis is relatively low. Power consumption of several cell types in commercial electro-refining service is in the range of 176 to 220 kWh per tonne of cathode deposited [1].

During the electrolysis the metals present in the anode which are of lower electrochemical activity than copper, chiefly silver, tellurium, gold and traces of platinum, are precipitated as slimes on the

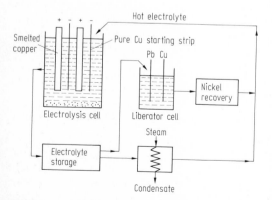

**Figure 11.8.** Diagram of a section of a electrolytic cell for the purification of copper

floor of the electrolysis cells as they are released from the anodes (Table 11.3). Any mercury, which is frequently present in the ore, is mostly volatilized in the hot smelting stages prior to electrorefining. Lead, which is dissolved from the anode as $Pb^{2+}$ is immediately precipitated as the insoluble sulfate, unlike nickel and many other metals more electrochemically active than copper the sulfates of which are soluble. The accumulated slime sludges are periodically pumped from the floor of the cells and dried preparatory to recovery of primarily the copper and the precious metals present. By a complex series of steps which include sulfuric acid acidification, roasting, copper electrowinning, silver cementation, fusions, and further electrowinning operations a precious metal alloy called Dore metal analyzing 8-9 % gold and 85-90 % silver is recovered from the residual copper, selenium, tellurium, and lead [15]. Frequently , the value of the precious metal content recovered from fire refined copper by electrorefining is sufficient to pay for a significant part of the costs of the electrolytic step. During normal cell operation the anode contaminants more electrochemically active than copper (and less than hydrogen), mainly arsenic, antimony, and bismuth, tend to accumulate gradually in the electrolyte. These are kept from depositing on the refined copper cathodes to any great extent partly by maintaining a much higher concentration of cupric ion in the electrolyte than the concentrations of these more active elements. Fortunately for the electrorefining process, cuprous oxide is generally the most significant impurity of fire refined copper. Since each mole of the cuprous oxide present in the anodes and exposed to the electrolyte puts a mole of cupric ion into solution by an auto-redox reaction (Equation 11.21),

$$Cu_2O + 2H^+ \rightarrow Cu^{2+} + Cu + H_2O \qquad (11.21)$$

without electrochemically removing a like amount onto the cathodes, the concentration of copper ion in the electrolyte rises more rapidly than the concentration of the other potential cathode contaminants. Incidentally, the elemental copper formed by Equation 11.21 is the chief source of the copper present in anode slimes. The electrolyte is continually recirculated through the electrolysis cells to minimize the build-up of impurities and avoid the negative effect on cathode purity that this would have. But when the copper content of the electrolyte reaches 40-50 g/L (from the initial 12-18 g/L), it must be purified before re-use. The concentrations of copper, arsenic, antimony, and bismuth as well as nickel, which is above hydrogen in electrochemical activity but whose sulfate is soluble in the electrolyte, must all be substantially decreased. A small portion (ca. 2-3 % of the total volume, per day) of the electrolyte stream is continuously bled from the main electrolyte stream for this purpose.

Most of the excess copper present in spent electrolyte is electrolytically removed from solution in liberator cells which are very similar to the electrolysis cells used for electrowinning. These have a pure copper cathode, on which copper is deposited (Equation 11.20) but a lead anode which reforms sulfuric acid from the hydrogen ions released from water (Equation 11.22) rather than contributing dissolved lead.

$$H_2O - 2e \rightarrow 2H^+ + \tfrac{1}{2}O_2 \qquad (11.22)$$

**Table 11.3** Analytical ranges of various products of the copper refining industry, U.S.A. and Japan[a]

|  | Anode (fire) refined) copper | Electro-refined copper | Dry, unprocessed anode slime |
|---|---|---|---|
| copper, % | 99.0 – 99.6 | 99.95 | 20 – 40 |
| oxygen, % | 0.1 – 0.3 | 0.02 – 0.04 | – |
| gold, g/tonne | 3 – 140 | 0.07 – 0.3 | 1 700 – 10 000 |
| silver, g/tonne | 70 – 3 400 | 2 – 20 | 34 000 – 300 000 |
| S, As, Sb, or Bi, % | 0.003 – 0.2 | 0.00001 – 0.0002 | 0.5 – 5.0 |
| lead, % | 0.01 – 0.15 | 0.0002 – 0.001 | 2 – 15 |
| nickel, % | 0.01 – 0.10 | 0.0001 – 0.002 | 0.1 – 2 |
| selenium, % | 0.01 – 0.06 | 0.0003 – 0.001 | 1 – 20 |
| tellurium, % | 0.001 – 0.02 | 0.0001 – 0.001 | 0.5 – 8 |

[a] Compiled from data of Kirk-Othmer [1] and Hofman [7].

Antimony and bismuth tend to precipitate on the bottom of the cell towards the later stages of this electrolysis, and arsenic reacts with the acid present under electrolytic conditions to form highly toxic arsine gas (boiling point $-55\,°C$; Equation 11.23) [23].

$$As + 3\,H^+ + 3\,e \rightarrow AsH_3\ (-0.54\ volts) \quad (11.23)$$

For this reason, liberator cells are normally hooded and well vented to avoid problems from the evolution of these gases. The "liberated" electrolyte is then further treated to remove nickel, either by crystallization as nickel sulfate or by dialysis, before it is returned to the electrolysis circuit for reuse [15].

New developments in the use of vigorous air agitation to boost cathode purities when electrorefining at very high (ca. $226\,A/m^2$) current densities [24] and procedures to improve electrorefining and electrowinning operations generally [25] have recently been published.

## 11.3 Fabrication and End Uses

For uses of pure copper, or critical alloys the 99.95 % copper cathodes obtained from electrolytic refining have to be washed, dried, and remelted before being cast into stock shapes useful for fabrication. The cast ingots can then be rolled into sheets (or foil) or hot or cold drawn into wire or pipe. Working the metal in these ways hardens and strengthens (e.g. raises the tensile yield point from $2,110\text{-}2,800\,kg/cm^2$ to $4,200\text{-}4,900\,kg/cm^2$ [7] which, when continued extensively can result in cracks and fractures in the metal. But by heating to a temperature of $500\text{-}700\,°C$ followed by quenching (rapid chilling) the metal is made malleable again, or annealed [7]. The annealed stock may then be worked again and sold either in a work-hardened state or annealed, depending on the requirements of the application.

The high electrical conductivity of copper, $0.586 \times 10^6$ mho x cm at $20\,°C$, or about 95 % of that of the most conducting metal, silver, has led to the large scale use of copper in wiring and electrical machinery of all types. More than half of the metal produced in Japan, the U.S.A., and Western Europe is used in electrical or electronic applications [26]. Even though aluminum possesses only some 62 % of the conductivity of copper, because of its light weight and relatively lower cost it has become competitive with copper for some electrical applications.

The relative thermal conductivities of the metals quite closely parallel electrical conductivities, so that again copper, with a conductivity of 0.934 cal x $cm^{-1}$ x $sec^{-1}$ x $K^{-1}$ at $20\,°C$ (3.98 W · $cm^{-1}$ x $K^{-1}$ at 300 K) emerges as the best low cost metal for efficient heat transfer [27]. This property has led to its employment in industrial heat exchangers as well as in food processing and domestic cookware. Cookware is usually, though not always, lined with stainless steel to decrease the possibility that copper might be taken up by or discolour food. While copper is an essential trace element in man [28] and the average American intake has been established to be 2 to 5 mg daily, it has also been known to cause mild toxic symptoms from ingestion of citrus juices that have been in long term contact with, or cooked in copper vessels [29]. Copper and its salts have a relatively low mammalian toxicity, although instances of acute copper poisoning have been described (from 1-12 g of copper sulfate). Chronic toxicity is known to lead to liver cirrhosis and nerve, brain, and kidney damage (Wilson's disease) [28].

Excellent resistance to corrosion both promotes the use of copper in some of the applications mentioned above and favours its use as an expensive, though extremely long-lived roofing material. This property has also led to the use of copper in all kinds of high quality piping functions as well as for gutters and drains.

Its malleability and corrosion resistance are also factors in the extensive use of copper as a coinage metal, although undoubtedly here its beauty is an important contributing factor. The naturally attractive red colour is also the stimulus for its use in many types of decorative interior fittings.

About one quarter of the copper produced goes directly into copper alloys, in the less critical alloying applications primarily as fire refined copper, although some electrolytically refined material is also used for alloying. The two principal alloy groups are the brasses, which comprise copper alloyed primarily with zinc, and the bronzes which are mainly copper/tin alloys. Both of these alloy groups are stronger than copper, one motivation for their use, but there are other reasons which favour their choice. The brasses are less expensive than pure copper, hence their employment in such uses as cartridge cases for ammunition. While the brasses are less corrosion resistant than copper this property is still sufficiently good for brasses to be used in automobile radiators where its good heat transfer capabilities and higher strength as well as lower cost favour this choice. The zinc content of brasses ranges all the way from 5 %, in gilding metal, through 30 % and 40 % in cartridge brass and Münz metal, respectively, which are the brass alloys containing the highest proportion of zinc [27]. Some brasses also contain 1-3 % lead which improves machinability.

The tin content of bronzes is lower, ranging from 1.25 %, to 8 % in the phosphorbronzes which also contain 0.25 % phosphorus, and confer not only higher strength but also better corrosion resistance in the alloy than pure annealed copper [27]. These alloys are therefore ideally suited to marine exposure in such applications as ships' propellers and other marine exterior fittings as well as for water lines carrying salt water and the like. A low coefficient of friction between bronze and steel [30], as well as the superior strength of bronze over copper have led to applications of bronze as a bearing metal and as the casting metal for gears.

Two special copper alloys, beryllium copper and nickel silver, although only used on a small scale, contribute valuable metal properties for special uses. Two percent beryllium added to copper gives greater fatigue resistance to the metal and confers a non-sparking (on impact) quality to tools made of this alloy, important when working in flammable or explosive atmospheres. Nickel silver, a copper/nickel/zinc alloy with an appearance very like silver, is important as the strong base metal for silver plated tableware.

Copper salts such as copper sulfate, while not consuming a large proportion of the copper produced are nevertheless of some importance as agricultural fungicides. In this application the salt is employed as a spray or a dust. Care should be taken to ensure that inhalation contact by applicators is minimized [28]. Copper salts are also used to correct very occasional local soil deficiencies in this element. Copper sulfate itself is also an excellent algicide in swimming-pools and fish rearing ponds.

## 11.4 Emission Control Practice Related to Copper Processing

### 11.4.1 Mining and Concentration

Problems at the mining stage, whether by open pit or underground methods, center around overburden and waste rock disposal and potential contamination of surface water courses from disposal of mine drainage waters. Waste rock disposal in mined-out areas is being practised where feasible but because the crushed material does not pack as densely as the bedrock from which it was derived this procedure still generally requires some supplementary surface disposal. By skilful placement and profiling followed by plantings to encourage re-establishment of flora, dry land surface disposal can be accomplished in an aesthetically acceptable manner.

Percolate from such dry land disposal sites as well as mine drainage waters often contain traces of dissolved copper and other metals such as iron (Table 11.4), and is also usually low in pH all arising from bacterial action on the sulfidic minerals present [31] (e.g. Equation 11.24).

$$2\,CuS + 3\,O_2 + 2\,H_2O \xrightarrow{\text{bacteria}} 2\,Cu^{2+} + 2\,H_2SO_4 \quad (11.24)$$

Copper may be recovered from such waste streams by cementation, or by cementation plus liming, both for the purpose of metal recovery from such

**Table 11.4** Partial analysis of typical copper mine and smelter waters [32]

|  | Ion concentrations in mg/L | | | | | |
|---|---|---|---|---|---|---|
|  | $Cu^{2+}$ | $Fe^{2+}$ | Fe, total | $Zn^{2+}$ | $H_2SO_4$ | pH |
| Mine waters: | | | | | | |
|   Typical | 970 | 8 350 | 10 500 | 4 000 | – | 1.4 |
|   S. American | 120 | 2 000 | 4 000 | 1 000 | – | 1.4 |
|   Canadian | 156 | – | 176 | 328 | < 10 | 3.8 |
| Typical wash water | 100 | > 5 | 10 | 24 | 100 | 2.8 |
| Electrolytic | 240 | – | 3 | < 1 | 760 | 2.0 |
| Copper refinery | 446 | – | 4.5 | 43 | 1 810 | 1.5 |

waste streams and to decrease the toxicity risks on final stream discharge (Equations 11.25—11.28).

Cementation:

$$Cu^{2+} + Fe \rightarrow Fe^{2+} + Cu \text{ ("cement copper")} \tag{11.25}$$

Liming:

$$Cu^{2+} + Ca(OH)_2 \rightarrow Cu(OH)_2 \downarrow + Ca^{2+} \tag{11.26}$$

$$Fe^{2+} + Ca(OH)_2 \rightarrow Fe(OH)_2 \downarrow + Ca^{2+} \tag{11.27}$$

$$2\,Fe^{2+} + O_2 + 2\,H_2O + Ca(OH)_2 \rightarrow$$
$$2\,Fe(OH)_3 \downarrow + Ca^{2+} \tag{11.28}$$

Cementation, which involves the passage of the solution containing copper ions over a large surface area of scrap iron followed by settling to remove the finally divided copper so precipitated, is quite effective for copper removal. However, subsequent liming is also necessary, both to remove most of the iron dissolved in this way and to raise the pH to more normal values before discharge. More recently, four stage solvent extraction in the presence of proprietary complexing agents has been tested and found to be capable of decreasing $Cu^{2+}$ concentrations from about 400 to about 6 mg/L [32].

It is possible to recirculate a large fraction of the water used for froth flotation by decantation of a clear supernatant liquor from a settling lagoon or thickener overflow [33]. But, because of the high sensitivity of froth flotation to low concentrations of surface active agents, the water throughput of a froth flotation plant cannot operate as a completely closed circuit since build-up of certain dissolved impurities in the water could markedly affect flotation efficiencies. Traces of metals in the flotation waters effluent, some suspended and some dissolved, are captured prior to discharge by liming and settling as outlined for mine drainage waters or tailings percolate.

Other types of water effluent problems can arise from froth flotation operations. For example, with multimineral flotation when separate concentrates are required of each of say copper, lead, and zinc, copper would normally be floated first, then lead, and finally zinc sulfide. The method of effecting this separation requires addition of sodium cyanide as a depressant. The waste water stream from this process contains 50 mg/L or more of dissolved cyanide which requires detoxification before discharge. Calcium hypochlorite has been found to be effective to neutralize the cyanide by oxidation to cyanate (Equation 11.29).

$$2\,NaCN + Ca(ClO)_2 \rightarrow 2\,NaCNO + CaCl_2 \tag{11.29}$$

When treatment is undertaken on large scale, however, chlorine addition has been found to be both more convenient and more economical [34] (Equations 11.30-11.31).

$$Cl_2 + H_2O \rightarrow HCl + HClO \tag{11.30}$$

$$2\,Ca(OH)_2 + 2\,HCl + 2\,HClO \rightarrow$$
$$CaCl_2 + Ca(ClO)_2 \tag{11.31}$$

If the chlorine addition is undertaken prior to pH reduction with lime the reaction sequence is thought to involve intermediate formation of cyanogen chloride which is decomposed to cyanate on addition of base (Equations 11.32, 11.33).

$$NaCN + Cl_2 \rightarrow CNCl + NaCl \tag{11.32}$$

$$2\,CNCl + 2\,Ca(OH)_2 \rightarrow$$
$$Ca(CNO)_2 + CaCl_2 + 2\,H_2O \tag{11.33}$$

Using the latter waste treatment sequence it has been found that feed values of 68 mg/L cyanide and 42 mg/L copper are simultaneously decreased to concentrations of 0.1 mg/L or thereabouts, for both contaminants [34]. To achieve this level of control required a treatment rate of 3.4 kg of chlorine per kg of cyanide in the effluent, and a 15-30 minute residence time for destruction of the cyanide only. About three to four times this residence time was required for effective neutralization of copper cyanide, in particular. Ideally, the proportion of oxidant added should be sufficient to convert cyanate to carbon dioxide and nitrogen and thus avoid the possibility of reversion of cyanate back to cyanide after discharge. Ozonation has also been found to be effective for the neutralization of cyanide in waste waters [35].

Disposal of the large volumes of finely pulverized gangue separated from mineral concentrates may also be made to underground or strip mined areas, but again disposal volumes in excess of the containment capacities are a problem. Dry land disposal, followed by aesthetic contouring and re-establishment of plant life is one measure which may be adopted but this method requires monitoring and installation of control measures for any dissolved metals which may be picked up by water percolate through the tailings. No uncontrolled drainage directly to a small lake or a water course is allowed by pollution control agencies (e.g. reference 36). Direct tailings disposal is being practised to large lakes [34] and to ocean inlets [37], when these are close to the mine site, but the long term acceptability of these practices is yet to be conclusively determined [38, 39].

Solution mining can also generate complex recovery problems, particularly with in-place leaching methods. On completion of the mining step the fractured deposit requires extensive flushing to remove residual components of the leaching solutions used [40]. Deep well injection, or various solution concentrating procedures are used for ultimate disposal of waste solutions from these flushing operations.

## 11.4.2 Smelter Operation

Of significant concern in the smelting of copper is control of the associated dust and sulfur dioxide produced. Upwards of a tonne of sulfur dioxide is produced for every tonne of copper obtained from sulfidic minerals and in the early days of the industry this sulfur dioxide was simply allowed to dissipate in the air around a smelter [5]. The effect on local air quality, even with these early, relatively small scale works was so severe that stack discharge adopted to obtain mixing of the smelter gases with larger volumes of air to obtain decreased ground level sulfur dioxide concentrations. Stack heights gradually increased over the years to exploit dilution effects for larger volumes of sulfur dioxide, culminating in the construction of the world's tallest stack, 380 m high, by the International Nickel Company (Inco) at Sudbury for this purpose, in 1970 [41, 42]. At the same time, some smelters adopted sulfur dioxide containment and utilization for sulfuric acid production as a means to decrease discharge problems. In the process the sulfuric acid byproduct obtained was sold to obtain a process credit [43]. Containment and sulfur dioxide utilization have now been adopted by Inco [44], as well as more generally by the industry, as the limited atmospheric capacity for dispersion of enormous quantities of sulfur dioxide worldwide has become increasingly recognized. This desirable approach is now enforced by the Environmental Protection Agency in the U.S.A. with a requirement that 95 % (or better) of the sulfur in the ores being processed by a smelter there must now be contained during smelting [45].

The potential quantity of dust discharged per tonne of ore processed varies significantly with the particular stage of processing, although the crushing and dry grinding stages as well as final rotary drying which involve dispersion of finely divided ore particles in air are likely to contribute most to dust losses. Conducting the final stages of ore grinding wet, of course, avoids dust problems at this stage as well as providing improved means of control of the final particle size obtained from the operation. The potential for dust losses at each subsequent stage is further dependent on the fineness of the charge(s) delivered to that stage, the degree of agitation (particularly if gas phase) needed to achieve the process requirements of that stage, and the average specific gravity of the feed. For all stages the proportion of copper present, which is about 10 % in roaster dusts, 25 % in reverberatory furnace dusts, and 45-55 % in converter dusts, is sufficiently high to provide economic material recovery as well as emission control incentives to capture these [1].

Simple gravity settling of dust in a gravity settling chamber or in settling chambers with wire baffle grids have been used for many years with some 84 % mass containment success being obtained with copper blast furnace discharges [7]. Sometimes the dust settling and gas carrying functions are combined in a single large cross sectional area (slow gas flow) balloon flue [11] (Figure 11.9). The tighter more recent emission control requirements as well as an interest in more trouble-free acid plant operation, however, have generally required employment of electrostatic precipitation and/or scrubbers as further dust control stages for smelter flue gas treatment [1, 21]. Electrostatic precipitation, which has a control efficiency of about 98 % by mass on the particle size range 0.1-100 $\mu$m gives the method a capability to control at

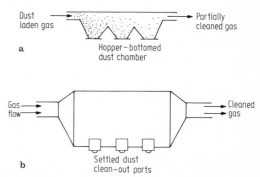

**Figure 11.9.** Operating details of a. gravity settling chamber and b. section of a balloon-flue for removal of the coarser dusts from copper smelter operation

least part of the fume component as well as the larger particle size dusts contained in smelter discharges [46]. Fumes consist of very finely divided solid material which remains in suspension in air which may have been formed from sublimation, or from condensation and solidification of such components as oxides of antimony, arsenic, lead, zinc and sulfates. Water, sulfuric acid, and mercury may also have been condensed and absorbed onto the fume particles, depending on the constituents present in the ore feed to the smelter. Dusts containing these elements are collected and specially treated for recovery of the components of interest or concern, e.g. mercury by sulfatization [47, 48]. Baghouses are also efficient for dust control in this application [11], as long as the design bag fabric is compatible with acid gas contact and measures are taken to ensure that flue gases temperatures are not in excess of the working range of the fabric. The additional blower energy required to force gases through a filter fabric is offset to a significant extent by the electrical saving which would otherwise be required to maintain the charge on the elements of an electrostatic precipitator.

The problem with sulfur dioxide containment in connection with operation of a copper smelter is that many components of the smelter produce sulfur dioxide concentrations of 1 or 2 % whereas the normal minimum economic concentration required by most types of sulfur dioxide conversion processes (e.g. for sulfuric acid or sulfur production) is 3.5 to 4 % [11]. Thus, the economic solution to containment problems rests with either modification of smelter processes to obtain higher sulfur dioxide concentrations directly, or with methods to capture sulfur dioxide at relatively low concentrations by procedures which subsequently allow regeneration of a high concentration sulfur dioxide gas stream for efficient conversion. There are also the throw-away approaches to sulfur dioxide containment, lime or limestone wet scrubbing, which are worth considering by at least small scale smelters [49] (Equations 11.34-11.36).

$$Ca(OH)_2 + H_2O + SO_2 \rightarrow CaSO_3 + 2\,H_2O$$
$$(11.34)$$

$$CaSO_3 + \tfrac{1}{2}\,O_2 + 2\,H_2O \rightarrow CaSO_4 \cdot 2\,H_2O$$
$$(11.35)$$

$$CaCO_3 + H_2O + SO_2 \rightarrow CaSO_3 + CO_2 + H_2O$$
$$(11.36)$$

The gypsum-calcium sulfite product of this control system occasionally causes blockages and other problems in the scrubber slurry handling system and the insoluble product is generally only suitable for landfill. This choice thus adds an operating cost to the smelter without conributing any byproduct credit and hence is adopted only as an expedient required to meet emission control objectives.

Altered smelter operations which serve to raise sulfur dioxide concentrations in the flue gases include switching from chain grate and conventional multiple hearth type roasters to fluidized bed types where the air to hot sulfide ore particle contact is much better, and there is less accompanying entrained air which is not actually serving to oxidize sulfur. In the extreme case this change alone can bring about a rise from 1 % to 12 % sulfur dioxide in the "off gases" produced by the roaster, quite a favourable concentration for all sulfur dioxide conversion processes [11].

Fluid bed, autogenous roasting of the type just described elegantly solves the sulfur dioxide containment problem for concentrates of high sulfur ores where this roasting method is most feasible. But for ore concentrates of lower sulfur content, particularly when the sulfur content is less than the copper content, external fuel combustion is required to provide sufficient heat for adequate roasting of the charge. This addition of the external combustion gases to the roaster gases serves to dilute the sulfur dioxide produced to less than economic concentrations. This disadvantage has led to the abandonment of roasting as a separate smelter operation, particularly in locations which have a low sulfur to copper ratio in the concentrate [1].

Continuous smelting processes [50] or flash smelting processes using oxygen-enriched air [51] are being employed to combine two or three smelting stages, roasting, reverberatory smelting, and/or converting, into one [52]. In this way sulfur removal from the charge to the extent desired is obtained with good control and high sulfur dioxide concentrations are obtained in the exit gases produced. For instance Outokumpu furnace designs using air feed for sulfur combustion are capable of producing exhaust gases containing 14 % sulfur dioxide [51]. Use of a top blown rotary converter (TBRC) also promises to streamline copper production by merging smelting stages and by allowing the use of oxygen-enriched air for blowing [50, 53, 54]. Use of oxygen-enriched air is not possible in converters because of tuyere (air nozzle) damage [55]. If oxygen-enriched air is used for sulfur combustion it is possible to obtain flue gases containing 75 to 80 % sulfur dioxide [51, 55], sufficiently high to practise direct sulfur dioxide recovery, if desired. These joint developments are being adopted by several new smelters, which are in the best position to capitalize on these improvements [56-59].

The various enriched sulfur dioxide streams, when combined with lower sulfur dioxide flue gases from other smelter areas, can yield a blended gas stream of 4 to 6 % sulfur dioxide suitable for acid production. Smelter acid is not as pure as the acid produced from sulfur combustion, so that it does not command as high a price without some clean-up. Nevertheless the smelter product is quite suitable directly in the form obtained for many uses, e.g. fertilizer phosphate production, so that a reasonable by-product credit is possible by this sulfur dioxide conversion choice. It has been estimated that copper smelters which exercise this route for sulfur dioxide control produce some four tonnes of sulfuric acid for each tonne of copper [55]. Of the total sulfuric acid produced in the U.S.A. and Canada, smelter sources represented more than 6 % of the U.S. total in 1965 and more than 60 % of the Canadian total for 1976 [1,60].

Sulfur dioxide containment and subsequent reduction to a sulfur product has also been tested in the smelter context with sufficiently promising initial results to have prompted further study on a pilot plant scale [11]. The American Smelting and Refining Company's process to achieve this involved combustion of the cooled and cleaned smelter gases with methane (Figure 11.10; Equation 11.37).

$$4\,SO_2 + 2\,CH_4 + air \xrightarrow{1250\,°C} S + H_2S + COS \\ + SO_2 + 2\,CO_2 + H_2O + N_2 \qquad (11.37)$$
(unbalanced)

The carbonyl sulfide and some of the residual sulfur dioxide were then reacted at 450 °C in the presence of a bauxite catalyst to obtain further conversion of sulfur dioxide to sulfur vapour (Equation 11.38).

$$2\,COS + SO_2 \rightarrow 2\,CO_2 + 3\,S \qquad (11.38)$$

After electrostatic precipitation of products of these two processes, the residual hydrogen sulfide and sulfur dioxide, present in about a 2:1 ratio, are reacted in the presence of a Claus-type activated alumina catalyst to further raise the sulfur yield (Equation 11.39).

$$2\,H_2S + SO_2 \rightarrow 3\,S + 2\,H_2O \qquad (11.39)$$

Actual sulfur recoveries reported were about 90 %, but based on experience with Claus plant operational efficiencies it was predicted that with the addition of a second stage of Claus conversion the recoveries could be boosted to about 95 % [11].

The U.S. Bureau of Mines has also developed and tested to a pilot scale a sulfur dioxide to sulfur conversion process involving initial absorption of

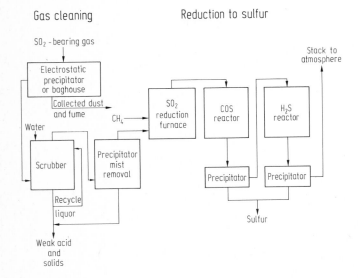

Figure 11.10. Flow sheet of the American Smelting and Mining Company's process for conversion of sulfur dioxide to sulfur [11]

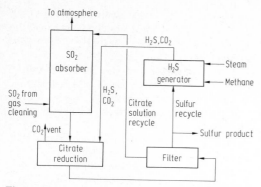

**Figure 11.11.** The citrate process for sulfur dioxide capture and conversion to elemental sulfur developed by the U.S. Bureau of Mines [11]

sulfur dioxide in an aqueous solution of citric acid, $HOC(CH_2CO_2H)_2CO_2H$, and sodium citrate [61, 49, 62] (Figure 11.11). Contacting this solution with hydrogen sulfide in a second reactor vessel then reduces the contained sulfur dioxide to elemental sulfur, which is readily filtered from the resulting slurry (Equations 11.40-11.42).

$$SO_2 + H_2O \rightarrow H_2SO_3 \qquad (11.40)$$

$$H_2SO_3 + H_3Cit \rightarrow H_2SO_3 \cdot H_3Cit \qquad (11.41)$$

$$H_2SO_3 \cdot H_3Cit + 2\,H_2S \rightarrow 3\,S + H_3Cit + 3\,H_2O \qquad (11.42)$$

The filtrate is returned to the absorber for re-use. A part of the product sulfur is reacted with methane and steam in the presence of alumina to produce the hydrogen sulfide required for reduction of the citrate complex (Equation 11.43).

$$4\,S + CH_4 + 2\,H_2O \rightarrow 4\,H_2S + CO_2 \qquad (11.43)$$

This sequence of reactions has been found to be over 90 % efficient for sulfur dioxide containment and conversion to sulfur for reverberatory furnace flue gases containing 1.5 % sulfur dioxide [11].

## 11.5 Hydrometallurgical Copper Recovery

### 11.5.1 Basic Principles

Hydrometallurgy, or the application of aqueous solutions for metal recovery from ores, has been practised for copper recovery in relatively straightforward sequences for many years. The original

impetus for solution methods for copper extraction, particularly before the development of froth flotation concentration technology, was the existence of large ore bodies of low copper grade which were not amenable to economic working using conventional smelting methods. Hydrometallurgy also holds a complementary position to concentrator technologies for the scavenging of copper from concentrator tailings, since froth flotation is poor at floating copper oxides [33] which are readily extractable by appropriate aqueous solutions. Most recently, hydrometallurgical copper recovery methods have been refined to be effective with concentrates of more complex ores with the primary objective of eliminating sulfur dioxide discharge problems resulting from conventional smelting methods. Thus, while all hydrometallurgical methods essentially have a low polluting potential, some of the methods developed most recently have deliberately selected chemistry which directly yields elemental sulfur (rather than sulfur dioxide) from the sulfur minerals present in the ores, as a further step towards clean processing.

Hydrometallurgical copper recovery can be conveniently considered in two stages, the extraction stage, where the various forms of copper in the ore are placed into solution, usually as cupric ion, and the recovery stage, where dissolved copper is regained as solid, pure, or nearly pure copper metal ready for fabricating or final smelting stages.

The solvent most commonly employed is 5 to 10 % sulfuric acid in water, such as diluted smelter acid, which can provide this at low cost. This has a solvent capability for a number of copper minerals, not just the oxide (Equations 11.44-11.47).

$$CuO + H_2SO_4 \rightarrow CuSO_4 + H_2O \qquad (11.44)$$
$$Cu_2O + 2\,H_2SO_4 + \tfrac{1}{2}O_2 \rightarrow 2\,CuSO_4 + 2\,H_2O \qquad (11.45)$$

$$CuSiO_3 \cdot 2\,H_2O + H_2SO_4 \rightarrow \\ CuSO_4 + SiO_2 + 3\,H_2O \qquad (11.46)$$

$$Cu_2(OH)_2CO_3 + 2\,H_2SO_4 \rightarrow \\ 2\,CuSO_4 + CO_2 + 3\,H_2O \qquad (11.47)$$

These reactions, and in particular the carbonate solvating ones, deplete the sulfuric acid content of the extraction solvent at the extraction stage. Later metal recovery from the solutions, however, returns an equivalent acid strength to the water. Also, in the presence of air and adequate time for bacterial oxidation to take place, *Thiobacillus ferro-oxidans* converts sulfide minerals to sulfates

[31], thus in effect contributing to the net acid content of the solvent when metal is recovered from the solution.

Extraction of metal may be accomplished by leaching in place, by heap or percolation leaching, and by leaching with mechanical agitation in separate tanks. Leaching in place requires prior fracturing of the ore by explosives to provide passages for percolation of the leach solutions. Copper-containing percolate is recovered for recirculation and eventual metal recovery by pumping this solution from lower galleries of the in-place leaching zone. This method is as effective with low grade zones of an ore body loosened in place as with ores which have been passed through crushing stages and then returned to the mine for disposal.

Heap and percolation leaching are used with low grade ores crushed to an average 1 cm particle diameter specifically for the purpose. Heap percolation is practised on large scale, 100,000 tonne by 35 m high flattened top ore heaps constructed so as to allow even distribution of liquor and air throughout the heap. A series of vats containing the crushed ore is employed for percolation leaching, which relies on either alternate filling and emptying cycles or continuous slow flow through the vat series to gradually put the metal into solution. Percolation leaching, by ensuring thorough contact of the percolating solution with the whole of the crushed ore and as practised for many years at Chuquicamata, Chile [1], thus provides better control of extraction conditions than heap leaching, of long-lived importance to the Rio-Tinto Company, Spain [6], but does so at a higher capital cost. Agitation leaching is the extraction method of choice with very finely divided ores which may have already been partially processed and would not readily lend themselves to percolation methods because of more difficult liquor passage. This material, which also has a tendency towards local channelling of extraction liquors in a percolation mode, is sufficiently fine that it is fairly readily kept in aqueous suspension. Thus, for agitation leaching the finely ground pulp is placed with a sufficient quantity of leach solution to obtain a very fluid mass, in vats equipped with air-lift (a pachuca) [21] or mechanical agitation to maintain the whole mass in suspension. On completion of the extraction phase, which is orders of magnitude more rapid than either in-place, or heap/percolation leaching methods, the liquor containing the dissolved metal values is separated from the gangue as the overflow from a thickener. The slurry underflow from the thickener is filtered to collect residual copper-containing liquor and washed before it is discarded.

Once the copper (and any other metals of interest) have been obtained in solution, the difficult and usually time-consuming part of the process is over. Copper is recovered from these solutions by any of a variety of methods.

Cementation, or precipitation of the copper from solution by passage over scrap iron, during an approximately one hour contact time, produces a finely divided form of the metal analyzing ca. 90 % Cu (dry) by replacement of the dissolved copper by iron. Theoretically this should only require 0.879 kg of iron for each kilogram of copper obtained (Equation 11.25) but normal experience is an iron consumption of 1.3 to 3 kg [33, 63]. In the U.S.A. in recent years this procedure for the recovery of copper from aqueous solutions has grown in importance to the point where some 10 % of U.S. production in 1973 was from cement copper, mostly by the Kennecott Copper Corp. and Anaconda Co. [63].

Copper may also be recovered from leach solutions electrolytically. This process, called electrowinning, requires use of an insoluble anode such as hard lead, comparable to the liberator cell used for liquor purification in copper electro-refining. Consequently, there are net electrochemical reactions involved in the electrowinning (Equations 11.20, 11.22), as opposed to the situation with electro-refining, so that operating voltages of about 1.7 V are required for this step. This results in an electrical power consumption for electrowinning of about 2.8 kWh/kg of copper, much higher than the approximately 0.2 kWh/kg required for electro-refining.

## 11.5.2 New Hydrometallurgy Developments

A more selective copper recovery from leach solutions may be obtained by solvent extraction using a dissolved, copper selective complexer such as an hydroxyoxime (Figure 11.12), in the organic phase. For example, complexation of two moles of this selective additive to cupric ion confers a marked preference of the copper (II) species for the organic, rather than the aqueous phase, allowing it to be efficiently captured from a relatively dilute aqueous solution [32, 64]. Note from the figure

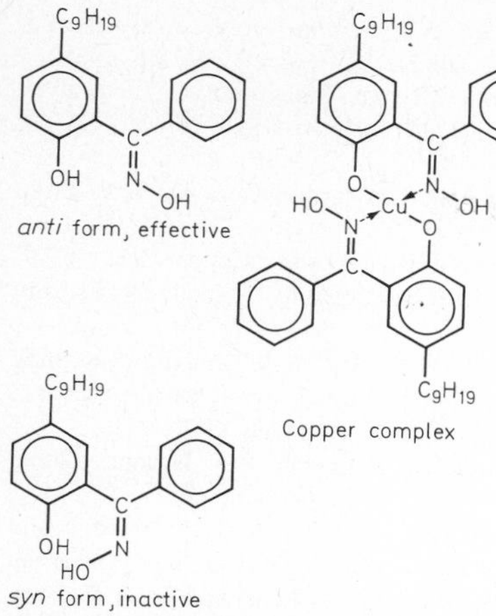

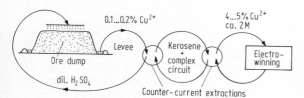

**Figure 11.12.** Structure of the two forms of the copper selective complexer, LIX65N, and the kerosine soluble copper complex formed on contact of the anti form with cupric ion [78]

that the metal complex is stabilized by the formation of two six-membered rings which include the metal ion, stabilizing and favouring the complexed, over the uncomplexed state. After separation of the metal rich organic phase from the original aqueous scavenging solution the copper may be back-extracted to a fresh aqueous acid phase in preparation for electrowinning or other methods of recovery of the metal values [65, 66] (Figure 11.13). In this situation final copper recovery from the aqueous solution is greatly facilitated by the higher concentration and purer state of the cupric ion, generally well compensating for the cost of the extraction step. Other ways in which the metal values from the primary leach solutions may be concentrated preparatory to metal recovery include ion exchange techniques and adsorption/desorption from activated carbon [67].

Anaconda's announcement of pilot plant development of a sulfuric acid based leaching process employs a number of innovations [68]. Concentrates are leached with ca. 90 % sulfuric acid (much more concentrated than usually used for

leaching) to yield a solution of dissolved metal sulfates and releasing sulfide sulfur in elemental form, in the process reducing a portion of the sulfuric acid to sulfur dioxide and water. Purification of the solution, followed by treatment with hydrogen cyanide and sulfur dioxide then yields a solid phase of high purity cuprous cyanide, all accomplished at atmospheric pressure. High purity copper, 99.99 % Cu, is obtained on reduction of the separated cuprous cyanide with hydrogen (Equation 11.48).

$$CuCN + \tfrac{1}{2} H_2 \rightarrow Cu + HCN \qquad (11.48)$$

The new Sherritt-Cominco copper process also introduces some hydrometallurgical wrinkles into sulfuric acid leaching procedures to make the method applicable to chalcopyrite ($CuFeS_2$) ores [69]. After thermal rearrangement and reduction of the concentrate with hydrogen it is leached first with sulfuric acid to dissolve out the iron [70] (Equation 11.49).

$$FeS + H_2SO_4 \rightarrow FeSO_4 + H_2S \qquad (11.49)$$

The hydrogen sulfide is recovered for processing to sulfur and the iron-containing solution is separated for iron precipitation. The copper, which remains undissolved in the concentrate up to this stage as $Cu_5FeS_4$, is then separately dissolved out of the concentrate with sulfuric acid under oxidizing conditions (Equation 11.50).

$$Cu_5FeS_4 + 6 H_2SO_4 + 3 O_2 \rightarrow$$
$$5 CuSO_4 + FeSO_4 + 4 S + 6 H_2O \qquad (11.50)$$

After purification of this copper-containing solution, metallic copper is obtained either by electrowinning or hydrogen reduction [71]. Claims of at least 98 % copper recovery, gold and silver capture at least equivalent to the extent of conventional smelting, recovery of 60 to 90 % of the sulfur in elemental form as well as recovery of other metal values in saleable from have been substantiated with pilot plant data [72, 73].

Another hydrometallurgical variant particularly appropriate for chalcopyritic ores but also adaptable to porphyry copper and oxides, is the

**Figure 11.13.** Counter current solvent extraction, using a copper selective complexer in the organic phase used to raise the concentration and purity of copper ion solutions fed to electrowinning

Cyprus Metallurgical Processes Corporation's ("Cymet") chloride leaching process, and adaptations of this [74]. In brief, copper concentrates are first leached with an aqueous ferric chloride solution. Cupric ($Cu^{2+}$) ion in the ore reacts with the chloride leach medium to form cuprous ion and thus dissolves as cuprous chloride (CuCl). After a series of several more steps cuprous chloride is crystallized from the solution in pure form [75]. Final metal recovery is obtained by hot reduction of the cuprous chloride with hydrogen (Equation 11.51).

$$CuCl + \frac{1}{2}H_2 \rightarrow Cu^0 + HCl \qquad (11.51)$$

This sequence of reactions has been successfully tested with 680,000 kg of concentrate analyzing about 28 % copper, 30 % iron and 30 % sulfur, with sulfur recovery ultimately effected in elemental form.

Ammonia complexation has also been employed, primarily for nickel-copper separations [21] but also for copper recovery, when this is from native copper or oxide ores [1]. The dissolving process is thought to involve formation of soluble cuprous ammonium carbonate (Equations 11.52, 11.53).

$$Cu + Cu(NH_3)_4CO_3 \rightarrow Cu_2(NH_3)_4CO_3 \quad (11.52)$$

$$2\,Cu_2(NH_3)_4CO_3 + 4(NH_4)_2CO_3 + O_2 \rightarrow$$
$$4\,Cu(NH_3)_4CO_3 + 4\,H_2O + 2\,CO \qquad (11.53)$$

Most interesting new developments relate to the combining of the technology of electrolytic metal recovery from heap leached solutions [76] with the actual copper extraction process. In laboratory scale tests, where the finely ground (2-3 $\mu m$) copper ore was both leached and electrolytically deposited in a single electrochemical cell, an "electroslurried" product of 99.9 % copper was obtained [77]. While this purity is not up to normal electrical conductor standards, it is very respectable for a single stage electro-hydrometallurgical process capable of recovering 98 % of the copper present in the ore feed.

# Relevant Bibliography

1. Flotation Agents and Processes: Technology and Applications, M.W. Ranney, editor, Noyes Data Corp., Park Ridge, N.J., 1980
2. K.N. Subramanian et al, Production of Copper by a Pressure Leaching, Precipitation, Electrowinning Process, Can. Min. and Metallurg. Bull. 71, 132, Sept. 1978
3. Copper Refining Technology Reduces Costs, Energy, and Pollution, Ind. Eng. 10, 48, Jan. 1978
4. C.R. Bayliss, Modern Techniques in Electrolytic Refining of Copper, Electron and Power 22, 773, Nov.-Dec. 1976
5. R. Wood, Copper Extraction Goes On Site, Proc. Eng. 90, Sept. 1976
6. F. Habashi and B.I. Yostas, Copper from Chalcopyrite by Direct Reduction, J. Metals 29, 11, July 1977
7. G.J. Snell and M.C. Sze, New Oxidative Leaching Process Uses Silver to Enhance Copper Recovery, Eng. and Min. J. 178, 100, Oct. 1977
8. Copper in the Environment, J.O. Nriagu, editor, Wiley, New York, 1979, two volumes
9. International Symposium on Copper in Soils and Plants, Murdoch University, 1981, J.F. Lonergan and A.D. Robson, editors, Academic Press, Sydney, 1981
10. P.A. Spear and R.C. Pierce, Copper in the Aquatic Environment: Chemistry, Distribution, and Toxicology, National Research Council, Ottawa, 1979
11. Jan Leja, Surface Chemistry of Froth Flotation, Plenum Publishing Corp., New York, 1981
12. D.H. Pulling and F.A. Garner, Application of a Continuous Technique to Secondary Copper Smelting, Can. Mining and Metallurg. Bull. 70, 122, Dec. 1977

# References

1. Kirk-Othmer Encyclopedia of Chemical Technology, 2nd edition, John Wiley and Sons, New York, 1965, volume 6, page 131
2. H.H. Coghlan, Notes on the Prehistoric Metallurgy of Copper and Bronze in the Old World, The University Press, Oxford, 1951
3. F.S. Taylor, A History of Industrial Chemistry, Arno Press, New York, 1972, page 37
4. R. Chadwick, Chem. In Brit. 17 (8), 369, Aug. 1981
5. J.H. Beynon and D. Betteridge, Chem. In Brit. 15 (7), 340, July 1979
6. D. Avery, Not on Queen Victoria's Birthday; the Story of Rio Tinto Mines, Collins, London, 1974
7. H.O. Hofman, Metallurgy of Copper, McGraw-Hill, New York, 1914
8. C.J.S. Warrington and R.V.V. Nichols, A History of Chemistry in Canada, Pitman and Sons, Toronto, 1949, page 15
9. Statistical Yearbook, 1952, United Nations, New York, 1953, and earlier editions
10. Minerals Yearbook, Vol. 1, Metals and Minerals, 1980, U.S. Bureau of Mines, Washington, 1981, page 261
11. Control of Sulfur Oxide Emissions in Copper, Lead and Zinc Smelting, Information Circular 8,527, U.S. Bureau of Mines, Washington, 1971

12. J. Leja, J. Chem. Educ. *49* (3), 157, March 1972

13. Froth Flotation, 50th Aniversary Volume, D.W. Fuerstenau, editor, Am. Inst. Mining, Metal., and Petroleum Engineers, Inc., New York, 1962

14. A. Sutulov, Copper Porphyries, University of Utah Printing Services, Salt Lake City, Utah, 1974. Cited by reference 15

15. Kirk-Othmer Encyclopedia of Chemical Technology, 3rd edition, John Wiley and Sons, New York, 1979, volume 6, page 819

16. E.J. Mahne, Chem. In Can. *23* (3), 32, March 1971

17. T. Young, Proc. Roy. Soc. (London), *95*, 65 (1805). Cited by A.M. Gaudin, Principles of Mineral Dressing, McGraw-Hill, New York, 1939

18. V.I. Klassen and V.A. Mokrousov, An Introduction to the Theory of Flotation, translated by J. Leja and G.W. Poling, 2nd edition, Butterworths, London, 1963

19. Mineral Processing, A. Roberts, editor, Proceedings of the Sixth International Congress, Cannes, May-June, 1963. Pergamon Press, Oxford, 1965

20. Flotation Agents: A Broader Mix, Can. Chem. Proc. *57* (4), 33, April 1973

21. J.R. Boldt, Jr. and P. Queneau, The Winning of Nickel, Longmans of Canada, Ltd., Toronto, 1967

22. Kirk-Othmer Encyclopedia of Chemical Technology, 3rd edition, John Wiley and Sons, New York, 1978, volume 10, page 523

23. F.A. Cotton and G. Wilkinson, Advanced Inorganic Chemistry, A Comprehensive Text, 3rd edition, Interscience, New York, 1972, page 373

24. W.W. Harvey, M.R. Randlett, and K.I. Bangerskis, J. Metals *30*, 32, July 1978

25. A.S. Gendron, R.R. Matthews, and W.C. Wilson, Can. Min. and Metallurg. Bull. *70B*, 166, Aug. 1977

26. W. Gluschke, J. Shaw, and B. Varon, Copper: The Next Fifteen Years, D. Reidel (The United Nations), Dordrecht (Holland), 1979, page 30

27. Properties of Copper Metals, Canadian Copper and Brass Assoc., Toronto, 1967

28. Patty's Industrial Hygiene and Toxicology, 3rd edition, G.D. Clayton and F.E. Clayton, editors, John Wiley and Sons, New York, 1981, volume 2 A, page 1,620

29. McGraw-Hill Encyclopedia of Science and Technology, McGraw-Hill, New York, 1982, volume 3, page 641

30. Handbook of Chemistry and Physics, 51st edition, R.C. West, editor, Chem. Rubber Publ. Co., 1970, page F-16

31. Biological Leaching, A New Method of Metal Recovery, B.C. Research, Vancouver, ca. 1972

32. G. Barthel, J. Metals *30*, 7, July 1978

33. W.A. Triggs and A.M. Laird, Mining Mag. *124* (6), 438, June 1971

34. A.G. Eccles, Can. Mining and Metall. Bull. *70* (785), 141, Sept. 1977

35. G.I. Mathieu, Ozonation for Destruction of Cyanide in Canadian Gold Mill Effluents — A Preliminary Evaluation, Canmet Report 77-11, Dept. of Supply and Services, Ottawa, 1977

36. Pollution Control Objectives for the Mining, Mine-milling, and Smelting Industries of British Columbia, Department of the Environment, Queen's Printer, Victoria 1976

37. C.A. Pelletier, Marine Tailings Disposal at Island Copper Mine, Rupert Inlet, B.C., 79th Ann. Meeting, Can. Inst. of Mining, Ottawa, April 16-20, 1977

38. M.J.R. Clark, A Preliminary Review of Buttle Lake Water Quality, Report No. 80-2, Waste Management Branch, Ministry of the Environment, Victoria, B.C., 1980, 132 pp

39. M.J.R. Clark and T.O. Morrison, Impact of the Westmin Resources Ltd. Mining Operation on Buttle Lake and the Campbell River Watershed, Waste Management Branch, Ministry of the Environment, Victoria, B.C., 1982, 960 pp

40. Solution Mining Poses Environmental Challenges, Chem. Eng. News *60* (14), 31, April 5, 1982

41. L. Hodges, Environmental Pollution, Holt, Rinehart, and Winston, Inc., New York, 1973, p. 271

42. Law Makers Say, Can. Chem. Proc. *55*, 47, April 1971

43. S.J. Williamson, Fundamentals of Air Pollution, Addison-Wesley, Reading (Massachusetts), 1973, page 191

44. Inco Entering Acid Production, Can. Chem. Proc. *65* (2), 10, March 27, 1981

45. A.V. Colucci, Sulfur Oxides: Current Status of Knowledge, EPRI EA-316, Final Report, Electric Power Research Institute, Palo Alto, California, Dec. 1976. Cited by reference 15

46. Air Pollution Handbook, P.L. Magill, F.R. Holden, and C. Ackley, editors, McGraw-Hill, New York, 1956

47. J. Rastas, E. Nyholm, and J. Kangas, Eng. Min. J. *172*, 123 (1971)

48. J. Kangas, E. Nyholm and J. Rastas, Chem. Eng. *78E*, 55 (1971)

49. Cleaning Up $SO_2$, Chem. Eng. *80* (9), AAA, April 16, 1973

50. F.C. Price, Chem. Eng. *80* (9), RR-ZZ, April 16, 1973

51. D.G. Treilhard, Chem. Eng. *80* (9), P-Z, April 16, 1973

52. Copper Smelting Is Going Continuous, Can. Chem. Proc. *60* (4), 21, April 1976

53. Successful Use Of TBRC, Chem. in Can. *26*, 9, May 1974

54. Copper Refining Via Rotary Smelting, Can. Chem. Proc. *60* (1), 6, Jan. 1976

55. T.J. Chessell, Can. Chem. Proc. *63*, 21, March 21, 1979

56. M.P. Amsden, R.M. Sweetin, and D.G. Treilhard, J. Metals *30*, 16, July 1978

57. New Copper Recipe, Dravo Review, 2, Fall 1978
58. Afton, New Canadian Copper Mine, World Mining *31* (4), 42, April 1978
59. Afton Mine On Stream, Mining Mag. 2, Jan. 1978
60. P.D. Nolan, Can. Chem. Proc. *61* (4), 35, April 1977
61. Citric Acid Used, Chem. Eng. News *49*, 31, June 14, 1971
62. F.S. Chalmers, Hydroc. Proc. *53* (4), 75, April 1974
63. Copper Industry Uses Much Scrap Iron, Env. Sci. and Techn. *7* (2), 100, Feb. 1973
64. D.S. Flett, Chem. and Ind. (London) 706, Sept. 3, 1977
65. M.A. Hughes, Chem. and Ind. (London) 1,042, Dec. 20, 1975
66. M.C. Kuhn, Mining Eng. *29*, 79, (1977)
67. F. Habashi, Chem. Eng. News *60* (6), 46, Feb. 8, 1982
68. Chemical Route to Copper, Chem. Eng. *77* (8), 64, April 20, 1970
69. C-S Process Offers Unique Steps, Can. Chem. Proc. *63*, 25, March 21, 1979
70. D.E.G. Maschmeyer, P. Kawalka, E.F.G. Milner, and G.M. Swinkels, J. Metals *30*, 27, July 1978
71. G.M. Swinkels and R.M.G.S. Berezowsky, Can. Min. Metallurg. Bull. *71*, 105, Feb. 1978
72. P. Kawulka, C.R. Kirby, and G.L. Bolton, Can. Min. Metallurg. Bull. *71*, 122, Feb. 1978
73. D.E.G. Maschmeyer, E.F.G. Milner, and B.M. Parekh, Can. Min. Metallurg. Bull. *71*, 131, Feb. 1978
74. New Refining Processes for Copper, Chem. Eng. News *55* (37), 7, Sept. 12, 1977
75. Cyprus Reveals Details, Eng. Mining J. *178*, 27, Nov. 1977
76. C.C. Simpson, Jr., J. Metals *29*, 6, July 1977
77. Lower Energy process for Copper Refining, Chem. Eng. News *58* (23), 7, June 9, 1980
78. 1979/80 Statistical Yearbook, United Nations, New York, 1981, and earlier
79. P.J. Bailes, C. Hanson, and M.A. Hughes, Chem. Eng. *83 C*, 86, Aug. 30, 1976

# 12 Production of Iron and Steel

## 12.1 Early History and Development

Iron, at 5 % of the earth's crust, is the most abundant metal after aluminum and the fourth most abundant element present in the surface rocks of the earth [1]. Iron is also the pre-eminently useful metal of our society for the manufacture of machinery of all types and for constructional purposes. The general wide utility of iron, on its own quite a versatile metal, and of an increasingly wide variety of steels, iron alloys which are being tailored to meet a range of demanding applications, combine to make the tonnage of iron produced each year easily exceed the combined annual production of all the other metals used by man.

Iron, like aluminum, is a metal which is relatively easily oxidized and so is seldom found in nature in metallic form, except in recent meteorites. In fact sideritic (iron-containing) meteorites were probably the origin of the first iron worked by early man. It is also possible that lumps of iron-rich ores placed around large very hot campfires were reduced to a spongy type of iron mixed with slag. Slag could have been mostly beaten out from these lumps and simple tools fashioned in the same way from this quite malleable form of iron. Because of the low likelihood that meteoritic iron would be found to the same extent as lumps of native copper in many areas, historians believe that the iron age followed the copper and bronze ages in prehistoric stages of metallurgical development since the stone age [2]. Some parts of the world without any significant surface exposures of copper mineral however, could have experienced direct development from the stone to the iron ages. In either case the use of iron was certainly known by the Egyptians dating back to 3 to 4,000 B.C., both for tools and simple jewellery [2, 3].

The methods used by the early small scale smelters of the ore were probably simpler versions of the methods used in Europe before about 1750. At this time a furnace was built of a fire-resisting clay and fired by burning charcoal raised to smelting temperatures by a pair of continuously operated foot bellows [4, 5]. Selected ore, of 40 % or more iron, was then added in lumps a few at a time and the charcoal fired heating continued for several hours. The furnace was then allowed to cool, the front was broken open and the lumps of spongy iron mixed with slag removed. Hammering was used to drive out most of the slag and to shape the residual iron. Direct reduction of iron ore as accomplished by these methods, in the absence of liquefaction, produced a metallic iron which was low in carbon and hence ductile enough to be fashioned by beating.

These small scale operations permitted the production of only a few kilograms of iron from each firing. The development of the forerunners of the modern blast furnace in rudimentary form is believed to have occurred in Belgium in about 1400. These larger units, although still fired by charcoal, could produce several hundred kilograms every 24 hours but also required mechanization to provide sufficient blast which was usually provided by water power.

But because of the greater cost and organization required for blast furnace construction and operation, large blast furnaces grew alongside the earlier procedures and only very gradually displaced the early smaller scale operations.

To provide the charcoal required for both firing and the reduction process took enormous quantities of wood and caused rapid depletion of forest stands near smelters, both in Europe and North America [4, 6]. Depletion of charcoal resources, and the increasing availability of coal led to many attempts to fire blast furnaces directly with coal, with uniformly poor results. The sulfur content of the coal contributed to a high sulfur content in the iron which made it brittle when hot, an undesirable property. Gradually it was learned that prior coking of the coal could be used to obtain equivalent quality of iron from coke of coal origin as could be obtained using wood charcoal. Only subsequently was it learned that sulfur had been the cause of the early problems with iron quality, and that coking got rid of this problem by getting rid of the sulfur before it contacted the ore. As coke came into more general use and became available in quantity the size of blast furnaces being built also became greater, developments which coincided with the invention of the steam engine which was ideally suited to provide the increasing demand of power for the blast.

Impurities such as sand, clay etc. present in iron ore were combined with lime produced by the addition of either limestone or oyster shells into the furnace charge [4, 6]. The slag so formed melted at the blast temperatures, and being lighter than the melt of iron, floated on top. Slag and iron were withdrawn separately, from upper and lower tapholes, at appropriate intervals.

From these early and longer term developments the iron and steel industry has grown to the point where world production of iron ore, based on iron content, exceeded half a billion tonnes in 1978 (Table 12.1). While the major producers of ore are also, in general, major producers of pig iron, this is by no means universally the case. Sweden, Australia, Brazil, Canada, and India each mine and process iron ore corresponding to 8, 7, 4, 2, and 2 times the mass of pig iron they produce, respectively, the rest of the ore production being exported mainly to other industrialized countries. Of this last group, Japan and West Germany are currently importing nearly all (99.5 %, and > 98 %) of their iron ore requirements, a trend which in recent years appears to be towards a complete dependence on imported ores. The U.S.A. depends on imported ores for about $^1/_3$ of her requirements, a proportion which has been fairly steady, at least for the last ten years.

Steel, derived both from new pig iron produced from ores as well as from recycled steel obtained from structural shapes, old automobiles, ships and the like, is a better defined material of construction for most uses. World steel production currently hovers around $1\,^1/_3$ times world pig iron production, a reflection of the recycled metal content. Again, we have some countries, like China and the U.S.A., whose national steel consumption exceeds even their very large scale of production by about $^1/_3$ and about $^1/_5$, respectively (Table 12.2). France and the United Kingdom each produce slightly more steel than they consume, whereas Italy and West Germany each produce close to $1\,^1/_3$ times their annual consumption. And Japan, even with a very large per capita consumption, produces some $1\,^2/_3$ times her annual requirements. This is undoubtedly a reflection of the growth in export significance of shipbuilding, automobile manufacture, and heavy engineering generally to the Japanese economy, as compared to those of the other large scale steel producers. Even though some 55 nations

**Table 12.1** Production of pig iron and iron ore by the major world producers, given in thousands of metric tonnes [a]

|  | 1950 | 1960 | 1970 | | 1978 | |
|---|---|---|---|---|---|---|
|  | Pig iron | Pig iron | Ore, Fe content | Pig iron | Ore, Fe content | Pig iron |
| Australia | 1 116 | 2 922 | 28 676 | 5 769 | 54 739 | 7 096 |
| Brazil | 729 | 1 750 | 24 739 | 4 296 | 45 133 | 10 530 |
| Canada | 2 266 | 4 025 | 29 187 | 8 424 | 24 367 | 10 578 |
| China | 2 392 [b] | 27 500 | 24 000 | 22 000 | 32 500 | 34 790 |
| W. Germany | 9 511 | 25 739 | 1 773 | 33 897 | 510 | 30 417 |
| India | 1 708 | 4 260 | 19 654 | 7 118 | 24 472 | 9 718 |
| Japan | 2 299 | 12 341 | 916 | 69 714 | 361 | 88 102 |
| Sweden | 837 | 1 627 | 19 804 | 2 842 | 21 487 | 2 363 |
| U.S.A. | 60 217 | 62 202 | 53 808 | 85 141 | 50 512 | 79 542 |
| U.S.S.R. | 25 000 [c] | 46 800 | 106 058 | 85 933 | 133 894 | 110 702 |
| Other | 6 825 | 69 434 | 112 685 | 114 666 | 142 225 | 136 462 |
| World Totals | 112 900 | 258 600 | 421 300 | 439 800 | 530 200 | 520 300 |

[a] Countries producing more than 20 million tonnes per year of iron ore (Fe content basis) or 29 million tonnes per year of pig iron in 1978 included. Data from U.N. Statistical Yearbooks.
[b] Value for 1944, published data closest to 1950.
[c] Value for 1952, published data closest to 1950.

**Table 12.2** Production and consumption of steel by the major steel producing countries, in thousands of metric tonnes [a]

| | 1970 | | 1978 | | |
| | Production | Total consumption | Production | Total consumption | Consumption, kg/capita |
|---|---|---|---|---|---|
| China | 18 000 | 22 504 | 31 780 | 42 706 | 46 |
| France | 23 773 | 23 236 | 22 841 | 19 199 | 360 |
| W. Germany | 45 040 | 40 601 | 41 253 | 32 209 | 525 |
| Italy | 17 277 | 21 113 | 24 283 | 18 822 | 332 |
| Japan | 93 322 | 69 882 | 102 105 | 61 507 | 535 |
| U.S.S.R. | 115 889 | 110 234 | 151 453 | 155 000 [b] | 591 [b] |
| U. Kingdom | 28 316 | 25 539 | 20 311 | 19 429 | 348 |
| U.S.A. | 119 309 | 127 304 | 124 314 | 146 445 | 670 |
| Other | 132 974 | 140 287 | 185 860 | 209 883 | 1 to 756 [c] |
| World | 593 900 | 580 700 | 704 200 | 705 200 [d] | 166 |

[a] Countries producing more than 20 million tonnes of steel per year, from 1979/1980 Statistical Yearbook [34].
[b] Estimated by extrapolation, 1970 – 1976 period. 1978 not available.
[c] Range of kilograms per capita consumption in 1978
 e.g. Afghanistan, Burma and Ethiopia with 1 kg per capita, Czechoslovakia with 756 kg per capita.
 See text for explanations.
[d] Includes estimated consumption value for U.S.S.R. for 1978.

were reporting steel production on some scale in 1970, the top eight countries produced more than $^3/_4$ (77.6 %) of the world's steel in that year. Large scale steel production appears to be becoming less exclusive since by 1978 the same eight countries produced only about 70 % of the world's steel.

## 12.2 Reduction of Iron Ores to Iron

### 12.2.1 Direct Reduction

A relatively large number of naturally occurring minerals contain iron but only a small number are rich enough in iron and occur in extensive enough deposits to be classed as ores. The important ores in North America are the oxides magnetite ($Fe_3O_4$, 72.4 % iron), hematite ($Fe_2O_3$, 70.0 % iron), and limonite, a hydrated iron oxide ($2 Fe_2O_3 \times 3 H_2O$, 59.8 % iron). In Europe, including the U.K., an iron carbonate siderite ($FeCO_3$, 48.2 % iron) is also an important ore for pig iron production, though this ore is of less significance in North America. The sulfide minerals pyrite ($FeS_2$) and chalcopyrite ($CuFeS_2$) are more important for their sulfur content than for their iron content. While the iron content of deposits of these minerals is never quite as high as the calculated percentages given

for the pure mineral, many commercial ores are mined with an iron content in the 50-60 % range and some deposits are worked with iron compositions of as high as 66-68 % [7].

If the ore is contained in a hard rock matrix it is mined from underground or surface deposits by drilling and blasting followed by recovery with power shovels. Many deposits are sufficiently loosely consolidated that direct recovery with power shovels is possible, by either open cut or open pit operation. Both types of ore are also frequently beneficiated or cleaned of extraneous material at the mine site. Beneficiation may involve any of rough sorting, crushing, screening, magnetic separation (magnetites only), washing (for clay or sand removal) and drying, all designed to improve the grade of the ore prior to shipping or smelting.

Direct reduction processes of various kinds, similar in nature to prehistoric reduction methods in that at no time during the process is the iron content entirely fused (melted), are still used on a small scale to produce spongy and powdered iron products. In practice all direct reduction methods involve contacting of a solid or gaseous reductant, at temperatures of 1,000 to 1,200 °C, with a good analysis magnetite or hematite ore to effect reduction without melting [8]. Both solid reductants such as powdered coal and coke, and gaseous reductants such as natural gas and hydrogen, are being used. The advantage of direct reduction methods for iron production is that the iron product obtained is low in carbon and can therefore be used for steelmaking directly, without requiring carbon removal as a preliminary step. For this reason the direct reduction product is occasionally referred to as "synthetic scrap", since it is possible to use this type of iron to replace part or all of the scrap component of steelmaking. Catalyst applications also provide a small but important market for both the spongy and the iron powder forms of directly reduced iron. The very high porosities and specific surface areas make these forms of iron far more valuable than iron filings or turnings for catalyst applications. Direct reduction processes have been of interest for years and continue to be practised on a small scale for special purposes. Nevertheless, because of various feed composition and processing constraints required by these methods blast furnace reduction of iron ores is, and has always been the dominant method used to obtain metallic iron from its ores. Direct reduction methods are, however, particularly in favour in parts of the world where there is abundant natural gas but negligible coal reserves, such as in Iran, Qatar, and Venezuela [9]. They are also of value for small scale iron making operations which feed their product directly to an electric furnace for steelmaking.

### 12.2.2 Blast Furnace Reduction of Iron Ores to Pig Iron

Reduction of iron ore to the metal in a blast furnace requires the reduced metal to be melted in the process, the key physical difference between direct reduction and blast furnace reduction. Two to three tonnes of iron ore are required for each tonne of iron produced. The main impurity in iron ore is generally silica. To avoid contaminating the iron with silica, limestone or dolomite ($CaCO_3$ + $MgCO_3$) is also added to combine with the silica and to provide a lithophilic phase to dissolve up any other acidic oxides which do not react with lime (Equations 12.1-12.3).

$$CaCO_3 \rightarrow CaO + CO_2 \tag{12.1}$$

$$CaO + SiO_2 \rightarrow CaSiO_3 \text{ (slag)} \tag{12.2}$$

$$CaO + Al_2O_3 \rightarrow Ca(AlO_2)_2 \tag{12.3}$$

In this way a separate molten slag phase is formed

from the ore impurities which can be easily separated from the molten iron. In addition some 550-1,000 kg of coke is used per tonne of iron produced, to provide both much of the heat and the reducing agents necessary for the formation of pig iron.

The whole reduction process is carried out in a blast furnace consisting of large, almost cylindrical stack up to 10 m in diameter and 70 m high [8], which is tapered to a larger diameter at the bottom to avoid bridging (jamming) of ore as it moves down the stack (Figure 12.1). The outer shell is constructed of heavy (ca. 2 cm thick) rivetted or welded steel plates, which is then lined with silica-containing (upper region) or carbon (hearth area) refractory bricks to make the structure resistant to the intense heat. This unit functions as a heterogeneous counter-current reactor in which the solid reactants are charged in at the top, and the firing air is blown in at the bottom. The ore, limestone, and coke, in lumps to permit movement of gases through the charges, are added in alternating layers to the top of the furnace using a skip cart on rails. A double bell arrangement at the top of the furnace prevents gas loss from the furnace during the charging process by only opening one bell at a time during the addition steps. The remaining raw material, the firing air, is preheated to about 850 °C in regenerative stoves and then blown through water-cooled steel tuyeres (or nozzles) at the hearth area of the furnace.

Initial warming of a newly lined blast furnace is conducted gradually, over several days, to avoid any unnecessary thermal stresses. Fueled first by wood or oil and then by coke, the approach to normal operating temperature by the furnace is gauged from peep holes through the tuyeres. As these temperatures are reached ore and limestone are blended with the coke additions through the double bell arrangement at the top of the furnace. Heat, provided by primary combustion in the oxidizing zone near the hot air blast of the tuyeres is sufficient to cause carbon dioxide and water vapour (from the air) reduction (Equations 12.4-12.6).

Primary combustion and reductant formation:

$$C + O_2 \rightarrow CO_2 \tag{12.4}$$

$$CO_2 + C \text{ (white hot)} \rightarrow 2 CO \tag{12.5}$$

$$H_2O + C \text{ (white hot)} \rightarrow CO + H_2 \tag{12.6}$$

Carbon monoxide is the primary agent responsible

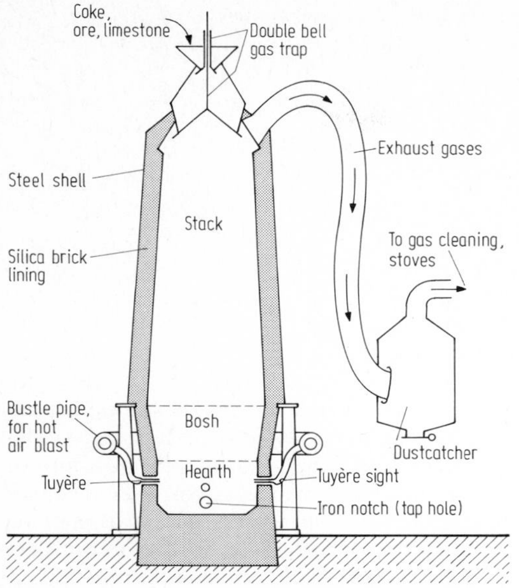


**Figure 12.1.** Diagram of the operating details of a blast furnace for the production of pig iron

for reduction of the iron oxides in the ore, assisted by carbon and to a small extent by hydrogen (Equations 12.7-12.12).

Primary reducing reactions:

ca. 500 °C $3\,Fe_2O_3 + CO \rightarrow 2\,Fe_3O_4 + CO_2$ (12.7)

ca. 850 °C $Fe_3O_4 + CO \rightarrow 3\,FeO + CO_2$ (12.8)

ca. 900 °C $FeO + CO \rightarrow Fe + CO_2$ (12.9)

ca. 900 °C $Fe_2O_3 + 3\,H_2 \rightarrow 2\,Fe + 3\,H_2O$ (12.10)

The solid-solid or liquid-solid reactions between ferric and ferrous oxides and carbon also occur at temperatures above about 500 °C [7].

$$Fe_2O_3 + C \rightarrow 2\,FeO + CO \quad (12.11)$$

$$FeO + C \rightarrow Fe + CO \quad (12.12)$$

The iron, with its melting point depressed to about 1,100 °C by dissolved carbon (melting point of pure iron is 1,535 °C), drips through the charge and collects on the hearth. Being the denser fluid product of the blast furnace, it forms a lower phase as it accumulates on the hearth. Molten pig iron is periodically withdrawn from this accumulation by drilling or knocking out a clay plug placed in a lower tap hole in the side of the hearth.

The slagging reactions (Equations 12.1-12.3) start to occur as the charge moves down the furnace and the limestone or dolomite used as a flux reaches temperatures of about 900 °C and higher. As carbon dioxide loss occurs the lime (and magnesia, if present) start to react with and dissolve other oxide impurities of the charge into the slag layer. It is also possible to reduce silica and phosphate by the same kinds of reactions used to reduce iron which, in the process, contaminates the pig iron (Equations 12.13, 12.14).

$$SiO_2 + 2\,C \rightarrow Si + 2\,CO \quad (12.13)$$

$$P_2O_5 + 5\,C \rightarrow 2\,P + 5\,CO \quad (12.14)$$

The presence of the basic slagging ingredients minimizes, but does not eliminate these reductions from occurring. The molten slag, which is less dense than iron, floats on top of the iron layer and is periodically withdrawn in this form through an upper tap hole in the hearth of the blast furnace. About 0.5-1.5 tonnes of slag are produced for each tonne of pig iron, the low end of the range being obtained with high grade ores while using a coke with a low ash content for reduction. Thus utilization of this byproduct of pig iron production is a significant part of normal blast furnace operation. Slag is used for road aggregate (ca. 50-60 % of the total), in crushed form as a stone component of asphalt, and in granulated form (by running the melt into water) for use as an ingredient in the manufacture of cement (ca. 10-20 % of the total). It is also used to a small extent (ca. 5 %) for the preparation of mineral wool ("rock wool") insulation materials. The approximate chemical composition of blast furnace slag is given in Table 12.3.

The carbon monoxide produced in the furnace is only partially consumed on its passage through the granular charge (Table 12.3). But capture of this important fuel gas for energy recovery, by the double bell closure at the top of the furnace, is only a relatively recent (ca. 1840) innovation in the long history of iron making. After gas collection from the upper part of the furnace the exit gas is passed to a dust catcher for removal of most of the entrained dust and is then used, partly to fuel three or four regenerative stoves and partly for other purposes. These regenerative stoves, which are designed to have a large thermal mass from an extensive internal firebrick checker-work of narrow passages, are first heated by blast furnace gas. Then, when the temperature of the brick checkerwork of one stove has been raised sufficiently, the blast furnace gas is switched to fuel the burner of,

**Table 12.3** Approximate composition ranges of the product of blast furnace operation [a]

| Iron | | Slag | | Exit gases | |
|---|---|---|---|---|---|
| Component | % by wgt | Component | % by wgt | Component | % by wgt |
| iron | 90 – 94 | lime | 30 – 48 | nitrogen | 58 – 60 |
| carbon | 3 – 4.5 | silica | 28 – 42 | carbon monoxide | 27 – 30 |
| silicon | 1 – 2.5 | alumina | 10 – 20 | carbon dioxide | 8 – 14 |
| phosphorus | 0.04 – 2.5 | magnesia (MgO) | 2 – 14 | hydrogen | 0.4 – 3 |
| manganese | 0.4 – 0.7 | sulfur | 1 – 3 | water vapour | 0.4 – 1.0 |
| sulfur | 0.02 – 0.2 | iron oxide (FeO) | 0.3 – 2 | | |
| | | manganese oxide, (MnO) | 0.2 – 1.5 | | |

[a] Compiled from data of Dearden [3] and Kirk-Othmer [7].

and heat another stove. At this point the preheated stove is used to heat the air blast being fed to the tuyeres in the hearth area of the blast furnace. By a fuel gas and air switching arrangement like this blast air is continuously preheated using the fuel value of the furnace exit gases, and in so doing decreasing the overall coke consumption to significantly below that required before gas collection was practised.

Originally the molten iron product from the blast furnace, as it was drained from the lower notch on the hearth was led to a series of shallow trenches shaped in sand. As the molten iron moved along each long trench, it filled a series of short smaller trenches terminated by roughly rectangular hollows each of which, in cooling, eventually formed a 20-25 lb ingot of iron. This product became known as "pig iron" by the appearance parallels between this manual method or forming ingots to a sow suckling her young. Today pig iron ingots are made on a continuous pig-casting machine, which consists of an endless chain with metal moulds and into which the molten iron is ladled. A water spray, as the moulds with molten metal move through it, quickly chills the melt allowing pig production at the rate of two tonnes per minute or more. A lime or similar wash, sprayed into the insides of the emptied moulds as they return to the ladle for refilling, prevents the pig iron from sticking to the moulds.

Blast furnace production of iron allows the hot, newly reduced iron to trickle through the bed of heated coke to the hearth. Since carbon is significantly soluble in molten iron, pig iron usually contains from 3-4.5 % carbon. It also contains smaller percentages of other reduced elements such as silicon, phosphorus, manganese etc. generated by the same reducing processes as yielded the iron

(Table 12.3). Because of these impurities, but primarily because of the effect of the high carbon content on the iron crystal structure, the blast furnace product is brittle, hard, and possesses relatively low tensile strength. Hence the crude pig iron product of the blast furnace, in this form, is little used directly (Table 12.4).

However, after remelting in a cupola furnace and some minor composition adjustments pig iron is used to make various types of cast irons, with improved properties. Molten gray cast iron, for instance, is poured into moulds to make automotive engine blocks. In this application the hardness and graphite content of this material contribute to its desirable wear resistance. But the cast irons, a generic term for a group of iron casting products including pearlitic gray cast iron, white cast iron, chilled iron, nodular iron etc., together comprise an important field of iron metallurgy. Their properties, appropriately varied by the adjustment of microstructural constituents and crystal structures, can cover a wide range of possibilities. The corrosion resistance of some cast irons make these useful for the fabrication of certain types of chemical reactors. The low cost of the raw material means that the wall thickness of a reactor may be made sufficiently thick that a low rate of corrosion can be accommodated without perforation. Construction of low speed flywheels from cast iron and its use as a ballast in ships are also applications which depend on low cost and high mass density for their success. But by far the majority of pig iron produced moves directly, either still in molten state or in ingot form, to one or other of the various steelmaking processes.

A number of operating advantages are obtained from pig iron production in very large furnaces, one aspect of which is the lowered coke consump-

**Table 12.4** Approximate material balance data for a modern blast furnace producing 600 tonnes of pig iron per day [a]

| Raw Materials | | Product and Byproducts | |
|---|---|---|---|
| Beneficiated iron ore (50 – 60% Fe) | 1 000 –  1 200 tonnes | Pig iron | 600 tonnes |
| Hard coke | 500 –   600 tonnes | Slag | 300 –   400 tonnes |
| Limestone (or dolomite) | 200 –   300 tonnes | Dust | 60 –    80 tonnes |
| Blast air, > 800 °C | 2 000 –  2 400 tonnes | Exhaust gases | 2 800 – 3 400 tonnes |
| and 550 kPa (80 psi) | $1.5 - 1.9 \times 10^6$ m$^{3}$ [b] | | $2.2 - 2.6 \times 10^6$ m$^{3}$ [b] |
| Water; as steam | 20 –    40 tonnes | | |
| Water, cooling | 14 000 – 18 000 tonnes | | |
| Electricity | 2 400 – 28 600 kWh | | |

[a] Data compiled from [3, 8, 10, 11].
[b] Volumes at standard conditions i.e. 15 °C and 1 atmosphere.

tion from the lower furnace heat losses resulting from the smaller surface to volume ratio. This factor, together with more extensive use of richer, beneficiated ores which also may be partially pre-fluxed, use of all charges in a more uniform size range (by prior agglomeration) plus additions of oxygen and steam to the blast air, have combined to nearly halve the coke consumption per tonne of pig iron produced in American practice, since 1950 [10]. The addition of steam to the blast has been a growing practice, serving to refine the control of the blast temperature as well as providing additional reducing gas at the bosh zone via Equation 12.6 [8]. Occasionally fuel oil or natural gas are used to provide additonal combustion and reducing capacity to a blast furnace, in which case the coke requirement is decreased correspondingly. Recent energy balance data incorporating many of these variations is given by Kirk-Othmer [11].

An idea of the scale of operation of a typical modern blast furnace, incorporating some of these innovations, may be gained from consulting Table 12.4, where a range of input-output data for an intermediate size furnace capable of producing 600 tonnes of pig iron per day is given. The largest

blast furnaces now operating in the U.S.A. and Japan are capable of producing 10,000 tonnes of pig iron per day [11], more than 16 times this capacity. The total U.S. iron production in 1979 of some 80 million tonnes was produced by 168 large and small furnaces, mostly operating in the Pittsburgh and Chicago areas.

## 12.3 The Making of Mild, and Carbon Steels

Pig irons each contain 6 to 8 % of total impurities which give the product brittle, low strength characteristics. Broadly speaking, pig irons may all be classified into two groups, those that are low in phosphorus and those that are high (Table 12.5). Mild steels and high carbon steels are made by significantly decreasing the concentrations of the impurities present in pig irons and controlling the final concentration of carbon. These products, in contrast to the properties of either of the types of pig irons, are considerably stronger materials and in addition may have other specific desirable properties contributed to them by more careful

**Table 12.5** Percentage of impurities present in two typical pig irons which are representative of the product obtained from the two main types of iron ores, compared to the composition range of common steels [a]

| Element | No. 1, % | No. 2, % | Common Steels, % |
|---|---|---|---|
| Carbon | 3.5 – 4.0 | 3.0 – 3.6 | 0.1  – 0.74 |
| Silicon | 2.0 – 2.5 | 0.6 – 1.0 | 0.01 – 0.21 |
| Sulfur | < 0.040 | 0.080 – 0.10 | 0.028 – 0.055 |
| Phosphorus | < 0.040 | 1.8  – 2.5 | 0.005 – 0.054 |
| Manganese | 0.75 | 1.0 – 2.5 | 0.35 – 0.84 |

[a] Compiled from G. R. Bashworth [35] and J. A. Allen [12].

control of composition and by heat or other types of subsequent treatments. As a result of this improvement in properties more than 90 % of all pig iron produced, as ingots, hot metal, or molten metal (particularly at integrated operations where the steelmaking is closely associated with blast furnace operation), proceeds directly to various steelmaking procedures. Also, over 90 % of all steels made, the "common steels" of Table 12.5, are carbon steels. Thus, the bulk of the following discussion is centered on this important group of products. Mild steels, the product used for the vast majority of structural and engineering purposes contains 0.15 to 0.25 % carbon. High carbon steels, defined as steels containing 0.6 to 0.7 % carbon, possess important heat treating qualities which can permit them to hold a sharp "edge". Mild steels are not usually heat treated.

The processes used in making common steels may be classified on a chemical basis using the composition of the refractory linings used for the vessels used for containment, and the primary composition of the slag. Combined with this classification the processes used may be further subdivided in accordance with the steelmaking technology applied, that is whether they are based on pneumatic, open hearth, or electric furnace procedures.

Acidic steelmaking employs silica ($SiO_2$) refractory linings in the equipment used, and generates silica rich slags via additions of sand for removal of impurities. Silica, being a non-metallic oxide, allows removal of carbon, manganese, and silicon during the metallurgical steps but cannot achieve removal of sulfur or phosphorus. Thus, pig irons rich in sulfur, or phosphorus, or both elements cannot be satisfactorily processed under these conditions. Basic steelmaking procedures use magnesia (MgO) refractory linings and limestone or dolomite added to the charge as a slag former. Equipment with a basic lining is not only capable of removing carbon, manganese and silicon, but is also capable of removing phosphorus and sulfur by capture of their oxides in the basic slag. Addition of basic slag formers to an acid lined vessel or vice versa is never practised because by doing this lining life is drastically shortened by their direct interaction.

The technological classification is based on the method used to effect removal of pig iron impurities. All pneumatic methods primarily rely on the heat content of the molten charge plus the heat of the refining reactions themselves to provide the heat required by the process. The purification is effected by blowing air or oxygen through or onto the molten charge. Open hearth methods all rely on the combustion of an external fuel, usually natural gas or oil, to provide the heat required for the process. Hence, stoves and their regenerative heating capabilities (as with blast furnace gas) are important for open hearth operation. Electric furnace methods derive heat requirements from electric arc or resistance sources, and as such permit control of the atmosphere above the melt independently of the heating process. This capability has important consequences, particularly for the production of special alloy steels.

### 12.3.1 Pneumatic Steelmaking: The Bessemer Process

The earliest of the pneumatic methods of steelmaking, the Bessemer process, was devised and patented in England in 1855 by Henry Bessemer [12]. Originally prompted by an interest in improving the quality and strength of cast iron gun barrels to obtain greater range from lighter guns, this process was developed from the perception by Bessemer that the carbon in molten iron could be burned out by air blowing. In this way it was expected that this method would raise the speed and decrease the cost of the current crucible method of steelmaking [13].

Bessemer steel is made in a Bessemer converter, a pear-shaped vessel with a double bottom, lined with silica and capable of being tilted on its axis. A typical converter can hold 15 to 25 tonnes of metal at one time (Figure 12.2) which is charged with molten pig iron while the converter is lying on its side. The air pressure to the perforated inside base of the unit is then turned on and the vessel gradually raised to the upright position so that the air is sparged through the molten pig iron via the inner perforated floor of the converter. Blowing is continued for a period of 15 to 20 minutes until the operator, from observing the colour and luminosity change of the flame at the mouth of the converter, judges the refining process to be complete. Carbon, as well as silicon, manganese, and some iron are removed, largely via reactions 12.15 to 12.19.

$$2\,Fe + O_2 \rightarrow 2\,FeO \tag{12.15}$$

$$2\,Mn + O_2 \rightarrow 2\,MnO \tag{12.16}$$

$$Si + O_2 \rightarrow SiO_2 \tag{12.17}$$

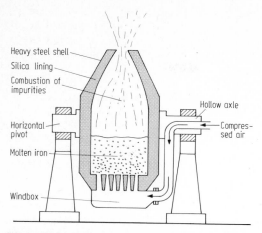

**Figure 12.2.** Cut-away view of the main operating details of a Bessemer converter during a "blow". It is filled and emptied by tilting onto its side, while the air flow is off

$$2\,C + O_2 \rightarrow 2\,CO \qquad\qquad (12.18)$$

$$C + O_2 \rightarrow CO_2 \qquad\qquad (12.19)$$

The involatile products of these reactions form an acidic slag ($SiO_2$ x FeO; $SiO_2$ x MnO etc.) which forms a separate lighter phase on top of the molten iron. These exothermic reactions occurring during the blow are sufficient to raise the temperature of the converted metal from about 1,350 °C to around 1,600 °C or higher, adequate to keep the residual iron molten. This temperature rise is necessary since the melting point of the iron phase rises as the concentration of impurities is decreased by the conversion process. While impurities are present the melting point is depressed significantly.

The Bessemer process is fast, one converter being capable of producing nearly a tonne of steel per minute as compared to 10 to 14 days for smaller quantities per unit via the crucible methods current at the time the Bessemer process was developed [13]. The product is of acceptable quality for some uses, at least when the pig iron is derived from low phosphorus ores and is smelted in blast furnaces using low sulfur cokes. However, with high phosphorus pig irons the decarbonized product is "cold short" (brittle at ambient temperatures) making it next to useless for the vast majority of applications of the steel [5]. Another difficulty is that if the ore or the smelting process produces a pig iron which is high in sulfur (e.g. from sulfur-containing cokes) the Bessemer product is "hot short", or brittle

when hot. This made hot shaping or working of the product impossible, further adding to steelmaking difficulties with this product. And the speed of Bessemer steelmaking is both an advantage and a disadvantage, in that there is no time during the process for product sampling, analysis, and composition adjustment. Thus, much of the product quality depends on the skill and judgement of the operator. Overblowing, resulting in total carbon removal and some iron oxide and excessive iron nitride formation may be partially corrected by blending the blown product with a small amount of hot melt pig iron and mixing well [8], and other minor adjustments may be made in the ladle receiving the Bessemer product. But removal of the nitrogen, even with strong deoxidizers such as aluminum, is never easy, and yet is essential to obtain good stamping quality steels (i.e steels destined for shaping into auto or appliance body panels, etc.). The combination of these stringent raw material requirements and the short composition adjustment time thus has restricted production of acid Bessemer steels (produced in converters with acidic linings) to low carbon mild steels. However, steelmaking comprising these chemical and technological elements is still being practised on a small scale, in the U.K. and other European countries at least into the 1960's, as an element in the range of methods employed for producing low cost steels [12].

*Basic Bessemer, or Thomas Process*

The problem of phosphorus or sulfur removal from pig irons high in these elements was solved by the use of a basic refractory lining in the converter, such as magnesia (MgO) or lime (CaO), rather than an acidic lining. When this was accompanied by the addition of limestone or dolomite with the molten charge of pig iron, phosphorus and sulfur constituents could be relatively efficiently removed in the basic slag layer so formed. These improvements, which were developed and announced in London in 1878 by the cousins S.G. and P.G. Thomas, were adopted to some extent in the U.K. steel industry for the processing of moderate phosphorus content ores. But for the steel makers of Lorraine, a district of high phosphorus iron ore deposits in NE France, the Thomas process, as this came to be known, was enthusiastically adopted to achieve phosphorus removal and obtain the resulting improved steel quality [12].

Phosphorus removal in this process is thought to occur primarily as the oxide, whereas elemental sulfur is thought to form a slag constituent by direct replacement of the oxygen of calcium oxide (Equations 12.1, 12.20-12.22).

$$4P + 5O_2 \rightarrow 2P_2O_5 \tag{12.20}$$

$$P_2O_5 + 4CaO \rightarrow 4CaO \cdot P_2O_5 \text{ (slag)} \tag{12.21}$$

$$2S_{(Fe)} + 2CaO_{(slag)} \rightarrow 2CaS_{(slag)} + O_{2(Fe)} \tag{12.22}$$

An important attraction of the acidic Bessemer process and of the basic Bessemer, or Thomas, process is the rapid steelmaking capability, about a tonne a minute per converter, that these methods provide. But the acidic Bessemer process has difficulties in dealing with phosphorus and sulfur in pig irons and both processes tend to cause some nitriding, absorption of nitrogen from the blowing air by the hot metal (Table 12.6). Increasing presence of nitrogen in a finished steel increases its susceptibility to strain aging, or embrittlement under stress, a feature which makes this product unsuitable for any shaping or finishing operations, such as wire drawing, which are to be conducted on the cold metal. These disadvantages, plus the inability to handle significant quantities of scrap metal or an initial solid iron charge have decreased the relative importance of both of these pneumatic processes in modern steelmaking to 10 % or less of the total [12]. Variants of these methods such as using oxygen-enriched air, oxygen and steam, or oxygen and carbon dioxide (pure oxygen blowing introducing too severe a thermal stress on a Bessemer converter), all introduced to decrease the nitriding tendency, have been adopted to a varying, though small, extent [12]. A combination of one or other of these pneumatic processes as a preliminary to an open hearth furnace (duplex steelmaking), or as a preliminary to both open hearth and electric furnaces (triplex steelmaking), has led to achievement of a decreased nitrogen content of the product

**Table 12.6** Nitrogen content of blast furnace metal (pig iron) and various steels [a]

| Iron product | Range of nitrogen content, % |
| --- | --- |
| Pig iron | 0.002 – 0.006 |
| Bessemer steels | 0.010 – 0.020 |
| Duplex steels [b] | 0.005 – 0.008 |
| Open hearth steels | 0.004 – 0.007 |

[a] Compiled from [8, 12].
[b] See text for details.

and improved steelmaking flexibility, features which have extended the utility and practice of these pneumatic processes [8]. Details of the associated processes in duplex and triplex steelmaking are given later.

### 12.3.2 The Open Hearth Process

Generally improved furnace combustion efficiency, regardless of the application, was the primary objective of F. Siemens, who was from Germany but conducted most of his combustion trials and developments in Britain. It was he who applied regenerative methods to recover waste heat from a combustion process by passing the combustion gases through a chamber containing an open, firebrick chequerwork. When the chequerwork reached nearly the combustion gas temperature, the hot gas flow was diverted to another chamber of firebrick chequerwork to heat this. At the same time the air flow to the combustion process was directed through the preheated chamber of hot chequerwork in order to preheat the combustion air. This air preheat not only raised the completeness and efficiency of the combustion process itself with whatever fuel was being used, but it also enabled much higher ultimate temperatures to be reached than was possible in the absence of the air preheat. The two sets of firebrick chequerwork were operated on swing cycles such that as the air preheat chequerwork cooled down the air stream was redirected to the other, already hot chequerwork, and the hot combustion gas stream was returned to reheat the initial, now cool chequerwork. In this way a continuous heat recovery from the spent combustion gases was obtained. This procedure sustained the thermal efficiency advantage described for the first cycle. With these developments alone, by 1857 Siemens was able to claim fuel savings of some 70 to 80 % from that previously consumed. Industries such as glassmaking employed furnaces of this type and were among the first technical groups to adopt these improvements [5].

These combustion developments were first tested for steelmaking in 1863 in France, under license from Siemens, by a father and son team, E. and P. Martin [5]. Siemens himself tested the furnace for this purpose in Wales, three years later, and both operations were pronounced a success. These were the origins of the open hearth process often referred to as the Siemens-Martin process, after the joint contributions of these inventors.

The open hearth furnace, using the Siemens regenerative principle to raise maximum combustion temperatures to the region of 1,650 °C [10] and at the same time conserve fuel, possesses a heating capability independent of the heat content of impurity burning reactions of the iron charge. Because of this, the charge placed in the open hearth furnace may be any proportion of scrap to hot metal up to 100 % scrap since the furnace has the capability to melt this. Usual practice, however, is to use roughly a 50:50 mix of scrap to melt [14].

The actual steelmaking components of an open hearth furnace consist of a rectangular heating zone with a floor containing a long shallow depression of 200 to 500 tonnes capacity into which the charge is placed (Figure 12.3). Gas or oil burners at each end are provided with preheated combustion air via hot chequerwork, and the flames so generated are deflected down onto the charge by a convex-down, reverberatory roof. So a combination of radiant heating and hot gas contact with the charge transfers sufficient heat to melt it. To protect the structural components of the depression holding the charge from the intensely high temperatures and also to aid in removal of phosphorus and sulfur impurities from the charge, the depression is normally lined with a magnesia brick refractory.

The procedure for steelmaking in an open hearth furnace is to place the solid components of the charge, the limestone, any ore requirement, and rusty scrap steel into the depression, cold. The scrap is preferably rusty to provide oxygen which will assist in impurity removal. Iron ore (e.g. $Fe_2O_3$), may also be added to provide oxygen. The fuel, usually natural gas, coal gas, or oil, is then ignited at the burners and this inital charge is heated until it just starts to melt. During this process some oxidation also inevitably occurs as a result of the necessarily oxidizing atmosphere of the furnace.

At these temperatures carbon dioxide is chemically oxidizing towards iron, and causes some additonal formation of ferrous oxides, apart from that present in the initial charge. Once the initial solid components are melted, the blast furnace melt, at 1,300-1,350 °C, is added causing initiation of the main carbon removing reaction (Equation 12.23).

$$2\,C_{(Fe)} + O_{2\,(Fe)} \rightarrow CO_{(gas)} \tag{12.23}$$

Oxygen, from the scrap, or the ore, or dissolved in the molten iron from air, reacts with the carbon present in the blast furnace melt to form carbon monoxide. As carbon monoxide bubbles up through the charge it serves to stir the fluid mass and aids in the rise of silica, manganese oxide, phosphorus oxides etc. to form a slag layer. While these reactions are going on, limestone decarbonization also occurs and the carbon dioxide formed further contributes to the carbon monoxide formed by its conversion with dissolved carbon (Equations 12.1, 12.24).

$$CO_{2\,(Fe)} + C_{(Fe)} \rightarrow 2\,CO_{(gas)} \tag{12.24}$$

Thus, the slag formed on the open hearth charge is quite frothy. It is pushed off the pool of molten steel beneath it with a type of hoe, once or twice during the process. Blast furnace metal melts at about 1,130 °C, whereas pure iron, more closely representative of steels, melts at 1,535 °C. For this reason, as carbon removal from the blast furnace melt proceeds the temperature of the charge needs to be raised to avoid solidification of the charge and effectively stopping the steelmaking process until remelting is obtained. Since the bulk of heat transfer in the open hearth process occurs via direct contact of hot combustion gases with the surface of the charge, it is important that the frothy slag layer not be allowed to get too thick or heating may become ineffective.

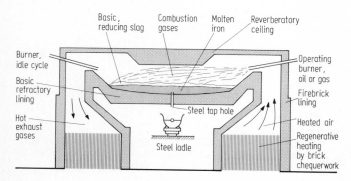

**Figure 12.3.** Operating details of an open hearth (Siemens-Martin) furnace for steel production

When a satisfactory composition of a particular charge is obtained the product is tapped from a hole in the base of the depression and cast into large ingots ready for final shaping. Open hearth processing is quite slow, 6 to 12 hours normally being required per batch, about half the time for charging. But this does provide plenty of time for sampling, analysis, and adjustment of the carbon content by additions of further scrap or hot blast furnace metal. Some alloys of up to 10 % of another metal are also possible in an open hearth process, which in this way provides greater product flexibility than possible with the pneumatic steelmaking processses.

Open hearth processes employing silica refractory linings are also known [10, 8]. However, since this type of lining does not provide phosphorus or sulfur removal capability it is usually reserved for open hearth operation on solely scrap iron charges, or on the occasional high grade iron ores free of these elements. In this way the independent heating capability of an open hearth furnace frees it from necessarily being associated with a blast furnace operation, and allows steels to be made locally entirely from scrap.

### 12.3.3 Electric Furnace Steel

Early development of practical furnaces heated by an electric arc at around 1876 again owed much of their development to the contributions to this area by F. Siemens, the pioneer of conventionally fueled regenerative furnaces. Again it was others who first utilized these principles in steel processing. Stassano, in Italy, succeeded in melting steel scrap in 1899 using this furnace design, with an arc struck between closely spaced carbon electrodes, although he was unable to reduce iron ores by this method [3]. It was the designs of Paul Heroult, working in France at about the same time, that were more successful and became the forerunners of electric furnace steelmaking processes used to the present day [3, 10].

Heroult, having gained experience from his development of the electrolytic method for reduction of aluminum, abandoned the indirect arc electric furnace designs, in which the arc was struck between opposed carbon electrodes which carried the power, and which were in vogue at the time. Instead he used direct arc designs which still employed the electrodes to carry power to the heating zone of the furnace. But the heating effect was obtained by two arcs, one struck between one electrode and the surface of the metal and the other between the other electrode and the metal. The metal itself served as the power conductor between the electrodes, thus producing two arcs in series from the same current flow (Figure 12.4). The double or multiple arcs in direct contact with the metal provided by the direct arc design gave much improved heat transfer efficiency and contributed significantly to improved electrical efficiency per mass of metal melted, over the earlier indirect arc designs. This heating principle also forms the basis of the design of all modern steelmaking arc furnaces [3, 15].

Even though electric power is relatively more expensive than natural gas, oil, or coal gas as a source of energy, its higher efficiency of utilization in the arc furnace significantly makes up for this. It is estimated that an open hearth furnace, even with regenerators, is only able to utilize 15 to 20 % of the heat energy in the fuel supplied to it for steelmaking whereas the parallel value for the electric furnace is of the order of 55 to 60 % [3].

Electric furnace steelmaking, like the open hearth process, has a means of charge heating independent of a hot metal charge or the combustion of impurities, and thus for fifty years or so since 1900 has been primarily used for the remelting of clean scrap, particularly of valuable alloys. But since about 1950 electric furnace steelmaking has been much more widely adopted because it permits control of both the heat input *and* the atmosphere above the charge. It does not need a moreorless oxidizing atmosphere as required for both the pneumatic and open hearth processes, but can be provided with an appropriate atmosphere for the

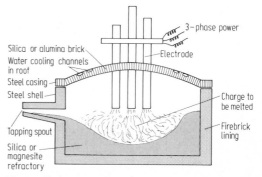

**Figure 12.4.** Schematic design of a direct arc electric furnace for steelmaking. The whole unit may be tilted to the side for emptying the finished steel melt

type of steel being made, a particularly important factor for some of the alloy steels. Special versions of arc furnaces are also used to permit melting under vacuum for products destined for particularly demanding applications where small amounts of dissolved gases in the steel could contribute to less than ultimate properties.

For an electric furnace with basic refractory, while the power is off, a small amount of limestone is charged initially to provide basic slag forming components. This is followed by the remainder of the charge consisting entirely of scrap steel for simple refabrication, or scrap steel plus alloying metal(s) if special alloys are being produced. If carbon steels are being made, a small amount of ore, plus scrap to provide oxygen and a significant component of solid or molten pig iron will be added. For furnaces producing specialized alloys the total charge volumes may be only 1 to 5 tonnes although electric furnaces accepting charges as large as 150 tonnes are in common use [15]. On completion of the charging step the arcs are struck and heating begins, very soon producing an equilibrium reducing atmosphere of carbon monoxide and carbon dioxide in the vapour space above the charge. A period of 2 to 3 hours is required for the usual charge to melt completely, after which samples can be taken to check for composition.

If the carbon content is still too high for carbon steel or mild steel preparation, further oxygen can be added using a water-cooled lance discharging the gas close to the surface of the melt (about 1 to 1.5 m above) while the arc is off. The initial slag containing silica, MnO, FeO, $P_2O_5$, and calcium sulfide in a lime base, what is called an oxidizing slag, is skimmed off. Then just before pouring the steel it is "killed" (deoxidized) by addition of one or more star-shaped pieces of aluminum to the melt, depending on the oxygen content determined by analysis. The stronger oxidizing tendencies of aluminum effectively combines with the oxygen of any residual iron oxides, forming further slag and decreasing the incidence of imperfections in castings made from the melt (Equation 12.25).

$$2\,Al + 3\,FeO \rightarrow Al_2O_3 + 3\,Fe \qquad (12.25)$$

For the preparation of alloy steels such as with

**Table 12.7** Raw materials required to make one tonne of ingot by the various steelmaking processes [a]. Amounts actually consumed in the process are in kilograms unless otherwise stated.

| Raw material | Basic Bessemer process | Hot metal, fixed open hearth furnace | Electric arc process | | Basic LD oxygen converter |
|---|---|---|---|---|---|
| | | | Common steels | Special alloy steels | |
| Iron: | | | | | |
| molten furnace | 1 050 – 1 150 | 550 | nil | nil | 850 – 1 025 |
| cold pig | nil | slight | some [b] | 0 – 75 | nil |
| Scrap iron (or steel) | 0 – 80 | 525 | 1 075 | 1 000 – 1 100 | 100 – 300 |
| Oxidizing ore or millscale [c] | low to 60 | 100 | 12.5 | 2 – 18 | 7 – 75 |
| Fluxes, as limestone etc. | 110 – 180 | 200 | 55 | 25 – 75 | 60 – 80 |
| Finishing alloys | ca. 10 | 14 | 10 | various | 6 – 7 |
| Oxygen, m$^3$ (15 °C, 1 atmos) | nil | varies | 150 ft$^3$ | varies | 1 700 – 2 000 |
| Primary fuel, kJ | | | | | |
| Natural, or coal gas, (BTU) | nil | 4.2 – 5.3 × 10$^6$ (4 – 5 × 10$^6$) | nil | nil | nil |
| Electricity, kWh | nil | nil | 500 – 550 | 650 – 750 | nil |

[a] Adapted from data in [12].
[b] Some used occasionally.
[c] Millscale is the iron oxide (mostly $Fe_3O_4$) "crust" removed from steel billets during rolling operations.

chromium, vanadium, tungsten, or manganese, most of the alloying metal is in place at the start of the melt. To prevent oxidation of the alloying metal a reducing slag containing burned lime (CaO) and powdered carbon is used. This practice provides both the protection of the reducing atmosphere of carbon monoxide and carbon dioxide in the vapour space above the metal, as well as the protective layer on the metal. On completion of melting a sample is taken and the analysis made to determine whether any final adjustment of alloying metal concentration is necessary before pouring.

One important class of steel alloys, the stainless steels, require an electric furnace for their preparation. The corrosion resistance of this class of alloy, conferred by the presence of about 18 % chromium plus 8 % nickel in the iron, is a valuable feature of its use in industrial vessels, the food industry and for interior and exterior fittings.

The stainless steels are only one example of the present day versatility of the electric furnace in steelmaking. The electric furnace has a high flexibility to accept solid, molten, or a mix of these types of charges, to operate with acidic or basic linings depending on need, and to provide an oxidizing, reducing, or neutral (inert) atmosphere. The combined appeal of these attributes, plus its ability to operate under clean conditions and under a partial vacuum for steels destined for critical service have led to the rapid adoption of electric furnace methods as a component of steelmaking practice worldwide, since about 1950 [12, 15, 16].

Electric furnaces employing induction heating have also been used sporadically for steelmaking [15]. In this procedure high frequency electric power of 1,000 to 2,000 cycles per second or more is passed through a water cooled copper coil (conductors are actually heavy walled copper pipe) surrounding a non-magnetic, non-conducting crucible made entirely of refractory. In this way, the charge placed in the crucible serves as both the "core" and the single turn secondary "winding" of a step down transformer. The power flow and eddy currents generated in the charge as a result of this placement brings about rapid melting and permits any composition adjustments to be made. Electrical consumption is about the same as for an electric arc furnace, 660-880 kWh per tonne of steel, but the capital cost of an induction furnace averages two to three times that of an arc furnace, mainly for the motor-alternator and condensers required to provide the high frequency AC current [3]. But for steel requiring clean conditions and for specialty steels, induction furnaces fill an important need [12].

## 12.4 Recent Developments: Oxygen Steelmaking

The use of essentially pure oxygen for the oxidation of iron impurities, together with the advantages that this practice entails, became a commercial possibility as the availability of oxygen on a tonnage scale in the 1930's became feasible. Early development of this process, originally called the "Linzer Düsenverfahren", took place in Linz and Donawitz in Austria during the 30's and 40's [5]. Either the original process name or the places of development led to the identification of this development as the LD process [12]. The first two oxygen converter plants using this principle became operational in Austria in 1952 and 1953.

The LD process, also referred to as basic oxygen steelmaking (BOS, also used to refer to the product, basic oxygen steel), carried out in a basic oxygen furnace (BOF), are all terms used synonymously to refer to these developments. The vessel used for these processes is similar in shape and capacity to the Bessemer converter, but is lined with a basic refractory. From a specially designed and water cooled nozzle located in the neck of the vessel a substantially pure stream of oxygen is directed onto the surface of the mostly molten blast furnace metal charge. Impurities in the hot metal combine rapidly with the high oxygen content atmosphere with which they come into contact. The resulting oxides then combine with lime, or limestone which is also added to the charge, to form a slag layer to permit removal of the impurities.

Oxygen steelmaking is rapid, like the older pneumatic processes, tap to tap times normally being in the range of 18 to 20 minutes [12], and the product is not subject to nitriding since nitrogen gas contact is avoided. Also, because the heat production from impurity oxidation is roughly equivalent to Bessemer expectations and there is no heat loss to nitrogen of air, it is possible to incorporate up to 30 % by weight of scrap steel or special high grade iron ores into the 70 % or so of molten blast furnace iron which usually comprises the bulk of the charge in the LD process [14]. This ability to accept a proportion of solid charge with the blast furnace melt contributes flexibility to the process.

To do all this requires some 65 m³ of oxygen per tonne of steel (about 2,100 ft³/ton, at 16°C, 1 bar) [12], a net raw material cost to the process. But in return for this raw material cost, time from charging to tapping the steel product for the larger LD furnaces of 200 and 300 tonnes capacity is still only about twice as long as for the smaller LD units, or about 40 minutes [12, 17]. In comparison to the 10 to 12 hour turnaround time for a single charge to a 300 tonne capacity open hearth furnace the cycle time for an oxygen steelmaking using variants of the equipment described here are also practised [4, 10, 12]. In total, oxygen steelmaking methods were used to produce about one half of the steel of the U.K. and the U.S.A., about 70 % of the steel in West Germany, and about 80 % of the total in Japan, in 1974 [17], and these proportions of the total are likely to grow still further.

## 12.5 Iron and Steels, and their Properties and Uses

The wrought iron of historical importance, produced by reducing the iron ore without melting the iron component and then beating out most of the ferrous silicate slag, was virtually pure iron containing traces of carbon (0.02—0.04 %) and strings of residual slag (1.0—2.0 % FeSiO₃). The low carbon content was a result of reduction without fusion and contributed the good malleability, ductility and the associated good cold working properties that are synonymous with wrought iron. Nineteenth century iron works could form sheets and bars by cold rolling of blooms (balls of reduced

iron from which much of the slag has been removed by mechanical hammering) of wrought iron [4]. Cold rolling hardens and generally strengthens the iron, a process called work-hardening, which is a desirable procedure for some end uses (Table 12.8). If, however, the iron shapes are required to be ductile for subsequent shaping etc., the cold rolled shapes must first be annealed, or heated to a high temperature and then allowed to cool slowly to resoften the metal for continued cold forming. Wrought iron is still produced on a small scale by pouring molten Bessemer iron into slag that is molten but below the melting point of the iron, and by other similar methods [7]. The spongy mass which forms is then mechanically treated, in this way forming a wrought iron very similar in properties to the earlier hand worked material. Residual slag stringers left in the wrought iron contribute to the toughness of the iron in this form, and also make a small contribution to corrosion resistance. The ductility, weldability and toughness of wrought iron continue to make it valuable for many of the same uses as the original hand-produced material, for example pipe, boiler tubes, rivets, heavy chains and hooks, and ornamental iron work.

### 12.5.1 Iron-Carbon Alloys: Steels, and Cast Iron

One of the most important characteristics of iron products which enables a wide difference in spectrum of properties to be achieved, is the concentration of carbon present. Ordinary steels are iron-carbon alloys, which are simply referred to as steel, and are so important that this alloy comprises

**Table 12.8 a** Percentage composition of various forms of iron and carbon steels[a]

| Element | A Ingot iron | B Wrought iron | C Gray cast iron | D Carbon steels SAE 1020 | E Ultra-high-strength steel, 300-M |
|---|---|---|---|---|---|
| Iron | 99.9+ | 97.5[b] | 94.3 | 99.1 | 94.0 |
| Carbon | | | 3.4 | 0.2 | 0.43 |
| Silicon | | | 1.8 | 0.25 | 1.6 |
| Manganese | | | 0.5 | 0.45 | 0.80 |
| Nickel | | | | | 1.85 |
| Chromium | | | | | 0.85 |
| Molybdenum | | | | | 0.38 |
| Vanadium | | | | | 0.08 |

[a] Compiled and recalculated from data in [21 and 22].
[b] Residue consists of slag.

**Table 12.8 b** Selected properties of various forms of iron and carbon steels [a]

| Material, form, condition | | Yield [c] strength MPa | Tensile [c] strength MPa | Elongation [d] in 5 cm % | Hardness [e] Brinell | Density g/cm³ | Melting point °C |
|---|---|---|---|---|---|---|---|
| A | annealed | 130 | 260 | 45 | 67 | 7.86 | 1 530 |
|   | hot rolled | 200 | 310 | 26 | 90 | | |
| B | hot rolled | 210 | 330 | 30 | 100 | 7.70 | 1 510 |
| C | as cast | — | 170 | 0.5 | 180 | 7.20 | 1 230 |
| D | annealed | 260 | 450 | 30 | 130 | 7.86 | 1 515 |
|   | hot rolled | 290 | 470 | 32 | 135 | | |
|   | hardened [f] | 430 | 620 | 25 | 179 | | |
| E | hardened [g] | 1 650 | 2 000 | 10 | 535 | 7.84 | 1 504 |

[a] Compiled and recalculated from data in [21 and 22].
[b] Residue consits of slag.
[c] Megapascals convertible to pounds per square inch by multiplying by 145. Yield strengths corresponds to the stresses causing permanent deformation of the metal (i.e. beyond the elastic limit). Tensile strengths correspond to stresses at the point of ultimate metal failure.
[d] A measure of ductility.
[e] The Brinell hardness number is based on the identation of the surface under test by a ball of specific diameter, usually 10 mm, when pressed into the test surface. It is usually expressed in kg per square mm, thus small numbers represent low hardness.
[f] Hardened by water quench followed by 540 °C temper.
[g] Hardened by oil quench followed by 320 °C temper.

more than 98 % of all iron alloys produced. In most iron-carbon alloys the carbon is present as iron carbide, $Fe_3C$, also called cementite. Since the carbon content of cementite is only 6.69 %, a small change in the carbon content of an iron causes a large change in the concentration of the cementite present in the iron. Cementite is soluble in molten iron, one of the reasons why carbon is accumulated in the product of the blast furnace process for reduction of iron ores. This is an advantage, since the melting point of the iron-cementite mixture is depressed to about 1,100 °C permitting accumulation of the product in the base of the furnace at temperatures considerably below the 1,535 °C melting point of pure iron. At the same time it is a disadvantage since, in the process, rather more cementite is dissolved in the molten iron than is wanted for most uses.

If the percentage of carbon in molten iron is 1.7 % or less, which amounts to 25.4 % or less cementite, it remains dissolved in the iron when solidified by quick cooling (e.g. when quenched in water) to form only one type of crystal. The crystals of this solid solution have cementite evenly distributed in the crystals of $\gamma$-iron and yields an extremely hard, brittle product called martensite. If the iron containing from about 0.82-1.7 % carbon is al-

lowed to cool more slowly, cementite can separate from the iron and the result is an intercrystalline structure of cementite and pearlite. The formation of pearlite, which consists of alternate layers of ferrite (pure iron) and cementite, contributes significantly softer, more malleable and tougher properties to the iron than obtained from a quenched product. As the amount of carbon in the iron is brought down below about 0.82 %, slow cooling allows ferrite plus pearlite (interlayered ferrite and cementite) phases to separate, again giving a relatively soft, tough product. More detail concerning these phase changes and their effect on properties of carbon steels is available from the more extensive accounts in The Making, Shaping and Treating of Steel [8] and the McGraw-Hill Encyclopedia of Science and Technology [10].

Iron-carbon alloys, commonly referred to as carbon steels or simply steel, are roughly classified in accordance with their carbon content, a reflection of the significant influence of carbon content on their properties. Thus mild steels normally contain about 0.25 % carbon, medium steels about 0.45 % carbon, and high carbon steels about 0.7 to 1.7 % carbon. Both medium and high carbon steels are commonly tempered (heat treated) to optimize their properties for the particular function they are intended to perform, although any of the carbon steels are amenable to property modification by heat treating [18, 19]. In fact, that is the general definition of a steel, an iron-carbon alloy that contains less than 1.7 % carbon and is amenable to hardening by quenching (Table 12.8). As a rule, the strength and hardenability of a steel is raised at the same time as the toughness and ductility are decreased, with an increase in the carbon content of a steel. A recent user classification system has been proposed [20].

To obtain optimum properties from heat treating the finished steel article is first hardened by heating to red heat followed by quenching by dipping in water, brine, or oil any of which may also be cold or warm depending on the degree of hardness desired. The martensite structure obtained in the iron by this treatment is very hard and a tool at this stage will hold a cutting edge well, but it is also rather brittle and in this condition can fracture quite easily. It will also be under severe internal stress because of the rapid cooling rate, which for a small article will tend to increase its susceptibility to fracture under shock, and for a large part may actually cause it to crack as a result of the quenching alone. For this reason large parts will be oil or air quenched to reduce this risk. To gain the desired degree of toughness the quenched article is tempered (or "drawn down") by reheating to a particular controlled temperature according to the end use desired in an air or oil bath, and then allowed to cool slowly. This combined treatment gives a cutting product or spring the desired combination of hardness (sharp edge retention) and toughness (resistance to shattering under sudden physical shock).

If a carbon steel is required in a ductile state either for intermediate cold forming (e.g. auto body panels) or machining, or in its final use form then it is annealed. To accomplish this the finished part is heated to above the iron phase transition temperature, about 650 °C, and then allowed to cool very slowly. Annealing not only accomplishes a softening and ductility improving effect on the steel (Table 12.9) but also achieves stress relief. Large coldformed or welded industrial vessels or ships' components will have stresses induced in the structure from the fabrication methods used, which may be removed by an annealing procedure such as the one outlined. In this way the service life and reliability of the unit may both be greatly improved.

When the carbon content of the iron rises to above 1.7 % ( > 25.4 % cementite) it is classed as a cast iron. At this concentration the proportion of cementite becomes so high that it is no longer completely soluble even in solid $\gamma$-iron and forms a separate cementite phase on hardening from the melt. Cast iron may be made by mixing molten pig iron with molten steel scrap, or by melting steel scrap in direct contact with coke in a cupola furnace (virtually a small version of a blast furnace). It will ordinarily contain 2.0-4.0 % carbon, only a little less than pig iron, together with 0.5-3.0 % silicon, 0.5-1.0 % manganese and traces of phosphorus and sulfur. The hard brittle nature of many cast irons, hot or cold, makes these impossible to shape by rolling or forging (hammering while hot) the solidified metal, so that casting followed by machining if required are the usual methods of fabrication. Fortunately, as in the original blast furnace reduction of iron from its ores, the presence of the cementite and other impurities depresses the melting point of the iron to the 1,100-1,200 °C range making both the melting process and the casting step easier than it would be for pure iron.

It is possible to produce a moderately tough cast iron by special alloying techniques, and by control of the microstructure through the selection of the

Table 12.9 Tempering (draw down) temperatures for some particular uses of carbon steels [a]

| End use | Reheat temperature °C | Colour of of Oxide layer |
|---|---|---|
| razor blades, twist drills | 232 | pale straw yellow |
| axes, taps and dies | 255 | brown-yellow |
| cutlery | 277 | purple |
| watch springs | 288 | bright blue |
| chisels | 300 | dark blue |

[a] Compiled from [3, 36] and other miscellaneous sources.

appropriate cooling cycle. And some versions of cast iron may be converted to malleable iron by annealing for a period of several days. The long anneal causes dissociation of the cementite ($Fe_3C$), the cause of the hardness, into ferrite (iron) and spherical zones of graphite making the product relatively soft and significantly tougher than the material as originally cast. With these special irons primary fabrication by casting can be followed by other shaping techniques.

### 12.5.2 Alloy Steels

Mild steels and high carbon steels may be fabricated and heat treated in ways to achieve most combinations of properties which may be required. But plain carbon steels suffer several serious drawbacks, primarily in the areas of strength, toughness, and hardness for some particular applications, and in corrosion resistance. For example, heat treating of very large machine parts made of carbon steels can only achieve very shallow surface hardening. Also, heat treating of large parts introduces severe internal stresses and strains which causes distortion of the part from the originally machined shape. In small parts this distortion may be acceptable or compensated for, but it may be sufficiently great with large parts to introduce fitting problems and a tendency to crack. With a situation such as this, a drastic quench cannot be used and therefore optimum hardness of the carbon steel is not achievable. Alloying with elements other than carbon can enable achievement of hardenability, a term reserved to mean hardness at depth, with a less severe (such as an air) quench. Addition of molybdenum, for example, not only achieves hardenability but at the same time or separately from the heat treating step contributes other desirable properties (Table 12.10).

Table 12.10 gives examples of the kinds of properties achievable with several alloying elements to demonstrate the range of variation possible. Many hundreds of alloy steels are available, including other elements than those listed, to tailor-make a steel to suit a particular application [8, 21, 22]. For example, nickel-chromium combinations confer stainlessness, as outlined in the table, more than twenty different commercial variations being available to achieve specific objectives. The stainless steels, from their general utility, probably represent the largest single class of alloy steels [23]. Aluminum may be added in trace amounts to tie up

oxygen and avoid the deleterious properties induced by the presence of small amounts of ferrous oxide. A small percentage of manganese is useful to tie up any residual traces of sulfur as well as for the other functions of this alloying metal given in the table. While ferrous sulfide introduces brittleness in the steel, making hot working impossible, manganese sulfide does not interfere in this way which thus permits normal forging operations to be carried out without defects being introduced. Boron, too, is a relatively recently discovered alloying element, a few hundredths of a percent contributing low cost hardenability. Coating developments with hard carbides such as titanium promise to increase the life of cutting tools made with the new high-speed-tool-steels [24].

## 12.6 Emission Control in the Production of Iron and Steel

For an industry which involves large scale handling of sometimes finely divided solid materials, the operation of large scale combustion processes for coking, ore reduction to iron, and for subsequent steelmaking operations, the production of iron and steel necessarily requires attention to emission controls. Containment in this instance is necessary not only for aesthetic reasons but also for maintenance of public health in associated built-up areas, for industrial hygiene requirements of plant operators and also for better product yields from the raw materials entering the processes used.

It should be remembered that many pollutional aggravations of the industry have been greatly decreased by technological changes which generally were introduced as a means of improving operating efficiency but which, in addition, contributed significantly reduced levels of emission per tonne of iron or steel produced. For instance, blast furnace top gas containment measures were introduced to conserve the carbon monoxide fueling capacity of these gases for preheating of blast air. But at the same time this measure served to greatly decrease both particulate and gaseous pollutant discharges from blast furnace operations. Since the top gas containment measures were not adopted primarily with emission control in mind, until very recently blast furnace operating procedures allowed both some intentional bleeding (pressure relief), and required no particular care to control accidental gas

**Table 12.10** Composition and applications of some common examples of classes of alloy steels [a]

| Alloy steels | Percentages of alloying elements | Properties achieved | Typical uses |
|---|---|---|---|
| Chromium | Cr 0.6 – 1.2 C 0.2 – 0.5 | improved strength, toughness, heat resistance | improved castings |
| | Cr 16 – 25, Ni 3.5 – 20 | combination confers stainlessness, heat resistance, strength | stainless steels: chemical, medical food processing equipment |
| | Cr > 12%, Ni > 8% | as above plus hardenability | stainless cutlery |
| Copper | Cu 0.10, C ca. 0.2 | improved corrosion resistance | copper-bearing steels, bridges exterior steelwork |
| Nickel | Ni 6 – 9, C 0.2 – 0.5 | low temperature toughness | snowmobile parts icebreakers, armour plate (6.5% Ni) |
| | Ni 30 – 45% | low coefficient of thermal expansion i.e. $0.8 \times 10^{-6}$ as opposed to 10 to $20 \times 10^{-6}$ for most other steels | Invar: graded seals to glass, length standards (surveying), compensated pendulums and balance wheels |
| | Ni 60, Cr 16 | heat and electrical resistance | Nichrome heating elements |
| Manganese | Mn 10 – 15, C ca. 1 | tough, acquires hardness on repeated impact | rock crusher jaws, power shovel bucket teeth, burglar-proof safes |
| Molybdenum | Mo 5 – 9 | hardenability, high temperature strength | high speed tools, steam boilers and pipes, jet engine turbines |
| Silicon | Si 3 – 5 | low hysteresis and eddy current losses, increased magnetic permeability | electric power transformer plates |
| | Si 13 – 15, C 0.8 | superior hardness acid resistance | e.g. Duriron (14.5% Si) industrial valves, piping |
| Tungsten | W 5, C 0.5 | assists in retaining hardness while hot | Tungsten steel, drill bits |
| | W 17, Cr 10, V 0.3, C 0.7 | same as above, remains hard and strong even when hot | high speed steel, high speed cutting and machine tools |
| Vanadium | V 0.2 – 0.5 + Cr ca. 1 | high strength, toughness, heat treatability | gears, driveshafts, axles, ball bearings, mechanics tools |

[a] Compiled from data of [13, 21, 22, 37].

losses containing around 35 g/m³ (15 grains/ft³) particulate, as well as gaseous pollutants [25]. More recently even these relief and accidental discharges practices are now being regulated by requiring passage through at least two levels of control devices before discharge. In this way the industry generally has progressed from no containment to 90 + % containment to both its own and society's significant benefit. In some cases containment levels as high as 98 + % have been achieved with further mutual gains. A similar progression, with similar mutual benefits, could be outlined for the adoption of the BOP (basic oxygen process) for steelmaking. Since the air and water emission control and waste disposal problems of the raw material preparation area, blast furnace operations, and steelmaking and fabrication areas have many overlapping aspects, emission control aspects will be treated by the emission control area rather than process area.

### 12.6.1 Air Pollutants and Control Measures

Aerial discharges from facilities for producing iron and steel may be conveniently considered in sequence, from raw material preparation to the fabrication of finished products. Thus, coke plants and sintering plants used to prepare the blast furnace fuel and ore components come first, followed by the various steelmaking procedures such as open hearth, basic Bessemer, electric and basic oxygen processes.

In the preparation of raw materials as blast furnace feed, the coke ovens used to prepare a dense, desulfurized metallurgical coke from coking grade coals produce a significant mass of particulate discharge in the absence of control measures. A battery of coke ovens used to process 4,500 tonnes of coking coal per day, and to yield about 3,200 tonnes of coke is estimated to generate a solid discharge of about 0.1 % of the mass of coal processed, or about 4,500 kg of particulate per day [26]. This particulate emission occurs at all stages of coke preparation, including charging, pushing, and quenching stages of the operation as well as from leaks in the ovens themselves, thus making control measures difficult. No estimates of current feasible particulate emission control efficiencies are available.

While the coking process does produce a blast furnace coke feed substantially free of sulfur, the gaseous product, coke oven gas, has a sulfur gas content of 900-1,100 g/m$^3$ (at 15 °C, 1 bar) [25]. The principal sulfurcontaining constituent of coke oven gas is hydrogen sulfide, which may be removed either by the vacuum carbonate or the Stretford processes. Sulfur gas removal efficiencies of the Koppers Company's vacuum carbonate process have been found to be about 90 %, whereas the Stretford process is capable of maintaining a 99 % sulfur gas containment, both efficiencies while

operating on coke oven gas streams [25]. In general, the final choice of desulfurization process appears to rest on the most feasible sulfur gas product desired, contact process sulfuric acid primarily being preferred for the former process, and Claus plant technology for conversion to elemental sulfur for the latter (see Chapter 7). Condensible hydrocarbons such as benzene (and other aromatics) and phenols have always been recovered more or less efficiently [27].

Sintering machines are used to prepare larger nodules from beneficiated iron ore fines and iron oxide containing dusts recycled from particulate emission control equipment. Uncontrolled operation, it is estimated, would discharge particulate at the rate of about 0.3 % of the mass of sinter produced or about 2,700 kg from a machine producing 900 tonnes of sinter per day [26]. Control measures relying primarily on cyclones are estimated to decrease the particulate emission to about one quarter of the uncontrolled levels. With the introduction of sintering machines as an integral part of iron reduction complexes, however, it also became possible to utilize the sintering machine as a roaster to enable sulfur removal from sulfur-containing iron ores. While this procedure has the advantage of producing an ore feed amenable to blast furnace reduction from materials which could not be used previously, it also produces a new waste gas stream from the sintering machine which contains sulfur dioxide. No concentration or mass emission rate data for sulfur dioxide in sinter plant exhaust gas is available, since this has not normally been recovered [14]. But a mathematical model presenting some means of estimation of these quantities has been described [14].

Particle loadings of the top gases leaving a blast furnace are very dependent on the particle size distribution of the raw materials charged to the furnace. With no particular care taken to eliminate fines from the ore or flux, dust loss rates of the order of 200 kg per tonne of pig iron produced, corresponding to a particulate loading in the off gases of about 2 g/m$^3$, are reported [14]. Screening of the charge to remove fines roughly halves this loss rate and the particulate loadings to about 100 kg/tonne and 1 g/m$^3$, and at the same time improves blast furnace operation by providing for a freer flow of gases through the charge (cf. Table 12.4). But the greatest reduction in dust carryover in the top gases, to the range of 15-20 kg/tonne of pig iron produced, is obtained by prior briquetting, pelletizing [28], or nodulizing the charge before placement in the blast furnace. These size refinement techniques are often carried out with the flux and coke also incorporated into the lumps charged, which further improves blast furnace gas flow.

All the measures described above amount to methods to decrease dust entrainment during normal blast furnace operation are, in effect, pre-combustion or pre-reduction methodologies. In addition to these dust *avoidance* techniques, modern blast furnaces also usually require three stages of gas cleaning for the top gases before these are employed as a fuel for blast stove heating and other energy requirements. A dust catcher situated close to the furnace captures the coarse particulate (see Figure 12.1) and then by either low and high energy scrubbers, or by a low energy scrubber followed by electrostatic precipitation the residual mid-size range and aerosol particles are removed [11, 26]. By these measures the actual particulate loss rate after final combustion of the blast furnace top gases and eventual discharge amounts to some 1 to 18 kg per tonne of pig iron produced. Magnetic filtration has been proposed as a safer method than electrostatic precipitation for iron particulate removal from combustible blast furnace gases, to avoid explosion risks [29].

All the steelmaking procedures produce exhaust gases having a moderate particle loading before treatment, but nowhere near as severe as the particulate loadings obtained from blast furnace operation. For instance, nominal dust loadings of 0.5-1.4 g/m$^3$ are experienced during the charging phase (loading ore, flux, scrap etc.) of an open hearth furnace [27]. Higher dust loss rates, of 4.6-7.0 g/m$^3$, are experienced when an oxygen lancing (of basic oxygen process steelmaking utility) stage is employed in open hearth operation [27]. Originally, open hearth furnaces were operated without emission control devices. However, present practice is to use either electrostatic precipitation, for which 90-97 % particulate control efficiencies are reported [14], or a wet scrubber for particulate control [14, 25, 27]. A wet scrubber also gives efficient particulate control but this produces a waste water stream that requires treatment. The fine particles captured from the waste water stream by chemical coagulation and settling followed by neutralization of the supernatant liquor, which is normally acidic from dissolved sulfur dioxide, before discharge [27, 30]. When a scrubber is employed as part of the system used to clean up blast furnace top gases, the waste liquor is normally alkaline from absorbed limestone. Hence when this is the control system used, the waste liquor may be usefully employed to neutralize the acidic scrubber liquors from associated steelmaking operations.

For Bessemer, electric arc, and basic oxygen furnaces pollutant containment is practised via a hood with connecting ductwork placed over the opening of the vessel during active steelmaking phases. In these situations emission control is exercised by electrostatic precipitation or scrubbers, as is usual with open hearth furnaces, or with a baghouse [25]. Particulate collection efficiencies of 80-90 % are cited for the former two methods [14]. More detailed control information is available relating to the use of baghouses for the control of electric arc furnace emissions. For direct arc furnaces of 4-15 tonnes capacity, particle loadings were found to be 0.8-1.2 g/m$^3$ and particle size distributions generally in the range of 50-70 % (by weight) below 5 $\mu$m and less than 15 % above 40 $\mu$m [26]. Under these conditions, using Orlon fabric bags, particulate mass control efficiencies of 98 to 99 % were reported. Gas temperatures were kept below about 80 °C to avoid fabric damage.

Ingot reheating prior to hot rolling or forging the final products has only the associated normal types of combustion processes to provide reheat energies. Hence there is little of air emission control concern that is particular to the iron and steel industry for this area of operations.

## 12.6.2 Water Pollutants and Control Measures

Preparation of blast furnace coke involves the indirect heating of metallurgical coal to 1,000-1,100 °C, in the absence of air, in a battery of refractory brick-lined coke ovens. The overall procedures used at such an operation are collectively referred to as a "byproduct coke plant" from the association of byproduct recovery with the coke formation process. Byproduct recovery was not feasible with some of the early coking processes because of the coking configurations used. In a byproduct coke operation indirect heating of the coal charge is continued until pyrolysis is complete and all of the volatile matter has been vaporized, a process which takes 16 to 24 hours. Then the residual lumps of coke, still hot, are pushed out of the oven through a quenching shower of water and into a rail car for final shipment. About 700 kg of coke plus a number of volatile products are recovered from each tonne of metallurgical coal heated. More details on the coking process itself are available in the texts by Kent [31] and Shreve and Brink [32]. The bulk of water usage in a coking operation, about 42 m³/tonne (10,000 U.S. gal/ton) of coke produced, is for indirect cooling purposes, mainly for indirect cooling of the volatiles stream to collect coal tar and other condensible components of interest. This water stream, which is warmed but is otherwise uncontaminated, only requires thermal loading considerations on discharge. Water vapour, lost from the coal on heating, is also condensed at this stage of the process. This aqueous condensate stream, which will also contain ammonia, sulfides, hydrocarbons, phenols etc. (see Tables 12.11, 12.12) all condensed from the coal

volatiles, is one stream requiring treatment before discharge both from a chemical recovery standpoint and for emission control.

About 2 m³ of water per tonne of coke (ca. 500 U.S. gal/ton) is used for quenching purposes, about a third of which is vaporized in the coke cooling process [25]. The vaporized fraction is discharged as steam but the residual water, which will contain suspended coke fines ("coke breeze") plus low concentrations of cyanides and phenols, requires treatment before discharge. Coke fines are removed by settling and when recovered are blended into a mixture of ore and limestone for briquetting as blast furnace charge, or are used as fuel for sintering [31]. Recycling much of the supernatant water minimizes the volume requiring treatment.

The aqueous condensate stream is the most complex to deal with. Phase separation produces a water phase and an oily tarry phase, each of which is then treated separately [30]. Phenol is kept in the aqueous phase during ammonia recovery by converting it to the less volatile sodium salt (Equation 12.26).

$$PhOH + NaOH \rightarrow NaOPh + H_2O \qquad (12.26)$$

Ammonia is then removed from the aqueous phase by distillation and either recovered by scrubbing with sulfuric acid to produce ammonium sulfate, which is sold as a fertilizer constituent, or it is discarded by discharge to air. Economical re-acidification of the aqueous ammonia recovery residue with flue gas, and subsequent distillation allows recovery of both phenol and sodium carbonate (Equations 12.27, 12.28) [30].

**Table 12.11** Typical characteristics of the waste water stream from the ammonia still of a battery of by-product coke ovens [a]

| Parameter | Significance, ppm by weight |
|---|---|
| 5 day BOD, 20 °C | 3 974 |
| Suspended solids: | |
|     Volatile | 153 |
|     Total | 356 |
| Nitrogen: | |
|     As ammonia, $NH_3$ | 187 |
|     Organic and $NH_3$ | 281 |
| Phenol | 2 057 |
| Cyanide | 110 |
| pH | 8.9 |

[a] Selected data from [38].

**Table 12.12** Approximate coke and volatile product yields in kilograms, from one tonne of metallurgical coal [a]

| Solid and condensed liquids | | Gaseous products | |
|---|---|---|---|
| coke, with coke breeze | 750.0 | ammonia | 3.5 |
| light oil | 0.5 | hydrogen sulfide | 3 |
| heavy creosote | 4.5 | carbon dioxide | 9 |
| pitches | 24 | nitrogen | 4 |
| benzene | 7.5 | hydrogen | 16 |
| toluene | 3.0 | carbon monoxide | 45 |
| xylene | 0.8 | methane | 65 |
| naphthelene oil | 4.0 | ethane | 5.5 |
| anthracene oil | 7.0 | ethylene | 10 |
| other organics | 1.3 | other gases | 5 |
| aqueous liquor | 60 | Total gases | 143.5 |
| | | (ca. 305 m³ at 15 °C, 1 bar) | |

[a] Data selected and calculated from that of [31].

$$CO_2 + H_2O \rightarrow H_2CO_3 \qquad (12.27)$$

$$2\,NaOPh + H_2CO_3 \rightarrow Na_2CO_3 + PhOH \quad (12.28)$$

Cyanide may either be detoxified to cyanate by oxidation with chlorine or hypochlorite (see Chapter 11), or the oxidation may be conducted by air in the presence of sulfur dioxide and copper ion at pH 9 or 10 [33]. Depending on the cyanide neutralization technology employed, BOD reduction to 80 to 90 % of the original values will be achieved via either biological waste treatment or alkaline chlorination methods before discharge of this stream [25].

The separated organic phase is commonly used as fuel. Otherwise it will be put through a sequence of washes and stills for chemical recovery, or it may be segregated into a portion destined for each appropriate end use area at some coking locations [31]. Details of some of the gaseous and fluid products recoverable as the byproducts from a byproduct coke plant are given in Table 12.12. About one third of the coke oven gas is burned to provide the heat for coal carbonization and usually the rest of this component is also burned, but for other energy needs within the steel plant.

Blast furnace water requirements are primarily for water jacket cooling of the critical tuyere and bosh areas of the furnace. Otherwise the severe operating temperatures in this region would cause softening or melting of the bronze tuyere castings and also the steel furnace shell. Like other indirect cooling applications, the primary concern on discharge of this water would be to minimize thermal loadings. Water is also frequently used for scrubbing of blast furnace top gases, and less often for fracturing (breaking up) of hot slag. These latter water requirements, where practised, would require settling and appropriate chemical neutralization before discharge.

The steelmaking techniques all have a cooling water requirement of some 12 to 15 m$^3$/tonne (ca. 3,000 to 3,700 U.S. gal/ton) as well as a much smaller requirement of about 80-100 L/tonne (ca. 20-25 U.S. gal/ton) for scrubbing of exhaust gases from the various processes [30]. Indirect cooling water disposal requirements are similar to those from any other source. Treatment of scrubber effluent from this source will generally be similar to the procedures used with blast furnace scrubber effluent, except that these steelmaking waste streams will usually be acidic whereas blast furnace effluent is normally alkaline. After particulate re-

moval by settling, pH neutralization can be economically effected by stream blending, particularly if the steel mill is operated in close proximity to supporting blast furnaces.

Water effluent problems of hot and cold steel rolling mills differ from those of other areas of the complex. Even though the primary water requirement in a hot rolling mill is for cooling of the heavy iron or steel rolls used for shaping the steel, the cooling function here is by direct spray onto the rollers. Thus, the water picks up small particles of iron, scale (mostly iron oxide), and lubricating oil in its passage across the face of the rollers, which are retained in suspension as the water reaches mill collection channels. Some additional water, used in high pressure jets to flush scale from a steel sheet as it forms, is also collected in the same waste water system. A large tank, called a "scale pit", serves the combined functions of scale settling and oil creaming to stimultaneously remove the bulk of both of these contaminants before discharge of the water stream (Figure 12.5). A system of paddles on endless chains is used to move settled scale out of the scale pit, up an inclined plane to drain off most of the fluid, and thence to a collection for incorporation into blast furnace feed [27]. Creamed oil collected on the water surface of the scale pit is removed by suction pumping through a skimmer floating on the fluid surface [25]. If the cleaned exit water from the scale pit still does not meet local effluent guidelines it can be subjected to further chemical polishing and/or biological treatment steps before discharge.

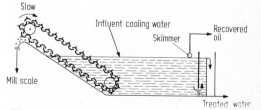

**Figure 12.5.** Schematic diagram of a scale pit showing how simultaneous oil and scale removal is achieved from steel rolling mill waste waters

# Relevant Bibliography

1. R.B. Elver, Direct Iron Process and their Prospects in Eastern Canada, Dept. of Mines and Technical Surveys, Ottawa, 1960
2. K.M. Leopold, Approaches to Controlling Air Contamination at the Foundry, Mod. Casting 67, 64, Sept. 1977
3. H.B.H. Cooper and W.J. Green, Energy Consumption Requirements for Air Pollution Control at an Iron Foundry, J. Air Pollut. Control Assoc. 28, 545, May 1978
4. Foundry Monitors Waste Water, Foundry Manag. and Tech. 195, 70, Jan. 1977
5. G.G. Cillie, Use of Water in the Iron and Steel Industry, J. of Metal 11, 35, (1971)
6. Removal of Phenol and Thiocyanate from Coke Plant Effluents at Dofasco, Proc. 16th Ont. Indust. Waste Conf., Ontario, 1969
7. F.C. Lauer, Solvent Extraction Process for Phenols Recovery from Coke Plant Aqueous Waste, Iron and Steel Eng. 46, 99, (1969)
8. Air and Water Quality Control at Stelco's Hilton Works, Iron and Steel Eng. 53, 75, Nov., 1976
9. J.H. Waugh and R.J. Triscori, Consider all the Options for a Sinter Plant Air Pollution Control System, Iron and Steel Eng. 54, 36, March 1977
10. N.T. Stephens, J.M. Hughes, B.W. Owen, and W.O. Warwick, Emissions Control and Ambient Air Quality at a Secondary Steel Production Facility, J. Air Pollut. Control Assoc. 27, 61, Jan. 1977
11. D. Marchand, Possible Improvement to Dust Collection in Electric Steel Plants and Summary of all Planned and Existing Systems in the Federal Republic of Germany, Ironmaking and Steelmaking 3 (4), 221, (Discussion p. 230), 1976
12. Recycling Steel Automotive Bumpers, Industrial Finish. 52, 40, Nov. 1977

# References

1. B. Mason, Principles of Geochemistry, 3rd edition, John Wiley and Sons, New York, 1966, page 46
2. J. Newton Friend, Iron in Antiquity, Charles Griffin and Co., London, 1926, page 27
3. J. Dearden, Iron and Steel Today, 2nd edition, Oxford Univ. Press, Oxford, 1956, page 19
4. W.K.V. Gale, Iron and Steel, Longmans, Green and Co., London, 1969
5. W.K.V. Gale, The British Iron and Steel Industry, David and Charles, Newton Abbot, 1967
6. J.B. Pearse, A Concise History of the Iron Manufacture of the American Colonies up to the Revolution and of Pennsylvania until the Present Time, Burt Franklin, New York, 1970, reprint of the 1876 edition
7. Kirk-Othmer Encyclopedia of Chemical Technology, Interscience, New York, 1952, volume 8
8. The Making, Shaping and Treating of Steel, 9th edition, H.E. McGannon, editor, United States Steel, Pittsburgh, 1971
9. 1980 Minerals Yearbook, volume 1, Bureau of Mines, U.S. Dept. of the Interior, Washington 1981, page 405
10. McGraw-Hill Encyclopedia of Science and Technology, McGraw-Hill, New York, 1971, volume 13, page 98
11. Kirk-Othmer Encyclopedia of Chemical Technology, 3rd edition, Wiley Interscience, New York, 1981, volume 13, page 743
12. J.A. Allen, Studies in Innovation in the Steel and Chemical Industries, A.M. Kelley, New York 1968
13. D.A. Fisher, The Epic of Steel, Harper and Row, New York, 1963
14. C.S. Russell and W.J. Vaughan, Steel Production: Processes and Residuals, Johns Hopkins University Press, Baltimore, 1976
15. Kirk-Othmer Encyclopedia of Chemical Technology, 2nd edition, Interscience, New York, 1969, volume 18, page 715
16. G.E. Wittur, Primary Iron and Steel in Canada, Department of Energy, Mines and Resources, Ottawa, 1968
17. M. Finniston, Chem. Ind. (London), 501, June 19, 1976
18. Potentials for More High-strength, Low-alloy Steels in Autos, Metal Prog. 109 (2), 26, Feb. 1976
19. M.S. Rashid, Science 208, 862, May 1980
20. J.T. Sponzilli, C.H. Sperry, J.L. Lytell, Jr., Metal Prog. 109 (2), 32, Feb.
21. Chemical Engineers Handbook, 4th edition, R.H. Perry, editor, McGraw-Hill, New York, 1969
22. Lange's Handbook of Chemistry, 10th edition, N.A. Lange, editor, McGraw-Hill, New York, 1969
23. W.A. Luce, Ind. Eng. Chem. 49 (9) part II, 1643, Sept. 1957
24. R.F. Bunshah and A.H. Shabaik, Research/Devel. 26 (6), 46, June 1975
25. The Steel Industry and the Environment, J. Szekely, editor, Marcel Dekker, New York, 1973
26. Handbook of Environmental Control, volume I, Air Pollution, CRC Press, Cleveland, Ohio, 1972
27. Industrial Pollution Control Handbook, H.F. Lund, editor, McGraw-Hill, New York, 1971
28. Waste Iron-bearing Fumes, Chem. Eng. News 55 (16), 17, April 18, 1977
29. Magnetic Filter May Cut Pollution, Chemecology 11, July 1980
30. N.L. Nemerow, Industrial Water Pollution, Origins, Characteristics, Treatment, Addison Wesley, Reading, Mass. 1978
31. Riegel's Handbook of Industrial Chemistry, 7th edition, J.A. Kent, editor, Van Nostrand Reinhold Co., New York, 1974

32. R.N. Shreve and J.A. Brink, Jr., Chemical Process Industries, 4th edition, McGraw-Hill, New York, 1977

33. Process Destroys Cyanide, Can. Chem. Proc. *66* (6), 8, Sept. 10, 1982

34. 1979/80 Statistical Yearbook, United Nations, New York, 1981

35. G.R. Bashworth, The Manufacture of Iron and Steel, 3rd edition, Chapman and Hall, London, 1964, cited by J.A. Allen, Studies in Innovation in the Steel and Chemical Industries, A.M. Kelley, New York, 1968

39. E. Oberg, F.D. Jones and H.L. Horton, Machinery's Handbook, 20th edition, Industrial Press, Inc., New York, 1975

37. H.H. Sisler, C.A. Vanderwerf, and A.W. Davidson, General Chemistry, a Systematic Approach, MacMillan Co., New York 1949

38. American Iron and Steel Institute, Annual Statistical Report, 1949. Cited by reference 30

# 13 Production of Pulp and Paper

## 13.1 Background and Distribution of the Industry

The production of wood pulp, and paper from this, is a primary industry in the sense of its utilization of a wood primary resource as its chief raw material, but it is also a secondary industry in the sense that it consumes large quantities of bulk inorganic chemicals such as chlorine, sodium hydroxide, pigments etc. produced by the primary chemical, or commodity chemical industry. In fact the pulp and paper business area alone consumes close to 10 % of the inorganic chemicals produced in the countries in which it is a dominant industry, thus not only contributing directly to the employment and business activity in these areas, but also indirectly by its purchase from these primary areas of chemicals production.

The development of a large scale pulp and paper industry is generally favoured by the availability of an extensive domestic wood supply, e.g. Brazil, Canada, Finland, Sweden, U.S.A., and the U.S.S.R. (Table 13.1). Japan is alone among the very large scale pulp producers in relying to a significant extent on imported pulp logs for her pulp and paper industry. Norway on the other hand, with a large forest potential, does not appear to have exploited this to the same extent as other nations with this resource. Brazil, a relative newcomer to the pulp producing fraternity, is one of the areas most committed to the "farming" of wood on a continuous harvest basis, using fast growing varieties of trees. The extent of exploitation of the available wood supply by all the major pulp producing countries has aroused concern about the inability of natural processes to continue to provide wood at the rate that it is being cut. In most cases this has resulted in better organized care and harvesting of wood, and increased encouragement of renewal by reforestation practices which are more effective than the natural processes.

Financially, an active pulp and paper industry provides a significant fraction of the gross national product of the countries where this is a major business area, in Canada amounting to some $ 8 billion annually. Of twenty manufacturing categories in Canada, the annual value of pulp and paper production is only exceeded by the food and beverage, and the transportation equipment sectors at $ 25.4 billion and $ 19.9 billion respectively, and about

**Table 13.1** Major world producers of wood pulp, thousands of metric tonnes [a]

|  | 1950 | 1960 | 1970 | Chemical 1979 | Mechanical 1979 | % of 1979 world totals |
|---|---|---|---|---|---|---|
| Brazil | 139 | 453 | 811 | 2 320 | 155 | 2.1 |
| Canada | 7 686 | 10 397 | 16 609 | 11 734 | 7 588 | 16.1 |
| China |  |  | 1 220 | 1 233 | 796 | 1.7 |
| Finland | 1 913 | 3 700 | 6 471 | 4 600 | 2 238 | 5.7 |
| France | 500 | 1 139 | 1 787 | 1 361 | 453 | 1.5 |
| Japan | 748 | 3 524 | 8 801 | 7 902 | 1 757 | 8.0 |
| Norway | 1 073 | 1 654 | 2 182 | 920 | 491 | 1.2 |
| Sweden | 3 448 | 5 611 | 8 142 | 6 906 | 1 981 | 7.4 |
| U.S.A. | 13 471 | 22 966 | 39 304 | 39 763 | 3 977 | 36.4 |
| U.S.S.R. | 2 046 | 3 213 | 6 679 | 6 741 | 1 840 | 7.2 |
| W. Germany | 914 | 1 452 | 1 732 | 664 | 1 136 | 1.5 |
| Other | 2 682 | 5 480 | 11 034 | 9 241 | 4 195 | 11.2 |
| World | 34 620 | 59 580 | 104 772 | 93 385 | 26 607 | 100.0 |

[a] Includes totals of both chemical and mechanical pulp where undifferentiated. Pulp weight specified on an air dry basis, i.e. 10% moisture. Data compiled from [28, 109].

**Table 13.2** Growth in exports of wood pulp, by countries exporting more than 500 000 tonnes in 1979 [a]

| | Thousands of metric tonnes | | | |
| | 1950 | 1960 | 1970 | 1979 |
|---|---|---|---|---|
| Brazil | 0 | — | 40 | 582 |
| Canada | 1 675 | 2 360 | 5 063 | 7 090 |
| Finland | 1 056 | 1 587 | 2 057 | 1 865 |
| Norway | 546 | 802 | 981 | 573 |
| South Africa | 0 | 74 | 278 | 529 |
| Sweden | 2 091 | 2 931 | 3 762 | 3 519 |
| U.S.A. | 87 | 1 036 | 2 808 | 2 662 |
| U.S.S.R. | n.a. | 257 | 448 | 680 |
| Other | 243 | 632 | 1 410 | 2 673 |
| Total | 5 698 | 9 679 | 16 847 | 20 173 |

[a] From [28].    n.a. = not available.

equalled by the value of petroleum production. What is perhaps of equivalent importance to a national economy, because of its ability to earn international exchange, is the value of exports in this commodity area. This factor is particularly significant to countries which both produce pulp and paper on a very large scale and have a relatively small population and hence consumption, for example Canada, Finland, Norway, and Sweden. Brazil is likely, very shortly, to join this group.

In paper consumption the largest producers, U.S.A. and Canada, are also the largest consumers, at least on a per capita basis (Table 13.3), the U.S. alone consuming one half of the total world production. They, and Japan, the runner-up after the U.S. on a consumption rate by country basis, all still show a trend of increasing annual newsprint consumption of 5 % to 30 % over the nine year period. Denmark, the Netherlands, Sweden, and the United Kingdom, all nations known to be great newspaper subscribers, all lie in the mid range for per capita newsprint consumption, the last two showing a 36 and a 14 % decrease in per capita consumption over the same period. Among the smaller scale newsprint users per capita consumption trends ranged from a 100 % increase for China, to a stabilized consumption for India, Iraq, and Ethiopia, to decreases of 58 % and more for Malawi and Uganda.

Early papermaking was accomplished one sheet at

**Table 13.3** Annual consumption of newsprint by a selection of countries both by country, in thousands of metric tonnes (Mg) and per capita [109]

| | 1970 | | 1978 | |
| | Total | Per capita | Total | Per capita |
| | Mg | kg | Mg | kg |
|---|---|---|---|---|
| U.S.A. | 8 924 | 43.6 | 9 919 | 45.4 |
| Canada | 657 | 30.8 | 943 | 40.1 |
| Australia | 449 | 35.9 | 472 | 33.0 |
| Denmark | 149 | 30.2 | 169 | 33.2 |
| Netherlands | 379 | 29.1 | 412 | 29.5 |
| Sweden | 343 | 42.7 | 227 | 27.4 |
| United Kingdom | 1 544 | 27.8 | 1 331 | 23.8 |
| Japan | 1 973 | 18.9 | 2 445 | 21.3 |
| West Germany | 1 077 | 17.7 | 1 166 | 19.0 |
| France | 606 | 11.9 | 611 | 11.5 |
| U.S.S.R. | 928 | 3.8 | 1 147 | 4.4 |
| China | 571 | 0.74 | 1 315 | 1.5 |
| India | 182 | 0.33 | 217 | 0.33 |
| Iraq | 3.1 | 0.33 | 3.8 | 0.31 |
| Sudan | 2.7 | 0.21 | 1.8 | 0.11 |
| Afghanistan | — | — | 1.2 | 0.057 |
| Ethiopia | 0.90 | 0.036 | 1.1 | 0.036 |
| Malawi | 0.20 | 0.046 | 0.10 | 0.019 |
| Uganda | 1.0 | 0.10 | 0.20 | 0.016 |
| Other | 3 773.1 | — | 3 740.8 | — |
| Total, world mean | 21 563 | 5.86 | 24 123 | 5.7 |

a time, by hand, by methods which originated in China at about 105 A.D. From that time till about 1760 papermaking remained a batch operation, single sheets being prepared by dipping a mould (or screen) into a vat holding a suspension of fiber in water. After draining and pressing of the excess water the sheets were dried in the open air. N.L. Robert, in 1761, took the first steps towards developing a machine for papermaking on a continuous basis [1]. But it was not until 1822 that a Fourdrinier paper machine, initially operated in France, was first described as a commercially successful, continuous papermaking machine. Thus, papermaking by anything like the methods used today is a relatively recent activity.

A suspension of pulp in water is the common raw material for virtually all types of papermaking. The pulp is mostly cellulose and is derived from wood fibers, separated from whole wood by one means or another. But many other sources are used. Cotton and linen rags (as offcuts from clothing manufacture) which are virtually 100 % cellulose are still used to produce small quantities of the very highest quality papers. Where wood may be scarce, any of straw, cornstalks, hemp, jute or bagasse (sugar cane fiber) may also be employed as raw materials [2]. Papermaking from bagasse has to be able to cope with associated carbon particles left on the stems when cane field foliage is burned just prior to sugar harvest, and with the high proportion of pith. Esparto grass, a tough dryland cover, and bamboo [3] are also used. The very rapid growth of bamboo allows "cropping" for papermaking and the tough internodes of bamboo may be incorporated into the process by prior crushing between steel rollers. Modern chemical pulping processes are also capable of accommodating many other types of annual plants for paper pulp preparation which permits the use of quite a range of renewable raw materials as pulp sources. But by far the dominant raw material used worldwide for papermaking is wood, so the detailed discussion which follows will center on the methods used with this source of pulp.

## 13.2 Wood Composition and Preparation for Pulping

### 13.2.1. Wood Composition and Morphology

The woody part of a tree serves to provide both mechanical support to hold the form of the tree upright for optimum exposure of the leaves to sunlight and air, and as a conduit to carry water and trace nutrients from the roots to the leaves and photosynthetic products to where needed. Hollow, interconnected fibers composed mostly of cellulose and oriented along the axis of the tree serve both of these functions (Figure 13.1).

Cellulose, the main component of the conduit fibers is a polymer of $\alpha$-D-glucose (Figure 13.2). The initial step in its formation is the photosynthetic reaction, producing glucose from water and carbon dioxide (Equation 13.1).

$$6\,H_2O + 6\,CO_2 \underset{\text{sunlight}}{\overset{\text{chlorophyll}}{\rightarrow}} C_6H_{12}O_6 + 6\,O_2 \tag{13.1}$$

Glucose in solution is carried in the sap to provide both energy for plant metabolism, by the reverse of the photosynthetic reaction, and to the cambium layer of the wood, where the new growth occurs. The cambium layer which comprises the interface between wood and bark (the "sapwood"), is where new fibers form by processes which deposit insoluble cellulose from the soluble glucose and cellobiose, a disaccharide of glucose. Rings of fibers which form in the spring during a period of rapid growth are relatively thin-walled and have a large lumen (bore, or duct). These are collectively referred to as springwood. In summer, when the growth is slowed, fibers are formed with thicker walls and smaller lumens, which are collectively referred to as summerwood. It is these differentiating growth features which give rise to the characteristic annual rings of wood, and which also allow longer term historical climatic variations to be estimated from cross sections of old specimens. The core of the trunk or limb inside the cambium growth layer is dead wood and is termed heart wood.

Cellulose is formed by elimination of water from adjacent $\alpha$-anomers of the two cyclic forms of glucose (Equation 13.2),

$$n\,C_6H_{12}O_6 \rightarrow (C_6H_{10}O_5)_n + n\,H_2O \tag{13.2}$$
$$n = \text{about } 10\,000$$

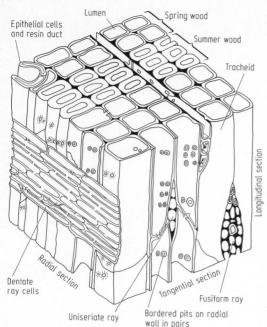

Lumen
Epithelial cells
and resin duct
Spring wood
Summer wood
Tracheid
Longitudinal section
Radial section
Dentate
ray cells
Tangential section
Fusiform ray
Uniseriate ray
Bordered pits on radial
wall in pairs

**Figure 13.1.** Enlarged cubic section of a typical softwood
[11] Reprinted courtesy John Wiley and Sons

ments, on marked dilution with water. Cellulose
does not melt on heating and is relatively stable to
heat, not decomposing until it reaches 260-
270 °C. It is also relatively stable towards oxidation, all in all a set of properties highly suitable for
the raw material of a product with such wide application. Chemically speaking, cellulose is the
same from one species to the next, although there
are morphological differences between the cellulosic fibers from different species.

Lignin, which functions primarily as an interfiber
bonding agent to hold the wood structure together
[4], is also an important component of wood. It is a
three dimensional, crosslinked polymer of heterogeneous structure composed of predominantly n-
propylbenzene units joined by a variety of ether (C-
O-C) and carbon-carbon (C-C) links [5] (Figure
13.3). Lignin is also biosynthetically derived from
glucose, like cellulose although a little more re-
motely [6, 7]. Because of the high aromatic-
phenolic content in the polymer, lignins are all
more or less brown in colour, and are relatively
readily oxidized. The molecular weight of natural
lignins, *in situ,* is not well defined, and even in the
natural state probably consists of polymers comprising a range of molecular weights. Molecular
weights determined on isolated lignins mostly
range from 1,000 to 20,000, depending to a significant extent on the origin of the lignin and the
method used to determine the molecular weight
[8].

Lignin is relatively stable to most aqueous mineral
acids but is soluble in hot aqueous base and hot
aqueous bisulfite ion ($HSO_3^-$). It has low
softening and melting points, e.g. native spruce lignin softens at 80-90 °C (wet), 120 °C (dry) and
melts at 140-150 °C [8, 9]. Higher temperatures
than this cause carbonization. The lignins from different species of tree, and even from the same
species growing in different locations are chemically significantly different from one another, particularly when comparing softwoods and hardwoods [4, 8].

Less dominant components of wood are the hemicelluloses, so called because at first it was thought
that this component consisted of lower molecular
weight intermediates enroute to the biosynthesis of
cellulose. However, the hemicelluloses are soluble
in aqueous alkali, which was one of the early
means for differentiating them from cellulose. This
property is also a clue to the amorphous nature of
hemicelluloses, which are branched polysac-

hence a "dehydro" polymer. Formation of cellulose from the α-anomer is what gives the
polymer its characteristic linear form which favours strong intermolecular hydrogen bonding between adjacent polymer chains. With some 10,000
glucose units (molecular weights about 1.5 million)
per chain [4] individual molecules can be up to
0.012 mm in length. This high molecular weight
and the close intermolecular association makes fibers composed of cellulose both strong and insoluble in water. Both factors also confer cellulose
with significant resistance to hydrolysis because of
the relative inaccessibility of the acetal links between adjacent glucose units. The hydrolytic stability, strength, and insolubility are necessary features of the structural role of cellulose in the plant
as well as being important for papermaking and
paper properties.

Chemically cellulose is thus a two dimensional
high molecular weight polymer with no conjugation, hence white (colourless) in colour. It has no
taste or odour and is insoluble and relatively stable
in water or aqueous alkali. It is, however, soluble
in strong aqueous mineral acids but may be regenerated, generally to lower molecular weight frag-

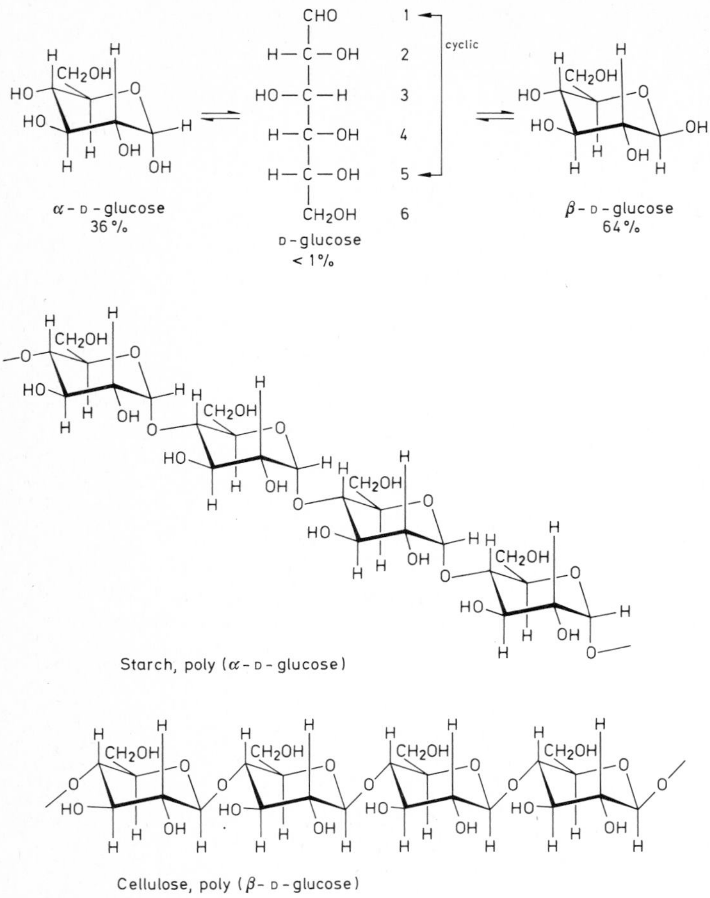

**Figure 13.2.** Linear and stereochemical representations of the cyclic forms of glucose, and their relationship to the two regular dehydropolymers of glucose

charides of a variety of sugars, all of lower molecular weight than cellulose. None have simple trivial names. The name of the particular hemicellulose is derived from the sugar monomer from which the polymer is composed, for example xylose forms xylans, galactose forms galactans, mannose forms mannans, etc. As with lignin, the chemical structure and composition of the hemicellulosic fraction of wood also differs from species to species in woods, grasses, and annual plants. For example, with hardwoods it is the xylans which predominate and with softwoods glucomannans generally comprise the main hemicellulose component. But the exact hemicellulosic content of any particular type of wood is very heterogeneous, and there is even significant evidence that the xylan

from a softwood is not identical with the xylan from a hardwood [2]. To add to these complications, isolation of a particular hemicellulosic fraction from wood inevitably modifies it somewhat, so that it is very difficult to determine what the precise composition of the hemicellulose in the wood was. Thus the structure of hemicelluloses, much like lignins, remains rather poorly defined as they occur in the natural state.

The amorphous nature and lower molecular weight make hemicellulose an important component of papermaking, probably by its ability to improve interfiber bonding. But there is inevitably some hemicellulose loss from the pulp, particularly with some chemical pulping processes, because hemicellulose is less stable to acid and alkali than cellulose. The

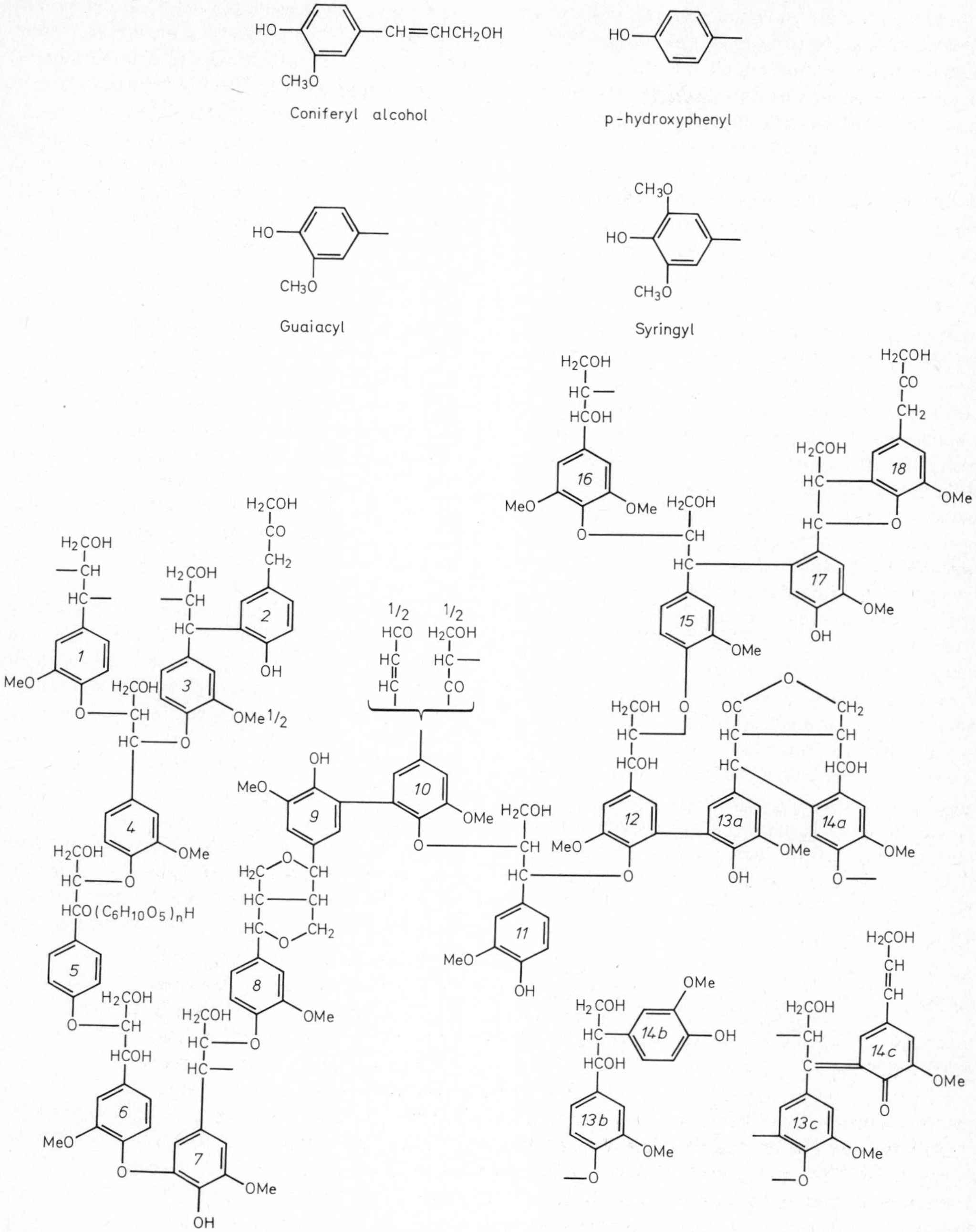

**Figure 13.3.** Structure and identity of decomposition fragments and an approximation of the structure of a softwood (spruce) lignin. Reprinted from [4, 579], courtesy Marcel Dekker

sum of the cellulose and hemicellulose components in wood or pulp are collectively referred to as holocellulose, because of the high value placed on the hemicellulose component in this application. However, for some types of chemical pulps destined for end uses requiring pure cellulose such as for the manufacture of cellophane or cellulose acetate (the so-called "dissolving pulps"), the presence of hemicelluloses causes subsequent problems with filtration etc. Thus, when pulping to produce a dissolving pulp, the process is modified to remove as much of the hemicellulose as possible.

In addition to these more dominant components of dry wood, there are a group of miscellaneous non-polymeric, or very low molecular weight polymeric, materials which are soluble in cold water and are relatively easily separated from the wood. These are collectively referred to as extractives and comprise from 3-8 % by weight of dry wood [2]. Despite their relatively lesser importance in paper the extractives do cause processing problems with some species of wood, and hence one has to be aware of their presence. As examples, the presence or formation of fatty acids from some species of wood during pulping can cause foaming problems, either during pulping or on waste disposal, or both. Fatty acids, and resin acids formed from oxidized terpenes, can also cause corrosion problems.

The proportion of all of these components found in wood varies from species to species, and with specimens of the same species grown under different conditions. However, in general hardwoods (from deciduous trees) contain more cellulose and less lignin than softwoods (coniferous trees; Table 13.4). Also keep in mind that the chemical nature

of the lignin, hemicelluloses, and miscellaneous components will also differ significantly from one species to another, which will affect the details of the optimum pulping method to be used for any particular species. The objective of all pulping processes destined to paper is to produce a fiber suspension in water suitable for papermaking from a wood raw material by fiber separation by largely physical work, or by chemical treatment, or by a combination of these methods.

## 13.2.2 Wood Preparation for Pulping

After wood has been harvested, today employing extensive mechanization [10], and brought in to the pulp mill by water, rail, or road it has to be prepared for pulping. Bark removal, or debarking, is a preliminary step common to all pulping processes. Even though leaving the bark in place as the wood is pulped would tend to slightly increase the fiber yield, at least for some woods and some pulping processes, it would contribute an undesirable highly coloured, nonfibrous constituent to the pulp with most species of wood.

With smaller logs a drum debarker is commonly used [10]. This consists of a steel cylinder 3-4 m in diameter and 15-25 m long, placed on a horizontal or near horizontal axis and capable of slow rotation on its axis. It is fitted with welded, log-tumbling baffles inside and has longitudinal slots cut into the periphery to allow bark fragments to fall out or be flushed out with water. Slow rotation of the drum containing a load of logs removes the bark by a combination of abrasion and impact attrition, after which the debarked logs are removed from the outlet gate. Drum debarkers may be operated dry, producing a dry bark byproduct of higher fuel

Table 13.4 Typical approximate percentage compositions of dry wood and cotton, on an extractive-free basis [a]

| Constituent | Softwood e.g. fir | Hardwood e.g. aspen | Cotton |
|---|---|---|---|
| holocellulose | | | |
| cellulose | 42 | 42 – 48 | 99 |
| hemicellulose | 27 | 27 | 1 |
| total holocellulose | 69 | 75 | 100 |
| lignin | 29 | 15 – 20 | 0 |
| pectin, starch, ash, etc. | 2 | 4 | 0 |
| total | 100 | 100 | 100 |

[a] Analysis after thorough extraction with water and non-polar organic solvent, and drying. Compiled from data in [4, 11].

value to the pulp mill, but sometimes slowing the debarking step as a result. Or they may be operated wet giving faster debarking, particularly when using hot water with cold or frozen logs, but resulting in a bark of lower fuel value and requiring a drying step before combustion. Wet debarking also produces a waste water stream.

The bark of very large trees is normally removed in one or other type of in-line debarker. A single log at a time is moved slowly, in an axial direction, while at the same time being rotated slowly on its axis either past rapidly rotating slack chains which serve to beat the bark off, or in a "hydraulic debarker" past a water jet operating at 70 bar (1,000 psi) or higher and directed to the wood/bark interface. The energy of the water jet literally rips the bark off [11]. A dry system also in common use is another version of inline debarker for large logs in which the log is moved axially, without rotation. At the same time a large stout ring fitted with spring-loaded knives pointing radially inwards is revolved around the log in a lathe-like fashion, removing the bark as the log moves throgh.

To obtain an idea of yields through these initial stages, when the bark is removed from a cord or $128\,\mathrm{ft}^3$ ($3.625\,\mathrm{m}^3$) of wood, 70 to 80 $\mathrm{ft}^3$ (1.98-$2.27\,\mathrm{m}^3$) of debarked solid wood is normally obtained [10], a yield of 55-63 %. The volume measure used by the pulp and paper industry to specify this particular product is the cunit (c unit) meaning 100 $\mathrm{ft}^3$ (equivalent to 2.83 $\mathrm{m}^3$) of bark-free wood, i.e. the raw material suitable for papermaking.

Once the bark is removed, small logs which are to be reduced to a pulp suspension in water by grinding (one form of mechanical pulping) proceed directly to the stone groundwood operation. Very large logs which cannot be accommodated by the stone groundwood equipment directly will be sawn to manageable block sizes before grinding. Water channels are frequently used to convey both types of wood to grinding operations by floating. While this particular part (stone grinding) of wood processing utilizes short lengths of whole logs, or sawn rectangular blocks of wood as the common form of wood entry for processing (see later), all of the other methods of pulp production, including other mechanical, as well as chemical methods, require a wood chip feed.

Thus, chipping, or the process of reducing a log to chips of about 2.5 x 2.5 x 0.5 cm thick, is a common preliminary to all other pulping methods except stone groundwood. Wood chips are a form of

the raw material which is more convenient and uniform for solids transport within the mill complex by conveyor belt or pneumatic delivery systems than are whole logs. Also a chip medium is more amenable to direct physical conversion to fibers because of the much easier penetration of heat, energy, and moisture through these thin sections of wood. Chemical pulping methods require a chip format for the wood in order to obtain easier, faster, and more uniform penetration of chemicals when digesting under pressure.

A chipper in this service consists of a 2-4 m diameter heavy steel disk fitted with 5 or 6 radially mounted steel knives, each of which protrude about 2.5 cm (the chip length desired) from one face of the disk [10,12]. A spout, aimed at about 30° from the axis of rotation, directs the log to be chipped towards the rotating disk, end on and at the ideal angle to minimize the rotational energy required to drive the chipper. In the chipping process, this angular feed produces a splayed angular edge cut to the chip which optimizes the ease of liquor penetration when pulped. As the chips are cut, slots through the steel disk ahead of each of the knives allow the chips to pass straight through the disk, aided by blowing vanes attached to the opposite face of the chipper from the knives.

A chipper of the type described, driven by a 500 hp (ca. 800 kW) electric motor is capable of reducing a 0.6 m diameter x 8 m log to chips in less than 30 seconds and could produce something like 50 tonnes of chips per hour [12]. More recent models obtain much greater production rates by using 2,000 - 2,500 hp (1,600 - 2,000 kW) drive motors [10].

Two stage chipping proposals have also been made to obtain chips with less compression damage [13]. The first stage of the proposal involved removal of thin slices off a log with the cut taking place along the axis of the log (or fiber) in much the same way as the "peeling" of a log in a lathe for veneer or plywood production. Chopping of these longitudinal slices into short lengths is then accomplished in a conventional chipper with only slightly higher power cost, overall, than required for conventional chipping, and achieving greater chip uniformity.

Whatever method of chipping is used care has to be taken to ensure an optimum quality and size consistency of the chips produced, both of importance in the yield and the strength of the pulp produced. Fines have to be screened out and cooked under different conditions than regular sized chips to obtain optimum pulp yields, and oversize material has to be further processed for use. Blade sharpness, speed of disk rotation, weight of log(s) in the spout, and whether the butt, or the top of the log is fed in first, all significantly affect chip consistency and quality [2].

## 13.3 Mechanical Pulping of Wood

The wood pulps produced which rely solely on physical work on the log, without the use of chemicals, to reduce the log to a fiber suspension in water, are collectively referred to as mechanical pulps. By the nature of the method used to produce it, the pulp consists of a mixture of all of the insoluble constituents present in the original wood, less a small amount that is degraded during the pulping step which is lost in soluble form in the water phase. Thus the yield, or the weight of pulp obtained from the weight of wood pulped is normally high, 90 % or better, and the pulp is appropriate for the manufacture of low cost papers. With care in operation, and particularly while pulping favourable species of trees such as white spruce, the pulp obtained is light in colour, although not completely white. The whiteness and strength of the low cost papers made from mechanical pulps may both be improved somewhat by blending a proportion of fully bleached, long fiber chemical pulp into this before papermaking.

Relatively coarse mechanical pulps, with only 10-15 % incorporation of bleached chemical pulp to provide sufficient strength, are used in the preparation of hanging stock (wallpapers). Newsprint, which is used for newspapers, inexpensive magazines, paperback books and other low cost publishing requirements employs a blend of 15-35 % chemical pulp stocks into the mechanical pulp, to obtain an economically thin sheet of sufficient strength to avoid breakage in modern high speed printing machines. And about 10 % or so of mechanical pulps, particularly of stone groundwood types, may be blended into the pulps used for some fine papers as a means of improving smoothness and opacity.

### 13.3.1 Stone Groundwood

Whole, debarked logs or sawn wood blocks with their axes parallel to the axis of the stone (Figure 13.4), are pressed against a large grindstone rotat-

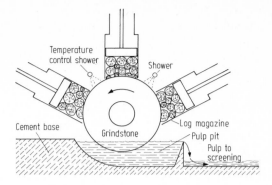

Three-pocket hydraulic grinder

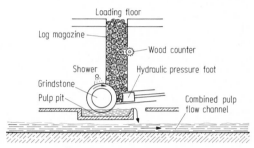

Hydraulic magazine grinder

**Figure 13.4.** Grindstone and log feed configurations used for the production of stone groundwood pulps from whole logs or blocks

ing in a pit of water, to produce a suspension of 3-4 % groundwood pulp in water [11]. The axis of the log is kept parallel to the axis of the stone to obtain as long a fiber from the wood as possible and at the same time to minimize the grinding power requirement. Friction between the log and the stone produces heat which softens the lignin allowing cellulose fibers to be torn out with minimum fiber damage. Grinding a log end first into the stone or under cold conditions would produce more of a wood flour of little papermaking utility rather than a fibrous pulp, and would cost more power to achieve. To optimize the thermal softening of lignin many stone groundwood mills now use a system of showers with flow control and a pit water level below the base of the stone to obtain higher stone and wood temperatures and achieve better regulation of these [10]

Usually, stone dimensions range from 1.8 to 2 m in diameter by 1 to 1.5 m wide. They are made of synthetic grits such as aluminum oxide or silicon carbide, and embedded in a ceramic matrix for a long working life under the high centrifugal forces involved under hot, wet conditions. To drive a working stone of this size at the required 250-300 rpm (peripheral speeds of 25-30 m/sec) requires 4,000 to 5,000 hp (ca. 3,200 - 4,000 kW), usually provided by large electric motors. Grinding pressure, or the force with which the logs are gravity fed or hydraulically pressed to the stone is the means used to regulate both stone speeds and motor power requirement. Mechanical power, in the range of 1,600 to 1,750 hp hours (1,200 - 1,300 kWh) per tonne of pulp for newsprint grades [2], is the chief production cost for groundwood pulping. For hanging stock the power requirement is slightly less than this, and for book papers slightly more.

The last step in stone groundwood production is a screening of the fiber suspension in water to remove shives, or fiber bundles (small splinters) from the stock where individual fibers have not been fully dispersed in the water. If paper is made from stock containing shives it will have low strength properties in the area around each shive because of poor interfiber cohesion, making the sheet susceptible to tear either on the papermaking machine itself or during subsequent printing. The separated shives are either discarded (burned when dry) or fed in a water suspension to a refiner for defiberization.

### 13.3.2 Chip Refiner Groundwood

It became possible to produce a pulp from chips [10], as well as from whole logs, when the process of refiner mechanical pulping was developed to accomplish this during the 1950's. This convenient method of producing mechanical pulp was rapidly adopted as a complement to stone grinding by many newsprint mills, but did not, however, completely displace stone grinding as a mechanical pulping procedure because of the differing qualities of the pulp produced by the two methods [14].

The refiner, a machine used to convert a chip to a fiber suspension in water, consists of two stout steel disks of about 1 m diameter, which are fitted with hardened steel bars and depressions on one face of each (Figure 13.5). The barred faces of the two disks are placed together, and the spacing between the bars of the facing disks is adjustable to allow control of the fineness of the fiber suspension produced. One disk may be stationary and the other rotated, or the disks may both be rotated in op-

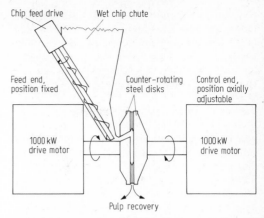

**Figure 13.5.** Operating details of a refiner used to reduce a wood chip feed to pulp

posite directions, in either case by powerful electric drives of up to 10,000 hp (ca. 8,000 kW). Wet chips, together with some additional water, are fed to the center of the space between the two refiner disks while they are rotating. The adjacent faces of the two disks are each slightly concave so that the chips are able to drop into the wider space existing between the disks, near their centers. Friction between the chips and the disks produces heat and steam which softens the lignin and starts the defiberization process. As the chip fragments get smaller they are carried by the continuing flow of water and fresh chips towards the periphery of the disks, where the spacing between the disks is less, and the bar/depression frequency is higher (Figure 13.5). In this way the chips are rapidly reduced to a separated, individual fiber suspension in water, as the fiber fragments reach the periphery of the disks. Peripheral speeds of refiner disks may be as high as 160 m/sec, significantly greater than the usual peripheral speeds of grinders [11].

To obtain a well-processed pulp for newsprint manufacture usually requires passage of the chips and the partially refined pulp through two or three refiners operating in series [15]. Paper produced from this type of pulp is generally stronger, from the longer average fiber lengths produced, than paper produced from stone grinding of the same wood [11, 16]. At the same time, and for the same reason, it also generally has a higher surface roughness and lower opacity than stone groundwood newsprint, although these factors are affected by refiner adjustments which increase or decrease the power requirement to obtain the pulp (Table 13.5). Chip refining, however, generally has a somewhat higher power cost for pulp production, say about 80 to 100 hp days/air dry tonne for spruce, as opposed to about 80 to 90 when stone grinding [10].

The above reasons, apart from the "paid for" aspects of the average stone groundwood pulping operation, are responsible for the continuance of stone grinding as a component of the mechanical pulping operations of many newsprint mills. Many newsprint mills use blends of refiner and stone groundwood pulps with a proportion of chemical pulp to raise the strength. However, it is possible to produce a good newsprint solely from refiner groundwood pulps because of the superior fiber lengths obtained by this method of mechanical

pulping [17]. One further advantage of refiner pulping methods is the capability it provides of producing a useful pulp from planer shavings or sawdust, options not possible with stone grinding. While it is not possible to make a useful paper from sawdust pulp alone, it can be readily blended with more conventional pulps to produce a sheet meeting printer's specifications.

### 13.3.3  Thermomechanical Pulp (TMP)

As more experience was gained, it was realized that stone or refiner mechanical pulping at higher temperatures, achieved by operating with a lower and carefully controlled proportion of water to logs or chips, decreased the power input required per tonne of pulp produced. Carrying this realization a step further, the idea of subjecting chips to a short

**Table 13.5** Comparison of properties of papers produced from spruce, stone or refiner groundwood pulps at similar and dissimilar grinder work inputs and some related terminology [10, 110]

| | Stone groundwood | Refiner groundwoods | |
| --- | --- | --- | --- |
| | | Moderate work | Higher work |
| Power (hp days/air dry ton) | 76 | 75.3 | 89 |
| Can. standard freeness[a] (mL) | 115 | 164 | 101 |
| Burst factor[b] | 12.1 | 12.6 | 18.2 |
| Tear factor[c] | 56.7 | 87.8 | 86.0 |
| Breaking length[d], m | 2 920 | 2 680 | 3 200 |
| Brightness[e] | 61.0 | 59.0 | 60.5 |
| Opacity[f], % | 95.0 | 92.5 | 95.0 |

[a] **Canadian standard freeness** (CSF) is a measure of the ease of water flow through a standard perforated plate, specified in mL, from a one liter sample of pulp at 20 °C and 0.3% consistency (weight fraction of pulp in the suspension). Drained water from the sample passes into a dual outlet funnel, one outlet higher than the other. The amount of water collected from the upper "overflow" outlet, in mL, is recorded as the CSF. The rate of water removal on the paper machine of a "free" pulp will be faster than the rate of water removal from a "slow", or slow draining pulp.

[b] **Burst Factor** The pressure developed by a fluid, in pounds per square inch, required to split open a sample of the paper when it covers a circular hole of 1.2 inches diameter and is protected by a rubber diaphragm, is defined as burst (lb/in$^2$, or g/cm$^2$). When burst is divided by the basis weight, e.g. in g/m$^2$, it is called burst factor.

[c] **Tear** corresponds to the tensile force in grams required to continue a tear that has already been started in a single sheet of paper. Corresponds to the "internal tear resistance" as opposed to "edge tear resistance" which is frequently a much higher value. Tear, divided by basis weight, is the tear factor.

[d] **Breaking Strength** is a hypothetical value, calculated from tensile strength measurements, to give a relative measure of the length of a strip of paper which is just self supporting, when it is held vertically.

[e] **Brightness** is a measure of overall reflectivity of a paper to white light, or whiteness. It is based on a scale of 100 for pure magnesium oxide and 8 for carbon black, when the reflectance of a beam of light, of wavelength 457 mμ, is compared at angle of 45° from the axis of the incident light.

[f] **Opacity** is a measure of sheet "see-through". It is obtained by dividing the reflectance of the sheet when backed by a black body by the reflectance of the same sheet when backed by a white body having an effective reflectance of 89% (absolute), and expressed as percent.

period of preheating before refining was developed with the expectation that this might both decrease refining power requirements and produce a longer-fibered pulp potentially able to yield a stronger paper than the product from conventional refining. Combinations of these features were the benefits achieved by the first commercial installations which began operation in the early 1970's.

Chips, shavings, or sawdust are first washed to remove any grit or other potentially damaging contaminants, and then digested with steam at 120 to 130 °C, 170-315 kPa (25 to 45 psi) for 2-3 minutes [18, 19, 20]. While still under pressure, and with only a 60-75 % water content (much less than required for conventional chip refining), the fragmented wood is refined and the pressure then released, producing a fluffed up, bulky pulp. After addition of a small amount of water and a further stage of refining at atmospheric pressure a pulp is obtained which has longer average fiber lengths, higher strength (from less damaged fiber) and a lower fines content than either stone groundwood or conventional chip refiner groundwood pulps.

The favourable strength properties of TMP allow it to be used to replace part or all of the chemical pulp component in a newsprint furnish, the replacement rate ranging from 2 to 4 tonnes for every 1 tonne reduction in chemical pulp content [21]. Since the yield from wood for TMP, about 90 %, is nearly twice that of chemical pulps, chemical pulp replacement by TMP in newsprint very definitely obtains a saving in wood requirements for newsprint production. The savings in wood generally far surpass the additional refining energy requirement for thermomechanical pulping of some 1,000 kWh per tonne more than that required for conventional refining or stone grinding. For coniferous wood TMP requires about 2,000 - 2,500 and stone grinding about 1,200 - 1,650 kWh per air dry tonne [19,21]. A further advantage of TMP over at least chemical pulps is more straightforward effluent treatment [22, 21]

## 13.4 Chemical Pulping Processes

Wood processes involving the use of chemicals for either softening or removal of much or all of the lignin, all have the objective of assisting the process of fiber separation from the wood. In the simpler processes of pulping employing chemicals all that is involved is a chemical presoak of whole

(small) logs or chips prior to the use of conventional mechanical pulping methods of defiberization to produce a fiber suspension in water. These presoak methods are usually reserved for the mechanical pulping of hardwoods and are referred to as chemi-mechanical pulping, not as true chemical pulping processes. At the other extreme the more rigorous chemical pulping methods remove all, or nearly all of the lignin from wood chips leaving cellulose fiber as intact as possible. In this case a fiber suspension in water is obtained directly, with no or negligible mechanical work required for fiber separation. Despite the selectivity aim of chemical pulping methods to soften and dissolve lignin alone, without affecting the cellulose, there is inevitably some cellulose chain shortening and some end group conversion to carboxyl that occurs at the same time. Nevertheless, since about 1930 the range of chemical pulping techniques that have been developed are adequate to produce satisfactory pulps for papermaking from virtually any species of tree as well as from grasses and annual plants. Thus, a much wider range of raw materials can be handled by chemical pulping methods than is possible by straight mechanical pulping procedures.

### 13.4.1 Chemimechanical Pulping

Direct grinding or refining of hardwoods (from deciduous trees) tends to give pulps with a short average fiber length which produces a relatively low strength sheet on the paper machine. This is particularly the case on mechanical pulping of *hard* woods from deciduous trees both since the fibers are shorter and finer, in the wood, and since there is a greater tendency for fibers to be fragmented, instead of whole, as they are torn out from the denser wood structures.

However, if the wood is given a presoak in a hot solution of sodium hydroxide it is possible to obtain a more useful pulp with less fiber damage [11]. Temperatures may be at only slightly above ambient up to 80 °C, depending on the species to be pulped and treating time, and small amounts of sulfur dioxide may be added to the liquor to decrease the tendency of the wood to darken under the aerated alkaline conditions. Whole logs or blocks may require a 4 to 10 hour presoak prior to grinding whereas chips, for refining, only need a 10 to 15 minute contact time. In this way, pulp yields of 85-90 % are obtained with properties parallel to those expected for groundwood pulps generally, and from species of tree not otherwise amenable to mechanical pulping. At the same time the chemical pretreatment also significantly decreases the power input required to reduce the wood to fibers [11]. Even mixed blends of soft, and hardwood chips have been found amenable to chemical pretreatment before mechanical pulping and still yield useful pulps [21]. Where feasible this approach simplifies newsprint product for mills using a mixture of coniferous and deciduous woods.

### 13.4.2 Semichemical Pulping

The essence of semichemical pulping methods, as the name implies is that chips are subjected to a more intensive chemical treatment than a simple presoak but not as severe as for chemical pulping, to cause partial chemical delignification [2, 11]. Intermediate intensity chemical treatment is achieved by a number of means, among them decreased ratios of chemicals to wood, and decreased cooking times and/or temperatures in combination with the use of chemicals which pulp at near neutral pH rather than under highly acid or highly alkaline conditions. Following the cook, the chips are refined at much lower power inputs per tonne of pulp than required for straight mechanical pulping, because of the prior chemical softening and partial solubilization of the lignin. Since there is only partial lignin solubilization during the pulping process relatively high yields of pulp of 65 to 80 % are obtained.

There are at least half a dozen variants of semichemical pulping which have been practised, and some of these, namely neutral sulfite semichemical (NSSC) pulping, softwood bisulfite high yield pulping, and softwood sulfate pulping for linerboard production are still common [2, 11]. The second and third of these procedures are quite closely related to straight chemical pulping which will be discussed shortly. But NSSC pulping has a number of differentiating features of interest. Hence this procedure, which is primarily of value in the pulping of hardwoods, will be discussed here as an example of a semichemical pulping procedure.

The digestion liquor used for NSSC pulping essentially consists of sodium sulfite and sodium carbonate dissolved in water, but at the proportions and the pH used actually comprises a mixture of sodium sulfite and sodium bicarbonate. The make-up sodium carbonate is either purchased or made

from aqueous solutions of low grade sodium hydroxide by contacting this with flue gas (Equations 13.1, 13.2).

$$CO_2 + H_2O \rightarrow H_2CO_3 \tag{13.1}$$

$$H_2CO_3 + 2\,NaOH \rightarrow Na_2CO_3 + 2\,H_2O \tag{13.2}$$

Burning of elemental sulfur is the first stage of sodium sulfite preparation. Repeated contact of the sulfur dioxide so formed with the aqueous sodium carbonate in a gas-liquid absorption unit (sulfiting tower, Figure 13.6) produces the approximate proportion of sodium sulfite and sodium bicarbonate (about 1 mol: 2 mol) required (Equations 13.3, 13.4).

$$SO_2 + H_2O \rightleftharpoons H_2SO_3 \tag{13.3}$$

$$H_2SO_3 + 2\,Na_2CO_3 \rightarrow Na_2SO_3 + 2\,NaHCO_3 \tag{13.4}$$

For practical purposes liquor recycle to the sulfiting tower is continued until the pH of the underflow reaches 8.5, at which time a portion of this stream may be continually bled off for use as pulping digestion liquor [2]. In this situation the actual proportion of salts present in solution is about 82 % $Na_2SO_3$, 4 % $NaHSO_3$, and 14 % $NaHCO_3$. Occasionally, for some species of wood or with differing pulping objectives liquor of about pH 8 may be used, achieved by using more prolonged contact between the sulfur dioxide and sodium carbonate solutions (Equation 13.5).

$$H_2SO_3 + 3\,Na_2CO_3 + SO_2 \rightarrow$$
$$2\,Na_2SO_3 + 2\,NaHCO_3 + CO_2 \tag{13.5}$$

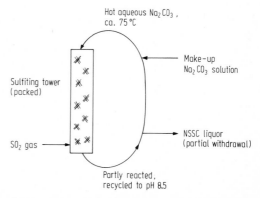

**Figure 13.6.** Preparation of neutral sulfite semichemical pulping liquor via counter-current contacting of a sulfur dioxide gas stream with aqueous sodium carbonate

Care has to be taken to ensure that the digestion liquor does not become too acid, otherwise cellulose degradation during the pulping step becomes more significant.

For NSSC pulp production the effective chemical concentrations are about 120 g/L of sodium sulfite and about 20 g/L of sodium carbonate. Chips to be pulped are cooked in this two buffer system at temperatures of 140-170 °C for 1-4 hours depending on both the type of wood being pulped and the pulping objective [2]. High grade pulps destined for bleaching will be pulped at ratios of sodium sulfite to wood of about 15-20:100 by weight, significantly more than used for coarser pulps. They also demand higher temperatures and/or longer times, and give correspondingly lower yields of finished pulp, because of the greater degree of delignification achieved. Coarser pulps, destined for board production (cardboard, or box liners) and the like, will normally, by pulping chips at sulfite to wood ratios of about 1:10, be subjected to milder conditions and for shorter times. In this way pulps are obtained at higher yields that at the same time retain considerable individual fiber stiffness, a property important for the manufacture of products such as corrugating medium. The dual buffer system that these pulping liquor compositions comprise exert their pulping action by both sulfonation of lignin present in the chips [10] and probably by fragmentation by reduction as well [23], both processes yielding soluble lower molecular weight lignin fragments from the original insoluble high molecular weight lignin.

The cooked pulp is filtered from the pulping liquor, washed, and then refined to complete the defiberization which has intentionally been only partially completed by the chemical treatment step in order to maximize the yield. A high yield of the clean, refined pulp is obtained, which is light in colour, but which still contains some lignin. Hence, this product is generally unsuitable for high grade papers. But the light colour of the product minimizes bleaching costs for incorporation of this product into intermediate or coarser paper grades. There is also no significant smell problem associated with the pulping step, as there is with some other chemically-based pulping processes.

However, one problem with NSSC pulping is the difficulty of recycle or disposal of the waste (spent) pulping liquor. It can be concentrated and burned but this produces sulfate, not sulfite. If there is a kraft pulping operation at the same site, or in close proximity to the semichemical pulping interests, the sodium sulfate from the NSSC component may be usefully employed to provide both sodium ion and sulfate make-up chemical for the kraft mill. In fact, if the kraft pulp production rate is 3 or more times the NSSC production rate, the kraft process is capable of profitably using all of the spent liquor of the NSSC component [2]. If, however, there is no nearby sulfate consumer, it is possible to obtain sulfite from sulfate by a complex series of steps but these procedures add significantly to the costs and thus makes recycle less attractive.

Alternatively the spent NSSC liquor may simply be discharged. But untreated, this liquor has a very high B.O.D. and several inorganic and organic toxic constituents so that its disposal to fresh water streams and lakes has not been practised for some time. Disposal in the open ocean, at depth, in areas of vigorous currents could probably be accomplished acceptably, but since mills fronting on the sea are usually on sheltered inlets or fjords the same kinds of impact will be realized on marine discharge as on fresh water discharge, in this situation. Details of other expedients possible are discussed with emission control aspects of sulfite pulping.

### 13.4.3  Chemical Pulping

Wood chips supply the fiber raw material for the vast majority of chemical pulping operations, as for chemimechanical and semichemical methods. The difference between chemical pulping, and these other pulping methods which also employ chemicals and already described, is in the fact that here the chemical treatment step is severe enough to dissolve all, or much of the lignin originally present in the wood. The chemicals employed operate at low or high pH (not near neutral), higher ratios of chemicals to wood are used, and the cooking conditions are generally more severe than for the processes employing chemicals discussed thus far. Thus, chemical pulping methods require little or no mechanical work on the treated chip, subsequent to the chemical treatment step, in order to obtain a fiber suspension in water from the original wood.

Photomicrographs of chemical pulps show that very little fiber damage is incurred in the pulping step. Consequently papers made from chemical pulps, particularly from kraft chemical pulps, are strong. Since there is not only lignin loss, but for some chemical pulping processes (for instance, the acid sulfite process) hemicellulose is largely lost as well, the yields of pulp from straight chemical processes are in the 40 to 60 % range, lower than resulting from the process employing milder chemical treatment and much lower than achieved by the straight mechanical procedures. But because the product from chemical pulping processes consists of almost pure holocellulose, or cellulose, it is more or less readily bleached to high brightness (whiteness) levels, and this brightness and the strength properties already mentioned are both much more stable. These factors make papers produced from chemical pulps both more pleasing in appearance and more durable than papers produced from other types of pulps, but it also makes them more expensive because of the increased wood and chemical cost required for their manufacture. The various chemical pulping processes in use may be conveniently classified according to the pH of the digestion liquor used to carry out the pulping.

### 13.4.3.1 Acid Sulfite Pulping

After the discovery of the softening action of sulfurous acid on wood, by Benjamin Tilghman in the U.S. in about 1857, and without any significant understanding of the chemistry involved this idea was developed experimentally as a chemical method for wood pulping by 1867 [10]. It was adopted on a commercial scale in the ensuing few years, both in the U.S. and elsewhere, and rapidly became the dominant chemical pulping method used worldwide, until about 1937 when the volume of kraft pulp produced first exceeded the volume of sulfite pulp [2].

Digestion liquor for acid sulfite pulping consists of a solution of sulfur dioxide in water, the sulfur dioxide being obtained by sulfur combustion or by roasting of pyrite [10], and containing added lime. The lime component is added to the liquor by passage of the cold sulfur dioxide solution down one or more Jenssen towers packed with limestone, in a manner very similar to the sulfiting towers used to prepare NSSC pulping liquor (Equations 13.6, 13.7).

$$SO_2 + H_2O \rightleftarrows H_2SO_3 \qquad (13.6)$$

$$2\,H_2SO_3 + CaCO_3 \rightarrow Ca(HSO_3)_2 + H_2O + CO_2 \qquad (13.7)$$

Since calcium sulfite is not very soluble in water, the last stage(s) of liquor preparation involve absorption of sulfur dioxide into the cold liquor under pressure. By these means an eventual ratio of about 1 mole of calcium hydrogen sulfite (calcium bisulfite) to a total of 4 to 5 moles of (sulfur dioxide + sulfurous acid) is obtained, and at this stage represents an aqueous solution containing a net 6.5 to 9 % dissolved sulfur dioxide in all forms [10].

The terminology generally used to describe the various forms of sulfur dioxide present calls the total dissolved (sulfur dioxide + sulfurous acid) "free SO_2", and the sulfur dioxide tied up in calcium hydrogen sulfite as "combined SO_2" (Equation 13.8).

$$
\begin{array}{ccl}
SO_2,\ H_2SO_3 & Ca(HSO_3)_2 & (13.8) \\
\text{``free SO}_2\text{''} & \text{``combined SO}_2\text{''} & \\
4-5 \quad : & 1 & \text{mole ratio of free} \\
& & \text{to combined SO}_2
\end{array}
$$

However, the effective sulfur dioxide accessible for pulping is a composite of these two forms of sulfur dioxide on the basis that the "available SO_2" = free SO_2 + 1/2 (combined SO_2). The reason for this situation is that, as sulfur dioxide is consumed or vented from the digestion vessel during pulping, the equilibrium represented by equation 13.9 is displaced to the right,

$$Ca(HSO_3)_2 \rightarrow CaSO_3 + H_2O + SO_2 \qquad (13.9)$$

and in the process releasing one half of the combined SO_2 to the solution to become available for pulping.

Conditions required for sulfite pulping vary with the species of wood being pulped as well as with the degree of delignification and the chemical to wood ratio being used, but normally lie in the 8-12 hour range at 110-145 °C, frequently following a 2 hour liquor penetration period. Longer times will usually be used at the lower temperatures, and shorter times, to avoid undue cellulose degradation, at the higher temperatures. Typical chemical to wood ratios may be estimated from the typical usage rate on the West Coast of North America, 182,000 L (50,000 U.S. gal) of liquor containing 8.80 % total SO_2, 1.10 % combined SO_2 at 70 °C; to 36 tonnes of bone dry wood [10]. Of course it is not necessary that the chips be dry for digestion, but the moisture content must be known, roughly, in order to have a relatively consistent ratio of active chemicals to the wood content of the chips. The combination of the heat, and the aqueous low pH conditions brings about hydrolytic fragmentation and sulfonation of the lignins as well as formation of sugar sulfonic and aldonic acids, chemical changes which contribute significantly to the dissolving of these smaller, more polar fragments in water. Sulfuric acid concentrations in the pulping system are minimized, during the initial liquor preparation, by combustion of sulfur or pyrite in a deficiency of air and rapid, post-combustion chilling of the gases produced, and secondly by keeping the digester air content to a minimum at the time of loading. Without these precautions the higher acidities which might result can cause a measurable reduction in pulp yield from aggravated hydrolysis of cellulose. The cellulose hydrolytic action of sulfonic acids, strong acids which are known to form during the digestion is thought to be controlled by the sufficient concentration of weaker, buffering acids also present in the pulping system.

Acid sulfite pulping produces a light-coloured easily bleached pulp. After washing and screening and even without bleaching it has a brightness compatible with direct blending into mechanical pulps for newsprint production. With two or three bleaching stages high brightness bond papers may be produced from these pulps. However, because of the less than optimum paper strength properties obtained and because chemical recovery procedures are expensive (see references given, Relevant Bibliography), acid sulfite pulping has become much less important in recent years for the production of chemical pulps in competition with the kraft process. It is still of value for the manufacture of pulps for some specialty uses such as tissues and printing papers [24], and for the production of dissolving pulp for conversion to a range of useful cellulose derivatives employed as reconstituted fibers and thermoplastic materials. Yield advantages and lower bleaching requirements than necessary for kraft pulps promise a continued place for sulfite-based pulping processes for an appropriate spectrum of applications [25, 26].

### 13.4.3.2 Other Sulfite-based Pulping Processes

Acid sulfite pulping, originally developed around low-cost calcium ion at present suffers from a number of disadvantages relative to alternate liquor components employing other cations. Among these, calcium sulfite is virtually insoluble, introducing difficulties during the preparation of digestion liquor, chemical recovery is sufficiently

complex that it is uneconomic, and even disposal of the waste liquor has attendant difficulties (Section 13.5). Because of these disadvantages, most modern sulfite mills have switched from calcium-based to magnesium, sodium, or ammonium based systems which circumvent many of these difficulties.

With magnesium, sodium, or ammonium based systems the bisulfite and sulfite salts are all soluble at all proportions in the presence of sulfurous acid. Even magnesium sulfite, with a solubility of about 1.25 g/100 mL, cold, is about 160 times as soluble as calcium sulfite at the same temperature and its sulubility *increases* with temperature. So liquor preparation with these more recently employed sulfite salts is easier, whether for acid sulfite pulping or for pulping under less acid (bisulfite pulping) or neutral (NSSC) conditions. And, in fact, sulfite pulping has also been experimentally tested under alkaline conditions. For ammonium based systems, ammonium hydroxide is contacted with a sulfur dioxide gas stream for liquor preparation. Magnesium based systems contact a magnesium hydroxide slurry with the sulfur dioxide gas stream, whereas sodium based systems normally employ sodium carbonate lumps in a sulfiting tower, in a method similar to that used for NSSC liquor preparation, as the source of the sodium ion although low cost sodium hydroxide may also be used for this purpose.

To give an idea of the versatility of operating pH possible with ammonium, magnesium, and sodium-based sulfite pulping systems, below are outlined examples of the results of operating with various ratios of sodium carbonate to sulfite. The analogous results may be achieved with ammonium and magnesium ions, at least with pH's up to the bisulfite range. Sodium ion is the only one useful throughout the whole pH range to even monosulfite and alkaline sulfite pulping liquor preparation (Table 13.6). With sodium carbonate as the base it is possible to use carbonate to sulfurous acid ratios ranging from 1:3 where the active pulping species are bisulfite ($HSO_3^-$) and sulfurous acid, corresponding to the conditions employed in acid sulfite pulping, to 1:2 where the principal pulping species is primarily bisulfite, such as used in bisulfite pulping [11]. At the 1:1 to 2:1 range of ratios of carbonate to sulfurous acid, sulfite and bicarbonate are the active pulping principals, as used in NSSC pulping. And when one reaches the 3:1 to 4:1 range of ratios, sulfite and carbonate are the domi-

**Table 13.6** Chemical pulping processes classified according to operating pH, and end uses of the pulps obtained.

| Pulping process | Liquor pH at 20 °C | Base options | Appropriate end uses |
|---|---|---|---|
| Acid sulfite | 1 – 2 | $Ca^{2+}$, $Mg^{2+}$ | 1. Chemical conversion to derivatives of cellulose, e.g. viscous rayon, cellulose acetate. |
| | | $Na^+$, $NH_4^+$ | 2. Filler pulps for paper requiring little strength, e.g. writing papers. Provides bulk, cushioning properties. |
| Bisulfite | 3 – 5 | $Mg^{2+}$, $Na^+$, | 1. Unbleached – in newsprint (25 – 30%, rest groundwood) |
| | | $NH_4^+$ | 2. Bleached – fine papers (up to 100%) |
| Neutral sulfite semichemical [a] | 6 – 8 | $Na^+$, $NH_4^+$ | 1. Fluting board – provides stiffness, crush resistance to corrugated cardboard |
| | | | 2. Coarse wrapping papers |
| Monosulfite | 8 – 10 | $Na^+$, $NH_4^+$ | As for neutral sulfite |
| Alkaline sulfite [b] (Experimental) | 11 – 13 | $Na^+$ | 1. Unbleached – box liner-board, other packaging materials |
| | | | 2. Bleached – high quality papers |
| Soda | 11 – 13 | $Na^+$ | Outmoded for chemical pulping, see text. |
| Kraft | 11 – 13 | $Na^+$ (+ $Na_2S$) | 1. Unbleached – packaging materials: boxboard, sack and bag papers |
| | | | 2. Semibleached – newsprint (25 – 35%) |
| | | | 3. Bleached – high quality papers, all types |
| Prehydrolysis kraft | 11 – 13 | $Na^+$ ($Na_2S$) | Chemical conversion |

[a] Not generally used as a full chemical pulping process, but may also be used for the production of fine papers. Details of this process are included here to demonstrate the pulping versatility of sulfite ion over the full range of pH conditions.
[b] Only at the laboratory stage at present, but see [27].

**Table 13.7** Ratios of sodium carbonate to sulfurous acid [a] required for sulfite pulping at various pH's

| Process | Required mole ratio of | | Active pulping entities present |
|---|---|---|---|
| | $Na_2CO_3$: | $H_2SO_3$ | |
| Acid sulfite, | 1 | 3 (and excess) | $H_2SO_3$, $HSO_3^-$ |
| pH ca. 1 – 2 | | | |
| Bisulfite, | 1 | 2 | $HSO_3^-$ |
| pH ca. 3 – 5 | | | |
| NSSC [b], | 1 | 1 | $SO_3^{2-}$ |
| pH ca. 6 – 8 | 3 | 2 | $SO_3^{2-}$, $HCO_3^-$ |
| NSSC and monosulfite, | 2 | 1 | $SO_3^{2-}$, $2 HCO_3^-$ |
| pH ca. 8 – 10 | | | |
| Alkaline sulfite [c], | 3 | 1 | $SO_3^{2-}$, $2 CO_3^{2-}$ |
| pH 11 – 13 | 4 | 1 | $SO_3^{2-}$, $3 CO_3^{2-}$ |

[a] The existence of sulfurous acid ($H_2SO_3$) is somewhat hypothetical, but is a convenient concept for considerations required here. Sulfite salts are common.
[b] Neutral sulfite semichemical pulping.
[c] Experimental in nature, at present [27].

nant anions present, important in the experimental alkaline sulfite pulping method [27] (Table 13.7). Acid sulfite and bisulfite pulping processes yield light coloured pulps, suitable for direct blending with groundwood pulps for newsprint production. These are also pulps which may be bleached easily for fine paper production, and require low input of beating energy to obtain good sheet formation and strength. They may also be readily purified with solutions of sodium hydroxide for the preparation of dissolving pulps [2]. But to offset these advantages, long cooking times are required, and either disposal, or recovery of the chemicals from the spent liquor present problems.

### 13.4.3.3 Alkaline Pulping: The Soda, and Kraft (Sulfate) Processes

The older soda process which involves the pulping of chips by cooking with aqueous sodium hydroxide at 160-170°C, was originally devised by Watt and Burgess in England in 1851. It requires long cooking times and gives poor yields of a dark pulp still having a significant lignin content, even with hardwoods which give the most favourable results. Also the strength properties of soda pulps are second rate. Any "soda" mills still operating generally employ a small amount of sulfide in the pulping liquor [2], which is not strictly soda pulping, but should be classed as a modified process bearing more than a superficial resemblance to kraft pulping. Thus, at present virtually all alkaline pulping, worldwide, is conducted by the kraft process.

The kraft, or sulfate, process is today the pre-eminent chemical pulping procedure. In Canada alone, 80 % of all the chemical pulp produced is by this process, and worldwide some 85 % of the total is via this route [28]. The current prominence of this pulping procedure warrants a rather fuller discussion of the details of this process than is devoted to the other chemical pulping methods.

This process, in which wood chips are cooked in a pulping liquor consisting of a solution of sodium hydroxide and sodium sulfide in water (so-called "white liquor"), was originally developed by Dahl, in Germany, in 1879 [4]. The presence of the sodium sulfide accelerates the pulping rate so that kraft pulping requires a shorter cooking time than soda pulping. The sulfide also serves to stabilize the cellulose somewhat, so that the pulps produced suffer minimal fiber damage and hence make a strong paper sheet. This high strength pulp product is the origin of the name given to the process, "kraft" being German for "strength". Factors which contribute to the greater strength of pulps made by this process include not only the presence of the sodium sulfide and the shorter cooking time (and consequently less cellulose degradation) but also the lower ratio of chemicals to wood required to obtain effective pulping. A further advantage of kraft pulping lies in its ability to handle hardwoods or softwoods, even of resinous or pitchy species, since the liquor composition is such that the resins are dissolved. Resinous species are more difficult to handle by any of the acidic sulfite pulping processes.

A feature of the kraft process vital to its continued succes is the integral, well-tested chemical recovery system. The digestion liquor for each batch of chips to be pulped is predominantly obtained from the chemicals recovered from the spent liquor of previous digestions, and will have the approximate composition given in Table 13.8. For a pulping context it is usual to specify all of the components present in the digestion liquor on a "Na$_2$O

**Table 13.8** Typical analysis of white liquor used for kraft pulping [a]

| Component | Concentrations, as | | Function |
|---|---|---|---|
| | Na$_2$O equivalent g/L | Molar M/L | |
| NaOH | 73.0 | 2.34 | primary pulping agent |
| Na$_2$S | 31.4 | 0.51 | accelerates pulping, raises yield of cellulose and hemicellulose by protective action. |
| Na$_2$CO$_3$ | 18.2 | 0.29 | negligible. Presence incidental as a result of chemical recovery system. |
| Na$_2$SO$_4$, Na$_2$SO$_3$, Na$_2$S$_2$O$_3$ | ca.  1.0 | — | none. Traces present incidentally. |

[a] Data from Kirk-Othmer [11]. Liquor with the composition given here will normally be diluted to a net sodium hydroxide concentration of about 50 g/L Na$_2$O [10]

equivalent'' basis, which has the effect of putting the active constituents on the same sodium ion content basis. Thus, the actual concentration of sodium hydroxide present for a 73 g/L, $Na_2O$ equivalent is given by Equation 13.10.

$$73 \text{ g/L (Na}_2\text{O equiv.)} \div 62 \text{ g/mol (Na}_2\text{O)} \times$$
$$2 \text{ mol/mol (NaOH from Na}_2\text{O)}$$
$$= 94.2 \text{ g/L NaOH} \qquad (13.10)$$

By a similar calculation the actual concentration of sodium sulfide present works out to 39.5 g/L.
For pulping purposes, the "effective alkali" present is somewhat higher than the actual sodium hydroxide concentration present in the white liquor because of the existence of a hydrolytic equilibrium between sodium sulfide and sodium hydrogen sulfide (Equation 13.11).

$$Na_2S + H_2O \leftrightharpoons NaOH + NaHS \qquad (13.11)$$

Thus, the effective alkali comprises the concentration of sodium hydroxide present, plus one half of the concentration of sodium sulfide present, when both are expressed as $Na_2O$, or for the typical analysis of Table 13.8, about 88-89 g/L $Na_2O$. Specified in molar terms, since the molar concentration of sodium hydroxide derived from sodium sulfide is equivalent to the initial molar concentration of sodium sulfide, the effective alkali concentration of the white liquor of Table 13.8 would be 2.85 M.
"Sulfidity" is also an important parameter relating to the liquor composition for kraft pulping. It simply refers to the concentration of sodium sulfide divided by the concentrations of sodium sulfide plus sodium hydroxide the "active alkali", as opposed to "effective alkali", all expressed in terms of their $Na_2O$ equivalents. Equation 13.12 gives the method of determining sulfidity, expressed as a percentage.

$$\% \text{ Sulfidity} = (Na_2S)/(Na_2S + NaOH) \times 100 \qquad (13.12)$$

The desirable range for sulfidity is 15 to 35 %. Values lower than this tend to give incomplete pulping and weaker pulps, akin to the properties of pulp produced by the soda process, and higher values tend to give sluggish delignification [10]. Sulfidities above this range also tend to raise potential sulfur losses in both the pulping and chemical recovery segments of kraft pulping.

*Kraft Pulping*

The cook is carried out in a digester, a steel or stainless steel cylindrical pressure vessel about 3 m in diameter and 16 m high, or about the height of a 5 storey building, usually fitted with external heat exchangers. For a digester this size 50 to 60 tonnes of chips per batch (commonly containing 50 % moisture) are loaded from the top, while steam is blown through the opening of the digester. Steam blowing serves to both displace much of the air from the digester during the loading process and on white liquor addition, it also aids in liquor penetration into the chip from the combined air displacement, warming and wetting actions on the wood. Any knots, or uncooked chips screened from a previous digestion will also be added at this stage. At the time of white liquor addition, sufficient "weak black liquor" is also added to bring the content of $NaOH + Na_2S$ present in the digestion liquor to the concentration range of about 50 g/L $Na_2O$ equivalent, and to a liquor to wood ratio of about 16 kg of effective alkali for each 100 kg of dry wood. The heavy steel lid is bolted down and heating begun by pumping liquor from the digester, through the external steam-heated heat exchangers, and back to the digester. A common heating schedule is to take $1\frac{1}{2}$ hours to bring the temperature up to about 170 °C and 620 kPa (kilopascals, i.e. six atmospheres pressure, 90 psig), after which it is held at this temperature for a further $1\frac{1}{2}$ hours to complete the digestion step. Chemical to wood ratios and digestion times and temperatures will be varied somewhat in particular situations depending on the species of wood being pulped, and the degree of delignification desired for the product.
After the digestion is completed a valve at the base of the digester is opened while it is still at operating temperature and pressure, causing a very rapid discharge of the contents of the digester into a blow tank. The fast pressure drop experienced by the pulp on discharge assists in blowing the fiber bundles apart, aiding in the production of a pulp from the chips.
Cooked pulp plus weak black liquor, spent white liquor which has picked up lignin and some hydrolyzed cellulose and hemicellulose etc., proceeds to screens which serve to remove any uncooked fragments and knots from the pulp. The bulk of the weak black liquor is separated from the cooked pulp in a decker, basically a filter which captures the pulp on a screen and removes the bulk of the liquor by suction. The pulp is then rinsed in washers, units very like drum filters, equipped with 2 or 3 fresh water fed shower tubes which run

parallel to the axis of the drum and which rinse off any residual weak black liquor from the pulp. At this stage the kraft process product, a fairly clean pulp, though still dark brown in colour from residual lignin etc. (i.e. the raw material for "kraft" paper bags etc.), is obtained separated from the weak black liquor which contains all of the spent chemical constituents of the original white liquor plus the components of the wood which were solubilized during the course of the pulping process.

The stages just described, from the point of loading chips into a cylindrical digestion vessel, comprises *batch* pulping of wood. To maintain a continuous stream of cooked pulp for subsequent stages requires say six to ten of these digesters, operating in parallel in a staggered loading sequence, and each discharging 5 to 7 loads of 12 to 15 tonnes (about $1/4$ of the weight of wet wood charged to each digester) of cooked pulp from each load, per 24 hours. Or all the stages described up to this point may also be conducted continuously in a single, more complex digestion unit performing all these functions in turn on a mass of chips which moves steadily through the unit. One such commercially available version is the Kamyr continuous digester (Figure 13.7). Many pulp mills use a combination of batch and continuous digesters to obtain maximum flexibility in terms of the types of wood being pulped, and to obtain the desired degree of pulp dilignification proportioned as required for the end use objectives of the pulp.

### Kraft Cyclic Chemical Recovery Process

The objectives of chemical recovery from the weak black liquor separated from the pulp after kraft pulping are threefold. Recovery and regeneration of most of the chemicals employed in the original

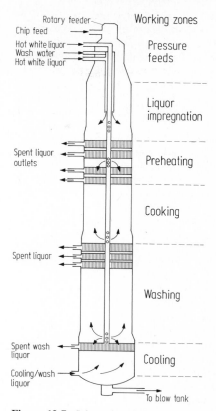

**Figure 13.7.** Schematic of the operating details of a Kamyr continuous digester

digestion liquor is a primary objective but a further objective is also to generate a significant fraction of the total steam requirements of the pulp mill from combustion of the dissolved lignin residues and

**Plate 13.1.** Pulp washer for unbleached kraft pulp. Four shower heads above the rotating drum of washer distribute fresh water onto surface of pulp while reduced pressure in the interior of drum pulls the rinse water through pulp

other organic constituents (comprising nearly half the original dry weight of the wood) contained in the weak black liquor. Not only is energy recovered by burning the organic constituents present in the spent liquor, but the third objective, avoidance of the significant polluting potential on untreated discharge of spent pulping liquor into a stream or lake, is also achieved. The environmentally-acceptable alternative to this, treatment of the spent liquor by artificial lagoon aeration, would be a net cost to the process rather than a net credit, as provided by energy recovery.

Initially the weak black liquor is contacted with air, either by sparging air through a tank of the liquor, or by causing thin films of black liquor to fall through an air chamber. The objective of this step is to decrease the losses of foul-smelling volatile sulfur compounds which would otherwise occur during subsequent evaporation or combustion steps (Equations 13.13 - 13.15).

Volatile sulfide loss:
$$H_2O + CO_2 \text{ (from combustion)} \leftrightarrows H_2CO_3 \tag{13.13}$$
$$Na_2S + H_2CO_3 \rightarrow Na_2CO_3 + H_2S \uparrow \tag{13.14}$$

$$2\,NaSCH_3 + H_2CO_3 \rightarrow Na_2CO_3 + 2\,CH_3SH \tag{13.15}$$
$$\text{methyl mercaptan}$$

By oxidizing sodium sulfide to sodium thiosulfate, and methyl mercaptan (methanethiol) to dimethyldisulfide, the first component is made virtually involatile, and the volatility of the second is greatly decreased (Equations 13.16, 13.17).

Black liquor oxidation:
$$2\,Na_2S + 2\,O_2 + H_2O \rightarrow Na_2S_2O_3 + 2\,NaOH$$
$$\Delta H = -900 \text{ kJ } (-215 \text{ kcal}) \tag{13.16}$$

$$2\,CH_3SNa + \tfrac{1}{2}O_2 + H_2O \rightarrow$$
$$CH_3SSCH_3 + 2\,NaOH \tag{13.17}$$

After oxidation of the weak black liquor, of about 15-18 % total solids content by weight, it is concentrated. Concentration before combustion is necessary to increase the energy recovery possible on combustion of the organic constituents. Otherwise a large proportion of the energy derived from ultimate combustion is consumed in evaporating the high water content of the liquor. Evaporation to 50-55 % solids is usually carried out in 3 to 6 stages of multiple effect evaporators. By operating under

reduced pressure, and by using the steam obtained from the first and each of the subsequent stages to heat the liquor of the following stage, 5 to 5 $^1/_2$ kg of water removal can be achieved for each kg of steam used for heating the first stage in a sextuple-effect evaporator. The last stage of black liquor evaporation, from 50-55 % solids to about 65-70 % solids, is conducted by direct contact of the black liquor with hot flue gases from the recovery boiler (Figure 13.8). Increased surface area for evaporation and increased evaporation rates are obtained by slowly rotating parallel discs which are half immersed in the liquor and half exposed to the hot flue gases. In this manner direct contact evaporation also accomplishes a moderate scrubbing effect on the flue gases, at the same time consuming the bulk of the residual sodium hydroxide (Equations 13.18, 13.19),

$$2\,NaOH + CO_2 \rightarrow Na_2CO_3 + H_2O \tag{13.18}$$
$$2\,NaOH + SO_2 \rightarrow Na_2SO_3 + H_2O \tag{13.19}$$

and heat is transferred from the flue gases to the black liquor so that it leaves the last stage of evaporation hot. At this point it is referred to as strong black liquor. Addition of make-up chemicals as either sodium sulfate ("salt cake") or sulfur as necessary, is carried out at this stage to replace any losses of sodium and sulfur occurring elsewhere in the process. In fact, the reason "sulfate process" is synonymous for the "kraft process" is because the primary make-up chemical requirement of the kraft process is sodium sulfate, although sodium sulfate is not actually an active constituent of kraft pulping liquor.

The objectives of black liquor combustion in the furnace of the recovery boiler are complex. These include combustion of the organic components present in the strong black liquor to generate heat and in turn to produce steam, and to recover the inorganic chemicals originally present in the black liquor, in *reduced* form, that is as sodium sulfide rather than sodium sulfate. Preheated (about 120 °C) strong black liquor is first sprayed into the hot, drying zone of the recovery furnace (Figure 13.9). The residual water in the black liquor practically instantaneously flashes off, producing a char which adheres to the walls of the furnace. As the thickness of the layers of carbonized char on the walls builds up, pieces fall off and land on the hearth. At the hearth conditions are chemically reducing, maintained by providing inadequate air for total

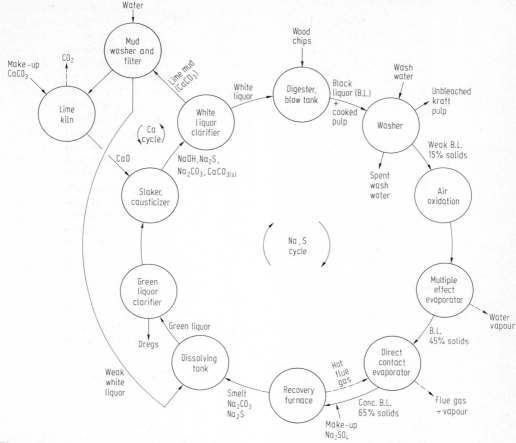

**Figure 13.8.** Cyclic process for recovery of chemicals and energy from the spent pulping liquor of the kraft process

combustion, and temperatures of ca. 1,000 °C prevail, sufficiently high to provide the energy required for the endothermic nature of many of the reducing reactions. Here, sodium sulfate and sodium thiosulfate, produced at the black liquor oxidation stage, are reduced to sodium sulfide, and any remaining sodium hydroxide is converted to sodium carbonate (Equations 13.20 -13.25).

Examples of recovery furnace reducing zone reactions:

$$2C + O_2 \rightarrow 2CO \tag{13.20}$$

$$C + H_2O \rightarrow CO + H_2 \tag{13.21}$$

$$Na_2SO_4 + 4CO \rightarrow Na_2S + 4CO_2 \tag{13.22}$$

$$Na_2SO_4 + 2C \rightarrow Na_2S + 2CO_2 \tag{13.23}$$

$$Na_2S_2O_3 + 4CO + 2NaOH \rightarrow \\ 2Na_2S + 4CO_2 + H_2O \tag{13.24}$$

$$Na_2CO_3 \rightarrow Na_2O + CO_2 \text{ (slight)} \tag{13.25}$$

The hearth is sloping so that as the sodium sulfide and sodium carbonate (melting points 950 °C and 851 °C) accumulate in the char on the hearth the mixture, called smelt, trickles out, to be dissolved immediately in lime mud wash water for the first stage in fresh liquor preparation. Dissolving of the molten smelt in water is a noisy procedure, not only from the large temperature differential but also from the exothermic nature of the solution process as well [29].

At the upper parts of the recovery furnace second-

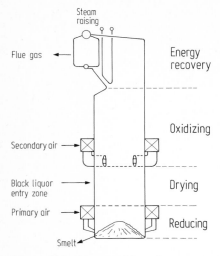

**Figure 13.9.** Diagram of the main components of a kraft recovery boiler, for strong black liquor combustion. Reprinted from Canadian Chemical Processing, November 1970, courtesy of Southam Business Publications Ltd

ary air is admitted to obtain an oxidizing atmosphere, and the temperatures are still sufficiently high that any residual reduced compounds present in the gases are oxidized (Equations 13.26 - 13.29)

Recovery furnace oxidizing zone reactions:
$$2\,CO + O_2 \rightarrow 2\,CO_2 \tag{13.26}$$

$$2\,H_2 + O_2 \rightarrow 2\,H_2O \text{ (minor)} \tag{13.27}$$

$$2\,H_2S + 3\,O_2 \rightarrow 2\,SO_2 + 2\,H_2O \text{ (traces)} \tag{13.28}$$

$$CH_3SH + 3\,O_2 \rightarrow SO_2 + CO_2 + 2\,H_2O \text{ (traces)} \tag{13.29}$$

As a result, gas temperatures rise to the neighbourhood of $1{,}250\,°C$ in this region of the furnace, from which heat recovery is practised via steam superheating tubes and water tubes to generate high and low pressure steam. After heat recovery the flue gases are first passed through the direct contact evaporator, for additional heat utilization, and through electrostatic precipitators, for particulate removal, before delivery to the stack for discharge. The electrostatic precipitator catch is usually returned to the black liquor stream for recycling.

Very great care is required for normal recovery furnace operations to minimize risk of explosions. Some of the potential hazards are from such operating errors as spraying in too weak black liquor causing quantities of water, from incomplete flash evaporation, to land on the red hot hearth supporting char and smelt. Or an explosive gas composition may be generated in the furnace by incorrect proportioning of combustion air relative to the combustible organics in the black liquor. In the corrosive gas atmosphere prevailing in a recovery furnace a water tube can spring a leak, with the same effect as spraying in a too dilute black liquor. Fortunately, the shell of the recovery furnace is stoutly built so that the minor bangs resulting are alarming but generally not damaging.

When smelt is dissolved in lime mud wash water the resulting solution is referred to as green liquor, from the colour imparted to it from suspended inorganic insoluble matter, or "dregs", present. The dregs, which consist mostly of carbon plus other inorganic insolubles such as iron, are settled out to leave a supernatant clarified green liquor which consists essentially of a solution of sodium carbonate and sodium sulfide in water.

For white liquor preparation, clarified green liquor is fed to a causticizer which forms sodium carbonate on the addition of lime, leaving the sodium sulfide unchanged (Equation 13.30).

$$Na_2CO_3 + Ca(OH)_2 + Na_2S \rightarrow$$
$$2\,NaOH + CaCO_3 \downarrow + Na_2S \tag{13.30}$$

The calcium carbonate (lime mud) is settled out from this mixture, filtered, and rinsed with water which is then subsequently used for dissolving smelt. The supernatant liquor is now a solution of sodium hydroxide, sodium sulfide, and small amounts of unreacted sodium carbonate of very nearly the correct composition required for a new batch of white liquor. Calcium carbonate in the lime mud is recycled by first calcining to decarbonate, followed by slaking to regenerate the calcium hydroxide required for a subsequent causticization (Equations 13.31, 13.32).

$$CaCO_3 \xrightarrow[1200\,°C]{} CaO + CO_2 \tag{13.31}$$

$$CaO + H_2O \rightarrow Ca(OH)_2 \tag{13.32}$$

Several important byproducts may be derived and are recovered from many kraft pulping operations, among which are kraft turpentine, tall oil, and dimethyl sulfoxide. Kraft turpentine is condensed from digester relief gas while pulping of coniferous species, pine for example yielding about 10 kg per tonne of pulp [2]. Kraft turpentine contains a number of diterpenes such as $\alpha$, and $\beta$-pinene, camphene, dipentene etc. similar to the composition of conventional turpentine as obtained from the dry distillation of pine wood [30] (e.g. naval stores industry). In its crude state as recovered it also contains water, volatile acids, and reduced sulfides including methyl mercaptan and hydrogen sulfide in particular. By careful washing and fractional distillation of this crude product several useful terpenes are recovered [30].

If tall oil recovery is to be effected, black liquor from softwood pulping is taken out at an intermediate stage of the multiple effect evaporation when it contains about 30 % total solids, and is allowed to stand [31, 32]. Tall oil recovery is generally uneconomic from spruce, or hardwoods. On standing, soaps (sodium salts of acids present) cream to the top of the vessel and are skimmed off, and the residual black liquor is returned to the evaporators. The soap yield, which can range from 10 to 200 kg/tonne of pulp [11] (or even higher for pine), is then acidified and the free fatty acids and resin acids obtained are separated by distillation. The fatty acids recovered consist mainly of oleic and linoleic acids and are employed in soap manufacture and as drying oil components of paints and varnishes [33]. Resin acids, which comprise primarily terpene acids such as abietic acid and its positional and reductive variants, are mainly employed in paper sizing.

Dimethyl sulfoxide $((CH_3)_2SO$; b.p. 189 °C) is made from dimethyl sulfide recovered by condensation from kraft digester relief gases. This useful water miscible industrial solvent is produced by air oxidation of the sulfide in the presence of catalytic concentrations of nitrogen oxides.

# 13.5 Bleaching of Wood Pulps

Mechanical and chemical (in particular kraft) pulps are normally more or less brown as initially obtained. The colour is due in part to residual lignin or chemically modified lignin and miscellaneous colour bodies present in the pulp. The objective of bleaching is to obtain the maximum increase in brightness possible with a minimum yield loss, consistent with the pulp end use and bleaching cost. Brightness is a term applied to the whiteness of a pulp, usually determined from a small paper sample or "handsheet" prepared by hand from a pulp sample. It is measured in a spectrophotometer

at a 45° angle to the incident beam of blue light of 457 nm, and is based on an arbitrary scale of 8 for the brightness of carbon black, and 100 for the brightness of pure magnesium oxide. On this basis, brightnesses of about 55 to 60 are common for unbleached mechanical pulps, which may be raised to the 65-70 range by bleaching. Unbleached sulfite pulps will have a brightness of about 60, and kraft pulps substantially less, but both may be raised to brightnesses in the 80 range, about the brightness of ordinary bond papers, with the appropriate number of bleaching stages. Very high brightnesses of 85 to 90 are achievable with chemical pulps using one or two additional bleaching stages. Because the bleaching procedures for the high lignin content mechanical, and chemi-mechanical pulps differs significantly from the procedures used with chemical pulps the two groups will be discussed separately. Neutral sulfite semichemical pulps are not normally bleached.

## 13.5.1 Bleaching of Mechanical and Chemimechanical Pulps

Mechanical pulps, since they comprise the whole wood put into a pulp suspension in water, have approximately the same lignin content as the wood from which they were derived and an as-pulped brightness usually in the 50 to 60 range. Either a chemically oxidizing or a chemically reducing system, or occasionally both are used with these pulps to gain a 10-12 point of improvement in brightness. The systems used are lignin-retaining bleaching methods, since the objective of these is to decolourize lignin without significant solubilization in order to maximize the yield of bleached pulp obtained.

Oxidizing systems are based on peroxide, and may use solutions of either sodium peroxide or hydrogen peroxide in water, usually with added disodium ethylenediamine-tetra acetic acid ($Na_2$ EDTA) to suppress metal ion catalyzed spontaneous decomposition of the peroxide. Capital P is the appropriate symbol used to designate a peroxide stage in bleaching. The active decolourizing agent is probably hydroperoxide anion, $HOO^-$, formed by an equilibrium between hydrogen peroxide and sodium hydroxide (Equations 13.33, 13.34).

$$H_2O_2 + NaOH \rightarrow HOO^- + Na^+ + H_2O \quad (13.33)$$
$$H_2O + Na_2O_2 \rightarrow HOO^- + 2Na^+ + OH^- \quad (13.34)$$

Peroxide induced lignin decolourization is achieved by hydroperoxide ion attack of the carbonyl carbon of quinones and ketones which form a part of long conjugated chains (the origin of a large part of the colour) in the lignin, thereby interrupting the conjugation and reducing the intensity and wavelength of the contributing colour [34, 35] (e.g. Equations 13.35, 13.36).

In this way the external, or peripheral colour-inducing functionalities are oxidized and decolourized over a bleaching period of 2-3 hours while the bulk of the lignin macromolecule remains intact so that a brightness improvement of 8 to 10 points is possible and yet the yield loss is kept low. Unfortunately peroxide-based bleaching systems tend to be expensive relative to other alternatives for groundwoods, and hence this method is generally only used for pulps destined for tissue papers, paper napkins, and some specialty papers rather than for newsprint applications.

Sodium bisulfite ($NaHSO_3$) was once used to some extent for reductive bleaching of mechanical pulps, but its action is slow and ineffective so that its use has been largely discontinued. Present reductive bleach systems for groundwoods employ either zinc or sodium hydrosulfite ($ZnS_2O_4$; or $Na_2S_2O_4$; sometimes referred to as dithionites) dissolved in water. The former system, developed in the mid 50's [10], was simply and inexpensively instituted by reacting powdered zinc with commonly available pulp mill chemicals (Equation 13.37).

$$Zn + 2\,H_2SO_3 \rightarrow ZnS_2O_4 + 2\,H_2O \qquad (13.37)$$

By treating pulp with this powerful reducing agent direct brightness improvements of 10 to 12 points are possible, particularly if care is taken to exclude oxygen. Otherwise a part of the active reducing capacity is wasted from the very rapid (probably only diffusion limited) reaction of zinc hydrosulfite with oxygen (Equation 13.38).

$$2\,ZnS_2O_4 + 2\,O_2 + 2\,H_2O \rightarrow$$
$$Zn(HSO_4)_2 + Zn(HSO_3)_2 \qquad (13.38)$$

But the disposal of waste groundwood bleach liquor from this particular bleaching reagent has posed toxicity problems for shellfish in particular [36], so that many mills have now switched to sodium hydrosulfite which is equally effective (although more expensive) and avoids this problem. Sodium hydrosulfite is normally made at the mill from purchased sodium borohydride (Equation 13.39).

$$NaBH_4 + 8\,NaHSO_3 \rightarrow$$
$$NaBO_2 + 4\,Na_2S_2O_4 + 6\,H_2O \qquad (13.39)$$

If both groundwood bleaching methods are to be used in sequence, a sulfur dioxide solution will be used as a stop bath for the pulp, after the peroxide stage, to ensure that any residual oxidants are kept out of the hydrosulfite stage. Combination bleaching will usually be conducted in this sequence not the reverse, which achieves less good results [10]. Brightness improvements of up to 16 points are possible in this way, better than possible with either bleach system alone but not quite a total additive effect. Because of the residual lignin still present in these pulps the brightness is not very stable, as observed by the significant darkening of a newspaper by 10 points or more if it is left exposed to sunlight for an afternoon, compared to another section of the same newspaper kept out of the sun.

$$(13.35)$$

$$+ \quad RCOOH \qquad (13.36)$$

## 13.5.2 Bleaching of Chemical Pulps

Chemical pulps destined for high brightness papers are cooked under such conditions that a relatively low residual lignin content is retained. Bleaching of these pulps involves selective solubilization and washing out of lignin in the early stages, causing some yield loss, followed by decolourization of any lignin residues present in the cellulose fiber which result in smaller yield losses. Thus, the more stages of bleaching required to obtain the required degree of brightness in the finished pulp, the greater yield loss is experienced.

The ease of bleaching of a chemical pulp is related to its initial colour and its lignin content. To establish this a preliminary permanganate number (also kappa, or K-number) test, specified as the number of mL of 0.10 N potassium permanganate which is decolourized by 1 g of dry pulp at 25 °C in 5 minutes [2], is run on the pulp to determine its bleachability. Easy bleaching pulps give permanganate numbers of 6 to 10 or less, whereas pulps which give permanganate numbers of 20 or more are generally classified as unbleachable.

The chlorination, or C stage of bleaching, which is conducted at a pH of about 2 by a solution of chlorine in water, serves to relatively rapidly chlorinate and demethoxylate lignin via the action of both dissolved chlorine and of undissociated hypochlorous acid [37] (Equation 13.40).

$$Cl_2 + H_2O \rightarrow HCl + HOCl \qquad (13.40)$$

A slower oxidation of lignin also occurs, serving in combination with the first two processes to yield more soluble fragments of lignin than the higher molecular weight forms originally present. Chlorination thus serves to accomplish a significant amount of lignin removal, though a kraft pulp does not appear to be a great deal brighter after this stage. Cellulose is not chlorinated and is more or less stable under these conditions, although there is some oxidative degradation, resulting mainly from the action of the undissociated hypochlorous acid inevitably present.

A caustic extraction, E, stage is the means by which a solution of about 0.5-0.7 % sodium hydroxide in water is used at 50-60 °C to remove solubilized lignin produced by either C, or H (hypochlorite) stages.The high pH conditions serve to convert any phenolic hydroxyls to their sodium salts, raising their solubility, and at the same time converts any residual fatty acids and rosin acids

present to salts, aiding in their removal by washing. Some hemicellulose removal is also experienced at this stage, even though this component is a desirable one for papermaking. Hemicellulose losses are minimized by keeping the caustic solutions dilute, but may be maximized, when required, such as for the preparation of dissolving pulps free of hemicellulose, by using 1 to 5 M (4 to 20 %) *cold* aqueous sodium hydroxide [10].

Hypochlorite, H, bleaching employs a solution of either calcium or sodium hypochlorite in water, made at the pulp mill by adding chlorine to a solution of lime (as a suspension) or sodium hydroxide in water (Equations 13.41, 13.42).

$$2\,Ca(OH)_2 + 2\,Cl_2 \rightarrow$$
$$Ca(OCl)_2 + CaCl_2 + 2\,H_2O \qquad (13.41)$$
$$2\,NaOH + Cl_2 \rightarrow NaOCl + NaCl + H_2O$$
$$\qquad (13.42)$$

The base used is employed in excess to ensure that the pH is kept at, or below 12, both to ensure that hypochlorite, $OCl^-$ (and not hypochlorous acid, HOCl), is the bleach active species (see Figure 13.10) and to stabilize the hypochlorite ion for this purpose. Both the C and H stages are designed to operate at pH's well clear of neutral to minimize the cellulose degrading action of undissociated hypochlorous acid. Hypochlorite ion is a less powerful oxidant than chlorine (or hypochlorous acid) and hence this reagent is an appropriately milder treatment suitable for later bleach stages. It primarily serves to oxidize and dissolve the chlorinated residues from the chlorination stage, the so-

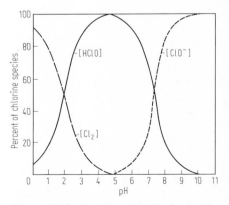

**Figure 13.10.** Variation with pH of proportions of chlorine, hypochlorous acid, and hypochlorite ion in 0.05 M chlorine water at 25 °C [37]

called "refractory material", which was not dissolved in the E stage. Some cellulose degradation is also experienced here.

A solution of chlorine dioxide, $ClO_2$, in water, referred to as a D bleaching stage, provides good bleaching action with less loss in pulp viscosity (i.e. less cellulose degradation) than some of the other procedures just described. Even solutions of chlorine dioxide are unstable so that this is usually made right at the mill site, for example by treating an aqueous solution of sodium chlorate with sulfur dioxide (Equation 13.43).

$$2\,NaClO_3 + SO_2 \rightarrow 2\,ClO_2 + Na_2SO_4 \qquad (13.43)$$

The byproduct sodium sulfate produced may be used to provide a part of the chemical make-up requirements for a kraft pulp mill. It is a milder oxidizing agent than hypochlorite and may be employed either on its own, as a separate bleach stage, or in combination with chlorine to replace a part of the stronger acting chlorine [38]. Chlorine dioxide may be economically produced at the same time as chlorine by employing sodium chloride or hydrochloric acid in the generators (Equations 13.44, 13.45).

$$2\,H_2SO_4 + 2\,NaClO_3 + 2\,NaCl \rightarrow$$
$$2\,ClO_2 + Cl_2 + 2\,Na_2SO_4 + 2\,H_2O \qquad (13.44)$$

$$4\,HCl + 2\,NaClO_3 \rightarrow 2\,ClO_2 + Cl_2 +$$
$$2\,NaCl\ (recycled) + 2\,H_2O \qquad (13.45)$$

Combination usage both gives pulps of higher strength than obtained using chlorine alone in the first stage [39], and greatly facilitates recovery of bleach plant effluents [40].

A peroxide, P, stage is also occasionally used as a polishing bleach for chemical pulps, using the same basis as already discussed [41]. Peroxide has also been tested for initial bleaching of softwood kraft pulps after an acid pretreatment, in the process achieving a significant reduction in waste loads but costing substantially more than conventional bleach sequences [42].

Oxygen bleaching, an O stage, has also become a part of the bleach sequence of many pulp mills. Delignification and decolourization are effected by treating the chemical pulp at medium to high consistencies (10 - 15 %, or about 30 % pulp by weight) in a dilute solution of sodium hydroxide, with up to 5 bars pressure of oxygen [43, 44]. Addition of a small amount of a magnesium or manganese salt serves to stabilize the cellulose

against severe degradation. One of the principal motivations for an oxygen bleaching step since it can replace a C E or C E H combination, is the decrease in waste load and toxicities possible by this means. It does, however, require rather different bleach plant equipment to put into practice.

Easy bleaching (low permanganate number) sulfite pulps can normally be bleached to a brightness of 80 or so with a single H stage, and to 85 or higher with a C E H sequence. Three to five bleaching stages, e.g. C E H, C E H D, C E H E D sequences are required for harder bleaching sulfite pulps [2]. To contrast with sulfite pulps, the normally dark kraft pulps can only be bleached to the 40 to 50 brightness range with a single hypochlorite stage. For semibleached grades of brightness in the 70 to 80 range requires at least three or four bleach stages, e.g. C E D, C E H D, are required for kraft pulps. A typical progression of pulp, chemicals, and water for a semibleached kraft pulp is given in Figure 13.11. To obtain high, or "super" brightness kraft pulps of 85-90 brightness levels requires five to seven stages of bleaching. Sequences such as C E D E D, C E H D E D, and C E H E D P are commonly used, requiring a period of 18 to 20 hours for passage of the pulp through the bleach plant.

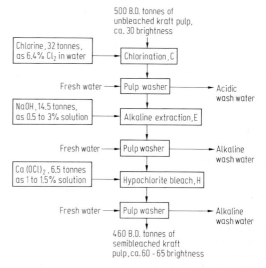

**Figure 13.11.** Progression through a typical C-E-H bleach sequence starting with 500 B.D. (bone dry) tonnes of unbleached kraft pulp

# 13.6  Market Pulp, and Papermaking

Bleached or unbleached chemical pulps may progress along either of two paths. It may either be fed as a dilute water suspension into a fairly coarse wire mesh to form a thick pulp mat, which is then dried to about 10 % moisture (roughly in equilibrium with atmospheric moisture), cut into sheets of about 0.6 x 1.2 m, and wrapped and baled for shipment as 232 kg (500 lb) units of market pulp. Or it may be fed in diluted form directly onto the Fourdrinier wire of a paper machine, with the intention of producing one of the wide variety of trade papers. The sheets produced for market pulp are deliberately made thick and only loosely consolidated to facilitate repulping by stirring with water, at the destination point, for eventual papermaking. Because mechanical pulps are much less colour stable and would not as readily take to repulping without a loss in sheet properties, these are nearly always converted to the finished product, newsprint or paper "boards" of various types, at the pulping site.

With any pulp source, eventual papermaking requires the appropriate single, or a blend of pulps to be prepared into a furnish, which is a dilute pulp suspension in water ready to be fed to a paper machine. Before papermaking, chemical pulps may be beaten (mechanically worked in units similar to refiners) to a varying extent depending on the properties desired in the finished sheet. Typical compositions of the slurry of about 2 % pulp in water that comprises the furnish for a newsprint might be a blend of 60-70 % groundwood (mechanical) + 30-40 % sulfite, semibleached, or bleached kraft pulps to provide strength. Bag papers and board, such as linerboard for corrugated board manufacture would normally use 100 % unbleached kraft pulp, hence the term "kraft paper". Board destined for use as corrugating medium would normally use a higher yield stiff pulp fiber such as produced by the NSSC process. Bond or book papers would use fully bleached sulfite, rag, or kraft pulps or blends of these.

The furnish is fed to the headbox of a paper machine (Figure 13.12), a large complex operation of the order of a city block in length. Sufficient additional water (white water) is added to bring the pulp concentration in the headbox down to about 0.5 %. White water comprises the water removed from the fibers as the paper sheet is formed, which is continuously recycled. This low concentration of pulp in water is necessary to obtain smooth formation, or evenness of fiber distribution as sheet formation occurs. The diluted furnish is fed through an adjustable horizontal slot (the slice) in the headbox onto an endless wire screen (the Fourdrinier wire) which is moving rapidly away from the headbox. Direct drainage plus suction boxes draw off more than 90 % of the water while the sheet is still on the wire, enough that the delicate 20 % solids sheet may be passed on to the press section of the paper machine, where the sheet is now supported on felts. By pressing the damp sheet into the drying felts the pulp content is nearly doubled, giving the sheet enough strength to carry it through the drying section. Here, 30 to 50 steam heated rolls rapidly evaporate the remaining water to give a sheet containing only 5 to 8 % moisture as it emerges from the dryer. The last function of the paper machine carried out on the still warm paper sheet is a smoothing one, performed as the sheet passes through the calendar stack. Here, heavy smooth steel rollers serve to decrease surface roughness and largely eliminate any residual "sidedness" of the sheet that occurred as it formed on the Fourdrinier wire [45]. Any adjustment of the final paper roll width, from the 7 to 10 m width produced by the typical paper machine is carried out as a separate step on a slitter-rewinder, using the 20 tonne rolls of paper obtained directly from the paper machine. The final step is a heavy paper wrapping, applied both to protect the contents from physical damage during shipment, as well as to avoid any change in the uniformity of the moisture content of the product.

Paper is an extraordinarily interesting commodity in the wide variations possible in all its properties, depending on the end use objectives. While superficially it appears to be uniform, it is actually significantly anisotropic. Even the best papers are usually sided, that is they have a wire side which is rougher, and for a newsprint will have a higher proportion of chemical fibers than will the smoother felt (or top) side of the sheet. The sheet will have a machine direction and cross direction to it, an anisotropy feature which can be adjusted somewhat. Tensile and folding strengths are greater in the machine direction, which corresponds to the direction of orientation of the majority of fibers, than in the cross direction [46]. This property is considered in printing, for instance, since books lie better with the pages having the ma-

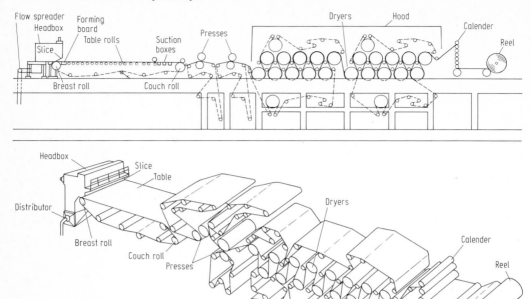

**Figure 13.12.** Principal components of a
Fourdrinier paper machine [11]. (Beloit Corp.)

chine direction up and down the page. There are all
sorts of other variable properties such as density,
thickness, porosity, smoothness, stiffness, strength
etc., as well as variations on the way each of these
properties are measured which is beyond the scope
of the present discussion [47].

## 13.7 Potential Pulping Emissions and Control Measures

Each of the pulping methods, mechanical, chemi-
mechanical (presoak), semichemical, and the kraft
and sulfite chemical processes produce different
types of wastes, so that for any consideration of
emission control measures each should be considered separately. Since mechanical, and the kraft
chemically-based pulping methods dominate the
world sources of pulp (Table 13.1) the discussion
will center on the considerations required for these
two processes, although many of the measures discussed will also be applicable to parallel problems
of other pulping methods. Pollution control
problems of the pulp and paper industry have been
relatively recently catalogued and surveyed [48,
49] and details of current research in the area reviewed [50] together with considerations of how

these considerations apply in a wide range of case
histories [102].

Wastes generated from wood preparation for pulping do not generally pose a major problem. Hydraulic debarker effluent amounting to some 14
m$^3$/tonne (3700 U.S. gal/tonne) is relatively clean,
containing only about 10 ppm suspended solids as
wood fines and grit and about 5 mg/L BOD
(biochemical oxygen demand) [49]. Primary clarification is required to decrease the suspended
solids, while the BOD level is regarded as inconsequential [48]. Toxic rosin acids flushed from the
bark/wood interface of some species are usually
dealt with by simple dilution with other mill effluents. Many mills have settled on drum or in-line
dry debarking systems, particularly when an inad-
equacy of water supply adds to the generally minor
problems of waste water treatment. For either type
of debarking system the bark itself is normally
burned in a bark fueled (hog fuel) boiler and energy
is recovered from the process as steam.

### 13.7.1 Effluents of Mechanical Pulping

Mechanical pulping processes, because of their nature, do not present any direct air emission problems. However, air emission problems are possible

at the point of electric power generation, if this is from combustion sources as it frequently is for pulp mill site generation facilities. Control of these potential emission sources is discussed later.

There is, however, an overall water usage estimated to be about 21 m³ per tonne (5,500 U.S. gal/tonne) of newsprint required not only for such applications as log or block transport and wet grinding operations but also for sheet preparation on the paper machine itself. The types and approximate levels of impurities present in the waste water stream are given in Table 13.9.

The philosophy of pulp mill emission control, particularly with valuable fiber constituents which may be lost in waste streams, is to practise fiber recovery as much as economically feasible before employing other measures to minimize waste impact on discharge. Recovery is obtained by sedimentation and/or flotation methods. Sedimentation, either in a pond, which is periodically drained and the accumulated settled material cleaned out, or via a clarifier (which operates in a similar manner to a thickener) produces a high water content largely combustible waste also containing a certain amount of grit. This 90-95 % water content waste is dewatered to 50-60 % solids in a filter or centrifuge and then burned [52], accomplishing both disposal and energy recovery.

Flotation methods allow recovery of air avid fibers as a thick mat which forms on top of an aeration cell. This mat, which is largely free of grit, may be returned to the pulp working stream as a useful papermaking constituent. Any problems with slime growths in the aeration cell, which can adversely affect the value of the recovered fiber are controlled by alternation of conditions such as pH or temperature, or by chlorination [49]. Although organomercurial slime control agents have been ex-

tensively used for this purpose this practice has now been discontinued [53], at least in Canada, Sweden, and the U.S.A.

After grit removal and fiber recovery, waste water streams from mechanical pulping are treated in aeration lagoons for reduction of the biological oxygen demand (BOD). By this means it has been found practical to achieve 40-75 % decrease of the BOD in 4 days and 90 % or better decrease in 7 days using a surface to volume ratio of about 1 hectare (1-2 acres) per million gallons per day of waste water flow [54]. Lagoon treatment with an adequacy of floating mechanical aerators, rather than activated sludge treatment, has been found to be the optimum solution from both a cost and efficiency standpoint for BOD reduction for this type of water stream [54]. The large volume and good mixing provided by this system can tolerate the shock loadings, which frequently occur in pulp mill operations, well. Added phosphate and nitrogen nutrient requirements are less, and sludge production is lower, but the land requirement is greater so that land cost may be a significant factor in the final choice of the method to be used.

## 13.7.2 Control of Emissions from Kraft Pulping

Kraft or sulfate pulping, involving as it does both vigorous chemical reactions under forcing conditions for lignin solubilization and a series of combustion and solution reactions for regeneration of pulping chemicals from spent liquor, has a good deal more complex emission control requirements than mechanical pulping procedures. These may be conveniently considered in the separate areas of air pollution control and water pollution control.

### 13.7.2.1 Air Pollution Problems and Control

Air pollution problems center around control of the formation and discharge of reduced sulfur compounds, which cause severe odour problems (Table 13.10), and also from the discharge of particulates. Reduced sulfur compounds, in particular methyl mercaptan and dimethylsulfide, arise from the action of hydrosulfide and methyl sulfide anions on the methoxyl groups present in the lignin of wood (Equations 13.46 - 13.48).

$$HS^- + Lignin\text{-}O\text{-}CH_3 \rightarrow Lignin\text{-}O^- + CH_3SH$$
$$(13.46)$$
$$CH_3SH + NaOH \rightarrow CH_3S^- + Na^+ + H_2O$$
$$(13.47)$$

**Table 13.9** Typical characteristics of the waste water produced by a groundwood pulping operation [a]

| Component | Value, ppm |
|---|---|
| Total solids | 1 160 |
| Suspended solids | 600 |
| Ash of above | 60 |
| Dissolved solids | 560 |
| Ash of above | 240 |
| BOD, 5 day | 250 |

[a] Data from [49].

**Table 13.10** Reduced sulfides contributing to the odour control problems in kraft pulping [a]

| Compound | Boiling point, °C | Henry's law constant at 38 °C [b] | Odour threshold, ppb | Industrial hygiene requirement, ppm |
|---|---|---|---|---|
| hydrogen sulfide, $H_2S$ | -62 | 700 | 5 | 10 |
| methyl mercaptan, $CH_3SH$ (methane thiol) | 6 | 180 | 5 | 10 |
| dimethylsulfide, $CH_3SCH_3$ | 38 | 50 | ca. 10 | – |
| dimethyldisulfide, $CH_3SSCH_3$ | 116 | 30 | ca. 100 | – |

[a] Compiled from [55, 111 and 112].
[b] Equilibrium mole fraction in the gas phase divided by the mole fraction in a water phase at 38 °C. If base is present the equilibria of the first two compounds will affect this ratio. $H_2S$, $k_1=2.1 \times 10^{-7}$, $k_2=< 10 - 14$; $CH_3SH$, $k=4.3 \times 10^{-11}$ for aqueous solutions at 100 °C.

$$CH_3S^- + Lignin\text{-}O\text{-}CH_3 \rightarrow$$
$$Lignin\text{-}O^- + CH_3SCH_3 \qquad (13.48)$$

In this way, about 3 % of the total sulfide loading to the digester of some 80-100 kg per air dry tonne of pulp is converted to methyl mercaptan and dimethylsulfide. Raising the sulfidity, the pulping temperature or the time, or when pulping hardwoods, all tend to be factors tending to produce proportionately more of these volatile sulfides [55]. Hardwood pulping, because of the much higher methoxyl content of hardwood lignins, nearly doubles the amount of volatile sulfides generated relative to that expected from the pulping of softwoods under the same conditions. Of the remaining sulfide in the liquor about 6 to 8 % becomes bound to lignin, mostly in soluble but invo-

latile forms, and the residue is predominantly converted to a variety of inorganic forms of oxidized, dissolved sulfur.

Escape of the volatile sulfides can occur from the venting of excess digester pressure, which occurs at intervals during the digestion, at the blow, when the contents of the digester are released to the blow tank, and during the last stage of black liquor evaporation in the direct contact evaporator, when there is rapid passage of a gas stream past the hot concentrated liquor. These, and more minor loss points are detailed in Table 13.11. Control measures for reduced sulfides center on sulfide containment from the major loss points, the direct contact evaporator and digester relief and blow gases. Black liquor oxidation, now almost universally

**Table 13.11** Total reduced sulfide (TRS) emissions from the chemical recovery operations of a kraft pulp mill in kg/S per air dry tonne of pulp produced [a]

| Source | Total reduced sulfide emissions, kgS/tonne | |
|---|---|---|
| | no controls | With controls |
| Direct contact evaporator | 7 – 10 | 0.05 – 1 |
| Recovery furnace | 0.1 – 1 | 0.05 – 1 |
| Digester and evaporators, non-condensible gas | 1.0 – 1.5 | 0.0 – 0.3 |
| Black liquor oxidation | 0.05 – 0.2 | 0.05 – 0.2 |
| Pulp washer hoods | 0.05 – 0.2 | 0.05 – 0.1 |
| Dissolving tank vents | 0.05 – 0.2 | 0.05 – 0.1 |
| Lime kilns | 0.05 – 1 | 0.05 – 0.1 |
| Totals | 8.3 – 14.1 | 0.03 – 2.8 |

[a] Compiled from data of [55].

practised (Figure 13.8), serves to decrease the volatile sulfide losses on direct contact evaporation to generally acceptable levels [56], although it has been found that for optimum benefit the time interval between oxidation and evaporation should be kept short [57]. Otherwise sugar-based reductants still present in the liquor cause re-formation of volatile sulfides on standing. Indirect black liquor evaporation for all stages, including the last one, has also been advocated by some Swedish mills.

Digester relief and blow gases are chilled, removing a "foul condensate" and the uncondensed gas fraction is moved on to a gas accumulator, which serves to smooth out the very wide range of gas flow rates from these sources. Venting this accumulator at a steady rate through a shower of weak black liquor having some reserve alkalinity serves to effectively capture at least hydrogen sulfide and methyl mercaptan. Any residual dimethylsulfide and dimethyldisulfide not captured may be destroyed by venting this into the combustion air stream of the lime kiln. Novel and effective alternative scrubbing techniques have also been described [58, 59].

Sulfide emissions may of course also be by-passed by a change of the pulping process to avoid the use of sulfide in the pulping liquor altogether [60]. Both the old soda process and nitric acid pulping [61] suffer from the fact that the pulp produced has inferior strength properties, but the employment of anthraquinone in conjunction with basically a soda pulping procedure promises to put this procedure on a much more competitive basis [62, 63, 64]. Particulate emissions can amount to as much as

0.25 tonne per tonne of pulp produced, particularly for an older pulp mill operating without control devices, as outlined in Table 13.12. Control measures involve installation of electrostatic precipitators or wet scrubbers, or occasionally both, on the recovery furnace flue gas discharge. These measures can effect containment of 95 % or better of the potential particulate discharge [58, 59]. The precipitator catch is returned to the black liquor stream for eventual chemical recovery from the inorganic chemicals present. Fumes lost from the dissolving tank vent are being captured by demister pads and small low energy scrubbers feeding the captured material to the green liquor circuit. Lime kiln dust containment is now generally via wet scrubbers, the scrubbate being used for slaking or other recycle functions in the lime circuit [65].

### 13.7.2.2 Water Pollution Control Practices

The principal waste water producing sectors of a large fully bleached kraft mill are summarized in Table 13.13 giving ranges of values both to cover normal operating variations as well as to reasonably represent a range of ages of pulp mills. Heat loadings through the discharge of hot waste water streams are now seldom a problem since heat exchange recovery is quite generally practised to reduce energy costs. Also the total volumes of water consumed are so large, 75 to 115 $m^3$ per tonne of unbleached kraft pulp and 150 to 208 $m^3$ per tonne for fully bleached kraft pulp, that discharge of the blended waste water stream is negligibly if any higher in temperature than source water.

**Table 13.12** Particulate emission loads from the chemical recovery operations of a kraft pulp mill, per air dry (10% moisture) tonne of pulp produced [a]

| Source control | Particulate discharge, kg/tonne pulp | | |
| --- | --- | --- | --- |
| | Particulate composition | No controls | With controls |
| Recovery furnace | $Na_2SO_4$, $Na_2CO_3$, $NaCl$ [a], etc. | 100 – 200 | 1 – 10 |
| Lime kiln | CaO dust | 10 – 20 | 0.4 – 2 |
| Dissolving tank vent | $Na_2S$, $Na_2CO_3$ $Na_2SO_4$, $NaCl$ [a] | 3 – 4 | 0.4 – 1 |
| Totals | | 113 – 224 | 1.3 – 13 |

[a] Salt build-up will affect coastal mills using tidewater transport and storage of logs, particularly when good particulate emission control is in place (see text).

**Table 13.13** Aqueous effluent characteristics and volumes per tonne of pulp (TP) from a fully bleached kraft pulp mill producing market pulp [a]

| Water quality parameter | Debarking | Pulping, screening | Chemical recovery | Causti-cizing | Bleaching 1st E | Bleaching All other | Paper Machine |
|---|---|---|---|---|---|---|---|
| Volume [b] m$^3$/TP | 7.5 – 15.0 | 22.7 – 45.4 | 3.8 – 13.2 | 1.9 – 7.5 | 19.0 – 38.0 | 56.8 – 94.6 | 1.9 – 5.7 |
| Suspended Solids kg/TP | 10 – 15 fiber, grit | 5 – 15 fiber | 0.5 – 1 | 1 – 10 lime | 1 – 2 fiber | 1 – 3 | 0.5 – 1 |
| pH | 6 – 8 | 6 – 9 | 6 – 9 | 8 – 10 | 9 – 11 | 2 – 3 | 5 – 6 |
| BOD$_5$, kg/TP | 3 – 4 | 6 – 15 | 1 – 2 | 1 – 2 | 5 – 10 | 10 – 12 | 0.5 |
| Colour APHA units | 300 – 700 | 1 000 – 1 500 | 200 – 400 | 400 – 600 | 20 000 – 30 000 | 1 000 – 1 500 | low |
| Foam [c] | occas. | occas. | occas. | none | yes, low | yes, high | none |
| Toxicity, Acute bio-assay | some | 2nd most | none [d] | none [d] | most toxic | 4th most | none |

[a] Compiled from personal contacts and the data of [48, 49 and 54].
[b] Converted to m$^3$ per tonne of pulp, by multiplication of U.S. gal values by 3.785 × 10$^{-3}$.
[c] Occas., short for occasionally. Low, and high refer to the relative ease of a foam being generated with these particular effluents.
[d] Effluent streams marked as having no toxicity are only thus if no condensates are added to these streams. Condensate addition contributes several toxic aromatic compounds to the stream.

Water quality parameters which may be affected by kraft process effluents, depending on the point of origin, include the process waste streams which are high in suspended and dissolved solids, highly acid or alkaline, and may have a high BOD. Other features which, without treatment, will have a significant impact on receiving waters include intensely coloured streams, the intensity of which is measured by comparison with an American Public Health Association (APHA) standard absorbance produced by a cobalt/platinum complex in water, or streams which have a tendency to produce a foam or that are highly toxic to fish, both effects produced by dissolved constituents present. Useful general reviews of waste water treatment for the pulp and paper industry have been published recently [66, 50]. One factor to bear in mind in considering the summary of waste water data tabulated is that the chemical recovery part of kraft pulping has already served to decrease potential effluent loadings of about 1,000 kg of BOD and 300 kg of dissolved inorganic chemicals per tonne of pulp which would have had to be coped with in the absence of chemical recovery, far greater than the pollutant loadings of even the worst streams for which data are listed.

Whether or not treatment is desirable or required for the particular component streams outlined depends on firstly, the degree to which the water quality parameters of the effluent stream depart from the normal situation in the receiving body of water, and secondly the gross effluent volume of any stream showing a significant variation of one or more parameters. One factor which should also be borne in mind in considering the treatment measures to adopt is that, while a low volume stream showing a gross deviation from normal in one or other water quality parameter may, if the volume is small enough, be adequately diluted on blending with 190 m$^3$ of process water to no longer pose a problem, it is very frequently easier to devise a treatment measure to improve the quality of a differentiated than a blended stream.

On the bases outlined above the main problem streams from a kraft pulping operation are those from the hydraulic debarker, the pulping and screening area, causticization, and both bleachery streams. Chemical recovery process waste waters are relatively benign, as long as relief gas or evaporator foul condensates are kept segregated from them [67]. Paper or pulp machine waters are generally clean as long as types of specialty papers involving use of other chemicals, dyes, pigments, or fillers are not being produced.

Suspended solids removal from the waste streams so affected is generally by use of a clarifier on each of these streams, while still segregated, to remove 90-95 % of settleable suspended solids [68]. Combustible sludges obtained are dewatered by filtration or centrifugation to 40-50 % solids or higher and then burned, or other means of disposal may be adopted [52]. Lime circuit clarifier sediments can be simply filtered and returned to the lime kiln for burning.

pH deviations from normal can be largely corrected by the blending of streams of opposing pH's as far as possible, *after* other required treatment steps. Final adjustment may be necessary through reagent addition, usually using 90+ % sulfuric acid for correction of excess alkalinity, and lime or 50 % sodium hydroxide for correction of excess acidity. An automated operating system has been described by Bruley [54].

Reduction of the biochemical oxygen demand (BOD) is desirable, particularly for waste streams where the loading is 3 to 4 kg per tonne of pulp, or more. In concentration terms, BOD's of 75 to 300 ppm (mg $O_2$/L) are common in kraft pulping effluents [49]. Biological waste treatment in artificially aerated waste lagoons, after adjustment of the water conditions to a pH in the 6.5 - 9 range, the temperature to 12-32 °C, and addition of ammonium phosphate to provide nitrogen and phosphorus nutrients, accomplishes accelerated BOD reduction, as outlined in more detail earlier. Use of oxygen, rather than air for treatment has also been advocated in this application [69, 70].

The highly intense brown colour, produced by ionized phenolics and condensed phenolics found primarily in the first caustic extraction or E stage, may be dealt with in a number of ways [71]. Liming greatly decreases the intensity of the colour by precipitating a lime/organics complex [72, 73], effective but producing combined organic-inorganic sludge which poses disposal problems in the attempt to not discard the lime. Activated carbon has also been used for colour adsorption [74], but carbon regeneration is difficult and not complete. It is however possible to operate using mill generated carbon [75]. Soil infiltration has been tested as a separate measure and found to be effective [76] but its applicability depends on the particular soil type and land availability features of the mill site. Reverse osmosis [77, 78] and ozone treatment [79] have also been suggested. The method of choice in any particular situation will depend on local factors.

The toxic properties of effluent streams may be contributed by sulfides, such as contained in the foul condensates of digester relief gas or the first stage of multiple effect evaporation, by long chain unsaturated fatty acids and the like [80, 81] released from wood components during kraft pulping, or by chlorinated phenols produced in bleaching steps [81, 82]. Foul condensates may be detoxified by air or steam stripping of the condensate, plus disposing of the volatiles so obtained by capture in black liquor, or combustion in the recovery boiler or incinerator. Normal biodegradiation procedures are usually sufficient to deal with the toxicants released from wood, but the chlorinated phenols are substantially resistant to short term biodegradation [83, 84]. Use of the foaming tendencies of these same effluents, however, to generate a head of foam in which the toxicants are concentrated has been found to be an effective method to remove these toxic compounds [85-87]. If liming is practised for colour removal, this also serves to decrease toxicities via adsorption onto the sludges as they come out of solution, and as is also feasible by adsorption onto activated carbon [74].

Changing the basic pulping technology used, such as by the utilization of oxygen or air pulping [88] and/or bleaching [89-91] could serve to avoid many of these problems. Total recycle of bleachery effluents as well as of pulping spent liquor has also been found to be a viable alternative [39, 92, 93], particularly in mill locations with a marginally adequate water supply.

Foam control can be achieved by adding commercial anti-foaming agents but many of these are also toxic, which could upset the operation of a biobasin or contribute a toxicant to the effluent. Foams can, of course, be stimulated to form in a plant treatment unit to achieve other objectives, then skimmed and burned or otherwise disposed of, in the process simultaneously getting rid of the foam and the toxicants (Figure 13.13). Some mills operating in cold continental climates find that a 0.5-1 m head of foam on a biopond can actually be an aid to maintaining optimum pond temperatures for continued BOD reduction under winter operating conditions. Otherwise, floating plastic foam mats may be required to keep pond temperatures high enough to maintain biochemical and biological activity.

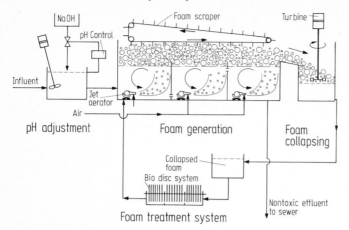

**Figure 13.13.** Proposed scheme for a foam production and separation device for the removal of both toxicants and foamable constituents from kraft process waste water streams [84]. Reprinted courtesy of Wheatlands Journals

## 13.7.3 Environmental Impacts of Papermaking and Paper Recycling

Waste water streams from paper machines while making papers composed entirely of wood pulp or while preparing market pulps generally do not pose serious waste loads on discharge. But while producing filled, coated, sized or coloured papers, whether coloured by pigments or dyes, the clay, coating materials, pigments etc. present in the discharge fraction of the whitewater have to be substantially contained or removed before water discharge. Many of these waste stream components are not only undesirable from their impacts on receiving waters but are also highly visible constituents of a waste stream. The volume of waste water required to be treated from this source is minimized by operating the paper machine using 90 % or higher water recycle [94, 95]. This measure also achieves a substantial direct saving of fillers, dyes, or the expensive titanium dioxide based coatings by capture of these constituents into the paper sheet as it is formed. Treatment measures for the rejected fraction of recycled water, necessary to avoid build-up of resinous components, dissolved solids, or the establishment of microbiological growths involves settling for waste streams containing primarily suspended solids, as already described. Activated carbon adsorption [96, 97] or one of the other colour removal techniques already described under pulping waste water emission control bay be employed for colour removal.

Fiber re-use by repulping of printers offcuts, computer and office paper wastes, as well as the paper component of municipal garbage is also an excellent way to decrease the overall impact of papermaking operations [98, 99]. Wood requirements may be greatly reduced, though not eliminated, since it is usual to blend virgin fiber with recycled stock. Also energy and chemical utilization are greatly decreased [100, 101] as are the waste loadings to air or water. However, recycling does not eliminate external impacts since significant waste loads to water are generated from the repulping, de-inking, washing, and simple bleach stages required [102].

Waste paper stock destined for recycling is normally segregated into grades corresponding to whether originally produced by chemical or mechanical methods. The separate stocks are then repulped in hot dilute sodium hydroxide, sodium phosphate or sodium silicate in water, and cleaned via a series of raggers, cyclones and screens. Froth flotation is used to remove most of the ink, via a pigment rich froth [103, 104]. After a further screening and wash, the stock is bleached with one or two stages of sodium peroxide, hypochlorite, or a chlorine dioxide rich mix of chlorine and chlorine dioxide [105, 106]. On completion of these steps there is perhaps a 15-20 % mass loss in reprocessing of chemical pulps and a 20-25 % loss from groundwood pulps [102, 107]. The product pulp may be incorporated into pulps prepared from new fiber, at varying proportions, to produce a sheet with very little difference in final properties from a sheet produced entirely from new fiber [108]. Fine papers and newsprint may be made from the cleaner grades of recycle fiber, and liner boards, corrugating medium and asphalt shingle stock are prepared from the less clean grades.

# Relevant Bibliography

1. Pulp and Paper: Chemistry and Chemical Technology, 3rd edition, J.P. Casey, editor, John Wiley and Sons, New York, 1979-1981, volumes 1-3
2. Kirk-Othmer Encyclopedia of Chemical Technology, 3rd edition, John Wiley and Sons, New York, 1982, volume 19
3. E. Horntvedt, A Sodium-base Sulfite Recovery Process Based on Pyrolysis, TAPPI 53(11), 2147, Nov. 1970
4. Pulp production: Sprucing up Old Technology, Chem. Eng. News 58(21), 26, May 26, 1980
5. J.P. Hanson, No-sulfur Pulping Pushing Out NSSC Process at Corrugating Medium Mills, Pulp and Paper 52(3), 116, March 1978
6. P. Uronen, Computer Control in the Recovery Area of the Kraft Process, TAPPI 61(11), 57, Nov. 1978
7. Is Cheap Recovery in Sight for Sulfite?, Can. Chem. Proc. 61(11), 32, Nov. 1977
8. Can Wetox Handle Spent Sulfite Liquor?, Can. Chem. Proc. 61(11), 20, Nov. 1977
9. D.W. Emmerson, Synthetic Papers and Oxygen Pulping Developments, Chem. In Can. 25, 21, Sept. 1973
10. J.C. Mueller, Fermentation Utilization of Spent Sulphite Liquor, A Review and Proposal, Pulp and Paper Mag. of Can. 71(22), T488, Nov. 20, 1970
11. W. Flaig, Slow Releasing Nitrogen Fertilizer from the Waste Product, Lignin Sulphonates, Chem. Ind. (London), 553, June 16, 1973
12. Lignin Conversion Process Shows Promise, Chem. Eng. News 58(44), 35, Nov. 3, 1980
13. J.J. Garceau, S.N. Lo, and L. Marchildon, A Cost Evaluation of Alternative Sulphite Spent Liquor Strategies, Pulp and Paper Can. 77(10), T174, Oct. 1976
14. D. Craig and C.D. Logan, Method of Producing Vanillin and Other Useful Products from Ligno-sulfonic Acid Compounds, Can. Pats. 615,552 and 615,553 to the Ontario Paper Co., Feb. 28, 1961
15. W.M. Hearon and C.F. Lo, Separating Phenols from Alkaline Pulping Spent Liquors, U.S. Pat. 4,208,350 to Boise Cascade Corp., June 17, 1980
16. Abstract Bulletin of the Institute of Paper Chemistry, Appleton, Wisconsin, volume 53, 1983, and earlier

# References

1. R.H. Clapperton, The Paper-making Machine, Pergamon, Toronto, 1967
2. J.P. Casey, Pulp and Paper Chemistry and Chemical Technology, 2nd ed., Interscience, New York, 1960, volume 1: Pulping and Bleaching
3. U. Mai Aung, Pulp and Paper Mag. of Can. 62(3), T230, March 1961
4. H.I. Bolker, Natural and Synthetic Polymers, An Introduction, Marcel Dekker, New York, 1974
5. E. Adler, Ind. Eng. Chem. 49(9), 1377, Sept. 1957
6. K. Freudenberg and A.C. Neish, Constitution and Biosynthesis of Lignin, Springer-Verlag, New York, 1968
7. Lignins, Occurrence, Formation, Structure, Reactions, K.V. Sarkanen and C.H. Ludwig, editors, Wiley-Interscience, Toronto, 1971
8. F.E. Brauns, The Chemistry of Lignin, Academic Press, New York, 1952
9. D.A.I. Goring, Tech. Section Proc., 49th Annual Meeting, Can. Pulp and Paper Assoc., Montreal 64, T-517, Jan. 1963
10. Handbook of Pulp and Paper Technology, 2nd edition, K.W. Britt, editor, Van Nostrand Reinhold, New York, 1970
11. Kirk-Othmer Encyclopedia of Chemical Technology, 2nd edition, John Wiley, New York, 1968, volume 16, page 680
12. D.J. MacLaurin and A.M. Van Allen, Paper Trade J., 1, Nov. 25, 1948
13. K.C. Logan, O.J.K. Sepall, J.L. Chollet and T.B. Little, Pulp and Paper Mag. of Can. 61(11), T515, Nov. 1960
14. O.L. Forgacs, Technical Assoc. Proceedings, Can. Pulp and Paper Assoc., 49th Annual Meeting, Montreal, 64, T-89, Jan. 1963
15. W.F. Holzer, J.T. Henderson, W.B. West, and K.F. Byington, Tech. Section Proceedings, Can. Pulp and Paper Assoc. 63, T-192, 1962
16. R.M. Dorland, D.A. Holder, R.A. Leask, and J.W. McKinney, Tech. Section Proceedings, Can. Pulp and Paper Assoc. 63, T-443, 1962
17. M.T. Neill and L.R. Beath, Tech. Assoc. Proceedings, Can. Pulp and Paper Assoc., 49th Annual Meeting, Montreal, 64, T-299, Jan. 1963
18. J.C. Gauss and D.J. Wachowiak, TAPPI 62(9), 27, Sept. 1979
19. Pulpers like TMP, Can. Chem. Proc. 61(11), 26, Nov. 1977
20. Mechanical Pulps, Can. Chem. Proc. 57(12), 44, Dec. 1973
21. Developments and Trends in Thermomechanical Pulping, Defibrator, Stockholm, 1980
22. Domtar is Trying TMP, Can. Chem. Proc. 58(3), 10, March 1974
23. F.A. Cotton and G. Wilkinson, Advanced Inorganic Chemistry, a Comprehensive Text, 3rd edition, Interscience, New York, 1972, page 545
24. T. Fossum, TAPPI 65(10), 69, Oct. 1982
25. A. Wong, TAPPI 65(10), 11, Oct. 1982
26. J.E. Laine, J. Turunen, and R.H. Rasanen, TAPPI 62(5), 65, May 1979
27. Alkaline Sulfite May Challenge Kraft, Can. Chem. Proc. 54(11), 56, Oct. 1970
28. Reference Tables, 35th edition, Canadian Pulp and Paper Association, Montreal, 1981

29. C.E. Rogers, H.P. Markant, and H.N. Dluehosh, Pulp and Paper Mag. of Canada *62*(C), T101, (1961)

30. J. Drew and G.D. Pylant, Jr., TAPPI *49*(10), 430, Oct. 1966

31. Crude Tall Oil Recovery, Tech. Section Proc., 47th Ann. Meeting, Can. Pulp and Paper Assoc., Montreal, Jan. 1961

32. Tall Oil, Can. Chem. Proc. *59*(1), 38, Jan. 1975

33. Tall Oil Fatty Acids, Chem. Eng. News *57*(40), 10, Oct.10, 1979

34. K. Kratzl, P.K. Claus, A. Hruschka, and F.W. Vierhapper, Cellul. Chem. and Tech. *12*, 445 (1978)

35. C.W. Bailey and C.W. Dence, TAPPI *52*(3), 491 (1969)

36. D.V. Ellis, P. Gee, and S. Cross, Water Poll. Res. J. of Canada *15*(4), 303 (1981)

37. T.P. Nevell, Chem. Ind. (London), 253, 15 March, 1975

38. A. Teder and D. Tormund, TAPPI *61*(12), 59, Dec. 1978

39. H. Rapson, C.B. Anderson, and D. Reeve, Pulp and Paper Mag. Can. *78*(6), T137, June 1977

40. W.H. Rapson, Pulp and Paper Mag. Can. *68*(12), T-635, (1967)

41. R.R. Kindron, TAPPI *62*(9), 67, Sept. 1979

42. M. Ruhanen and H.S. Dugal, TAPPI *65*(9), 107, Sept. 1982

43. L. Nasman and G. Annergren, TAPPI *63*(4), 105, April 1980

44. O. Samuelson and L.A. Sjoberg, TAPPI *62*(12), 47, Dec. 1979

45. B.I. Howe and J.E. Lambert, Pulp and Paper Mag. of Can. *62*(No.C) T139, (1961)

46. J.W. Swanson and E.J. Stone, Tech. Section Proc., Can. Pulp and Paper Assoc. *63*, T-251 (1962)

47. J.P. Casey, Pulp and Paper Chemistry and Chemical Technology, 2nd edition, Interscience, New York, 1961, volume II, Papermaking

48. Industrial Pollution Control Handbook, H.F. Lund, editor, McGraw-Hill, New York, 1971, page 18-1

49. N.L. Nemerow, Industrial Water Pollution, Addison-Wesley, Reading, Mass., 1978, page 439

50. G.W. Gove and I. Gellman, J. Water Poll'n Control Fed. *46*(6), 1235, June 1974

51. L. Allan, E. Kaufman, and J. Underwood, Paper Profits, Pollution in the Pulp and Paper Industry, MIT Press, Cambridge, Mass. 1972

52. T.R. Aspitarte, A.S. Rosenfield, B.C. Smale, and H.R. Amberg, Methods for Pulp and Paper Mill Sludge Utilization and Disposal, U.S. Env. Prot. Agency, Washington, 1973

53. C.T. Charlesbois, Three Perspectives on Mercury in Canada: Medical, Technical, and Economic, Science Council of Canada, Ottawa, 1976

54. A.J. Bruley, The Basic Technology of the Pulp and Paper Industry and its Waste Reduction Practices, EPS 6-WP-74-3, Environment Canada, Ottawa, 1974

55. K.V. Sarkanen, B.F. Hrutfiord, L.N. Johanson, and H.S. Gardner, TAPPI *53*(5), 766, May 1970

56. F.E. Murray, Pulp and Paper Mag. of Can. *72*, 57 (1971)

57. J.M. Bentvelzen, W.T. McKean, and J.S. Gratzl, TAPPI *59*(1), 130, Jan. 1976

58. A.J. Teller and L.L. Bebchick, Pulp and Paper Mag. of Can. *81*(12), T358, Dec., 1980

59. A. Wawer, Paper Trade Journal, 36, Dec.1, 1976

60. L.A. Cox and H.E. Worster, TAPPI *54*(11) 1890, Nov. 1971

61. W. Beazley, Paper Industry *39*(7), 596, Oct. 1957

62. H.H. Holton and F.L. Chapman, TAPPI *60*(11), 121, Nov. 1977

63. M.B. Hocking, H.I. Bolker, and B.I. Fleming, Can. J. Chem. *58*, 1983 (1980)

64. J. Blair, TAPPI *63*(5), 125, May 1980

65. P. Bruce, Pulp and Paper *52*(3), 180, March 1978

66. Proceedings of Seminars on Water Pollution Abatement Technology in the Pulp and Paper Industry, Report EPS 3-WP-76-4, Environment Canada and the Canadian Pulp and Paper Association, Ottawa, 1976

67. B.R. Blackwell, W.B. Mackay, F.E. Murray, and W.K. Oldham, TAPPI *62*(10), 33, Oct. 1979

68. E.G.H. Lee, Pulp and Paper Mag. of Can. *78*(9), 69, Sept. 1977

69. R.R. Peterson, J. Water Poll. Contr. Fed. *47*(9), 2317, Sept. 1975

70. M.V. Nelson, TAPPI *63*(3), 61, March 1980

71. R.J. Rush and E.E. Shannon, Review of Colour Removal Technology in the Pulp and Paper Industry, Report EPS 3-WP-76-5, Environment Canada, Ottawa, 1976

72. M. Gould, Chem. Eng. *78*, 55 (1971)

73. D.J. Bennett, C.W. Dence, F.L. Kung, P. Luner, and M. Ota, TAPPI *54*, 2019 (1971)

74. A. Wong, T. Tenn, J. Dorice, and S. Prahacs, Pulp and Paper Mag. of Can. *78*(4), T74, April 1977

75. New Scrubbing and Effluent Treatment Processes use Activated Carbon, Chem. In Can. *28*(11), 19, Dec. 1976

76. J.C. Mueller, B.C. Guidelines, B.C. Research, Vancouver, Feb. 1981

77. H. Beder and W.J. Gillespie, TAPPI *53*(5), 883, May 1970

78. H. Lundahl and I. Mansson, TAPPI *63*(4), 97, April 1980

79. Oxidation with Ozone, Can. Chem. Proc. *61*(5), 4, May 1977

80. J.M. Leach and A.N. Thakore, Prog. Water Tech. *9*, 787 (1977)

81. C.C. Walden and T.E. Howard, TAPPI *60*(1), 122, Jan. 1977

82. A.N. Thakore and A.C. Oehlschlager, Can. J. Chem. *55*, 3298 (1977)

83. J.C. Mueller and C.C. Walden, J. Water Poll. Control Fed. *48*(3), 502, March 1976

84. J.M. Leach, J.C. Mueller, and C.C. Walden, Proc. Biochem. *10*(1), 7, Jan./Feb. 1976

85. K.S. Ng, J.C. Mueller, and C.C. Walden, J. Water Poll. Control Fed. *48*(3), 458, March 1976

86. K.S. Ng, J.C. Mueller, and C.C. Walden, Can. J. Chem. Eng. *55*, 439, Aug. 1977

87. K.S. Ng, J.C. Mueller, and C.C. Walden, Pulp and Paper Mag. of Can. *75*(7), T263, July 1974

88. C. de Choudens and P. Monzie, Pulp and Paper Mag. of Can. *78*(6), T118, June 1977

89. Oxygen Treating, Can. Chem. Proc. *55*(4), 52, April 1971

90. W.L. Carpenter, W.T. McKean, H.F. Berger, and I. Gellman, TAPPI *59*(11), 81, Nov. 1976

91. A. Wong, M. Le Bourhis, R. Wostradowski, and S. Prahacs, Pulp and Paper Mag. of Can. *79*(7), T235, July 1978

92. D.W. Reeve, G. Rowlandson, J.D. Kramer, and W.H. Rapson, TAPPI *62*(8), 51, Aug. 1979

93. Great Lakes Ok's Effluent-free Idea, Can. Chem. Proc. *62*(10/11), 24, Oct./Nov. 1978

94. R.J. Spangenburg, Pulp and Paper *47*, 46 (1973)

95. L.C. Aldrich and R.L. Janes, TAPPI *56*, 92 (1973)

96. P.B. Dejohn and R.A. Hutchins, J. Amer. Asso. Text. Chem. and Colorists *8*(4), 34169, April 1976

97. Activated Carbon, Can. Chem. Proc. *58*(12), 24, Dec. 1974

98. P.E. Gishler, Recycling of Waste Paper, Research Council of Alberta, Edmonton, 1972

99. Recycling of Paper, Can. Chem. Proc. *55*(11), 58, Nov. 1971

100. J.R. Fooks and C.H. Langford, Envir. Letters *6*(3), 205, (1974)

101. J.R. Fooks and C.H. Langford, Chem. In Can. *27*, 16, Jan. 1975

102. T.E. Rattray, Paper Recycling in Canada, Environment Canada, Ottawa, Feb. 1974, ms., 22 pp.

103. L.H. Pfalzer, TAPPI *65*(10), 73, Oct. 1982

104. L. Pfalzer, TAPPI *62*(7), 27, July 1979

105. W.H. Rapson and C.B. Anderson, Bleaching Pulp Recovered from Municipal Garbage, Chem. Pulping and Bleaching Conf., Vancouver, Nov. 1971

106. J.P. Casey, Pulp and Paper Chemistry and Chemical Technology, 2nd edition, Interscience, New York, 1961, volume 2

107. D.W. Duncan, TAPPI *62*(7), 31, July 1979

108. De-inking Plant, Chem. In Can. *33*, Jan. 1981

109. 1979/80 Statistical Yearbook, United Nations, New York, 1981

110. J.P. Casey, Pulp and Paper Chemistry and Chemical Technology, Interscience, New York, 1961, volume 3

111. M.A. Karnovski, J. Chem. Educ. *52*, 490, Aug. 1975

112. R. Cederlov, M.L. Edfors, L. Friberg, and T. Lindval, TAPPI *48*(7), 405, July 1965

# 14 Fermentation Processes

## 14.1 Introduction and General Principles

Strictly speaking, fermentation is the process of anaerobic breakdown or fragmentation of organic compounds by the metabolic processes of micro-organisms. However, fermentation processes can be considered, in a more general way, to relate to the chemical changes of a substrate accomplished by selected micro-organisms or extracts of micro-organisms, to yield a useful product. This less specific definition includes micro-biological processes carried out under anaerobic (fermentative, or in the absence of air), aerobic (respiratory), and enzymatic (via extracts) conditions.

Historically, microbes of one kind or another have been used by man for centuries for the preparation of a variety of foods and beverages. Each fermentation product probably arose from what was originally an accidental discovery, for example grapes or grape juice fermenting on storage or mould growth on curds (milk solids). These were recognized as changes which increased the menu variety and the storage stability of the foodstuff, hence the appeal of the products of these processes. It is easy to speculate, then, that each accidental discovery led to the conducting of experiments aimed at being able to obtain the same change in the character of the food intentionally and at will. From probable early beginnings of this kind, yeasts have been used in the arts of wine, beer, and spirits production, as well as for the *in situ* production of the carbon dioxide used for leavening of bread and other bakery goods. Bacteria have been employed for the preparation of a variety of cheeses, sauerkraut, and yoghurt. Moulds too, still have a significant utility in the preparation of certain cheeses. However, it is only since about 1900 that fermentation processes have been employed on any sort of scale for the production of industrial products, in addition to foods.

The impetus for industrial utilization of micro-organisms is based on several factors favouring their use. Most importantly, a fermentation process is very often the only feasible way to obtain a complex, or a highly specific chemical transformation of the raw material. Also a chemical change conducted enzymatically (via micro-organisms or extracts) has a far lower energy barrier incurred by the process than a chemical synthesis route to the same change, meaning that the process may very often be carried out at ambient, or near ambient conditions. Under these conditions the enzymatic processes are $10^9$ to $10^{12}$ times as rapid as the corresponding chemical route to the same product [1, 2]. Another feature which is of particular importance with the production of some complex biologicals is that many enzymes are totally specific and stereoselective for the product of interest.

Under appropriate conditions the rate of reproduction of yeasts and bacteria is extremely rapid, of the order of minutes, so that there is little delay in building up the number of individuals present to maximize the conversion rate of raw material (Table 14.1). Their small size, and thus large surface to volume ratio, is also favourably disposed to permit relatively rapid diffusion of substrate into the cell, and metabolized product out. But utilization of micro-organisms for chemical change does introduce a difference in the profile of the rate of appearance of product. With a conventional batch reaction using only chemical starting materials, if all of the reactants are present in the reactor at the start of the process an initial maximum rate is observed, which gradually tapers to lower rates. In contrast to this, chemical conversions using microbes generally start relatively slowly while the number of organisms builds up during a reproductive phase from the small number present on initial inoculation (Figure 14.1). Following this lag phase or induction period, the maximum conversion rate of substrate to product is observed during the interval BC. The interval CD corresponds to the period when there is no further increase in numbers of organisms and the concentration of available substrate has decreased sufficiently to limit its availability to organisms, slowing its rate of metabolic conversion to product.

For large scale fermentations, in particular, the proportion of time during which a fermenter is

**Table 14.1** Salient details of the four principal groups of micro-organism useful in industrial fermentations [a]

| Class, complexity, and size | Reproduction | Occurrence, (common uses) |
|---|---|---|
| **Bacteria** | | |
| simple, single chromosome (prokaryote), usually single-celled organisms, 1 – 6 μm diam. spherical (coccus) or rod-like (Bacterium) in shape | asexually, by binary fission (simple cell division) | air, water, soil (sour milk, cheeses, yoghurt) |
| **Yeasts** | | |
| two or more chromosomes (eukaryote), single-celled near spherical, 5 – 12 μm diameter | usually asexual by "budding", under some conditions sexual (via spore cells) | air, water soil (beer, wine, spirits, bread) |
| **Actinomycetes** | | |
| multicellular organisms intermediate in complexity between bacteria and moulds, filamentous bacteria whose mycelium is readily fragmented | see bacteria/moulds | normally inhabit soil (antibiotics) |
| **Moulds (funguses)** | | |
| two or more chromosomes (eukaryote), readily visible multicellular filaments | varies, generally asexually via wind transmitted spores, and sexually via ascospore formation | air, water, soil (cheeses, antibiotics) |

[a] Compiled from [10, 11, 13 and 153].

producing under these high rate conditions should be large relative to the total fermenter operating time to minimize operating costs. That is, the time

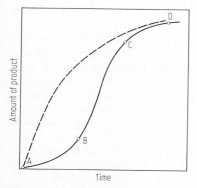

**Figure 14.1.** A qualitative comparison of production curves via a conventional chemical process (----) and a fermentation process (—)

period, BC, during which maximum production is observed should be large relative to the total of the remaining intervals that the fermenter is occupied, as given by Equation 14.1.

If interval BC >> interval (AB + CD + DA), then batch operation is generally favoured.

(14.1)

In this expression, the interval DA corresponds to the turnaround time, or time required to empty the fermenter, clean and sterilize it, refill it with fresh sterile substrate, and introduce a 'starter' of the appropriate organism. If this situation does in fact hold, then batch operation of the fermentation process is generally quite commercially acceptable [3]. If however, BC, the interval for maximum production rate, is short relative to the total fermenter operating time (Equation 14.2),

If interval BC < interval (AB + CD + DA), then continuous operation is generally favoured.

(14.2)

then in general there is a strong incentive to try to develop a continuous, rather than batch mode of operation for the process. A continuous mode of operation would be particularly favoured in this situation for very large scale fermentations. In a continuous mode, which is generally accomplished by staging the fermentations in several vessels rather than in a single one, the use of the high fermentation rate continuously in one or more of the vessels thus obtains the most efficient overall rate of conversion of substrate to product.

Whilst there are many problems attached to operation of fermentations on a continuous basis, not the least of which are the potential for genetic variation with time of the micro-organism itself and the maintenance of the ideal conditions for growth [4], there are nevertheless several fermentation products which are produced on a commercial scale using a continuous mode of operation. Baker's yeast [5], yeast from petroleum and petroleum products, and vinegar are all presently in large scale production using continuous culture methods [3, 6]. Even in the highly market sensitive brewing [3, 7] and wine making [8, 9] industries, which have traditionally operated exclusively by batch fermentation and are still dominated by this mode of operation, continuous mode fermentations have now been proved to be feasible. Of course waste water treatment plants also operate in a continuous mode. Waste water treatment is also an example of a fermentative process, but one which employs a wide variety of micro-organisms to accomplish the collective changes desired, a non-specific conversion.

The requirements for producing a specific product via an industrial process using fermentation are, first and foremost, a culture of a specific micro-organism that produces the desired end product [10, 11]. This is usually the dominant product, such as during the anaerobic production of ethanol (ca. 95 % selective) from sugar by yeasts, in which case the process is termed homofermentative. On the other hand fermentations which produce several products, such as the anaerobic production of butanol (60-70 %), acetone (20-30 %), and ethanol (ca. 10 %) from sugars by *Clostridium acetobutylicum* [12], may also be useful if all the products are recoverable and saleable. In this case the process is termed heterofermentative. Other fermentation requirements are that the microbe must be relatively easy to culture, and to preserve in a dormant state for reference and stock renewal

purposes. It should also be able to be easily maintained in an active state having consistent genetic characteristics. Raw materials appropriate for conversion must be economic for the end product desired, i.e. sugars and starches for beverage alcohol, low cost starches or cellulosic substrates for fuel alcohol, etc. Acceptable to good yields should be produced, based on the relative prices of substrate and end product. The fermentation should proceed rapidly under conditions (pH, temperature, pressure, nutrients, etc.) that can be relatively easily provided. Finally, it should be possible to recover the product of the fermentation in a straightforward manner.

Many of the requirements summarized above are quite general ones for any viable industrial process. The key difference in these requirements is the participation of the micro-organism in the process, the locating, selection, and improvement of which is primarily the province of microbiologists [13]. Cultures of the desired strains are maintained in dormant state and supplemented under scrupulous conditions. Portions of the cultures can then be taken, and provided with ideal propagating conditions to build up the numbers of organisms available to provide a 'starter' for inoculation to the medium of a full scale fermenter. The medium, or growth environment, will consist of a solution of the primary substrate to be converted, usually in water. It will also have small amounts of nutrients, such as ammonium phosphate, to provide a source of nitrogen and phosphorus, and traces of other minerals such as may be required for favourable microbial growth and reproduction. After pH adjustment, and sterilization of the medium to avoid interference from any competing organisms, the temperature will be brought to the appropriate range and the starter added.

One remaining consideration affecting the mode of operation of industrial fermentations concerns whether it is an aerobic or anaerobic process which is required. Some micro-organisms can only function under aerobic conditions (in a respiratory mode) and are referred to as obligate aerobes. Others, obligate anaerobes, can only function in the absence of air (in a strict fermentative sense). Many species of these obligate anaerobes are so sensitive to air that they are killed on air exposure. Yet a third group are capable of functioning under either anaerobic or aerobic conditions and are referred to as facultative micro-organisms. The metabolic processes and end products of facultative organisms may be significantly altered depending on whether the conditions provided are anaerobic or aerobic. Thus, both the range of substrates which may be utilized, and the product(s) obtained may be changed depending on the metabolic mode forced onto the organism employed by the conditions provided. Yeasts, for example, in the presence of air ferment simple sugars to carbon dioxide and water, yet under anaerobic conditions produce ethanol instead.

Details of the particular type of industrial fermentation vessel to be used will be determined, among other things, by whether the fermentation is to be conducted aerobically or anaerobically. It should be noted that even anaerobic fermentations, such as are used for alcohol production, can be conducted in open-topped fermenters when these are located inside a building. In this situation, the top of the fermenting medium is protected from any significant contact with the air by the blanket of co-produced carbon dioxide, which is heavier than air. But generally closed top fermenters are preferred because of the reduced risk of contamination of the process by unwanted organisms.

## 14.2 Brewing of Beer

Brewing, a term generally reserved for the large or small scale manufacture of beer, has been practised for more than 8000 years. Certainly beers, beverages produced from a fermented brew of cereal(s) and hops, have been produced at least since the dawn of recorded history and probably well before that [14-17]. For many countries, dominated by nations with a temperate climate, brewing represents the sector of the fermentation industry with the highest volume of product and the highest value of annual sales (Table 14.2). In keeping with this, beer production also occupies a larger volume of total fermenter capacity than any other specific fermentation product, for example a gross of 128,000 $m^3$ for the U.K. [1]. Beer production, even on a global basis, thus represents the most important sector of beverage alcohol production in terms of both volume and value of product. In the U.K. the only sector exceeding brewing, in terms of total fermentative capacity, is the 2.8 million cubic meters utilized in total for waste water treatment.

Beer, being a perishable product, is largely consumed in the country of origin. On this basis Australia, Czechoslovakia, and West Germany show the highest per capita production [18], producing over 140 L per person per year (Table 14.2). In contrast to this, the countries with a low level of production average out at less than 5 L per capita. The manufacture of beer is an example of an anaerobic fermentation process which uses varieties of yeast to produce a beverage of low alcohol content from basically a solution of simple sugars in water. To avoid problems of corrosion, or product contamination from metals, all brewery processing is conducted in glass-lined or epoxy resin-lined steel, or stainless steel, except perhaps for the cereal cooker where copper is occasionally used to obtain more efficient heat transfer. Three types of beer are recognized, based on their respective alcohol content, specified by volume. Light beers are those containing less than about 4 % ethanol. Regular beers run around 5 % alcohol, and the malt liquors have an alcohol content in the 5.5 to 6.5 % range.

**Table 14.2** Annual production of beer by the major world producers [a]

| | 1968 Totals million L | 1978 Totals million L | 1978 Liters per capita[c] |
|---|---|---|---|
| Australia | 1 408 | 2 001 | 148 |
| Belgium | 1 189 | 1 340 | 139 |
| Canada | 1 506 | 2 129 | 93 |
| Czechoslovakia | 2 010 | 2 206 | 154 |
| France | 1 996 | 2 278 | 43 |
| GDR | 1 501 | 2 230 | 131 |
| FRG | 7 373 | 8 792 | 145 |
| Japan | 2 428 | 4 421 | 40 |
| Mexico | 1 252 | 2 257 | 47 |
| Netherlands | 685 | 1 465 | 112 |
| Poland | 945 | 1 138 | 35 |
| Spain | 1 026 | 1 735 [b] | 51 |
| U.S.S.R. | 3 830 | 6 414 | 24 |
| United Kingdom | 5 143 | 6 642 | 120 |
| U.S.A. | 13 790 | 18 000 [b] | 89 |
| Yugoslavia | 475 | 1 001 | 49 |
| Other | 10 125 | 16 431 | 4.9 |
| World | 56 682 | 80 480 | 18.5 |

[a] All countries producing more than ten billion ($10^{10}$) liters annually in 1978, from Statistical Yearbooks [154].
[b] Estimated.
[c] Calculated from 1978 production divided by 1979 population.

In outline the aqueous fermentation substrate for beer is made by brewing cereals with hops, which provide the characteristic bitter flavour component of the product. Barley is the chief cereal ingredient which is first malted, both to give enzymes required to break down starches to simple sugars and also to contribute an important component of the flavour. But barley, being a fairly expensive source of starch, is usually supplemented by the addition of starch adjuncts such as corn (maize), oats, millet, wheat, or rice [19]. The adjunct makes a relatively small direct contribution to the flavour but has a greater effect from the dilution of the strong malt flavour that it replaces. Various strains of yeasts are used in the brewing industry which are selected by the industry as a whole for their efficient conversion of glucose to ethanol, and further developed by individual brewers to achieve the particular product qualities desired.

## 14.2.1 Details of the Steps Involved in Brewing

Initially the barley component of the brew is malted, which usually involves three steps. In the first step, cleaned, graded whole barley grains are soaked in cold water until the desired moisture content, 44 to 48 %, has been reached. Then, in the second stage, the grains are spread out to germinate. During the germination step, which requires a period of 5 to 8 days, the enzymes diastase and maltase (among others) are formed by the sprouting seeds for the primary purpose of mobilizing the starch reserves in the seed for growth utilization. Diastase promotes the hydrolytic conversion of starch, a polysaccharide, into the disaccharide, maltose. This hydrolysis is accomplished relatively easily since starch is a polymer (ca. 2,000 units) of the alpha anomer of glucose, in which the glucosidic links are relatively accessible to water. Maltase in turn rapidly hydrolyzes maltose, a disaccharide of two glucose units, to glucose itself. Once these enzymes have formed in the germinating barley it is heat-killed and dried in a current of hot air. This product, 'light malt', is ready for incorporation into a brew. Or, as a third step in malting it may also be cooked or roasted in a kiln forming a dark or 'black malt', to contribute additional colour and stronger flavour elements to particular brews of beer in addition to its primary function of providing starch-hydrolyzing enzymes. Careful control of temperature, humidity, and air flow is required

through all these malting stages to produce an acceptable product.

The considerable time and space required for operating the malting step, and the degree of industry standardization of this process has led to the virtual take over of this step today by large, centrally located malting companies. Dried malt of various grades is then shipped from the central processor on order, as required by the various brewers.

Mashing, the next step in the process, requires the grinding of the dried malt and the starch adjunct (malt surrogate), which may consist of up to 25 to 30 % of the total starch. Some of the ground malt and the whole of the adjunct are then mixed with water and boiled for half an hour in a cereal cooker to convert insoluble to soluble starch (Equations 14.3, 14.4).

$$\text{insoluble starch} \rightarrow \text{soluble starch} \quad (14.3)$$
$$\text{(intact cell membranes)} \quad \text{(ruptured cells)}$$

$$(C_6H_{10}O_5)_n \rightarrow (C_6H_{10}O_5)_n \quad (14.4)$$
$$\text{intracellular} \quad \text{extracellular}$$

Meanwhile, in the mash tun (tub, or mixer), the remainder of the malt is suspended in water and warmed to about 50 °C. The soluble starch in water from the cereal cooker, while still hot, is then added to the mash tun containing the warmed malt in water to produce a resultant temperatures of 70-75 °C. At this temperature (*not* higher or enzymes are destroyed) about half an hour is provided for hydrolysis of the soluble starches to maltose and glucose (Equations 14.5, 14.6).

$$\text{soluble starches} \xrightarrow[\substack{\text{amylases,} \\ \text{diastase}}]{\text{hydrol.}} \text{disaccharide}$$
$$\text{(extracellular)}$$

$$\xrightarrow[\text{maltase}]{\text{hydrol.}} \text{monosaccharide} \quad (14.5)$$
$$\text{(simple sugar)}$$

$$(C_6H_{10}O_5)_n \xrightarrow{^n/_2\,H_2O} {^n/_2}\,(C_{12}H_{22}O_{11}) \xrightarrow{^n/_2\,H_2O}$$
$$\text{starch} \qquad\qquad \text{maltose}$$
$$n\,C_6H_{12}O_6 \quad (14.6)$$
$$\text{glucose}$$

It is estimated that under these conditions the mashing step effectively converts 75 to 80 % of the starches present to simple sugars [20]. Higher mashing temperatures tend to give lower proportions of sugars to dextrins, and lower mashing temperatures tend to give higher proportions of sugars

100 glucose units) and undissolved fragments of grain, malt, and husks (Table 14.3). The hot wort is first lautered, a coarse filtration step, by passing it into a vessel with a perforated false bottom (lauter tun) which retains any undissolved grain, malt, and husk fragments. Filtered wort is then boiled for a period of 2 to 3 hours, hops being added after the first hour, usually in several portions. Hops, which consist of the bracteole or cone of a vine *Humulus lupus,* provide the characteristic bitter flavour component of a beer contributed by the several humulones and related compounds [7, 22-24] (Equation 14.7) which are extracted from the hops during this boiling step.

**Plate 14.1.** Stainless steel, filter-press style wort cooler for beer production

to dextrins [7, 21]. This step is sometimes referred to as saccharification, literally sugar formation.

The product of the mashing stage, called a wort, consists of varying proportions of maltose and glucose, flavour constituents from the malt, plus dextrins (incompletely hydrolyzed starches of 30-

humulone

lupulone

(14.7)

This boiling period not only extracts the hoppy flavour constituents, but also serves to destroy residual enzymes, coagulates undesirable proteins, and on completion provides a sterile medium for subsequent fermentation. After boiling, the hot wort is allowed to settle to remove the coagulated proteins while it is kept covered to avoid contamination by any unwanted organisms. Subsequently it is cooled through a plate cooler (heat exchange to water) to 8-11 °C, the product at this stage being referred to as hopped wort.

Before, or immediately after the yeast is added for fermentation the hopped wort is oxygenated briefly. This assists in starting the fermentation process more rapidly once the brew is placed under anaerobic conditions. A period of 6 to 7 days depending on the particular brew being processed, is required to complete the fermentation once the yeast is added. During the active phase this is ac-

**Table 14.3** Breakdown of the principal carbohydrate constituents of a normal wort used for making beer [a]

| Component | Concentration g/L |
|---|---|
| maltose | 57.7 |
| tri- to nonasaccharides | 22.8 |
| higher carbohydrates | 10.8 |
| Total carbohydrates (91.0% of extract) | 101.1 |
| Fermentable carbohydrates (69.7% of extract) | 77.4 |

[a] Data selected from the more detailed listing given in [7].

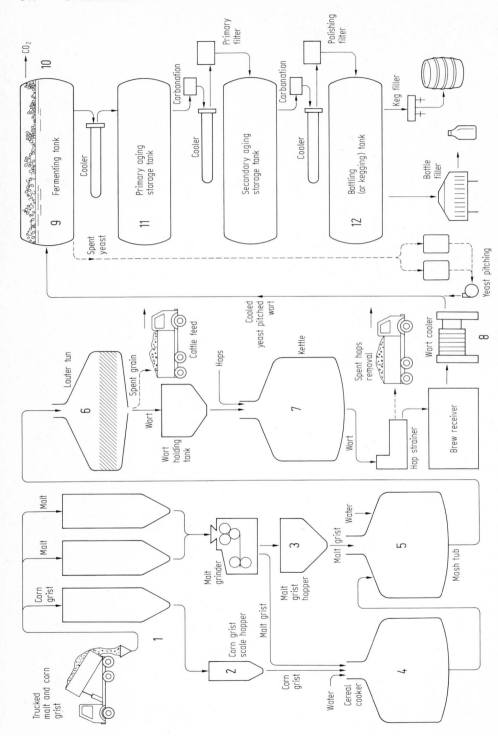

**Figure 14.2.**  The principal steps of beer production in a modern, large scale brewery [20]. Reprinted courtesy of Southam Business Publications, Ltd.

complished by a yeast count in the fermenting wort of about 12 million/mL, originally added as a starter of yeast cells in a propagating suspension of about 1 % of the volume of hopped wort to be fermented.

Heat evolution of 74.5 to 93.3 kJ, depending on the source of combustion data used [25, 26], is evolved for each mole of glucose fermented to ethanol and carbon dioxide (Equation 14.8).

$$C_6H_{12}O_6 \xrightarrow[\text{(yeast)}]{\text{zymase}} 2\,CH_3CH_2OH + 2\,CO_2 \quad (14.8)$$

$$\Delta H = -74.5 \text{ to } -93.3 \text{ kJ } (-17.8 \text{ to } -22.3 \text{ kcal})$$

The heat evolution on maltose fermentation, which involves a preliminary mildly exothermic hydrolytic step to glucose before alcohol formation (Equation 14.9), can be approximated by multiplying the figure taken for glucose by two. A closer estimate may be obtained by adding the further 15.9 kJ/mole required for maltose hydrolysis [26] to this number.

$$C_{12}H_{22}O_{11} + H_2O \xrightarrow{\text{maltase}} 2\,C_6H_{12}O_6 \quad (14.9)$$

$$\Delta H = -15.9 \text{ kJ } (-3.8 \text{ kcal})$$

Fermentation temperatures of 15 to 18 °C are desirable for beer production. Above this range some flavour elements are lost and head (froth) formation of the finished beer is poor. Therefore, brewers normally use refrigeration to maintain fermentation temperatures in this optimum range.

Yeasts used in brewing (and for winemaking and spirits production) are homofermentative, that is, under anaerobic conditions they are about 95 % selective for ethanol as the end product of glucose metabolism [27]. Thus Equations 14.8 and 14.9 represent a quite close approximation of the overall chemical change accomplished by fermentation, even though they do not do justice to the quite complex sequence of biochemical steps required to achieve these net chemical changes [13]. Using these equations it is possible to calculate the approximate alcohol concentration to be expected on fermentation of any given initial concentration of fermentable sugars. During the fermentation step, virtually all of the fermentable sugars present, comprising about 2/3 of the original carbohydrate mashed, is converted to ethanol and carbon dioxide (Table 14.3). Tri-, and higher polysaccharides which are not fermentable by yeast remain as a dissolved carbohydrate fraction in the beer. Most breweries collect the carbon dioxide formed during

fermentation and clean it after compression by passage through two (or more) stages of active carbon adsorption followed by a water wash. The 99.99 % pure carbon dioxide is then liquefied for storage for later use as needed in packaging, inert gas blanketing and the like. In continuous mode brewing systems part of the carbon dioxide produced is used to keep the yeast and the wort agitated and moving during the fermentation step [7].

After fermentation the brew is separated from the yeast by centrifuging, or by decantation through an outlet raised slightly above the bottom of the fermenter, which leaves nearly all of the yeast behind as a 'heel' in the fermenter. Each fermentation produces enough yeast for the pitching (starting) of five further brews of equivalent size [18]. After careful checking for strain purity and freedom from mutations the middle portions of this yeast will be reused, sometimes for as long as three months or 8 to 12 fermentations, before it is arbitrarily discarded and a starter freshly prepared from pure culture stock used to re-initiate the whole process. The decanted fermented brew is filtered and the opalescent product at this stage is referred to as 'green' beer.

Green beer requires aging at temperatures as near 0 °C as possible for a period of 2 to 6 weeks, usually with at least one intermediate decantation plus filtration. During this period some further protein settling occurs, clarifying the product, and a certain amount of flavour mellowing (maturation) is achieved to produce a more palatable product. The aged product, with an approximate analysis as given in Table 14.4, is now ready for final packaging and marketing.

Finished beer is marketed in two forms, bottled (or canned) and draught. The proportions of each sold varies widely depending on market area. In Canada over 95 % bottled beer is sold in Quebec and only about 80 % bottled in Alberta. Prior to packaging of either form of beer in North America, the mature product is usually given a final polishing filtration to yield the product of high brilliance (clarity) demanded by the market. In Europe some opalescence (haze) is generally accepted by customers, so that the final product is not usually filtered before packaging. The product is carbonated to the extent of about 0.4 % carbon dioxide by weight (2.6 to 2.9 % by volume) [28], using a part of the purified carbon dioxide produced by fermentation. It is then loaded into bottles previously flushed out with carbon dioxide if permitted by the bottling

**Table 14.4** Composition by weight and other properties of a typical finished beer [a]

|  | Composition [b] | Per 341 mL bottle[c], approx. |
|---|---|---|
| Alcohol content, by weight | 3.8 – 3.9% | 11.3 g |
| by volume | ca. 5.2% | – |
| Degree of fermentation | 66% | – |
| Density | 1.003 – 1.007 g/mL | – |
| Total unfermented carbohydrates | 4.0 – 4.6% | 12.6 g |
| fermentable sugars (as maltose) | 0.90% | 2.6 g |
| Water | 92% | 270 g |
| Proteins (as N, × 6.25) | 0.30 – 0.50% | 1.3 g |
| Humulones | ppm range | – |
| Minerals, ppm |  |  |
| calcium (as Ca) | 40 | 15 mg |
| phosphorus (as P) | 150 – 250 | 93 mg |
| Vitamins ppb |  |  |
| thiamin (vitamin B-1) | 20 – 90 | 19 μg |
| riboflavin (vitamin B-2) | 200 – 1 300 | 270 μg |
| pantothenic acid (vitamin B-5) | 1 000 | 340 μg |
| pyridoxin (vitamin B-6) | 400 – 800 | 205 μg |
| nicotinamide | 10 000 | 3 400 μg |
| inositol | 29 000 | 9 900 μg |
| biotin (vitamin H) | 10 | 3 μg |
| Other properties |  |  |
| pH | 4.1 – 4.4 | – |
| food energy (as alcohol, carbohydrates, proteins) | 1 780 – 1 970 kJ/L (425 – 470 kcal/L) | 607 – 672 kJ (145 – 160 kcal) |

[a] Data compiled from [20, 155 and 156].
[b] By weight, except where otherwise stated.
[c] Equivalent to a 12 ounce bottle.

equipment used. Following bottling, the shelf life of the bottled product is improved to about 3-4 months by pasteurizing for 6 to 10 minutes at 60 °C. After pasteurization a simple residual 'head' check determines the quality of the seal obtained on capping, and any imperfects are discarded. Optimum product shelf life is obtained under cool, dark storage conditions.

Draught beer is produced from the same brew, but has a lower level of carbonation, 2.2 to 2.4 % carbon dioxide by volume. In Canada it is packaged in aluminum or stainless steel 'half barrels' of 12 $^1/_2$ Imperial gallons capacity, and is not pasteurized [29]. Thus, draught beer has a much shorter shelf life than bottled beer, about 21 days with refrigerated storage, so that a brewer apportions his newly-brewed product appropriately to ensure that all that is packaged as draught is consumed within this period. The incentive for the brewer to use the draught package lies in the very much lower packaging cost in this form. In return,

consumers enjoy a flavour unaffected by pasteurization and less affected by the higher carbonation levels required for bottled beer. Even for a large scale brewery-bottling plant, the cost of bottling very often exceeds all the other costs of brewing.

### 14.2.2 Product Variety and Quality

Apart from the small difference in the product due to the method of packaging, bottled or draught, there are many other factors which influence the quality and type of product emerging from the brewing process. To start with, the water quality and dissolved ion content are highly important, both for the chemistry of the starch conversion step and in the flavour of the finished beer [7]. Some very soft waters may actually require addition of small amounts of calcium chloride and/or sulfuric acid, or sometimes gypsum ($CaSO_4$ × $2 H_2O$) to produce an adequacy of dissolved ions and a sufficiently low pH for a good product.

A breakdown of a conventional proportioning of

**Table 14.5** Major ingredients and appropriate current costs of brewing, per Canadian barrel and per hectoliter of beer [a, b]

|  | Per Canadian barrel | Per hectoliter (100 L) |
|---|---|---|
| brewer's malt | 34 – 37 lb | 11 – 12 kg |
| malt adjunct(s) | 12 – 14 lb | 4 – 5 kg |
| hops | 0.27 – 1.5 lb | 0.090 – 0.50 kg |
| yeast | ca. 1 lb | ca. 0.135 kg |
| Production costs: |  |  |
| labour | $ 4.50 | North |
| power, steam | $ 4.00 | American |
| bottling | $ 6.00 | cost. estim. |

[a] Canadian barrel of beer = 25 Imperial gallons (or 30.023 U.S. gallons ≈ 113.6 L), U.S. beer barrel = 31 U.S. gallons ≈ 117.3 L, English beer barrel = 36 Imperial gallons ≈ 163.6 L.
One dozen small bottles = 0.90 Imperial gallons.
One hectoliter (European marketing unit) = 100 L = 22.00 Imperial gallons.
[b] Ranges calculated from data of [11 and 157].

the main ingredients used in brewing is given in Table 14.5. Lager beers, which would follow the proportions tabulated fairly closely, have two distinctions. First they are produced by a bottom-fermenting strain of yeast, *Saccharomyces carlsbergensis,* named after its discovery and isolation by E.C. Hansen at the Carlsberg Institute, Copenhagen, in about 1870 [18], or a variety of *Saccharomyces uvarum.* Bottom fermentating yeasts settle out or sink in the green beer on completion of fermentation. Two or more subsequent settlings, or rests while chilled, of the decanted lager beer from the fermentation tank is sufficient to remove much of the yeast, at the same time as other undesirable protein constituents ('lagern' being German for 'to rest').

Lagers comprise more than 95 % of the brew sold in the U.S.A., but just over 40 % of the beer sales in Canada, and less than 5 % of the beer sold in the U.K. [29]. More than 50 % of the beer sold in Canada and nearly 90 % of that sold in the U.K. is an ale, the basic difference being that it is produced by a top-fermenting yeast, *Saccharomyces cerevisiae,* or a related strain. Not only is the yeast skimmed from the top of the green beer, on completion of fermentation, but fermentation is carried out at 15 to 20 °C, slightly higher than required for lagers, and therefore requiring somewhat shorter times. Ales are also hopped at a higher rate

than lagers [28, 30], giving the product a more bitter hoppy flavour component.

Other brew variants are pilsner, a light beer of 3.4 to 3.8 % alcohol content and a little higher hop content, originally produced in Pilsen, Bohemia, and bock beer, a heavy, darker, sweet product of lower hop content produced particularly in spring by German breweries. Porters and stouts are dark, heavy, strongly-flavoured English beers produced using more hops and using malts which have been baked (roasted or black malt), and little or no malt adjuncts. Porters and stouts, produced by top fermentation, are also sold in Canada but together comprise less than 1 % of the Canadian brewing market [29].

The perishability of beer is attributable to its low alcohol content, which is inadequate to prevent attack by micro-organisms, and to the unstable nature of many of its constituents. Thus, bottled and draught beers are at their best immediately after packaging. Optimum keeping qualities are obtained by ensuring minimum oxygen content [31] and aseptic conditions during aging and packaging, by keeping the product under cool conditions to slow any chemical changes, and by storage of the product in the dark or subdued light to avoid the generation of 'light struck flavours' [32].

### 14.2.3 Brewing Emissions and Controls

Air emission problems of breweries are not serious in terms of mass emission rates of long-lived air pollutants. The main concern of air emission control efforts is with containment of process odours. Water scrubbers are used on brew house stacks, to eliminate this potential problem area [20, 33]. If spent yeasts and spent grains are dried on site, the resulting odours generated are controlled by installation of afterburners on the dryer vent stacks [20, 33]. The high contact temperatures of the odorous vapours in the presence of excess oxygen eliminates the odours pyrolytically.

Waste water streams having a high BOD and high suspended solids are the main aqueous emission problem areas. It is estimated that the equivalent of about 9 kg of BOD loading results from each 1,000 L of beer (90 lb BOD/1,000 Imp. gal.) produced in a commercial brewery [20], with a breakdown as to origin roughly as given in Table 14.6. Beer itself has a BOD in the 50,000 to 90,000 range and a COD some 20 to 30 % higher than this [34], an oxygen demand problem which has to be dealt with when any spillage enters the sewerage system.

**Table 14.6** Characteristics of typical aqueous brewery waste
streams [a]

| Origin | Biochem. oxygen demand, mg/L | Suspended solids, mg/L |
|---|---|---|
| Beer spillage | 90 000 | 4 000 |
| Washings from process units | 200 – 7 000 | 100 – 2 000 |
| Cleaning solutions | 1 000 | 100 |
| Screen and press liquor [b] | 15 000 | 20 000 |
| Wort trubs [c] | 50 000 | 28 000 |
| Spent yeast | 150 000 | 800 |
| Precipitates from clarification | 60 000 | 100 |

[a] Data obtained from [34].
[b] From the pressing of wet spent grains prior to drying.
[c] Consists of haze, which forms either during wort boiling or cooling, which is removed before fermentation.

The heaviest BOD loading for sewage treatment is contributed by the spent yeast, beer spillage, and trub (accumulated clarification precipitates) streams. Where the local sewerage regulations permit, some breweries simply pay a treatment surcharge to the local authority, and discharge the brewery liquid wastes to the municipal sewerage system. The surcharge is required by the municipality to offset the higher treatment costs imposed on the sewage treatment plant by these wastes, which have BOD's several orders of magnitude higher than normal domestic sewage. This arrangement then allows direct discharge of these waste streams blended with other aqueous wastes of the operation, but otherwise without treatment by the brewery [20]. Or the brewer can deal with each of these high impact waste streams individually to avoid imposing the waste load on the sewerage system.

The spent grains and yeasts, for instance, can beneficially be recovered from these waste streams and sold as cattle feed supplements [20, 34]. If the feed usage can be accomplished promptly, to very local agricultural interests, these wastes can be sold wet. This is advantageous in two ways, since it avoids the formation of the high BOD 'screen and press liquor' stream altogether [34], as well as avoiding the capital and energy costs involved in drying. However, these waste materials are highly perishable. So if local regular disposal in a short time frame is not possible, pressing and drying of spent grains and yeasts is necessary to improve keeping qualities. This step also reduces shipping costs, but nevertheless it remains a more costly alternative to direct disposal. If no feasible feed markets exist, these wastes are dried and then incinerated to accomplish some energy recovery with final disposal.

For the beer spillage, washings etc., components of brewery waste water streams, which have a lower BOD but which are still high enough to have a substantial sewerage impact, a variety of methods are employed to decrease the final treatment load. Conventional activated sludge methods are used by some breweries to reduce simultaneously the BOD and suspended solids content of these streams to values acceptable to municipal sewerage systems without a treatment surcharge [20, 34]. With added ammonia used as a nitrogen source, BOD reductions of as high as 95 % and suspended solids reduction of 97 % have been reported [35], using this method. Other variations on this theme such as rotating biological contactors [36] (RBC's, or 'biodisk units'), plastic media-filled bio-towers [34, 37], equivalent to trickling filters, but using multiple liquor recycle to achieve effective reduction of these very high BOD's (so-called high rate trickling filters), and even the UNOX process developed by Union Carbide [34], have all been used by the brewing industry to accomplish the same objectives. Tests were conducted with the UNOX system, for example, in which pure oxygen was used to aerate a combined influent stream consisting of a mixture of brewery wastes and domestic sewage of about 1,000 mg/L BOD, and 400 mg/L suspended solids. Both BOD and suspended solids were decreased by about 95 % during these tests [34].

# 14.3 Winemaking

The art of winemaking, like brewing, was developed well before the beginnings of recorded history. It is easy to speculate, again, that these skills were acquired by experimentation following the accidental fermentation of stored fruit or fruit juices by natural yeasts which were originally present on the surface of the fruit. Several theories exist. The theory that grape cultivation spread parallel with culture from East to West is most widely accepted. It is thought that viticulture originated in the lands south of the Caspian Sea since signs of grape cultivation and winemaking have been discovered in Mesopotamia dating back 4,500 years. Certainly winemaking and aging skills were well known to the Greeks, some 2,500 years ago [38], and very gradually spread from there, and probably other centers, in the following centuries.

From a chemical standpoint winemaking is similar to brewing since the essential alcohol-forming step still involves an anaerobic fermentation using yeast. But there are also some significant differences. The key distinction is that winemaking starts with a fruit, usually grape, which already contains sugar, the fermentation substrate. Thus many of the complicated preliminary steps required in brewing to produce a sugar solution for fermentation from starch-based raw materials, are unnecessary for winemaking. This greatly simplifies the early steps of the winemaking process.

A number of varieties of grapes form the dominant winemaking raw material but many fruits other than the grape, such as apples, loganberries, raspberries, strawberries, etc. are also used. Apart from the vested incentive to produce an appealing beverage product, there is also an economic value-added incentive to winemaking from grapes on a large scale. To illustrate this, using Canadian mid-1973 prices, grapes sold wholesale at some 190 $/tonne and retailed fresh at something like 660 $/tonne [39]. To contrast with these figures, the roughly 600 to 800 L of wine obtainable from the tonne of grapes (ca. 160 Imp. gal/ton grapes) was worth some $ 1,600, a coarse measure of the profitability, or the value-added incentive. Consequently, half or more of the grapes grown go into winemaking. In Canada, in the mid 1970's, the proportion going into winemaking was over 75 %. There are many differences in the types of grapes grown, depending on the suitability of climate, micro-climate, and soil types in an area, as well as the type of wine to be produced. Also the sugar content of any particular variety of grape, a most important parameter for winemaking, will vary significantly depending on the locale in which it is grown, and weather variations, year by year for this area. These variable factors affecting the raw material in turn greatly affect the type and quality of finished wine produced. They are also significant developmental factors in the wine industry of any particular country (Table 14.7). It is possible to ship the fruit some distance from a grapegrowing to a wine-producing area, such as may be necessary to supplement the sugar content of a local juice by importing high sugar content grapes. However, because of the perishability and relatively low value of raw grapes this is only practised to a limited extent. Thus, winemaking is more or less restricted to countries which are able to grow their own suitable varieties of grapes, which reduces the number of countries engaged in winemaking on a commercial scale to about $^1/_3$ of the number of countries which produce beer [40].

Another significant feature of these combined natural factors is the limitation placed on wine production volume by the volume of available fruit. In 1980 this aspect restricted world wine production to about $34 \times 10^9$ L as compared to a world beer production for 1978 of some 2 $^1/_2$ times this figure. It also serves to impose quite wide, weather-dependent fluctuations, year by year, on the volume of wine produced. For example, France alone produced over $8 \times 10^9$ L, almost a quarter of the world total, in 1973, and yet poor growing conditions forced a drop of this volume to barely $5 \times 10^9$ L in 1977. With the influence of this major component, total world production variations are also significant, $35.4 \times 10^9$ L versus $28.2 \times 10^9$ L for the years mentioned above, but show a smaller fractional variation than the figures for France alone. Beer production, which is not raw material limited in the same way as wine, shows far smaller annual variations in production volume. Other statistical features of the major wine producers are the unsteady, but nevertheless definite upward trends in production volumes shown by the majority of countries of Table 14.7. But some, like Algeria, Austria, and West Germany appear to be experiencing significant downtrends in annual wine production despite an overall increase in world production levels.

**Table 14.7** Wine production by countries producing more than 200 million liters in 1980 [a, b]

| | 1960 Totals million L | 1970 Totals, million L | 1980 Totals million L | Per capita, L |
|---|---|---|---|---|
| Algeria | 1 585 | 869 | 260 | 13.2 |
| Argentina | 1 675 | 1 836 | 2 300 | 84.2 |
| Australia | 154 | 287 | 414 | 28.3 |
| Austria | 808 | 310 | 268 | 35.7 |
| Brazil | 142 [c] | 190 | 275 | 2.2 |
| Bulgaria | 405 [c] | 409 | 445 | 50.2 |
| Chile | 485 | 401 | 570 | 51.4 |
| France | 6 311 | 7 540 | 7 155 | 133.2 |
| W. Germany | 684 | 1 012 | 399 | 6.5 |
| Greece | 290 | 453 | 440 | 46.3 |
| Hungary | 296 | 438 | 565 | 52.8 |
| Italy | 5 534 | 6 887 | 7 900 | 138.5 |
| Portugal | 1 146 | 1 150 | 943 | 95.0 |
| Romania | 588 [c] | 449 | 895 | 40.2 |
| S. Africa | 287 | 424 | 630 | 21.5 |
| Spain | 2 126 | 2 501 | 4 243 | 113.4 |
| U.S.A. | 1 033 | 969 | 1 729 | 7.6 |
| U.S.S.R. | 777 | 2 685 | 2 940 | 11.0 |
| Yugoslavia | 335 | 548 | 678 | 30.3 |
| Other | (361) [c] | 841 | 872 | 0.3 |
| World total | 24 300 | 30 199 | 33 921 | 7.8 |

[a] From Statistical Yearbook [40] and The World Almanac [158] for 1980 population estimates.
[b] To convert liters to Imperial gallons, divide by 4.55; to U.S. gallons divide by 3.79. A common international unit of wine production volume is the hectoliter, equivalent to 100 L.
[c] Averages for 1961 – 1965 period, 1960 not available. These estimates have not been included in the world total figure for 1960, hence the negative entry under "Other".

## 14.3.1 Classification of Wines

A feature of wines which distinguishes this alcoholic beverage area from beer is the much wider range of product variety and alcohol content achievable with this product. The 'natural' wines, as a class, include all wines in which the alcohol content is derived entirely by natural fermentation. The alcohol content of natural wines is thus limited to the range of about 7 to 14 % alcohol, by volume. The upper end of this concentration range is only attainable by using special varieties of wine yeasts. These wines are also commonly called table, or dinner wines. All, however, have a higher alcohol content than even the malt liquor types of beer.

Most natural wines are sold 'still', that is with no residual carbonation or effervescence. However, some are carbonated or 'sparkling' varieties, which effervesce on opening from excess dissolved carbon dioxide present in the wine. Originally, this carbonation was accomplished by bottling the wine prior to complete fermentation so that the carbon dioxide produced during a secondary fermentation generated a pressure in the bottle. This procedure is still used by some wineries but because of its high cost is reserved for only a very few of the finest quality wines. Good sparkling wines are also made by fermentation in large pressure tanks. The larger scale of operation. lowered labour requirement, and decreased risk of breakage using this method gives a lower cost product. More often today the carbonation is applied artificially after fermentation, using either pure, commercial carbon dioxide or a part of the carbon dioxide collected during fermentation after it has been purified. This is applied to the chilled wine just prior to and during bottling. As a group, the sparkling wines are classified according to their alcohol and carbon dioxide content [39]. Thus, crackling wines are those carbonated to less than 2 bars pressure and containing 7 to 13 $\frac{1}{2}$ % alcohol, by volume. Sparkling wines all contain about 7 % alcohol, and may be carbonated to a varying extent. Champagnes may have an alcohol content in the 7 to 13 $\frac{1}{2}$ % range and be carbonated to the extent of 2 to 6 bars pressure.

An alcohol content of about 14 % is the highest which can ordinarily be achieved directly by fermentation. Concentrations of up to about 18 % by volume can be obtained under highly favourable conditions by using some special strains of wine yeasts. At these high alcohol concentrations further sugar conversion by the yeast is inhibited. Nevertheless it is possible to produce a wine of higher alcohol content, in the 14 to 20 % range (by volume), by the addition of brandy (distilled from wine) or fermentation alcohol to a natural wine. So fermentations to higher than about 14 % alcohol are seldom used. Wines where the natural alcohol content is supplemented by addition are referred to as 'fortified', or dessert wines, and include the varieties commonly known as sherries and ports. One of the original motivations for fortification of these wines was to improve the keeping qualities for shipment and export, by making the beverage completely unattractive for further yeast action. Production volumes of the fortified wines are similar, in total, to the volumes of natural wines produced.

Both natural and fortified wines may be further classified according to their residual sugar content. Dry wines have 1 % or lower residual sugar content, i.e. as little unfermented sugar present in the finished wine as the vintner is able to achieve. This residual sugar usually consists of only unfermentable pentoses and the like [41]. Red natural wines are obtained by fermenting the juice of special varieties of grapes in the presence of the skins to extract some of the colour, or occasionally by heating the must (juice for fermentation) in the presence of the skins before fermentation. These wines tend to be drier, on the whole, than the white varieties of wines which are fermented in the absence of skins. Sweet wines, which are more often whites, are those which contain from 2.5 to 10 % residual sugars. These are produced by starting with an initial sugar content which is higher than can be completely fermented by the yeast. Occasionally, however, the sweeter character is obtained by a stopped fermentation, by removing the yeast from the fermentation at the appropriate stage desired.

## 14.3.2 Principal Steps of Winemaking

Grapes are harvested in the fall, the exact timing depending on the grape variety, current weather conditions and grape sugar content. On arrival at the winery the green stems are separated from the grapes, and then the fruit is crushed. It is necessary to minimize the contact of fruit or juice to air at this stage, to prevent oxidation which can affect the flavour. If at all possible the time from picking the fruit to crushing is kept to less than 24 hours. If a longer keeping time is necessary, such as when shipping grapes or for temporary storage, a nitrogen or carbon dioxide atmosphere can be maintained above the fruit preferably at temperatures just above freezing to suppress microbiological action, or the fruit may be treated with sulfur dioxide to prevent oxidation. For production of white wines the juice is only left in contact with the skins very briefly at the juicing stage, after which it is separated. But red grapes, when destined for red wines, are crushed and kept with the juice for fermentation. By fermentation in the presence of skins (pomace), much of the colour is extracted by the alcohol which forms in the fermenting mixture. This colour extraction is also promoted by the agitation induced by carbon dioxide evolution during the fermentation. Sometimes it is also assisted by occasional pumping of the fermenting must over the skins, which tend to rise to form a thick mat at the top of the fermenter.

Large numbers of natural yeasts (and other microorganisms) are found adhering to the fruit in the field [27]. Originally it was these yeasts which were used to carry out the fermentation of juice to wine, with somewhat variable results. More often now this initial juice, or must, is sterilized and inactivated by addition of 50 to 200 mg/L available sulfur dioxide to the must before fermentation [42]. Commercially, either liquefied sulfur dioxide itself or a 10 % solution of sodium or potassium metabisulfite ($K_2S_2O_5$) may be used for this purpose. The 10 % metabisulfite solution, equivalent to about 5 % available sulfur dioxide [43] (Equations 14.10, 14.11), is the more convenient form for small scale use.

$$Na_2S_2O_5 + H_2O \rightarrow 2\,NaHSO_3 \qquad (14.10)$$

$$NaHSO_3 + H_2O \rightarrow NaOH + SO_2 \qquad (14.11)$$

**Table 14.8** Range of concentrations of the principal constituents of must and dry wine, in percent by weight[a, b]

| Component | Must | | Wine | |
|---|---|---|---|---|
| Water | 70 | – 85 | 80 | – 90 |
| Total carbohydrates | 15 | – 25 | 0.1 | – 0.3 |
| glucose | 8 | – 13 | 0.1 | – 0.5 |
| fructose | 7 | – 12 | 0.05 | – 0.10 |
| pentose | 0.08 – | 0.20 | 0.08 | – 0.20 |
| Alcohols | | | | |
| ethanol | trace | | 8 | – 15 |
| methanol | 0.0 | | 0.01 | – 0.02[c] |
| C 3 and higher | 0.0 | | 0.008 – | 0.012 |
| glycerol | 0.0 | | 0.30 | – 1.40 |
| Total organic acids | 0.3 – | 1.5 | 0.3 | – 1.1 |
| d-tartaric | 0.2 – | 1.0 | 0.1 | – 0.6 |
| l-malic | 0.1 – | 0.8 | 0.0 | – 0.6 |
| l-citric | 0.01 – | 0.05 | 0.0 | – 0.05 |
| succinic | 0.0 | | 0.05 | – 0.15 |
| lactic | 0.0 | | 0.1 | – 0.5 |
| acetic | 0.00 – | 0.02 | 0.03 | – 0.05 |
| pH | < 3.3 – | 3.6 | | |
| Tannins | 0.01 – | 0.10 | 0.01 | – 0.30 |
| Aldehydes | trace | | 0.001 – | 0.050 |
| nitrogenous compounds | 0.03 – | 0.17 | 0.01 | – 0.09 |
| Inorganic compounds | 0.3 – | 0.5 | 0.15 | – 0.40 |

[a] Selected components from the comprehensive listings of reference.
[b] Composition of the starting grape is about 80 – 90% pulp and juice, 10 to 20% seeds, skin, and stems [27].
[c] Methanol is not a product of the fermentation but arises from the hydrolysis of naturally occurring pectins [6].

Sulfur dioxide sterilization not only inhibits the activity of the wild yeasts present, but also serves as a chemically reducing preservative, preventing oxidation of the fresh juice, and lowers the pH, producing conditions which will tend to favour the activity of the tolerant yeasts added by the winemaker, over the wild yeasts present. If boiling were used for sterilization in winemaking, as it is in brewing, much of the fruity subtlety of the product would be lost in the process, and these additional advantages would not be obtained.

Before fermentation is begun, the must is analyzed for sugar and total acid content. Adjustments will be made to the concentrations of these components with added sweeter juices or sucrose [27], and tartaric or citric acids if necessary. The product at this stage of the process will have an analysis roughly as given in Table 14.8.

For fermentation the must is allowed to stand for a minimum of 2 hours, and sometimes up to 24 hours or so, for the antiseptic action of the sulfur dioxide to take effect, and for the sulfur dioxide content to be decreased somewhat by forming combined sulfur dioxide and by breathing [6]. If carried out at low temperatures this time also allows heavy sediments to settle (fruit particles, seeds, sand, etc.) from which the juice should be separated. The cleaner fermentation obtained by doing this gives a better preservation of the natural fruitiness and aromatic values in the product.

After this standing period a starter of 1 to 2 % of the must volume of the appropriate variety of yeast is added and fermentation started at close to ambient conditions, about 20 °C (Equations 14.8, 14.9). Generally, fermentation temperatures for winemaking are far less critical than for the fermentation of wort to a beer, so that many winemakers simply allow the process to take its course. But for better control of flavour and to avoid problems with product loss from foaming, chilled water is used by many vintners to control white wine fermentations to within 2 or 3° of 15 °C, a process which takes 1 to 2 weeks to complete [39, 27]. However, for high quality wines from good, sound fruit close temperature control in the 3 to 10 °C range is required, which permits only 1 to 1 ¹/₂ % sugar conversion over 24 hours. At these temperatures the fermentation may take up to 6 weeks but produces a wine with better varietal character (reflecting the species of grape fermented), higher residual sugar, and higher volatile acidity and glycerin.

Red wines are fermented at about 20 °C in the presence of the skins for three to six days, depending on the intensity of colour (anthocyanins) and flavour (tannins) desired. The partially fermented must is then decanted and pressed from the skins, and a secondary slower fermentation carried out to the extent required. Acid and glycerin formation, small but important components of both white and red wines, tend to increase towards the end of the fermentation under the influence of the inhibiting ethanol (Equation 14.12).

$$2\,C_6H_{12}O_6 + H_2O \xrightarrow{\text{yeast}} CH_3CH_2OH +$$
$$CH_3COOH + 2\,CH_2OHCHOHCH_2OH +$$
$$2\,CO_2 \qquad\qquad (14.12)$$

Completion of fermentation for either type produces a 'young' table wine which is still not a palatable product as it stands.

Sulfur dioxide addition is required as soon as fermentation is stopped and must be maintained at or above 25 ppm of free $SO_2$ throughout the subsequent clarification steps to prevent oxidation, to assist in biological stabilization, and to bind aldehydes in the new wine. The binding of aldehydes (mostly acetaldehyde) performs a taste-improving function on the wine.

Clarification is needed after fewrmentation, primarily to remove the spent yeasts in the wine to prevent their autolysis in the product, which would adversely affect the flavour (Figure 14.3). It also serves to remove any other suspended material affecting the clarity. Simple settling and racking (decantation), repeated several times, accomplishes much of this. Occasionally, however, it is necessary to add fining agents, such as isinglas (a fish protein), gelatin, a bentonite clay, or a highly refined diatomaceous earth, to promote coagulation and precipitation of colloidal matter in the young wine [44].

Prolonged chilling to near freezing to promote tartrate crystallization, or ion exchange methods are being employed by some wineries to remove excess tartrates [45], particularly before the bottling of young red wines [6]. This practice avoids the formmation of haze from crystallization of the residual tartrate in the bottle, which can occur particularly when the wine is chilled or aged in the bottle.

A final polishing step in clarification is filtration, used primarily to ensure that no yeast or other micro-organism remains in the wine during the subsequent aging steps and to produce a brilliantly clear product [46]. At this stage the wine will have an analysis roughly in the ranges quoted in Table 14.8.

Aging completes some of the objectives of the clarification step but more importantly accomplishes a maturation of many of the flavour components of the young wine to give a less harsh, more palatable product. Colloquially what is achieved is a mar-

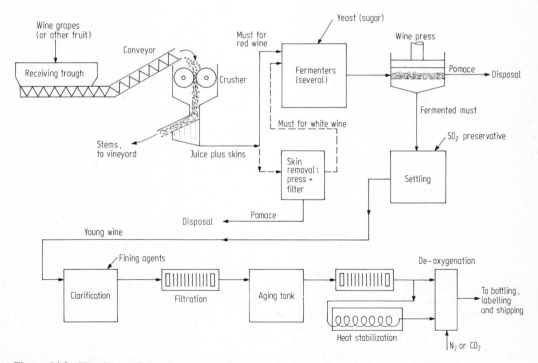

**Figure 14.3.** Flowsheet outlining the sequence of steps required to produce wine on a commercial scale

rying of flavours, and a development of bouquet, which we know happen through a complicated variety of chemical steps. Probably the most important of these is esterification, the combining to some extent of the acids and alcohols present to form esters which contribute more fruity, less harsh flavour and 'nose' components than initially contributed by the ester components present in the young wine (Equation 14.13).

$$RCOOH + R'CH_2OH \xrightarrow{H^+} RCOOCH_2R' + H_2O \quad (14.13)$$

Of the esters formed ethyl acetate is the most likely to form, because of the relatively higher concentrations of ethanol and acetic acid than other ester forming constituents present. It is also thought to be the most important from the point of view of the progress of maturation [6]. Malic to lactic acid conversion (Equation 14.14) is also an important part of the aging process, particularly for red wines [47].

$$HOOC-CHOH-CH_2-COOH \xrightarrow{\text{Lactobacillaceae}}$$
malic acid
$$HOOC-CHOH-CH_3 + CO_2 \quad (14.14)$$
lactic acid

Several North American white wines are bottled and marketed after only about 4 months aging following fermentation, and having undergone 2 or 3 rackings before bottling [11]. Many are marketed in under a year [6]. Red wines are usually aged longer than this, a minimum of 24 months being normal [48] and some as long as 4 to 6 years before bottling. Sherries and ports, the fortified wines, are generally aged 3 years or longer for full maturation and smoothness to develop before marketing. Sherries are subjected, during this aging process, to a partial oxidation with air, and a 'baking' at about 50 °C for one or two months to produce the characteristic sherry flavour.

Prior to bottling, most commercial wines are stabilized and blended. Stabilization, by one or more of the following techniques, is conducted to improve the keeping qualities of a wine. Filtration through a small pore filter removes yeast cells that might continue to ferment residual sugar when the wine is in the bottle, particularly with sweet wines. Filtration also prevents lysis from contributing intracellular amino acids etc. from ruptured yeast cells to the wine. Pasteurization, for 2 minutes at

86 °C, accomplishes similar objectives and may be used in series with filtration or on its own as a stabilizing step, although it does not remove killed microorganisms from the wine. By this stage it is assumed that the number of organisms remaining is small. Pasteurization also coagulates some proteins, thereby assisting in their removal. Gentle sparging of the wine with carbon dioxide or nitrogen helps to preserve it from oxidation on storage, by removing most of the oxygen that might have been dissolved in the wine from contact with air during the previous processing stages. Residual sulfur dioxide, present at an average level of about 200 mg/L (as combined, and free) in Canadian wines, and at about twice this concentration in many European varieties [39], also provides further protection against wine oxidation or secondary fermentation on extended storage.

Final blending of the products of different fermentations based on similar or different grapes is conducted to achieve more or less consistent product flavour characteristics in a particular brand name of wine. Once the stabilized blend is bottled, particularly when closed with polyethylene or polyethylene-lined caps, there is negligible further change on storage. Bottled wine, when kept in a cool, dark place, is a relatively stable product even on prolonged keeping. This stability is assisted by the practices of filtration and pasteurization ('hot bottling') described, and by the approximately 150-200 mg/L of total residual sulfur dioxide present [39]. Government regulations in Canada [39] and the U.S.A. [42] permit up to 350 mg/L total sulfur dioxide in natural or dessert wines, still below the obtrusive level for the average palate. However, the sensitive palate of a trained wine taster can taste concentrations of free (as opposed to 'combined') sulfur dioxide above about 25 mg/L. Combined sulfur dioxide is present mostly as its bisulfite addition compound with acetaldehyde [49]. In the U.K. sodium or potassium sorbate, and ascorbic or erythorbic (isoascorbic) acid addition is now also permitted [50] to absorb oxygen, in the process helping to maintain a free $SO_2$ level for longer periods of time. In the presence of these compounds less sulfur dioxide is needed to maintain the required free $SO_2$ level for good keeping qualities, thus improving the taste.

Related to product quality control, not keeping properties, the presence of ubiquitous asbestos fibers in all types of finished wines [51, 52] is now being minimized through better final product sampling and tests, and by avoidance of asbestos-containing filtering materials. The quite wide dispersion of asbestos fiber in the atmosphere [53-55] means that harvested fruit of all kinds, not only grapes, will inevitably contain a measurable adhering asbestos fiber count.

### 14.3.3 Utilization and Disposal of Winery Operating Wastes

Stems and pomace (skins and seeds), which together amount to 10 to 20 % of the weight of grapes crushed, represent a significant disposal problem, particularly for wineries producing upwards of 36 million liters (ca. 8 million Imp. gallons) of wine per year. Some wineries return this component of their operations directly to the soil of the vineyards, as organic fertilizer [48]. At one time the pomace component of these wastes was dehydrated for use as a component of cattle feed, the seeds being separated out first for oil recovery [6, 56], but both procedures have now declined in importance [6]. This may be partly because of the low nutritional value of fermented pomace, and partly because of storage difficulties. Pomace which is heap stored can start to ferment, and if fed to cattle in this state can be fatal. It appears that South Africa has made the most comprehensive use of winery byproducts; alcohol, tartrates, cattle feed, methane, and fertilizer all being produced from pomace [57].

Probably the most prevalent values to be recovered from pomace before disposal, because of the simplicity of recovery by water extraction, are the sugars from the pomace of white wines, and the residual sugars plus alcohol from the pomace of red wines [6]. After encouraging completion of the fermentation of the residual sugars present in the combined extracts, ethanol, useful for fortifying brandies, is recovered from both types of fermented extracts by distillation.

Fresh lees, the accumulated spent yeast from wine fermentations may be usefully used for alcohol recovery by distillation of the wine separated with the yeast. Or some wineries pressure filter the lees for further wine recovery and produce a dry yeast cake [46]. The solid residue may then be dried and sold as a cattle feed supplement.

Monopotassium tartrate recovery may be practised from pomace, lees, or most economically from the virtually pure monopotassium tartrate obtained from specific tartrate removal steps. In Canada and the U.S.A. monopotassium tartrate from this source provides most of the commercially marketed tartrates for use in such diverse products as baking powder, cream of tartar and the like.

A poorly fermented batch of wine can sometimes be made substantially palatable by chemical means, depending on the cause of the problem. It can then be blended with other batches of wine and the blend marketed as an acceptable product. If it is not possible to correct its faults, however, it may be simply distilled for recovery of alcohol which may then be used for wine fortification. The aqueous residue, now significantly lower in BOD, is then discarded.

An 'off' wine may also be converted to wine vinegar by acetification (Equation 14.15), a process that is sufficiently profitable that at least one California winery has specialized in producing wine vinegar, rather than wine [6].

$$CH_3CH_2OH + O_2 \xrightarrow[aceti]{Aceto-bacter} CH_3COOH + H_2O \tag{14.15}$$

Vinegar production, as mentioned earlier, represents one of the fermentation processes that is practised both on a batch basis, literally 'in the barrel', and in continuous flow modes where the acetifying wine is continuously recirculated over a bacterial support. With care and appropriate local circumstances for byproduct sale and disposal, few of the byproducts and waste materials of a winery operation need to be wasted.

Dual aerated lagoons operated in series have been found to be an effective way of dealing with aqueous wastes not amenable to any of these recovery procedures, or for dealing with the aqueous residuals from these [58]. Better than 95 % BOD removal has been reported for influent BOD's in the range of from 131 to 5,000 mg/L when retention times averaging 30 days for the first lagoon and 100 days for the second were used.

## 14.4 Beverage Spirits

Spirits represent a class of alcoholic beverage which generally have an alcohol content of 20 % or more, above the range achievable solely by fermentation. Thus the spirits classification includes those beverages where distillation is necessary to raise the alcohol content above that achievable by fermentation and where the distillate forms an essential and significant part of the final product.

To be able to produce beverage spirits required a knowledge of both fermentation processes and of the art of distillation. Thus the preparation of spiritous liquors is historically much more recent than the arts of beer or wine production. Informed estimates place the earliest successful attempts at producing a distilled beverage from wine as occurring in Greece or Egypt some 1800 years ago, and probably first accomplished by alchemists [59]. But the original discoveries did not become widespread because of the secretiveness of the early practitioners, so that it was perhaps a thousand years later before distilled beverages became at all widely available.

A synonym for whisky, the Gaelic "uisge-beatha", and the names attached to other national distilled beverages such as the French "eau de vie", and the Russian "vodka" all can be liberally translated as "water of life". These give some idea of the early recognition of the significant stimulant effects of the beverage. All of these terms came into use in the last 800 to 900 years.

## 14.4.1 Specifying the Alcohol Content of Spirits

Much of the English-speaking world uses the proof system on a degrees Sikes scale, which originated in the U.K. in 1816, for specifing the alcohol content of beverage spirits [60]. In this system 100 % alcohol is equivalent to 175.2° proof. Proof alcohol, synonymous with 100° proof on this scale, is thus equivalent to 57.1 % alcohol by volume. In Canada and the U.K. proof spirit (100° proof) is specified in statutes on a density basis as being exactly twelve-thirteenths the weight of an equal volume of distilled water at 51°F [59] (ca. 10.5°C). Rumour has it that the early basis for deciding the alcohol content of spirits lay in the test that gunpowder moistened with an equal amount of proof (or higher alcohol content) spirits still ignited when touched with a flint or lighted taper [60]. When moistened with underproof spirits it would not.

Under this proof system, spirits that are 20° overproof, which is the same as saying 120° Sikes or 120 proof, corresponds to a beverage which contains 120/175.2 x 100 %, or 68.5 % alcohol by volume (Table 14.9). Similarly, spirits that are 70° proof is the same as say that it is 30° Sikes or 30° underproof and contains 70/175.2 x 100 %, or 40 % (39.95 %) alcohol by volume.

The most convenient way of determining ethanol concentration is by measuring the density and temperature of the solution and relating this to concentration by the use of appropriate tables. In turn the concentration determined in this way can then be related to the British proof system, or to the Sikes scale for specification on the bottle.

The density system can be used to determine alcohol concentration for any country. But for marketing purposes the U.S.A. uses a simpler proof system based on 100 % alcohol being equivalent to 200° proof. Fractions of this concentration, by volume, are specified by simply multiplying the concentration, in percent by volume, by two (Table 14.9). In France, alcohol concentrations in spirits are simply specified as percent alcohol by volume, sometimes referred to as the Gay-Lussac system. Percent alcohol by *weight* is used in Germany.

The excise taxes levied on distilled spirits production is based on the volume of proof spirits produced, which as outlined above corresponds to 57.1 % ethanol (or ethyl alcohol) by volume in Canada and the U.K., and 50.0 % alcohol by volume in the U.S.A. Since the concentration basis of proof alcohol for taxation in the former areas is higher and the Imperial gallon is equivalent to 1.20 U.S. gallons, the alcohol content of one British 'proof gallons' is equivalent to the alcohol content of 1.37 U.S. "proof gallons". This summarizes the trade and tax basis for specifying beverage alcohol concentrations for many producing areas.

## 14.4.2 Steps in Spirits Production

The initial steps for spirits production parallel those described for beer or wine in that initially one starts with either a cereal source for starch to be hydrolyzed to sugar, or an inexpensive source of sugar itself. Thus a variety of grains, or molasses are common raw materials. There are many features of the various types of spirits produced which are parallel. Whisky has been for a long time, and still is the dominant product of the distilled beverage class (Table 14.10), although there has been a significant increase in the proportions of this class represented by vodka, cordials, and brandies in the U.S.A. over the last twenty years (Table 14.10). Because of the continued dominance of whisky in this class, the discussion of distilled beverage production will be centered on this product. Generally the choice of the main cereal used for the production of spirits, or whisky in particular, is based on the lowest cereal cost, on a starch content

**Table 14.9** A comparison of the systems for specifying the alcohol content of beverages.

| Percent Alcohol by volume [a] | Approximate density at 15 °C [b] | Canadian/ British proof [c, d] | Scale, degrees Sikes [d] | Equivalent U.S.A. proof |
|---|---|---|---|---|
| 100 | 0.7939 | 75.2 | 175.2 | 200 |
| 97 | 0.8079 | 70 | 170 | 194 |
| 91 | 0.8306 | 60 | 160 | 182 |
| 86 | 0.8465 | 50 | 150 | 172 |
| 80 | 0.8639 | 40 | 140 | 160 |
| 74 | 0.8799 | 30 | 130 | 148 |
| 69 | 0.8925 | 20 | 120 | 138 |
| 63 | 0.9068 | 10 | 110 | 126 |
| 57.1 | 0.9199 | proof [e] | 100 | 114.2 |
| 51 | 0.9325 | 10 | 90 | 102 |
| 50 | 0.9344 | 12.5 | 87.5 | 100 (proof) |
| 46 | 0.9419 | 20 | 80 | 92 |
| 40 | 0.9519 | 30 | 70 | 80 |
| 34 | 0.9604 | 40 | 60 | 68 |
| 29 | 0.9666 | 50 | 50 | 58 |
| 23 | 0.9730 | 60 | 40 | 46 |
| 17 | 0.9791 | 70 | 30 | 34 |
| f | f | 80 | 20 | f |
|  |  | 90 | 10 |  |
| 0 | 1.000 | 100 | 0 | 0 |

[a] System used in France, sometimes referred to as the Gay-Lussac system. In Germany, alcohol concentrations are specified on the basis of percent by *weight*.
[b] Related to the density of water, also at 15 °C. Values will deviate slightly from this with sugar content, and at differing temperatures. Taken from [159].
[c] Values above "proof" are quoted as "overproof", values below proof as "underproof" to keep the units unambiguous.
Not exact, except where three significant figures quoted. Values rounded downward, see note b above. Canadian and British practice sometimes uses these proof numbers for specifying alcohol concentrations, without using the prefixes "over-" or "underproof". Value on this scale is obtained by multiplying the percent alcohol by volume by 1.752.
[e] Proof alcohol, in the British system, or 57.1% alcohol, by volume.
[f] Correspondence between the different scales becomes inaccurate in this range.

basis, since the sugar ultimately available for fermentation depends on the amount of starch hydrolyzed. Small amounts of the appropriate cereal, for example rye for rye whisky, corn for bourbon, may still be employed to provide the desired flavour element in the final beverage. However, the low cost starch source will still provide most of the sugars required. Rye is used, and extensively by distillers in the areas where it is plentiful [60]. But many other starch sources such as corn, wheat, millet, rice, and potatoes also provide supple-

mentary sources of starch for spirits production, depending on local cost and availability. Millet, with a high starch content of 65-68 % and a low cost in Western Canada, is a grain frequently used there. Small amounts of rye are also sometimes used in hydrolyzed form specifically for favourable yeast propagation (ca. 2 % of the total starch input) and not primarily to provide the fermentable sugars used.
Whatever grain is employed, high quality and an accurate knowledge of the water content at the time

**Table 14.10** Recent trends in the U.S. consumption of classes of distilled beverages (spirits) [a]

| Class of spirits | Percent of U.S. consumption, by class and year | | |
|---|---|---|---|
| | 1960 | 1966 | 1976 |
| whiskies [b] | 74.4 [b] | 68.2 [b] | 53.5 [b] |
| gin | 9.3 | 10.5 | 9.8 |
| vodka | 7.8 | 10.4 | 18.0 |
| cordials (liqueurs) | 3.8 | 4.3 | 6.8 |
| brandy | 2.6 | 3.2 | 3.9 |
| rum | 1.6 | 2.2 | 4.2 |
| Other | 0.5 | 1.2 | 3.8 |
| | 100.0 | 100.0 | 100.0 |
| Total volumes: | | | |
| $L \times 10^6$ | 888.3 | 1 169.2 | 1 621.8 |
| wine gallons [c] $\times 10^6$ | 234.7 | 308.9 | 428.5 |

[a] Selected from data of [160].
[b] Includes blends, straights, bonds, Scotch, Canadian, and other types of whiskies. In 1949, consumption of all types of whiskies, totalled 87.8% of all distilled beverage consumption, in the U.S.A.
[c] The wine gallon is a strictly volume measure, in the U.S.A. eqivalent to a U.S. gallon (3.7054 L), in Canada and the U.K. equivalent to an Imperial gallon (4.546 L).

of purchase is vital because of the close checks maintained by most governments on distillery operations. Every bushel (British: 36.37 L, U.S.: 35.24 L) of grain shipped to a distillery has to be accounted for in terms of the proof gallons produced. For any grain the standard "distillers bushel" is 56 lbs. (ca. 25.4 kg) [59]. Since this is measured on a mass, and not on the usual volume basis, an unsuspected high water content can cost the distiller not only the lost raw material but also the excise tax on the proof gallons not obtained, for which he is still liable based on the quota of grain received. Any discards caused by any mould or disease not recognized on delivery can also cost lost production and tax liability.

With a cereal start, the preliminary steps prior to fermentation are very similar to the first steps of beer production, to obtain a sterile mash ready for inoculation. The mash is cooled to about 20 °C at which point the yeast starter, which is propagated from a master culture in 4 to 5 stages to a volume of about 2 to 3 % of the mash volume to be fermented, is added. Fermentation temperatures are far less critical for spirits production than for beer, since the important flavour elements are not af-

fected by temperature in the same way. Since only a few minutes at 50 °C is enough to kill the yeast [6], however, sufficient cooling is applied to keep the closed fermenter temperature below about 38 °C to avoid this. The higher fermentation temperatures used here speeds up the process significantly to complete the fermentation in a period of only about 3 days instead of the 6 or 7 days required for the refrigerated fermentation of beer. Under these conditions, by the time 50 % of the sugar in the mash has been converted there is no further yeast propagation. From this point on the remaining sugar to alcohol conversion is accomplished by metabolism by the yeast population already present. The product of the mash fermentation is a "beer" containing 7.5 to 8 % alcohol by volume. This name does not imply any beverage qualities to the product at this stage. It is merely a reflection of the similarities of the methods used to obtain it. Beverage spirits are obtained from this fermentation product by various distillation, blending, and aging procedures.

### 14.4.3 Distillation of "Beers"

The theoretical basis of the success of distillation as a method for enrichment of the alcohol content of the aqueous solutions of alcohol produced by fermentation lies in the application of Raoult's Law. This law quantitatively relates vapour composition to liquid composition, for mixtures of ideal liquids, A and B, according to Equation 14.16.

$$P_A = X_A P^o{}_A \qquad (14.16)$$

where:
$P_A$ = partial pressure of the vapour of liquid A, in the vapour mixture
$X_A$ = the mol fraction of A in the liquid phase, and
$P^o{}_A$ = the vapour pressure of pure liquid A, at the same temperature

While few real liquids behave quite like the ideal liquids required for strict adherence to Raoult's Law, many mixtures of real liquids, including ethanol/water mixtures, do approximately follow this relationship on heating (Figure 14.4). Water has a vapour pressure of 327 mm Hg at 78.3 °C. Ethanol, with a boiling point of 78.3 °C, has a vapour pressure of 760 mm Hg at this temperature and consequently forms a higher mol fraction in the vapour space above a heated ethanol/water mixture, than it comprises in the liquid phase. Condensation of the alcohol enriched vapour mixture

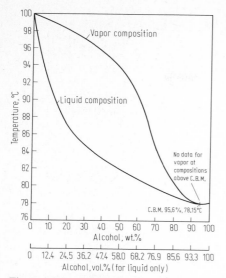

**Figure 14.4.** Liquid vapour composition diagram for ethanol-water mixtures [11]. Reprinted courtesy McGraw-Hill

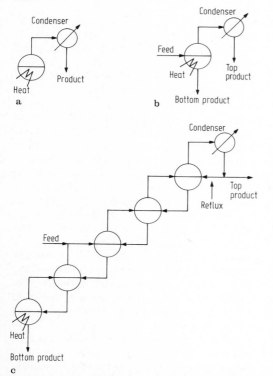

**Figure 14.5.** The relationship of batch (simple) distillation to single and multistage continuous distillation [63]. Reprinted courtesy M.L. Burstall

obtained in this way produces a solution of ethanol in water again, but now enriched in the concentration of ethanol over that present in the initially heated mixture.

In a laboratory batch distillation the process described above may be carried out very easily, but this only achieves a limited (by the liquid-vapour composition diagram) improvement in concentration of ethanol obtained with each repetition of the simple distillation step (Figure 14.5 a). Also, as the distillation proceeds, the concentration of alcohol in the distilling vessel becomes depleted. Consequently there is also a gradual depletion in the alcohol concentration obtained in the vapour and condensate, later in the distillation. Despite these problems related to batch distillation of whisky "beers", many small distilleries still use essentially this batch method to raise the alcohol concentrations to the requirement of the product [61].

It is also possible to operate a distillation in a continuous mode, which yields a steady state concentration of alcohol in the product. When an incoming beers stream is fed continuously to a distilling vessel, and a vapour stream and a distilled (alcohol depleted) liquid stream are continuously removed from the vessel, the mode of operation is referred to as continuous distillation. Continuous distillation gives a further processing advantage over batch distillation by simplifying quality control, because of the predictable and steady alcohol concentrations obtained from both product streams (Figure 14.5 b). The compositions of both the distillate and the residue streams, within the limits dictated by the liquid-vapour composition diagram, will depend primarily on the beers feed rate and the heat input to the system.

With a feed beers stream consisting of 7 to 8 % alcohol, the potential improvement in alcohol concentration achievable with one stage of efficient distillation is significant, ideally close to 55 % alcohol in the distillate. This figure may be estimated by drawing a horizontal line from the 8 % alcohol point of the liquid composition line to the vapour composition line. To allow more latitude in congener (flavour component) addition, blending, and the alcohol concentrations that are taken for aging, most distillers strive for higher alcohol concentrations than this. This may be achieved either by several repetitions of simple distillations, or by several stages of continuous distillation (Figure 14.5 c).

Continuous distillation is the only way that the

scale of beers processing by the average North American distillery, which ferments over 25 tonnes of grain per day, can be conveniently handled. To carry this out in several physically separated units, as depicted in Figure 14.5 c, would introduce formidable heat and material balance problems to obtain smooth operation. Because of this, continuous separations of this kind are almost invariably carried out in a single tube or column in which a series of perforated plates or bubble cap plates serve as separate stages of distillation (Figure 14.6). Using this arrangement, hot vapours from each stage heat the next higher stage as they partly condense, so that only a single source of heat at the bottom of the column is sufficient to warm all the higher plates. As excess higher boiling fluid builds up on a plate, it overflows into a "downcomer" tube which carries it down to the next lower plate. Any more volatile, lower boiling alcohol present in the fluid on a plate has a tendency to be vaporized and carried up the distilling column in vapour form as the fluid on the plate becomes heated by the upward moving vapours passing through it. In these ways·the liquid on each plate gradually comes to, and stays in equilibrium with the hot vapours continuously being forced through the liquid on the plate. Plate choice in any particular application is a function of the variables the particular distilling column is required to meet. Perforated plate columns are the most efficient for good liquid/vapour contact but are rather sensitive to distillation rates. Too low a distillation rate and the plates become dry from drainage through the perforations as well as via the downcomer. Too high a distillation rate and a perforated plate column has a tendency to flood (become overloaded with fluid) and hence lose distillation efficiency. A bubble cap column, in contrast, can handle a wider range of distillation rates for a given column without the drying out or flooding problems of a perforated plate column. But a bubble cap column provides less efficient liquid/vapour contact than a perforated plate column, and also costs more, for a given distilling capacity [62]. More often than not, however, bubblecap plates are specified at the time of column design when it is desired to obtain the principal advantage of this system, flexibility in operating rates [63]. For a large scale spirits producer fermenting over about 40 tonnes/day, the main beers distilling column can be shorter if the separation task is split among several columns, and in this application four columns are generally adequate [64]. Splitting

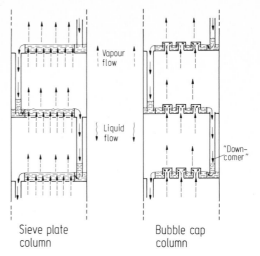

**Figure 14.6.** Horizontal sections of a sieve plate (left) and bubble cap (right) distilling columns

the distilling duty in this way enables the three subsidiary columns, the selective distillation, the aldehyde, and the fusel oil columns to be sized significantly smaller than the main whisky separating column which has the job of alcohol enrichment of the whole of the beers stream (Figure 14.7).

In essence the task accomplished by this group of columns allows efficient continuous production of a consistent product. The first, and largest column, the beer distilling or whisky separating column, produces an overhead stream containing 55 to 60 % alcohol together with the remaining volatiles, mainly aldehydes and fusel oils. This stream of several components is collectively referred to as the "high wines" stream of the distillery. At one time, with only some subsequent adsorption of impurities on charcoal, this stream comprised the whisky product of early distilleries which was sent to aging [60]. The bottom stream from this column consists of stillage, which is basically water from which all the alcohol has been stripped, but which will still contain some unfermented sugars, dextrins and the like.

In the next step, conducted in a smaller selective distillation column, a neutral spirits stream (essentially 95 % ethanol, i.e. the azeotropic composition) plus roughly separated aldehydes (overhead) and fusel oils (bottom product) streams are separated from the high wines stream of the beers col-

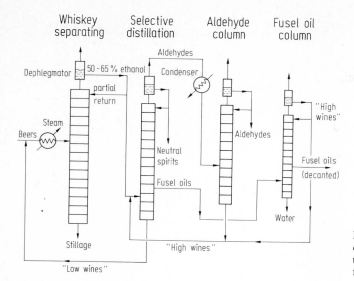

**Figure 14.7.** A suitable configuration of columns which can be used by a distillery for spirits production. Simplified from [64]

umn. The last two columns in the sequence function to recover high wines streams from the aldehydes and the fusel oils streams. The more concentrated streams of aldehydes and fusel oils obtained represent about 8 % and 7 % of the high wines stream, respectively [60].

### 14.4.4 Distinct Distilled Beverage Products

The aldehydes and fusel oils contained in these product streams of the distillation are also referred to as congeners or flavour components. Most of the aldehyde content is acetaldehyde, although other aldehydes are also present [59]. The fusel oils (from the German "fuselöl" meaning "bad spirits") are thought to arise via yeast fermentation of amino acids present in the original mash [65]. Collectively they include amyl, and iso-amyl alcohols (ca. 50 %) and n-butanol (ca. 20 %) with the residue consisting of fatty acids and several other higher alcohols [64, 66] (Equation 14.17).

$$CH_3CH_2CH_2CH_2CH_2OH \qquad (14.17)$$
amyl alcohol

$$(CH_3)_2CHCH_2CH_2OH \qquad CH_3CH_2CH_2CH_2OH$$
iso-amyl alcohol \qquad n-butanol

Once the congeners are separated from the high wines stream, varying proportions of each may be added back to the neutral spirits fraction as required for the type of product being made. In this way they give the distiller much greater latitude

and control of the type of whisky flavour he can obtain in the final product than was possible for the early distillers, when there was little separation of the congener fractions from the high wines stream. Many of the whisky products of the distiller start out as simple blends of various combinations and proportions of the neutral spirits, aldehydes ("heads"), and fusel oils ("tails") streams, together with enough water to drop the alcohol content to the range required by statute. Bourbon-type whiskies generally have the highest concentration of congeners, followed by Scotch whiskies, though it can be seen that the ratio of aldehydes and fusel oils differs between these two types (Table 14.11). Thus, the relationship of congener proportions to final flavour achieved is a complex one, and bears little correlation to gas chromatographic analysis of the components present. Canadian blended whiskies, in contrast, are generally lower in congeners than either the bourbon or Scotch counterparts, only some $1/2$ to 1 % of the heads and tails streams being added to neutral spirits before aging to produce these.

The newly mixed solution of neutral spirits, congeners, and sufficient water to bring the mixture to 60 % alcohol is not a finished whisky. It still must be aged, or matured, to induce a mellowing and smoothing of the initial blend to produce a more palatable product. Maturation is a complex art that requires a skilful mix of measures taken to maximize product improvement, and at the same time

**Table 14.11** Congener content of some widely selling types of whiskies [a, b]

| Component [b] | American blended whisky | Bourbon whisky | Canadian blended whisky | Scotch blended whisky |
|---|---|---|---|---|
| fusel oil | 83 | 203 | 58 | 143 |
| total acid | | | | |
|   (as acetic acid) | 30 | 69 | 20 | 15 |
| esters | | | | |
|   (as ethyl acetate) | 17 | 56 | 14 | 17 |
| aldehyde | | | | |
|   (as acetalaldehyde) | 2.7 | 6.8 | 2.9 | 4.5 |
| furfural | 0.33 | 0.45 | 0.11 | 0.11 |
| total solids | 112 | 180 | 97 | 127 |
| tannins | 21 | 52 | 18 | 8 |
| total congeners, wt/vol % | 0.116 | 0.292 | 0.085 | 0.160 |

[a] a Selected examples from a tabulation of [64].
[b] Grams per 100 L at 100° proof, density of about 0.914 g/mL, at 20 °C. Thus, multiplying the numbers given by 10 gives a close approximation of respective concentrations in mg/L, and by 9.14 to give mg/kg.

to stay within government regulatory guidelines. During the process, water and alcohol diffuse out through the wood, and air diffuses in, the air very gradually oxidizing the aldehydes and other congeners present. A certain amount of flavouring and colouring is also picked up from the charred lining of the barrels, which at the same time adsorbs a certain amount of the fusel oil component from the mix [61]. These aging processes are carried out in charred oak barrels that are stored at a constant temperature as near 13-17 °C as possible, and a relative humidity of about 65 % for a period ranging from 4 to 18 years. If the conditions are cooler than this, the product doesn't mellow, and if warmer it causes greater than acceptable annual evaporative losses, something the distiller tries to avoid. The complex congener concentration trends during the aging process have been determined and tabulated [64]. Lower humidity levels or frequent variations in temperature in the storage warehouses also tend to increase evaporative losses and hence tend to be factors which increase the cost of the aging process.

In Canada, the government requires a distiller to demonstrate production of at least 2 1/2 proof gallons (2 1/2 Imp. gal. of 57.1 % alcohol by volume) for each bushel of grain raw material mashed. If this is not achieved, the distiller is still taxed on this basis, hence the strong emphasis on pure, high quality raw materials by the distiller. In practice a distiller can generally produce 3 1/2 to 4 proof gallons (Canadian) per bushel, which is still taxed at the same rate but allows some tax margin for error. The government also allows for an aging loss of 3 % per year. If the loss rate exceeds this, the distiller is still liable for taxation at the rate of the product volume he should have had with a 3 % per year loss rate. This is one of the strong incentives, apart from product quality, for the care taken in providing the proper aging conditions.

After aging for the required length of time the product is bottled, sometimes after blending the contents of a number of barrels with differing flavour nuances to yield a whisky with consistent flavour characteristics appropriate for the particular brand. When taking into account the average annual loss rates, it is simple to see why an 8 or 12 year old whisky costs more per bottle, than a 4 year whisky, since there may easily be 10 to 20 % less whisky to bottle after these aging periods. This factor is apart from the inventory and inflation costs of keeping a valuable product stored for any length of time. Once bottled there is very little, if any change in the product because of the impermeability of the glass. There is also no stability problem since whiskies, and the spirits class generally, have a sufficiently high alcohol content to inhibit or kill any micro-organisms which may be present.

Vodka, presently second in sales volume to whiskies in the U.S.A., consists of straight neutral spirits (95 % alcohol) diluted with pure water to the marketing concentration required. Ideally there should be no residual congeners in a first class vodka, which is ensured by some producers by passing the neutral spirits through activated charcoal, or by redistillation before water addition. Perhaps the growth in the proportion of vodka in the distilled spirits market can be attributed to the fact that it is a beverage without significant aroma, character, or taste. For this reason, it readily lends itself to the preparation of many types of mixed drinks, punches, and homemade cordials.

Gin is produced entirely from neutral spirits too (i.e. without the fermentation congeners), but has an aromatic flavour element derived from a final distillation over juniper berries or coriander (a member of the carrot family) seeds.

Brandies have the special feature that they are prepared by distillation of a wine, generally a grape wine, rather than a "beer". The great variety in flavour and qualities of the various brandies originates from the different types of wines distilled, whether the distillation was batch or continuous, and aging and blending details [6, 61].

Rums are a distinctive beverage of this class in that they consist of the distillate from fermented molasses or other solution of sugar cane juice origin. So-called light or heavy rums are obtained depending on the sugar cane product that is fermented, and the details of distillation and maturation process used.

Cordials (liqueurs) may be made from any of the distilled spirits already described, blended in combination with various fruit juices, flowers, botanicals, and miscellaneous other flavouring materials [67]. Maceration, percolation, and distillation techniques are all methods which are used to place the blend of flavour components into the spirits base used for solution. Over the years many generic names have come into common use to enable recognition of the principal flavour elements, such as Creme de Cacao (chocolate), Creme de Menthe (peppermint), Triple Sec (citrus peel), Anisette (aniseed), etc. Also certain brand names have, by tradition, established a well-known flavour association with them such as for example Benedictine (aromatic plant flavours) and Drambuie (whisky flavour). Considering the wide range of flavour types produced in the cordial class alone, the range of consumer-acceptable flavour variations possible from the distilled beverage class far exceeds that possible from either of the other two classes of alcoholic beverage, the beers and the wines. But a producer of distilled beverages, in turn, has to face a greater technological and investment input and far closer government regulation than the other two classes, to produce a competitive consumer product for market.

### 14.4.5 Environmental Aspects of Distillery Operations

Similarities of raw materials and many of the procedures of distillery operations to those of brewing and winemaking mean that many of the emission control problems are similar. Much of the discussion of those sections is also appropriate here.

The higher fermentation temperatures and rates used for whisky production, over those used for beer or wine production, undoubtedly increases the vaporization rate of organics from the fermenting mash. Not surprisingly, one distillery, while producing at the rate of 5.14 proof gallons (U.S.) per bushel of grain (equiv. to 552 L 50 % alcohol/$m^3$ grain), had an average ethyl alcohol vapour emission rate corresponding to 182 g/$m^3$ of grain input during the course of the fermentation [68]. Vapours of other organics were also lost, prominent among these being ethyl acetate averaging 0.59 g/$m^3$, isoamyl alcohol at 0.17 g/$m^3$, and isobutyl alcohol at 0.051 g/$m^3$, all expressed per cubic meter of grain input. The study citing these data used activated charcoal for vapour capture and analysis. This system, if appropriately scaled up, could also be used for control and recovery of these organics, if necessary for the locale of operation.

In the area of aqueous waste generation a distinguishing aspect of distillery operations which has to be dealt with in addition to the usual types of waste streams generated by breweries and wineries, is the stillage or "distillery slops" stream. Beer and wine, as products, still contain virtually all of the unfermented original ingredients present, as well as the fermentation products. Distillery operations, however, since they are primarily interested in the alcohol and congener fractions produced by fermentation, are left with a residual waste water stream from the distillation containing 2 to 7 % total solids [69]. This consists of mostly organic matter comprising starches, dextrins, and unfermented sugars with beers, or unfermented sugars with wines. There also will be a significant nitrogen content present in this stream when the distillery has fed the yeast content (lees) with the beers or wines stream to the still. Biochemical oxygen demands of 1,500 to 5,000 ppm have been cited for stillage in the absence of lees [6], and in the 7,000 to 50,000 range for lees stillage [6, 69]. Mass emission rates of 0.60 kg/tonne BOD and 0.90 kg/tonne COD, both expressed per tonne of grapes crushed, have been cited for a winery operating with stills. These figures probably represent very conservative emission rates as compared to distillery operations producing solely brandies from newly fermented wines, or whiskies.

As with breweries, if the distillery is operating near the sewerage system of a large center of population, liquid wastes can generally be discharged directly to this system. If the distillery waste waters can be diluted with a much larger volume of lower BOD domestic sewage, the composite stream will usually be amenable to normal methods of sewage treatment. But this disposal method will also generally require the payment of a surcharge to compensate for the additional treatment costs so imposed [6].

Stillage is being tested as the organic feed to an anaerobic digestion unit, where cultures of bacteria are used to convert the carbon content of the wastes into methane [70]. Preliminary projections, through combustion of the methane produced rather than using purchased oil for the distillery's energy requirements, give an estimated 530 m$^3$ (116,000 Imp. gallon) annual oil saving for this distillery.

If the distillery is operating in a remote area, with plenty of available land, stillage may be discharged to an aerated pond for treatment or may be used for irrigation [6, 69]. Decomposition and integration of the waste into the soil are accelerated by discing or light harrowing at regular intervals.

An alternative to these measures is to recover dried yeast and dried spent grains from the stillage, generally for incorporation into cattle feed. The lower BOD values of the residual weaker stillage remaining may then be more economically treated for BOD removal than the waste streams which contain the unsegregated spent yeast and spent grains. Treatment in a trickling filter, in which 60 to 90 % BOD reductions have been observed, or in a biofiltration plant employing two aeration units in series which has demonstrated similar levels of improvement, are effective methods of dealing with this waste stream [69].

**Table 14.12** Trends in the U.S. supply of industrial ethanol from fermentation versus synthetic sources [a]

| | Industrial ethanol, million L of 95% | | | Percent by fermentation |
|---|---|---|---|---|
| | Fermentation [b] | Synthetic [c] | Total | |
| 1930 | 364 | 3.4 | 367.4 | 99 |
| 1935 | 325 | 35 | 360 | 90 |
| 1940 | 364 | 122 | 486 | 75 |
| 1945 | 669 | 223 | 892 | 75 |
| 1950 | 232 | 394 | 626 | 37 |
| 1955 | 168 | 652 | 820 | 26 |
| 1960 | 95 | 939 | 1 034 | 10 |
| 1965 | 229 [d] | 1 126 | 1 455 | 16 |
| 1970 | 241 [d] | 1 164 | 1 305 | 18 |
| 1975 | 316 [d] | 850 | 1 166 | 27 |
| 1980 | 246 | 574 | 810 | 30 |
| 1982 | – | – | 720 [e] | – |

[a] Compiled from [115, 161 – 163].
[b] Fermented from molasses, grain, sulfite liquors, cellulose etc.
[c] From ethyl sulfate, and directly from ethylene, the latter predominating since 1970.
[d] Includes beverage alcohol, therefore higher than the straight industrial product volume.
[e] Preliminary estimate.

# 14.5 Industrial Ethyl Alcohol

Before 1945, most of the supply of ethyl alcohol for industrial solvent or feedstock uses was derived from fermentation sources (Table 14.12). Since this time, the reliability and low cost of petrochemical routes to the product caused a rapid replacement of fermentation sources by synthetic routes in the U.S.A. The early petrochemical sources were based on the hydrolysis of ethyl sulfate, but more lately this has been dominated by processes based on the direct gas phase hydration of ethylene [71] (Equation 14.18-14.20).

$$CH_2=CH_2 + H_2SO_4 \rightarrow CH_3CH_2OSO_3H \quad (14.18)$$

$$\begin{aligned} CH_3CH_2OSO_3H + H_2O \rightarrow \\ CH_3CH_2OH + H_2SO_4 \end{aligned} \quad (14.19)$$

$$CH_2=CH_2 + H_2O \xrightarrow{H_3PO_4} CH_3CH_2OH \quad (14.20)$$

Since 1975 however, increased costs and scarcity of the petroleum resource appear to have caused a reversal of this trend and have renewed the interest by producers in fermentation routes to ethanol [72]. Several recent announcements of new construction plans for large fermentation units based on a variety of substrates strengthen the indirect evidence of the reversal of this trend [73-77], and at least one distillery formerly used for whisky production has been converted to industrial alcohol production [78].

Starches, and fruit, or sugarcane sugars can be, and are used for industrial alcohol production in a similar manner to that used for the beverage products [79-81]. Industrial end uses, however, decrease the quality requirement on the raw materials which may be used [82]. For instance it is possible to conceive of mobile, trailer-mounted fermentation distillation units which can usefully convert off-specification fruit and starch crops to fuel grade ethanol on a contract basis. Also the lower quality requirement means that many sources of fermentable sugars that are inappropriate for beverage alcohol production, such as waste sulfite liquors, may be profitably tapped for industrial alcohol production [79, 83]. As much as 80 L of ethanol per tonne of wood pulp produced is possible from waste sulfite liquor under some conditions [84]. The inadvertent cellulose hydrolysis to sugars obtained during sulfite pulping may be deliberately promoted with heat and hydrochloric (40 %, at 20 °C), or sulfuric (0.5 %, at 130 °C) acids in what

have been known as the Bergius and the Scholler processes (Equation 14.21).

$$(C_6H_{10}O_5)_n + n\,H_2O \xrightarrow{H^+} n\,C_6H_{12}O_6 \quad (14.21)$$
cellulose               glucose

Chemical hydrolysis of cellulose has not, however, been as economically attractive as some more recently developed enzymatic routes [85-87]. But the door that the cellulose hydrolytic route in general opens to the use of all kinds of farm wastes, straw, sawdust, waste (and non-recyclable) paper etc. to produce sugars for fermentation ethanol provides significantly widened raw material prospects for alcohol production.

All fermentation processes to alcohol strive to obtain a 6 % or higher concentration of ethanol in water from the fermentation step, the minimum which is presently economically attractive to recover. Lower alcohol concentrations than this cause the energy costs for distillation to become disproportionately high. From the fermented solution of ethanol in water, procedures similar to those outlined for beverage spirits production yield a 95 % alcohol, 5 % water solution corresponding to the azeotropic composition [88]. This is the maximum concentration of ethanol which can be achieved directly by distillation from low concentrations of ethanol in water, but is a product already suitable for many industrial uses.

For the 10 % alcohol-gasoline blends, the formula for the "gasohol" proposals for automotive use, as well as for many of the more critical industrial applications, anhydrous ethanol is required. Anhydrous ethanol is miscible with gasoline, but the presence of even traces of water in the alcohol or the hydrocarbon component causes a water-rich phase to separate from the composite fuel [89]. Small amounts of 95 % ethanol may be conveniently dried by addition of the correct amount of quicklime (calcium oxide, Equation 14.22).

$$CaO + H_2O \rightarrow Ca(OH)_2 \quad (14.22)$$

No stable hydrate is formed so that a mole of calcium oxide is required for each mole of water to be removed [90]. Since both lime components are slightly soluble in the ethanol the dried alcohol still requires distillation to obtain a high purity product. Azeotropic distillation is the most economical method to dry large volumes of ethanol. A third component, usually benzene, is added to the alcohol/water binary azeotrope already obtained from the dilute solutions of ethanol in water. The boiling point of an azeotropic composition is always below that of the component with the lowest boiling point, and the vapour pressures of each of the components present, at the boiling point, are additive. Therefore on distillation of the 95 % ethanol in the presence of added benzene the ternary azeotrope, of water, alcohol, and benzene, with a boiling point of 64.6 °C, forms the overhead product from the first distilling column (Table 14.13). Thus, if sufficient benzene is added to remove all of the water from the ethanol/water azeotrope in this way, the bottom product from the distilling column A, is pure, dry ethanol (Figure 14.8). A slight excess of benzene is not critical to this result. Traces of toxic residual benzene remain in this alcohol product making it unsuitable as a beverage component, even though it was originally produced by fermentation.

Condensation of the top product from column A, which is the ternary azeotrope, then allows two phases to separate. The upper phase is rich in benzene (84.5 %) with only a trace of water present (1.0 %). The lower phase of the separator is rich in ethanol and water. Return of the upper phase di-

**Table 14.13** Boiling points and compositions of azeotropes appropriate to the drying of ethanol [a]

| Components and % composition | Boiling point, °C |
|---|---|
| water/ethanol/benzene | 64.6 |
| 7.4    18.5    74.1 | |
| benzene/ethanol | 67.8 |
| 67.6    32.4 | |
| benzene/water | 69.4 |
| 91.1    8.9 | |
| ethanol/ water | 78.2 |
| 95.6    4.4 | |
| pure ethanol | 78.5 |
| pure benzene | 80.1 |
| pure water | 100.0 |

[a] Data from [164].

rectly to an upper plate of the first column amounts to a short recycle time for much of the benzene added to this column.

The lower phase from the separator is redistilled in a second column, column B, recovering ternary azeotrope as the overhead product and water-rich ethanol as the bottom product. A third distilling column, column C, completes the recovery of ethanol

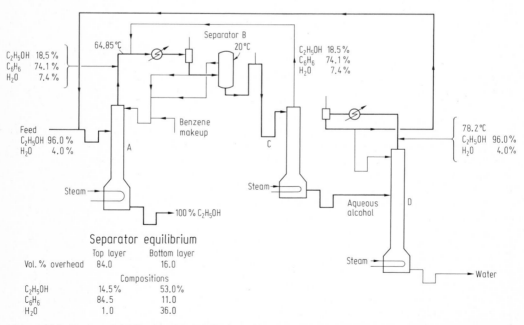

**Figure 14.8.** Drying of 95 % ethanol by distillation with benzene. Reprinted from Perry, pp. 13-45, 13-50; CE, 67 (10), 129 (1960) courtesy McGraw-Hill

as the ethanol/water azeotrope top product of this column, from the now water-rich ethanol stream of the second column. The bottom product of the third column is the water originally present in the ethanol/water azeotrope feed to column A, now entirely freed of solvents.

Columns B and C are smaller than A since the distilling volume to be accommodated is less at these stages. Apart from distilling energy costs the only additional raw material cost of drying ethanol by this method is for the benzene make-up (to replace process losses) amounting to about 0.2 % of the volume of ethanol produced. In part this is lost in the dried ethanol, as evidenced by a strong benzene absorption in ultraviolet spectra taken of ordinary absolute alcohol. Thus, if one wishes to obtain an ultraviolet spectrum of a water stable compound free of this benzene absorption, it is better to use 95 % ethanol as the solvent, to avoid this interference.

# 14.6   Aerobic Fermentations

Here the term fermentation is used in its more general sense, that is to refer to a chemical conversion accomplished by micro-organisms, rather than in its strict sense of a biological conversion carried out in the absence of air. Most micro-organisms are aerobes or they are facultative, i.e. they are capable of functioning under either aerobic or anaerobic conditions. Therefore, to accomplish fermentations under aerobic conditions gives far greater scope to the types of chemical change possible than if one is limited to anaerobic transformations. But the wide range of organisms and the range of appropriate conditions to which aerobes can adapt, mean that the risk of contamination by stray organisms is much greater than with conversions operating under anaerobic conditions. Also the very need of the aerobic system to be constantly replenished with oxygen or air provides greater risk of exposure of the process to organisms entrained in this aeration air. Nevertheless, as the technology for aerobic fermentations was developed, this also increased the number of options for new products accessible by this route. Yeasts can function in an aerobic mode, as has already been mentioned and as is operated commercially for bakers' yeast production, for instance. This fermentation mode is fine when increased cell mass is the primary objective. But the end products of aerobic fermentation of sugar by yeast are carbon dioxide and water, not very useful from a chemical conversion standpoint.

## 14.6.1  Operating Details of Aerobic Fermentation

Neither oxygen nor air is very soluble in water so that without taking special measures to increase the rate of exchange of dissolved gases from a medium to and from air this rapidly becomes the limiting factor for aerobic conversions. In the laboratory, the rate of air exchange per unit volume of liquid is increased by using only a thin layer of the liquid fermentation medium in the bottom of an Erlenmeyer flask. Air exchange between the inside of the flask and the outside air is obtained using a tight, sterile cotton plug which also serves as a filter barrier to intrusion by outside organisms. Frequently the flask is also gently and continuously shaken to further boost the air/liquid gas exchange rate. These relatively simple methods function well for preliminary tests and process exploration, but are inappropriate for larger scale operation.

For larger scale production use, shallow pans made of non-corrosible metal or plastic and placed in portable racks, may still be used in some situations to provide efficient liquid-air contact. When placed in a sterile room under closely controlled temperature and humidity and with frequent sterile air changes, this method functions well. The same efficient air-liquid contact may be obtained from a loosely packed, moist, solid medium which is allowed to be permeated with air. For instance bran is used as the substrate for enzyme production in this manner, by *Aspergillus oryzal* [10], and wine vinegar (acetic acid) is produced by trickling wine (basically ethanol in water) over a short tower of wood chips which are used as the support to provide air to *Acetobacter aceti* [6] (Equation 14.15). In fact by law, vinegar is a solution of 3 to 5 % acetic acid in water that *must* be produced by fermentation, and not by petrochemical methods.

However, the most common procedure for large scale aerobic fermentions is to conduct these in stirred deep tanks of the liquid medium [91]. Efficient air exchange is obtained by continuously pumping sterile air into the liquid column in the tank through crossed or coiled perforated pipe placed at the bottom of the tank, at rates up to one volume of air per unit of medium volume per minute [92, 93]. This method does promote aerobic growth throughout the medium accomplishing a large volume of production or conversion in a limited space. But problems relating to the maintenance of sterility and occasionally from foam formation have to be dealt with to avoid losses of medium or product [4, 94]. Overall, however, submerged aerobic fermentations of this type represent the most common method used by industry for aerobic conversions.

## 14.6.2 Important Aerobic Fermentations to Single Cell Protein

Despite not knowing the detailed metabolic pathways of many aerobic fermentative conversions, several important products today are obtained largely, if not exclusively, via the operation of aerobic fermentation processes. Of course the activated sludge, trickling filter, and aerated lagoon modes of industrial and domestic waste treatment are also very large, non-specific versions of aerobic fermentations.

Probably the largest scale, specific product aerobic fermentations are those now used to produce single cell protein (SCP) or yeast biomass from hydrocarbons [95, 96]. Here, an aerated yeast suspension in water containing trace nutrients is used to generate yeast biomass from gas oil, a mixture of straight chain hydrocarbons in the $C_{12}$ to $C_{20}$ range [3, 97] (Equation 14.23).

$$\begin{array}{cccc} \text{yeast} + \text{water} + \text{gas oil} + \text{air} \rightarrow \text{yeast biomass} \\ \text{(with} & \text{(}C_{12} \text{ to} & \text{(with} \\ \text{nutrients)} & C_{20}\text{)} & \text{ammonia)} & (14.23) \end{array}$$

The yeast produced by continuous culture techniques is separated from the liquid medium and solvent washed by centrifugation or filtration techniques. After drying, a protein supplement containing 65 to 68 % protein and suitable for addition to animal feeds is obtained. This protein content compares very favourably with that of dry fish meal, which contains about 65 %, and dry skim milk powder, with about 32 %. Processes using essentially this operating procedure have been run on the thousands of tonnes per year scale in the U.K., France, and Italy [97], but regulatory problems with facilities operating on unpurified gas oil feedstocks have caused some shutdowns [98-100]. Nevertheless, because of the cell mass doubling time of 2.5 to 3 hours and the very efficient carbon conversion to protein of this technology, these developments deserve to be considered further. If

problems involved with human consumption of SCP can be resolved this certainly represents a far more direct and less energy intensive route to protein than via meat or fish, and possibly even less energy intensive than plant proteins produced using high technology (chemical fertilizer, tractors, herbicides, pesticides) agricultural methods.

It is also possible to convert methane, the lowest cost paraffin on a carbon content basis, to a mixed cell biomass in a similar type of fermenter [101, 102]. But because of difficulties with obtaining sufficiently high concentrations of methane in the water phase, and the associated pressure required to keep it there, this system is not generally favoured [103].

Methanol, relatively easily derived from methane and also readily purified, avoids some of the contamination problems of the higher, n-paraffin carbon sources. At the same time methanol is easily put into aqueous solution at any desired concentration. This is the basis of the U.K.'s Imperial Chemical Industries process to protein from methanol using *Methylophilus methylotrophus,* on a scale of 30,000 to 50,000 tonnes per year [104, 105] (Figure 14.9). Recombinant DNA technology was employed to raise the efficiency of methanol conversion by this organism [106, 107]. The dry product, trade named Pruteen, contains 72 % protein and is highly suitable for animal feed supplementation.

## 14.6.3 Examples of Aerobic Fermentation of Hydrocarbon Substrates

Aerobic micro-organisms are capable of highly useful selective oxidative transformations on specific hydrocarbon substrates which would at best be less efficient and some would be impossible by straight chemical means [108]. For one example strains of *Pseudomonas desmolytica* can carry out the specific oxidation of p-cymene to cumic acid [109], a process difficult to achieve by ordinary chemical means (Equation 14.24).

Some steroid transformations which are difficult to achieve by ordinary synthetic means may also be accomplished readily by specific microbiological methods [110, 111]. Salicylic acid may be obtained in a 94 % yield, on a weight for weight basis, by the action of *Pseudomonas aeruginosa* on naphthalene [112] (Equation 14.25).

Acetylation of the salicylic acid with acetic anhydride or ketene can then nearly quantitatively

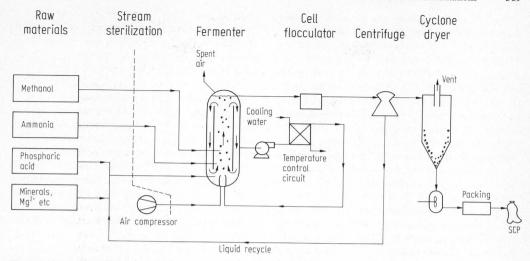

**Figure 14.9.** Details of the ICI single cell protein (SCP) process based on methanol [91]

p-cymene → cumic acid    (14.24)

naphthalene → salicylic acid → "ASA"    (14.25)

convert this to acetylsalicylic acid (ASA, or Aspirin), one of the most widely sold, non-prescription analgesics.

### 14.6.4 Amino Acids by Fermentation

As the pressures on the available food supply become greater, the limited supply of available first class (animal) protein is going to place an increasing dependence by man on plant (vegetable) sources of protein for this necessary component of our diets. Most of the first class, animal-derived protein (meat, fish, eggs, milk, cheese, etc.) has a food demand to produce it of 2 to 20 times the weight of protein obtained. This factor makes these an expensive source of protein in relation to the amount of arable land required to support their production, as compared to obtaining the same protein more directly from plant sources.

The problem with using plant sources of protein to provide a large fraction of our protein requirements is that ordinary plant sources do not provide the proper balance of the amino acids required for normal human nutrition [113, 114]. Some of the deficient amino acids may be synthesized by the body. But some, the "essential amino acids", must be acquired through the diet to obtain proper human growth and nutrition. Commercial production of amino acids has, thus far, been predominantly for animal feed supplementation. However, it can be seen from the comments above that to have a supply of synthetically-produced essential amino acids available to enrich cereals with those in which they are deficient could provide a nutritional advantage. Cereals enriched in this way could greatly extend the suitability of plant crops to provide most, if not all, of the human protein requirement. Thus, much of the corn, oats, wheat and other crops now grown for animal feed could, instead, be diverted with appropriate amino acid supplementation to human consumption. In this way more people could be fed from the same area of arable land than would be possible with a heavy dietary requirement of first class (i.e. animal) protein.

At the moment glutamic acid, a non-essential amino acid, is produced on the scale of about 270,000 tonnes per year [115], mostly in Japan. This is the largest volume of any commercially produced amino acid. Common starting materials for fermentation production of glutamic acid are glucose, molasses, corn or wheat glutens, or Steffen's waste liquor (from beet sugar mills) plus trace nutrients [3, 79], although it is also possible to use purified fractions of n-paraffins ($C_{12}$ to $C_{20}$) or acetic acid as substrates [115] (Equation 14.26).

$$\text{glucose, or} \rightarrow \underset{\text{glutamic acid}}{HOOCCH_2CH_2CH(NH_2)COOH}$$
$$+$$
$$\rightarrow \underset{\text{zwitterionic form}}{HOOCCH_2CH_2CH(NH_3)COO^-} \quad (14.26)$$

Glutamic acid, as the end product of this fermentation, results from a tinkering of the normal cell metabolism of *Micrococcus glutamicus* in such a way that glutamic acid, as it is formed, "leaks" through the cell membrane and accumulates in the aqueous medium [115]. Simple treatment of the glutamic acid product with sodium carbonate then yields monosodium glutamate (MSG). MSG is not produced from a nutritional interest but from an incentive from the large market for this product as a flavour enhancer. But the experience gained by the fermentative production of glutamic acid on a commercial scale has served to pave the way for the commercial production of essential amino acids.

When monosodium glutamate is produced by a petrochemical route [3] (Equation 14.27), the dl product obtained by this process is only about half as effective as the fermentative product on a weight basis.

$$CH_2{=}CH{-}CN \underset{\text{reaction}}{\overset{oxo}{\rightarrow}} OHC{-}CH_2{-}CH_2{-}CN$$

$$\underset{\text{2. } NH_3}{\overset{\text{1. HCN}}{\rightarrow}} NC{-}CH(NH_2){-}CH_2{-}CH_2{-}CN$$
$$\underset{\text{lysis}}{\overset{\text{hydro-}}{\rightarrow}} HOOC{-}CH(NH_2){-}CH_2{-}CH_2{-}COOH$$
$$(14.27)$$

For this, and other economic reasons the petrochemical route is not as attractive as the fermentative route, so is not practised commercially. Of the two essential amino acids which are generally deficient in cereals, lysine and methionine, only lysine is currently produced by fermentation to any significant extent [115]. Again a carbohydrate source is a suitable substrate for cultures of *Corynebacterium glutamicum* to produce lysine (Equation 14.28).

$$\underset{\text{source}}{\overset{\text{carbohydrate}}{\rightarrow}}$$
$$H_2N{-}CH_2CH_2CH_2CH_2CH(NH_2){-}COOH \rightarrow$$
$$\text{lysine} \quad +$$
$$H_2N{-}CH_2CH_2CH_2CH_2CH(NH_3){-}COO^-$$
$$\text{zwitterionic form} \quad (14.28)$$

As with glutamic acid production, some modification of the normal metabolic pathway of *C. glutamicum*, imposed by the fermentation conditions, is necessary in order to accumulate lysine in the medium [115]. At present, worldwide production of lysine by fermentation routes totals some 36,000 tonnes annually, most of which (some 95 %) is by Japan [116]. About 9,000 tonnes of racemic lysine (dl mixture) per year is also produced synthetically from petrochemical sources. One interesting synthetic suggestion for lysine, which does not appear to have been taken up, is to start with caprolactam (or 6-amino-caproic acid), a starting material readily available on a tonnage scale [117].

As yet methionine, the other amino acid in which cereals are deficient, is only produced from pentaerythritol or acrolein by synthetic means, not by fermentation [115, 118] (e.g. Equation 14.29). However, it is probably only a matter of time before the production of this and other essential amino acids yield to production via biological methods. This will come either from a detailed understanding of the metabolic processes of more types of suitable micro-organisms or by genetic engineering techniques [119-122] to generate specific new types of micro-organisms with the required synthesis characteristics.

$$CH_3SH + CH_2{=}CHCHO \rightarrow CH_3SCH_2CH_2CHO \xrightarrow[\text{2. }(NH_4)_2CO_3]{\text{1. NaCN}} \tag{14.29}$$

$$CH_3SCH_2CH_2\overset{\;\;-\;\;+}{\underset{\underset{CONH_2}{|}}{C}}HNHCOONH_4 \quad CH_3SCH_2CH_2\overset{|}{\underset{\underset{NH-CO}{|}}{C}}H{-\!\!-}CO{\Big\rangle}NH \rightarrow CH_3SCH_2CH_2\overset{}{\underset{\underset{NH_2}{|}}{C}}HCOOH$$

dl-methionine

Plant genetic experiments presently being undertaken with corn [123] (maize) and with rice [124] may also help to improve the nutritional properties of these widely grown cereal crops, and in so doing help to close the amino acid nutritional gap which exists between plant sources of proteins and first class proteins [114, 125]. However, this will still not detract from the many other uses of amino acids, such as for chelating agents, as dietary supplements, as building blocks for specialty proteins, and a small but important use as components of intravenous feeding solutions [126]. All of these uses will continue to provide incentives for large scale biochemical and chemical synthesis of amino acids.

### 14.6.5 Other Important Aerobic Fermentation Products

Since the original development and commercialization of penicillin production [127] in about 1943, the antibiotics as a class and the penicillins in particular are still dominant in sales value in the aerobic fermentation products area [128]. The submerged culture methods used to prepare penicillin itself, which sold for 330 $/g in 1943, were sufficiently improved and refined that by 1970 the price had dropped to about 2 $/g [10] (Equation 14.30).

$$RCONH{-}\overset{H}{\underset{\underset{O}{\overset{|}{C}}}{\overset{|}{C}}}{-}\overset{S}{\overset{\diagup\;\diagdown}{CH}}\quad C(CH_3)_2$$
$$C{-}N{-\!\!-}CH(COOH)$$

R = PhCH$_2$–
"Penicillin"
benzylpenicillin,
or Penicillin G

R = Ph-O-CH$_2$–
Phenoxymethyl
penicillin, or
Penicillin V

R = Ph–CH(NH$_2$)–
Ampicillin
(semi-synthetic)                              (14.30)

Details of the more complex production methods and process flowsheets are given by Shreve and Brink [11]. Despite the decreased cost of penicillin and of the related antibiotics such as streptomycin, neomycin, the tetracyclines, etc. as their production methods were streamlined, the medical benefits of these drugs are so well established that the value of U.S. production of drugs in this class now approaches a billion dollars annually [128].

Many vitamins too, are produced by fermentation methods, although some are also produced by chemical synthesis procedures. Nevertheless those that are produced fermentatively represent another important class of submerged aerobic culture products. The total value of prescription vitamins alone that were marketed in the U.S.A. in 1980 exceeded $ 130 million [128]. In 1959, during the early development of the commercial fermentative product vitamin B-12, it sold for 139 $/g [10]. Again through processing improvements, the price of this product has been brought down dramatically so that it now sells for 8 $/g [129]. Considering only the two B vitamins, riboflavin (vitamin B-2), presently selling at 56 $/kg, and vitamin B-12, these alone now account for U.S. sales of more than $ 20 million, annually.

Citric acid is the largest scale pure organic acid produced by a microbiological process. Present world production is estimated to be well in excess of 200,000 tonnes per year [130, 131]. It is produced by *Aspergillus niger* from a medium containing sugars, obtained generally from low cost cane or beet molasses. Occasionally, however, a pure sugar solution may also be used. The stoichiometry of this process was originally established by Currie [132] (Equations 14.31, 14.32).

$$C_6H_{12}O_6 + {}^3\!/_2\,O_2 \rightarrow$$
glucose
$$HOOCCH_2{-}COH(COOH){-}CH_2{-}COOH$$
citric acid
$$+\,2\,H_2O \tag{14.31}$$

$$C_{12}H_{22}O_{11} + H_2O + 3\,O_2 \rightarrow$$
sucrose

$$2\,HOOCCH_2-COH(COOH)-CH_2COOH$$
citric acid

$$+ 4\,H_2O \hspace{3cm} (14.32)$$

To obtain citric acid as the metabolic product again requires interfering with the normal Krebs tricarboxylic acid cycle in such a way that citric acid metabolism is blocked. Various methods may be used to accomplish this [11]. Most commonly the metabolic control is achieved by careful regulation of concentrations of trace metals available as co-enzymes to the various enzyme pathways used by *A. niger,* so that some of these are rendered ineffective [130] (are blocked). The isolated product is useful as a synthetic flavouring agent, particularly for all kinds of fruit flavours. It is also a valuable chelating agent, and may shortly be in large scale demand as a component of an aqueous sulfur dioxide scrubbing solution. No synthetic method of citric acid production yet devised has been able to compete effectively with this microbiological method.

Several other organic acids are also made commercially on a large scale by microbiological techniques. Some of the better known examples of these are acetic acid (vinegar) via the *Acetobacter aceti* metabolism of alcohol in water, and lactic acid (Equation 14.33), produced by *Streptococcus lactis* fermentation of hexose sugars [10].

$$CH_3CH(OH)COOH \hspace{0.5cm} HOCH_2(CHOH)_4COOH$$
lactic acid                 gluconic acid

$$H_2C=C(COOH)-CH_2COOH \hspace{1.5cm} (14.33)$$
itaconic acid

Common sugar sources for lactic acid are molasses, hydrolyzed corn, and whey, a milk byproduct of cheese production. Gluconic and itaconic acids are also microbiological oxidation products, obtained by methods similar to those used for citric acid [10]. Gluconic acid is formed, via simple oxidative fermentation, by *Aspergillus niger* or *Acetobacter suboxydans*, of the glucose aldehyde group.

## 14.6.6 Soluble, and Immobilized Enzymes

Further development of a branch of the fermentation industry has pursued the isolation of enzymes, which are the catalytically active proteins of a cell. In the presence of the appropriate co-enzyme, these proteins are the active principle of the cell which serve to carry out the chemical changes from which the cell gains energy. Rennet, a milk curdling enzyme used in cheesemaking, and the amylases, used in food processing and for laundry stain removal are among the earliest examples of isolated enzymes which have been used commercially in the absence of host cells [133]. The essence of the value of enzyme use is that the enormous rate enhancements obtained by microbiological means can be achieved in the absence of the organism, if the enzyme responsible is stable and can be isolated [134]. A direct result of this utility is a predicted increase in value of the world enzyme market from $300 million in 1979 to $480 million in 1985 [135].

The advantage of using isolated enzymes rather than whole cells to carry out a chemical change is that the process does not have to cater to the idiosyncrasies of living cells. However enzymes, too, are sensitive entities so that care has to be exercised in the conditions under which they are used, in order to maintain catalytic activity. Very often the problem here is that use of the isolated enzyme is frequently a more expensive proposition, on a single use basis, than propagated cells. There are other disadvantages to their use too, such as the need to remove enzyme from the product once the change desired is completed. Or, alternatively, long reaction times may be necessary if it is decided to use a low concentration of an expensive enzyme to decrease enzyme cost [136].

Some of these difficulties of enzyme utilization may be decreased or eliminated, and the full potential of even expensive enzymes realized by fixing, or immobilizing the enzyme in some way. The solution of substrate for conversion is then passed through a bed or tube of the immobilized enzyme, and the product(s) collected as the solution emerges after contact with, and conversion by, the enzyme. A number of methods are being tested for enzyme immobilization, the actual method used in any particular case depending on the operating details of the enzyme system employed and the nature of the solvent to be used as the substrate carrier, which is usually water [137, 138]. Enzyme, or inactivated cells, may be encapsulated in a film, or encased in a gel which is permeable to both substrate and product, but not to enzyme [136]. Porous glasses, or insoluble polymers such as a derivatized cellulose may be used as a support onto which enzyme is adsorbed [139]. Pendant functional groups of a polymer, such as those of the ion exchange resins, can be used either to ionically bind the enzyme to the resin active sites or to convalently bond the enzyme to the resin [140, 141]. Resin may also be used to form the backbone of a polymer chain using a linking bifunctional monomer such as glutaraldehyde to react with enzyme sites that do not affect its catalytic activity [138, 141].

Hollow fiber technology may also be usefully wedded to immobilized enzyme technology by making the wall of a hollow fiber out of a gel in which the enzyme is already immobilized [138]. The finished fiber may then be used simply as a large exposed surface area for enzyme activity, packed in small chopped pieces inside a larger tube. Or it may be used as a membrane into which substrate solution permeates from the outside, under a pressure differential. Passage through the tube wall allows contact of the substrate with the gelled enzyme, which converts substrate to product. Then, under the influence of the differential pressure across the wall of the tube, the product would flow through to the inside of the tube, eventually to be collected at a multi-tube header.

Immobilized enzymes have already been successfully tested for sugar conversion to ethanol [142]. Conversion of the lactose of whey to glucose and galactose is another confirmed application of immobilized enzymes. Preliminary tests have prompted predictions of 300 day operating cycles [143], when processing 6,500 L/day (1,800 U.S. gal/day) of whey [144]. This application also produces a saleable commodity from a whey analyzing about 93.5 % water, 5 % lactose, 1 % protein, and 0.5 % ash, which was an increasingly aggravating waste disposal problem for cheese makers [144]. In doing so, it produces two monosaccharides of higher sweetening power and therefore greater food processing value than lactose, from the lactose-containing waste.

Enzymes, in both soluble and immobilized forms, are also increasingly being employed in the food processing industry, commonly to convert low sweetening power sugars into more potent sweeteners. As examples, glucose is isomerized to fructose, which is 2 1/2 times as sweet, by an immobilized glucose isomerase enzyme in a fixed or a fluidized bed [145]. Also a high fructose corn syrup, to be produced from corn using two stages of immobilized enzymes, is likely shortly to become a commercial reality. Starch hydrolysis to glucose would be accomplished by an immobilized

glucoamylase in the first stage, in 6 minutes, followed by glucose isomerization to fructose, as outlined above [145].

This power of enzyme systems for specific interconversions of sugars has also been most usefully exploited as an additional fresh milk processing step, at present tested only in Milan. This process converts the normal lactose milk sugar, to which some people react unfavourably, into glucose and galactose which are much better tolerated [144]. Milk treated in this way can now be used as a beverage by those who could only tolerate milk substitutes previously.

Amino acid production too, may extensively benefit from this immobilized enzyme technology [140]. Optically active D(-)-$\alpha$-phenylglycine is one current example which has been confirmed, but more are likely to follow.

The promise of fixed enzymes to process many food and waste products is only just beginning to be realized [4, 146]. The prediction is that combining this technology with the developing field of synthetic enzyme preparations [147-152] will lead to a wide range of feasible, clean, and specific chemical conversions never before thought to be possible.

# Relevant Bibliography

1. Fermentation Advances, D. Perlman, editor, Academic Press, New York, 1969
2. Annual Reports on Fermentation Processes, D. Perlman, editor, Academic Press, New York. Volume 1 issued in 1977, of a continuing series
3. J.E. Bailey and D.F. Ollis, Biochemical Engineering Fundamentals, McGraw-Hill, New York, 1977
4. S. Aiba, A.E. Humphrey, and N.F. Millis, Biochemical Engineering, Academic Press, New York, 1977
5. Industrial Aspects of Biochemistry, Volume 30, Parts I and II, B. Spencer, editor, North-Holland Publishing Co., Amsterdam, 1974
6. G.L. Solomons, Materials and Methods in Fermentation, Academic Press, New York, 1969
7. Chemistry of Winemaking, ACS Advances in Chemistry Series No. 137, American Chemical Society, Washington, 1974
8. Wine Production Technology in the United States, ACS Symposium Series No. 145, American Chemical Society, Washington, 1981
9. 10. Applied Biochemistry and Bioengineering, L.B. Wingard, Jr., E. Katzir, and L. Goldstein, editors, Academic Press, New York, ca. 1978

Volume 1, Immobilized Enzyme Principles, 1976
Volume 2, Enzyme Technology, 1979
11. 12, 13. Advances in Biochemical Engineering, T.K. Ghose, A. Fiechter, and N. Blakebrough, editors, Springer-Verlag, Berlin
Volume 10, Immobilized Enzymes I, 1978
Volume 11, Microbiology, Theory and Application, 1979
Volume 12, Immobilized Enzymes II, 1979
14. Immobilized Enzymes in Food and Microbial Processes, A.C. Olson and C.L. Cooney, editors, Plenum Press, New York, 1974
15. P.A. Hahn, Chemicals from Fermentation, Doubleday, Garden City, New York, 1968
16. C.S. Pederson, Microbiology of Food Fermentations, 2nd edition, Avi Publishing Co., Westport, Connecticut, ca. 1979

# References

1. P. Dunnill, Chem. Ind. (London), 204, April 4, 1981
2. F.J.E. Bailey and D.F. Ollis, Biochemical Engineering Fundamentals, McGraw-Hill, New York, 1977
3. L.M. Miall, Royal Inst. of Chem. Reviews 3 (2), 135, Oct. 1970
4. E.L. Gaden, Jr., Scientific American 245 (3), 180, Sept. 1981
5. Fermentation Unit Runs Nonstop for 75 Days, Chem. Eng. News 58 (22), 27, June 2, 1980
6. M.A. Amerine, H.W. Berg, and W.V. Cruess, The Technology of Wine Making, 2nd edition, Avi Publishing Co., Westport, Conn., 1967
7. Kirk-Othmer Encyclopedia of Chemical Technology, 3rd edition, John Wiley and Sons, New York, 1978, volume 3, page 692
8. A. Weisenberger, Food Eng. 50 (6), 136, June 1978
9. G. Anton and A.M. Vicente, Food Eng. 51 (2), 76, Feb. 1979
10. Riegel's Handbook of Industrial Chemistry, 7th edition, J.A. Kent, editor, Van Nostrand/Reinhold, New York, 1974, page 156
11. R.N. Shreve and J.A. Brink, Jr., Chemical Process Industries, 4th edition, McGraw-Hill, New York, 1977, page 525
12. W.N. McCutchan and R.J. Hickey in Industrial Fermentations, L.A. Underkofler and R.J. Hickey, editors, Chemical Publishing Co., New York, 1954, volume 1, page 347
13. H.J. Phaff, Scientific American 245 (3), 76, Sept. 1981
14. C. Dagleish, La Recherche 11 (110), 434, April 1980

15. M.E. Chavez, Liquor: The Servant of Man, Little, Brown and Co., Boston, 1965. Cited by F.K.E. Imrie, Chem. Ind. (London) 584, July 17, 1976
16. H.S. Corran, A History of Brewing, David and Charles Publishing Co., Newton Abbott, U.K., 1975
17. F.W. Salem, Beer, Its History and Economic Value as a National Beverage, Arno Press, New York, 1972, reprint of 1880 edition
18. A.H. Rose, Scientific American 245 (3), 126, Sept. 1981
19. D.W. McClure, J. Chem. Educ. 53 (2), 70 (1976)
20. T. Lom, Can. Chem. Proc. 58 (12), 14, Dec. 1974
21. M.A. Pozen, Enzymes in Brewing, Lefax, Philadelphia, Pa., ca. 1975
22. L.O. Spetsig, On the Isohumulones of Beer, Proc. European Brewery Conv. Congress, Stockholm, 1965, page 416
23. M.W. Brenner, C. Vigilante, and J.L. Owades, Proc. Amer. Soc. Brewing Chemists, 48, (1956)
24. R. Stevens, Chem. Rev. 67, 19 (1967)
25. M.S. Kharasch, Bureau of Standards J. of Research 2, 359 (1929), cited by Handbook of Chemistry and Physics, 39th edition, C.D. Hodgman, editor, Chemical Rubber Publ. Co., Cleveland, Ohio, 1957, and by Lange's Handbook of Chemistry, 10th edition, N.A. Lange, editor, McGraw-Hill, New York, 1969
26. J.F. Thorpe and M.A. Whiteley, Thorpe's Dictionary of Applied Chemistry, 4th edition, Longman's Green and Company, London, 1941, volume 5, page 8
27. M.A. Amerine, Scientific American 211 (2), 46, Aug. 1964
28. R.I. Tenney, The Brewing Industry, in Industrial Fermentations, L.A. Underkofler and R.J. Hickey, Chemical Publishing Co. Inc., New York, 1954, volume 1, page 172
29. Brewing in Canada, Brewers Association of Canada, Ltd., Ottawa, 1965
30. F.H. Thorp and W.K. Lewis, Outlines of Industrial Chemistry, The MacMillan Company, New York, 1917, page 444
31. Z. Valyi and G.E.A. Van Gheluwe, Tech. Quart. Master Brewers' Assoc. of America 3, 184 (1974)
32. A. Vogler and H. Kunkely, J. Chem. Educ. 59 (1), 25, Jan. 1982
33. Avoiding an Environmental Hangover, Can. Chem. Proc. 64 (2), 29, March 1980
34. S. Beszedits, Water and Polln Control 115 (10), 10, Oct. 1977
35. H.M. Malin, Jr., Envir. Science and Tech. 6, 504 (1972)
36. B.O. Botuk, N.G. Dimitrevskii, Y.P. Vasilenko, and L.K. Kryzhanovskii, Ferment. Spirit. Prom (USSR) 3, 22 (1973). Chem. Abstr. 79, 23294 (1973)

37. J.F. Walker, Water Polln Man., 174 (1972), cited by Water Polln Abstr. (U.K.) *45*, 2953 (1972)

38. H.W. Allen, A History of Wine, Horizon Press, New York, 1961

39. L. Morisset-Blais, Can. Chem. Proc. *57*(12), 31, Dec. 1973

40. Statistical Yearbook, United Nations, New York, 1981, and earlier years

41. P. Duncan and B. Acton, Progressive Wine-making, Amateur Winemaker, Andover, Hants, 1967

42. Kirk-Othmer Encyclopedia of Chemical Technology, 2nd edition, John Wiley and Sons, Inc., New York, 1970, volume 22, page 307

43. F.A. Cotton and G. Wilkinson, Advanced Inorganic Chemistry, 3rd edition, Interscience, New York, 1972, page 448

44. L.G. Saywell, Clarification of Wine, Lefax, Philadelphia, Penn., June 1963

45. Tramp Ions Can Mar Wine, Can. Chem. Proc. *60*(5), 31, May 1976

46. L. Trauberman, Food Eng. *39*, 83, Nov. 1967

47. K. Dimick, M. Burns, G. Burns, and E. Montoya, Spectra Physics Chrom. Rev. *9*(1), 2, March 1983

48. W.H. Detwiler, Chem. Eng. *75*, 110, Oct. 21, 1968

49. M.A. Amerine, H.W. Berg, and W.V. Cruess, The Technology of Wine Making, 3rd edition, The Avi Publishing Co., Westport, Conn., 1972

50. P. Shaw, Sunday Times Magazine, 13, Feb. 6, 1977

51. Test: Asbestos in Wine, Canadian Consumer *7*(3), 44, June 1977

52. H.M. and R. Pontefract, Nature *232*, 332, (1971)

53. E.J. Chatfield and H. Pullan, Can. Research and Devel. *7*(6), 23, Nov.-Dec. 1974

54. C.T. Charlebois, Chem. In Can. *30*(3), 19, March 1978

55. I.J. Selikoff and D.H.K. Lee, Asbestos and Disease, Academic Press, New York, 1978

56. E. Klinger, Analyst *46*, 138 (1921), from Farbenzeit *26*, 6, (1920)

57. A. Agostini, Wynboer *32*(395), 13, 1964. Cited by reference 6

58. T.J. Tofflemire, S.E. Smith, C.W. Taylor, A.C. Rice, and A.L. Hartsig, Water and Wastes Eng. F-3, Nov. 1970

59. Kirk-Othmer Encyclopedia of Chemical Technology, 2nd edition, Interscience, New York, 1963, volume 1, page 501

60. W.F. Rannie, Canadian Whisky, The Product and the Industry, W.F. Rannie Publishing Co., Lincoln, Ontario, 1976

61. J.S. Swan and S.M. Burtles, Chem. Soc. Revs. *7*(2), 201, (1978)

62. Chemical Engineer's Handbook, 4th edition, R.H. Perry, editor, McGraw-Hill, New York, 1963

63. B.G. Reuben and M.L. Burstall, The Chemical Economy, Longman Group Ltd., London, 1973, page 428

64. Kirk-Othmer Encyclopedia of Chemical Technology, 3rd edition, John Wiley and Sons, New York, 1978, volume 3, page 830

65. C.R. Noller, Chemistry of Organic Compounds, 3rd edition, W.B. Saunders Company, Philadelphia, 1966

66. McGraw-Hill Encyclopedia of Science and Technology, 5th edition, McGraw-Hill, New York, 1982, volume 4, page 346

67. P.V. Price, Penguin Book of Spirits and Liqueurs, Handbook Series, Penguin Books Inc., New York, 1981

68. R.V. Carter and B. Linsky, Atmos. Envir. *8*(1), 57, Jan. 1974

69. N.L. Nemerow, Industrial Water Pollution, Origins, Characteristics, Treatment, Addison-Wesley, Reading, Mass., 1978

70. Methane Pilot for Distillery, Can. Chem. Proc. *66*(1), 8, Feb. 19, 1982

71. F.A. Lowenheim and M.K. Moran, Faith, Keyes, and Clark's Industrial Chemicals, 4th edition, Wiley-Interscience, New York, 1975

72. U.S. Output of Synthetic Ethanol to Cease, Chem. Eng. News *61*(18), 18, May 2, 1983

73. Ethanol Fuel from Wood, Can. Chem. Proc. *66*(1), 8, Feb. 19, 1982

74. USDA Process Converts Xylose to Ethanol, Chem. Eng. News *59*(25), 54, June 22, 1981

75. B. Orchard, Can. Chem. Proc. *64*(8), 55, Nov. 26, 1980

76. Ethanol for Gasoline Trucked to California, Chem. Eng. News *57*(26), 10, June 25, 1979

77. Ethanol Joint Venture Formed, Chem. Eng. News *58*(29), 16, July 21, 1980

78. National Distillers to Convert Plant for Gasohol, Chem. Eng. News *58*(19), 8, May 12, 1980

79. W.L. Faith, D.B. Keyes, and R.L. Clark, Industrial Chemicals, 3rd edition, John Wiley and Sons, New York, 1965

80. J.G. da Silva, G.E. Serra, J.R. Moreira, J.C. Concalves, and J. Goldemberg, Science *201*, 903, Sept. 8, 1978

81. W.H. Long, Chem. Eng. News *59*(36), 7, Sept. 7, 1981

82. Gasohol: Focus Shifts to Effect on Animal Feed, Chem. Eng. News *59*(3), 81, Jan. 19, 1981

83. Ethanol for Gasoline Trucked to California, Chem. Eng. News *57*(26), 10, June 25, 1979

84. C.R. Noller, Textbook of Organic Chemistry, 3rd edition, W.B. Saunders Company, Philadelphia, 1966

85. Ethanol Fuel from Wood, Can. Chem. Proc. *66*(1), 8, Feb. 19, 1982

86. USDA Process Converts Xylose to Ethanol, Chem. Eng. News 59 (25), 54, June 22, 1981

87. W. Worthy, Chem. Eng. News 59 (49), 35, Dec. 7, 1981

88. W.R. Oliver, R.J. Kempton, and H.A. Conner, J. Chem. Educ. 59 (1), 49, Jan. 1982

89. Microemulsions Stabilize Ethanol-diesel Oil Blends, Chem. Eng. News 58 (50), 29, Dec. 15, 1980

90. N.V. Sidgwick, The Chemical Elements and Their Compounds, Clarendon Press, Oxford, 1950, volume I, page 248

91. P. Prave and U. Faust, Chem. In Brit. 14 (11), 552, Nov. 1978

92. Y. Hirose, J. Nakamura, H. Okada, and K. Kinoshita, Agric. Biol. Chem. 29, 931, (1965)

93. B. Kristiansen, Chem. Ind. (London) 787, Oct. 21, 1978

94. A.P. Kouloheris, Chem. Eng. 77 (16), 143, July 27, 1970

95. D.D. MacLaren, Chemtech 5 (10), 594, Oct. 1975

96. G. Hamer and I.Y. Hamdan, Chem. Soc. Revs. 8 (1), 143 (1979)

97. C.A. Shacklady, New Scientist 43, 5, Sept. 25, 1969

98. Italy Confirms Petroprotein Stance, Chem. Eng. News 56 (27), 8, July 3, 1978

99. Health Issue Halts Petro-protein Plans, Chem. Eng. News 51 (10), 9, March 5, 1973

100. BP Contests Petroprotein Plant Health Issue, Chem. Eng. News 56 (12), 12, March 20, 1978

101. J.C. Mueller, Can. J. Microbiol. 15 (9), 1047, (1969)

102. J.C. Mueller and C.C. Walden, Can. Home Economics J. 9, April 1970

103. H. Dalton, Chem. Soc. Revs 8 (2), 297, (1979)

104. ICI to Scale Up Single Cell Protein Process, Chem. Eng. News 5 (42), 25, Oct. 11, 1976

105. ICI's Bugs Come On Stream, Chem. In Brit. 17 (2), 48, Feb. 1981

106. ICI Studies Single-cell-protein Source, Chem. Eng. News 58 (16), 26, April 21, 1980

107. Single Cell Makes Protein More Efficiently, Chem. Eng. News 58 (41), 17, Oct. 13, 1980

108. M.J. Johnson, Chem. and Ind. (London), 1532 (1964)

109. K. Yamada, S. Horiguchi, and J. Takehashi, Agric. Biol. Chem. 29, 943 (1965)

110. W. Charney, New Scientist, 10, Sept. 25, 1969

111. S.C. Beesch and G.M. Shull, Ind. Eng. Chem. 49 (9), 1491, Sept. 1957

112. T. Ishikura, H. Nishida, K. Tanno, M. Mujachi and A. Ozaki, Agric. Biol. Chem. 32, 12 (1968)

113. B. Harrow, and A. Mazur, Textbook of Biochemistry, W.B. Saunders Co., Philadelphia, 1958, page 345

114. M. Hamdy, Chemtech 4 (10), 616, Oct. 1974

115. D.E. Eveleigh, Scientific American 245 (3), 154, Sept. 1981

116. Lysine Set for Big Growth in South America, Chem. Eng. News 56 (2), 15, Jan. 9, 1978

117. J.H. Ottenheym and J.W. Gielkens, in Petrochemical Developments Handbook, Gulf Publishing, Houston, Texas, 1969, page 96

118. R.F. Goldstein and A.L. Waddams, The Petroleum Chemicals Industry, 3rd edition, E. and F.N. Spon Ltd., London, 1967, pages 345, 407

119. J.L. Fox, Chem. Eng. News 58 (11), 15, March 17, 1980

120. J.L. Fox, Chem. Eng. News 59 (14), 17, April 6, 1981

121. J.L. Fox, Chem. Eng. News 60 (13), 10, March 9, 1982

122. D.A. Hopwood, Scientific American 245 (3), 90, Sept. 1981

123. J.L. Fox, Chem. Eng. News 59 (49), 31, Dec. 7, 1981

124. High Protein Rice Developed at USDA, Chem. Eng. News 60 (8), 7, Feb. 22, 1982

125. A. Moriarty, Can. Res. 8 (5), 46, Sept.-Oct. 1975

126. Capacity Jumps for Amino Acids, Chem. Eng. News 61 (1), 18, Jan. 3, 1983

127. P. Craig, Chem. In Brit. 15 (8), 392, Aug. 1979

128. Y. Aharanowitz and G. Cohen, Scientific American 245 (3), 140, Sept. 1981

129. Chemical Marketing Reporter 220, 41, June 18, 1982

130. L.M. Miall, New Scientist 43, 8, Sept. 25, 1969

131. Kirk-Othmer Encyclopedia of Chemical Technology, 3rd edition, John Wiley and Sons, New York, 1980, volume 9, page 861

132. J.N. Currie, J. Biol. Chem. 31, 15 (1917). Chem. Abstr. 11, 2814 (1917)

133. J.L. Meers, Chem. In Brit. 12 (4), 115, April 1976

134. M.L. Sinnott, Chem. In Brit. 15 (6), 293, June 1978

135. Strong Growth Ahead for Industrial Enzymes, Chem. Eng. News 59 (3), 37, Jan. 19, 1981

136. C. Bucke and A. Wiseman, Chem. Ind. (London), 234, April 4, 1981

137. W.R. Vieth and K. Venkatasubramanian, Chemtech 3 (11), 677, Nov. 1973

138. Enzyme Utilization-A Sleeping Giant, Envir. Science and Tech. 7, 106 (1973)

139. W. Worthy, Chem. Eng. News 55, 23, May 30, 1977

140. C.J. Suckling, Chem. Soc. Revs. 6 (2), 215, 1977

141. W.R. Vieth and K. Venkatasubramanian, Chemtech 4 (1), 47, Jan. 1974

142. Ethanol Via Fixed Bed, Can. Chem. Proc. 65 (1), 18, Feb. 1981

143. Immobilized Enzyme Half-life Lengthened, Chem. Eng. News 55 (43), 16, Oct. 24, 1977

144. R. Greene, Chem. Eng. 85 B, 78, April 10, 1978

145. Enzymes Now Winning New Applications, Can. Chem. Proc. *61* (12), 20, Dec. 1977

146. D. Perlman, Chemtech *4* (4), 210, April 1974

147. R. Breslow, Science *218*, 532, Nov. 5, 1982

148. G. M. Whitesides and C-H. Wong, Aldrichimica Acta *16* (2), 27, 1983

149. C.J. Suckling and K.E. Suckling, Chem Soc. Revs. *3* (4), 387, 1974

150. Researchers Close In On Synthetic Enzyme, Chem. Eng. News *57* (15), 26, April 9, 1979

151. R. Breslow, Chem. In Brit. *19* (2), 126, Feb. 1983

152. J.L. Fox, Chem. Eng. News *61* (7), 33, Feb. 14, 1983

153. F.K.E. Imrie, Chem. Ind. (London), 584, July 17, 1976

154. 1979/80 Statistical Yearbook, United Nations, New York, 1981, and earlier years

155. The Great Canadian Beer Book, G. Donaldson and G. Lampert, editors, McClelland and Stewart, Toronto, 1975

156. G.-B. Ledoux, Protect Yourself (Ottawa) *7* (9), 19, Oct. 1979

157. Kirk-Othmer Encyclopedia of Chemical Technology, 3rd edition, John Wiley and Sons, New York, 1978, volume 3, page 692

158. The World Almanac and Book of Facts, Doubleday, New York, 1982

159. Lange's Handbook of Chemistry, 10th edition, N.A. Lange, editor, McGraw-Hill, New York, 1969, page 1183

160. Business Week, issues of Feb. 17, 1962 and March 21, 1977, cited by reference 64

161. Kirk-Othmer Encyclopedia of Chemical Technology, 3rd edition, John Wiley and Sons, New York, 1980, volume 9, 338

162. Outlook for Organic Solvents Remains Bleak, Chem. and Eng. News *55* (44), 10, Oct. 29, 1979

163. No Recovery in Sight for Major Solvents, Chem. and Eng. News *59* (45), 12, Nov. 9, 1981

164. Handbook of Chemistry and Physics, 56th edition, R.C. Weast, CRC Press, Cleveland, Ohio, 1975

# 15 Petroleum Production and Transport

## 15.1 Production of Conventional Petroleum

Early use of petroleum or mineral oil, as opposed to animal or plant oils, was by direct harvesting of the crude product from generally superficial surface seeps and springs. For example tar, obtained from the Pitch Lake (La Brea) area, Trinidad, has been used for the caulking of ships since the Middle Ages and is still marketed to the extent of about 142,000 tonnes per year [1, 2]. Tar from the Alberta tar sands was used in the 1700's by Cree Indians of the Athabasca river area to seal their canoes, as recorded by Peter Pond. Also a thick bituminous gum was collected from the soil surface in the vicinity of the St. Clair River, Southern Ontario [3], and from Guanoco Lake, Venezuela [4], and these too were marketed for a range of purposes.

It was a relatively small but important step to advance from opportunistic oil recovery and utilization from surface seeps of this kind to the placement of wells in or near these surface deposits in an organized attempt to try to increase production. Probably the first wells drilled in the Western World with the specific intent of oil recovery were placed during the period 1785 to 1849 in the Pechelbronn area of France [5]. None of these wells, which ranged in depth from 31 to 72 m, were prolific producers but they did serve to establish the practice of drilling for oil in Europe. In North America, "Colonel" Edwin H. Drake drilled a well to about 19 m (60 feet) in depth at Titusville, Pennsylvania in 1859, where he struck oil. This well initially produced oil at the rate of about 20 barrels per day. Some historical accounts of the development of the North American petroleum industry credit this event as being the first instance of crude oil production from a well [6, 7]. But more comprehensive accounts also describe the accomplishments of James M. Williams in Canada who, at about the same time, had already marketed some 680 m³ (150,000 gallons) of oil produced from his wells near Black Creek in the vicinity of Petrolia and Oil Springs in Southern Ontario [5, 8, 9]. Some of this production was also partially refined, on site. The first of Williams' wells, several of which were placed by digging rather than by drilling, produced oil at a depth of about 18 m (59 feet) in late 1857 to early 1858 [1, 3]. By 1860 each of Williams' five successful wells were producing on the average about 3.2 m³ (20 barrels) of oil a day netting an annual production rate for Ontario of some 1,600 m³ (10,000 barrels) in that year. U.S. production rapidly made up for the marginally later start in the production of crude oil with a gross of something more than 40,000 m³ (250,000 barrels) in the same year [1].

### 15.1.1 Modern Exploration and Drilling for Oil and Gas

Prospecting for petroleum (literally "rock oil") draws significantly on the present state of knowledge regarding its genesis. One requirement for oil formation is the presence, millions of years ago, of shallow seas having a rich microscopic and/or larger animal life and probably also plant life. For the effective formation of oil deposits this condition must have been accompanied by a large sediment source such as a turbid river. Thus, as organisms died or were trapped in the accumulating sediments anaerobic decomposition occurred with the assistance of bacteria. As further sediments accumulated the transformation process was further assisted by the contributions of heat (geothermal and decompositional) and possibly also pressure. Pressure of accumulating sediment eventually caused expression of the fluid and gaseous decomposition products which then migrated through a porous deposit to formations of lower pressure, generally upwards or sideways (horizontally) until the mobile products reached impermeable strata. If this type of hydrocarbon trapping occurred a petroleum reservoir or deposit was formed (Figure 15.1). Or there may have been no nonporous barrier between the formation zone and the surface, or a fault line may have subsequently permitted migration in which cases a large part of the hydrocarbon content either moved to a fresh entrapment zone or was lost to the atmosphere and surroundings by evapo-

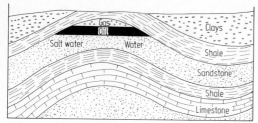

Anticlinal trap

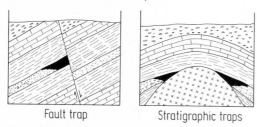

Fault trap          Stratigraphic traps

**Figure 15.1.** Three types of structural traps which can provide a situation favourable to the formation of oil accumulations. Other types of structural traps such as salt domes are also recognized

ration and weathering. Losses of the latter type probably combined with some further bacterial action are what have given rise to surface deposits of viscous, high molecular weight hydrocarbons such as occur at Pitch Lake, Trinidad and the other surface seeps and springs mentioned earlier.

While the harvesting and tapping of more or less self evident oil deposits was sufficient to provide for the early demand for petroleum products, the incentive very soon became strong to devise methods to explore areas without any such obvious signs, to improve the prospects of success on drilling. For a new area this initially involves large scale aerial strip photography, with significant overlap of adjacent frames and strips to allow stereoscopic viewing to determine the topography and geologic features of the area of interest. From study of these photographs, maps are made which include accurate placement of the main physical features of importance and also the principal aerially observable geological features of interest. Aerial surveys may be continued via gravity meter measurements, in which the small differences in the force of gravity from variations in rock distribution may be measured, and/or by magnetometer surveys, from which local variations in the earth's magnetic field are used to estimate the thickness of underlying sedimentary rock forma-

tions [10]. More recently, geochemical methods which rely on determining hydrocarbon gas content above surface soils are also being employed [11]. The presence of a hydrocarbon "halo" at the surface, from the diffusion of gases from an underground deposit, is taken as promising evidence of an accumulation.

More detailed actual ground surveys are conducted in the areas revealed to be of prime interest from the large scale methods. Surface study draws on a variety of procedures of which the first essential is geologic mapping, a step which may include fossil collecting and analysis. This is commonly followed by seismic surveys to try to establish reasonable evidence for the occurrence of an anticlinal structure or other equivalent suitable rock structure in which a petroleum reservoir may occur. Evidence of the presence of, for example, a thick sequence of marine sedimentary rock, including some rich in kerogen, would be taken as a promising sign that the area had a good possibility of hydrocarbons having been formed with which a reservoir could have been filled. Seismic study employs a small explosive charge or several charges to set off shock waves which are transmitted through the rock formations. Placement of 12 or more sensitive geophones (listening devices) in a pattern around the site of the explosive charge allows the travel time and frequency of the shock waves to be traced on a high speed chart. Study of these charts, in turn, provides geologists with sufficient shock wave intensities and times, including reflections from deep-seated formations and other data, to permit a reasonable estimate of the shape of the underlying formations to be assembled from an orderly series of these tests.

Test drilling will follow these preliminary procedures, if the site area still looks promising. During exploratory drilling of this kind, particularly in areas being newly explored, core samples (clean cylinders of rock) are taken using a special circular cutting drilling bit. The cores are carefully stored at the surface in the sequence in which they were obtained. Study of these permits determination of formations or strata at this drilling site which are parallel with those previously collected at other drilling sites. It also enables any fossils present to be identified and from a prior knowledge of these, the approximate age of the formations being penetrated may be determined. Stratum dating is an important feature of core analysis because of the good correspondence observed between formation age and the geological periods during which many oil deposits were laid down [12]. It is also possible, from core samples, to establish whether the underlying strata at the test site are appropriate for petroleum entrapment or containment or not. The presence of vugs, or holes in a limestone formation, for instance, signals the possibility of that formation serving as an oil reservoir. Evidence of a porous and permeable zone of the appropriate geological age overlain by an impermeable shale would also be regarded as promising.

Because of the high cost of coring in both time and money, use of this means of obtaining information about underlying strata is usually held to the minimum that is necessary in the drilling circumstances. Much subsurface geologic information, particularly when the geology of the area is known in general, is derived from a detailed study of the rock cuttings (chips) brought up with the flow of drilling mud from the operation of the regular rotary bit. More coring is done in areas which are being newly investigated, usually when the bit has just entered an oil-stained and/or porous zone as evidenced by the cuttings. Cuttings can also be used to a significant extent for hole to hole comparisons, again decreasing the extent to which coring has to be used for this purpose. Coring, however, remains the only method by which precise data on porosity and permeability of a formation can be obtained, from physical tests conducted at the surface on the comparatively larger volumes of rock provided by core samples.

Drilling success rates, that is wells drilled which show *any* evidence of the presence of oil or gas, are markedly improved with the sophistication of the initial exploration techniques used. Ratios of 1 success in 30 or more holes is about all that can be expected without the application of the methods outlined whereas success ratios of approximately 1 in 5 to 1 in 10 are achievable when using evidence gained from these combined procedures [13]. Since to drill and complete (bring into production) a typical oil well in an accessible region can cost upwards of $ $^1/_2$ million, and $ 10 million or more in frontier regions, it is important to assemble as much favourable evidence as feasible before drilling. For a well to be worth bringing into production it must be capable of a significant initial rate of oil or gas production. Otherwise the cost of the associated surface collection equipment becomes too high in relation to the anticipated value of the product, so the well is capped. The success rate of producing wells, compared to dry holes or non-producers, runs to about 1 in 9 or 10 wells completed in North America, hence the high exploration investment required by the oil industry.

Early drilled wells were made by a cable-tool system in which a heavy tool, about 10 cm in diameter by 1.5 to 2 m long and with a sharpened end, is alternately lifted and then let drop in the hole by a cable and winch system. At intervals the loosened material, suspended in a few cm of water at the base of the well, was lifted out with a bailer end to the cable system. The bailer consisted of a length of pipe fitted with a weighted valve at the lower end which opened when the pipe was lowered to the end of the hole, and which reclosed on lifting. Then the percussion drilling could be resumed. This method, which had been used by the Chinese for water and brine wells for more than a thousand years, was effective, simple, and inexpensive and permitted the reaching of depths as great as 1,100 m [5]. It was virtually the only method used to drill oil wells in the nineteenth century, and the predominant method for the first two decades of the twentieth century. But cable tool systems were slow relative to other developing methods, particularly in softer formations. They had a depth limitation too and, using this system, it was not possible to provide an effective safeguard for pressure containment in the event that the drill penetrated formations under high gas pressures.

Rotary drilling, which gradually became the dominant well drilling method in the first quarter of this century, enabled more rapid well completion and permitted working to depths as great as 8,000 m but required a greater equipment and labour investment to achieve this [5]. First, a larger derrick or support structure about 40 m high is required above ground to enable lifting, lowering, and guidance of the drill string and to act as a temporary storage rack for lengths of drill pipe during bit changes etc. (Figure 15.2). The actual drilling is accomplished by special types of bit, about 16 cm in diameter, threaded to the end of 9 m lengths of about 8 cm diameter special high strength steel pipe sufficient to reach the bottom of the hole from the surface. A lifting system from the top of the derrick allows control of the weight or force on the bit and hence the torque required to rotate it as the bit cuts into the rock formations being penetrated. Since three 9 m lengths (the working unit kept stacked inside the derrick) of even 12 cm diameter drill pipe weighs some 730 kg, sufficient pipe to reach a

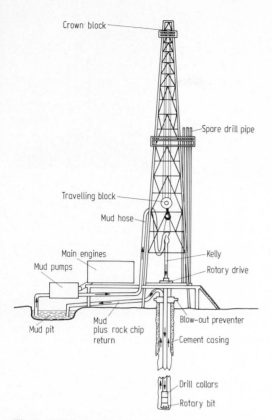

Crown block

Spare drill pipe

Travelling block

Mud hose

Main engines

Mud pumps

Kelly

Rotary drive

Mud pit

Mud plus rock chip return

Blow-out preventer

Cement casing

Drill collars

Rotary bit

**Figure 15.2.** Schematic diagram of a rotary drilling rig showing hoisting derrick and routing of drilling mud

2,000 m depth would weigh some 54 tonnes. Thus this control task using massive equipment is a delicate one, assigned to the head driller of each operating crew to ensure optimum drilling rates without breakage of the drill string from the application of excessive vertical pressure on the bit. A rotating table at the base of the derrick, driven by 2,000 to 3,000 kW of tandem diesel engines, is used to grip and rotate the pipe string from the top of the string at 100 to 250 rpm [13].

A specially formulated drilling mud pumped down the inside of the pipe string and returning between the pipe and the side of the hole, provides lubrication and cooling for the pipe string and drill bit. It also serves to carry rock chips up to the surface for removal from the mud by screening, assists in sealing any somewhat porous formations which are being penetrated and, most importantly, provides a safety feature in the event that a formation under high gas pressure is reached. In fact the density of

the drilling mud used will be deliberately increased by the addition of powdered barytes ($BaSO_4$; density $4.15 \, g/cm^3$), iron oxide, or the like when it is anticipated that a high pressure formation is likely to be encountered. This can be predicted, for instance, when further developing an oil reservoir with known high formation pressures which has already been tapped by exploratory wells.

Blowouts, or the loss of drilling fluid and sometimes the whole of the drill pipe and derrick as well when unanticipated high pressures are encountered, only rarely occur today. Warning signs such as a rapid unexplained increase in drilling mud volume, sometimes accompanied by a frothy appearance, are taken seriously and the drill mud exit stream flow rate may be restricted to increase the drilling mud pressures in the hole. Also modern drill rigs are equipped with blowout preventers, massive valves placed immediately under the derrick floor and cemented into the top of the hole, as well as additional valves which may be placed at one or two intermediate depths for deeper wells, which may be closed at a moment's notice on the threat of a blowout. In these ways both the hazards, and the significant hydrocarbon and formation pressure losses which can result from a blowout are generally avoided.

One further mode of drilling, in addition to the cable tool and rotary methods already described, is the turbo drill. This method, developed significantly by the Russians, uses much of the same equipment such as derrick, draw works (lifting system) and mud pumps of the rotary drilling method. But instead of imparting the turning force to the bit by rotating the string of pipe from the top of the string, in the manner of a conventional drill, this system uses a turbo-powered unit located in the drill string at the bottom of the hole and just above the drill bit itself to rotate the bit. The pipe string remains stationary, except for vertical movement during bit changes etc. Power is carried to the turbo unit and thence to the bit via the high pressure mud stream itself, flowing down the inside of the string of pipe. Since the drill string does not need to provide the rotating torque to the drill bit with this method, the pipe stresses are much less. For this reason turbo-drilling is of primary value for the making of very deep holes of 5,000 m or more.

## 15.1.2 Petroleum Production

Conventional petroleum accumulations can generally be classified into one or other of four types of

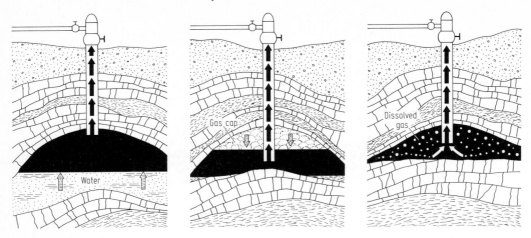

**Figure 15.3.** Diagrammatic representations of three of the four common types of oil reservoir: **a** Water drive; **b** Gas cap drive; **c** Dissolved gas drive [9]. Courtesy G.A. Purdy

oil reservoirs (Figure 15.3) [13]. Production from a water drive reservoir involves the movement of petroleum upward into the producing well by displacement of petroleum from the lower portions of the reservoir into the producing zone by the hydrostatic pressure of underlying water, also present in the formation. Conversely a gas cap drive reservoir relies on the pressure of a separate natural gas phase within the formation to move oil downward towards a producing well. In some reservoirs containing both oil and gas however, the two components are more or less homogeneously distributed throughout the formation, a production situation described as a dissolved gas drive. When this occurs, both components move into producing wells. And the fourth type of reservoir, where little or no production assistance is derived from the natural pressure of the formation contents, is described as a gravity drainage reservoir. In some petroleum producing situations one or other combination of these reservoir types may be encountered.

The fraction of oil-in-place which is recoverable from conventional petroleum reservoirs varies greatly with the reservoir type, the oil viscosity, formation pressure and the production rate and finesse employed. Water drive reservoirs, because of the positive displacement aspect, generally give the highest ultimate petroleum recoveries, or up to about 70 % of the oil-in-place [9, 14]. Estimates of ultimate recoveries possible from gas cap drive and dissolved gas drive types of reservoirs are generally much lower, in the 25 to 50 % range for the

former and 10-30 % for the latter. Recovery from gravity drainage reservoirs will also be low, generally towards the lower end of the recovery ranges of the two gas drive types of reservoirs.

When a new oil formation of one of the first three types is first tapped, the formation presures of up to 20 MPa (3,000 lb per square inch, 200 bars) or more which may be encountered are sufficient to force the oil being produced to the surface under formation pressure. Thus, artesian well fashion, the crude oil (plus associated gas) simply requires piping to a temporary holding area or to a wellhead separation manifold. In the early days of the oil industry, when competing interests were all vying for product from the same oil reservoir under little or no government regulation, it became a race to obtain as many producing wells into the deposit as soon as possible to obtain the maximum share of the product. At the time there was also little application for the co-produced natural gas so that any which did separate from the oil product during production was simply flared at the well site. While these practices may well have given one company a larger share of the product recovered from a particular oil field than another company, they also served to decrease markedly the ultimate total recovery of oil possible from the field, to the detriment of all producers. This yield decrease occurred as a net result of the premature loss of formation pressure, an increase in the average viscosity of residual formation oils still in place, and from discontinuities in fluid-filled voids in the rock. All of these production problems were a direct result of the rapid production rates, sometimes complicated by reservoir structural features.

For some newly developed oil fields and for most oil fields when they reach a mature stage of production, formation pressures are inadequate to raise the oil to the surface although they may be sufficient to raise the oil some distance towards the surface of a producing well. In these instances pumping, with specially designed versions of deep well lift pumps, is used to bring the oil to the surface. When gas is present with the oil, gas separators are placed in the well below the pumping zone to avoid the interference of vapour lock which would otherwise severely affect pumping efficiency.

## 15.1.3 Economizing Techniques

Today most oil producing areas have some framework of drilling and production regulations in place designed to put the development of oil fields and the production of petroleum on a more orderly and less wasteful basis. For instance in British Columbia regulations limit the number of wells which may be drilled per unit area at the surface, roughly corresponding to not more than one well per 4 hectares (10 acres), and which limit the production rate per well based on a formula which takes into account factors such as the unrestricted production rate for a well on initial testing, estimated reserves present in the particular oil field, product properties and the like [15]. Production rate of a particular well is controlled by installation of a choke, a plate with a hole of an appropriate size for the required production rate, or by appropriately adjusted valves installed at the well head. One estimate of an efficient production rate for a water drive reservoir is that the total annual production of all wells producing from the reservoir be not more than 3 to 5 % of estimated ultimate yield (not oil-in-place) of the reservoir [14].

Despite conservation measures such as these taken during the early life of mature oil fields, and sometimes even for new oil fields which have little or no formation pressure when initially brought into production, pumping has to be used to maintain or initiate petroleum production. This is one common way in which the producing life of an oil field, even though slowed, may be extended for some years.

Water flooding, by pumping water or water plus additives into the formation via one well, while producing displaced oil plus water from surrounding wells is a further method of improving the ultimate recovery of oil [16-18]. These methods, which are variously called secondary or tertiary oil recovery techniques depending on the stage of life of an oilfield in which they are employed, represent two variants of what are collectively referred to as "enhanced oil recovery" (EOR) methods [19-21].

Some recovery of reservoir pressure and a decrease in viscosity of the residual petroleum in the reservoir are both obtained by practising return of natural gas, or other petroleum gases, to the formation. High formation pressures contribute to the solubility of methane, ethane, propane etc. in the residual petroleum which brings about the decreased viscosity. Carbon dioxide injection is also used which accomplishes similar objectives as natural gas return [22-25]. Nitrogen has also been used for this purpose [23, 26].

An effective variant of the gas injection enhanced oil recovery procedure is the proposal to inject molasses or some other low cost fermentable material plus a species of bacterium such as *Clostridium acetobutylicum* into the formation [29]. Fermentation products such as acetone, butanol, and lower molecular weight acids which eventually comprise some 50 % of the dry weight of fermentable material originally injected, and carbon dioxide which comprises the balance, both serve to augment production. The combined solvent action of the former products and the pressurization obtained from the carbon dioxide have been reported to increase oil flow by 200 to 250 %.

Direct application of heat via *in situ* combustion [28, 29] or via superheated steam generation at the surface and injection [20, 21, 30] are other effective methods to boost production, either in mature oil fields or in "heavy oil" fields where even the initially discovered petroleum is quite viscous. While both formation heating methods achieved production rate improvements by viscosity reduction, the apparent simplicity of the in-situ combustion concept is somewhat offset by the difficult separation task of recovered oil from intimate admixture with an aqueous solution containing nitrogen oxides and sulfur oxides as well as other combustion products. Not only do the acids present in the aqueous phase contribute to the stability of the emulsions obtained from the producing wells of an *in situ* combustion project, but until thorough separation is achieved the mixture is highly corrosive to ordinary steel pipes and tanks.

Sometimes, for a particular well situation, it is decided that a measure is needed to increase the porosity of the formation in the producing zone, either to initiate production in a new oil field or to stimulate production in a mature one. This may be achieved hydraulically ("fracturing") by pumping a mixture of water and coarse sand into the well at a high flow rate and pressure. Under the impetus of pumps driven by 2,000 to 3,000 horsepower (1,500-2,250 kW), the injected high pressure stream of water and sand opens up cracks in the formation extending radially from the well being stimulated. When pumping is stopped the coarse sand present jams in the cracks and holds them open. Otherwise they would tend to close again from the weight of the overlying rock when the pressure from the injected water is removed. These propped open cracks now serve as channels through which oil may flow to the producing well. If the producing zone of a well is in a limestone formation, acidizing techniques may be used to stimulate production. Corrosion-inhibited hydrochloric acid is pumped down the well, which dissolves the limestone with which it comes into contact and in this way generates new channels for oil flow (Equation 15.1).

$$CaCO_3 + 2\,HCl \rightarrow CaCl_2 + H_2O + CO_2 \quad (15.1)$$
$$\text{soluble}$$

Explosives may also be used in various ways to augment oil flow. For old or new wells, any deposits of wax or bitumen accumulated in the casing at one time was removed and at the same time the producing zone enlarged by setting off a delicately placed charge of nitroglycerin in a "torpedo", a special device designed for this purpose. More often today casing cleaning functions are conducted with better control by using a combination of solvent and scraping tools. A more specialized well servicing unit, the "gun perforator" may be used to boost production of a sluggish producing zone. The unit is lowered to the required level where high powered, steel-piercing shells previously loaded into the device are fired. The paths cut into the rock by the high velocity shells provide new channels for enhanced oil flow. A gun perforator may also be lowered to a higher or lower section of the steel casing of the well, which was not originally brought into production but which was noted as a potential producing zone at the time the well was originally drilled and logged (formation details

recorded). Firing the shells at this new perforator position both penetrates the steel casing at this point and provides short flow channels in the rock at this new producing zone to improve net production from the well.

Of all the economizing techniques described water flooding is generally considered to be the most effective, particularly in sandstones [14]. Frequently, however, a combination of these techniques will provide the optimum improvement in production rates for any particular situation.

## 15.1.4 Supply Prospects of Conventional Petroleum

World crude oil production, and presumably consumption, has been growing relatively steadily during the last fifty years [31]. From production values of 177 million metric tonnes and 262 million metric tonnes in 1930 and 1940 for the early part of this period [32], this has grown to a world production figure of $3.09 \times 10^9$ metric tonnes for 1978 (Table 15.1). Although growth has continued, on the whole, throughout this period the rate of growth has varied widely. The rate of growth in terms of the doubling time, or the time in years to obtain double the inital annual production rate stood at 21 years for the 1930 to 1940 period. But the doubling time rapidly shrank from this value to 11.8 years for 1940 to 1950, 8.5 years for 1950 to 1960, and 8.7 years for 1960 to 1970, a trend which was certainly reason for concern about the projected life of the petroleum resource. In fact, Hubbert, in 1973, estimated that world petroeum production would peak at about the year 2000 [31]. However, since this estimate, the doubling time for the 1970 to 1978 production interval has shown a dramatic reversal of the earlier growth trend by increasing to a period of 28.1 years. Even though this decrease in the growth rate of world petroleum production (and presumably consumption) represents a slightly shorter time period than the earlier figures quoted, the trend change is so large that it was evidently a real one. It was the result of a number of factors but undoubtedly significant among these was the imposition and continuance of the oil embargo by the Organization of Petroleum Exporting Countries (OPEC) in 1974.

Perhaps the combination of circumstances leading to reduced oil production and consumption during the decade of the 70's was a useful exercise to the world. This brought to the notice of the major petroleum consumers the ultimately finite nature of

**Table 15.1** Trends in annual petroleum production by the present major world producers of crude oil [a]

| | Density g/cm$^3$ | Production, in millions of metric tonnes [b] | | | |
| --- | --- | --- | --- | --- | --- |
| | | 1950 | 1960 | 1970 | 1978 |
| Algeria | 0.80 | < 0.1 | 8.8 | 49.0 | 54.0 |
| Canada | 0.85 | 3.7 | 26.0 | 62.0 [c] | 64.3 [c] |
| China | 0.86 | – | 3.7 [d] | 23.9 | 104.1 |
| Indonesia | 0.85 | 6.4 | 20.6 | 42.6 | 80.5 |
| Iran | 0.86 | 32.3 | 52.2 | 191.3 | 262.8 |
| Iraq | 0.85 | 6.5 | 47.5 | 76.5 | 125.6 |
| Kuwait | 0.86 | 17.3 | 81.9 | 150.6 | 126.0 |
| Libya | 0.83 | – | – | 159.8 | 95.4 |
| Mexico | 0.89 | 10.3 | 14.4 | 21.5 | 60.8 |
| Nigeria | 0.85 | – | 0.9 | 54.2 | 94.9 |
| Saudi Arabia | 0.86 | 26.9 | 62.1 | 188.4 | 476.3 |
| U.S.S.R. | 0.86 | – | 147.9 | 353.0 | 572.5 |
| United Arab Emirates | 0.85 | – | – | 37.7 | 89.6 |
| United Kingdom | 0.86 | < 0.1 | ca. 0.1 | ca. 0.1 | 52.9 |
| U.S.A. | 0.85 | 270.1 | 348.0 | 475.3 | 429.2 |
| Venezuela | 0.90 | 78.2 | 152.4 | 194.3 | 113.5 |
| Other | | 33.3 | 90.3 | 196.6 | 284.2 |
| World total | | 485.0 | 1 056.8 | 2 276.8 | 3 086.6 |

[a] Data is given for those countries producing more than 50 million metric tonnes in 1978, obtained from [32]. Does not include natural gas or natural gas liquids.

[b] A close estimate of the production volume in petroleum industry "barrels" may be obtained by dividing the number of metric tonnes by the appropriate density figure and then multiplying by 6.2898. The standard petroleum industry barrel is equivalent to a volume of 0.159 m$^3$, or 35 (34.97) Imperial gallons, or 42 U.S. gallons.

[c] Includes synthetic crude oil produced from tar sands: 1.63 million metric in 1978. Very little was produced from tar sands in 1960, and none in 1950.

[d] Data for 1959.

present global deposits of petroleum, and in so doing accelerated conservation efforts and the process of development of other energy and chemical feedstock sources. While it is reasonable to believe that new deposits of petroleum are presently continuing to be laid down, the rate of geological development of these is undoubtedly several orders of magnitude slower than our present rate of petroleum consumption. The major oil producing contries of the world are able for a time to increase, or maintain their large production rates by continuing to explore for, discover, and develop new deposits to bring into production to make up for declining production from older oil fields. But if the production rate is raised a great deal, even a diligent search for new deposits is not able to keep up the oil reserves (proven potential oil production, from oil-in-place) to annual production rate ratio up to original values. For example Indonesia, Iran, Iraq, and Saudi Arabia all show continuing increases in production rates. At the same time all these countries show significant decreases in this reserves to annual production ratio with values in the 33 to 68 "year" range decreasing to the 24 to 35 "year" range even in the relatively short period from 1970 to 1978 (Table 15.2). The projected lives of the petroleum reserves of these countries is still quite good even at their present large production rates, and especially so when compared to their own relatively small domestic consumption rates.

Worldwide resource life projections of this kind are smaller but the decreasing life trends are also smaller, moving from a value of 29.7 "years" in

**Table 15.2** Trends in petroleum reserves and the anticipated reserve life at current production rates for the present major oil producing countries [a]

|  | 1970 | | 1974 | | 1978 | |
|---|---|---|---|---|---|---|
|  | Reserves, million tonnes | Ratio, reserves/ production | Reserves, million tonnes | Ratio, reserves/ production | Reserves, million tonnes | Ratio, reserves/ production |
| Algeria | 1 056 | 22.3 | 1 158 | 23.8 | 1 309 | 24.2 |
| Canada [b] | 1 157 | 19.1 | 895 | 12.0 | 791 | 12.8 |
| China | – | – | 2 024 | 31.1 | 2 738 | 26.3 |
| Indonesia | 1 367 | 32.5 | 1 614 | 23.7 | 1 070 | 13.3 |
| Iran | 8 204 | 42.8 | 9 315 | 30.9 | 6 148 | 23.4 |
| Iraq | 3 873 | 50.7 | 4 724 | 48.8 | 4 702 | 37.4 |
| Kuwait | 10 443 | 76.0 | 10 469 | 81.4 | 10 184 | 80.8 |
| Libya | 3 959 | 24.5 | 3 039 | 41.4 | 3 719 | 39.0 |
| Mexico | 779 | 35.5 | 433 | 14.6 | 3 884 | 63.8 |
| Nigeria | 757 | 14.0 | 2 655 | 23.8 | 1 678 | 17.7 |
| Saudi Arabia | 12 041 | 68.1 | 14 780 | 35.0 | 15 911 | 33.4 |
| U.S.S.R. | 7 930 | 22.5 | 6 607 | 14.4 | 7 990 | 14.0 |
| United Arab Erimates | 2 161 | 56.9 | 3 397 | 41.9 | 4 318 | 48.2 |
| United Kingdom | 132 | 1 590 [c] | 1 641 | 18 650 [c] | 1 393 | 26.3 |
| U.S.A. | 5 270 | 11.1 | 4 629 | 10.7 | 3 801 | 8.9 |
| Venezuela | 2 009 | 10.4 | 2 090 | 13.4 | 2 492 | 22.0 |
| Other | 6 562 |  | 6 060 |  | 5 559 |  |
| World | 67 700 | 29.7 | 75 530 | 27.0 | 77 687 | 25.2 |

[a] Values selected and calculated from data of Statistical Yearbooks [32]. Reserve data may be converted to petroleum industry barrels by dividing the number of metric tonnes by the appropriate density figure (obtained from Table 15.1) and the multiplying by 6.2898. The petroleum industry barrel is equivalent to 0.159 $m^3$, ca. 35 (34.972) Imperial gallons or 42 U.S. gallons (exactly).

[b] Does not include synthetic crude oil produced from the tar sands, nor the reserves represented by the tar sands.

[c] Anomalously high values for the ratios of reserves to production are obtained here during the early stages of development of a new large petroleum producing area, North Sea oil (see text).

1970 to 25.2 "years" in 1978, a reflection of averaging effects. It is the large scale consumers who are also producers, like the U.S.A., who have had difficulty in matching their production to their consumption rates that have tended to decrease the world resource life projections. For instance the U.S.A., in 1970, had a reserve to production ratio of about 11 "years", lower than world projections and much lower than the Middle Eastern oil producers. And this value has also shown a significant decreasing trend to a value of about 9 years for 1978. In fact the conventional petroleum production rates for Canada and the U.S.A., two large per capita consumers of petroleum products, both peaked some time ago, Canada in 1973 and the U.S. in 1970 (Figure 15.4). While new discoveries may bring about a reversal of the declining conventional petroleum production trends of these and other large scale consuming countries this is going

to be an increasingly costly and difficult exercise. As preliminary evidence of this, by 1975 already about 19 % of total world crude oil production was from offshore wells.

Some countries with a large petroleum consumption, most notably China, the United Kingdom, and the U.S.S.R., have been able to continue to increase their petroleum production from either diligent exploration, late resource development, or a combination of these factors (Table 15.2). The U.S.S.R. is in fact a net exporter of oil, mainly to other socialist bloc nations. Norway and the U.K., from new discoveries of natural gas and oil in the North Sea in the late 60's and early 70's, have both been able to dramatically increase their domestic petroleum production from prior negligible levels (Figure 15.4). Both countries were able to increase production by a factor of more than 600 from 1970 to 1978, a larger factor than shown by any other

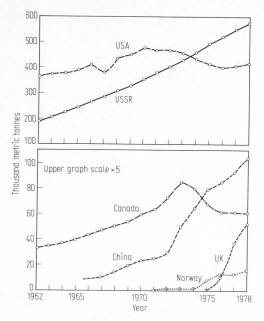

**Figure 15.4.** Trends in annual conventional petroleum production volumes for examples of countries where this appears to have peaked, and which have shown recent and sometimes dramatic growth

country and very much more than the factor of ca. 1.4 times shown by world production increases during the same period. When these few promising examples are viewed on a world scale, in which half the cumulative total oil produced to 1970 took from 1857 to 1960, a 103 year period, and the other half took only ten years, from 1960 to 1970, it puts these developments into proper perspective [31].

On balance then the world, and in particular the large scale petroleum consumers are increasingly exploring ways to supplement the convenient and efficient liquid fuel and chemical feedstock aspects of a conventional petroleum resource. Coal can and is being used in this way to some extent. From current projections the known reserves of hard coal have an estimated life based on present consumption rates of more than 200 years, considerably more than the projected life of petroleum resources [33]. But extraction of coal from deposits is generally more difficult and dangerous than extraction of conventional petroleum, equipment to consume it directly needs to be more complex, and efficient emission control is more involved than for consumption of the conventional petroleum resource.

Coal may be partially converted to a liquid fuel by anaerobic distillation (Table 12.12), or may be made to yield a higher return of liquid fuel by gasification followed by catalytic hydrogenation of the primary liquid gasification products using Fischer-Tropsch technology [34]. This technology was developed and operated on a large scale in Germany during the exigencies of war when the country was cut off from sources of conventional petroleum. The processes operated then were not economically viable and were dismantled at the end of the war. The plants operated today by the South African Coal, Oil, and Gas Company (Sasol) are also not directly profitable, but they continue to be run for strategic and political reasons [35, 36]. These operations do prove that liquid fuel from coal is technologically feasible. Second generation, more direct coal liquefaction plants may bring down the cost of liquid fuel from this source, but the technology to accomplish this has as yet only been tested on a small scale and full scale operation is of the order of a decade away [37-39].

Using technology that bears some resemblance to production of liquid fuels from coal, hydrogenated urban refuse has also been studied as a potential source for liquid fuel [40, 41]. Refuse hydrogenation appears to be too costly, and to provide too small a return of liquid fuel to justify the large capital cost, except perhaps for very large urban centers. In the latter situation escalating costs for the more traditional disposal methods, transport and sanitary landfill or incineration alternatives, can raise the incentive for construction of a refuse conversion plant. The lower capital and operating costs of more straightforward refuse pyrolytic plants, an alternative route to liquid fuels from municipal waste, have been found to be economically feasible by some cities [42]. A mixture of oils which is produced by such a unit is feasible to be used directly as a fuel for space heating combustion units, even if not for vehicles [43]. While the liquid fuel yield from a refuse pyrolysis unit is smaller than from a refuse hydrogenation process, the capital cost is also lower. The capital costs of either type of refuse treating unit are partially offset by the savings in the reduced purchases of conventional petroleum sources for heating, and by the decreased disposal costs for the sterile waste residue obtained, which amounts to only about one tenth of the original refuse volume. The gross volume of municipal waste available is large, but nowhere near large enough to provide more than a small fraction of our total liquid fuel requirements. Thus, though it may make a useful contribution, municipal waste can never become a significant contributor to the world liquid fuel supply.

It is also possible to produce liquid fuel, more specifically ethanol, by fermentation or other processing of carbohydrates. Elegant technology for this type of conversion already exists in the form of alcoholic beverage and spirits production, and this technology has been widened to utilize lower cost cellulosic sources of carbohydrate and enlarged in scale to increase the volumes of alcohol produced [44-46] (see also Chapter 14). Brazil, for instance has carried the development of this technology to the point where automobiles are being produced capable of operating entirely on ethanol. In general, however, this renewable fuel technology is being employed as a supplement to conventional automotive fuels, as a "gasohol" blend with them, and as an aid to conserving conventional petroleum supplies rather than as a means of entirely replacing them [47-49].

Probably the more important and more significant developments to extend the available supply of liquid fuels lie in the processes being introduced to recover oil from the tar sands, and in pilot plant projects involving oil shale pyrolysis experiments to liquid fuels.

## 15.2  Liquid Fuel from Non-conventional Geological Sources

### 15.2.1  Petroleum Recovery from Tar Sands

Major deposits of tar sands, also called bituminous sands or oil sands, occur in several areas of the world, and in total represent considerable hydrocarbon reserves (Table 15.3). The largest of these in terms of both areal extent and reserves in place occurs in Canada in the Fort McMurray area of Alberta. While the reserves in place for this deposit are estimated to be of the order of $120 \times 10^9$ metric tonnes it is thought that only about one quarter of this, or about $30 \times 10^9$ metric tonnes, is likely to be eventually recoverable [50, 51]. Nevertheless the potentially recoverable synthetic crude oil from the tar sands of this one locale represents an enormous hydrocarbon potential since even the recoverable oil estimate is equivalent to about 95 % of the estimated reserves in the Middle East, at present the area producing the largest quantity of conventional petroleum.

The second largest tar sand deposits occur in the Oficina-Tremblador area of Venezuela, and are also sufficiently large to have a significant development potential for the production of synthetic crude oil. Smaller deposits also occur in Malagasy, U.S.A., Albania, and Trinidad, all of which are less significant from a world reserve standpoint but still are important for the locales in which they occur. Other oil sand deposits are known, such as occur in the Lower Triassic Bjorne formation of Melville Island in the Canadian Archipelago [52, 53], but these are generally less well delineated as to extent.

The tar sands situated in Alberta, Canada, consist of deposits of oil-bearing sandstones. Surface exposures occur in parts of the Athabasca deposit, but much of the deposit lies 100 m or more beneath the surface. Oil present in these Lower Cretaceous sandstone deposits is very viscous and thick and partially oxidized so that it cannot be recovered by simple pumping alone. Hence, the use of the terms tar or bitumen to describe this heavy oil fraction, which averages about 12 % ranging up to as much as 18 % of the deposit by weight. Deposits with less than 2 or 3 % bitumen have been excluded from the reserve data given in Table 15.3.

The surface exposures and the near surface deposits (< 45 m of cover) of the Athabasca region may be profitably surface mined for bitumen recovery. These shallow deposits, which are accessible by surface strip-mining techniques, amount to a total of about 10 % of the Athabasca deposit. The remainder of the Alberta tar sands, which lie under 75 m or more of overburden, is uneconomical to surface mine and at the same time is too poorly consolidated for underground mining. These deeper deposits are yielding bitumen to the surface via various *in situ* techniques.

To obtain bitumen free of sand from the strip mined bituminous sands requires treatment with hot water and steam, together with a small amount of sodium hydroxide (as a surfactant), in a process originally devised by K.A. Clark [54-56]. In primary separation cells, the hot frothy bitumen rises to the surface and is skimmed off for further processing (Figure 15.5). The separated sand sinks to the bottom of the cell and is moved hydraulically to the tailings disposal area. By directing the discharge piping carrying the sand tailings onto the walls of the tailings pond, the walls of the pond are

**Table 15.3** Location and significance in millions of metric tonnes, of the known major tar sands deposits of the world [a]

| Location, and Name of deposit | Areal extent $km^2$ ($10^3$ ha) [b] | Overburden meters | Reserves in place [c] |
|---|---|---|---|
| **Canada, Alberta** | | | |
| Athabasca | 2 323 | 0 – 580 | 99 510 |
| Cold Lake | 486 | 210 – 790 | 8 188 |
| Peace River | 456 | 90 – 430 | 5 310 |
| Wabasca | 714 | 76 – 762 | 5 500 |
| Totals | 3 979 | | 118 508 |
| **Venezuela** | | | |
| Oficina-Tremblador | 2 323 | 0 – 900 | 31 790 |
| **Malagasy** | | | |
| Bemolanga | 39 | 0 – 30 | 278 |
| **U.S.A.** | | | |
| Utah deposits | 19.8 | 0 – 610 | 276 |
| California deposits | 2.7 | 0 – 180 | 34 |
| New Mexico, Santa Rosa | 1.9 | 0 – 12 | 9 |
| Kentucky, Asphalt | 2.8 | 2 – 15 | 8 |
| Total | 27.2 | | 327 |
| **Albania** | | | |
| Selenizza | 2.1 | shallow | 59 |
| **Trinidad** | | | |
| La Brea | 0.05 | 0 | 10 |

[a] Compiled from a considerable range in magnitude of data given in [50, 145].
[b] To convert to acres multiply by 2.47.
[c] To convert into units of petroleum industry barrels, since the density of these heavy oils is close to 1.0, multiply the number of tonnes given by 6.290.

continuously raised and made thicker by the precipitated sand. The tailings pond is a necessary feature of a hot water process tar sands extraction plant to retain waste waters. These waste waters, because they contain about 20 % of accumulated mineral fines plus 1-2 % bitumen in a very stable suspension, cannot be discharged without treatment [57, 58].

There is also a middle, less well differentiated fraction from the primary separation cell containing a mixture of water, clay fines, and bitumen. This "middlings" fraction is processed through scavenger cells in which more vigorous aeration and flotation techniques recover some further bitumen froth, which is eventually combined with the bitumen layer from the primary separation cell for further processing. It is the waste water stream from this cell which contributes much of the mineral fines fraction to the tailings pond. Effective treatment of the intractable tailings pond to reduce the accumulated volume of this waste stream has not, as yet, been undertaken.

The bitumen layer from these two separation processes, being a similar density to water and also quite viscous, contains some occluded water plus a small amount of mineral fines. By diluting this fraction with naphtha, in which the bitumen is soluble, the viscosity is decreased sufficiently to allow phase separation to take place. Phase separation, accelerated by centrifuging, produces separate streams of the waste water (plus some mineral fines) and the naphtha solution of bitumen. Flash distillation of the naphtha solution then yields the crude bitumen product and permits recovery of the naphtha for recycle.

Crude bitumen, as recovered from the tar sands in this way, is a black, very viscous (about 100 centipoise at 38 °C) rather intractable material as in-

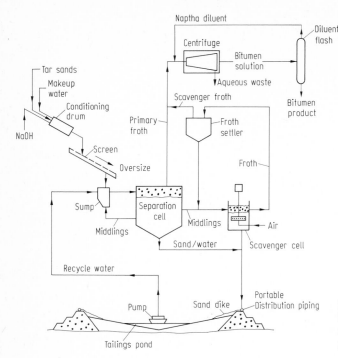

**Figure 15.5.** Schematic diagram of the hot water process for the extraction of bitumen from tar sands [63]. Reprinted with permission

itially obtained. This crude bitumen is not very useful directly, since its thermal softening point is too low in this form to make it useful as a component of road building materials. And with its high viscosity and relatively high pour point (minimum temperature at which it will still just pour) of about 10 °C it would have to be heated to move it via pipeline to another location for further processing, and this would be expensive. So, right at each extraction plant, hot crude bitumen obtained directly from the

**Table 15.4** Approximate breakdown of products on the delayed coking of bitumen recovered from the Alberta tar sands by hot water extraction [a]

| Component | %, by weight | Boiling range, °C | Density g/cm³ |
|---|---|---|---|
| Gases [b] | 8 | — | — |
| Naphtha | 12.7 | 95 – 190 | 0.775 |
| Kerosene | 15 | 190 – 260 | 0.832 |
| Gas oil + heavy fuel oil | 42.1 | 260 – 460 | 0.884 |
| Coke | 22.2 | > 460 | — |

[a] Compiled from [50, 146].
[b] Breakdown given in Table 15.5.

naphtha recovery unit is fed to a coker. Here, by applying heat at temperatures up to about 500 °C, gases and the volatile liquids present are distilled off and the heavier asphaltene fraction (Equation 15.2) consisting of higher molecular weight polyaromatics [59, 60], is cracked to more volatile hydrocarbons and coke (Table 15.4; Figure 15.6). Examples of hypothetical asphaltene skeletons [61, 62]:

Equation 15.3 gives an example of the kinds of chemical changes occurring.

$$C_{42}H_{56} \rightarrow 2\,C_8H_{18} + C_4H_{10} + C_2H_6 +$$

a typical asphaltene, octane, butane, ethane

$$C_2H_4 + 18\,C \qquad (15.3)$$

ethylene, coke

The gases obtained from volatilization and asphaltene pyrolysis comprise about 8 % by weight of bitumen coked, more than half of which consists of hydrogen and methane (Table 15.5). Thus, the composite gaseous product is a valuable fuel gas and can also be used for other purposes, such as the production of hydrogen. Volatile liquid products amount to some 68 % of the bitumen feed, leaving a residue of about 20-22 % coke. Overall mass bal-

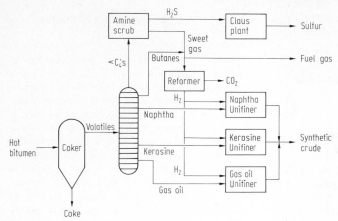

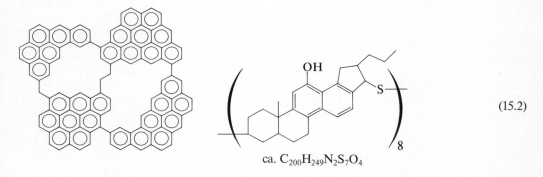

$$\text{ca. } C_{200}H_{249}N_2S_7O_4 \qquad (15.2)$$

**Figure 15.6.** Flow diagram for the purification and stabilization of the products of bitumen pyrolysis [63]. Reprinted with permission

ance estimates and a more detailed compilation of the properties of the products than is presented here, is available [63].

**Table 15.5** Composition of gas stream from delayed coker, before and after sweetening [a]

| Component | Mole fraction, % | |
|---|---|---|
| | Crude | Sweetened |
| hydrogen | 24.2 | 31.7 |
| carbon monoxide (+ N$_2$) | 0.9 | 1.7 |
| carbonyl sulfide | 0.1 | 0.0 |
| hydrogen sulfide | 15.2 | 0.0 |
| carbon dioxide | 1.2 | 0.0 |
| methane | 32.9 | 37.9 |
| ethane | 11.6 | 13.1 |
| ethylene | 2.3 | 2.6 |
| propane | 7.6 | 8.6 |
| propylene | 3.4 | 3.8 |
| butane | 0.6 | 0.6 |
| | 100.0 | 100.0 |

[a] From Bachmann and Stormont [146].

The coke residue, the least valuable of the coking products, is a result of the hydrogen short stoichiometry of the process. With a proximate analysis of carbon 80 %, volatiles 10 %, sulfur 6 %, and ash 4 %, the coke has a fuel value near that of high rank coals. It is burned in the site power plants to provide steam and electrical power for oil sands processing. However, the high sulfur content detracts from its wider utility as a fuel. Any coke in excess of the current fuel requirement has also been finely powdered and incorporated into the dyke walls as a measure to help trap any hydrocarbons present in water seepage through the wall. The crude liquid product, the primary objective of the coking step, contains alkenes and other unstable components resulting from the pyrolysis reactions and is also high in bound sulfur. Interaction of the unstable components with alkenes and air tends to cause accumulation of gummy material on storage and the presence of sulfur compounds is undesirable in fuels. High pressure hydrogenation (Unifining) is used on site to produce both more stable alkanes from alkenes, and volatile hydrogen sulfide from the sulfur bound to organic compounds (Equations 15.4, 15.5).

$$R-CH{=}CH_2 + H_2 \xrightarrow[\text{high pressure}]{\text{catalyst}} R-CH_2-CH_3$$

(15.4)

$$R_2S + 2\,H_2 \rightarrow 2\,RH + H_2S \qquad (15.5)$$

The hydrogen sulfide is now easily separated from the liquid hydrocarbon stream by distillation, and is then converted to elemental sulfur, another product of tar sands operations, via the Claus process. The composite, now stabilized liquid hydrocarbon stream comprises the synthetic crude oil product of tar sands extraction plants.

In this way the useful petroleum fractions are recovered from the surface, or near surface exposures of tar sand by the two presently operating hot water process extraction plants in Alberta. Together Suncor [64] (since 1968) and Syncrude [65] (since 1978) now produce of the order of 9 x $10^6$ metric tonnes of synthetic crude per year, and between them supply something like 10 % of Canada's current crude oil requirements. Other processes for bitumen recovery from minable sands, such as preliminary partial sand removal with the help of cold water, followed by direct coking of the whole of the bitumen/solid residue, and solvent extraction methods have both been tested [66, 67] but are apparently not sufficiently attractive as yet for commercial development.

Various measures are being developed or are actually undergoing small scale field trials for recovery of the 90 % of bitumen in place, which is in deposits too deep for surface mining. Among these experimental *in situ* methods being tested are formation heating using steam [68-70], *in situ* bitumen combustion [71-73], or electrical resistance heating [74] to increase bitumen temperatures sufficiently to make it flow and allow recovery of the hot accumulated bitumen by pumping. Water flooding with a small amount of polymer or sodium hydroxide additive content to enhance the interfacial interactions, in the presence of heat, has also demonstrated some positive results [75]. Mining integrated with one or more of the above methods is also under consideration as an *in situ* recovery method [76]. As one or more of these *in situ* combinations is refined to efficient full scale production the petroleum recovery potential of the Alberta tar sands should eventually be realized.

## 15.2.2 Petroleum from the Oil Shales

The world oil shale deposits, too, represent a considerable potential source of oil, distributed among a number of different countries (Table 15.6), but realizing this potential with oil shale is a significantly different proposition than with the tar sands. Firstly, the organic fraction present in the rock is not an oil or bitumen, but kerogen. Kerogen, which is a complex, high molecular weight ($> 3,000$), three dimensional polymeric solid, is intimately distributed through the rock. It is also insoluble in water or common organic solvents [77]. For example, less than 1 % of the organic constituent, consisting mainly of traces of bitumen commonly associated with the kerogen, is recovered on extended Soxhlet extraction of oil shale with boiling toluene. Thus, in contrast to the tar sands, the organic component of oil shales in place is not in fact an oil. However, on retorting the shale, a process amounting to vigorous heating in the absence of air to temperatures of the order of 500 °C, it may be made to yield from 21 L/tonne shale (5 U.S. gal/short ton) up to occasionally as high as 417 L/tonne (100 U.S. gal/ton) of a dark viscous oil (Tables 15.7, 15.8).

A second misnomer of the oil shale resource is used for the kerogen matrix, which is not a true shale but a marlstone, composed primarily of dolo-

Table 15.6 Estimates of the distribution of potential oil-in-place by country, from the major oil shale deposits of the world [a]

|  | Potential oil-in-place $10^9$ m$^3$ |
| --- | --- |
| Brazil | 127 |
| Canada | 8.0 |
| China | 4.5 |
| Sicily | 5.6 |
| The Congo | 15.9 |
| U.S.A. | 350 |
| U.S.S.R. | 16.7 |
| Other | 3.4 |
| Total | 531.1 |

[a] Corresponds to 1969 estimates from [147]. Data published in 1981 gives a world total potential of 332.8 $\times$ $10^{12}$ m$^3$, considerable more than quoted here, distributed by grade as 83%, 21 – 42 L/tonne shale; 16%, 42 – 104 L/tonne; and 1%, 104 – 417 L/tonne [80].

**Table 15.7** Percent elemental composition by weight of kerogen and the products from the retorting of oil shale [a]

|  | Raw Shale | Kerogen | Retorted Shale | Crude Oil |
|---|---|---|---|---|
| Organic carbon | 16.5 | 80.5 | 4.94 | 84.68 |
| Hydrogen | 2.15 | 10.3 | 0.27 | 11.27 |
| Nitrogen | 0.46 | 2.39 | 0.28 | 1.82 |
| Sulfur | 0.75 | 1.04 | 0.62 | 0.83 |
| Oxygen | – | 5.75 | – | – |

[a] Data from [78, 79].

**Table 15.8** Properties of typical crude shale oil, and composition of gases from a surface retorting of oil shale [a]

| Shale oil composite | | Typical composition of gases, mole % | |
|---|---|---|---|
| Gravity, °API [b] | 22 | Hydrogen | 22.4 |
| Pour point, °C | –1.0 | Carbon dioxide | 21.4 |
| Carbon, % | 84.7 | Carbon monoxide | 3.6 |
| Hydrogen, % | 11.3 | Methane | 15.2 |
| C : H ratio | 7.5 | Ethane | 10.3 |
| Nitrogen [c], % | 1.8 | Ethylene | 5.4 |
| Sulfur, % | 0.8 | Propane | 4.0 |
| Boiling ranges, °C | % | Propylene | 3.7 |
| Initial to 204 | 18 | Butane | 1.6 |
| 204 – 316 | 24 | Butenes | 2.7 |
| 316 – 482 | 34 | C$_5$, and higher | 5.4 |
| 482 and above | 24 | Hydrogen sulfide | 4.3 |
|  | 100.0 |  | 100.0 |

[a] From Atwood [78].
[b] °API gravity, short for degrees, American Petroleum Institute gravity is a density scale used to relate this property of different crude oils and distillate fractions.

$$°\text{API gravity} = \left(\frac{141.5}{\text{specific gravity } 60°/60\,°\text{F}}\right) - 131.5$$

Thus, water, with specific gravity of 1.00, has a °API of 10, crude oils run the range from about 5 to about 65, lubricating oils run about 26 to 35, and gasolines run about 60. The less dense the crude or distillate fraction, the higher the °API.
[c] Nitrogen-containing constituents are mainly pyridines, quinolines, pyrroles, and carbazoles [75].

mite ($CaCO_3$ x $MgCO_3$; ca. 32 %), calcite ($CaCO_3$; 16 %), quartz ($SiO_2$; 15 %), illite (a silica clay; 19 %), and an albite ($Na_2O$ x $Al_2O_3$ x 6 $SiO_2$; ca. 10 %). The remaining 8 % of the rock matrix is composed of smaller amounts of several other minerals [78, 79]. Because some of these carbonate constituents are susceptible to thermal decomposition, e.g. dolomite at 600-750 °C [80] (Equation 15.6),

$$CaCO_3 \cdot MgCO_3 \underset{> 600\,°C}{\rightarrow} CaO + MgO + 2\,CO_2 \qquad (15.6)$$

their presence influences the composition of the pyrolysis gases, particularly at higher retorting temperatures (Table 15.8).

The Fischer Assay, which is a standardized laboratory test in which an oil shale sample is retorted at 500 °C to determine its oil yield, provides a measure of the grade of oil shale being processed [81]. Commercial processes such as the Oil Shale Corporation's TOSCO II process give oil recoveries (synthetic crude or syncrude) of up to 100 % of the Fischer Assay [77]. Oil and gas for-

**Table 15.9** Effect of retorting temperature of Colorado oil shale on product distribution[a]

| Retorting conditions | Distillate hydrocarbon yield, volume percent[b] | | |
|---|---|---|---|
| | Saturates | Olefins | Aromatics |
| 537 °C | 18 | 57 | 25 |
| 649 °C | 7.5 | 39.5 | 53 |
| 760 °C | 0 | 2.5 | 97.5 |
| 871 °C | 0 | 0 | 100 |
| Simulated *in situ* | 41 | 37 | 22 |
| Actual *in situ* | 59 | 16 | 25 |

[a] Data selected from that of [80].
[b] Volume percents quoted for product test distiilation to a 300 °C boiling point.

mation from the kerogen occur by what is now believed to be two successive first order processes [80, 82] (Equations 15.7, 15.8),

$$kerogen \overset{k_1}{\to} bitumen + gas + coke \qquad (15.7)$$

$$bitumen \overset{k_2}{\to} oil + gas + coke \qquad (15.8)$$

to give a product distribution of about 70 % oil, 10 % gas and light oils, and 20 % coke which remains on the solid residue [81]. Thus, the observed products arise both directly from kerogen pyrolysis, and indirectly from kerogen via a bitumen intermediate formed from the kerogen. The actual product distribution is similar to that obtained on the coking of tar sands bitumen. Heat for the py-

rolysis is mostly provided from combustion of a part of the kerogen in the shale feed, and sometimes by burning a portion of the gases and/or the residual coke in the pyrolyzed shale (e.g. Figure 15.7). Higher overall thermal efficiencies are obtained for those processes which utilize the residual coke as well as pyrolysis gases for shale heating.

The key difference between the methods used for oil recovery from oil shales and that used for tar sands is in the methods used for separation of the organic constituent from the naturally occurring material. Oil shale processing requires the whole of the mined material to be heated up to the pyrolysis temperatures of 500 °C or more, whereas the hot water process for tar sands extraction requires the mined tar sand (plus process water) to be heated to only around 70-80 °C. It is only the extracted bitumen, comprising some 10-12 % of the mined mass, that has to be heated up to ca. 500 °C during the coking step to obtain synthetic crude. Because *all* the oil shale must be heated to pyrolysis temperatures to effect oil recovery, efficient heat transfer and reclamation from hot spent shale is more crucial to the commercial success of oil recovery from oil shales. This is achieved in the TOSCO II process by recirculating steel or ceramic balls (Figure 15.7). A number of variants have either been tested or are in the process of being tested to determine the most efficient method by which efficient heat transfer maey be achieved, and to confirm the quality of the synthetic crude obtained [81, 83-86]. One very recent procedure borrows

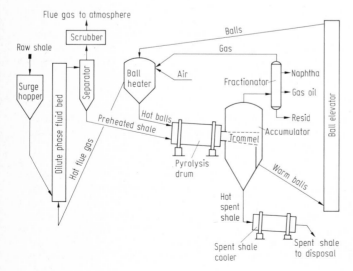

**Figure 15.7.** Diagram of the components of The Oil Shale Corporation's TOSCO II process for the retorting of mined oil shale for the production of synthetic crude [77]. Courtesy L. Yardumian

from coal gasification technology by using the hot spent shale itself to transfer heat to the incoming pulverized raw shale. In so doing this Lurgi-developed process obtains oil recoveries equivalent to 100 % of the Fischer Assay and at the same time achieves overall thermal efficiencies of 92.7 % [87].

Whatever the pyrolysis process used the crude black, viscous pyrolysis oil requires an upgrading step to make it suitable as a refinery feedstock. This is accomplished by high pressure hydrogenation in a manner very similar to the upgrading step used for the coking products in tar sands processing [88].

As with the deeper tar sands deposits, *in situ* methods are required to economically exploit the deeper oil shale deposits. The nature of the organic fraction of oil shales requires devising methods to conduct underground kerogen pyrolysis, permitting subsequent or concomitant recovery of the oil. The most promising methods to date utilize explosives to fracture the oil shale formation [75] to initially loosen the material sufficiently to permit underground combustion and flow of hot gases to take place [80]. Following the fracturing, air is injected in the presence of an ignition source located at the top of the loosened rubble. This heats the formation sufficiently to cause kerogen pyrolysis and oil generation [89]. As the oil forms, it trickles to the bottom of the fractured rubble from where it is recovered by conventional pumping.

At the moment high projected prices for oil obtained from oil shale have stalled the construction of commercial scale oil shale processing plants, although there have been many pilot scale test plants built. One recent venture, that of The Oil Shale Corporation (TOSCO) constructed near Grand Valley, Colorado, was designed to process 100 short tons of shale per day (ca. 90 metric tonnes, 36,000 U.S. gal/day) and was operated close to this scale of production at intervals from 1965 to 1967. Increases in the world price of conventional petroleum have re-awakened an interest in the development of U.S. oil shale deposits. Public announcements of commercial scale plants capable of producing $1,600 \, m^3$/day (10,000 barrels/day) to be constructed by 1982, to be enlarged to $8,000 \, m^3$/day (50,000 barrels/day) by 1990 have been made [90, 91].

# 15.3 Environmental Aspects of Petroleum Production

## 15.3.1 From the Exploratory Geology Phase

The significance of any environmental impact of petroleum production is to a great extent affected by the particular stage in the producing process. Land exploration by aerial surveying, or by the use of aerial or surface based geophysical techniques such as seismic, gravimetric, magnetic, or electrical, methods will have relatively little direct effect on the land surface. Except for any disturbances caused by the placement of seismic shots, only the travel requirements of a field crew will need to be considered. The locations where this travel requirement may pose a significant local impact, however, is particularly in tundra, permafrost, and other ecologically delicate regions. Summer passage of exploration teams in the Arctic, for instance, has been known to leave thermokarst line scars which remain and in some cases worsen in prominence for decades after the event because of the low temperatures, limited plant species diversity, and the short growing season [92, 93]. And these polar and subpolar areas, frequently referred to as "frontier" regions, are increasingly being explored for oil prospects as the possibilities for the discovery of new oil deposits in accessible regions become less likely. These exploratory impacts can be minimized, however, through use of proper vehicles such as air cushion hovercraft, helicopters, and vehicles such as the Rolligon which transmit very low load densities to the travelled surface by the use of wide, ultra low pressure pneumatic tires [94, 95]. Timing this activity to coincide with a frozen surface layer, particularly when a snow cover is present, helps to minimize impact on the tundra surface even though the colder weather may introduce other field work difficulties. Also there is some evidence that the vegetation in those thermokarst scars which do eventually heal is more profuse than in adjacent, undisturbed terrain.

Offshore exploration, whether this is conducted in the relatively shallow marine waters of a contiguous continental shelf or in large lakes, has a different set of risks attached to it. Sediments may be disturbed and the locations of any underwater installations such as power or telephone cables, and water or oil pipelines etc. has to be determined beforehand. The overlying water column however, does serve to minimize any direct visual impact of these activities.

## 15.3.2 Impacts of Exploratory and Production Drilling Activities

After the surface and seismic exploration phases have been conducted to gain information to assess the prospects of a drilling site, wells are drilled to test this information. These wells, termed exploratory or exploration wells, are placed with little or no foreknowledge of the exact nature of the conditions to be expected in the formations being penetrated. In this situation the greatest environmental risk faced by these operations arises from the threat of a "blowout" or an uncontrolled release of drilling fluid mixed with oil, gas and brine under high pressure, which can occur when the drill rig penetrates a high pressure reservoir. The on-site hazards of such an event comprise not only the mud and debris from the well itself but can occasionally also include the forcible ejection of the whole of the drill string (one or more km of heavy pipe). Also there is the associated risk of a well head explosion or fire from the release of flammable gases at the surface. A blowout may also have a lesser impact over a wider area of thousands of hectares, too, from the discharge of highly toxic hydrogen sulfide gas which frequently occurs in significant concentrations in well-head natural gas. Fortunately the bad smell of hydrogen sulfide serves as something of an early warning signal of its presence. However, relatively short exposures quickly dull the olefactory nerves giving a false indication of dissipation of the gas, a tendency which has to be guarded against. These collective risks in the drilling of exploratory wells have eled to them being termed "wildcats" by drillers, as recognition of their unpredictable nature.

Obviously, to be able to prevent a blowout has not only favourable environmental consequences, but also the avoidance of significant costs for the exploration company concerned in the form of less risk of injury to personnel or damage to equipment as incentives. The immediate large scale losses of the valuable resource and the longer term potential savings in ultimate hydrocarbon recovery from a reservoir in which it has been possible to maintain formation pressures are both other real benefits to blowout control. In the event that the blowout catches fire, the very high cost and personnel risk involved to put this out and gain control of production again gives blowout control a high priority in the drilling phase.

The main safeguards employed to avoid blowouts on penetration of a formation under high pressure, in addition to increasing the density of drilling mud and restricting its outflow, are at least a pair (usually) of blowout preventers. These consist of stout large bore valves, actuated by a pair of hydraulic rams. The upper valve, which is used for control when a drill string is in place, has semi-circular cutouts in the two faces of the valve corresponding to half the cross sectional area of the pipe used in the drill string. When activated the two halves close tightly around the drill string, controlling both movement of the pipe itself and formation pressures. There is also a lower blowout preventer, without the semi-circular cutouts, which may be used to close off the well when there is no drill string present. In deeper wells, when the size of the well casing (pipe lining) is decreased to a smaller size for the lower reaches of the well, an additional one or more blowout preventers will be installed at the junction. This measure both increases the possible safeguards the driller may apply and increases the capability to control very high pressures which are more likely to be encountered at depth. These blowout prevention measures have certainly increased the rarity of blowouts but they still do occur occasionally, despite the precautions. Shielding of potential ignition sources and enforcement of no smoking regulations helps reduce the risk of fire in the event of a blowout, or from the dissipation of gas captured in the drilling mud.

An unsuccessful wildcat is called a dry hole. A successful wildcat is called a discovery well, and when the drilling equipment is removed and it is brought into production there is very little impact on the surrounding area during the producing phase. Further wells, called appraisal wells, will then be drilled better to determine the size and extent of the initial discovery. These will have a much lower risk of the occurrence of a blowout because the depths at which formations under high pressure may be encountered and the pressures to be expected will already have been established from the data recorded on drilling the discovery well. Further wells, called development wells, may or may not be drilled following completion of the appraisal wells depending on how promising the results of the appraisal are.

Still on land but in tundra or permafrost areas exploratory drilling may have a longer term effect because of the lower natural recovery capability [96]. On completion of the wildcat wells in these areas greater care than usual is necessary for disposal of waste drilling mud etc. to avoid extensive slumping and terrain collapse from the introduction of a thaw susceptible area into the permafrost [97].

Carrying the problems outlined for land drilling into an offshore drilling context, whether this is undertaken in the relatively shallow seas of a continental shelf or in fresh water lakes, several complicating factors are introduced simultaneously. If drilling is pursued from a ship, maintenance of position, normally accomplished by radial placement of four (or more) anchors or sometimes by computer-controlled propellors, becomes a nontrivial exercise. Add to that the complicated motions of the ship in rough weather and it usually means that drilling has to be curtailed until the arrival of better conditions. Maintenance of position is no problem for sea-based drilling platform with ballasted stabilizing legs extending well into ocean depths not subject to wave action, or to the sea floor. Also drilling from artificial islands, constructed from material dredged from the surrounding area, is positionally stable but all of these situations are still highly susceptible to severe weather conditions. They are also methods which are only applicable to the more shallow exploratory areas. Add the problem of floating massive sea ice to any

of these offshore drilling methods and one can see why oil exploration in this frontier area is so expensive [98].

Blow-out control methods used for offshore drilling activities are still basically the same as used on land, but installation and confirmation of reliability are both less straightforward when the drilling is undertaken in water. But there is little doubt that in the event of the failure of blowout prevention equipment, which has occurred in several instances (e.g. Table 15.11), the immediate oil pollution risk to a much wider area is substantially greater than would be generated by a similar event occurring on land, because of the mobility of both the water and oil phases. Measures to bring the blowout under control are also much more difficult to exercise, because of poor visibility and accessibility problems at the underwater site.

### 15.3.3 Emission Problems of Synthetic Crude Production

The obtainment of synthetic crude from tar sands or oil shales is not without their own particular kinds of emission problems. One of the problem areas in the extraction part of the hot water process for tar sands extraction arises from the clay mineral fines which are present to a varying extent in the deposit, from less than 1 % to 15 % or more of the mined material. Presence of these mineral fines

**Plate 15.1.** Offshore drilling rig, the SEDCO-708 of the Southeastern Drilling Company, Dallas. Note men at railing for scale. (John McKay, Times-Colonist, Victoria, with permission)

interferes with efficient bitumen separation in the primary separation cell and requires the back-up scavenger cell to maintain bitumen recovery efficiencies. Selective mining could be used to avoid the problem by leaving high fines tar sands in the deposit. But this procedure would raise mining complexity and cost and would waste bitumen present in the fines, so is not practised. As a result of this, the spent water discharged from the scavenger cells still contains much of the mineral fines plus traces of bitumen. This is accumulated in large holding ponds, totalling $30 \, km^2$ (about 11 square miles) in extent for the larger of the two extracting plants which are presently operating [99].

Prolonged settling in the holding ponds allows a supernatant part of this waste water stream to be recycled into the extraction process. But gradually a sludge consisting of about 78 % water, 20 % mineral fines, and 1-2 % bitumen is accumulated in the holding pond. Ordinarily, the residual sand from the extraction process could be accommodated in the mined-out areas as mining proceeds. But to combine the extracted sand with the very large additional volume of water contained in the sludges makes containment of the gross residues of the extraction process in the mined out areas impossible with present technology [65]. However, work continues to try to find an economic method of sludge dewatering that will allow spoils replacement to take place [57, 58, 100-102]. Possibly spherical agglomeration methods, which are presently being pioneered for direct solvent extraction of bitumen from tar sands by the National Research Council of Canada, may also be usefully applied to this problem area [67].

Sulfur gases arising during synthetic crude production from the bitumen, and from the high pressure hydrogenation process for synthetic crude stabilization are captured in amine scrubbers, and are subsequently converted to sulfur via the Claus sequence of reactions. Of the order of 1,500 tonnes of sulfur is produced daily from these sources by the two presently operating hot water extraction plants [65]. However, there have occasionally been problems in maintaining the ambient air sulfur dioxide concentrations to below the ambient air guidelines set by the Alberta government, particularly during an inversion episode [103]. It has also been recommended that monitoring, and air quality modelling studies be continued to determine the interaction of the two plumes, those of Suncor and Syncrude, and obtain greater precision in the area

modelling studies [104]. Vanadium and nickel recovery from fly ash, on the scale of 1,600 and 3,900 tonnes per year, respectively, has also been considered [105].

Production of synthetic crude oil from oil shale has three main environmental problem areas. These center on a scarce water resource in the operating area (at least in the U.S.), disposal of spent shales and reclamation of the disposal areas, and sulfur gas containment. Even without a water requirement for extraction, the water needs are significant, for pretreating shale, for condensing the crude and upgraded shale oils, and for the operation of any scrubber based emission control devices. Thus, since the present pilot plants are operating in an arid area, provision of an adequate water supply is a major concern of prospective large scale shale oil producing units [81].

'Burned' shale, containing little in the way of plant nutrients and having a high salt content is not easily reclaimed [106], particularly because of the dry conditions and the large amounts involved. An oil shale industry producing a million barrels of oil per day would produce over 1.25 million tonnes of burned shale each day [107]. Since the areas potentially involved are large, and confounded by the aridity, this too is likely to be a costly side to shale oil production. However, at least reclamation efforts apparently will not be complicated by health problems from either the raw oil shale rock or from the spent shale [108]. Sulfur gas containment from either oil shale pyrolysis or oil upgrading can be accomplished using the largely existing technology already outlined for tar sands processing. Environmental aspects of oil shale processing have been recently reviewed [80].

Both *in situ* bitumen from tar sands and shale oil from *in situ* shale pyrolysis processes promise to have a lower environmental impact than the methods based on surface mining. There will, however, be surface water treatment requirements for both processes, from water recovered with bitumen from steam drive tar sands processing, and small volumes of formation water which will be mixed with the shale oil recovered by pumping from the lower reaches of the fractured deposit.

### 15.3.4 Petroleum Shipment

The major petroleum producing areas of the world do not coincide very closely with the major consuming areas, so that much ocean shipping activity today is involved with tanker movement of oil [109]. When the producing area is accessible to the consuming area by a land route, or for the transport of oil to or from an oil port to a refinery, a pipeline is normally used. And for smaller quantities, segregated products, or shorter distances, small tanker ships, tank trucks, or rail tank cars are generally used.

Since the volumes of oil shipped long distances are very large, and transportation costs per tonne of oil shipped these distances are at a minimum when using large ocean tankers or pipelines, more than half the present world's ocean cargo weight is now oil, and has been at least since the mid 60's [32]. More than 75 % of all this oil traffic is now crude oil [110], which means that small unit segregation for different products is only required for a relatively small proportion of the total tanker tonnage. Thus the appropriateness of very large shipping units and the increased shipping economy to be realized by the use of very large tankers has led to a rapid escalation in the size of the largest tankers used, from the 5,000 to 15,000 deadweight *tonnes* (dwt) size range in the 1939-45 period [13] to 85 vessels of over 320,000 deadweight tonnes operating in 1980 with more on order. In total, some 80 % of the tonnage of the world tanker fleet in 1980 consisted of vessels of over 72,000 dwt and several vessels were in regular operation of 430,000 dwt [109, 111].

An important aspect of this escalation in tanker size is the corresponding decrease in vessel maneuverability. Whereas the standard World War II tanker of 15,000 dwt had a loaded draft of about 9 m and a

length of 170 m, the loaded 430,000 dwt very large crude carrier (VLCC) requires a draft of about 27.5 m and has to steer a length of more than 400 m, both severe maneuverability restricting factors. In the event of a serious accident involving a VLCC the polluting potential also becomes enormous, but at the same time it should be remembered that the probable number of accidents for the same volume of oil transported goes down. It would take some thirty or more World War II vintage tankers to move the same volume of oil as one VLCC.

Bearing all of this in mind, therefore, it comes as a surprise to most people that only about 9 % of the world oil loss to the oceans is via accidental tanker spills, i.e. large point discharges (Table 15.10). The amount of oil this represents, ca. 100,000 to 270,000 tonnes per year, equates to about one major VLCC accident every two years or around ten accidents involving smaller vessels every year. Much larger quantities than this, from 530,000 to 670,000 tonnes per year are estimated to be discharged to the sea as a result of normal tanker operations. Direct discharge to the sea of water used for tank cleaning operations of tankers or used as sea water ballast, which is employed for vessel stability on the return trip, are the major tanker oil discharge practices which contribute to this total. Today, many tanker facilities are being equipped with shore lagoon storage facilities to accept large volumes of oily ballast and tank cleaning waste waters, developments stimulated by both environmental, and oil value motivations. Smaller capacity oil/water separation units are then used to recover the oil before the salt water is returned to the sea.

Various methods to recycle and reintegrate waste oils into the refining process to produce either the usual or particular petroleum products [112, 113], or at least for energy recovery [114, 115] are assisting to gradually decrease the very large amounts of oil lost to the oceans by poor used oil disposal practices. The stimulus here has come from both the severalfold increase in value of petroleum in the last decade, as well as from an increased recognition of the harm done to even the vast oceanic ecosystem [116-118] by continual discharge at this rate. Pipelines may also incur oil loss from a line rupture caused by corrosion, subsidence or land slip, or from seismic disturbances [119]. However losses from this source are usually kept small by regularly scheduled inspections of pipeline rights of way, and containment and

**Table 15.10** Estimates of direct petroleum losses to the world's oceans [a]

| Source | Thousands of metric tonnes | Percent of total |
|---|---|---|
| Waste oil disposal industrial, motor oils | 3 300 – 745 | 67 – 25 |
| Tankers, normal operations tank cleaning, ballasting | 530 – 725 | 11 – 25 |
| Other ships, normal operations cleaning, bilges | 500 – 635 | 10 – 22 |
| Refineries, petrochemical plants | 300 – 405 | 6 – 14 |
| Tanker, ship accidents, and other accidental spills | 200 – 300 | 4 – 9 |
| Offshire petroleum production, normal operations | 100 – 144 | 2 – 5 |
| Range of total | 4 930 – 2 954 | |

[a] Compiled from [148, 149].

cleanup operations are generally more straightforward than with ocean shipping accidents. One important precautionary recommendation in this regard is to dike any pipeline rights of way which run alongside a lake, stream, or maritime coast, at the time of construction [120]. This can be aesthetically blended into natural contours, and seeded for erosion protection to make the measure not unattractive. Rail and truck shipments probably account for even smaller losses than pipelines, overall.

More novel proposals for frontier oil shipment include dirigibles having a 450 tonne payload [121]. Helium as a lifting gas would be safer, but the use of methane for this purpose would allow the craft to serve a dual role, movement of oil and natural gas. Boeing officials conducted a preliminary feasibility study in 1972 in which 12-engined jet aircraft of 154 m wingspan carrying up to one million kilograms of cargo per trip were judged to be a competitive option [122]. Perhaps one factor in the non-adoption of these alternatives was consideration of the result of a serious accident.

### 15.3.5 Oil Recovery and Cleanup from Oil Spills

Major oil spills on land occur only rarely because of diking precautions taken around the tankage of tank farms at the time of construction, and regular inspection of pipelines and pipeline pumping stations. When spills do occur, usually as a result of some type of oil transfer or transport accident, temporary dikes of sand bags or earth are put in place by locally available equipment whenever possible to contain the spill promptly and decrease the size of the area affected. Diking may also serve to prevent oil migration to any local bodies of water and associated compounding of spill problems. Following containment, the best means of dealing with the spill is recovery by pumping, of as much of the spilled oil as possible. Unfortunately, there is often a significant delay between the time of a spill and the initiation of recovery from a lack of suitable recovery equipment at the site of the spill. Depending on the viscosity of the oil spilled (i.e. light naphtha, crude oil, or viscous residual fuel oil) and the time taken to recover as much of oil as possible from the spill containment area, a degree of the oil absorption into the underlying soil will occur [123]. The non-recoverable superficially absorbed oil can be cleaned up by burning, or by using detergents to-

gether with hot water and steam, but while these measures may serve to remove the *oil* from the site, the after effects from these clean-up procedures may make recovery of the site itself more prolonged from the clean-up measure than it would from the residual oil [124]. In fact experiments have shown that in some cases appropriate distribution of fertilizer over the spill-affected area sufficiently stimulates bacterial activity, from the abundance of nutritive elements and substrate so provided, that this procedure was recommended for optimum plant re-establishment and site recovery for some situations [125-127].

Any oil migrated to deeper areas of soil will tend to move further through zones where the soil has been previously disturbed, such as along utility and pipeline trenches and foundation lines. As it does so, it will also tend to seep through an increasingly wide cross sectional area of soil. If the water table is high in the region of the spill, the oil may reach it and contaminate ground water of the immediate area before soil adsorption slows migration [118]. If, however, the water table lies at significant depths the risk of oil contamination from a recovered surface spill is slight unless the spill is both large and oil recovery measures ineffective. Deep regions of disturbed soil within the spill area increase the likelihood of ground water contamination from the spill.

The philosophy of oil spill clean-up from water surfaces is virtually the same as for oil spills on land; that is first containment, then recovery of as much as is feasible, followed by clean-up of residues, which will occur particularly on fixtures and exposed beach areas. But on water, even under ideal conditions, each of these steps is more difficult to apply than on land. And oil spills at sea frequently occur under storm conditions which in many cases contribute to the occurrence of an offshore drilling or shipping accident in the first place. Containment and recovery under these conditions is many times more difficult than on land, and under some circumstances virtually impossible, at least until heavy weather subsides.

Floating booms may be used for containment, logs chained end to end serving moderately well in calm waters, although there is some seepage through the log connecting links. On waters with vigorous wave action logs are virtually useless, since wave action carries oil both over and under the logs and accelerates seepage through the joints. Thus, where wave action has to be contended with the

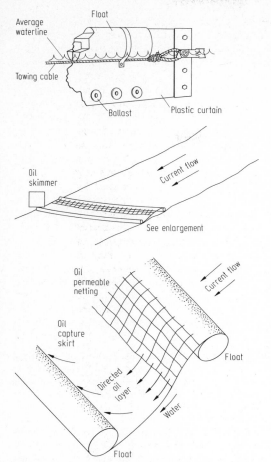

**Figure 15.8. a** Important components of an oil boom, especially constructed to contain an oil spill under open water conditions [128]. Reprinted courtesy Pollution Engineering. **b** A boom specifically designed to capture spills in moving water

only effective control of spreading is obtained through the use of an appropriately sized floating boom especially constructed for the purpose, such as the Slickbar (Figure 15.8a). Here, a weighted skirt extending well below the longitudinal floating members prevents oil flow under the boom, a high profile float prevents splashover, except from the largest waves, and interlocking connections between floating units prevents seepage between adjoining segments [128, 129]. In moving currents of water, such as occasioned by tides or to contain spills in rivers, a skimming boom is required to

avoid oil loss under or around the boom. This is the basis of the Steltner oil guide boom design [130] (Figure 15.8b).

Assuming it is possible to more or less effectively contain a spill the next priority is oil recovery. A number of devices have been developed to accomplish this, all having some feature to preferentially pick up oil, rather than water. Some are more effective with low viscosity oils and in relatively calmer seas while others perform better with very viscous oils and oil-soaked debris (Figure 15.9) [120, 124]. Surface activity features are employed either alone, or in combination, to obtain oil-water differentiation with the various procedures illustrated. For example, a steel belt is naturally oleophilic and if contacted into an oil spill at more than a critical angle, water will run off the belt while oil remains adhering to the surface. A flexible wiper can then strip the oil film off, into a receiver located immediately beneath the wiper (the Sandvik system). Open cell polyurethane foam, like many plastics, is naturally somewhat oleophilic and may be treated with agents to boost this oil preference relative to water. Moving of an open-celled foamed polyurethane belt in the manner of a roller towel through the oil layer of a contained spill serves to pick up mostly oil in the pores of the foam plastic. Subsequent passage of the belt through a pair of closely spaced rollers wrings the oil into a receiving vessel. This is the basis of the operating principle of the Slicklicker oil spill cleanup device.

All recovery operations also face the problem of storage of the recovered oil, once reclaimed from water. For an accident involving a very large crude carrier this is not a trivial problem since well over 100,000 tonnes of oil may be involved at one time (Table 15.11). Containment may be accomplished to barges, kept in reserve at high risk areas. Or it may be effected to air lifted high capacity rubber storage containers of Goodyear design [131], which may be dropped for use at the oil recovery area. When full these containers are built sturdily enough to withstand towing to an oil processing area, and may be re-used.

Destruction of a spill on water by burning is difficult, particularly with bunker fuels which are difficult to ignite at the best of times. The cooling effect of the water lying in direct contact with the layer of spilled oil requires that the oil layer be 4 mm thick or more before ignition is possible. Addition of gasoline or some other volatile fraction may still be

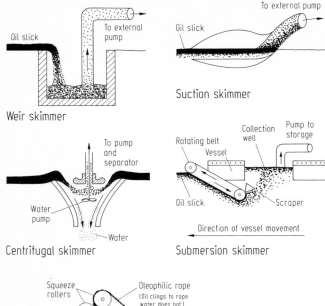

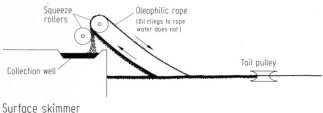

**Figure 15.9.** Illustrations of the operating principles of different types of recovery equipment for oil spills on water [118]. (Shell Canada)

required to accomplish ignition with heavy crude oil or residue fuel oil spills. However, in a situation where a wrecked tanker cannot be safely unloaded, burning may be a useful option to decrease the volume of oil loss to water.

Final stages of clean-up of beach and dock facilities will use combinations of straw, seaweed and specially made oil absorbing materials (e.g. Graboil, Sorboil etc.), chemical oil dispersants and steam to assist natural processes in removal of residues [120, 124].

Perhaps most important in encouraging the adoption of oil spill avoidance measures by all segments of the oil industry and in implementing spill clean-up procedures promptly by transporters and producers alike, is to provide authorities with the power to assess at least a part of the direct clean-up costs on the spiller, and a method to assign blame where this may be ambiguous. Cost assessments to the spiller encourage the adoption of shipping precautions such as double hulls, advanced navigation and propulsion and communication systems, and

well trained crews as important spill avoidance measures. It would also stimulate transporters and dock facilities jointly to purchase and maintain stocks of spill containment and clean-up equipment at high risk locations. A complicating factor in the setting up of emergency equipment pools and in the rapid deployment of clean-up crews in the event of a spill is the frequently complex arrangements re tanker ownership and registration, operating company and degree of training of crew, and oil ownership, control of each of which may reside with different companies.

The Intergovernmental Maritime Consultative Organization (IMCO), established in 1958 by the United Nations and with headquarters in London, serves in an advisory and consultative capacity in an attempt to decrease the incidence of deliberate and accidental oil discharges [132]. IMCO is not, however, empowered to enforce regulations and is not engaged in spill clean-up or compensation. Reimbursement is provided in instances where a national government can prove negligent operation

**Table 15.11** A list of some recent marine oil spill accidents and the extent of oil loss incurred from each [a]

| Location | Date | Type of Accident | Amount and type of oil lost [b] |
|---|---|---|---|
| Canada | | | |
| Alert Bay, British Columbia | Jan. 24, 1973 | Irish Stardust tanker, grounded | 1 100 barrels fuel oil |
| Chedabucto Bay, Nova Scotia | Feb. 4, 1970 | Arrow tanker grounded, split in two | 75 500 barrels Bunker C |
| Chile | | | |
| Straits of Magellam | Aug. 1974 | S. S. Metula tanker, grounded | 283 000 barrels crude |
| Ireland | | | |
| Bantry Bay | Oct. 21, 1974 | crude oil spill | 18 600 barrels crude |
| Japan | | | |
| Mizushima Refinery | Dec. 18, 1974 | Coastal refinery oil tank rupture | 292 700 barrels type not known |
| U.K. | | | |
| Seven Stones Reef, England | Mar. 18, 1967 | Torrey Canyon tanker grounded, broke in two | 700 000 Kuwait crude |
| North Sea | Apr. 23, 1977 | Ekofisk well blow-out | unknown volume crude |
| U.S.A. | | | |
| Nantucket Island, Mass. | Dec. 20, 1976 | Argo Merchant tanker grounded | 181 000 barrels No. 6 fuel oil |
| Santa Barbara, California | Jan. 28, 1969 | oil well blow-out | 79 285 barrels crude |
| Corpus Christi, Texas | Jan. 1977 | Esso Castellon tanker broke from mooring | 500 000 barrels type not known |

[a] Selected examples from the recent compilation of [150]. Detailed citations for many of these accidents are available in the reference given.

[b] Recalculated into the petroleum industry barrel of 0.159 m$^3$. Originally cited volume units were converted to barrels by using the following factors: tonnes $\times$ 6.2898; short tons $\times$ 5.720; Imperial gallons $\div$ 35; U.S. gallons $\div$ 42.

of a tanker, by the Tanker Owners Voluntary Agreement Concerning Liability for Oil Pollution Damage (TOVALOP), which is an insurance measure initiated by this group in 1969. The various limitations to compensation imposed by TOVALOP [132] still make it desirable for other national interests to maintain clean-up funds and equipment. For instance Canada, in 1972, initiated a 15 cents per short ton levy on all oil shipments into Canada by sea [133]. The accumulated proceeds of this fund are to be used to pay for the clean-up costs of oil spills.

Prompt clean-up and blame assignment are both assisted by rapid detection of spills and prompt analysis of the hydrocarbon and trace element content. As an oil spill ages and weathers both the hydrocarbon and trace element distributions become altered, making assignment as to origin of the spill more difficult [134]. Aerial surveillance and detection has permitted more rapid discovery of unreported spills and slicks [135]. Analytical matching procedures range from gas chromatographic methods, in which relative volatilities of hydrocarbon components are compared [136, 137], to infrared spectrophotometric methods, in which ratios of absorbances are compared [138, 139], to ultraviolet fluorescence and luminescence tests [140, 141], and combinations of these methods sometimes combined with others [142-144]. The methods tested to date can in many cases permit an infallible linking of an oil spill, to the source of oil present in the spill from samples taken of the spill and the various possible sources.

# Relevant Bibliography

1. J.J. Fitzgerald, Black Gold With Grit, Gray's Publishing Ltd., Sidney, British Columbia, 1978
2. R. Green and E. McCrary, Barge System Recovers Oil from Tankers, Chem. Eng. *85*, 90, June 5, 1978
3. F. Zurcher and M. Thuer, Rapid Weathering Processes of Fuel Oil in Natural Waters, Envir. Sci. and Tech. *12*, 838, July 1978
4. C.F. Logan, Aqua Dome Canopy Puts the Problem of Offshore Oil Leaks Under Cover, Offshore *37*, 111, Oct. 1977
5. J.H. Vandermeulen, Chedabucto Bay Spill: Arrow 1970, Oceanus *20*, 31, Fall 1977
6. J. Milgram, Clean-up of Oil Spills from Unprotected Waters, Oceanus *20*, 86, Fall 1977
7. E.J. Stephanides, Endless Loop Skims Oil from Water Surface, Design News *34*, 62, Feb. 6, 1978
8. M. Hura and J. Mittleman, High Capacity Oil-Water Separator, Naval Eng. J. *89*, 55, Dec. 1977
9. H.J. Lamp'l, Spill Control Systems and Standards, ASTM Stand. News *5*, 11, Nov. 1977
10. L. Kazmierczak, Spill Prevention and Control; a New Profession, ASTM Stand. News *5*, 19, Nov. 1977
11. 15th Oil Shale Symposium Proceedings, J.H. Gary, editor, Colorado School of Mines Press, Golden, Colorado, 1982
12. Chemistry of Asphaltenes, J.W. Bunger and N.C. Li, editors, Advances in Chemistry Series, No. 195, American Chemical Society, Washington, 1981
13. International Petroleum Encyclopedia, The Petroleum Publishing Co., Tulsa, Oklahoma, 1978. Published annually

# References

1. Our Petroleum Challenge, The New Era, The Petroleum Resources Communication Foundation, Calgary, ca. 1979
2. J.F. van Oss, Chemical Technology: An Encyclopedic Treatment, Barnes and Noble, New York, 1972, volume 4, page 34
3. R.M. McPherson and R.W. Ford, A History of the Chemical Industry in Lambton County, Chemical Institute of Canada, Sarnia, 1964
4. Encyclopaedia Brittannica, Macropaedia 1974, volume 18, page 712
5. J.E. Brantly, History of Oil Well Drilling, Gulf, Houston, Texas, 1971
6. P.H. Giddens, The Birth of the Oil Industry, Arno Press, New York, 1972, (reprint edition)
7. R.N. Shreve and J.A. Brink, Jr., Chemical Process Industries, 4th edition, McGraw-Hill, Toronto, 1977
8. The Canadian Petroleum Industry, by the Chemical Division, Shell Oil Co. of Can., Ltd., Ryerson Press, Toronto, 1956, page 31
9. G.A. Purdy, Petroleum, Prehistoric to Petrochemicals, McGraw-Hill, New York, 1958
10. S. McCutcheon, Science Dimension (Ottawa) *9*(6), 22, (1977)
11. Fresh Look at Geochemistry, Chem. Eng. News *59*(15), 60, April 13, 1981
12. E.J. Lynch, Formation Evaluation, Harper and Row, New York, 1962
13. The Petroleum Handbook, 4th edition, Shell International, London, 1959
14. Our Oil Resources, 2nd edition, L.M. Fanning, editor, McGraw-Hill, New York, 1950
15. Drilling and Production Regulations, Petroleum and Natural Gas Act, B.C. Dept. of Mines and Petroleum Resources, Queen's Printer, Victoria, B.C., 1969
16. E.V. Anderson, Chem. Eng. News *55*(4), 12, Jan. 24, 1977
17. Enhanced Oil Recovery, Chem. Eng. News *55*(19), 18, May 9, 1977
18. W. Worthy, Chem. Eng. News *57*(16), 40, April 16, 1979
19. Enhanced Recovery of Residual and Heavy Oils, 2nd edition, M.M. Schumacher, editor, Noyes Data Corp., Park Ridge, N.J., 1980
20. K.N. Jha, Chem. In Can. *34*(8), 19, Sept. 1982
21. C.E. Brown, Chem. Ind. (London) 875, Nov. 18, 1978
22. OTA Weighs Status of Enhanced Oil Recovery, Chem. Eng. News *56*(17), 6, April 24, 1978
23. H. McIntyre, Can. Chem. Proc. *66*(4), 38, June 1982
24. Alberta Stimulates Enhanced Oil Recovery, Can. Chem. Proc. *66*, 4, Nov. 1982
25. P.H. Abelson, Science *205*(4412), 21, Sept. 1979
26. Nitrogen Scheduled For Oil Recovery, Chem. Eng. News *58*(32), 7, Aug. 11, 1980
27. T.R. Jack, Enhanced Oil Recovery by Microbial Action, Second North American Chemical Congress, Las Vegas, Nevada, August 28, 1980. Also published in B.C. Guidelines, No. 9, B.C. Research, Vancouver, Sept. 1980
28. Cost Will Limit Use, Chem. Eng. News *59*(36), 60, Sept. 7, 1981
29. H. McIntyre, Can. Chem. Proc. *63*(8), 27, Nov. 14, 1979
30. Downhole Steam Generator, Chem. Eng. News *58*(37), 29, Sept. 15, 1980
31. M.K. Hubbert, Can. Mining and Metal. Bull. 37, July 1973
32. 1979/80 Statistical Yearbook, United Nations, New York, 1981, and earlier years
33. E.D. Griffith and A.W. Clarke, Scientific American *240*(1), 38, Jan. 1979

34. Liquid Fuels from Coal, Hydroc. Proc. *54* (5), 119, May 1975

35. M. Heylin, Chem. Eng. News *57* (38), 13, Sept. 17, 1979

36. O.H. Hammond and R.E. Baron, American Scientist *64*, 407, July-Aug., 1976

37. H.W. Habgood, Chem. In Can. *32* (10), 16, Nov. 1980

38. ADL Develops Coal Liquefaction Process, Chem. Eng. News *54* (28), 30, July 5, 1976

39. D.J. Hanson, Chem. Eng. News *59* (34), 13, Aug. 24, 1981

40. R.R. Grinstead, Environment *14* (4), 34, May 1972

41. Work Intensifies on Biomass, Chem. Eng. News *57* (40), 34, Oct. 1, 1979

42. M.B. Hocking, J. Envir. Systems *5* (3), 163, 1975

43. W.S. Sanner, C. Ortuglio, J.G. Walters, and D.E. Wolfson, Conversion of Municipal Refuse into Useful Materials by Pyrolysis, Report of Investigations 7428, U.S. Bureau of Mines, Washington, 1970

44. J.G. DaSilva, G.E. Serra, J.R. Moreira, J.C. Concalves and J. Goldemberg, Science *201*, 903, Sept. 8, 1978

45. N.V. Schwartz, Chem. In. Can. *29* (2), 27, Feb. 1977

46. Ethanol for Gasoline, Chem. Eng. News *57* (26), 10, June 25, 1979

47. D.A. O'Sullivan, Chem. Eng. News *57* (17), 11, April 23, 1979

48. B. Orchard, Can. Chem. Proc. *64* (8), 55, Nov. 26, 1980

49. Gasohol: Focus Shifts, Chem. Eng. News *59* (3), 81, Jan. 19, 1981

50. N. Berkovitz and J.G. Speight, Fuel *54*, 138, July 1975

51. L. Bridges, Science Affairs *9* (3), 23 (1976)

52. Geology of Economic Minerals of Canada, Part B. Geolog. Surv. of Can. Economic Geology, Report No. 1, R.J.W. Douglas, editor, Dept. of Energy, Mines, and Resources, Ottawa, 1968, page 589

53. H.P. Trottin and L.V. Hills, Geolog. Surv. of Can. Paper 66-34, Dept. of Energy, Mines and Resources, Ottawa, 1966

54. K.A. Clark, Can. Min. Metal. Bull. No. 212, 1385 (1930)

55. K.A. Clark, Nature. *127*, 199 (1931)

56. K.A. Clark, Trans. Can. Inst. Min. Met. *47*, 257, (1944)

57. M.B. Hocking, Fuel *56*, 334 (1977)

58. M.B. Hocking and G.W. Lee, Fuel *56*, 325 (1977)

59. M.L. Selucky, Y. Chu, T. Ruo, and O.P. Strausz, Fuel *56*, 369, Oct. 1977

60. T. Ignasiak, O.P. Strausz, and D.S. Montgomery, Fuel *56*, 359, Oct. 1977

61. J.G. Speight, Some Observations on the Chemical and Physical Structure of Bitumen, presented at the Symposium "Fossil Fuel Chemistry and Energy",

U.S. Bureau of Mines, Laramie, Wyoming, July 23-27, 1974

62. T. Ignasiak, A.V. Kemp-Jones, and O.P. Strausz, J. Org. Chem. *42*, 312 (1977)

63. M.B. Hocking, J. Chem. Educ. *54* (12), 725, Dec. 1977

64. Alberta's Oil Sands, Can. Chem. Proc. *56* (8), 10, Aug. 1972

65. G.R. Gray, Chem. In Can. *27* (1), 27, Jan 1975

66. The K.A. Clark Volume, M.A. Carrigy, editor, Research Council of Alberta, Edmonton, 1963

67. W. Campbell, Science Dimension (Ottawa), *8* (1), 10 (1976)

68. D.A. Redford, Chem. in Can. *28* (8), 20, Sept. 1976

69. Oil Sands R and D, Can. Chem. Proc. *62* (7), 30, July 1978

70. H. McIntyre, Can. Chem. Proc. *61* (12), 25, Dec. 1977

71. R.E. McRory, Oil Sands and Heavy Oils of Alberta, Alberta Energy and Natural Resources, Edmonton, 1982

72. Oil Sands Research, Chem. In Can. *30* (1), 4, Jan. 1978

73. J.C. Trantham and J.W. Marx, J. Petrol. Technology *18*, 109 (1966)

74. D.E. Towsonin, The Future of Heavy Crude and Tar Sand, Proc. of the 1st Unitar Conference, Edmonton, 1979, page 410

75. Shale Oil, Tar Sands and Related Fuel Sources, T.F. Yen, editor, Advances in Chemistry Series, No. 151, American Chemical Society, Washington, 1976

76. P.C. Quinn and G. Duncan, in "The Future of Heavy Crude and Tar Sand", Proc. of the 1st Unitar Conference, Edmonton, June 4, 1979

77. R.N. Hall and L.H. Yardumian, The Economics of Commercial Shale Production by the Tosco II Process, 61st Ann. Meeting, Am. Inst. Chem. Eng., Los Angeles, California, Dec. 5, 1968

78. M.T. Atwood, Chemtech *3*, 617, Oct. 1973

79. R.E. Bozak and M. Garcia, Jr., J. Chem. Educ. *53*, 154, March 1976

80. Kirk-Othmer Encyclopedia of Chemical Technology, 3rd edition, John Wiley and Sons, Ltd., New York, 1981, volume 16, page 333

81. D.L. Klass, Chemtech *5*, 1, Aug. 1975

82. Oil Shale, Tar Sands, and Related Materials, H.C. Stauffer, editor, ACS Symposium Series, No. 163, American Chemical Society, Washington, 1981

83. S. Katell and P. Wellman, Mining and Conversion of Oil Shale in a Gas Combustion Retort, Technical Progress Report 44, U.S. Bureau of Mines, Washington, 1971

84. Two New Shale Oil Processes, Chem. Eng. News *58* (12), 20, March 24, 1980

85. Three Australian Firms to Develop Oil Shale, Chem. Eng. News *58* (28), 26, July 14, 1980

86. Shell Shale Oil Retorting Process Promising, Chem. Eng. News *58* (37), 42, Sept. 15, 1980
87. Shale Oil Processes Ready, Chem. Eng. News *60* (15), 60, April 12, 1982
88. Process Upgrades Shale Oil to Usable Crude, Chem. Eng. News *56* (2), 33, Jan. 9, 1978
89. P.R. Tisot and H.W. Sohns, Structural Deformation of Green River Oil Shale as it Relates to In Situ Retorting, Report of Investigations 7576, U.S. Bureau of Mines, Washington, 1971
90. Energy Firms to Develop Oil Shale Plant, Chem. Eng. News *58* (27), 19, July 7, 1980
91. Production, Engine Tests Start for Shale Oil, Chem. Eng. News *58* (25), 39, June 23, 1980
92. W.E. Rickard, Jr. and C.W. Slaughter, J. Soil and Water Conserv. *28*, 263 (1973)
93. L.C. Bliss, Oil and Ecology of the Arctic, in: The Tundra Environment, Symposium of the Royal Society of Canada, Winnipeg, June 3, 1970, Trans. Roy, Soc. Can. 4th Series, VII, page 1. Cited by P.J. Webber and J.D. Ives, Envir. Conserv. *5*, 171, Autumn 1978
94. W.E. Rickard, Jr., and J. Brown, Envir. Cons. *1* (1), 55, Spring 1974
95. B. Henry, Science Dimension (Ottawa), *8* (1), 28 (1976)
96. E.D. Ershov, E.Z. Kuchokov, and D.V. Malinovskii, Moscow Univ. Geology Bulletin *34* (3), 60-68 (1979), cited by Bibliography and Index of Geology, *45* (1981)
97. H.M. French, Arctic *3* (4), 794, Dec. 1980
98. H.M. French, Can. Geogr. J. *96* (3), 46-51, July 1978
99. Facts and Figures, Syncrude, Fort McMurray, ca. 1978
100. G.J. Ewin, B.P. Erno and L.G. Hepler, Can. J. Chem. *59* (20), 2927 (1981)
101. T.E. Burchfield and L.G. Hepler, Fuel *58*, 745 (1979)
102. N.N. Bakshi, R.G. Gillies, and P. Khare, Env. Science and Techn. *9* (4), 363, April 1975
103. Sulfur Dioxide Concentrations, Chem. In Can. *28* (8), 17, Sept. 1976
104. S.B. Smith, Alberta Oil Sands Environmental Research Program 1975-1980, Summary Report, Environmental Summary Report, Environmental Consultants Ltd., Edmonton, 1981, page 39
105. Two Tar Sands Plants, Can. Chem. Proc. *57* (3), 52, Feb. 1973
106. W.D. Metz, Science *184*, 1271, June 1974
107. R.N. Heistand, Retorted Oil Shale Disposal Research, in Oil Shale, Tar Sands, and Related Materials, H.C. Stauffer, editor, ACS Symposium Series, No. 163, American Chemical Society, Washington, 1981
108. Shale Oil Materials Pose Few Health Problems, Chem. Eng. News *56* (41), 13, Oct. 9, 1978
109. World Oil and Gas in 1980, Shell Canada, Montreal, 1981
110. BP Statistical Review of World Energy 1981, The British Petroleum Company, London, 1982
111. J.R. Bright, Practical Technology Forecasting, 2nd edition, The Industrial Management Center, Inc., Austin, Texas, 1972, page 122
112. D.J. Skinner, Preliminary Review of Used Lubricating Oils in Canada, Environment Canada Report No. EPS 3-WP-74-4, Ottawa, June 1974
113. D. Bisson, Science Dimension (Ottawa), *7* (5), 8, (1975)
114. E. Baumgardner, Hydroc. Proc. *53* (5), 129, May 1974
115. N.J. Weinstein, Hydroc. Proc. *53* (12), 74, Dec. 1974
116. J.F. Payne, I. Martins, A. Rahimtula, Science *200*, 330, April 21, 1978
117. M. Blumer, H.L. Sanders, J.F. Grassle, G.R. Hampson, Environment *13* (2), 2, March 1971
118. Perspective on Oil Spills, Shell Canada, Montreal, 1981
119. R.O. Van Everdingen, Potential Interactions Between Pipelines and Terrain In a Northern Environment, National Hydrology Research Institute Paper No. 8, Ottawa, 1979
120. Report of the Task Force — Operation Oil, Cleanup of the Arrow Oil Spill in Chedabucto Bay, Information Canada, Ottawa, 1970, 4 volumes.
121. M. Cope, Weekend Magazine (Canada), p. 8, Nov. 4, 1972
122. Ultra Jumbo Jet Seen as Arctic Solution, Victoria Times, p. 23, May 19, 1972
123. D. Mackay, M.E. Charles and C.R. Phillips, The Physical Aspects of Crude Oil Spills on Northern Terrain, Information Canada, No. R72-9173, Ottawa, 1974
124. A. Southward, New Scientist, *79* (1112), 174, July 20, 1978
125. F.D. Cook and D.W.S. Westlake, Biodegradability of Northern Crude Oils, Information Canada, No. R72-8373, Ottawa, 1973
126. D. Parkinson, Effects of Oil Spillage on Northern Canadian Soils, Information Canada No. R72-12874, Ottawa, 1974
127. F.D. Cook and D.W.S. Westlake, Microbiological Degradation of Northern Crude Oils, Information Canada No. R72-12774, Ottawa, 1974
128. M.F. Smith, Pollution Eng. *2* (5), 24, Nov/Dec 1971
129. M.F. Smith, Envir. Contr. Sys. *142*, 32 (1971)
130. Oil-spill Boom, Water and Poll'n Contr. *112* (5), 13, May 1974
131. Rubber Containers Counter Threat of Oil Spills at Sea, Chem. Ecology 8, Feb. 1974
132. W.M. Ross, Oil Pollution as an International Problem, Western Geographical Series, Vol. 6,

University of Victoria, Victoria, British Columbia, 1973

133. Oil Spill Levy, Chem. In Can. *24*, 11, May 1972

134. G.W. Hodgson, M.T. Strosher, and E. Peak, Feasibility Study on a Method for the Detection of Oil Leaks into Water, Information Canada No. R72-9973, Ottawa, 1973

135. Progress in Aerial Detection, Chem. Eng. News *49* (25), June 21, 1971

136. G.A. Flanigan and G.M. Frame, Research/Development *28* (9), 28, Sept. 1977

137. F.K. Kawahara, Characterization and Identification of Certain Petroleum Products by means of Gas Chromatographic Analysis of Minor Components, Amer. Chem. Soc. Meeting, Los Angeles, April 1971

138. F.K. Kawahara and D.G. Ballinger, Ind. and Eng. Chem. Prod. Res. Develop. *9*, 553, Dec. 1970

139. F.K. Kawahara, Env. Sc. Techn. *3* (2), 150, Feb. 1969

140. Fingerprinting Oil Slicks, Water Polln. Control *110*, 31, Sept. 1972

141. P. John and I. Sontar, Chem. Brit. *17* (6), 278, June 1981

142. W.D. Johnson, F.K. Kawahara, L.E. Scarce, F.D. Fuller, and C. Risley, Jr., Proc. 11th Conf. Great Lakes Res., 1968, page 550

143. Chemical Tests Nail Oil Spill Suspect, Chem. Eng. News *53* (46), 7, Nov. 17, 1975

144. Publications and Articles Relating to the Chemical Analysis of Oil Pollution, U.S. Envir. Prot. Agency, Oil and Hazardous Materials Spills Branch, Edison, New Jersey, ca. 1977

145. L. Bridges, Science Affairs *9* (3), 4 (1976)

146. W.A. Bachman and D.H. Stormont, Oil and Gas *65*, *69* (1967)

147. Kirk-Othmer Encyclopedia of Chemical Technology, 2nd edition, John Wiley and Sons, New York, 1969, volume 18, page 1

148. L.T. Pryde, Environmental Chemistry, Cummings, The Philippines, 1973

149. Man's Impact on Terrestrial and Oceanic Ecosystems, W.H. Matthews, F.E. Smith, and E.D. Goldberg, editors, Mass. Inst. of Tech., Cambridge, 1971, cited by H.S. Stoker and S.L. Seager, Environmental Chemistry: Air and Water Pollution, 2nd edition, Scott Foresman, Glenview, Illinois, 1976

150. Judy Wigmore, Literature Review of Previous Oil Pollution Experience, Regional Program Report 78-24, Environmental Protection Service Canada, Vancouver, British Columbia, 1978

# 16 Petroleum Refining

## 16.1 Composition of Conventional Petroleum

It is commonly thought that crude oil from conventional oil wells is quite similar in composition, regardless of the source. This is not so. Both the physical characteristics and the composition vary widely, depending not only on the particular area of the world and the oil field from which it was obtained but also on the age, that is the number of years since production was initiated, in the oil field sampled.

Conventional crude oil, which ranges from green to brown to black in colour depending on the petroleum type and the mineral matter present, is a heterogeneous mixture of liquids, solids, and gases. Some components of the crude oil are soluble in others and some are not. Water may occur as a readily separated phase with the petroleum produced, as it generally occurs in the petroleum reservoir, or it may occur as an emulsion containing as much as 80-90 % water with the oil [1]. The pour point, used as one indicator of the low temperature viscosity or flow characteristics of an oil, is defined as being 3 °C (or 5 °F) above the setting temperature (maximum temperature at which no observable flow occurs) of the oil [2]. Pour points of some viscous conventional crudes can lie above 5 °C, while the pour points of the more fluid crudes lie well below −15 °C (Table 16.1)

The industry standard of bulk measurement for both crude oil and products was the barrel (abbreviated bbl) but now the metric tonne and the cubic meter are the most commonly used, especially in international trade. The metric tonne may be converted to the barrel volume unit by dividing the mass unit by the density (specific gravity) of the particular oil being measured and multiplying by 6.2898. For example 1 tonne of Canadian crude, of density 0.85 g/cm$^3$, would equate to 7.4 bbl of oil (Equation 16.1).

$$(1 \text{ tonne} \times 6.2898 \text{ bbl/m}^3) \div (0.85 \text{ tonne/m}^3)$$
$$= 7.40 \text{ bbl} \qquad (16.1)$$

In all bulk measurements of petroleum and especially of crude oil it is necessary to specify the gross concentration of solids and non-petroleum liquids present. This is normally stated as percent bottom sediment and water ( % BS & W), determined by simple centrifuging of a representative sample. Light petroleum gas is almost always found in solution in conventional crude oil and some wells are brought into production to produce solely natural gas. But even when the objective is to produce oil this may contain as much as 50 m$^3$/m$^3$ (ca. 300 ft$^3$/bbl) of dissolved gas, solution of the gas in the oil being assisted by the high formation pressures [1]. Gas is usually separated at well gathering stations by the controlled release of gas from solution. Depending on the producing temperature and pressure of the oil, gas may be released in several stages and flared, or may be compressed for pipeline transport or formation gas injection where oil, rather than gas, is the production required.

Formation water is produced along with the crude oil at rates averaging as much as 10 % by volume but with much higher volumes being experienced from oil fields using water flooding for enhanced recovery. The water (and its dissolved salts) is separated from the oil by simple gravity settling in field tank batteries following degassing, leaving the crude oil ready for delivery, often containing less than 0.5 % by volume sediment and water. This sediment and water contains salts, mainly sodium chloride, which can occur either dissolved, in the aqueous phase, or as a fine particulate suspension in the oil phase. Other entrained insoluble mineral matter such as sands and silts may also be present to the extent of 15-30 g/tonne oil (5-10 1b/1,000 bbl)[3] although the salt content alone can run as high as 300-1,500 g/tonne [4] (100-500 lb/1,000 bbl). This may seem to be an inconsequential amount. But when subsequent processing separates much of the oil content by distillation it concentrates the involatile mineral matter in the residue. This mineral matter, if not removed before distillation, can accumulate on heat transfer surfaces and in the process decrease thermal efficiency of the distillation. Thus, even these relatively small

**Table 16.1** Specifications of some examples of typical conventional and synthetic crude oils. Distillate and residue breakdowns are quoted in percent by weight.

| | Conventional crude oils [a] | | | Synthetic crude oils | |
|---|---|---|---|---|---|
| | Paraffin base oil (wax bearing) | Intermediate naphthene base oil (wax bearing) | Naphthene base oil (wax free) | From tar sands [b] | From oil shales [c] |
| Density, g/cm$^3$ [d] (specif. gravity) | 0.781 | 0.964 | 0.910 | 0.84 | 0.93 |
| Pour point, °C | < –15 | 5 | < –15 | –35 | 5 |
| Saybolt universal viscosity at 38 °C, seconds [e] | 34 | 4000 | 55 | 34 | 78 |
| Colour | green | brown-black | green | pale-yellow | black |
| Sulfur content, % | 0.10 | 3.84 | 0.14 | 0.03 | 0.72 |
| Distillation, | | | | | |
| 1st drop, °C | 34 | 138 | 157 | – | – |
| Fractions, % | | | | | |
| Gasoline + naphtha | 45.2 | 2.9 | 1.1 | 30 | 18 |
| Kerosene | 17.7 | 4.5 | 0.0 | 20 | 24 |
| Gas oil | 8.3 | 10.6 | 55.5 | 50 | 34 |
| Nonviscous lube | 9.8 | 8.6 | 14.2 | – | – |
| Medium lube | 3.8 | 6.7 | 4.7 | – | – |
| Viscous lube | 0.0 | 1.0 | 11.6 | – | – |
| Residue, % | 14.7 | 58.4 | 12.7 | – | 24 |
| Distillation loss, % | 0.9 | 1.9 | 0.2 | – | – |
| Carbon in residue, % | 1.1 | 18.2 | 4.5 | – | – |
| Carbon in crude, % | 0.2 | 10.6 | 0.6 | (22) [b] | – |

[a] Selected from [93].

[b] Data calculated from that for composite synthetic crude obtained after coking and Unifining of extracted bitumen, from [94]. Before Unifining (hydrogenation) mean density of the composite stream would be somewhat higher, and sulfur content would be about 3%. Proportions of distillate components are approximate, carbon content quoted is the code residue on pyrolysis of bitumen.

[c] Data calculated from that given for crude Fischer Assay Oil (the crude pyrolysate) from [95]. Proportions given on distillation are approximate. The percent residue quoted corresponds to the fraction of the pyrolysate having a boiling point higher than 482 °C.

[d] Density is not only a necessary property for the conversion of volume units such as m$^3$, bbl (barrels), or L to mass units such as tonnes, but is also a useful indicator of the composition of the crude oil (see text). °API is a petroleum industry density unit obtained from the specific gravity at 16°C: °API = (141.5/spec. grav.) – 131.5.

[e] Scale of viscosity, measured by the number of seconds required for a sample to pass through a standard orifice in a Saybolt viscosimeter, usually specified at 100 °F.

concentrations of non-dissolved material have to be considered when designing refinery process sequences.

The petroleum itself consists of a mixture of an enormous range of hydrocarbons and a number of heteroatom-substituted hydrocarbons. Even the natural gas, which mostly consists of methane, usually contains small amounts of 4 or 5 other hydrocarbon components and therefore is not uniform in composition. For instance, the paraffin hydro-carbons or crude oil, having a generic formula $C_nH_{2n+2}$, may be gases, liquids, or solids depending on their molecular weight (or the value of n). As examples of this group methane, $CH_4$, and the chief constituent of natural gas, has a boiling point of –164 °C and a melting point of –183 °C. Propane and butane, $C_3H_8$ and $C_4H_{10}$, are also both gases under ordinary conditions, but with boiling points of –42 °C and –1 °C respectively are both capable of fairly easy liquefaction. Butane, and higher (larger carbon number) members of this series occur not only in the straight chain form, referred to as the normal or "n" form but also in various branched chain structures of the same molecular formula but possessing different physical and chemical properties (e.g. Equation 16.2).

$$CH_3CH_2CH_2CH_3 \qquad CH_3\text{--}\overset{\overset{\displaystyle CH_3}{|}}{CH}\text{--}CH_3 \qquad (16.2)$$

n-butane (or butane)    isobutane
b.p. – 0.5 °C            b.p. – 12 °C

Pentanes, $C_5H_{12}$, with boiling points in the range of normal ambient conditions, are the constituents representing the approximate borderline between gases and liquids in the paraffin series (Equation 16.3).

As the carbon number gets larger in the paraffin series the number of possible structural isomers also gets larger so that hexane, for example, has five structural isomers, and heptane, six. All the eighteen isomers of octane have been isolated or synthesized, as have the thirty-five isomers of nonane [5]. Beyond this, little is established about the natural occurrence of the seventy-five possible structural isomers of decane ($C_{10}H_{22}$) or the over four thousand isomers possible with pentadecane ($C_{15}H_{32}$). Nevertheless, many of the possible par-

affin isomers are actually found in exhaustive separations of natural petroleum [6].

Normal octane, $C_8H_{18}$, with a melting point of $-57$ °C and boiling point of 126 °C, lies near the upper end of the liquid paraffinic constituents of gasoline. N-eicosane $C_{20}H_{42}$ $(CH_3(CH_2)_{18}CH_3)$, with a melting point of 36 °C and a boiling point of $>340$ °C is the first of the higher (longer carbon chain or larger molecular weight) paraffins which is isolated in solid form (a wax) under ordinary conditions. These larger saturated hydrocarbons occur dissolved in the lower molecular weight liquid hydrocarbons which comprise the bulk of light natural petroleums.

Saturated hydrocarbons also occur in petroleum in cyclic form, $C_nH_{2n}$ (if monocyclic). These cycloparaffins, referred to as naphthenes in the petroleum industry, occur primarily as five, six, and seven membered rings, with and without alkyl substituents, and also occasionally occurring as various combinations of two of these ring systems linked or fused together [6] (e.g. Equation 16.4).

Aromatic hydrocarbons, which occur to a varying extent in petroleum, have a higher ratio of carbon to hydrogen than any of the commonly occurring paraffins or naphthenes, corresponding to a molecular formula of $C_nH_{2n-6}$ if mononuclear (single ring only). Benzene, toluene and cumene

$$CH_3CH_2CH_2CH_2CH_3 \qquad CH_3\overset{\overset{\displaystyle CH_3}{|}}{CH}CH_2CH_3 \qquad CH_3\text{--}\overset{\overset{\displaystyle CH_3}{|}}{\underset{\underset{\displaystyle CH_3}{|}}{C}}\text{--}CH_3 \qquad (16.3)$$

n-pentane              isopentane             neopentane
b.p. 36 °C             b.p. 28 °C             b.p. 9.5 °C

cyclohexane            ethylcyclopentane      bicyclo[4.3.0]nonane                        (16.4)
b.p. 81 °C             b.p. 103.5 °C          (hexahydroindane) b.p. 161.08 °C

benzene       toluene       cumene            naphthalene                                 (16.5)
b.p. 80 °C    b.p. 111 °C   b.p. 152 °C       b.p. 218 °C
                                              m.p. 81 °C

are mononuclear examples of this series, and naphthalene, a dinuclear example (Equation 16.5).

These aromatics, as well as many others, have all been isolated from petroleum fractions [6]. The aromatic content of crude oils can vary widely but an aromatic content of a third or more of the total, as has been noted for some Borneo crudes [7], is not unusual. The density (or °API) of a crude oil is an indicator of the aromatic content since the high C:H ratio of aromatic components tend to make these the densest constituents present. This will be particularly true with polynuclear aromatic constituents since each additional ring further reduces the hydrogen count by two, increasing the already high C:H ratio and therefore also the density. While the bulk of the hydrocarbon content of crude oils is represented by the paraffins, naphthenes, and aromatics, small percentages of several other types of compound are also present. Olefinic hydrocarbons, unsaturated chain compounds having a carbon carbon double bond and a type formula $C_nH_{2n}$, also occur in natural petroleum but only to a very small extent since they are quite reactive compounds [8]. They are mentioned here, however, since they are formed to a significant extent by some refinery processes, particularly those involving cracking.

A wide variety of compounds which contain a heteroatom as well as carbon and hydrogen also occur in petroleum, but generally only to a limited extent.

Hydrogen sulfide and a variety of thiols, sulfides, and thiophenes are some examples of the sulfur compounds present [9] (e.g. Equation 16.6).

The sulfur content of petroleum containing these compounds is sometimes quite low but it can commonly be as high as 6 % of the total. The more stable oxygen derivatives of hydrocarbons such as paraffinic acids, ketones, and phenols (e.g. Equation 16.7), also occur in crude oils.

And nitrogen compounds, either on their own or complexed to a transition metal such as vanadium also occur in petroleum to a small extent. Pyridine, quinoline and many other heteroaromatics (e.g. Equation 16.8) have also been found, and the first two are also occasionally produced from petroleum. Benzonitrile has also been detected [9]. Much of the trace metal content of petroleums, in particular vanadium and nickel, is present in association with petroporphyrins, polycyclic pyrroles closely related in structure to hemes and chlorophylls. These represent examples of the more complex nitrogen heterocycles to be found in petroleum. In fact it is the presence of these particular heterocycles, with their complexed metal atoms, which contributes much to our present knowledge of the original biogenesis of petroleum hydrocarbons [10].

In the composite of molecular types found in a particular petroleum reservoir a light, fluid (low viscosity) crude is obtained if the proportion of low

$SCH_2CH_2CH_3$

$CH_3(CH_2)_5SH$

1-hexanethiol

1-thiapropyl-
benzene

2-methylbenzo-
(b) thiophene

(16.6)

$$CH_3(CH_2)_2\overset{\overset{\displaystyle CH_3}{|}}{C}HCH_2CO_2H$$

3-methylhexanoic acid

fluorenone

OH

$CH_3$

p-cresol

(16.7)

pyridine     quinoline     carbazole

CN

benzonitrile

(16.8)

molecular weight hydrocarbons (low carbon number, small molecules) to high molecular weight hydrocarbons is large. If, however, high molecular weight paraffins or polynuclear aromatics (asphaltenes) predominate, then one will have a viscous, high pour-point crude such as occurs in oil fields producing so-called "heavy crudes". The composition of crude oil produced from a particular oil field will vary somewhat too, with the stage of production. A higher proportion of lighter (lower molecular weight) hydrocarbons will generally predominate in the early stages of production. Later a somewhat larger proportion of higher molecular weight hydrocarbons would be obtained. The ratio of the percentage of paraffins, to the proportion of naphthenes and aromatics present will be relatively consistent from different producing wells of a particular oil field but can vary widely from one oilfield to another. With all these variables in mind, refineries seldom rely on crude oil from a single oil field for the whole of their production but will employ a selection of crudes depending on price, availability, operational processing equipment and proportions of the particular product mix being made at the time.

## 16.2 Pretreatment and Distillation

### 16.2.1 Crude Oil Desalting

Crude oils delivered to the refinery frequently contain significant quantities of water, sand, extraneous salts, etc. making desalting an important preliminary step before further processing. If these materials are not removed before the crude oil enters the system they can increase corrosion rates, and also the frequency of plugging (blockages) and the occurrence of scaling of refinery equipment such as heat exchangers. Two desalting methods are in common use, each of which requires the use of several simple processing units operating in series.

Chemical desalting is accomplished by addition of water equivalent to about 10 % of the volume of the oil to be treated plus sulfuric acid or sodium hydroxide as necessary for pH adjustment of the particular crude being processed. The pH adjustment is carried out just before the oil enters the desalting system charging pump (Figure 16.1). The acid or base addition may be sufficient by itself to cause rapid demulsification after mixing or a small con-

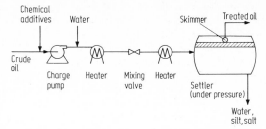

**Figure 16.1.** Flowsheet for the main steps involved in chemical desalting [11]

centration of a proprietary demulsifier such as Tretolite (a polyethylenimine) may have to be added with the water to assist the subsequent phase separation. In either case passage through the pump thoroughly mixes the constituents, further assisted by indirect heating to 65 to 180 °C (depending on oil viscosity) and by passage through a mixing valve [11].

After treatment, the mixture is passed through closed vessels under sufficient pressure to prevent any vapourization, where there is sufficient residence time to allow separation of the oil and aqueous phases. Oil is drawn from the top of the vessel and the aqueous phase containing the salts, sand, silt and other extraneous material is drawn from the bottom.

Desalting is also, and more frequently, carried out electrostatically [12]. As in chemical desalting, water is added to the oil stream (to dissolve suspended salts) and the stream passed through a mixing valve into the desalter where oil/water separation is principally achieved with a high potential electrostatic field instead of with demulsifying chemicals [13]. When appropriately practised the electrostatic field induces rapid coalescence and settling of water droplets together with any other aqueous-associated impurities. The water phase is drawn continuously from the bottom of the desalter vessel and passed to the refinery effluent treating plant. The oil phase is now ready for further processing. Desalting by one or other or a combination of these two procedures is now conducted as a routine preliminary step for most crudes processed in the U.S.A. [14].

### 16.2.2 Petroleum Distillation

The first stage of separation of useful components from crude oil is achieved by distillation, which accomplishes a rough grouping on the basis of boiling point differences. This was originally carried out on a batch basis, where the crude oil to be distilled was entirely in place in the distilling vessel at the start of heating. With a batch still the components of the crude are obtained progressively from this vessel in vapour form as heat is applied, lighter (lower boiling) components first. Mid-range and heavier, less volatile constituents are distilled later in the sequence, eventually leaving behind a viscous, high boiling point asphaltic residue in the distilling vessel. However, because batch distillations require inefficient staging of their sequences of filling, heating, stopping, pumping out the residue followed by refilling of the distilling vessel again, they tended only to be economic for distilla-

tive separations on a scale not exceeding about 300 bbl/day (ca. 12,000 U.S. gallons, 10,000 Imp. gallons, or 45 m$^3$ per day). This scale of production was only appropriate for local, very small distillate requirements from a simple refinery which also happened to be close to a petroleum producing area, not a very common situation.

Even a small modern refinery distills 2,000 to 10,000 bbl/day and the largest American refineries process 175,000 bbl (27,800 m$^3$) or more crude oil per day [15]. These are all scales of operation which demand continuous, rather than batch distillation to be accomplished efficiently and economically.

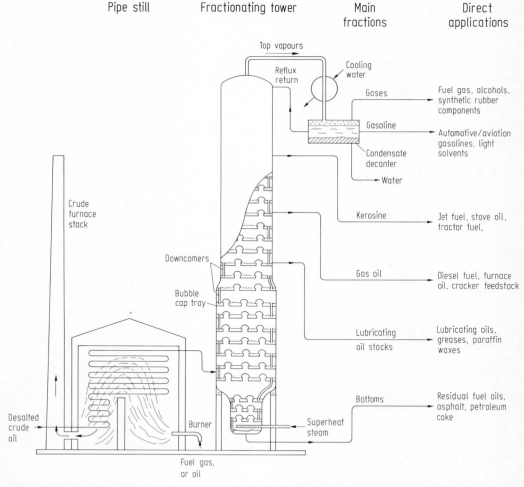

**Figure 16.2.** Partial cut-away view of continuous crude oil distillation via a pipe still and fractionating column. Frequently the heat content of the distillate fractions is employed to preheat the incoming crude oil (not shown here)

For continuous distillation the crude oil is first heated to 400-550 °C while it is continuously flowing through a pipe still, using natural gas, "light ends" (miscellaneous low-boiling hydrocarbons) or fuel oil for fuel. The heated crude is then passed into a fractionating tower near the bottom, the hot end of the tower. In the fractionating tower, a unit 2-3 m in diameter and 30-40 m high for a large modern refinery, the lower boiling components, as vapours, move up (Figure 16.2). As the vapour moves upward past each plate of the column it is forced, via the bubble caps ("bell caps") of that level, to pass through and come to thermal equilibrium with the liquid on that plate or tray. Hydrocarbons having a boiling point lower than the temperature of the liquid on the plate will continue to move up the column, in vapour form. Components of the heated crude having a boiling point higher or the same as the temperature of the liquid on the plate will tend to condense in the liquid on the plate. As liquid of similar boiling point accumulates on the plate it either overflows into the

projecting end of a "downcomer" pipe to the next lower plate, or is drawn off the plate and the column as one of the product streams of the crude distillation. The lower end of each downcomer is of sufficient length that it dips some distance into the liquid on the lower tray to prevent vapour movement *up* the downcomer.

In this way the lowest boiling components, the petroleum gases (mostly $C_3$ and $C_4$ hydrocarbons, and commonly referred to as liquefied petroleum gas, or LP gas) and frequently some water, are collected as vapours from the top of the fractionating column. The vapours pass through a dephlegmator (a partial condenser) which condenses naphtha, gasoline, and water vapour components out of this vapour stream, and allows passage of butane and lighter fractions through as vapours (Table 16.2).Water is phase-separated from the condensed liquids for removal in a unit outside the fractionating column, and most of the condensed naphtha/gasoline fraction is withdrawn from this as a product stream.

**Table 16.2** Atmospheric pressure boiling point ranges, and trends in properties and approximate composition of representative fractions of crude distillation [a]

| Fraction | Boiling range, °C | Density g/cm$^3$ | Viscosity[b] cSt at 38 °C | Range of n-paraffins[c] | No. of compounds | Sulfur, wgt. % |
|---|---|---|---|---|---|---|
| Gas | < 40 | 0.4 – 0.6[e] | – | C 1 – C 5 | 7 | 0.001 |
| Naphtha/ gasoline | 40 – 180 | 0.70 | low | C 6 – C 10 | > 100 | 0.011 |
| Kerosine | 180 – 230 | 0.79 | 1.1 | C 11, C 12 | >> 40 | 0.20 |
| Light gas oil | 230 – 320 | 0.85 | 3.7 | C 13 – C 17 | >> 12 | 1.40 |
| Heavy gas oil | 320 – 400 | 0.90 | 50 | C 18 – C 25 | >> 10 | 2.0 |
| Lubricant | 400 – 520 | 0.91 | 60 | C 26 – 38 | >> 7 | 3 |
| Residue | > 520 | 1.02 | solid[b] | > C 38 | – | 4 – 6 |

[a] Compiled from [1, 6 and 15]. The variable basis of the sources consulted necessarily makes the values given somewhat approximate, the uncertainty being roughly indicated by the number of significant figures quoted. The trends seen here are realistic.

[b] Viscosity in centiStokes is equivalent to units of mm$^2$/sec. For the lubricant fraction the value would be about 7 cSt at 99 °C and for the residue about 13 400 at 99 °C.

[c] Carbon numbers quoted are for normal (i.e. straight chain) paraffins. For branched paraffins the carbon numbers for the components included in this boiling range will tend to be somewhat higher, and for cyclic paraffins (naphthenes) and aromatics somewhat lower.

[d] Numbers given are number of individual compounds actually isolated from a sample in the boiling range given. This is preceded by a single carat if the compounds isolated comprised most of the sample of that boiling range, and a double carat if this comprised only a small fraction of the sample.

[e] A range since this is a composite of the densities at 15 °C of methane, 0.3; ethane, 0.37; propane, 0.51; isobutane, 0.56; butane, 0.58; and pentane/isopentane, 0.63, and is therefore significantly influenced by the properties of these hydrocarbons present.

A part of the phase-separated hydrocarbon stream is returned to the top plate of the fractionating column as a reflux for temperature control. This return of condensed light distillate to a top plate of the column is what controls the temperature of the cool end of the fractionating column to achieve the thermal gradient of 12-15 °C/m required. The bottom and hot end of the column is kept at about 500 °C via heat brought in through the entry of preheated crude oil, already mostly vaporized, and also occasionally by means of a reboiler operating at the bottom of the column indirectly heated by heat exchange fluid. To decrease the thermal gradient per meter of height, and sharpen (narrow the boiling range of) the fraction separated from crude oil the fractionating column will occasionally be split into two columns, each taking over one half of the fractionation duty of the single column they replace. One of these will normally operate over the lower temperature range of about 40 °C to 350-370 °C. The bottom, higher boiling point stream from this column, which includes all components boiling above about 370 °C, is then fed hot to a second column, which will normally operate under reduced pressure, to provide sharper fractions of the higher boiling constituents of the crude.

Whether the crude distillation is conducted in a single or dual column arrangement the principle of operation is the same. Hydrocarbon vapours plus a naphtha/gasoline fraction boiling in the 40-180 °C range is the top take-off product of the crude still. A kerosine/jet fuel fraction boiling at approximately 180-230 °C is taken off a plate further down the column, followed by fractions of light gas oil, b.p. 230-300 °C and heavy gas oil, b.p. 300-400 °C, from appropriate lower plates of the column (Figure 16.2). The highest boiling distillate (i.e. volatile) fraction from the crude still is the lubricating oil stream, boiling in the range 400-520 °C, which also contains much of the grease and wax yield of the crude oil. The residue or the bottom stream of the crude fractionating tower includes all the crude components not vapourized below about 520 °C, and consists of mostly asphalt and suspended petroleum coke. This is the highest distillation temperature normally used for crude distillation because at temperatures in this range the decomposition or break-down of the residual large hydrocarbon constituents begins to become significant.

Since the petroleum cuts (fractions) obtained from the crude distillation column(s) are quite wide in the boiling point ranges collected, many, if not all of these crude streams will be subjected to redistillation and/or stripping to sharpen (narrow down) the boiling point ranges. For instance the gas stream, composed entirely of components boiling at below 40 °C, is normally redistilled under pressure. Pressure distillation increases the boiling and condensation temperatures of the components present, and in this way enables condensation of many of the separated compounds from the original composite mixture of gases, and collection of these as liquids. Lower (and more expensive) refrigeration temperatures would be required to achieve condensation if distillation was conducted at atmospheric, rather than at higher pressures. Distillation under pressure also increases the difference between the boiling points of the individual components being distilled, which makes cleaner and sharper separations possible in fractionating columns of equivalent separation capability. Usually a separate column is used for the final separation of each component.

The naphtha/gasoline, kerosine, and higher boiling streams will usually be sharpened to narrower boiling ranges than obtained directly from the crude still by heating with steam in a reboiler to strip (remove by vaporization) excess volatiles. Alternatively the stream may actually be redistilled in a small column to separate it into two or three separate sharper fractions. The trends in density, viscosity, sulfur content etc. that are obtained as one proceeds from the low boiling point to higher boiling point fractions are quite compositionally informative (Table 16.2). For instance, as the carbon to hydrogen ratio increases a corresponding increase in density is observed. And the viscosity also increases regularly with an increase in molecular size. From this information it is also evident that much of the sulfur occurs in larger, rather than smaller molecules from the observed trend to increasing sulfur content with increasing boiling point. Much of the content of transition metals such as vanadium, nickel, and iron, also tends to remain with the high boiling and residual fractions of crude oil distillation [15], up to 250 ppm of vanadium being not uncommon in the residue [16].

Simple distillation, coupled with secondary distillation processes and stripping, enables the broad range of constituents present in crude petroleum to be separated into useful fractions consisting of hydrocarbon compounds of similar physical properties. Each fraction will contain from five up to over

100 different compounds (Table 16.2, 16.3). But the proportions of the constituents falling within these more or less arbitrary useful ranges of properties in different crude oils varies widely depending on the origin of the crude (Table 16.1). As little as 1 or 2 % of gasoline-type components is present in some crude oils while the minimum demand for gasoline, in Japan and Western Europe, is from 16 to 20 % of the petroleum refined (Table 16.4). In the U.S.A., the gasoline demand as a fraction of the crude oil refined runs to something like double these figures. This high a gasoline demand can be met from distillation alone of only a very few crude oils. Coupled to this there is also a seasonal swing in gasoline demand, particularly in northern countries. In Canada this seasonal swing amounts to a change in demand of from 27 % of refinery output in winter to about 35 % in summer, which corresponds to a 30 % jump in gasoline output required for summer production. Because of these poor matches between the proportions of constituents present in crude oil and separable by simple distillation, and the proportional demand of the main refinery product streams, some molecular modification has to be carried out on the product streams from crude distillation to achieve a better correspondence of product volumes to demand.

## 16.3 Molecular Modification for Gasoline Production

The naphtha/gasoline fraction from crude distillation, which is already composed of a large number of different hydrocarbons with physical properties in the right range, is still not a large enough fraction of the crude petroleum being refined to supply the demand. Thus, gasoline, as sold at the pump, contains not only desulfurized straight run gasoline, which is the component obtained directly from the distillation of crude, but also cracked gas oil, reformate, polymer gasoline, and occasionally natural gasoline as well. Natural gasoline, also referred to as casinghead gasoline, is the liquid fraction of $C_5$ and higher hydrocarbons which is condensed at the well-head (in producing field installations) from natural gas. It is often referred to as condensate, and is added to the crude oil before primary refinery distillation. The gasoline component of cracked gas oil comprises the $C_6$ to $C_{10}$ molecular fraction recovered from the partial thermal breakdown, or decomposition, of the larger molecules present in gas oil. Polymer gasoline, as the name would suggest, is produced by the fusing of smaller hydrocarbon molecules, primarily $C_3$ and $C_4$ hydrocarbons, to produce larger molecules of an appropriate size range for blending into gasoline. Reformate is straight run gasoline, desulfurized and upgraded to a higher octane rating by thermal or catalytic reforming. So by this combination of decreasing the molecular size range of crude oil components containing more than ten carbon atoms, and increasing the carbon number of primarily a $C_3$ and $C_4$ fraction of crude oil, the gasoline yield from the crude may be increased beyond that originally present to tailor the supply to suit the demand. The direct means of separating the first two of the four primary gasoline components from the crude oil have already been discussed. Details of the methods of production of the other two components follow.

**Table 16.3** Examples of some primary and secondary uses of the gas stream from petroleum refining streams

| Component | Direct uses | Primary products | Secondary products |
|---|---|---|---|
| methane | natural gas, heating | hydrogen | ammonia, methanol |
| ethane | natural gas | ethylene | polyethylene, ethanol, styrene, ethylene glycol |
| propane | bottled gas (liquified petroleum gas, LPG) | propylene | polypropylene, propylene glycol, cumene |
| butane | bottled gas, natural gasoline (winter blends) | butadiene, polymer gasoline | synthetic rubber, adiponitrile (nylons) |
| pentane | natural gasoline | gasoline | |

**Table 16.4** Principal uses and demand distribution for the primary liquid refinery
products [a]

| Refining stream | Components | Direct uses | Demand distrib'n, % by wgt | | |
|---|---|---|---|---|---|
| | | | Japan | U.S.A. | W.Europe |
| Gases | C 1 – C 5 | natural gas, LPG [b] etc. | 1 – 2 | 1 – 2 | 1 – 2 |
| Naphtha/gasoline | C 6 – C 10 | aviation gasolines, motor gasolines, light weed control oil, dry cleaning, and metal degreasing solvents | 20 | 42 – 45 | 16 – 21 |
| Kerosene | C 11, C 12 | diesel fuels, jet fuels, illuminating and stove oils light fuel oils | 8 – 9 | 9 – 11 | 4 – 8 |
| Light gas oil | C 13 – C 17 | gas turbine fuels, diesel fuel cracking feedstock, furnace oil | 7 – 8 | 20 – 22 | 33 – 35 |
| Heavy gas oil | C 18 – C 25 | cracking feedstock, fuel oils | 48 – 50 | 8 – 10 | 35 – 37 |
| Lubricant fraction | C 26 – C 38 | gear and machinery lubricating oils, cutting and heat-treating oils, lubricating greases, medicinal jelly, paraffin waxes | 2 – 3 | 1 – 2 | 1 – 2 |
| Residue | > C 38 | roofing, waterproofing and paving asphalts, residual fuel oils | – | 5 – 6 | 2 |
| Refinery consumption and losses | | | – | 20 – 22 | 11 – 13 |

[a] Compiled from data presented in [1, 15, 17, 96 – 98].
[b] Liquefied petroleum gas, i.e. propane, butane.

## 16.3.1  Thermal Cracking of Gas Oils

Cracking, as a general term applied to a refinery practice, is a process by which high temperatures and moderate pressures are used to break down, or decompose the larger molecules of gas oil into smaller ones suitable for incorporation into gasoline. The various cracking processes in regular use for production of gasoline constituents represent the most important of the petroleum modifying processes used by refineries today, with a capacity in North America exceeding 50 % of the current crude distillation capacity [17]. If the thermal decomposition is relatively mild, and is conducted on a residual feedstock to decrease its viscosity or to produce a heavy gas oil from it, the process is called viscosity-breaking, or "vis-breaking" for short. If the decomposition is severe or is continued for a sufficient time that coke is the residual product, the process is termed coking. If, however, the feedstock being treated is already in the naphtha/gasoline boiling range and is only subjected to mild decomposition to decrease the boiling point and increase the octane number slightly, then the process is referred to as thermal reforming [17]. These last three cracking variants are all related to, but far less important in terms of throughput volume than the processes which are used to crack gas oil directly to gasoline fractions.

There are two broad categories which can be used to classify all of the currently practised methods of cracking gas oil to gasoline. These are the straight thermal cracking, and the catalytic cracking processes. Thermal cracking, or the direct application of heat and pressure alone to modify gas oil refinery streams, was the original method by which the gasoline yield from a crude oil was increased. Today straight thermal cracking is only used in "simple" refineries where the range of accessible petroleum modification equipment is small and where the additional gasoline requirement, over and above that present in the crude, is not large. It involves initial formation of gas oil vapour, and then heating of this to the region of 500-600 °C under a pressure of about 3.5 x $10^6$ Pa (N/m²; or about 34 atmos) for a thermal contact time of the order of one minute. Temperatures this high are sufficiently energetic to cause homolysis (rupture to two radicals) of carbon bonds to form two highly reactive, but uncharged, radical fragments (Equation 16.9).

$$RC \overset{\displaystyle H \quad H}{\underset{\displaystyle H \quad H}{\vert \quad \vert}} : \overset{\displaystyle H \quad H}{\underset{\displaystyle H \quad H}{CCH_2CH_2R'}} \rightarrow RC\cdot \quad \cdot CCH_2CH_2R' \quad (16.9)$$

Each radical can then either attack another hydrocarbon molecule, by collision and abstraction of a hydrogen atom from the new molecule to form a new stable species from itself and another radical species from the subject of the collision (Equation 16.10).

$$RCH_2 + RCH_2CH_2CH_2CH_2CH_2CH_2CH_3 \rightarrow$$
$$RCH_3 + RCH_2CH_2CH_2CH_2\overset{\displaystyle \cdot}{C}HCH_2CH_3 \quad (16.10)$$

This process is thermodynamically favourable since a secondary (or internal) radical is more stable than a primary (terminal) one. Or, the fragment can undergo $\beta$-fission to lose ethylene and form a new smaller primary radical (Equation 16.11).

$$R'CH_2CH_2CH_2\cdot \rightarrow R'CH_3 + CH_2=CH_2 \quad (16.11)$$

It is this $\beta$-fission process that leads to a large part of the ethylene produced by thermal cracking. The hydrocarbon produced with an internal secondary radical center can also abstract a hydrogen atom from another hydrocarbon molecule, but this process will not be as favourable as the process which produced it since attacking and forming radicals will be of similar stability. Also there would be no reaction progress. But this radical center can undergo $\beta$-fission too, forming a smaller hydrocarbon radical and olefin products (Equation 16.12),

$$RCH_2CH_2CH_2CH_2\overset{\displaystyle \cdot}{C}HCH_2CH_3 \rightarrow$$
$$RCH_2CH_2\overset{\displaystyle \cdot}{C}H_2 + CH_2=CHCH_2CH_3 \quad (16.12)$$

a process which can continue (Equation 16.13).

$$RCH_2CH_2\overset{\displaystyle \cdot}{C}H_2 \rightarrow R\overset{\displaystyle \cdot}{C}H_2 + CH_2=CH_2 \quad (16.13)$$

Some of the larger olefins formed by these processes are useful gasoline constituents but the large amounts of ethylene also formed have to be used for other purposes. But there is one other process open to a terminal radical on a long hydrocarbon chain which can give rise to useful gasoline constituents and that is "backbiting". The radical on the terminal carbon can bend back on itself to pluck a hydrogen atom from the carbon atom six from the end of the chain, in so doing forming a more stable secondary radical from a primary one (Equation 16.14).

Progression of the resulting secondary radical center then gives heptene plus a primary radical fragment (Equation 16.15).

$$RCH_2CH_2\overset{\displaystyle \cdot}{C}HCH_2CH_2CH_2CH_2CH_3 \rightarrow$$
$$R\overset{\displaystyle \cdot}{C}H_2 +$$
$$CH_2=CHCH_2CH_2CH_2CH_2CH_3 \quad (16.15)$$

Thus, by the occurrence of many variations of processes such as these, thermal cracking produces a mixture of gaseous and liquid hydrocarbons of paraffinic (saturated) and olfinic (unsaturated) types, plus coke (Equation 16.16).

$$(16.14)$$

$$\text{gas oil} \quad 500-600\,^{\circ}C$$
$$\text{vapour} \qquad \rightarrow$$
$$\quad\quad \text{ca. 60 sec., 35 bars}$$
$$\quad\quad\quad \text{(34 atmos)}$$
$$\left\{\begin{array}{l} \text{— gases} \\ \text{— naphtha/gasoline} \\ \text{— cracked gas oil} \\ \quad \text{(heavier fractions)} \\ \text{— coke} \end{array}\right. \qquad (16.16)$$

The products, in particular the gasoline component, are separated from the mixture subsequently, by distillation (Figure 16.3). Of course the objective of the cracking process is to produce as much hydrocarbon in the $C_5$ to $C_9$ range appropriate for gasoline blends as possible (e.g. Equation 16.17).

$$C_{17}H_{36} \rightarrow \quad C_8H_{18} + C_9H_{18} \qquad (16.17)$$
$$\text{heptadecane \quad octane \quad nonene (olefinic)}$$

But larger paraffinic hydrocarbons cannot be cracked into two smaller paraffins because of the hydrogen-short nature of cracking stoichiometry. This is why the best that can be done is to obtain a paraffin plus an olefin. But the sensitivity of this process is such that at too high a temperature or too long a contact time the coke-forming reactions become more significant at the expense of hydrocarbons of the desired size range (e.g. Equation 16.18).

$$9\,C_{15}H_{32} \rightarrow \quad 16\,C_8H_{18} + 7\,C \qquad (16.18)$$
$$\text{pentadecane \quad octane}$$

This is one of the significant reasons for the care required in selection of cracking conditions to minimize the extent of coke formation.

## 16.3.2 Catalytic Cracking

During the late 1930's and early 1940's, it was discovered that when cracking to gasoline was conducted in the presence of a suitable catalyst the process was speeded up by a factor of from several hundred up to a thousand times the rates of straight thermal cracking (Table 16.5)[17]. This discovery firmly established the dominance of the "cat-cracking" route for the production of gasoline from gas oils.

Catalysts used in this process are of two types. Special acid-washed natural clays of particle diameters in the 2 to 400 $\mu$m range are used in fluidized bed versions of cracking (Figure 16.4) [18]. In these units the catalyst is kept suspended or "fluffed up" on an upward moving stream of hot gas oil vapours, which ensures both continuous exposure of all catalyst faces to the raw material and permits continuous turnover of catalyst material. Synthetic catalysts, prepared from a mixture of 85-90 % silica plus 10-15 % alumina, or from synthetic crystalline zeolites (molecular sieves) are also used [7, 18]. These synthetic catalyst types can either be made into a small particle size format suitable for use in a fluidized bed, or can be formed into 3 or 4 mm diameter pellets appropriate for crackers which use a moving bed mode of catalyst cycling.

A common feature of all of these heterogeneous materials is their acidic nature, i.e. they all function as active solid phase acids in the hot gas oil vapour stream. Synthetic silica/alumina catalyst composites, for example, have an acidity of 0.25 meq/g distributed over an active surface area of some 500 m$^2$/g. And it is this acidity that is one of the key features which distinguishes catalytic cracking from straight thermal cracking.

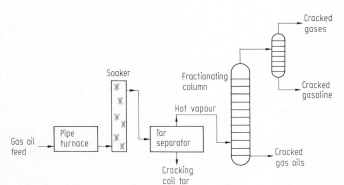

**Figure 16.3.** Tube and tank cracking unit for straight thermal cracking

**Table 16.5** Trends in the utilization of cracking processes for petroleum modification in the U.S.A. [a]

| Cracking process type | Cracking capacity as % of crude capacity | | | | | |
|---|---|---|---|---|---|---|
| | 1930 | 1940 | 1950 | 1960 | 1970 | 1979 |
| Thermal | 42 | 51 | 37 | 16 | 12 | n.a. |
| Catalytic | 0 | 3 | 23 | 36 | 37 | 33 |
| Hydrocracking | 0 | 0 | 0 | 0 | 5 | 5 |

[a] Data derived from [15 and 99].

Among the various possibilities open to acid catalyst action is the possibility that it may donate a proton to an olefin, generating a carbonium ion. The carbonium ion may simply lose a proton again but now from an internal position along the chain, which yields a different, more stable product (Equation 16.19).

$$RCH_2CH_2CH{=}CH_2 \rightarrow RCH_2CH_2\overset{+}{C}HCH_3 \rightarrow$$
$$RCH_2CH{=}CHCH_3 \qquad (16.19)$$

Or carbonium ion migration to a position further along the chain may occur, to move the cationic center to a position which then favours alkyl group migration towards a more stable carbonium ion, rather than olefin formation (Equation 16.20).

$$RCH_2CH_2\overset{+}{C}HCH_3 \rightarrow RCH_2\overset{+}{C}HCH_2CH_3 \rightarrow$$
$$\underset{\begin{array}{c}|\\RCH_2CH{-}\overset{+}{C}H_2\end{array}}{CH_3} \qquad (16.20)$$

Migration of a hydride ion from the branching carbon of the primary (terminal) carbonium ion so formed can then produce a relatively stable tertiary carbonium ion. An unchanged olefinic product may be formed from this by loss of a further proton (Equation 16.21).

$$\begin{array}{cc} CH_3 & CH_3 \\ | & | \\ RCH_2{-}\overset{+}{C}H{-}CH_2 \rightarrow & R{-}CH_2{-}\overset{+}{C}{-}CH_3 \rightarrow \end{array}$$
$$\underset{\begin{array}{c}|\\RCH_2{-}C{=}CH_2\end{array}}{CH_3} \qquad (16.21)$$

It is processes of these kinds that lead to the internal olefinic (alkene) and the branched paraffinic (alkane) and other branched olefinic products which are characteristic of catalytic cracking processes.

When a carbonium ion is formed some distance from the end of a large hydrocarbon molecule, ei-

ther by withdrawal of a hydride ion by another carbonium ion species in the gas phase or by an electron deficient area on the surface of the catalyst, it can form a neutral product by one of the processes just outlined. If so, the properties will change but there will be no significant decrease in average molecular size of the product. Or it can undergo $\beta$-fission via either of two possible routes each yielding two smaller fragments, one with a terminal carbonium ion and the other with a terminal olefin [7] (Equation 6.22).

$$\overset{1}{RCH_2CH_2CH_2}{-}CH_2\overset{+}{C}HCH_2\overset{2}{-}CH_2CH_2CH_2R'$$
Fission at: $\qquad (16.22)$
$$1. \rightarrow RCH_2CH_2\overset{+}{C}H_2 +$$
$$CH_2{=}CHCH_2CH_2CH_2CH_2R'$$

$$2. \rightarrow RCH_2CH_2CH_2CH_2CH{=}CH_2 +$$
$$\overset{+}{C}H_2CH_2CH_2R'$$

It is processes such as this which give the molecular size reduction activity of catalytic cracking processes.

These acid-catalyzed changes are brought about by passing hot gas oil vapours either through a mechanically circulated bed of the larger pellets, or upwards through a fluidized bed of more finely divided catalyst particles, the upflow of vapours providing the fluidizing action. To avoid loss of catalyst as the vapours of cracked gas oil leave the catalyst bed the vapours from fluidized bed crackers are passed through two cyclone separators in series. These serve to retain any entrained solid particles in the vapour stream from the catalyst bed and return these to the bed (Figure 16.4). The various products formed are then separated from the composite vapour stream on a distilling column.

Overall, catalytic cracking processes achieve a far higher cracking rate than is possible from straight thermal cracking. Furthermore this is achieved at somewhat lower temperatures and much lower pressures than is possible with thermal cracking

**Plate 16.1.** General view of the Chevron Refinery at Burnaby, B.C. Catalytic cracker with associated columns is shown in the center, reformer at the right hand side

(Table 16.6). Not the least of the benefits of catalytic cracking is that these operating advantages are obtained while at the same time obtaining a much higher proportion of gasoline-suitable products than obtained from straight thermal cracking, almost 50 % of the total. In fact modern fluid catalytic cracking units can give as much as 75 to 80 % gasoline.

Continuous decoking of the catalyst is necessary, at the high cracking rates of these systems, in order to maintain catalytic activity. This is accomplished by continuously moving a portion of the catalyst from the catalyst bed to regenerator, using pneumatic conveying or gravity (Figure 16.4). Here, the still hot catalyst particles are suspended in a current of air, which forms a second fluidized bed in the regenerator and effectively burns off much of the carbon (Table 16.5). Burning off

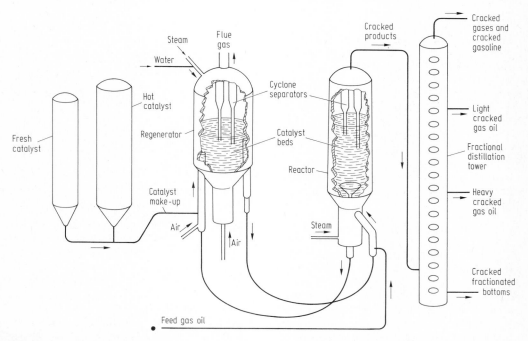

**Figure 16.4.** Main operating components of a fluid catalytic cracking unit (FCCU) [26]. Courtesy G.A. Purdy

**Table 16.6** A comparison of the conditions required for straight thermal and catalytic cracking of gas oil to gasoline [a]

|  | Straight thermal | Fluidized bed |
|---|---|---|
| Temperature, °C | 450 – 565 | 465 – 540 |
| Pressure [b], $N/m^2$ | $1.8 – 6.2 \times 10^6$ | $6.9 – 14.5 \times 10^4$ |
| (bars) | (14 – 60) | (0.7 – 1.4) |
| Contact time, sec | 40 – 300 | 100 – 300 [c] |
| Carbon removal | periodic shutdown | on spent catalyst, 0.5 – 2.6% C regenerated catalyst, 0.4 – 1.6% C |

[a] From [8 and 18].
[b] Above atmospheric pressure, i.e. gauge pressure, in pascals ($N/m^2$).
[c] Estimated from a space velocity range of 0.5 to 3.0. Space velocity here defined as (weight of oil feed per hour)/(weight of catalyst in reactor), since catalyst volume is not fixed in the fluid bed mode of operation.

carbon also serves to reheat the catalyst particles before these are returned to the endothermic processes going on in the cracker, this time accomplished using hot gas oil vapour as the driver. Thus, regeneration of catalyst not only allows a rapid recycle period, but also provides an efficient solid heat transfer medium between the regeneration unit and the cracker. This solids handling system for continuous catalyst regeneration is also a substantial petroleum engineering accomplishment since for a fluid bed cat cracker processing 100,000 barrels of gas oil per day the rate of catalyst movement between the cracker and the regenerator is at the rate of a boxcar load per minute, all without any moving parts [18]. The waste combustion gases from the regenerator exit through two cyclones in series to avoid catalyst losses, followed by heat exchange units for energy recovery as steam and frequently followed by an electrostatic precipitator for fine particulate emission control [18].

### 16.3.3 Polymer and Alkylate Gasoline

Much less significant than catalytic cracking in terms of the volume contributed to gasoline blends, the preparation of polymer gasoline by joining together two (or more) light hydrocarbons nonetheless forms an important part of integrated refinery operations. Even though the production of polymer gasoline comprises only 2-4 % of the volume of crude distilled, it enables the use of relatively large amounts of the $C_3$ and $C_4$ hydrocarbons formed as byproducts of cracking processes. Thus, it further incrementally increases the ultimate yield of gasoline possible from gas oil, and in the process generates a high quality component for gasoline blends.

Polymerization is carried out on a $C_3 + C_4$ hydrocarbon stream containing a high proportion of olefins, under heterogeneous conditions. Acid catalysis, usually phosphoric acid, is employed on a solid support. At pressures of $2.8-8.3 \times 10^6$ $N/m^2$ (27-82 atmospheres), high compared to the conditions used for cracking, and more moderate temperatures in the 176-224 °C range [18], carbon carbon bond formation is induced to occur under the combined conditions of acid catalysis and close molecular proximity. With isobutylene (2-methylpropene), for example, initial protonation gives t-butyl carbonium ion, which then adds to a further molecule of isobutylene to give a new branched eight carbon skeleton containing a different tertiary carbonium ion (Equation 16.23).

$$2\,CH_3\overset{\overset{\displaystyle CH_3}{|}}{C}{=}CH_2 \rightarrow CH_3\overset{\overset{\displaystyle CH_3}{|}}{C}{+} + CH_2{=}\overset{\overset{\displaystyle CH_3}{|}}{C}CH_3 \rightarrow$$

$$CH_3\overset{\overset{\displaystyle CH_3}{|}}{\underset{\underset{\displaystyle CH_3}{|}}{C}}CH_2\overset{\overset{\displaystyle CH_3}{|}}{C}CH_3 \overset{-H+}{\underset{+}{\rightarrow}} CH_3\overset{\overset{\displaystyle CH_3}{|}}{\underset{\underset{\displaystyle CH_3}{|}}{C}}CH_2\overset{\overset{\displaystyle CH_3}{|}}{C}{=}CH_2 \quad (16.23)$$

Loss of a proton back to the catalyst surface again releases a molecule of 2,4,4-trimethylpentene-1, a branched $C_8$ olefin, from the catalyst surface. Processes analogous to this can occur not only between two branched four-carbon olefins but also between straight chain olefins, propylene and $C_4$ olefins, and two propylene molecules. The carbonium ion formed by coupling of any two of these olefin units may also, occasionally, add a third olefin. The composite result of all of these processes is that a stream of $C_6$, $C_7$, and $C_8$ olefins (plus, of course, much still unreacted material) is obtained from the predominantly $C_3$ plus $C_4$ feed stream (Equation 16.24).

Polymerization:

$$C_3, C_4 \text{ olefins} \xrightarrow[\text{catalysis}]{\text{acid}} C_6, C_7, C_8 \text{ olefins} \quad (16.24)$$

Components appropriate for blending into gasoline will again be separated from this product stream by fractionation after polymerization.

While polymerization was the first refining process specifically designed to produce larger molecules appropriate for gasoline [17], from a predominately $C_3$, $C_4$ hydrocarbon fraction alkylation, a chemically similar process [7], is more prominent in importance today. Alkylation uses either concentrated sulfuric acid, employed for about 75 % of alkylation capacity, or hydrofluoric acid as liquid phase catalysts to form dimers or trimers from $C_3$ and $C_4$ olefins reacted with paraffinic hydrocarbons [19]. Because the yield of the alkylation process, based on the olefin feed, is about twice as high as the yield of the polymerization process, and the alkylation product is as good or better in terms of octane rating as the polymer product, alkylation is now a dominant synthesis route to superior gasoline components.

Alkylation conditions are quite mild with both common catalysts. Processes employing concentrated sulfuric acid operate at 2-12 °C (requiring refrigeration) and 4.1-4.8 x $10^6$ N/m$^2$ (4.1-4.8 atmospheres) and use a gas contact time of 5 minutes although commercial processes normally allow 15-30 minutes [17]. Catalysis with anhydrous hydrogen fluoride requires both somewhat higher temperatures, 25-45 °C, and correspondingly higher pressures than required by sulfuric acid, but functions as effectively at contact times of 20-40 seconds. Both catalysts operate by initial protonation of the double bond of an olefin. With isobutylene, for example, t-butylcarbonium ion is

formed. This t-butylcarbonium ion can add to a further molecule of isobutylene in the same manner as occurs during polymerization (Equation 16.25).

$$(CH_3)_3C+ + CH_2{=}C(CH_3)_2 \rightarrow$$
$$(CH_3)_3C\text{–}CH_2\overset{+}{\text{–}C}(CH_3)_2 \quad (16.25)$$

But rather than lose a proton, as would normally occur under polymerization conditions, the dimeric ion obtained abstracts a hydride ion (H–) from a neighbouring isobutane to give the saturated dimer 2,2,4-trimethylpentane, and regeneration of a tertiary butyl carbonium ion (Equation 16.26).

$$(CH_3)_3C\text{–}CH_2\overset{+}{\text{–}C}(CH_3)_3 + CH_3CH(CH_3)_2 \rightarrow$$
$$(CH_3)_3C\text{–}CH_2CH(CH_3)_2 + \overset{+}{C}(CH_3)_3 \quad (16.26)$$

The carbonium ion regenerated in this way can of course continue the process, a key feature being that under alkylation conditions this continuing active species is formed from *saturated* alkane, not olefin as required by polymerization. Different alkenes such as propylene, 1-butene, or the 2-butenes may also participate in the process of carbonium ion formation, in a manner analogous to the process of Equation 16.25, but neither n-butane nor n-pentane can replace an iso-alkane for the hydride transfer. An n-alkane is not capable of forming a sufficiently stabilized carbonium ion so that carbonium ion formation is thermodynamically unfavourable. Nevertheless, this is one advantage that alkylation possesses over polymerization as a route to gasoline constituents from light hydrocarbons alkanes (as long as they are branched) and alkenes are incorporated. Both alkylation and polymerization produce branched products, but the alkylation products are saturated (Table 16.5) whereas the polymerization products are alkenes.

### 16.3.4 Upgrading of Gasoline Components

Polymer and alkylate gasoline components have characteristically high octane numbers as obtained directly from the originating process because of the high degree of branching of these products (Table 16.7). Cat-cracked gasoline as a whole is also high octane (92 R.O.N. or better) mainly contributed by its high olefin content. In contrast straight run gasolines generally give low octane numbers because of their predominantly normal alkane composition. The octane number of a gasoline blend may be improved, within limits, by the addition of lead-

**Table 16.7** Motor octane numbers related to molecular weight and degree of branching of some representative examples of gasoline blend constituents [a]

| Component | Molecular weight | Octane number |
|---|---|---|
| n-butane | 58.1 | 89 |
| 1-butene | 56.1 | 92 |
| trans-2-butene | 56.1 | 95 |
| isobutane | 58.1 | 97 |
| n-pentane | 72.2 | 63 |
| isopentane | 72.2 | 90 |
| n-hexane | 86.2 | 26 |
| 2-methylhexane | 100.2 | 47 |
| 3-methylhexane | 100.2 | 55 |
| 2,2-dimethylpentane | 100.2 | 96 |
| 2,2,3-trimethylbutane | 100.2 | 101 |

[a] Dat selected from that of [100]. Motor Octane Numbers are determined under more rigorous test conditions than Research Octane Numbers, and hence are generally somewhat lower than the latter.

containing or other types of antiknock compounds. But with increasing incentives to reduce lead dispersion into the atmosphere there is a growing trend to upgrade the octane number of a gasoline by raising the octane number of the various hydrocarbon feedstocks being blended into it, rather than by lead-containing additives.

Catalytic thermal reforming, under a variety of trade names such as Platforming, Ultraforming, Renoforming etc. and using similar but somewhat milder conditions than catalytic cracking, accomplishes many changes towards improving octane ratings of normal (straight chain) alkane streams [20]. For example, naphthenes (cycloalkanes) are dehydrogenated to benzenes, and alkyl naphthenes undergo dehydroisomerization (e.g. Equations 16.27, 16.28).

In the same step alkanes are also isomerized to branched alkanes (e.g. Equation 16.29).

$$CH_3CH_2CH_2CH_2CH_2CH_3 \rightarrow$$

$$\begin{array}{c} CH_3 \\ | \\ CH_3\text{–}CH\text{–}CH(CH_3)_2 \end{array} \qquad (16.29)$$

Since aromatics such as benzene and toluene, and branched alkanes all possess higher octane numbers than their precursors, the octane number of the fuel is markedly improved by these structural modifications.

It should be noted that since sulfur is a poison for platinum-based catalysts, the feed for the catalytic reformer has to be essentially sulfur free. Sulfur is removed by passing the feedstock through a cobalt/molybdenum catalyst bed in the presence of hydrogen, normally generated from the catalytic reformer. Carbon-bound sulfur is converted to hydrogen sulfide (Equation 16.30). Hydrogen sulfide is easily separated from the other constituents after hydrogenation by stripping or fractionation. Hydrogen sulfide and mercaptans (thiols) may also be removed from refinery streams by washing with aqueous sodium hydroxide (lye treating, Equations 16.31, 16.32).

$$2\,NaOH + H_2S \rightarrow$$
$$Na_2S \text{ (water soluble)} + 2\,H_2O \qquad (16.31)$$

$$NaOH + RSH \rightarrow NaSR \text{ (water soluble)} + H_2O \qquad (16.32)$$

Sodium hydroxide may be regenerated, and mercaptan recovered from the alkaline scrubber effluent by steam stripping (Equation 16.33).

$$NaSR + H_2O \underset{steam}{\rightarrow} NaOH + RSH \uparrow \qquad (16.33)$$
$$\text{"mercaptan oil"}$$

$$(16.27)$$

$$(16.28)$$

$$(16.30)$$

Any mercaptans boiling below 80 °C are also readily dissolved in alkaline solutions (Equation 16.32). A common process for the removal of dissolved mercaptans of this kind, especially from catalytically cracked gasoline and liquefied petroleum gas, is the Universal Oil Products (UOP) Merox process which also uses caustic soda for extraction. In this process, however, the mercaptans are then oxidized to disulfides using air assisted by a metal complex catalyst dissolved in the caustic soda (e.g. Equation 16.34).

$$4\,NaSC_2H_5 + O_2 + 2\,H_2O \rightarrow$$

$$2\,C_2H_5SSC_2H_5 + 4\,NaOH \qquad (16.34)$$
diethyldisulfide

In this way the sodium hydroxide is regenerated for further use, and the disulfides formed, which are not soluble in sodium hydroxide, form an oily layer that can be removed. This is one way in which the sulfur content of gasolines is kept below the 0.1 % limit required for marketing [21].

## 16.3.5 Gasoline Blending

For optimum use of the various refinery fractions of the correct boiling range in gasoline, these must be blended to achieve the correct volatility and rate of combustion for the engines in which they are to be employed. More easy vaporization of a gasoline, as required under winter conditions, is obtained by increasing the proportions of butane and pentane in the blend. To decrease standing losses (vaporization from the fuel tank etc.) in summer the proportions of these more volatile constituents is decreased.

Control of the rate of combustion of a gasoline is necessary to obtain smooth engine operation over a range of loads and operating speeds. Rough engine operation, or knocking, is caused by too rapid combustion of the gasoline vapour/air mixture in the cylinder, a condition generally aggravated when the engine is operating under heavy load. Rough operation may also be the consequence of pre-ignition (too sensitive ignition) which might in turn be caused by hot carbon deposits on the cylinder head, by an overly hot exhaust valve, or by compression ignition of the fuel vapour/air mixture in the cylinder.

The components of commercial gasolines are always complex mixtures of hydrocarbons. Hence, these are best tested, individually or in already-mixed blends, in an operating engine to determine

their resistance to knock. An octane rating is then assigned to the particular fuel being used by comparison of its knock, or pre-ignition resistance as compared to the knock resistance of a standard fuel of known octane rating. Octane numbers have been defined for particular hydrocarbons. Pure 2,2,4-trimethylpentane, for example, which burns smoothly in a high compression engine, is assigned a value of 100. n-Heptane, which has a strong tendency to knock is assigned a value of zero. Octane numbers of other hydrocarbons with knock resistance values intermediate between these two are determined experimentally [22] and also can be obtained from suitable blends of the pure constituents. As a general rule, the octane number of a straight chain hydrocarbon is inversely proportional to molecular weight and directly proportional to the degree of branching or unsaturation (Table 16.7). Rather than using the expensive pure hydrocarbons for test purposes, in practice commercial blends of hydrocarbons of known standardized octane ratings are used for comparison with gasoline blends of current production to determine their market suitability, and make adjustments if necessary.

If the octane number of the blend is not high enough, it may be raised by increasing the proportion of high octane alkylate, catalytically reformed product, or aromatics in the blend [23, 24]. Or an antiknock compound may be added to accomplish inexpensively the same thing [25]. Tetraethyl or tetramethyl lead, with other additives, are still used to a significant extent to increase the octane number of "leaded" gasolines. For example, treatment with 1.8 mL of tetraethyl lead per Imperial gallon (4.546 L) can increase the octane number of a paraffinic gasoline from the 75 range up to about 85. But there is a decreasing rate of return on increasing the additive concentrations so that doubling the amount added to 3.6 mL/Imperial gallon, the maximum permitted by law because of the toxicity of these organic lead compounds, only increases the octane number from 85 to about 88. Gasolines with a high aromatic content have a better octane number response with tetramethyl lead than with tetraethyl lead, which is the additive normally used with high paraffinic and high naphthenic gasolines [26]. Both are thought to function by breaking up (shortening) the pre-flame chain reactions leading to combustion, thus decreasing the tendency of the fuel/air mixture to pre-ignite [26, 27]. The routine levels of tetraethyl lead addition in the U.S. used to be in the 0.48-0.83 mL/L range [28] (1.8-3.15 mL/U.S. gal; 2.2-3.8 mL/Imperial gallon) but by 1980, from environmental considerations [29, 30], this was to be reduced to the level of not more than 0.5 g of contained lead per U.S. gallon, even for leaded gasolines [31, 32].

In addition to the tetraethyl or tetramethyl lead, both types of anti-knock fluids also contain 1,2-dichloroethane and 1,2-dibromoethane (ca. 35 % of the fluid additive by weight), the halides of which react with the lead released on combustion to form lead bromide and lead chloride. These lead halides are volatile at the cylinder combustion temperatures of 800-900 °C, and leave the combustion chamber with the exhaust. Without these components, build-up of lead deposits in the combustion chamber can eventually interfere with proper engine operation. An identifying dye is also added to these anti-knock fluids so that when they are added to gasolines they may be immediately recognized as being leaded. This measure assists in ensuring that they are not unwittingly used for cleaning purposes or for combustion in heating units, end uses which pose unnecessary toxicity risks.

Toxicity considerations of the lead antiknock additives, as well as an interest in minimizing the increased refining costs to obtain gasoline components which have an intrinsically higher octane rating has stimulated the testing of numerous other organometallic compounds as antiknock additives [27]. Methylcyclopentadienylmanganese tricarbonyl (MMT), ferrocene, and other related compounds have been found to show antiknock activity, but these are not as compatible in this application as the lead alkyls [33, 34]. Many oxygen-containing compounds such as tertiary butanol, tertiary butyl acetate and in particular methyl tertiary butyl ether (MTBE) also have antiknock capabilities [35, 36]. But sufficient of these compounds needs to be added to the gasoline that they virtually qualify as a blending component rather than as an additive [27]. Fortunately, the blending of renewable fuel components such as fermentation ethanol into gasoline to decrease demands on non-renewable petroleum resources [37-39], also accomplishes octane improvement [15].

The final significant step to gasoline production after blending of the appropriate refinery streams and the addition of any necessary antiknock additives, is the addition of antioxidants. Mono-, di-, and tri-olefins such as are present particularly in catalytically-cracked gasoline components, are highly susceptible to gum formation on exposure to air. During shipping and final delivery it is virtually impossible to avoid air contact, which tends to peroxidize the labile allylic hydrogens of the olefins present, eventually giving rise to free radical centers. These radical centers then initiate polymerization to give the observed gums in the gasoline. However, addition of one, or a combination of p-phenylenediamine (PDA), N, N'di-sec-butyl-p-phenylenediamine (DBPDA, Equation 16.35) or 2.6-di-t-butylphenol as antioxidants effectively captures any radical centers which do arise before any polymerization occurs [17, 27]. The product of the antioxidant reaction is a stable free radical, too sluggish to initiate polymerization. Other less significant, though still important additives such as corrosion inhibitors, anti-icers, carburettor detergents, and intake valve deposit control additives may also be placed in particular gasoline blends. Thus gasoline, as an apparently simple though vital transportation fuel, actually represents a highly tailored smoothly performing and complex product of the petroleum refining operation.

## 16.4 Manufacture of Lubricating Oils

There are four main steps to producing a lubricating oil from the appropriate high boiling fractions separated on the crude distillation column. These are vacuum fractionation, to sharpen up the boiling point range of the fractions used, solvent dewaxing, to lower the pour point (gelling point) of lubricating oils, followed by decolourization and formulation steps.

### 16.4.1 Vacuum Fractionation

Redistillation of the appropriate lube oil fractions obtained from crude distillation is required to re-

$$H_2N-\!\!\left\langle\bigcirc\right\rangle\!\!-NH_2 \quad (CH_3)_2CHCH_2NH-\!\!\left\langle\bigcirc\right\rangle\!\!-N(H)CH_2CH(CH_3)_2$$

PDA                                                    DBPDA                                            (16.35)

move both "lighter ends" (lower molecular weight components such as gas oils) which are poor lubricants, and much of the high molecular weight asphaltic and wax constituents from the lube oil base stock. Because of the high boiling point of these oils, close to the temperatures at which cracking and pyrolysis occur, distillation is conducted under reduced pressures in the range 6 to 9 kPa [17] (0.06-0.09 atmospheres). At these pressures boiling points are decreased by 50 °C or so, sufficient to avoid thermal degradation of the oil on distillation.

Two to three steam ejectors operating in series with interstage condensers are normally sufficient to provide the pressure reduction necessary and are connected to the top end of the vacuum distilling column (Figure 16.5). Each take-off point on the main vacuum column is also passed through a stripper unit, a smaller version of the main column, to ensure thorough removal of volatiles before the lubricating oil proceeds to finishing stages. More volatile fractions are returned to the column.

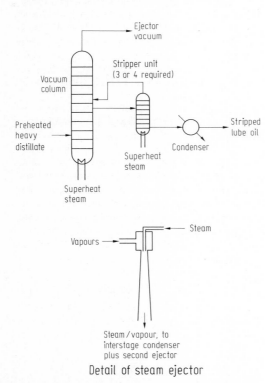

Detail of steam ejector

**Figure 16.5.** Simplified flowsheet of vacuum distillation for lubricating oil refining

## 16.4.2 Solvent Dewaxing

Waxes comprise mainly paraffinic hydrocarbons in the $C_{20}$ to $C_{30}$ molecular weight range. Their presence in lubricating oils tends to raise the pour point roughly proportional to the wax content. This is important since a lubricating oil with a high pour point makes a cold engine difficult or impossible to turn over or start, particularly under winter conditions. Thus, wax removal is a necessary part of lubricating oil manufacture. It is also one of the most complicated and expensive of lube oil stock preparation steps [40].

Solvents in use for wax removal include propane, which also serves as an auto-refrigerant by solvent evaporation, methyl ethyl ketone (MEK)-toluene mixtures, and methyl isobutyl ketone (MIBK) [41]. The last two systems are in widest use at present [40]. MEK-benzene was once a common solvent combination for dewaxing [26] but has now been all but abandoned from benzene toxicity considerations.

For dewaxing with an MEK-toluene mixed solvent, the solvent is mixed with 1 to 4 times its weight of vacuum-fractionated lube oil stock [17]. The intimate mixture is then chilled to somewhere in the −10 °C to −25 °C range, depending on the degree of dewaxing required, usually in two or more stages of heat exchange (Figure 16.6). Ammonia or propane evaporation is used as the refrigerant on the cold side of the heat exchangers. Scrapers operating continuously on the oil side of the exchanger prevent the accumulation of adhering wax on the chiller exchange surfaces. The emerging stream from the chillers is a slurry consisting of crystallized wax suspended in a mixture of dewaxed oil and solvent. While still cold, the wax is filtered from the oil/solvent mixture, and then washed with a small amount of fresh chilled solvent to recover the traces of oil still adhering to the wax. Eventual oil recovery from the oil/solvent filtrate is effected in a solvent stripper, which distills off the low boiling solvent mixture leaving behind a residue of dewaxed oil, the objective of this step in the lubricating oil refining process.

Waxes, in the 200 to 400 molecular weight range and in several melting point ranges, are recovered in finished form after several more processing stages [41]. The 50,000 tonne/year wax market in Canada of the early '70's broke down into applications by weight roughly as follows: packaging,

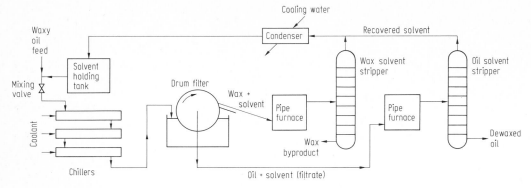

**Figure 16.6.** Flowsheet of the steps required for solvent dewaxing of a lubricating oil stock

32 %; tires and moulded rubber, 10 %; wire and cable insulation materials, 10 %; candles, 8 %; carbon paper, 8 %; and miscellaneous other uses, 32 % [42].

### 16.4.3 Lubricating Oil Decolourization

Lubricating oil stocks are much improved for their eventual end uses after vacuum fractionation and dewaxing steps but colour, caused by olefin and asphaltic residues, still must be removed to obtain long lubricant life. Originally these impurities were removed by the formation of the acid-soluble alkyl hydrogen sulfates and sulfonic acids (Equations 16.36-16.38),

$$RCH{=}CH_2 + HOSO_3H \rightarrow RCH(OSO_3H)CH_3 \tag{16.36}$$

$$2\,H_2SO_4 \rightarrow SO_3 + H_3O^+ + HSO_4^- \tag{16.37}$$

$$C_6H_6 + SO_3 \rightarrow C_6H_5SO_3H \tag{16.38}$$

obtained by treating the lubricating oil stocks with concentrated sulfuric acid [17]. Waste disposal problems, as well as other factors related to this decolourization process, have led to its gradual abandonment.

Today, clay adsorption or more commonly now hydrogen treating (catalytic hydrogenation) are used to accomplish decolourization [3]. Passage of oil through a clay coating on a continuous filter is quite effective, and a suitable method for smaller, less complex refineries. Larger refineries with a wider product range that can justify hydrogenation facilities will use catalytic hydrogenation. Reduction of olefins and aromatics to their saturated counterparts, paraffins and naphthenes, in

this way accomplishes both decolourization, by decreasing the number and extent of conjugated double bonded species present, and stabilization of the lubricating oil stocks towards oxidation.

The product at this stage is highly suitable for many industrial uses, and for straight lubricating functions. However, for all types of engine oil lubricating functions, where there are many demands other than lubrication which are placed on the oil, blending and several additives are required to meet these demands.

### 16.4.4 Formulation of Lubricating Oils

Both paraffinic and naphthenic (cycloparaffinic) stocks are used for formulation of lubricating oils, each with favourable characteristics for particular uses [40]. For instance, paraffinic stocks are generally preferred for their superior lubricating power and oxidation resistance. But naphthenic stocks have naturally lower pour points, that is they maintain flow characteristics at lower temperatures than paraffinics (Table 16.8) and possess a greater

**Table 16.8** A comparison of the properties of lubricating oil stocks[a].

|  | Paraffinic | Naphthenic |
| --- | --- | --- |
| Density, g/ml | 0.887 | 0.905 |
| Pour point, °C | 7 | –29 |
| Flash point, °C | 246 | 224 |
| Viscosity, cSt |  |  |
| 38 °C | 107 | 120 |
| 100 °C | 11.3 | 10.0 |
| Viscosity index | 100 | 57 |

[a] Calculated from the data of Gillespie et al. [40].

solvent power, features which tend to be more important for applications such as heat transfer, metal working, and fire-resistant hydraulic fluids [40]. Any significant content of residual aromatics in the lubricating base stock will have been removed before formulation by solvent extraction, using N-methylpyrrolidone, furfural, or less frequently today, phenol [3, 40] (Equation 16.39).

When an oil is used for lubrication of an internal combustion engine it not only has to reduce friction of moving parts but it has to accomplish this efficiently under a wide range of operating temperatures and loads. At the same time it performs heat transfer functions and has to deal effectively with carbon and dust particles, gum deposits, water and corrosive gases, and all of this without decomposing in the process.

Oil soluble detergents, added to the oil to the extent of 0.02 to 0.2 % by weight [1], capably handle many of these problems. Various sulfonates or phenolates are used, usually employed as their calcium salts, $CaX_2$ where $X$ = sulfonate or phenolate ligand, to confer oil solubility to the dispersant (e.g. Equation 16.40).

Occasionally calcium may be replaced by barium. Basic alkyl sulfonates, $Ca(OH)SO_3R$), which contribute both acidity neutralizing and dispersing activity to the oil, are also used [2]. All of these detergents, or dispersants effectively keep particulate impurities such as carbon, dust, or metal fines (e.g. lead, oxides, and salts) and water or acid droplets suspended in the oil, preventing deposition on or attack of critical moving parts. At the appropriate oil change interval the bulk of this suspended material is removed from the engine with the drained oil.

Mineral oils and in fact most liquids show a decrease in viscosity with an increase in temperature, i.e. an inverse relationship exists between the viscosity of an oil and its temperature. At higher operating temperatures the viscosity of the oil decreases (the oil becomes "thinner"). If this decrease of viscosity with increased temperature is too large, lubrication is impaired under hot conditions causing excessive engine wear and at the same time increasing oil consumption. Thus, it is important that this change in viscosity of an oil with temperature is kept within reasonable limits. This ratio of high to low temperature viscosity is an important property of a lubricating oil and is referred to as its viscosity index. In turn, the viscosity index is an empirical property determined by measuring the viscosities of an oil at 40 °C and at 100 °C and comparing these values with ASTM (American Society for Testing Materials) tables. Under this system Pennsylvania oils, which have a relatively small change in viscosity with temperature, are assigned a viscosity index of 100. U.S. Gulf Coast oils, with a relatively larger change in viscosity with temperature, are assigned a viscosity index of zero [3].

A few percent of a synthetic polymeric material, such as polymethyl methacrylate, polyisobutylene, or an ethylene-propylene copolymer added to an oil improves its viscosity index [1, 3]. In this way the viscosity of a low pour point oil, which remains fluid at low temperatures, although still reduced when hot, is still kept acceptably high because of the presence of the polymeric additive (Figure 16.7). The additive is thought to function by the individual polymer molecules assuming compact spherical shapes at low temperatures, because of low solvent power of the base oil under these conditions. At higher temperatures, when the oil is a better solvent for the polymer, the now well-solvated, extended polymer molecules undergo intermolecular intanglements which have the effect of partially offsetting the tendency of the base oil to "thin out" [43, 44] (Figure 16.8). Not only does this method of viscosity stabilization enable many

N-methylpyrrolidone            furfural            phenol            (16.39)

$$CH_3(CH_2)_{11}SO_3^-, \quad wax-\!\!\langle\bigcirc\rangle\!\!-SO_3^-, \quad CH_3(CH_2)_7-\!\!\langle\bigcirc\rangle\!\!-O^- \qquad (16.40)$$

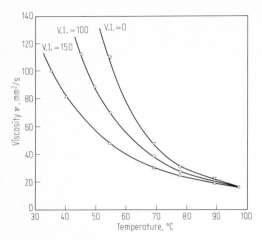

**Figure 16.7.** Relationship of viscosity indices to viscosity change with temperature at different polymer loadings [44]

base oils, which would be poorly suitable even as single grade engine oils on their own, to be used for engine lubrication. But it also permits the formulation of the increasingly popular multi-grade oils which may be used safely summer and winter, without requiring the traditional seasonal oil changes. In part, this is achieved because the same or similar additives also serve as efficient pour point depressants by decreasing the temperatures at which any residual wax may crystallize out, and cause the oil to gel [43, 45].

Rapidly moving engine parts tend to whip up a froth in the lubricant, as they operate, a tendency which is aggravated by the presence of the detergent additives in the oil. This can cause temporary failure of the lubricant circulating system in an engine from air entering the pump and causing a vapour lock. Small amounts, 1 to 10 ppm, of silicone antifoam agents prevent this [1]. Silicones are also effective lubricants in their own right, so that this additive does not adversely affect lubricity. Among

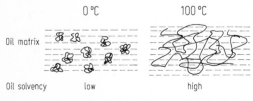

**Figure 16.8.** Mechanism of action of polymeric viscosity index improvers

many other additive materials which are formulated into lubricating oils are such ingredients as "oiliness" additives, polar compounds which tend to form a stable film on metal surfaces by plating out on them, and extreme-pressure additives which are commonly lead soaps [1]. Greases, which are solid to pasty lubricants used where less viscous materials would be forced out of or lost from the lubricating zone, consist of a thickening agent combined with a lubricating oil. Common thickening agents for greases are the soaps (metal salts of a fatty acid) of aluminum, barium, calcium, lithium, or sodium.

While lubricating oils represent only about 2 to 3 % of the total volume of crude oil processed, they are a high value component, and the nature of their only partially consumptive use encourages recycling of this petroleum fraction. Recycling also decreases the mutagenic burden on the environment that can occur from some disposal methods [46]. Almost half of the 850 million liters of lubricating oil sold in Canada each year, or about 360 million liters (80 million Imperial gallons), is potentially reclaimable, and yet in Canada less than 10 % of this is currently recycled [47, 48]. A re-refining scheme geared to downrate lubricating oil to a fuel oil after treatment, requires only removal of water and sludges in a relatively simple process [49]. But even this relatively low cost procedure is not used extensively at present. Full re-refining procedures utilizing six or more stages of upgrading do appear recently to be gaining favour, primarily because of the higher value of a lubricating oil stock as a product rather than a fuel oil [50, 51]. In 1974, capital costs for a 19 million liter (4.2 million Imp. gallon) per year re-refining plant to lube oil was estimated to be $0.9 to 1.4 million, with operating costs predicted to be around 3.9 to 5.8 cents per liter [52]. Areas with a high population density, which would tend to decrease collection costs, would thus appear to provide favourable situations for re-refining to be seriously considered.

### 16.4.5 Synthetic Engine Oils

Many special lubricant properties such as high viscosity indices, very low pour points, and high thermal stability accompanied by greater oxidation resistance, to an extent not possible with natural petroleum fractions, can all be achieved by the use of certain synthetic lubricating fluids [53]. Originally, developments along this direction were made using alkylated aromatics as the bulk constituent, but these had worse low temperature viscosity characteristics than refined petroleum stocks and were susceptible to oxidation. Olefin oligomers (low molecular weight polymers) which were developed subsequently were much better in these respects [54]. For automotive use, blends of olefin oligomer and high molecular weight esters such as di-2-ethylhexyl sebacate or trimethylolpropane triheptanoate are presently favoured from cost and performance considerations. These synthetic lubricant blends provide reduced wear, better engine cleanliness, longer oil change intervals, and slightly improved fuel economy [55], and achieves all this at lower oil consumption rates than experienced with conventional petroleum-based lubricants [54]. But the present much greater cost of this class of lubricant is likely to continue to limit their use to a fraction of the total automotive market, where their advantages outweigh their higher cost [56]. Much additional data relating to the various synthesized materials which may be used in this fashion, and details of their relevant properties have been published [3, 53, 57, 58].

## 16.5 Fuel Oils, Asphalts, and Pitches

The fuel oils, collectively, include a wide variety of petroleum end uses, from jet fuels at the more volatile low molecular weight end of the scale, through kerosene, stove oil, several grades of diesel fuel, to industrial heating fuel and ultimately to asphalt for road cements at the high molecular weight end. A convenient classification to consider this wide range of products is related to their distillate or residual origin, at the crude distillation step.

### 16.5.1 Distillate Fuel Oils

There is considerable overlap in the boiling ranges of the diesel, jet, kerosine, and light fuel oils represented by the distillate fuel oil class [1]. But all, if desulfurized, have the capability of burning relatively cleanly with little more than a trace of ash consisting of transition metal oxides (mainly vanadium).

Jet fuels, which amount to a blend of the higher boiling consistuents of the gasoline/naphtha fraction together with a proportion of the $C_{11}$, $C_{12}$ and higher components of which kerosine is primarily

composed, comprise the lowest boiling range of the fuel oil categories.

Kerosine is the lightest straight fuel oil in the distillate category, and has uses which range from lamp oils, to light stove oils and even to diesel fuels for use in Arctic service. Pour points of below $-50\,°C$ are the attraction for low temperature diesel operation, a use for which a small addition of lubricating oil stocks is also made to the fuel.

More conventional diesel fuel applications are served by those components of crude oil boiling in the range 190-385 °C, which is also referred to as No. 1 fuel oil. Different portions of this boiling range are selected for the particular class of diesel fuel being produced, based on volatility and pour point considerations. Diesel fuels also have higher flash points than gasoline or jet fuels, in the range of 40 to 50 °C [17], which is an important safety consideration.

Apart from the physical properties selected, as mentioned above, the most important quality of a diesel fuel is its cetane number. The cetane number is a measure of the ease of autoignition of a diesel fuel. Cetane (n-hexadecane, $C_{16}H_{34}$) itself performs very well in a diesel engine, meaning that there is negligible delay between the time of injection of the fuel into the cylinder and fuel combustion. Pure cetane is assigned a cetane number of 100. In contrast to this $\alpha$-methylnaphthalene ($CH_3C_{10}H_7$), which performs very poorly in a diesel engine, is assigned a cetane number of 0. Cetane numbers intermediate between these two numbers correspond to the percentage of cetane in a blend of these components [1].

Diesel fuels of low cetane number, the 25-40 range, are appropriate for operation of slow-speed engines such as are used for the propulsion of ships and some stationary engines. Medium and high speed engines require cetane numbers of about 40 and 50 to 65 respectively, for smooth operation [1]. Other operating features of the cetane rating system apply particularly at low temperatures when, within limits, the higher the cetane number the lower the operating temperature at which a diesel engine may be successfully started. It also relates to smoothness of engine operation, once it reaches operating temperature, and the formation of engine deposits and smoke. In general, but not invariably, the higher cetane number fuels tend to give the more favourable response for these qualities as well.

The higher boiling No. 2 and No. 3 distillate fuel

oils are largely used for space heating (homes and buildings), and can also form part of the fuel composition used in slow speed diesels. Also a part of these categories, in excess of demands, is employed as the solvent in the formulation of residual fuel oils.

### 16.5.2 Residual Fuel Oils, Asphalts and Pitches

As the name implies, these lower cost fuels are formulated from the residues of the refining and distillation processes. Pitchy residues of the crude distillation are dissolved in ("cut with") varying amounts of the residual heavy oil fractions remaining after catalytic cracking, so-called cat-cracked gas oil, or cycle oil (recycle stock). Thus, this group of fuels contain a relatively high concentration of all the less volatile and involatile impurities present in the original crude being distilled. These impurities include trace metals, heavy sulfur-containing compounds, asphalts, coke, and any residual incombustibles such as salt, silt etc. They still possess considerable fuel value however, in fact 5 to 10 % higher than the same volume of distillate fuel oils [59], but require combustion equipment that can cope with the higher viscosities and higher ignition temperatures required for these fuels.

Bunker A, or No. 4 fuel oil, so-called because of its use in the fueling (bunkering) of ships, is the lightest grade of residual fuel oil. It has the highest proportion of cracked gas oil to pitch, designed to stay fluid at ambient temperatures above freezing (pour point of ca. $-7\,°C$) [59] without difficulty. Because of its lower viscosity and hence greater versatility it is a residual fuel grade normally priced somewhat higher than the other available grades. But the price differential is still small enough to encourage the use of this fuel by particularly smaller combustion units operating under colder than average conditions, such as ships in Arctic service and the like. The use of especially low pour point ($-40\,°C$) No. 2 light diesel oil is often necessary in Arctic conditions for small ships, such as those of northern fishing fleets.

Bunker B, or No. 5 fuel oil has a lower proportion of cat-cracked gas oil to pitch than No. 4 fuel oil, but still remains relatively fluid at storage temperatures of ca. $10\,°C$. Combustion equipment however, would have to have some provision to heat the fuel prior to atomization, in order to obtain efficient combustion. Outside storage at temperatures much below $10\,°C$ would require insulated tanks fitted for heating, to enable pumping of the fuel under cold conditions.

Bunker C, or No. 6 fuel oil has the lowest proportion of cracked gas oil to pitch of all the residual fuels, and consequently storage tanks require heating even under ordinary ambient conditions, to enable the fuel to be pumped to the combustion unit and to assist in atomizing for burning. For large energy requirements this is not a difficult feature to arrange, and in return for this additional complication this fuel is the lowest in cost and the highest in fuel value among the fuel oils. The fuel value of No. 1 fuel oil is in the range 38,000 to 39,000 kJ/L (135,800-138,800 BTU/U.S. gal) whereas No. 6 fuel oil produces some 42,200 to 42,600 kJ/L (150,700-152,000 BTU/U.S. gal) on combustion.

A problem which occasionally arises, when there is a high density of thermal combustion sources in a particular area coupled with little air movement, such as during inversions, in that excessive concentrations of sulfur dioxide may accumulate in the ambient air of the area (see Section 16.6.1). The occurrence of this type of problem is of course aggravated by the combustion of high sulfur content residual fuels. However, the sulfur content of residual fuels may be decreased by using hydrodesulfurized cutting solvents, and by selecting the residuals from low sulfur crude oils for formation into residual fuels. The residues from high sulfur content crudes, if composed of appropriate components, could then go into the production of pitches and asphalts where the sulfur content is not subjected to combustion and discharge. A less desirable option for the residues from high sulfur content crudes is to formulate these into residual fuels to be marketed outside regulated areas, that is in areas where thermal combustion sources are smaller and less numerous. As a last resort, because of the high processing costs involved (much higher than for the desulfurization of distillate fractions), the formulated residual fuel itself may be hydrodesulfurized. When this measure of sulfur reduction is necessary a product price increase is generally necessary to offset the higher production costs.

### 16.5.3 Asphalts and Pitches

Usually the residue from crude distillation in North America will be processed to an asphalt product as much as possible, since as a component of residual fuel oils it generally fetches a lower price than as asphalt. This is because the residual fuel oil market has quite an extensive international supply network, hence ships and tankers can choose to load at centers offering the lowest cost bunkering and bulk fuels. And the price of the heavier grades of bunker fuel used in such large scale thermal operations as power generation has to be comparable to that of coal, in order to remain as a competitive energy source.

Asphalts are marketed in liquid, and low, medium, and high melting point solid grades, depending on the end use. The melting point and degree of hardness of an asphalt is affected by the completeness with which the volatile fractions have been recovered. If the melting point or hardness of a crude separated asphalt product is too low, it is oxidized by blowing air through the heated asphalt until the melting point is raised to the extent desired.

Low melting point asphalts are generally used in such applications as the waterproofing of flat, built-up roofs and the like, where the self-sealing qualities are an attraction. For sloping built-up roofs, higher melting asphalts are required. And for asphalt shingles, roll roofing and similar sheet roofing products the melting point has to be high enough to give an essentially non-sticky surface after fabrication, under ordinary ambient conditions. Road asphalts normally have not only a melting point specification, but also a penetration test ("pen test") and other requirements. The penetration test, conducted on an asphalt sample at 25 °C, is the extent to which the sample is penetrated by a steel ball or pin of standard dimensions while a force of 100 g is applied for 5 seconds. The result of this test is used as an important measure of the suitability of the asphalt for various paving applications. For application as road surfacing a narrow size distribution of crushed gravel is blended into the hot asphalt, which serves as the binder.

By the variety of processing procedures just outlined very little of the crude oil processed is wasted. From the high volatility dissolved gases present, all the way to the nearly involatile asphaltenes and the suspended carbon etc. present in the original oil, virtually all are converted to one or other of a host of useful products. Any combustible components that are not directly used towards the formulation of a product are used for the generation of the energy requirements of the refinery itself.

# 16.6 Refinery Emission Problems and Control

With the overall interest in producing useable products from all the feedstock components entering a refinery, and recognition of the fire/explosion or toxicity hazards to operating personnel in the event of losses to the atmosphere of many of the process streams, has meant that reasonable control over the loss of process streams has always been maintained. Nevertheless, there are particular components of refinery operations, and particular process streams where greater attention is necessary to ensure safe and environmentally-acceptable operation. For convenience of discussion these may be grouped into areas as they relate to atmospheric emission control, aqueous emission control, and waste disposal practices. Each is considered separately below.

## 16.6.1 Atmospheric Emission Control

Hydrocarbon vapour losses from refineries are estimated to amount to only about 5 to 10 % of the total hydrocarbons discharged to the atmosphere, from various air pollutant inventories [60-62]. But because these losses from each refinery location amount to a near-point source, control is still important to minimize local ambient air concentrations of hydrocarbons. Also losses of volatile sulfides, because of their intense odours, need attention to their control out of proportion to the masses lost [63].

Losses from pump seals, flanges connecting lengths of pipe, and operating vessels are best controlled by properly scheduled maintenance procedures [64]. Losses of methane or ethane from crude, or process stream preheat furnaces using these hydrocarbons as fuels used to occur occasionally, when there was inadequate air for complete combustion. These losses are now controlled by constantly maintaining a slight excess of combustion air, amounting to 1-2 % more oxygen than the theoretical requirement. A continuous oxygen analyzer is used to monitor the flue gases of these units and may be integrated into the system so as to control automatically the fuel-to-air ratio in this range. Or, in simpler systems the monitor may be wired to ring an alarm to alert operators to correct manually the fuel-to-air ratio, when the oxygen content of the flue gas deviates from set limits. Introduction of this type of flue gas combustion control system has also contributed a fringe benefit to refinery operation in that more efficient energy recovery is possible with the minimum, but still a slight excess of the air requirements for complete combustion, rather than with large excesses or deficiencies of air.

Potential hydrocarbon losses from the overpressuring of operating vessels is controlled via first, staged computer alerts and/or manual alarms to provide for correction of the condition. If the overpressure exceeds a second set point, pressure relief valves vent the contents of the vessel to a flare release system. The flare system provides a means of safely disposing of hydrocarbon vapours, in this way avoiding fire or explosion risks, by burning these at a non-hazardous point in the refinery. The details of coordination of good combustion conditions at the flare do not always work very well, however, occasionally generating large volumes of black smoke. Even though this is a less urgent problem than the upset causing the flaring it is still undesirable. More efficient modern flare designs, such as those of Indair [65] (Crauford Flares International), or John Zink solve these problems resulting from inefficient combustion via the use of the Coanda effect, or via steam injection at appropriate points in the flare streams to promote good fuel-air mixing, and hence smokeless combustion [66]. Use of steam for flaring has also been shown to reduce soot formation by the water gas reaction of carbon particles with steam [67] (Equation 16.41).

$$C + H_2O \rightarrow CO + H_2 \text{ (at high temperature)} \tag{16.41}$$

The carbon monoxide and hydrogen formed by this process both burn cleanly. Empirical and experimental data verify that a steam to hydrocarbon ratio of about 0.3:1 is necessary to obtain soot-free combustion. With current prices of crude oil and natural gas it is worth considering proposed containment measures for ordinary flared hydrocarbons for energy recovery or recycle [68].

Storage tanks holding the more volatile liquid products of refinery operation can lose a significant amount of the stored product as vapour, particularly when the emptying and filling cycle time for the tank is short. The storage tanks are always vented to the air to avoid build-up of vapour inside; on a hot day this can cause tank rupture from even a relatively small overpressuring. Or, conversely, the tank can be caused to collapse from at-

mospheric pressure acting on the outside of the tank during a cold night, which causes vapour condensation and internal pressure reduction. Losses of gasoline amounting to as much as 190,000 L per year from the venting of a single storage tank have been reported [14]. These losses may be controlled by fitting floating roofs to tanks used to store products of high vapour pressures, in this way eliminating the vapour space responsible for the losses from a fixed roof tank [69]. Or insoluble polyethylene spheres, two to three layers deep may be added to the surface of the liquid inside the tank to decrease losses from the vapour space by 90 % or more. Maintaining a slight and controlled positive pressure on storage tanks of this kind also helps to reduce vapour loss. These measures not only decrease emissions but are also generally cost effective, because of the value of the product so conserved. Floating roofs may also be used to decrease the explosion risk in fixed roof tanks that exists from the air/vapour mixture above volatile hydrocarbon products such as occurs during the storage of jet fuels and gasoline.

Crude oils all contain some sulfur, mostly in the form of thiols, sulfides, and hydrogen sulfide. The sulfur content, in combined form contained in these compounds, can range from about 6 % for light Arabian crudes, to 1-2 % for California crudes, down to as little as 0.4 % for some Canadian Peace River crudes [70]. Emission control of the volatile sulfides of crude oil during storage is also achieved using floating roof storage tanks or the equivalent. During refining, the sulfur content of gaseous hydrocarbon streams, which is mostly present as hydrogen sulfide, is removed by either scrubbing with organic bases (details, Chapter 7) or by physical absorption, for example by the Sulfinol process. Liquid process streams in the middle boiling point range require catalytic hydrogenation (Hydrofining or Hydrotreating) to convert the less volatile sulfides present to hydrogen sulfide and paraffinic hydrocarbons (Equation 16.42).

$$R_2S + 2H_2 \rightarrow 2RH + H_2S \tag{16.42}$$

Hydrogen sulfide, having a much lower boiling point than the liquid hydrocarbon streams, may then be readily separated from the hydrocarbon stream by stripping with air or steam. Separation of the sulfur-containing compounds present in higher boiling point streams, and from involatile residues requires similar chemistry to this, as a preliminary step, to convert high boiling sulfides to lower boiling hydrogen sulfide and thiols. Following the catalytic conversion the now relatively low boiling sulfur compounds may be separated by fractionation or stripping. But the problem with the residue stream in particular is that the wide variety of metallic and other impurities present make it difficult to provide suitably active catalysts for the chemical step, and even these are subject to relatively rapid deactivation by contamination. This problem makes it necessary to replace the catalyst relatively frequently, which makes desulfurization of refinery residue streams relatively more expensive than lighter streams. This is why residues are less often desulfurized. The sulfur content of a residual fuel oil may be minimized, without having to hydrodesulfurize the still residues, by thorough desulfurization of the catalytically cracked gas oil used as a diluent before it is blended with the residue.

Hydrogen sulfide accumulated from all of these desulfurization processes is converted to elemental sulfur using two or three stages of Claus reactors [71]. Any sulfur dioxide streams can also, with advantage, be diverted to the Claus plant where it can provide a part of the sulfur requirement for conversion to elemental sulfur. Claus plant operation is seldom a directly profitable operation for a refinery since the sulfur yield frequently amounts to only a few tonnes per day. But the improved ambient air quality achieved in the vicinity of the refinery, particularly from the elimination of the intense odours of reduced sulfur compounds, makes this measure worthwhile [72].

Carbon monoxide is a potential emission arising mainly from the catalyst regenerators of recent models of catalytic cracking units which run at about 620 °C, too low to obtain complete oxidation of carbon to carbon dioxide. Fortunately, carbon monoxide is a valuable industrial fuel. It is burned in a specially designed "CO-boiler" to obtain energy recovery in the form of steam and in the process discharges carbon dioxide, a more acceptable flue gas component. The most recent designs of catalyst regenerators now operate at somewhat higher temperatures than this which achieves complete conversion (and energy recovery) of the carbon monoxide to carbon dioxide, without the need for a CO boiler.

Control of catalyst particle losses from both the cracker and regenerator of fluid catalytic cracking units is achieved by two cyclones operating in series right inside each unit. This is usually followed by an electrostatic precipitator for fine particle control, working on the exhaust side of the catalyst regenerator [73].

Nitric oxide formation can occur to the extent of producing concentrations of 270 ppm or more at the hot metal surfaces of any of the refinery combustion units [74] (Equation 16.43).

$$N_2 + O_2 \quad \underset{\text{metal}}{\overset{\text{hot}}{\rightarrow}} \quad 2NO \qquad (16.43)$$

Its formation can be kept to a minimum by keeping the excess air supplied to combustion units to a minimum value for safe complete combustion [64]. Burner designs which produce a more diffuse flame front (large flame volume) achieve lower peak combustion temperatures which also help to decrease the formation of nitric oxide. Injection of ammonia into the flue gas, while it is still hot, has been demonstrated alone to decrease $NO_x$ concentrations to 80 to 120 ppm range, $^1/_3$ to $^1/_2$ of uncontrolled concentrations [74].

## 16.6.2 Aqueous Emission Control

Standards for aqueous effluent from refineries have been set, both in Canada and the U.S.A. (Table 16.9). For many of the more ordinary parameters such as total suspended solids, oil and grease, and ammonia nitrogen, American standards appear to be ten to twenty times more strict than Canadian federal standards. But in Canada each province sets its own emission regulations which are generally more restrictive than the federal standards [75]. Thus, the sulfide standard for the province of British Columbia is virtually the same as the U.S. standard, and BOD, ammonia nitrogen and total chromium, are only 3 to 5 times the U.S. standards. The standards for other provinces will differ from these. Perhaps these differences in standards are a reflection of the more numerous very large refineries operating in the U.S.A. than in Canada, or more particularly in British Columbia. Without the tighter U.S. standards the point source mass discharges of pollutants from major refineries could seriously tax if not swamp the assimilatory capacities of receiving bodies of water in their areas of operation.

Aqueous desalter effluent containing sediments, oil, dissolved salts, and sulfides is first treated by putting it through an American Petroleum Institute (API) separator for residual oil removal and recovery (Figure 16.9). This device is capable of ef-

**Table 16.9** Aqueous effluent standards for petroleum refining [a]

| Effluent component or property | Average of daily values for one month, max. permitted, g/m³ crude | | | Maximum values, any one day, g/m³ crude | |
|---|---|---|---|---|---|
| | U.S.A. | Canada [b] | B.C. [b, c] | U.S.A. | Canada [b] |
| Total susp. solids | 2.0 | 20.6 | 20 [d] | 2.4 | 42.8 |
| BOD | 2.0 | — | 6.6 | 2.5 | — |
| COD | 8.0 | — | — | 10.0 | — |
| Oil and grease | 0.4 | 8.6 | 1.7 | 0.5 | 21.4 |
| Phenols | 0.0060 | 0.86 | 0.066 | 0.012 | 2.1 |
| Sulfide | 0.035 | 0.29 | 0.031 | 0.055 | 1.4 |
| Ammonia nitrogen | 0.51 | 10.3 | 1.65 | 0.68 | 20.6 |
| Total chromium | 0.105 | — | 0.57 | 0.124 | — |
| pH | — | 6 – 9.5 | 6.5 – 8.5 | — | 6 – 9.5 |

[a] Derived from [75, 101 and 102].
[b] Recalculated from units of pounds per thousands barrels to grams per cubic meter of crude by multiplying by 2.855.
[c] Level A standards, test intervals vary from daily to monthly.
[d] Units of mg/L, not g/m³.

ficiently collecting any oil droplets larger than about 0.15 mm in diameter which are present in the desalter effluent. The skimmed creamed oil phase is routed to the refinery "slop oil" stream, joining any other off-specification liquid oil streams for reprocessing. Use of inclined plastic plates in an oil/water separator of this type greatly increases the speed and completeness of creaming, and reduces the average droplet size collected to about 0.06 mm [76]. After oil separation the water phase of the desalter is passed through a sour water stripper (Figure 16.10), where hydrogen sulfide is removed by blowing with low pressure, 140 kPa (ca. 130 °C, 20 lb/in²) steam [77, 78]. The separated hydrogen sulfide will be processed to ele-

mental sulfur in a Claus unit. The sweetened water obtained as the bottom stream from the stripper and now containing < 10 ppm hydrogen sulfide and < 50 ppm ammonia [79], will then normally be used to feed the incoming desalter water supply. In case of excess water build-up in the desalter circuit, it is bled from this point in the recycle to aerated bio-ponds etc. for further treatment before discharge.

Sour water (hydrogen sulfide in water), also obtained from a number of other points in the refinery operation such as from the aqueous condensate from the fractionation of the volatile products of a fluid catalytic cracker (FCC), hydrotreater, hydrocracker, or coker. In fact, in a refinery which is

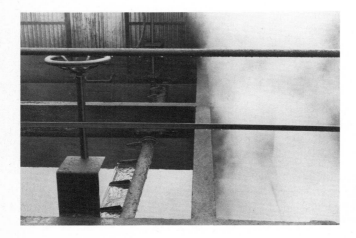

**Plate 16.2.** Top view of an API (American Petroleum Institute) separator showing skimmer with adjustment handwheel

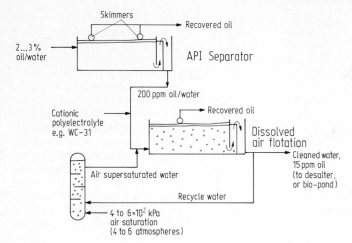

Skimmers

Recovered oil

2...3 % oil/water

API Separator

200 ppm oil/water

Cationic polyelectrolyte e.g. WC-31

Recovered oil

Dissolved air flotation

Air supersaturated water

Cleaned water, 15 ppm oil (to desalter, or bio-pond)

Recycle water

4 to 6×10² kPa air saturation (4 to 6 atmospheres)

**Figure 16.9.** Operating details of an American Petroleum Institute (API) separator, followed by a dissolved air flotation unit for oil/water separation

endeavouring to minimize water use, total sour waters may make up as much as 25 % of the total aqueous effluent discharged [77]. All of this will be treated for hydrogen sulfide (and ammonia) removal in manner similar to the procedure described for desalter effluent [78].

Site drainage waters from surface runoff invariably include some oil picked up from small spills and process leaks. These will also be routed through an API separator first, sometimes followed by a dissolved air separator (Figure 16.9), or the like [80]. Oil separated from this drainage water stream will join the slop oil circuit for recycling. During times of desalter operation under water short conditions, treated site drainage waters may be used for this purpose. Any water in excess of desalter requirements will proceed to an artificially aerated bio-pond for further BOD reduction before discharge [80, 81].

Phenols, which in many refineries are removed from hydrocarbon streams by treating with a solu-

tion of sodium hydroxide in water, produce a particularly intractable waste stream not easily amenable to BOD reduction in a bio-pond [82]. If feasible, this stream will be sold (or given) to a neighbouring petrochemical plant which will frequently be able to make use of the phenol content. Phenol is recovered from this stream by acidification, which may be accomplished economically by contacting with flue gases followed by solvent extraction [83] (Equation 16.44).

$$NaOPh + H_2CO_3 \rightarrow NaHCO_3 + PhOH \quad (16.44)$$

Caustic in water is also frequently used for removal of sulfides from certain liquid hydrocarbon refinery streams (Equations 16.45, 16.46).

$$2 NaOH + H_2S \rightarrow Na_2S + 2 H_2O \quad (16.45)$$

$$NaOH + CH_3SH \rightarrow NaSCH_3 + H_2O \quad (16.46)$$

Sodium sulfide, at least, is of value to pulp mills which use the kraft process for pulping and, if near by, provide one useful outlet for this waste product of caustic treating. However, if this disposal method is not feasible a "caustic oxidizer" may be required at the refinery to convert the sodium sulfide to the more environmentally acceptable sodium sulfate, for ultimate disposal by landfill (Equation 16.47).

$$Na_2S \text{ (aqueous)} + 2 O_2 \xrightarrow[heat]{} Na_2SO_4 \quad (16.47)$$

Or, alternatively, aqueous sodium sulfide may be acidified to produce a sour water stream (Equation 18.48),

$$Na_2S + H_2SO_4 \rightarrow Na_2SO_4 + H_2S \quad (16.48)$$

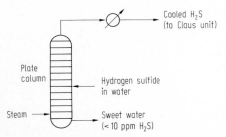

Cooled H₂S (to Claus unit)

Plate column

Hydrogen sulfide in water

Steam

Sweet water (< 10 ppm H₂S)

**Figure 16.10.** Details of hydrogen sulfide removal from water in a sour water stripper

which can then be routed to the sour water stripper for hydrogen sulfide recovery and conversion to sulfur. In this case the more economical method of acidification, flue gas contacting, cannot be used since this would liberate highly volatile hydrogen sulfide into the flue gases contacted.

When operating in areas with a restricted water supply, or on municipal water services, significant economies in water usage may be achieved by recycling once-used and warmed cooling water through direct or indirect cooling towers for reuse. Process cooling directly with air, by fan forcing against finned tubes is also a measure which can reduce overall refinery water requirements, particularly when operating in cold climates [71, 84].

The potential impact of inadvertent losses of some constituents of refinery liquid wastes has been recently reviewed by Cote [85], and a general summary of current developments of aqueous treatment processes has been presented by Baker [86].

### 16.6.3 Refinery Waste Disposal Practices

The components of most refinery liquid waste streams are recovered and re-used, whenever feasible [87]. But some of these, such as aqueous caustic phenolic or caustic sulfidic wastes, do not lend themselves readily to re-use in many refinery situations. Deep well disposal, incineration, or precipitation in some manner and landfilling of the separated solids are the measures used in these instances [87]. Preliminary concentration of brine streams by reverse osmosis can help decrease final disposal costs by decreasing the waste volume [88].

Sludges, such as those accumulated in the monoethanolamine or diethanolamine streams used for hydrogen sulfide removal by scrubbing, comprise another intractable waste material. Since these sludges are wholly combustible, most refineries dispose of these by incineration although a whole spectrum of other methods are used [87]. Some of those that do not practise incineration recycle sludges within the refinery operation. Others use landfilling (dry) or landfarming (discussed later) as disposal methods. Biotreatment or deepwell disposal are also used by some Canadian refineries, but to a more limited extent.

Spent acids are used where feasible to neutralize alkaline waste streams, or are neutralized with purchased lime, or caustic, and are then routed to the normal effluent treatment system for further clean-up before discharge [71, 87]. Waste solids from refinery operations include such materials as spent clay from decolourization of lubricating oils or waxes, sand used as a catalyst support, filter aid, or filter base, and exhausted Claus catalysts (primarily ferric oxide on alumina). These are disposed of by landfilling by the majority of Canadian refineries [87]. However, some use landfarming for disposal of these materials. One refinery recovers spent Claus catalysts for regeneration into new catalyst.

Landfarming, or the landspreading of oily sludges for degradation by soil micro-organisms, has been found to be an efficient method of disposal, providing proper conditions are maintained to achieve optimum oil decomposition rates within the soil. Examination of oil treated soils from a Texas test plot revealed that six genera of micro-organisms, *Pseudomonas* sp., *Flavobacterium* sp., *Nocardia* sp., *Corynebacterium* sp., *Arthrobacter* sp. and several unidentified yeasts, were dominant in the numbers of the micro-organism population present [87]. Oil degradation rates in soil have been demonstrated to be high, both in isolated boxed samples of soil, and from experimental active plots outside, particularly in warm weather (Tables 16.10, 16.11). Decomposition rates range from the neighbourhood of 150 kg oil carbon/hectare/day in the temperate climate of the area of Edmonton, Alberta [90], up to as high as 569 kg oil-C/ha/day for landspreading operations in Texas [89]. With proper application, cultivation at intervals, and addition of appropriate amounts of fertilizer, this method is not only used

**Table 16.10** Decomposition rates of accumulated sediments from a crude oil storage tank, in boxed soil tests [a]

| Elapsed time days | Application rate Oil content, % by wgt. | |
|---|---|---|
| | Medium | Heavy |
| 0 | 13 | 23 |
| 5 | 6 | 8 |
| 17 | 6.5 | 9 |
| 32 | 4.5 | 5 |
| 46 | 5 | 7 |
| 67 | 5 | 5.5 |
| 88 | 6.5 | 7 |

[a] From [103].

**Table 16.11** Decomposition rates of crude oil in the field, at Edmonton, Alberta [a]

| Elapsed time, days | Application rate oil content, kg/m$^2$ | | |
|---|---|---|---|
| | Light | Medium | Heavy |
| 0 | 5.5 | 13.7 | 24.9 |
| 43 | 3.7 | 10.7 | 17.2 |
| 340 | 2.1 | 8.5 | 11.5 |
| 460 | 2.0 | 7.2 | 11.0 |
| 735 | 1.5 | 4.2 | 7.8 |

[a] From [90].

for oily wastes, but also for other biodegradable waste products of refinery operation such as amine sludges and spent oily clay and sand [87], as already mentioned. If only light sludge loadings are applied it is said that the land can be returned to agricultural use after one season [84]. But it should also be noted that landfarming for sludge disposal, perhaps because of one or more of the following considerations, has been regulated against in some areas, for example the State of Minnesota. Among the recommendations of the Water Quality Programs Committee of the International Joint Commission (U.S.A. and Canada) it is proposed that piezometer wells be used to monitor groundwater pressure and quality at different depths , and that routine soil samples be taken for oil content and trace metal analysis from any operating landfarm disposal system [91]. And, on the retirement of land employed for landfarming disposal it is further recommended that these soils not be used for the growing of crops or for grazing because of the risk of contamination of foodstuff from metals or other materials accumulated in the soil [92].

## Relevant Bibliography

1. Petroleum Refinery Effluent Regulations and Guidelines, Water Pollution Directorate, Report EPS 1-WP-74-1, Ottawa, 1974 (plus updates of same)
2. Status Report on Abatement of Water Pollution from the Canadian Petroleum Refining Industry, Water Pollution Control Directorate, Report EPS 3-WP-76-11, Ottawa, 1976
3. Survey of Trace Substances in Canadian Petroleum Industry Effluents, by Can Test Ltd. and IEC International Environmental Consultants Ltd., Petroleum Association for Conservation of the Canadian Environment, Ottawa, 1981

4. F. Huang, and R. Elshout, Optimizing the Heat Recovery of Crude Units, Chem. Eng. Prog. 72, 68, July 1976
5. J.L. Allen, Evaluating a Waste-oil Reclamation Program, Plant Eng. 30, 255, April 29, 1976
6. Refining Developments, Special Report, Hydroc. Proc. 54, 93, Sept. 1975
7. D. Burris, Field Desalting, a Growing Producer Problem Worldwide, Pet. Eng. 46, 36, June 1974
8. R.G. Gantz, Sour Water Stripper Operations, Hydroc. Proc. 54, 85, May 1975
9. Urea Dewaxing, Edeleanu Gesellschaft. Hydroc. Proc. 51, 205, Sept. 1972
10. S.P. Ahuja, M. Derrien, and J.F. Le Page, Activity and Selectivity of Hydrotreating Catalysts, Ind. and Eng. Chem., Prod. Res. and Devel. 9, 272, Sept. 1970
11. J.G. Speight, The Chemistry and Technology of Petroleum, Marcel Dekker, New York, 1980
12. P.J. Bailes, Application of Solvent Extraction to Organic and Petrochemical Industries, Chem. Ind. (London), 724, Sept. 3, 1977

## References

1. Petroleum Products Handbook, V.B. Guthrie, editor, McGraw-Hill, New York, 1960
2. The Petroleum Handbook, 4th edition, Shell International Petroleum Company Ltd., London, 1959
3. Kirk-Othmer Encyclopedia of Chemical Technology, 3rd edition, Wiley-Interscience, New York, volume 14, page 477
4. Air Pollution Handbook, P.L. Magill, F.R. Holden, and C. Ackley, editors, McGraw-Hill, New York, 1956
5. C.R. Noller, Textbook of Organic Chemistry, Saunders, Philadelphia, 1966
6. F.D. Rossini, J. Chem. Educ. 37(11), 554, Nov. 1960
7. L.F. Fieser and M. Fieser, Advanced Organic Chemistry, Reinhold, New York, 1961
8. Riegel's Industrial Chemistry, 6th edition, J.A. Kent, editor, Reinhold, New York, 1962
9. G.C. Speers and E.V. Whitehead in: Organic Geochemistry, G. Eglinton and M.T.J. Murphy, editors, Longman-Springer Verlag, Berlin and London, 1969, p. 640
10. E. Eisima and J.W. Jurg in: Organic Geochemistry, G. Eglinton and M.T.J. Murphy, editors, Longman-Springer Verlag, Berlin, London, 1969, page 676
11. Chemical Desalting, Hydroc. Proc. 49(9), 235, Sept. 1970
12. Electrostatic Desalting, Hydroc. Proc. 49(9), 237, Sept. 1970

13. H.S. Bell, American Petroleum Refining, 4th edition, Van Nostrand Reinhold, 1959, cited by Riegel's Handbook of Industrial Chemistry, 7th edition, J.A. Kent, editor, Van Nostrand Reinhold, New York, 1974

14. Air Pollution Control Handbook, P.L. Magill, F.R. Holden, and C. Ackley, editors, McGraw-Hill, New York, 1956

15. Kirk-Othmer Encyclopedia of Chemical Technology, 3rd edition, Wiley Interscience, New York, 1982, volume 17, page 183

16. E.M. v. Z. Bakker and J.F. Jaworski, Effects of Vanadium in the Canadian Environment, National Research Council of Canada, Ottawa, 1980

17. Kirk-Othmer Encyclopedia of Chemical Technology, 2nd edition, Interscience, New York, 1968, volume 15, page 1

18. Riegel's Handbook of Industrial Chemistry, 7th edition, J.A. Kent, editor, Van Nostrand Reinhold, New York, 1974

19. L.F. Albright, Chem. Eng. 73, 143, Aug. 15, 1966, cited by Kirk-Othmer, 2nd edition, reference 17

20. S.R. Tennison, Chem. In Brit. 17(11), 536, Nov. 1981

21. R.N. Shreve, Chemical Process Industries, 3rd edition, McGraw-Hill, New York, 1967

22. T. Hutson, Jr. and R.S. Logan, Hydroc. Proc. 54(9), 107, Sept. 1975

23. G.C. Ray, J.W. Myers, and D.L. Ripley, Hydroc. Proc. 53, 141, Jan. 1974

24. Gasoline to Trigger Aromatics Shortage, Chem. Eng. News 59(20), 16, May 18, 1981

25. Unleaded Gasoline, New Sales For Oil Process, Chem. Eng. News 49(10), 14, March 8, 1971

26. G.A. Purdy, Petroleum Prehistoric to Petrochemicals, McGraw-Hill, Toronto, 1958

27. P. Polss, Hydroc. Proc. 52(2), 61, Feb. 1973

28. W. Hubis and R.O. Clark, Anal. Chem. 27, 1009, (1955)

29. EPA Opts for Stricter Lead-in-Gas Rules, Chem. Eng. News 60(32), 4, Aug. 9, 1982

30. E.J. Farkas, Water and Polln Control 110(6), 40, June 1972

31. Final Lead in Gasoline Rules Issued, Chem. Eng. News 57(38), 18, Sept. 17, 1979

32. EPA Mulls Over Lead-in-Gasoline Rules, Chem. Eng. News 60(18), 28, May 3, 1982

33. Lead Phasedown Delay Helps Aromatics, Chem. Eng. News 57(16), 4, April 16, 1979

34. Manganese Rather than Lead, Can. Chem. Proc. 56(6), 48, June 1972

35. MTBE Can be Made From Natural Gas, Chem. Eng. News 58(28), 26, July 14, 1980

36. Octane Improvement or BTX Extraction, Can. Chem. Proc. 64(6), 10, Sept. 3, 1980

37. DOE Endorses Use of Alcohol Fuels, Chem. Eng. News 57(29), 5, July 16, 1979

38. Gasoline Sparks Study on New Ethanol Plant, Chem. Eng. News 57(48), 8, Nov. 26, 1979

39. UN Workshop Urges Wider Use of Ethanol, Chem. Eng. News 57(17), 11, April 23, 1979

40. B. Gillespie, L.W. Manley, and C.J. Di Perna, ChemTech 8, 750, Dec. 1978

41. Esso Streamlines Lube Dewaxing, Can. Chem. Proc. 57(10), 40, Oct. 1973

42. L. Morriset-Blais, Can. Chem. Proc. 57(8), 48, Aug. 1973

43. I.S. Crighton, Chem. Ind. (London), 126, 16 Feb. 1974

44. H.G. Muller, Tribology International 11(3), 189, June 1978

45. R.R. McCoy and D.S. Taber, A Cold Look at Lubricants, SAE Transactions 80, paper 719716, 1971

46. J.F. Payne, I. Martins, and A. Rahimtula, Science 200, 329, April 21, 1978

47. D.J. Skinner and W.A. Neff, Preliminary Review of Used Lubricating Oils in Canada, Report EPS 3-WP-74-4, Environment Canada, Ottawa, 1974

48. D. Bisson, Science Dimension (Ottawa) 7(5), 8, (1975)

49. E. Baumgardner, Hydroc, Proc. 53(5), 129, May 1974

50. Waste Lubricant Recycling Booms, Can. Chem. Proc. 65(6), 16, Sept. 18, 1981

51. PROP on in BC, Can. Chem. Proc. 65(7), 6, Oct. 30, 1981

52. N.J. Weinstein, Hydroc. Proc. 53(12), 74, Dec. 1974

53. E.E. Klaus and E.J. Tewkesbury, Hydroc. Proc. 53(12), 67, Dec. 1974

54. D.B. Barton, J.A. Murphy, and K.W. Gardner, Synthesized Lubricants Provide Exceptional Extended Drain Passenger Car Performance, SAE Paper 780951, Nov. 13, 1978

55. C.E. Goldmann, A Synthesized Engine Oil Providing Fuel Economy Benefits, Paper SP-411, Soc. of Autom. Eng. Meeting, Milwaukee, Wisconsin, Sept. 13-16, 1976

56. Synthetic Lubricants Poised for Big Growth, Chem. Eng. News 58(13), 12, March 31, 1980

57. J.A.C. Krulish, H.V. Lowther, and B.J. Miller, An Update on Synthesized Engine Oil Technology, Paper 770634, Soc. of Autom. Eng. Meeting, Tulsa, Oklahoma, June 7-9, 1977

58. Synthetic Oils and Additives for Lubricants — Advances Since 1977, Chemical Technology Review No. 145, M.W. Ranney, editor, Noyes, New Jersey, 1980

59. Chemical Engineers Handbook, 4th edition, R.H. Perry, editor, McGraw-Hill, New York, 1963, page 9-6

60. L.S. Caretto, California Air Environment 5(1), 1, Fall 1974

61. The Clean Air Act Annual Report 1977-78, Ot-

tawa, 1979. Cited by M. Webb, The Canadian Environment: Data Book of Energy and Environmental Problems, W.B. Saunders, Toronto, 1980

62. G.J.K. Acres, Chem. Ind. (London), 905, Nov. 16, 1974

63. Odour Evaluation, Chem. In Can. *23* (10), 23, Oct. 1971

64. H.F. Elkin and R.A. Constable, Hydroc. Proc. *51* (10), 113, Oct. 1972

65. Gas in Your Pipe Stops Smoking, Indair Flare, Engineer *237*, 9, Sept., 27, 1973

66. W. Lauderback, Hydroc. Proc. *51*, 127, Jan. 1972

67. S.H. Tan, Flare System Design Simplified, in Waste Treatment and Flare Stack Design Handbook, Gulf Publishing Company, Houston, Texas, 1968, page 81

68. G.R. Kent, Hydroc. Proc. *51* (10), 121, Oct. 1972

69. Control of Atmospheric Emissions from Petroleum Storage Tanks, J. Air Polln. Control Assoc. *21* (5), 263, May 1971

70. C.M. McKinney, Hydroc. Proc. *51* (10), 117, Oct. 1972

71. W.J. Racine, Hydroc. Proc. *51* (3), 115, March 1972

72. J. Charlton, R. Sarteur, and J.M. Sharkey, Hydroc. Proc. *54* (5), 97, May 1975

73. C.S. Russell, Residuals Management In Industry, A Case Study of Petroleum Refining, Johns Hopkins University Press, Baltimore, Ohio, 1973

74. Pollution Control Process Demonstrated, Chem. Eng. News *56* (8), 23, Feb. 20, 1978

75. Pollution Control Objectives for The Chemical and Petroleum Industries of British Columbia, Water Resources Service, Queen's Printer, Victoria, 1977

76. J. Wardley-Smith, Prevention of Oil Pollution, Graham and Trotman Ltd., London, 1979

77. P.T. Budzik, Chem. in Can. *29* (3), 24, March 1977

78. R.J. Klett, Hydroc. Proc. *51* (10), 97, Oct. 1972

79. Fina Recovers $H_2S$ from Wastewaters, Can. Chem. Proc. *60* (10), 28, Oct. 1976

80. Waste Treatment and Flare Stack Design Handbook, Gulf Publishing Company, Houston, Texas, 1968

81. J.F. Ferrel and D.L. Ford, Hydroc. Proc. *51* (10), 101, Oct. 1972

82. J.C. Hovious, G.T. Waggy, and R.A. Conway, Identification and Control of Petrochemical Pollutants Inhibitory to Anaerobic Processes, EPA-R2-73-194, Washington, 1973

83. N.N. Li, W.S. Ho, and R.E. Terry, Extraction of Phenolic Compounds and Organic Acids by Liquid Membranes, 142nd Nat. Meeting, American Chem. Soc., Aug. 23-28, 1981

84. NPRA '74 Panel Views Processed, Hydroc. Proc. *54* (3), 127, March 1975

85. R.P. Cote, The Effects of Petroleum Refinery Liquid Wastes on Aquatic Life, with Special Emphasis

on the Canadian Environment, National Research Council, Ottawa, 1976

86. D.A. Baker, J. Water Polln Control Fed. *46* (6), 1298, June 1974

87. Canadian Petroleum Refining, Industry Waste Survey, Petroleum Association for Conservation of the Canadian Environment, Report No. 80-4, Ottawa, 1980

88. R.W. Newkirk and P.J. Schroeder, Hydroc. Proc. *51* (10), 103, Oct. 1972

89. C.B. Kincannon, Oily Waste Disposal by Soil Cultivation Process, EPA-R2-72-110, Washington, 1972. Cited by Beak Consultants Ltd., Landspreading of Sludges at Canadian Petroleum Facilities, Petroleum Association for Conservation of the Canadian Environment, Report No. 81-5A, Ottawa, 1981

90. The Reclamation of Agricultural Soils After Oil Spills, J.A. Toogood, editor, Dept. of Soil Science, The University of Alberta, AIP Publication No. M-77-11, Edmonton, 1977. Cited by Beak Consultants, see reference 89

91. A Review of the Pollution Abatement Programs Relating to the Petroleum Refinery Industry in the Great Lakes Basin, Great Lakes Water Quality Board, Windsor, Ontario, 1982

92. Manual for Landspreading of Petroleum Industry Sludges, Beak Consultants Ltd., Petroleum Association for Conservation of the Canadian Environment, Ottawa, 1981

93. E.C. Lane and E.L. Garton, Report of Investigations 3279, U.S. Bureau of Mines, Washington, 1935, Cited by Riegel's Handbook of Industrial Chemistry, J.A. Kent, editor, Van Nostrand Reinhold, New York, 1974, page 406

94. W.A. Bachman and D.H. Stormont, Oil and Gas J. *65*, 69 (1967)

95. M.T. Atwood, Chemtech *3*, 617, Oct. 1973

96. P.J. Garner, Chem. and Ind. (London), 131, Feb. 16, 1974

97. BP Statistical Review of the World Oil Industry — 1973, The British Petroleum Company Ltd., London, 1974

98. Esso Facts and Figures, Research and Analysis Division, Imperial Oil Ltd., Sarnia, 1981

99. Chemical and Process Technology Encyclopedia, D.M. Considine, editor, McGraw-Hill, New York, 1974

100. T. Hutson, Jr. and R.S. Logan, Hydroc. Proc. *54* (9), 107, Sept. 1975

101. D.R. Greenwood, G.L. Kingsbury and J.C. Cleland, A Handbook of Key Federal Regulations and Criteria for Multimedia Environmental Control, Env. Prot. Agency Report 600/7-79-175, Washington, August 1979. Cited by Kirk-Othmer Encyclopedia, 3rd edition, reference 15

102. Petroleum Refinery Effluent Regulations and Guidelines, Report EPS 1-WP-74-1, Water Pollu-

tion Control Directorate, Ottawa, 1974. From the Canada Gazette Part 2 of April-June 1982, which refers to the Consolidated Regulations of Canada, Vol. VII, c. 828, page 5225, Ottawa, 1978. These figures represent Canadian standards to date.

103. Industrial Oily Waste Control, W.K. Mann, H.B. Shortly, R.M. Skallerup, editors, American Petroleum Institute/American Society of Lubrication Engineers, New York and Baltimore, ca. 1970 (no date specified)

# Formulae and Conversion Factors

## Formulae

### Area

| | | Volume | |
|---|---|---|---|
| triangle | $\frac{1}{2}$ bh | cylinder | $\pi r^2 h$ |
| regular polygon* | $nl^2/(4 \tan\pi/n)$ | sphere | $\frac{4}{3}\pi r^3$ |
| circle | $\pi r^2$ | | |
| sphere | $4\pi r^2$ | | |

## Conversion Factors

| From | To | Factor |
|---|---|---|

### Length

| | | |
|---|---|---|
| miles | m | $1.609 \times 10^3$ |
| foot | m | 0.3048 |
| inches | m | $2.54 \times 10^{-2}$ |
| angstroms | m | $10^{-10}$ |

### Mass

| | | |
|---|---|---|
| ton, long (2240 lb) | kg | $1.015 \times 10^3$ |
| tonne, metric (1000 kg; 1 Mg) | lb | $2.204 \times 10^3$ |
| ton, short (2000 lb) | kg | $9.07 \times 10^2$ |
| pound (avdp) | kg | 0.4536 |
| ounce (avdp) | g | 28.35 |

### Area

| | | |
|---|---|---|
| square mile | acres | 640 |
| acres | ha(hectares) | 0.40469 |
| hectare | $m^2$ | $10^4$ |
| square foot | $m^2$ | 0.0929 |

### Volume

| | | |
|---|---|---|
| barrel (bbl) | $m^3$ | 0.159 |
| | Imp. gal | 34.971 |
| | U.S. gal (liq.) | 42 |
| cubic foot | $m^3$ | $2.832 \times 10^{-2}$ |
| Imperial gallon | L | 4.5460 |
| U.S. gallon (liq.) | L | 3.7854 |
| Imperial gallon | U.S. gal (liq.) | 1.20095 (i.e. $\frac{6}{5}$) |

### Pressure

| | | |
|---|---|---|
| bar | atmosphere | 0.98692 |
| | MPa | 0.1 |
| | Pa($Nm^{-2}$) | $10^5$ |
| atmosphere | bar | 1.01325 |
| | Pa (pascals) | $1.013 \times 10^5$ |
| | lb/$in^2$ | 14.696 |
| pound/$in^2$ | kPa | 6.89 |

### Energy

| | | |
|---|---|---|
| therm | MJ | 106 |
| British thermal unit (BTU) | kJ | 1.06 |
| calorie | J | 4.1840 |

---

* n = number of sides, $l$ = length of side

# Subject-Index

(Abbreviations: incl = includes; fig = figure; tbl = table)